LINEAR SYSTEMS

PRENTICE-HALL INFORMATION
AND SYSTEM SCIENCES SERIES
Thomas Kailath, *Editor*

LINEAR SYSTEMS

THOMAS KAILATH

Department of Electrical Engineering
Stanford University

PRENTICE-HALL, INC., *Englewood Cliffs, N.J.* *07632*

Library of Congress Cataloging in Publication Data

Kailath, Thomas.
   Linear systems.

   Includes bibliographies and index.
   1. System analysis. I. Title.
QA402.K295 1980   003   79-14928
ISBN 0-13-536961-4

Editorial/production supervision
   by Christopher Moffa and Lori Opre
Cover design by Lana Gigante
Manufacturing buyer: Gordon Osbourne

Printed in the United States of America

10 9 8 7 6 5 4 3

PRENTICE-HALL INTERNATIONAL, INC., *London*
PRENTICE-HALL OF AUSTRALIA PTY. LIMITED, *Sydney*
PRENTICE-HALL OF CANADA, LTD., *Toronto*
PRENTICE-HALL OF INDIA PRIVATE LIMITED, *New Delhi*
PRENTICE-HALL OF JAPAN, INC., *Tokyo*
PRENTICE-HALL OF SOUTHEAST ASIA PTE. LTD., *Singapore*
WHITEHALL BOOKS LIMITED, *Wellington, New Zealand*

for
SARAH
her
style and taste
judgment
love and support

CONTENTS

*Sections so marked throughout Contents may be skipped without loss of context.

CHAPTER 6 STATE-SPACE AND MATRIX-FRACTION DESCRIPTIONS OF MULTIVARIABLE SYSTEMS   345

CHAPTER 10 SOME FURTHER READING   632

APPENDIX: SOME FACTS FROM MATRIX THEORY   645

INDEX   671

PREFACE

Linear systems have been under study for a long time, and from several different points of view, in physics, mathematics, engineering, and many other fields. But the subject is such a fundamental and deep one that there is no doubt that linear systems will continue to be an object of study for as long as one can foresee. However, a particular feature of recent engineering studies, and the main focus of this book, is the emphasis on the structure of finite-dimensional linear systems. While such systems have been extensively studied, especially since the early 1930s, the frequency-domain techniques that were commonly used often did not specifically exploit the underlying finite dimensionality of the systems involved. Moreover, almost all this work was for single-input, single-output (or *scalar*) systems and did not seem to extend satisfactorily to the multi-input, multi-output (or *multivariable*) systems that became increasingly important in aerospace, process control, and econometric applications in the late 1950s. This fact, plus the importance of time-variant systems and time-domain characteristics in aerospace problems, led to a resurgence of interest, sparked by the work of Bellman and Kalman, in the state-space description of linear systems. This approach led naturally to more detailed examinations of the structure of finite-dimensional linear systems, or linear dynamical systems as they are often called, and to questions of redundancy, minimality, controllability, observability, etc. The papers [1]† and [2] give a good perspective of the situation around 1960. The state-space

---

†See the references following the Preface.

formulation led to some new proposals for system design and feedback compensation—pole-shifting controllers, quadratic regulator synthesis, state observers and estimators, noninteracting control, etc. But just as the state-space techniques were being codified into textbooks (References [3] and [4] nicely bracket the books of that period), Popov [5] and Rosenbrock [6] were showing how many of the scalar rational transfer function concepts could be naturally extended to matrix transfer functions and multivariable systems and how several questions could be more readily posed and solved in these terms. Since then these concepts have been effectively pursued by several researchers. By now, it seems to us, the main insight from this work is that transfer function (or high-order differential equation) descriptions and state-space (or first-order differential equation) descriptions are only two extremes of a whole spectrum of possible descriptions of finite-dimensional systems. We can work exclusively with one description or the other, but we can also easily translate results from one framework to the other, and, as expected, there are situations where a hybrid of the two extremes (using the so-called partial-state descriptions) is the most natural.

Our aim in this textbook is to take a beginning student, with some prior exposure to linear system analysis (elementary transform and matrix theory), through a motivated and integrated development of these new and fuller perspectives on linear system theory.

The detailed table of contents will provide a general idea of the scope of the book. Briefly, we start with scalar (single-input, single-output) systems and introduce the notions of state-space realizations, internal and external descriptions, controllability, observability, and their applications to minimal realizations, state-feedback controllers, and observers. While doing this, we also compare and contrast these state-space results with more classical transfer function ideas and gradually build up the awareness that equivalent results could have been obtained by working (carefully) with transfer function descriptions without reference to state variables or controllability or observability. The restriction to constant scalar systems in Chapters 1 to 5 allows one to gain this perspective at a fairly concrete and explicit level, so that the extension to multivariable systems can proceed more rapidly in the rest of the book (Chapters 6 to 9). Particular care was devoted to the selection and arrangement of topics in the scalar case, so that the parallel multivariable development is not only well motivated but, in the author's opinion, also quite insightful and powerful. Thus at many points the development reaches the frontiers of research (see also Chapter 10), equipping the reader for new studies and applications in the many fields where linear system theory can be important—e.g., in signal detection and estimation, system identification, process control, digital filtering, communication systems, and, generally speaking, the broad and exciting field of signal processing.

At Stanford, the material in the first five chapters and in Secs. 6.1, 6.2, and in Chapter 9 is covered in a 40 to 45-hour senior/first-year graduate course, with some of the sections indicated by asterisks being left for extra reading. Chapters 6 to 8 provide enough material for another 30-hour course for graduate students, with opportunities for further reading and development as term-paper projects. However, the material can be arranged in various ways, and I have tried to write in a way that will encourage browsing and self-study by students of various ages and backgrounds.

At this point, some explanation of the origins of this book may be helpful. For a variety of reasons, the state-space approach has been largely developed in control theory, and not in communication theory, where most of my own interests lie. In the mid-1960s, Schweppe [7] in the United States and Stratonovich and Sosulin [8] in the USSR began to show the usefulness of state-space methods in signal detection problems. Then Omura showed how the quadratic regulator control algorithm could be applied to certain feedback communication schemes [9]. These papers, and also the patient instruction of some of my early Ph.D. students, especially Jim Omura, Paul Frost, Roger Geesey, Ty Duncan, and B. Gopinath, gave me a greater appreciation of state-space theory and led me to introduce more of it into the Stanford linear systems course. However, it soon became clear that a deeper knowledge was necessary to really exploit the power of state-space methods. Also, the existing fashion in textbooks was largely oriented toward the background mathematics in differential equations and linear algebra, with less attention to the engineering significance and applications of the concepts peculiar to system theory. For example, much attention was devoted to Jordan forms, various ways of evaluating matrix exponentials, and numerous definitions of controllability and observability. The mathematics of all this was clear, but what the books did not really explain was why all this was useful to anyone—engineers or mathematicians.

It was the role of controllability in the pole-shifting problem for time-invariant systems (Chapter 3), and that of observability in the design of asymptotic observers (Chapter 4), that first gave me some meaningful indication of the value of these concepts. Then, as I examined the original research literature, I learned of their role in providing stability results for quadratic regulators [10] and optimum filters [11]. The significance of this stability is that the effect of numerical errors in computation, e.g., round-off errors, does not build up and destroy the calculation—obviously a very important practical consideration. It became clear that controllability and observability first arose as certain technical conditions to resolve existence and uniqueness conditions in certain optimal control and estimation problems. It was only somewhat later that Kalman isolated them and defined them via certain idealized problems [12], which for various reasons came to be overly emphasized in many treatments.

Moreover, as I began to obtain a better appreciation of the state-space point of view by applying it to various detection, estimation, and control problems, the pioneering and extensive studies of Rosenbrock [6] (and then

Popov, Forney, Wolovich, and others) clarified the power of the transfer function approach and the benefits to be gained by a better understanding of the relationships between it and the state-space approach. This book attempts a synthesis of the powerful new perspectives on linear system theory that are now available. The advantages of such a development are already to be seen in various areas, and I believe that a lot more will be done with it.

These background remarks also explain why the contents of this book do not quite follow the "traditional" (since 1963!) order of presentation found in most existing textbooks. One favorite topic in many of them is the time-domain solution of state-space equations. This is an interesting topic and can build well upon earlier knowledge of linear differential equations. However, I feel that the students' sense of accomplishment in mastering this material is somewhat illusory. First, if one really had to solve some equations, there are several readily available computer routines developed just for this purpose. But it is claimed that one should "understand" what one is computing. True, but this understanding comes from numerical analysis and not really from the pretty but particular mathematics learned in the linear systems course (see [13]). In fact, what is lost in dallying with this mathematics is the awareness that many of the things that can be done with state-space equations do not really need explicit time-domain solutions of the state equations. Therefore the solution of state-space equations has been deemphasized in this book. On the other hand, I have tried to show that the notion of explicit realizations of a given set of system equations (or transfer functions) can be a powerful aid in understanding and using linear systems. This theme first appears in Chapter 2 and continues to be developed throughout the book, e.g., in the exploration of multivariable systems (Secs. 6.4 and 6.5) halfway through the book, in the study of general differential systems in Chapter 8, and in the explanation of adjoints of time-variant systems in Chapter 9, and to a certain extent in the brief final Chapter 10. It may take time, and several readings, to adjust to the somewhat different perspectives of this book, and I can only offer my own experience as proof that it might be worthwhile.

While learning this subject and attempting to get some perspective on what was vital and what transient, I have found great help in going back to the original sources. For, as Robert Woodhouse [14] pointed out in 1810 (in the first book in English on the calculus of variations), "the Authors who write near the beginnings of science are, in general, the most instructive; they take the reader more along with them, show him the real difficulties, and, which is the main point, teach him the subject the way by which they themselves learnt it." Therefore, in these notes I have often made a special effort to point out the earliest papers on the different concepts and would encourage the active reader to pursue them independently. More generally, the references have been carefully selected for their significance, readability, and potential for further study and, in several cases, further independent investigation. Similarly, the exercises in this book are of various levels of difficulty and in several instances serve to complement and extend the material in the text. Therefore, all the exercises should at least be read along with each section, even if only a few are actually attempted.

I have also attempted to make the book reasonably self-contained, and every effort has been made to keep the proofs as simple and direct as possible. For example, things have been so arranged so that very little linear algebra is required either as a pre- or corequisite. What is really needed is some exposure to matrix manipulations and, more importantly, a recognition and acceptance by the student that, at this level, no course or textbook on linear algebra (or in fact any mathematical subject) can be a perfect or complete prerequisite for the material in any engineering course—there is no substitute for just buckling down to figure out many things for oneself (with liberal use of 1×1 or 2×2 matrices in the early stages). Of course some guidance is necessary, and therefore, in the Appendix and in Sec. 6.3, I have tried to collect the results from elementary algebra and polynomial matrix theory that are used in this book. However, they are not meant to be mastered before launching into the rest of the book—rather, the explicit references made in later sections to special results, such as determinantal and block matrix identities or the Cayley-Hamilton theorem or the Smith canonical form, are to be used as occasions for a more motivated study of the relevant special topics. Of course this may often be painful and slow, but my experience is that the student thereby achieves a better mastery of the material and, more important, a foretaste of the ability to pick out and learn enough about some special (mathematical) topic to try to resolve particular problems that he may encounter in his later work. The range of mathematics used in present-day engineering problems is so wide that one could spend all one's time taking "prerequisites"—especially since studying well-established material is so much easier than venturing out, even just a little, into some less well-defined territory.

Therefore in this book I have tried to subordinate the mathematical concepts to the system concepts—it is only too easy, and unfortunately only too common, for readers at this level to be led down very entertaining but ultimately deeply frustrating mathematical garden paths. My goal is not the presentation or development of "mathematical" system theory but an effort to introduce and use just as little mathematics as possible to explore some of its basic concepts. I try to follow an admonition of Joshua Chover [15]: "It is time to dispel a popular misconception. The goal of mathematics is *discovery*, not 'proof'." Or to make the point another way, I belong to the school that holds ideas and exposition to be more important than "mere" results [16, 4.B].

Finally, I should also caution that the unavoidable vagueness of real problems, arising from constraints of imperfect knowledge, nonmathematical performance specifications, economic constraints, flexible acceptability criteria, etc., means that solutions of the necessarily clean and specific mathematical problems of any theory can ultimately only serve as "guides" to the actual "resolution" of any engineering problem. Unfortunately this is a distinction that cannot really be conveyed by a textbook and is the reason good teachers (or engineers) can never be replaced by a book (or a computer program).

In this connection I should mention that, especially in the early chapters, the presentation is deliberately loosely organized, with emphasis on discussion and motivation rather than formal development. Several major themes are gradually developed in a spiral fashion, and readers should not expect to find all their questions answered the first time a topic is introduced. *Students will also find it helpful to frequently make up for themselves tables and charts of the major concepts, results and interrelations as they perceive them at various points in the course.* A continuous interplay between skills and knowledge must take place in any successful learning effort. As succinctly put by Edsger Dijkstra [17, p. 211], a scientific discipline is "not any odd collection of scraps of knowledge and an equally odd collection of skills" but "the skills must be able to improve the knowledge and the knowledge must be able to refine the skills." *Therefore, to really understand a subject one has ultimately to make a personal selection and resynthesis, modulated by one's own background and other knowledge, of the material in any given book or course.* My hope is that this book will provide enough material and opportunity for such an educational experience, via self-study and/or classroom instruction.

All this—

was for you, [dear reader].
I wanted to write a [book]
that you would understand.
For what good is it to me
if you can't understand it?
But you got to try hard—

Adapted from "January Morning" by William Carlos Williams†

REFERENCES

1. E. GILBERT, "Controllability and Observability in Multivariable Systems," *SIAM J. Control*, **1**, pp. 128–151, 1963.

2. R. E. KALMAN, "Mathematical Description of Linear Dynamical Systems," *SIAM J. Control*, **1**, pp. 152–192, 1963.

†From *Collected Earlier Poems* by William Carlos Williams, New Directions Publishing Corp., 1938. "January Morning" is published in adapted form (3 words changed), by permission of New Directions.

3. L. A. ZADEH and C. A. DESOER, *Linear System Theory—A State-Space Approach*, McGraw-Hill, New York, 1963.

4. C. T. CHEN, *Introduction to Linear System Theory*, Holt, Rinehart and Winston, New York, 1970.

5. V. M. POPOV, "Some Properties of Control Systems with Matrix Transfer Functions," in *Lecture Notes in Mathematics*, Vol. 144, Springer, Berlin, 1969, pp. 169–180.

6. H. H. ROSENBROCK, *State Space and Multivariable Theory*, Wiley, New York, 1970.

7. F. C. SCHWEPPE, "Evaluation of Likelihood Functions for Gaussian Signals," *IEEE Trans. Inf. Theory*, **IT-11**, pp. 61–70, July 1965.

8. R. L. STRATONOVICH and YU. G. SOSULIN, "Optimal Detection of a Markov Process in Noise," *Eng. Cybern.*, **6**, pp. 7–19, Oct. 1964 (trans. from Russian).

9. J. OMURA, "Optimum Linear Transmission of Analog Data for Channels with Feedback," *IEEE Trans. Inf. Theory*, **IT-14**, pp. 38–43, Jan. 1968. See also Ph.D. dissertation, Stanford University, Stanford, Calif., 1966.

10. R. E. KALMAN and R. KOEPCKE, "Optimal Synthesis of Linear Sampling Control Systems Using Generalized Performance Indexes," *Trans. ASME*, **80**, pp. 1820–1826, 1958.

11. R. E. KALMAN and R. S. BUCY, "New Results in Linear Filtering and Prediction Theory," *Trans. ASME Ser. D J. Basic Eng.*, **83**, pp. 95–107, Dec. 1961.

12. R. E. KALMAN, "On the General Theory of Control Systems," *Proceedings of the First IFAC Congress*, Vol. I, Butterworth's, London, 1960, pp. 481–493.

13. C. B. MOLER and C. VAN LOAN, "Nineteen Dubious Ways to Compute the Exponential of a Matrix," *SIAM Review*, pp. 801–836, Oct. 1978.

14. R. WOODHOUSE, *A History of the Calculus of Variations in the Eighteenth Century*, Cambridge University Press, London, 1810; reprint, Chelsea, New York, 1966.

15. J. CHOVER, *The Green Book of Calculus*, Benjamin, Reading, Mass., 1972.

16. B. PARLETT, "Progress in Numerical Analysis," *SIAM Review*, **20**, pp. 443–456, July 1978.

17. E. W. DIJKSTRA, *A Discipline of Programming*, Prentice-Hall, Englewood Cliffs, N.J., 1976.

ACKNOWLEDGMENTS

I have found the writing of this book to be surprisingly difficult. Perhaps the only things that have sustained me in this endeavour have been the flashes of pleasure in seeing so many different results fall so nicely into place, and the thought that I would at some stage have the opportunity to formally thank the many people who made it possible.

My first thanks go to my students, both in classes and outside, for their help in building my knowledge of system theory, and in helping in various ways to develop this book. I would particularly like to mention B. Dickinson, B. Friedlander, M. Gevers, B. Gopinath, S. Kung, B. Lévy, M. Morf, A. Segall, G. Sidhu, G. Verghese, and E. Verriest. M. Aref, K. Lashkari, and H. Lev-Ari assisted with the proofs and index. Sidhu, Segall, Dickinson, Morf, Friedlander, Verghese, and Lévy also taught the basic linear systems course at Stanford while these notes were under development. This and the Ph.D. studies of Dickinson, Morf, Kung, Verghese, and Lévy have provided the foundation for many sections of this book. This mention does not fully capture the measure of their assistance, especially so in the case of George Verghese, who provided boundless energy and selfless assistance in debating, revising, and developing many of what I judge to be the nicer features of this book.

Professors G. Franklin, M. Morf, L. Ljung, A. Bryson, M. Hellman, B. Anderson, and S. S. Rao also made helpful suggestions, and contributed examples and exercises, as they too taught from the notes for this book. Other friends made comments and suggestions on various portions of this material and it is a pleasure to again mention Brian Anderson, who valiantly reviewed several drafts, and John Baras, Stephen Barnett, Charlie Desoer, Patrick Dewilde, Eli Jury, Jorma Rissanen, Howard Rosenbrock, and Len Silverman. My secretary, Barbara McKee, must be even more relieved than they that the typing and retyping and constant revisions are finally over. It is a pleasure to thank her here for the many other contributions, well-known to my friends and colleagues, that she has made to my professional activities.

The productive environment at Stanford, developed by John Linvill and Ralph Smith, has fostered a fruitful interaction between graduate teaching and research that has been important to the development of this book. In this, I have also been aided by the consistent and unconstrained research support of the Mathematical Sciences Division of the Air Force of Scientific Research.

Of the many fine people whom I have been associated with at Prentice-Hall, I would especially like to thank John Davis, Hank Kennedy, and Paul Becker, and recently Lori Opre in production, and the Dai Nippon Printing Company in Japan, for their patience and assistance.

I gratefully remember G. M. Joshi and G. S. Krishnayya for their early influence and assistance, and am delighted to be able to thank Chandrasekhar Aiya, Bob Price, Paul Green, Bill Root, Jack Wozencraft, Bob Gallager, Jack Ziv, Lotfi Zadeh, Norm Abramson, Sol Golomb, Peter Whittle, Lou Stumpers, Dave Slepian, Allen Peterson, Manny Parzen, Moshe Zakai and Israel Gohberg for their help and guidance at various times and over many years. Space forbids explicit mention of many others, in this country and abroad, with whom I have enjoyed technical interactions in diverse fields.

A perceptive author (B. G. Raja Rao in *The Serpent and the Rope*) has written that all books are autobiographical. As the earlier remarks indicate, I am especially conscious of that, and would like to also express my appreciation to my large family, so happily extended by marriage, and especially to my parents, who have moulded us through faith, discipline and example.

My wife, Sarah, has been the best thing that ever happened to me. Completion of this book may be the best thing that ever happened to her, and to our children, Ann, Paul, and Priya.

> *"The lines have fallen unto me in pleasant*
> *places; yea, I have a goodly heritage"*

THOMAS KAILATH

Stanford, Calif.,

BACKGROUND MATERIAL 1

1.0 INTRODUCTION

This chapter contains a review of some of the basic definitions and background mathematics that will be used in this book. Most readers will be familiar with this material in one form or the other, and therefore no attempt is made at completeness here. However, we do try to provide some motivation for the various concepts and also to elaborate some aspects that are not always covered.

In Sec. 1.1 we use a very simple example to illustrate some of the difficulties encountered when attempting too formal a study of linear systems. An axiomatic treatment has some allure here, for both teacher and student, but we hope to illustrate that such a task is futile at this level. Our attitude is that there are enough new and interesting engineering and mathematical concepts to learn that we can forego long debates about the "proper" definitions of things we are comfortable with already—like linearity, causality, and time invariance.

In Sec. 1.2 we describe the main properties of unilateral Laplace transforms, which will be used to solve differential equations. The value of the so-called $\mathcal{L}_-$ transforms is motivated by several examples, and a handy but not too widely known generalization of the initial-value theorem is given.

In Sec. 1.3 we discuss the family of impulsive functions and their use in representing more complicated functions and in deriving input-output relations for linear systems.

1

Finally in Sec. 1.4 we make some introductory remarks on the use of matrices in this book. A more detailed presentation of matrix theory is given in an appendix.

It is not necessary to thoroughly digest the material in this chapter or the appendix before going on to Chapter 2. A quick reading will serve to fix our basic definitions and notations. This should suffice until occasions arise in later chapters where more detailed reference to this background material may be necessary.

1.1   SOME SUBTLETIES IN THE DEFINITION OF LINEARITY

A general approach to system theory may start with the observation that one method of representing a system, whether linear or nonlinear, time invariant or time variant, etc., is by a table of all the possible inputs to the system and the possible responses (outputs) for these inputs. However, only very rarely will such a crude technique be useful. One major reason for paying so much attention to linear systems is that for such systems the table of input-output pairs can be drastically abbreviated.

A system L is generally said to be linear if whenever an input u_1 yields an output $L(u_1)$ and an input u_2 yields output $L(u_2)$ we also have

$$L(c_1 u_1 + c_2 u_2) = c_1 L(u_1) + c_2 L(u_2) \tag{1}$$

where

$$c_1, c_2 = \text{arbitrary real or complex numbers} \tag{2}$$

Implicit in the above is the requirement that the space of possible inputs be closed under linear combination; i.e., $c_1 u_1 + c_2 u_2$ must belong to the space if u_1 and u_2 do.

To see how these properties of a linear system help condense the input-output pair description, let us assume that the input space consists of all periodic functions (with period T) that have a Fourier series representation. Then any input function can be written as

$$u(t) = \sum_{n=-\infty}^{\infty} \tilde{u}_n e^{j2\pi nt/T}, \qquad \text{where } \tilde{u}_n = \frac{1}{T} \int_0^T u(t) e^{-j2\pi nt/T} \, dt$$

and we shall have

$$L[u(t)] = L(\sum_n \tilde{u}_n e^{j2\pi nt/T}) = \sum_n \tilde{u}_n L(e^{j2\pi nt/T}) \tag{3}$$

This shows that to specify the system we need to know only its responses to the countably infinite set of inputs $\{e^{j2\pi nt/T}, n = 0, \pm 1, \ldots\}$, while quite clearly the space of all allowable inputs is many orders of infinity "larger." And, of course, in some instances we need tabulate even less, because it may be possible to give an explicit formula for calculating $L(e^{j2\pi nt/T})$ no matter what the value of n is.

However, even at this early stage, we can see the need for greater care in our calculations. For example, we must note that the additivity property (1) does not extend, without further assumptions, to infinite summations, so in general

$$L(\sum_1^\infty u_i) \neq \sum_1^\infty L(u_i) \tag{4}$$

We can easily see why care must be used in assuming equality in (4). If we set, say,

$$\sum_1^\infty u_i = \lim_{N\to\infty} \sum_1^N u_i = \lim_{N\to\infty} S_N \tag{5}$$

then equality in (4) would require

$$L(\lim_{N\to\infty} S_N) = \lim_{N\to\infty} L(S_N) \tag{6}$$

Even the most unsophisticated reader will agree that the required interchange of operations in (6) can be justified only if the system $L(\,\cdot\,)$ is "sufficiently smooth"; i.e., linearity as defined by (1) and (2) is not enough to ensure (6) without further qualifications. The standard example of such difficulties is a sort of mathematical pathology (see Exercises 1.1-2 and 1.1-3) and is usually justly ignored. However the following simple example, discovered by Adam Shefi (a Stanford student in 1966), is more arresting.

Example 1.1-1.   A Linear System?

Consider a system defined as follows. The inputs to the system are restricted to being piecewise continuous functions, bounded over $(-\infty, \infty)$ and having no more than a finite number of simple jump discontinuities in a finite time. The system response at any time t is defined to be the algebraic sum of the jumps that are present in the input signal from $-\infty$ up to the (present) time t.

The space of inputs is clearly a linear space, and moreover it is easy to see that the system obeys (1) and (2),

$$L(c_1 u_1 + c_2 u_2) = c_1 L(u_1) + c_2 L(u_2)$$

However, the system has some surprising properties. For example, the response to a step input is a step. Therefore we would think that the impulse response is a delta function and correspondingly that the frequency response (Laplace transform of the impulse response) is unity. However, the response to $\cos t$ or $\sin t$ or $e^{j\omega t}$ is zero because these functions have no discontinuities! Then, if we examine it more closely, we see that the response to an impulse is similarly ambiguous. If we use rectangular pulses to approximate an input delta function, the response will be equal to the input; however, if we use triangular pulses, the response will be zero. Similarly, the response to any *continuous* approximation to a step function is zero, no matter how good the approximation.

Clearly the system we have defined is not "smooth" or "continuous"—"small" changes in the input may give "big" changes in the output.   ∎

Now, to be able to regard a system as continuous, so that we may, for example, freely use superposition, Laplace and Fourier analysis, etc., we must first have a way of defining "small" and "big" in the input and output spaces. Unfortunately, this is a bigger task than is sometimes believed. The problem is that there are many ways of measuring size or distance, e.g., by appropriate *norms* or *metrics*, which are terms the reader may have encountered. With these norms the associated linear spaces can be regarded as (in increasing order of generality) Hilbert spaces, Banach spaces, or metric spaces, which are also terms the reader may encounter in the early pages of some introductory engineering texts. A discussion of Hilbert spaces, etc., has some intrinsic value of course, but unfortunately once we embark on such a route it will soon turn out that even more general concepts of distance will be needed. Thus, to "rigorously" include the operation of differentiation (to enable us for example to think of the derivative of a step function as an impulse), no simple metric will suffice, and the more abstract topological notions of *neighborhoods* and *open sets* have to be introduced.

To dally with this question a bit longer, we might ask, for example, what we mean by an infinite sum? Some reflection shows that this is actually a matter of definition,† and there can be many definitions. The usual one is a limit of partial sums:

$$\sum_1^\infty u_i = \lim_{N\to\infty} S_N, \qquad S_N = \sum_1^N u_i$$

However, this definition may not always be meaningful. For example, when

$$u_i = (-1)^{i+1}$$

the sequence S_N clearly does not have a limit. Therefore other definitions of *infinite sum* have been introduced, and one that is useful here is the Abel sum:

$$\sum_1^\infty u_i = \lim_{r\to1} \lim_{N\to\infty} \sum_1^N u_i r^{i-1}, \qquad 0 < r < 1$$

†This does need thought. As Hardy [1, Sec. 1.3] points out, a natural reaction is to ask "What is $1 - 1 + 1 - \cdots$?" rather than "How shall we *define* $1 - 1 + 1 - \cdots$?". It was not really until the time of Cauchy that even the greatest mathematicians recognized that a collection of mathematical symbols does not have a meaning until one has been assigned to it by definition.

For the sequence $u_i = (-1)^{i+1}$, we can check that the Abel sum is $\frac{1}{2}$. The same value is obtained for the so-called Cesàro $(C, 1)$ sum, defined as

$$\sum_1^\infty u_i = \lim_{N \to \infty} \frac{S_1 + \cdots + S_N}{N}$$

However, there are of course many other possibilities [1, pp. 73, 77, 85]. It is known that Guido Grandi, a professor of mathematics at the University of Pisa, argued that the two "evaluations"

$$0 = (1 - 1) + (1 - 1) + \cdots$$

and

$$\frac{1}{2} = \frac{1}{1 + x}\bigg|_{x=1} = 1 - x + x^2 + \cdots \bigg|_{x=1}$$

showed that he had proved the world could be created out of nothing! Other "arguments" can be used to give any rational number; Leibniz used a probabilistic argument to obtain the value $\frac{1}{2}$, while Lagrange favored the value $\frac{3}{5}$. The point is that different conventions are useful for different purposes! And "engineering reality" and "physical significance" are not good enough shibboleths to enable us to honestly say that one particular definition is the most useful (see [2] for an illustration of this fact in the context of "causality").

Therefore, in this book we shall not start out with a whole flock of definitions of linearity, causality, time invariance, ranges, domains, convergence, etc. We shall concentrate instead on a class of finite-dimensional linear systems described by ordinary linear differential or difference equations, for which we assume the reader already knows what linearity, causality, etc., mean. In other words, we shall not at all attempt an axiomatic treatment but shall try instead to learn more about the properties of systems that we have long considered as linear, time invariant, etc. Of course, we shall be dealing with idealized mathematical models, and we shall need a lot of mathematical arguments and proofs, but we shall aim to be, roughly speaking, "logically consistent" rather than "mathematically rigorous."

Therefore, the reader could now proceed to Chapter 2, after perhaps a quick glance at the background material on transforms and matrices in the Appendix. For a final look at "fine points" of definition, some exercises are given below, of which at least Exercises 1.1-1 and 1.1-7 should be attempted.

Exercises

1.1-1.   *Does superposition always work?*

Consider the linear time-invariant system shown in the figure. Time will be taken to be discrete and to assume only integral values, $k = 0, \pm 1, \pm 2, \ldots$. Assume that the system is at "rest."

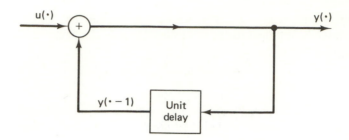

a. Determine the response of the system to a unit (discrete-time) impulse $u(k) = \delta(k)$, where $\delta(k) = 0$, $k \neq 0$, $\delta(0) = 1$.

b. Let $\phi(\cdot)$ be such that $\phi(0) = 1$, $\phi(-1) = -1$, and $\phi(k) = 0$ for all $k \neq \{0, -1\}$. It seems reasonable that we can represent $\delta(\cdot)$ by a linear combination of shifted $\phi(\cdot)$ functions as $\delta(\cdot) = -\sum_{l=1}^{\infty} \phi(\cdot - l)$. Find the response of the system to $\phi(\cdot - l)$ and to $-\sum_{l=1}^{\infty} \phi(\cdot - l)$ and compare with the result in part a.

c. It also seems reasonable to represent $\delta(\cdot)$ as $\delta(\cdot) = \sum_{0}^{\infty} \phi(\cdot + l)$. Find the response to $\phi(\cdot + l)$ and to $\sum_{0}^{\infty} \phi(\cdot + l)$. Compare with the result in parts a and b. Try to find an explanation for the difference.

1.1-2. *Classical Example of a Nonsmooth Linear System.*

From the property (a *part* of the definition of a linear system)

$$L(u_1 + u_2) = L(u_1) + L(u_2)$$

deduce that

a. $L(nu_1 + mu_2) = nL(u_1) + mL(u_2)$, where n, m are positive integers.

b. Extend the above to the case where n, m may be positive or negative.

c. Extend to the case where n, m may be any *rational* (ratio of two integers) number. However, we cannot extend to the case of real numbers n, m without also assuming that L is well behaved; see [3, pp. 115–118]. The argument, though worth reading once, is somewhat sophisticated, requiring use of the axiom of choice and of Hamel bases. The example in the text is much simpler.

1.1-3. *A Basic Functional Equation.*

Suppose $f(t)$ is a real-valued function of t, $-\infty < t < \infty$, such that

$$f(t + s) = f(t) + f(s), \qquad \text{all } t, s$$

It is well known that the only "well-behaved" solution is the linear one $f(t) = ct$. Show that this is true when $f(\cdot)$ is integrable over any finite interval. *Hint:* Show first that

$$I = \int_{0}^{t+s} f(x)\, dx - \int_{0}^{t} f(x)\, dx - \int_{0}^{s} f(x)\, dx$$

$$= tf(s) = sf(t)$$

Therefore when $t \neq 0$, $f(t)/t =$ a constant, say c. When $t = 0$, $f(t + s) = f(t) + f(s)$ implies $f(0) = 0$, so that $f(t) = ct$ also holds for $t = 0$. This elegant proof is due to Shapiro [4].

1.1-4. *Systems with Nonlinear Elements need not be Nonlinear!*

Is the following system linear? The input $u(t)$ is zero for $t \leq 0$. The initial voltage, $V_c(0)$, on the capacitor is zero. The output is $i(\cdot)$. The nonlinear elements are "ideal" diodes, which provide perfect transmission in the direction of the arrows and complete rejection in the reverse direction. This exercise shows a simple, though annoying, "subtlety" that has to be taken into account when giving a general definition of a linear system: This is the concept of "zero-state" linearity. [This example is due to D. Cargille, a student at the University of California at Berkeley (cited in [6, p. 140]).]

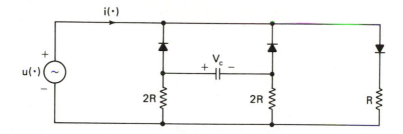

1.1-5.

Consider a system in the so-called *state-variable* or *state-space* form:

$$\dot{x}(t) = A(t)x(t) + B(t)u(t), \qquad x(t_0) = x_0$$
$$y(t) = C(t)x(t) + D(t)u(t), \qquad t \geq t_0$$

Is this system linear?

1.1-6.

A system has the property that its response to any input is zero. Is this system linear? Give a state-variable description of such a system. Can you say anything more about its linearity?

1.1-7.

Consider a system (a series *RL* circuit) where the (current) response $y(\cdot)$ is defined by the differential equation

$$\dot{y}(t) + y(t) = u(t), \qquad y(0-) = 1$$

a. Find the response of this system to an impulse input.
b. Find the response to a unit step input.
c. Can the response to the impulse be obtained by differentiating the response to the step?

1.2 UNILATERAL LAPLACE TRANSFORMS AND A GENERALIZED INITIAL-VALUE THEOREM

The unilateral Laplace transform (L.T.) will be used to solve some simple differential equations, and we shall therefore review this application via some examples. This material will undoubtedly be familiar to many readers, with perhaps the only novelty being a generalized form of the initial-value theorem, useful to determine the response to impulsive inputs. One may also note that we introduce the so-called $\mathcal{L}_-$ transform as being convenient for problems where time is the independent variable (as it generally is in this book).

The unilateral L.T. of a function $x(\cdot)$ is usually defined as

$$X_+(s) = \mathcal{L}_+[x(t)] = \int_{0+}^{\infty} x(t)e^{-st}\,dt \tag{1}$$

where s is a complex variable, $s = \sigma + j2\pi f$.

If $X_+(s)$ "exists" for some s, say $s_0 = \sigma_0 + j2\pi f_0$, then it exists for all s such that Re $s \geq \sigma_0$. The smallest value of σ_0, say α, for which (1) exists is called the *abscissa of convergence*, and $\{s: \text{Re } s \geq \alpha\}$ is called the *region of existence* (or *region of definition*) of $X_+(s)$.

Examples: The abscissa of convergence for $\exp bt$, b real, is b. For $\exp -t^2$ it is $-\infty$, while for $\exp t^2$ it is $+\infty$ (therefore $\exp t^2$ has no L.T.).

The original time function can be recovered from $X_+(s)$ by the contour integral

$$\mathcal{L}_+^{-1}[X_+(s)] \triangleq \frac{1}{2\pi j} \int_{c-j\infty}^{c+j\infty} X_+(s)e^{st}\,ds = x(t), \qquad t \geq 0+ \tag{2}†$$

where c is any real number greater than the abscissa of convergence, α. (At points of discontinuity, the contour integral gives the average of the right- and left-hand values.) An important property is that of *uniqueness*: $\mathcal{L}_+[x(t)] = \mathcal{L}_+[y(t)]$ on some line $s = \sigma + j\omega$ in their common regions of existence implies that $x(t) = y(t)$, $t \geq 0+$ (almost everywhere). In other words, $X_+(s) = 0$ can be true only for $x(\cdot)$ such that $x(t) = 0$, $t > 0+$.

A list of commonly used properties and examples is given in Table 1.2-1. In the table, we use certain standard conventions. Thus

$\dot{x}(t) \triangleq dx(t)/dt$

$1(t) \triangleq$ the Heaviside step function (also called the unit step function) with values $1(t) = 0$, $t < 0$, $1(0) = \frac{1}{2}$, $1(t) = 1$, $t > 0$

$\delta(t) \triangleq$ the Dirac delta function (also called the unit impulse function), further discussed in Sec. 1.3

†The symbol $\triangleq$ denotes "equal by definition."

An Application: The unilateral L.T. was first introduced into electrical engineering by J. R. Carson† as a "rigorous" substitute for Oliver Heaviside's "operational" (or "symbolic") methods for solving ordinary differential equations.

Table 1.2-1   Some properties and examples of unilateral Laplace transforms of functions defined for $t \geq 0+$

$x(t), t \geq 0+$	$X_+(s)$
$\dot{x}(t)$	$sX_+(s) - x(0+)$
$e^{-at}x(t)$	$X_+(s + a)$
$tx(t)$	$-dX_+(s)/ds$
$x(t - a)1(t - a)$	$e^{-as}X_+(s)$
$x(t/a)$	$a\,X_+(as)$
$\delta(t)$	0
$1(t)$	$1/s$
$t^{n-1}/(n - 1)!$	$1/s^n$
e^{-at}	$1/(s + a)$
$e^{-at}\sin bt$	$b/[(s + a)^2 + b^2]$
$e^{-at}\cos bt$	$(s + a)/[(s + a)^2 + b^2]$

Note: For the $\mathcal{L}_-$ transform, we just replace 0+ by 0− wherever it occurs in the above formulas and set $\mathcal{L}_-[\delta(t)] = 1$.

We shall illustrate the method by a simple example, solving for the current $y(\cdot)$ in a series RL circuit ($R = L = 1$) driven by a unit step voltage. We assume that the current at time $0+$ is y_0. The problem is therefore to solve the differential equation

$$\dot{y}(t) + y(t) = 1(t), \qquad y(0+) = y_0$$

Taking L.T.s of both sides of this equation, we have

$$[sY_+(s) - y(0+)] + Y_+(s) = \frac{1}{s}$$

Therefore

$$Y_+(s) = \frac{1}{s(s + 1)} + \frac{y(0+)}{s + 1}$$

$$= \frac{1}{s} - \frac{1}{s + 1} + \frac{y_0}{s + 1}$$

†Carson's monograph (1926) has been reissued in paperback form [5]. Despite its age, it still compares favorably with many more recent books on the Laplace transform and is therefore worth reading for more than just historical interest. However, note that Carson's definition of the L.T. is slightly different from ours. Carson, to preserve the similarity to Heaviside's operational calculus, uses an extra factor s in the definition [i.e., Carson defines $X(s) = s \int_{0+}^{\infty} x(t)e^{-st}\, dt$]. In addition, still following Heaviside, he uses p rather than s.

And, by inverse transformation, we obtain

$$y(t) = 1 + (y_0 - 1)e^{-t} \qquad \text{for } t \geq 0+$$

The Need for the $\mathcal{L}_-$ Transform.   In this problem one might argue that it is unreasonable to claim to know a priori the value $y(0+)$, which is the value of the current in the circuit just *after* the excitation has been turned on. It is more reasonable to assume that we know $y(0-)$, the value of the current just *before* the excitation is turned on. In general, knowing only $y(0-)$, we would have to compute $y(0+)$ (from the differential equation) before being able to use the $\mathcal{L}_+$ transform. This computation may or may not be easy, depending on the type of excitation. In the present problem it is easy: We claim that $y(0+) = y(0-)$. If not, $y(\cdot)$ would have a discontinuity at the origin and $\dot{y}(\cdot)$ would not be defined (would be infinite) there—but the right-hand side of the equation is defined at the origin. [More physically, we could argue that a (finite) voltage cannot change the current through the inductor instantaneously.] In other problems the computations may be more difficult; see the exercises.

Another approach that avoids supplementary calculations is to use an $\mathcal{L}_-$ transform,

$$\mathcal{L}_-[x(t)] = X_-(s) = \int_{0-}^{\infty} x(t)e^{-st}\, dt \tag{3}$$

with the inversion formula

$$\mathcal{L}_-^{-1}[X_-(s)] \triangleq \frac{1}{2\pi j} \int_{c-j\infty}^{c+j\infty} X_-(s)e^{st}\, ds = x(t), \qquad t > 0- \tag{4}$$

The formulas with this transform can be simply obtained from those for the $\mathcal{L}_+$ transform by replacing $0+$ by $0-$, *with one exception*. This is the so-called *initial-value* theorem, which we shall examine presently.

The major reason for our use of $\mathcal{L}_-$ transforms is that this enables us to handle initial conditions and impulsive inputs at the origin in a direct manner. Let us first examine the question of impulsive inputs.

The Skeleton-in-the-Closet of $\mathcal{L}_+$ Transforms.   From the formulas

$$\mathcal{L}_+[1(t)] = s^{-1} = \mathcal{L}_-[1(t)] \quad \text{and} \quad \mathcal{L}_+[\dot{x}] = s(\mathcal{L}_+ x) - x(0+)$$

we obtain

$$\mathcal{L}_+[\delta(t)] = s \cdot s^{-1} - 1 = 0 = \mathcal{L}_+[0]$$

This may be a surprise, though it is consistent with the direct calculation

$$\mathcal{L}_+[\delta(t)] = \int_{0+}^{\infty} \delta(t)e^{-st}\, dt = 0$$

The expected formula (and the one found in many textbooks that use $\mathcal{L}_+$ transforms) for the Laplace transform of $\delta(t)$ is 1. But this actually follows from the use of the $\mathcal{L}_-$ formulas,

$$\mathcal{L}_-[\delta(t)] = \mathcal{L}_-[d1(t)/dt] = s \cdot s^{-1} - 0 = 1$$

which also agrees with the direct evaluation

$$\mathcal{L}_-[\delta(t)] = \int_{0-}^{\infty} \delta(t)e^{-st} \, dt = 1$$

More detailed discussions of the $\mathcal{L}_-$ transform can be found in many places, e.g., Zadeh and Desoer [6, App. B] and Doetsch [7, pp. 107–108, 131–138].

Example 1.2-1.

Let us consider the solution of

$$\dot{x}(t) + 2x(t) = \delta(t), \qquad t \geq 0-, x(0-) = 1$$

by several methods.

1. *Classical.* For $t \geq 0+$, the equation is $\dot{x} + 2x = 0$, which has the solution $x(t) = x(0+)e^{-2t}$, $t \geq 0+$. Now we are not given $x(0+)$ but only $x(0-) = 1$. However, $x(\cdot)$ must have a unit discontinuity at the origin so that the $\dot{x}(\cdot)$ term on the left-hand side can match the $\delta(\cdot)$ on the right-hand side. Therefore, $x(0+) - x(0-) = 1$, from which we obtain $x(0+) = 2$.

2. $\mathcal{L}_+$ *Transform.* Taking $\mathcal{L}_+$ transforms, we get

$$sX_+(s) - x(0+) + 2X_+(s) = 0$$

or

$$X_+(s) = \frac{x(0+)}{s+2}, \qquad x(t) = x(0+)e^{-2t}, \qquad t \geq 0+$$

But now $x(0+)$ is unavailable and has to be found by the argument used in method 1.

3. $\mathcal{L}_-$ *Transform.* Taking $\mathcal{L}_-$ transforms, we get

$$sX_-(s) - x(0-) + 2X_-(s) = 1$$

or

$$X_-(s) = \frac{1}{s+2} + \frac{x(0-)}{s+2}$$

which yields

$$x(t) = e^{-2t} + e^{-2t}, \qquad t > 0-$$

There is no fuss about initial conditions here. ∎

In summary, note that both the $\mathscr{L}_-$ and $\mathscr{L}_+$ transforms give the same answer but that, clearly, the first is more direct, at least in problems where the initial conditions are most naturally given at $0-$. (Of course, not all problems are of this type, especially when we are dealing not with functions of time but say with functions of space, e.g., a mass distribution on a line.)

We shall conclude with an examination of the initial-value theorem.

An Initial-Value Theorem for $\mathscr{L}_-$ Transforms.   The $\mathscr{L}_-$ transform generally has the same properties as the $\mathscr{L}_+$ transform with values at $0-$ being used in place of values at $0+$. The most important exception is the initial-value theorem for $\mathscr{L}_-$ transforms. This theorem can be stated in varying degrees of generality, but for our purposes the following will usually suffice.

Let $x(t)$ be continuous for $t > 0+$, with possibly a step discontinuity at the origin, $x(0+) \neq x(0-)$, and also such that $X_-(s)$ is a rational function of s. Then

$$\lim_{s \to \infty} sX_-(s) = x(0+) \neq x(0-) \quad \text{in general} \tag{5}$$

A proof is outlined in Exercise 1.2-5; however, it is not hard to see heuristically why $x(0+)$ rather than $x(0-)$ arises in the initial-value theorem, for note that

$$sX_-(s) = \int_{0-}^{\infty} x(t)se^{-st}\, dt$$

and as s increases se^{-st} gets more and more concentrated toward the origin but always from the right; in the limit se^{-st} goes over into a delta function, and the integral reduces to $x(0+)$ (see also Exercise 1.2-5).

In some problems of interest to us, $x(\cdot)$ will have impulsive functions at the origin. In this case, the above theorem fails, and in fact $\lim_{s \to \infty} sX_-(s)$ turns out to be infinite. However, a simple modification allows us to handle such cases.

A Generalized Initial-Value Theorem for $\mathscr{L}_-$ Transforms.   Let $x(t)$ be continuous for $t > 0$ but possibly with $x(0+) \neq x(0-)$ and also possibly containing impulsive functions at the origin. Thus the transform $X_-(s)$ can have the form

$$X_-(s) = a_0 + a_1s + \cdots + a_ps^p + X_p(s) \tag{6}$$

where $X_p(s)$ is a *strictly proper* rational fraction. Then

$$x(0+) = \lim_{s \to \infty} sX_p(s) \tag{7}$$

The basis for a proof of this result is outlined in Exercise 1.2-5.

We may note that

$$\mathcal{L}_-^{-1}[s^k] = \delta^{(k)}$$

where $\delta^{(k)}$ denotes the kth derivative of the delta function. These objects may not be so familiar as delta functions, but since we often use such things in system theory (see, e.g., Sec. 2.3.2), we shall consider them briefly in the next section.

Exercises

1.2-1.

Solve the equations
a. $\ddot{x} + 3\dot{x} + 2x = \delta$, $x(0-) = 1 = \dot{x}(0-)$,
b. $\ddot{x} + \dot{x} = \delta^{(1)}$, $x(0-) = \dot{x}(0-) = \ddot{x}(0-) = 1$
by the classical method and by the $\mathcal{L}_-$ transform method.

1.2-2.

Solve the equations
a. $\ddot{x} - x = 2\delta + 3\delta^{(1)}$, $\dot{x}(0-) = 1 = x(0-)$,
b. $\ddot{x} + 2\dot{x} + x = \delta + 2\delta^{(1)}$, $\dot{x}(0-) = 2$, $x(0-) = 1$.

1.2-3.

Find $y(0+)$ and $\dot{y}(0+)$ for each of the differential equations
a. $2\ddot{y} + 4\dot{y} + 10y = \delta$, $y(0-) = 1 = \dot{y}(0-)$,
b. $\ddot{y} + y = 2\delta + \delta^{(1)}$, $y(0-) = 1$, $\dot{y}(0-) = 2$.
In both cases, determine the required answer by *both* time-domain and frequency-domain methods. [Do not calculate the complete solution $y(\cdot)$; it is not required.]

1.2-4.

Find the initial-value formula required to calculate $\dot{x}(0+)$ from knowledge of $X_-(s)$.

1.2-5.

Suppose that $x(0+) \neq x(0-)$ but that $x(\cdot)$ is continuous for all $t \geq 0+$. Then prove that $\lim_{s \to \infty} sX_-(s) = x(0+)$ by the following steps:
a. Note that $\mathcal{L}_-[\dot{x}(t)] = sX_-(s) - x(0-)$.
b. Evaluate $\lim_{s \to \infty} \mathcal{L}_-[\dot{x}(t)]$ as

$$\lim_{s \to \infty} \left[\int_{0-}^{0+} \dot{x}(t)e^{-st}\, dt + \int_{0+}^{\infty} \dot{x}(t)e^{-st}\, dt \right]$$

[A rigorous proof can be found in Sec. 8.6 of A. Zemanian, *Distribution Theory and Transform Analysis*, McGraw-Hill, New York, 1965.]
c. What happens if there is no discontinuity at 0; i.e., $x(0+) = x(0-)$?
d. Suppose $x(\cdot) = \delta(\cdot) + x_p(\cdot)$, where $x_p(\cdot)$ is as in (1·2·6). Then $x(0+) = x_p(0+)$. Use this observation to justify the generalized theorem quoted in the text.

*1.3 IMPULSIVE FUNCTIONS, SIGNAL REPRESENTATIONS, AND INPUT-OUTPUT RELATIONS

The concepts of impulsive forces, that is, very large forces acting for very short durations, are probably familiar to most of us from elementary dynamics and statics. For example, the forces that arise at the impact of a bat and ball or of two billiard balls are impulsive forces. In statics, concentrated point loads are impulsive loads.

In electrical engineering, impulsive functions arise, among other places, in the rapid charging of a capacitor by a current source. We shall examine this problem in some detail in order to try to get a better feeling for the nature and representation of such functions.

Consider a unit capacitor being charged by a current source. We would like to deposit Q units of charge on the capacitor "as rapidly as possible." But what does this mean? One answer is zero time (instantaneously), but this is clearly a physically unverifiable requirement. In any physical problem there will be a limit, set by the resolving power of our measuring instruments, on how small an interval of time can be distinguished. This smallest interval, say ϵ_0, will define the word *instantaneous* for our problem. In view of this limitation, any problem that is "physically reasonable" must have a solution that does not depend critically † on the fine structure of events that occur in a time less than ϵ_0.

We shall explore this in our example by studying the response to a wide variety of charging current sources of essential duration ϵ_0. To begin with, let us make the current $i(\cdot)$ a short intense pulse, zero everywhere except in $(0, \epsilon_0)$:

$$i(t) = Qp_{\epsilon_0}(t) = \begin{cases} \dfrac{Q}{\epsilon_0}, & 0 < t < \epsilon_0 \\ 0, & \text{elsewhere} \end{cases} \tag{1}$$

The resulting charge on the capacitor will be

$$q(t) = \int_0^t i(\tau)\,d\tau = \begin{cases} \left(\dfrac{Q}{\epsilon_0}\right)t, & 0 \leq t \leq \epsilon_0 \\ 0, & \text{elsewhere} \end{cases}$$

The time functions $i(\cdot)$ and $q(\cdot)$ are shown in Fig. 1.3-1(b).

However, the *rectangular* shape of the pulse function $p_{\epsilon_0}(\cdot)$ in the interval of time ϵ_0 should not be important. Thus, let us consider a current pulse of triangular shape as shown in Fig. 1.3-1(c). Here again, direct calculation shows that a charge Q is

---

*Sections so marked may be skipped without loss of context.

†If the problem is such that the solution does depend on such fine structure, either the problem is improperly posed or any solution that we find will be subject to "unobservable" forces beyond our control. In such situations no science is possible, and we shall take no further interest in them; however, things may be different in quantum mechanics.

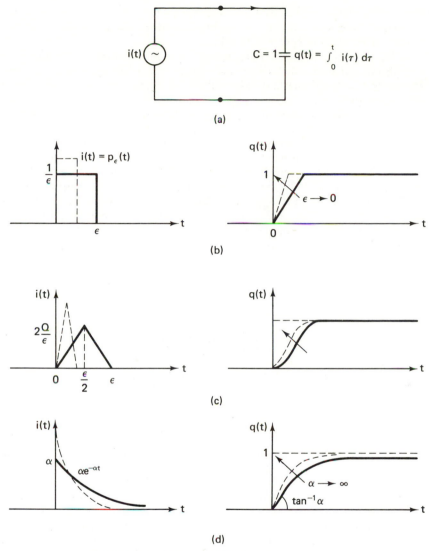

Figure 1.3-1.   Different ways of charging a capacitor "instantaneously."

deposited in time ϵ_0, even though the manner in which the charge rises from zero to Q is different from that in Fig. 1.3-1(b). This difference is unobservable and therefore irrelevant; what is relevant is that, insofar as our measurements can determine, both the rectangular and the triangular current pulses yield indistinguishable responses; viz., they both charge the capacitor "instantaneously." In fact it should be clear that all pulses of duration less than or equal to ϵ_0 and of area Q will be indistinguishable in this sense. We have used a rather strict definition of duration above. Even this is not necessary, as is shown by the pulse in Fig. 1.3-1(d); here, for large α, an exponential pulse of essential duration $\epsilon_0 = 1/\alpha$ delivers essentially Q units of charge in essentially a time ϵ_0.

What we need to do now is to try to find a way of exploiting this indifference to simplify our mathematical description of "instantaneous" events. Thus for a mathematical description (at least as usually understood), we need to specify our input and output time functions for all time, and it is cumbersome, even though it is the whole truth, to describe our charging current pulses above as "of arbitrary shape but having area Q for $t < \epsilon_0$ and being zero (or essentially zero) for $t \geq \epsilon_0$." To see what can be done, let us imagine shrinking the durations of the pulses in Fig. 1.3-1 to values less than ϵ_0. Insofar as our measurements are concerned nothing has changed, but "mathematically" the capacitor will be charging up more rapidly than before. In fact, we could let the pulse duration go to zero, and mathematically the capacitor would be charged in zero time. The current pulse that does this zero-time charging will be a rather peculiar object, for which it will be convenient to use a special symbol, $Q\delta(\cdot)$. From the examples of Fig. 1.3-1 it would appear that $Q\delta(\cdot)$ is characterized by the properties

$$Q\delta(t) = \begin{cases} 0, & \text{for } t \neq 0 \\ \text{undefined (it could be } \infty \text{ or zero),} & \text{for } t = 0 \end{cases} \tag{2}$$

but

$$Q \int_{-\infty}^{\infty} \delta(t)\, dt = Q \tag{3}$$

If we assume that the current source is $Q\delta(\cdot)$, the charge on the capacitor will be [by property (3)]

$$q(t) = Q \int_{-\infty}^{\infty} \delta(t)dt = Q, \qquad t > 0 \tag{4}$$

Since our smallest distinguishable time unit is ϵ_0, we cannot verify whether the charge Q is established in zero time as in (4). In fact, we know that charging as in (4) is physically impossible, but since the charge Q has been established in "our" zero time ϵ_0, it does not matter from a physical point of view whether or not we use the description (4). But from the point of view of a compact description and simple calculation it clearly does matter. Although in this example the gain in simplicity by using the symbol $\delta(\cdot)$ is not very great, we can easily give examples where it is considerable (see Exercises 1.3-1 and 1.3-2).

"Derivatives" of Discontinuous Functions and Delta Functions.   The introduction of the delta function enables us to give a meaning to the derivative of a function at a point of discontinuity. We note from Fig. 1.3-1(b) that $Q\delta(\cdot)$ can be regarded as the derivative of a step function. This derivative relationship is certainly true (in the conventional sense) as long as $\epsilon \neq 0$, and the use of the symbol $\delta(\cdot)$ allows us to extend it to the case where $\epsilon = 0$:

$$\delta(t) = \lim_{\epsilon \to 0} p_\epsilon(t) = \lim_{\epsilon \to 0} \frac{1(t) - 1(t - \epsilon)}{\epsilon} = \frac{d}{dt} 1(t) \tag{5}$$

Having gotten this far, we can go further and, using the same ideas, introduce a derivative for the delta function and in fact also second- and higher-order derivatives. To introduce the derivative, say $\delta^{(1)}(\cdot)$, we write

$$\delta^{(1)}(t) = d\delta(t)/dt = \lim_{\epsilon \to 0} \frac{\delta(t) - \delta(t - \epsilon)}{\epsilon} \qquad (6)$$

A pictorial interpretation using sequences is provided by Fig. 1.3-2. As $\epsilon \to 0$, the function in Fig. 1.3-2(a) tends toward the unit impulse $\delta(\cdot)$, while the function in Fig. 1.3-2(b) approaches what can be interpreted as its derivative. Since for very small ϵ the function in Fig. 1.3-2(b) consists of two narrow but strong pulses of opposite sign, the function in the limit is called the *unit doublet*. It is an idealization of a dipole in electromagnetic theory and of a couple (torque) in mechanics.

This procedure can be similarly extended to define $\delta^{(2)}(\cdot)$, $\delta^{(3)}(\cdot)$, and so on:

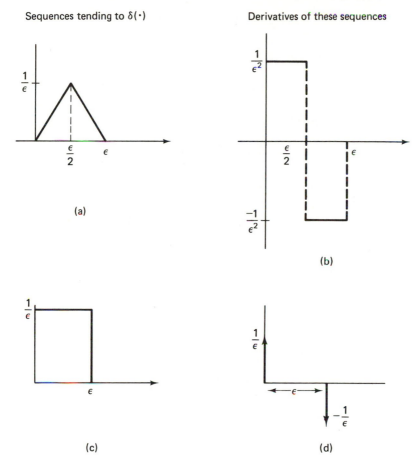

Figure 1.3-2.   Sequences for $\delta^{(1)}$. Note the convention used in (d) to represent a delta function.

We merely start with smoother and smoother approximating sequences for $\delta(\cdot)$. The derivatives obtained by the above procedures may be called *generalized derivatives* since the conventional derivatives are not defined.

The family of functions $\{\delta^{(k)}(\cdot),\ k = 0,\ 1,\ \ldots\}$ will be called the family of impulsive functions.

Some Properties. Impulsive functions cannot be treated as ordinary functions; if we did so, we would, for example, have to regard $\delta(f)$ and $\delta(\omega)$, $\omega = 2\pi f$, as the same functions because their values are the same everywhere. They are indeed different; see Exercise 1.3-3. The properties of impulsive functions (like their definitions) have to be set up by considering suitable sequences of approximating pulses for the impulsive functions. Many readers will be familiar with these procedures and with various properties. We shall give only a few examples.

1. *The Sifting Property:* Let

$$p_\epsilon(t) = \frac{1(t) - 1(t - \epsilon)}{\epsilon}$$

If $u(\cdot)$ is a continuous function, then it is clear that

$$\lim_{\epsilon \to 0} \int u(t)p_\epsilon(t)\, dt = u(0)$$

We write this symbolically as

$$\int u(t)\delta(t)\, dt = u(0) \tag{7}$$

If $p_\epsilon^{(1)}(\cdot)$ is the generalized derivative of $p_\epsilon(\cdot)$, we note that, if $u^{(1)}(0)$ exists,

$$\lim_{\epsilon \to 0} \int u(t)p_\epsilon^{(1)}(t)\, dt = \lim_{\epsilon \to 0} \int u(t)\frac{\delta(t) - \delta(t - \epsilon)}{\epsilon}\, dt$$

$$= \lim_{\epsilon \to 0} \frac{u(0) - u(0 + \epsilon)}{\epsilon} = -u^{(1)}(0)$$

Again we can write this relation symbolically as

$$\int u(t)\delta^{(1)}(t)\, dt = -u^{(1)}(0) \tag{8}\dagger$$

Similarly, if $u^{(n)}(0)$ exists, we can obtain

$$\int u(t)\delta^{(n)}(t)\, dt = (-1)^n u^{(n)}(0) \tag{9}$$

†If we set $u(t) = t$, we obtain

$$\int t\delta^{(1)}(t)\, dt = -1$$

which provides additional support for the interpretation of $\delta^{(1)}(\cdot)$ as a doublet (dipole, torque).

and, somewhat more generally, if $u^{(n)}(t_0)$ exists, we can obtain

$$\int u(t)\delta^{(n)}(t - t_0)\, dt = (-1)^n u^{(n)}(t_0) \tag{10}$$

Relation (10) shows how the impulsive functions *pick* out, or *sift* out, the values of $u(\cdot)$ or its derivatives at the point where the impulsive functions are located.

2. *Even and Odd Functions:* It will be convenient to *define* $\delta^{(n)}(\cdot)$ as an even function if n is even and as an odd function if n is odd. Therefore, we shall write

$$\delta(t - t_0) = \delta(t_0 - t) \quad \text{but} \quad \delta^{(1)}(t - t_0) = -\delta^{(1)}(t_0 - t) \tag{11}$$

3. *Convolution†:* It is easy to see (graphically or analytically) that

$$\lim_{\epsilon \to 0} p_\epsilon(\cdot) * u(\cdot) = \lim_{\epsilon \to 0} \int p_\epsilon(\tau) u(\cdot - \tau)\, d\tau = u(\cdot)$$

Again we shall write this relation symbolically as

$$\delta(\cdot) * u(\cdot) = \int \delta(\tau) u(\cdot - \tau)\, d\tau = u(\cdot) \tag{12}$$

Similarly, we can obtain

$$\delta^{(n)}(\cdot - t) * u(\cdot) = u(\cdot - t) \tag{13}$$

Note that there is no $(-1)^n$ term in convolution formula (13); this can be reconciled with the sifting formula (10) by using our definition of even and odd impulsive functions. For example, we have

$$\delta^{(1)} * u = \int u(\tau)\,\delta^{(1)}(\cdot - \tau)\, d\tau = -\int \delta^{(1)}(\tau - \cdot) u(\tau)\, d\tau = -[-u^{(1)}(\cdot)] = u^{(1)}(\cdot)$$

We note also that while sifting is a *local* property, involving only the value $t = t_0$, convolution is a *global* property, involving all values of t. A result of this fact is that for the convolution relation (13), unlike the sifting relation (10), no smoothness restrictions on the function $u(\cdot)$ are necessary. Thus in relation (12), $u(\cdot)$ may be a delta function, so that we have the formula

$$\delta(\cdot) * \delta(\cdot) = \delta(\cdot) \tag{14a}$$

or $u(\cdot)$ may be a derivative of $\delta(\cdot)$, yielding

$$\delta(\cdot) * \delta^{(n)}(\cdot) = \delta^{(n)}(\cdot) \tag{14b}$$

†We assume that the reader is familiar with the convolution of two functions, defined as $a(t) * b(t) = \int_{-\infty}^{\infty} a(\tau) b(t - \tau)\, d\tau = \int_{-\infty}^{\infty} a(t - \tau) b(\tau)\, d\tau$ [see also Eq. (26) below].

We can justify such formulas by returning to the approximating sequences defining $\delta(\cdot)$ and $\delta^{(n)}(\cdot)$.

4. *Laplace Transforms:* We have

$$\mathcal{L}_-[\delta(t)] = 1, \qquad \mathcal{L}_-[\delta^{(k)}(t)] = s^k \tag{15}$$

5. *Multiplication of Impulsive Functions Is Not Meaningful:* While the introduction of impulsive functions simplifies many calculations, we pay a price for these benefits. Thus the product of two impulsive functions cannot be defined in any satisfactory way. For a simple illustration, let us consider trying to define the product $\delta(\cdot) \cdot \delta(\cdot)$ via the sequence of rectangular pulses $p_\epsilon(\cdot)$. We would have

$$\delta(\cdot) \cdot \delta(\cdot) = \lim_{\epsilon \to 0} p_\epsilon(\cdot) \cdot p_\epsilon(\cdot)$$

which tends toward infinity as $\epsilon \to 0$. Attempts via other sequences and via other methods do not fare much better. In this case the idealization of the pulse to a delta function is not permitted, and we have to go through with the tedious exact calculations for every particular pulse shape.

6. *Generalized (or Distributional) Equality and Limits:* To always manipulate impulsive functions in terms of sequences of elementary pulses becomes tedious after a while. The situation is analogous to always manipulating real numbers in terms of sequences of rational numbers. For real numbers, we instead obtain certain defining properties and rules and use them for future manipulations. For impulsive functions, we can do roughly the same, as first shown by Laurent Schwartz in his theory of *generalized functions*, which he called *distributions*. Simple presentations of this theory can be found in [8] and [9]. We shall not need to study this, but it will still be useful to note some of its aspects.

The basic concept of the theory is that two expressions, say $A(\cdot)$ and $B(\cdot)$, that involve impulsive functions are said to be equal in the *generalized (or distributional) sense* if

$$\int A(t)\phi(t) \, dt = \int B(t)\phi(t) \, dt$$

for all $\phi(\cdot)$ such that both integrals have a meaning for all $\phi(\cdot)$ in some specified class. The functions $\phi(\cdot)$ are called *testing functions*.

For example, the equality

$$t\delta^{(1)}(t) = -\delta(t)$$

holds for all testing functions $\phi(\cdot)$ that are differentiable at zero:

$$\int_{-\infty}^{\infty} \phi(t)t\dot{\delta}(t) \, dt = -[\phi(t) + \dot{\phi}(t)t]_{t=0} = -\phi(0) = -\int \phi(t) \, \delta(t) \, dt$$

This notion of equality can also be used with nonimpulsive functions and turns out to be much closer in concept to our physical notions of equality than many of the more conventional mathematical definitions. (Although we shall not pursue it here, the reader may wish to consider the statement

$$\lim_{\omega_0 \to \infty} \cos \omega_0 t = 0$$

which is important in Fourier analysis but is not easy to explain in any but a distributional sense.)

Signal Representations.   Formula (12), which we repeat here,

$$u(t) = \int_{-\infty}^{\infty} u(\tau)\delta(t - \tau) \, d\tau, \qquad -\infty < t < \infty \tag{16}$$

can be interpreted as a resolution of the function $u(\cdot)$ into components $u(\tau)\delta(\cdot - \tau)$. To better understand this, note that we can make a stepwise approximation to $u(\cdot)$,

$$u(\cdot) \simeq \sum_n u(n\Delta)p_\Delta(\cdot - n\Delta)\Delta \tag{17}$$

where

$$p_\Delta(t - n\Delta) = \begin{cases} \dfrac{1}{\Delta}, & n\Delta \le t < (n+1)\Delta \\ 0, & \text{elsewhere} \end{cases}$$

Now if we set $\tau = n\Delta$ and take $\Delta \to 0$, $n \to \infty$, then $p_\Delta(\cdot - n\Delta) \to \delta(\cdot - \tau)$, and we obtain formula (16). A similar approach can be used to express $u(\cdot)$ in terms of other functions, e.g., step functions.

A more powerful method is the following. Since by (16) we know how to write $u(\cdot)$ as a combination of delta functions, we can represent $u(\cdot)$ as a combination of other functions, provided we should first find how to write delta functions in terms of these other functions. We then just substitute into (16). An important case is representation in terms of exponential functions. It is known that we can write

$$\delta(t - \tau) = \int_{-\infty}^{\infty} \exp j2\pi f(t - \tau) \, df \tag{18}$$

by which we mean, for example, that

$$\lim_{W \to \infty} \int_{-W}^{W} \exp j2\pi f(t - \tau) \, df = \lim_{W \to \infty} \frac{\sin 2\pi w(t - \tau)}{\pi(t - \tau)}$$

$$= \delta(t - \tau)$$

Therefore we can write

$$u(t) = \int_{-\infty}^{\infty} u(\tau)\delta(t - \tau) \, d\tau$$

$$= \int_{-\infty}^{\infty} e^{j2\pi ft}U(f) \, df \tag{19a}$$

where

$$U(f) \triangleq \int_{-\infty}^{\infty} u(\tau)e^{-j2\pi ft}\, d\tau \qquad (19b)$$

Equations (19a) and (19b) define a *Fourier transform pair* whenever the integrals "exist." As explained before, this is a purposely vague term, since it depends on our notions of convergence and equality.

Other decompositions of $\delta(t - \tau)$ give us other "transform pairs" (with the same qualifications on existence). Thus from

$$\delta(t - \tau) = \frac{1}{2\pi j} \int_{c-j\infty}^{c+j\infty} e^{s(t-\tau)} d\tau \qquad (20)$$

we get *Laplace transforms*. From

$$\delta(t) = \frac{1}{\pi t} * \frac{-1}{\pi t} \qquad (21)$$

we get *Hilbert transforms*. From

$$\delta_{nm} = \oint z^n z^{-m} \frac{1}{z} \frac{dz}{2\pi j} \qquad (22)$$

we get the *zee transform* pair,

$$u(n) = \sum u(m)\delta_{nm} = \oint z^n U(z) \frac{1}{z} \frac{dz}{2\pi j} \qquad (23a)$$

where

$$U(z) = \sum u(m)z^{-m} \qquad (23b)$$

And so on. These different representations have different domains of usefulness, as is discussed in books on transform theory. We shall only need some knowledge of Laplace transforms (as discussed in Sec. 1.2) and zee transforms (to be further discussed in Sec. 2.3.3 and Exercise 2.3-25).

Input-Output Relations. A major reason for studying the representation of arbitrary functions as combinations of "simpler" functions is to facilitate the analysis of the response of a linear system to an arbitrary input. Most readers will already be familiar with this concept and its applications, and therefore we shall give only an informal review, which can be omitted without loss of continuity.

Suppose that

$$h(t, \tau) = \text{the response of a linear system at time } t \text{ to a unit}$$
$$\text{delta function input at time } \tau, \ \delta(t - \tau)$$

Then

$$h(t, n\Delta)u(n\Delta)\Delta = \text{the response to } u(n\Delta)\Delta\delta(t - n\Delta)$$
$$\sum_n h(t, n\Delta)u(n\Delta)\Delta = \text{the response to } \sum_n u(n\Delta)\delta(t - n\Delta)\Delta$$

and in the limit as $n \longrightarrow \infty$, $\Delta \longrightarrow 0$, $n\Delta = \tau$, we can write

$$\int_{-\infty}^{\infty} h(t, \tau)u(\tau)\, d\tau = \text{the response at time } t \text{ to an input } u(\cdot) \qquad (24)$$

The function $h(\cdot, \cdot)$ is called the *impulse response* of the system, and the integral in (24) the *superposition integral*. Its significance is that we can often determine the impulse response fairly accurately by simple measurements. However, the fact that the impulse response is a function of two variables, the input time and the response time, makes for some complication. This is reduced for *time-invariant* systems, which have the defining property

$$h(t, \tau) = h(t + \alpha, \tau + \alpha) \qquad \text{for all } \alpha \qquad (25)$$

In other words, a system is time invariant if *delaying the input* by α seconds merely *delays the response* by α seconds. A consequence of this fact is that the response at any time t to an impulse input at time τ depends only on the *elapsed* time, $t - \tau$, rather than both $t - \tau$ and τ (or t and τ) as for time-variant systems. Mathematically, we see that by choosing $\alpha = -\tau$ in (25) we find

$$h(t, \tau) = h(t - \tau, 0)$$

Therefore the impulse response for a time-invariant system is actually a function of *one* variable rather than of *two*. By what is called "an abuse of notation," we shall write

$$h(t - \tau, 0) = h(t - \tau)$$

for convenience. Note that now

$$h(t) = \text{the response at time } t \text{ to a unit impulse at time } 0$$

Not only does the assumption of time invariance simplify the problem of measuring the impulse response, but the superposition integral (24) can be written as a *convolution integral*,

$$y(t) = \int_{-\infty}^{\infty} h(t - \tau)u(\tau)\, d\tau \qquad (26a)$$

or, by a change of variables, as

$$y(t) = \int_{-\infty}^{\infty} h(\tau)u(t - \tau)\, d\tau \qquad (26b)$$

We may note that the integral in (26b) can also be interpreted as

$$y(t) = \sum h(n\Delta)u(t - n\Delta)\Delta$$

which represents the response at time t as a *weighted* combination of inputs $u(t - n\Delta)\Delta$, with weights $h(n\Delta)$. Therefore the *impulse response* function $h(\cdot)$ is often also called the *weighting function*.

Suppose now that

$$u(t) = 0, \qquad t < 0 \tag{27}$$

Then we shall have

$$y(t) = \int_0^\infty h(t - \tau)u(\tau)\, d\tau = \int_{-\infty}^t h(\tau)u(t - \tau)\, d\tau$$

If the linear time-invariant system is *causal*, then there should be no response before an input is applied. In particular, we should have

$$h(t) = 0, \qquad t < 0$$

Therefore, for a causal time-invariant system with inputs obeying (27) we can write

$$y(t) = \int_0^t h(t - \tau)u(\tau)\, d\tau, \qquad t \geq 0 \tag{28a}$$

$$= \int_0^t h(\tau)u(t - \tau)\, d\tau, \qquad t \geq 0 \tag{28b}$$

The second integral in (28) expresses the fact that the weighting function of a causal system gives *zero* weight to *future* inputs. (The term *future* is actually also a matter of definition. We have used the "usual" definitions here, and they will suffice for our purposes. But it is possible to set up conventions where weighting functions of "causal" systems are not zero for negative arguments. We refer the curious reader to [2] for examples and further discussion.)

We also note that even though only causal systems can be implemented in real time, noncausal systems are still of interest. For one thing, the independent variable may not always be time, as, for example, in optics. Second, even when time is the independent variable, noncausal systems are useful in providing benchmarks of performance; the ideal band-pass filter is a classic example. Furthermore, as the reader may know, noncausal systems may often be satisfactorily approximated by causal systems plus a delay. Finally, noncausal systems may arise in the course of analysis, when decomposing or recombining causal systems.

Of course there is no essential reason for restricting ourselves to impulsive inputs; depending on the circumstances, it may be preferable to use steps, ramps, exponentials, Bessel functions, etc. The underlying principles are the same.

Transfer Functions.   Thus, suppose we consider exponential inputs

$$e^{st}, \qquad -\infty < t < \infty, \qquad s \text{ complex}$$

The response of a system with impulse response $h(t, \tau)$ to such an input will be

$$\int_{-\infty}^\infty h(t, \tau)e^{s\tau}\, d\tau = e^{st} \int_{-\infty}^\infty h(t, \tau)e^{-s(t-\tau)}\, d\tau$$

$$= e^{st}H(s, t), \text{ say}$$

We can work with this function, but it is a very important fact that for time-invariant systems $H(s, t)$ is independent of t, since

$$H(s, t) = \int_{-\infty}^{\infty} h(t - \tau)e^{-s(t-\tau)}\, d\tau = \int_{-\infty}^{\infty} h(\tau)e^{-s\tau}\, d\tau = H(s), \text{ say}$$

This is of course just the familiar fact that the response of a linear time-invariant system to an exponential input is also exponential, but possibly with a different amplitude and phase. The function $H(s)$ is called the *transfer function* of the system. Note that

$$H(s) = \frac{\text{response to } e^{st}}{e^{st}}$$

Since

$$u(t) = \int_{c-j\infty}^{c+j\infty} U(s)e^{st}\, \frac{ds}{2\pi j}$$

the response $y(\cdot)$ to a general input $u(\cdot)$ can be written as

$$y(t) = \int_{c-j\infty}^{c+j\infty} Y(s)e^{st}\, \frac{ds}{2\pi j}$$

where

$$Y(s) = H(s)U(s)$$

This is the familiar Laplace transform formula showing that convolution in the time domain is replaced by multiplication in the frequency domain.

However, many readers are already aware that the above discussion has to be qualified in various ways. For a given system, it is not generally true that the response to e^{st} is of the same form for all s—there may be regions of s (in the complex plane) for which the response is infinite. For example, consider a system with

$$h(t) = e^{-\alpha t} \cdot 1(t)$$

Then

$$H(s) = \int_{0}^{\infty} e^{-\alpha t}e^{-st}\, dt$$

$$= \begin{cases} \dfrac{1}{s + \alpha} & \text{for Re } s > -\alpha \\[2mm] \infty & \text{for Re } s < -\alpha \end{cases}$$

Therefore, in general, a *region of definition* has to be specified along with a transfer function $H(s)$. Good discussions of this problem can be found in Siebert [10] and Lathi [11] and in several other texts. In this book, we need not worry about this point because we shall restrict ourselves (with rare exceptions) to causal systems. In this case, some *right half plane* is always a region of definition, and this need not be explicitly specified in the "usual" calculations with unilateral Laplace transforms.

Exercises

1.3-1.

Consider an *RL* circuit as shown. Its differential equation is $\dot{y}(t) + \alpha y(t)$ $= u(t)$, $t \geq 0-$, where $u(\cdot)$ is as shown. We are given $y(0) = y_0$.

a. Find the response of the circuit to $u(\cdot)$, and show that by suitable choice of a we can make $y(t) = 0$, $t > \epsilon$. Can ϵ be made arbitrarily small?

b. What happens as $\epsilon \longrightarrow 0$ with $a = b/\epsilon$?

c. Compute directly the response to $u(\cdot) = b\delta(\cdot)$, and compare to the solution in part a.

d. What happens if $\epsilon \longrightarrow 0$ for fixed a?

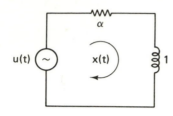   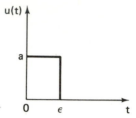

1.3-2.

a. Calculate the response, $y(\cdot)$, of a series *RL* circuit ($R = 1$, $L = 1$) to voltage inputs of the forms shown in the figure. Assume that the initial current $y(0)$ in the circuit is zero.

b. Find the limits of the solutions as $\epsilon \longrightarrow 0$. What do you observe?

c. Repeat part a but with $y(0) = 1$.

d. Use the concepts of linearity and your knowledge of the response to a delta function to obtain the responses in parts a and c to inputs (a) and (b) in the figure.

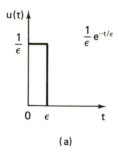   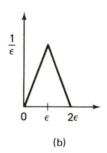   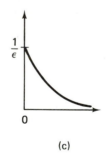

(a)   (b)   (c)

1.3-3.

Show by working with sequences that

a. $f(t)\delta(t) = f(0)\delta(t)$,

b. $f(t)\delta^{(1)}(t) = f(0)\delta^{(1)}(t) - f^{(1)}(0)\delta(t)$,

c. $t^3\delta^{(2)}(t) = 0$,

d. $\delta(at + b) = \delta(t + ba^{-1})/|a|$.

Also verify these equalities by using test functions.

1.3-4.   An Interesting Fourier Series

a. Show that

$$\sum_{-\infty}^{\infty} \delta(t - n) = \sum_{-\infty}^{\infty} e^{j2\pi nt} = 1 + 2 \sum_{1}^{\infty} \cos 2\pi nt$$

b. Show that

$$\sum_{-\infty}^{\infty} \delta^{(1)}(t - n) = -\sum_{1}^{\infty} 4\pi n \sin 2\pi nt$$

Lighthill [9, p. 68] points out that this series causes a lot of trouble in the classical theories of Fourier series. The classical "summability" methods give zero for the sum when $t \neq n$ (which is correct) but then also give zero when $t = n$, on the grounds that every term in the series vanishes at $t = n$. This approach leads to the unpleasant conclusion that there is no unique Fourier series expansion, because there exist series whose "sum" is zero everywhere.

1.3-5.

Let $u(t) = 1$, $t < 0$, $u(t) = e^t$, $t \geq 0$, and let $v(t) = \exp - 3|t|$, $-\infty < t < \infty$. Show that the convolution of $u(\cdot)$ and $v(\cdot)$ is $w(t) = \dfrac{\exp - 3t}{12} + (3/4) \exp t$ for $t \geq 0$ and $w(t) = (\exp 3t)/6 + \frac{2}{3}$ for $t < 0$.

1.3-6.

A voltage $e^{-3t} \cdot 1(t + T)$, $T > 0$, drives a series RC circuit ($R = 1 = C$). Show that the resulting current becomes infinite as $T \rightarrow \infty$. Why is the current not proportional to e^{-3t}?

1.3-7.

Evaluate the expression $1(t) * \delta^{(1)}(t) * 1(-t)$ as

a. $[1(t) * \delta^{(1)}(t)] * 1(-t)$,
b. $1(t) * [\delta^{(1)}(t) * 1(-t)]$,
c. $[1(t) * 1(-t)] * \delta^{(1)}(t)$.

What do you learn from this calculation?

1.4   SOME REMARKS ON THE USE OF MATRICES

It will soon become clear that matrix notation will be much used in this book, even if for nothing more than the compactness it affords. However, we shall actually find matrix theory much more useful than just as a notational device. Consequently, various results from matrix theory will be needed from time to time. Despite the fact that almost all students at this stage have had some exposure to matrix theory, it seems that the pressure of the constant use of matrices demanded in this book tends to make people confused about even simple things. The few remarks below are meant not so much to teach matrix theory as to provide a rather informal guide to its use. In the Appendix we have tried to collect all the matrix formulas and results that we shall need in the first part of this book. In the second part we shall need some properties of polynomial and rational matrices, and these will be developed in Secs. 6.3 and 6.5. We shall not explicitly use any results from linear algebra. However, this subject is especially useful in providing an overview and a geometrical picture of various results and calculations, and therefore we shall make a brief survey of some relevant parts in Sec. 5.2, showing how they illuminate some earlier calculations. We should also note that problems of actual numerical computation—conditioning, accuracy, and stability—are not treated; that would be a course in itself.

Perhaps the reason people have occasional difficulty in going from operations with numbers to operations with matrices is that some rules go through to matrices while others do not, and certain rules are applicable only to some matrices but not to all. Therefore, it is very important in this book, where matrices are used a lot, to try to keep the following questions in mind until you find they are superfluous:

1. What does this symbol represent? (For example, what are its dimensions? Is it a column vector, a row vector, or an $n \times m$ matrix? Is it a square matrix? Is it singular or not? And so on.)

2. Is the operation that we are going to do with this matrix permitted? Here are some basic rules and facts to help you:

1. Matrix multiplication is allowed only when the number of columns of the first matrix is equal to the number of rows of the second matrix: ($m \times n$ matrix) $\times$ ($n \times p$ matrix) = $m \times p$ matrix.

2. Given an $m \times n$ matrix A, only $n \times m$ matrices can multiply A *both* from the left and the right, and the result is an $n \times n$ or an $m \times m$ matrix, respectively; therefore, clearly $AB \neq BA$ (where B is $n \times m$).

3. Even if A is $n \times n$ and B is $n \times n$, so that both AB and BA are $n \times n$, still $AB \neq BA$ in general. One must always preserve the order in matrix operations.

4. If A is $n \times n$, A is said to be *singular* if det $A = 0$ and *nonsingular* if det $A \neq 0$.

5. If A is *nonsingular*, there exists a *matrix inverse* A^{-1} such that $AA^{-1} = A^{-1}A = I$, the identity matrix. For square $A, B, (AB)^{-1} = B^{-1}A^{-1}$. (Prove this!)

6. If A is *nonsingular* (note that when we say nonsingular we implicitly also say square) and if $AB = AC$ (B and C square or not), then by multiplying the last relation by A^{-1} *from the left*, we get $A^{-1}AB = A^{-1}AC$, so that $B = C$ (note that when you multiply, you must specify from which side and keep the order). Similarly, $BA = CA$ and A nonsingular implies $B = C$. Cases that do *not* imply $B = C$: $AB = AC$ and A is singular; $AB = AC$ and A is not square; $AB = CA$. Never *cancel* a matrix: Always multiply by its inverse, if this exists!

7. The *transpose* of a matrix $A = [a_{ij}]$ is a matrix A' such that $[A']_{ij} = [a_{ji}]$. That is, A' is obtained from A by interchanging rows and columns. We have the properties $(A + B)' = A' + B'$, $(AB)' = B'A'$.

8. *Transposition and inverses*: We shall write A^{-T} to denote the transpose of A^{-1}. We have the following properties: $(A')^{-1} = (A^{-1})' = A^{-T}$ and $(AB)^{-T} = A^{-T}B^{-T}$.

9. *Pre- and postmultiplication by rows and columns*: Note that premultiplying a matrix A by a row vector $[\alpha_1, \alpha_2, \ldots, \alpha_n]$ gives us a new row vector that is a weighted combination, weighted by the $\{\alpha_i\}$, of the rows of the original matrix. Similarly, postmultiplying by a column gives a weighted combination of the columns of the original matrix.

10. As should already be clear from the above, *we shall not use any special device* (e.g., underlining or boldface) *to distinguish matrices from scalars*; in cases of doubt, the reader can use context and the rules of compatibility to clarify the situation.

Also please reread the remarks in the introduction about there being no substitute for individual work in familiarizing yourself with matrix notation and matrix manipulation. When confused about matrix operations or when trying to "guess" or to "see how to prove" matrix results, try things out with 2×2 and/or diagonal matrices.

It will be a useful review of what we shall need of matrix theory to read *quickly* through the Appendix at this point, with initial emphasis on Exercises A.1, A.10–A.19, A.27–A.31, A.34–A.37, and A.42–A.50. A more detailed study of specific results can be undertaken when they are cited in later sections.

REFERENCES

1. G. H. HARDY, *Divergent Series*, Oxford University Press, London, 1949.

2. T. KAILATH and D. L. DUTTWEILER, "An RKHS Approach to Detection and Estimation Problems, Pt. III: Generalized Innovations Representations and a Likelihood-Ratio Formula," *IEEE Trans. Inf. Theory*, vol. IT-18, no. 6, pp. 730–745, November 1972.

3. E. KAMKE, *Theory of Sets*, Dover, New York, 1950.

4. H. N. SHAPIRO, "A Micronote on a Linear Functional Equation," *Am. Math. Mon.*, vol. 80, no. 9, p. 1041, November 1973.

5. J. R. CARSON, *Electric Circuit Theory and Operational Calculus*, Chelsea Publishing Co., New York, 1926 (2nd ed., 1953).

6. L. A. ZADEH and C. DESOER, *Linear Systems—A State-Space Approach*, McGraw-Hill, New York, 1963.

7.   G. DOETSCH, *Introduction to the Theory and Application of the Laplace Trans-formation*, Springer-Verlag, New York, Inc., New York, 1974.

8.   L. SCHWARTZ, *Methods of Mathematical Physics*, Addison-Wesley, Reading, Mass., 1964.

9.   M. J. LIGHTHILL, *Fourier Analysis and Generalized Functions*, Cambridge University Press, London, 1959.

10. W. M. SIEBERT, "Signals in Linear Time-Invariant Systems," in *Lectures in Communication System Theory* (E. J. Baghdady, ed.), McGraw-Hill, New York, 1961.

11. B. P. LATHI, *Signals, Systems and Communication*, Wiley, New York, 1965.

STATE-SPACE DESCRIPTIONS— 2
SOME BASIC CONCEPTS

2.0 INTRODUCTION AND OUTLINE

This chapter will introduce several fundamental notions and definitions. In a sense, most of the rest of the book consists of applications of the basic concepts of this rather long chapter.

An introduction to the major themes of this chapter is provided by considering a particular way of stabilizing a given system. This will be followed by an outline of the rest of the chapter.

Analysis of Stabilization by Pole Cancellation. For purposes of illustration we shall set up an extremely simple problem, which will help to highlight the main issues. The reader will recognize that in practice the same difficulties may arise in a less transparent fashion.

Consider a system with the transfer function

$$H_f(s) = \frac{1}{s - 1}$$

which is unstable† because of the pole in the right-half s plane. To stabilize it, we can precede $H_f(s)$ with a *compensator* (Fig. 2.0-1),

---

†We might ask why anyone would produce such a system. The point is that systems are designed under various constraints, of which stability may not initially be the most important one. In such cases, *compensation* is used to maintain *small-signal stability* about some desired *unstable operating point*.

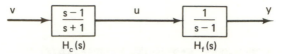

Figure 2.0-1.   Series or cascade compensation.

$$H_c(s) = \frac{s-1}{s+1}$$

to get an overall transfer function,

$$H_f(s)H_c(s) = \frac{1}{s-1}\frac{s-1}{s+1}$$

This is a nice outcome, but unfortunately this technique will not work: After a while the system will tend to *burn out* or *saturate*. To see why, let us first set up an *analog-computer simulation* of the cascaded system, for example, the one shown in Fig. 2.0-2. The reader can readily verify that these simulations

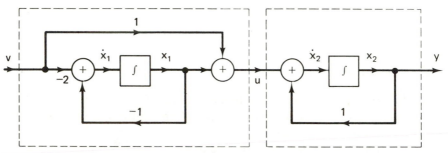

Figure 2.0-2.   Analog-computer simulation of the cascade system of Fig. 2.0-1.

have the desired transfer functions. To analyze the behavior of this realization, it is natural to consider the evolution of the major variables in the realization, which are clearly the integrator outputs, say $x_1(\cdot)$ and $x_2(\cdot)$. From Fig. 2.0.2, we can write the equations,

$$\begin{bmatrix} \dot{x}_1 \\ \dot{x}_2 \end{bmatrix} = \begin{bmatrix} -1 & 0 \\ 1 & 1 \end{bmatrix}\begin{bmatrix} x_1 \\ x_2 \end{bmatrix} + \begin{bmatrix} -2 \\ 1 \end{bmatrix}v, \qquad \begin{bmatrix} x_1(0) \\ x_2(0) \end{bmatrix} = \begin{bmatrix} x_{10} \\ x_{20} \end{bmatrix}$$

and

$$y = [0 \quad 1][x_1 \quad x_2]'$$

There are general methods of solving such so-called *state-space* equations (cf. Sec. 2.5), but here it will suffice to proceed as follows: The first equation is

$$\dot{x}_1 = -x_1 - 2v, \qquad x_1(0) = x_{10}$$

which yields $x_1(t) = e^{-t}x_{10} - 2e^{-t} * v$ (* denotes convolution). The second equation can then be solved by using Laplace transforms to give

$$Y(s) = X_2(s) = \frac{x_{20}}{s-1} + \frac{x_{10}}{(s-1)(s+1)} + \frac{V(s)}{s+1}$$

or

$$y(t) = x_2(t) = e^t x_{20} + \tfrac{1}{2}(e^t - e^{-t})x_{10} + e^{-t} * v$$

Therefore the overall transfer function, *which has to be calculated with zero initial conditions*, is $1/(s+1)$ as expected. Note, however, that unless the initial conditions can always be kept zero, $y(\cdot)$ will grow without bound. Because of stray voltages, it will be difficult to keep $x_{10} = 0 = x_{20}$, and therefore the above method of stabilizing the system is unsatisfactory. A reason often given is that the method relies on exact cancellation, which is difficult to ensure because of variations in component values, etc. However, the difficulty goes much deeper than this. The fact is that there is still a problem even with *perfect* cancellation, because almost all initial conditions (all except those for which $x_{10} = -2x_{20}$) will *excite* the unstable *mode of oscillation*.†

Some further insight into the difficulty is obtained by studying the situation (Fig. 2.0-3) in which $H_c(s)$ follows $H_f(s)$ rather than preceding it as in Fig. 2.0-1. From a transfer function point of view, this is completely equivalent to the configuration of Fig. 2.0-1. However, let us study a simulation of this rearrangement, as shown in Fig. 2.0-4. The state equations are easily

Figure 2.0-3. Compensator following the given system.

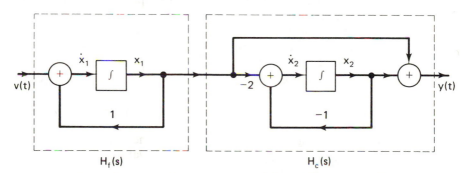

Figure 2.0-4. Analog-computer simulation of the system in Fig. 2.0-3.

---

†We shall use the term *mode of oscillation* (or often just *mode*) in a somewhat broad sense, relying on the reader's intuitive feeling for such a concept. More formal definitions can be given (cf. Sec. 2.5.2), but they are not really needed here.

seen to be

$$\begin{bmatrix} \dot{x}_1 \\ \dot{x}_2 \end{bmatrix} = \begin{bmatrix} 1 & 0 \\ -2 & -1 \end{bmatrix} \begin{bmatrix} x_1 \\ x_2 \end{bmatrix} + \begin{bmatrix} 1 \\ 0 \end{bmatrix} v$$

$$y = [1 \quad 1][x_1 \quad x_2]'$$

and we find that

$$y(t) = (x_{10} + x_{20})e^{-t} + e^{-t} * v(t)$$

Now the system is stable as far as $y(\cdot)$ goes, even if the initial conditions are nonzero. However, the realization in Fig. 2.0-2 is still internally unstable because $x_1(\cdot)$ and $x_2(\cdot)$ will have terms in them that will grow as e^t, and thus after a while the realization will saturate or burn out.

These examples raise many questions, but the main conclusion to be drawn here is that the *internal behavior* of a realization may be more complicated than is indicated by the *external behavior*. The internal behavior is determined by the *natural frequencies* of the (undriven) realization, which in our case are $s = +1, -1$. However, because of cancellation, not all the corresponding modes of oscillation will appear in the overall transfer function. Or, to put it another way, since the transfer function is defined under zero initial conditions, it may not display all the modes of the actual realization of the system. For a complete analysis, we shall need to have good ways of keeping track of all the modes, those explicitly displayed by the transfer function and also the "hidden" ones. It is possible to do this by careful bookkeeping with the transfer function calculations, but actually it was state-equation analysis as in the above examples that first clarified these and related questions. For example, it would be nice to have a simple explanation of the difference in behavior of the simulations in Figs. 2.0-2 and 2.0-4. In a sense to be made precise in Sec. 2.4, the explanation is that the unstable mode e^t in Fig. 2.0-2 is *observable* [it appears in the output], but it is not *controllable* [the external input $v(\cdot)$ cannot directly affect it], while in Fig. 2.0-4 the mode e^t is controllable but not observable. In Sec. 2.4 (Example 2.4-4) we shall also see how this difference in behavior can be predicted without any explicit calculations.

Before entering into a detailed study of these points, it seems reasonable to examine some methods of setting up simulations (analog-computer realizations) of general transfer functions (or of the underlying differential equation representations). Apart from its direct utility in simulation, the study of this problem will provide a natural introduction to the use of state-space descriptions and will highlight their manifold nonuniqueness, as we shall see in Secs. 2.1 and 2.2. This fact has important consequences for the analysis of systems, and we shall see our first examples of this in Sec. 2.3, where we shall introduce the important concepts of controllability and observability of state-space realizations, first for continuous-time and then for discrete-time systems. In Sec. 2.4, we shall then return better equipped for a deeper study of the differences between internal and external descriptions. In Sec. 2.5 we shall for the first time discuss (briefly) the explicit solution of state-space equations. The final section will provide a glimpse of stability theory.

This is a long chapter, but it is a basic one and well worth an investment of time. The reader will find that further studies in system theory, as, for example, in Chapters 6 and 8 of this book, will be much aided thereby.

2.1 SOME CANONICAL REALIZATIONS

Several important realizations of a given lumped dynamical system can be nicely introduced by considering how to simulate a given continuous-time transfer function on an analog computer or a discrete-time transfer function by a digital filter (without amplitude quantization). The two problems are very similar, but for various reasons we start with the continuous-time case; the discussion of the discrete-time case in Sec. 2.3.3 will then serve as a useful review, besides introducing some new insights. Many examples of the value of having concrete realizations in mind will appear throughout this book.

2.1.1 Some Remarks on Analog Computers

To see what might be done to simulate a continuous-time system, consider a system described by the differential equation

$$\ddot{y} + e^t y = t^2$$

If we could get a good differentiator, our first attempt at analog simulation might look like the circuit in Fig. 2.1-1. However, even ignoring questions relating to the availability of the functions t^2 and $\exp \pm t$, a fundamental weakness of this scheme is that in practice all signals are corrupted by noise. When such a signal is differentiated, the derivative of the usually rapidly varying *noise* will "drown out" the derivative of the signal.

For systems described by a general differential equation of the form

$$\frac{d^n y}{dt^n} = F\left(\frac{d^{n-1} y}{dt^{n-1}}, \ldots, \frac{dy}{dt}, y, u, t\right)$$

William Thomson, Lord Kelvin, made the important suggestion [1] that an integrating device be used as the basic building block for a simulation, to be obtained as follows: Assuming that $d^n y/dt^n$ is somehow directly available, integrate this signal successively n times, using integrating devices. The outputs of the integrators are the variables $\{d^{n-1} y/dt^{n-1}, \ldots, dy/dt, y\}$, which are used as the inputs to a device designed to deliver as output the quantity

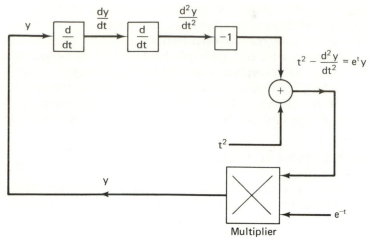

Figure 2.1-1.   Unrealistic implementation.

$F(d^{n-1}y/dt^{n-1}, \ldots, dy/dt, y, u, t)$. Finally we observe that this quantity is equal to $d^n y/dt^n$, so that we can close the loop.

These steps are summarized in Fig. 2.1-2, where it is also shown that each integrator must be provided with the initial value of its output. These initial values, which have to be given or calculated, can be introduced into the

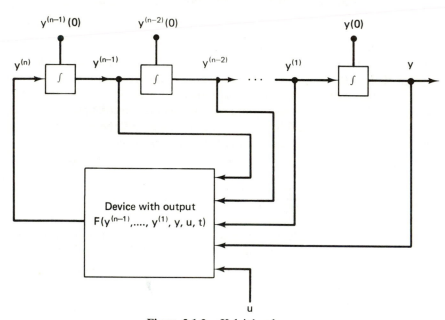

Figure 2.1-2.   Kelvin's scheme.

circuit in a variety of ways [2]–[4]. We postpone to Sec. 2.3 certain important questions connected with initial conditions; for the rest of this section we shall assume *zero initial conditions*.

To implement these ideas we need integrators, function generators, adders, and multipliers. Kelvin's scheme spurred a search for devices and techniques to achieve these desired operations with adequate accuracy. The lack of these delayed effective exploitation of his ideas until the 1930s, when Vannevar Bush built a highly precise electromechanical *differential analyzer*. As is to be expected, the differential analyzers or analog computers of today are purely electronic, with the development of cheap *op amps* (operational amplifiers) being a major technological milestone. The simulations can also be carried out (approximately) using digital circuitry—data registers, memory devices, and arithmetic units; we say approximately because of the error incurred when representing coefficients and signals by a finite number of bits. However, we shall not pursue these matters here (see [5]–[8]) but shall discuss certain analog simulations in terms of standard block diagram representations. Our goal is more to develop a tool for understanding systems than it is to learn about system simulation.

2.1.2   Four Canonical Realizations

We have to go beyond Kelvin's method when we have linear differential equations that contain derivatives of the forcing function, e.g., the equation

$$\ddot{y} + 5y = u + 2\dot{u}$$

If we try to apply Kelvin's technique, we meet the difficulty that

$$\ddot{y} = F(\dot{y}, y, u, t) = -5y + u + 2\dot{u}$$

cannot be generated from just y and u without using a differentiator. Several different methods have been proposed to get around this problem, some of which we shall describe here. Basically, we shall present four different "special" or *canonical* simulations (realizations) called the controller, controllability, observer, and observability canonical forms. Some other canonical forms will also be given in later sections (especially Secs. 2.1.3, 2.3.3, and 2.4.2), and it must be emphasized that there is an infinite number of different realizations. However, we shall soon describe applications (Secs. 2.3, 3.2, and 4.1) in which, just as with coordinate systems in electromagnetic theory, one or another of these canonical forms is most useful or natural.

We should stress that the methods of this section are chosen for their motivational content; to actually obtain these realizations in any given problem the matrix formulas given in Sec. 2.2, Eqs. (2.2-2)–(2.2-11), will provide the most direct route. Therefore, *one should go quickly through this section on a first reading*—the different names and realizations will be better assimilated as one progresses through Sec. 2.2 and especially Sec. 2.3.

A First Approach: Using Superposition. For linear systems one can exploit the idea of superposition. Consider the third-order system (the higher-order case being more difficult only in notation)

$$\dddot{y} + a_1\ddot{y} + a_2\dot{y} + a_3y = b_3u + b_2\dot{u} + b_1\ddot{u} \tag{1}$$

First, we examine the related system

$$\dddot{\xi} + a_1\ddot{\xi} + a_2\dot{\xi} + a_3\xi = u \tag{2}$$

Then, by linearity,

$$y = b_3\xi + b_2\dot{\xi} + b_1\ddot{\xi} \tag{3}$$

However, an important assumption is implicit here—namely that the initial conditions for (1) and (2) are all zero. Otherwise, superposition will not apply, as can be readily checked (also see Exercises 1.1-5 to 1.1-7). The problem of nonzero initial conditions will be treated in Sec. 2.3, though many readers might wish to try to do this on their own for the simple examples treated in this section.

In any case, returning to our problem, we can now proceed as follows. First simulate (2) using Kelvin's method. This gives us a system with input u and output ξ. Next, use differentiators to obtain y from ξ according to Eq. (3). This procedure results in the configuration shown in Fig. 2.1-3.

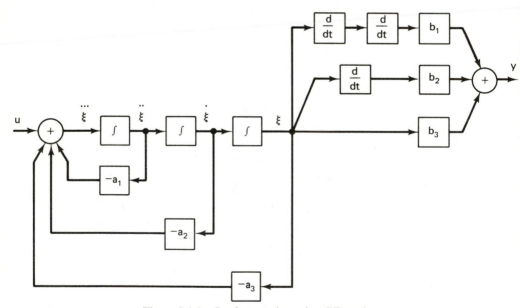

Figure 2.1-3. Implementation using differentiators.

Now we employ a simple trick to eliminate the differentiators by noticing that we have in effect integrators and differentiators in series. Hence, we simply move the lines with differentiators over the requisite number of integrators, thereby eliminating the need for differentiation. Thus we get the completed simulation as shown in Fig. 2.1-4. (Alternatively, just observe that

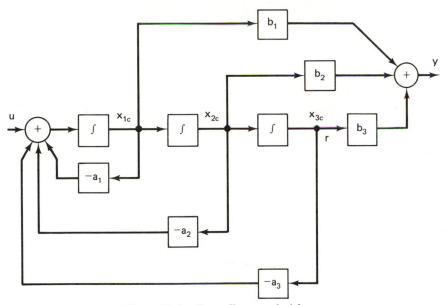

Figure 2.1-4.   Controller canonical form.

if ξ appears at the output of the last integrator, then the input to this integrator must be $\dot{\xi}$ and the input to the preceding integrator must be $\ddot{\xi}$. From them we just form the desired combination $b_3\xi + b_2\dot{\xi} + b_1\ddot{\xi}$.)

For reasons to be discussed later, this simulation or realization will be called the *controller canonical form* or *realization*. [We may note that there is no agreed convention in the literature as to this name or the others that we shall introduce presently. Therefore, at the moment, the reader will just have to remember our designations, though as he proceeds certain cues provided by some applications in which each of these forms is the most "natural" will help in properly identifying the various forms (see Secs. 2.3, 3.2, 3.5, and 4.1).]

A Different Realization.   Let us first look at the above procedure in a slightly different way. Recalling our assumption of *zero initial conditions* we can take Laplace transforms in (1) to get:

$$(s^3 + a_1 s^2 + a_2 s + a_3) Y(s) = (b_1 s^2 + b_2 s + b_3) U(s)$$

or

$$Y(s) = \frac{b_1 s^2 + b_2 s + b_3}{s^3 + a_1 s^2 + a_2 s + a_3} U(s) = \frac{b(s)}{a(s)} U(s) \tag{4}$$

Now what we did above was effectively to write

$$Y(s) = b(s)a^{-1}(s)U(s) = b(s)\xi(s), \text{ say,} \tag{5}$$

and first implement the simpler equation

$$a(s)\xi(s) = U(s)$$

On the other hand, it is very natural to investigate what happens if we write things in the reverse order,

$$Y(s) = a^{-1}(s)b(s)U(s) \tag{6}$$

This amounts to defining $m(\cdot) = b_3 u(\cdot) + b_2 \dot{u}(\cdot) + b_1 \ddot{u}(\cdot)$ and then implementing the equation

$$a(s)Y(s) = M(s) \tag{7}$$

which will give us the diagram shown in Fig. 2.1-5. Now we try to eliminate the differentiators by moving the input lines over the integrators. As a first step we obtain Fig. 2.1-6. At this point the going gets rougher, the main

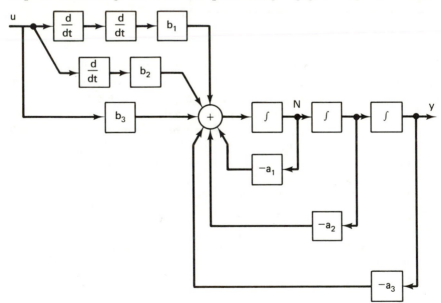

Figure 2.1-5. Intermediate step in realizing (5).

difficulty being the fact that if we move a line over the node marked N in
Fig. 2.1-6, then the quantities in the feedback lines are affected. However, by
some careful bookkeeping, it is not hard to show that we can obtain the form
shown in Fig. 2.1-7, where

$$\beta_1 = b_1$$
$$\beta_2 = b_2 - a_1 b_1 \tag{8a}$$
$$\beta_3 = b_3 - a_1 b_2 - a_2 b_1 + a_1^2 b_1$$

This form is called the *observability canonical form*, for reasons that will
become clearer in Sec. 2.3. Formula (8a) for the coefficients is somewhat
forbidding, especially when we consider general transfer functions. However,
more compact expressions can be given: For example, the reader can check
that

$$\begin{bmatrix} \beta_1 \\ \beta_2 \\ \beta_3 \end{bmatrix} = \begin{bmatrix} 1 & 0 & 0 \\ a_1 & 1 & 0 \\ a_2 & a_1 & 1 \end{bmatrix}^{-1} \begin{bmatrix} b_1 \\ b_2 \\ b_3 \end{bmatrix} \tag{8b}$$

Somewhat more explicitly, let

$$\frac{b(s)}{a(s)} = H(s) = \sum_1^\infty h_i s^{-t} \tag{8c}$$

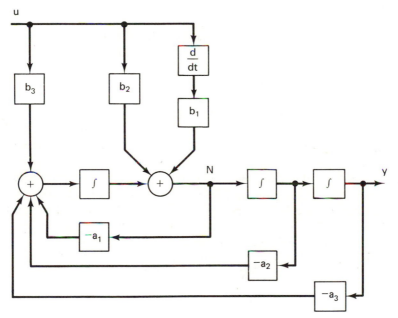

Figure 2.1-6.   Another intermediate step in realizing (5).

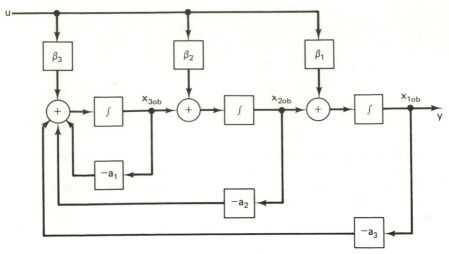

Figure 2.1-7.   Observability canonical form.

Then it can be shown† that

$$\beta_i = h_i, \qquad i = 1, \ldots, n \tag{8d}$$

The $\{h_i\}$ are known as the *Markov parameters* of the system and will be often encountered in later sections, especially Sec. 2.3.3.

A Different Method—The Observer Canonical Form.   The difficulties in obtaining the *observability form* can be eliminated by realizing $1/a(s)$, or equivalently (2), by a method somewhat different from Kelvin's. Let us write (2) as

$$\xi = s^{-3}u - a_1 s^{-1}\xi - a_2 s^{-2}\xi - a_3 s^{-3}\xi \tag{9}$$

Noting that $1/s^k$ is simply the operation of integration k times in succession, we can interpret this equation as saying that ξ can be obtained by integrating u thrice and then adding to it $-a_1$ times the integral of ξ, etc. This leads to a realization of $1/a(s)$ as shown in Fig. 2.1-8. Now it is easy, compared to Fig. 2.1-6, to compensate for the differentiators in $b(s)U(s)$ to obtain the form shown in Fig. 2.1-9. This is known, for reasons that will become clear later (Sec. 4.1), as the *observer canonical form*.

It is important to note that the gains involved in the controller and observer forms are directly the coefficients of the differential equation.

---

†The reader should try to do this now. Several different proofs of (8b)–(8d) will become evident as we progress (see, for example, Exercise 2.2-15 and the discussion of Figs. 2.3-1 and 2.3-2 in Sec. 2.3.3).

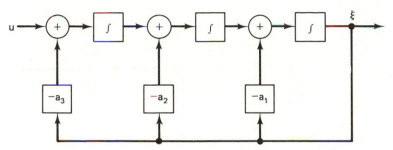

Figure 2.1-8.   Alternative (observer-form) realization of $1/a(s)$.

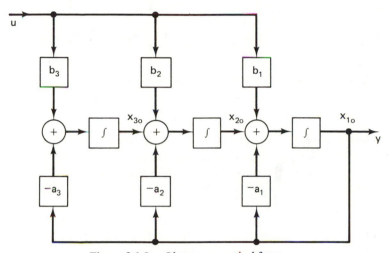

Figure 2.1-9.   Observer canonical form.

The Controllability Canonical Form.   To complete our discussions, we should write $Y(s)$ as

$$Y(s) = b(s)\left[\frac{1}{a(s)}U(s)\right]$$

and see what happens when we first realize $1/a(s)$ as in Fig. 2.1-8. Now we shall have a *dual* situation to that of Fig. 2.1-6, with derivatives appearing in output rather than in input leads. However, by a systematic and simple procedure we can move these leads back, one by one, to obtain the form shown in Fig. 2.1-10, where the $\{\beta_i\}$ have the same values as in Fig. 2.1-7; i.e., they are given by (8). This form is called the *controllability form* and will be encountered soon in Sec. 2.3.2.

Some General Remarks.   We have tried to give a motivated development of four important canonical forms. These forms will reappear often in this course, and the reader will gradually develop various cues for remembering them.

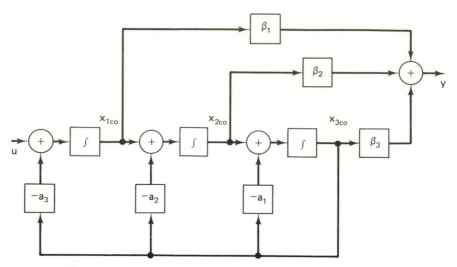

Figure 2.1-10. Controllability canonical form. This is *dual* to the form shown in Fig. 2.1-7.

One set of cues will be provided in the next section when we obtain compact matrix descriptions of these canonical forms. A mnemonic feature to note here is that in the *controller* and *controllability forms* the input either enters each integrator directly or by way of some previous integrations (but not other linear operations). Therefore every state variable is *controllable* in a sense that will be made precise in later sections, beginning with Sec. 2.3.2 On the other hand, the output in these forms is a linear combination of integrator outputs. Because of possible cancellations not every integrator output may appear in $y(\cdot)$. Therefore every state might not be *observable*, again in a sense that will be made precise in later sections, first in Sec. 2.3.1. Parallel statements may be made for the *observer* and *observability forms* if we interchange the roles of inputs and outputs. There is a certain *duality* phenomenon here, which we shall pursue more explicitly in Sec. 2.2. In Sec. 2.3, we shall see more fully why these just-mentioned properties are associated with *control* and *observation*, respectively.

Finally, to summarize the derivations of the four forms, the basic fact we should remember is that $1/a(s)$ can be realized in two fundamental ways—as in Fig. 2.1-3 or 2.1-8. Then if we write

$$Y(s) = b(s)[a^{-1}(s)U(s)]$$

these two forms for $1/a(s)$ yield the controller (Fig. 2.1-4) and controllability (Fig. 2.1-10) forms, respectively. On the other hand, if we write

$$Y(s) = a^{-1}(s)[b(s)U(s)]$$

they yield the observability (Fig. 2.1-7) and observer (Fig. 2.1-9) forms, respectively.

These various comments and mnemonics will be reinforced as the reader proceeds through the book; in particular, the remarks of the previous paragraph will reappear with more dramatic impact in our study of multi-input, multi-output systems (cf. Sec. 6.4).

We should also emphasize that we have talked here about realizations of differential equations with zero initial conditions. Interested readers should try to see how nonzero initial conditions $\{y(0), \dot{y}(0), \ldots\}$ can be incorporated into the above realizations; they will find this easiest to do for the observability form of Fig. 2.1-7 and just a little less so for the observer form of Fig. 2.1-9; however, the problem gets more complicated for the controller and controllability forms. These preliminary explorations will prepare the reader well for the more formal solutions in Sec. 2.3.1.

2.1.3   Parallel and Cascade Realizations

We have now discussed four different ways of simulating a system described by its input-output differential equation. But there are many other ways—in fact, a noncountable infinity—of obtaining analog-computer simulations. However, we shall not pursue this topic much further here because our aim is not the study of analog-computer simulations as such. In particular, we shall not examine the question of which forms are numerically least sensitive to small changes in parameter values. Such questions are addressed in more specialized books, where it is shown that (*where possible*) a new form, the so-called *sum* or *parallel* or *diagonal* form, is perhaps best from a sensitivity point of view. Certain related *product* or *cascade* forms are also important in this regard; see, e.g., [5]–[9]. We shall discuss these forms briefly, partly because the diagonal form, though not always achievable, is useful as a conceptual tool to check out new results and conjectures (see Sec. 2.4.1).

Sum or Parallel Realizations.   If we make a partial fraction expansion

$$\frac{b(s)}{a(s)} = \sum_{1}^{n} \frac{g_i}{s - \lambda_i}$$

each term on the right can be easily realized and $H(s)$ is then obtained as a parallel combination of these elementary realizations (Fig. 2.1-11). This is sometimes called a *diagonal-form representation.*† [*Notice the compact form used for simplicity of representation in the* (b_2, c_2) *line; this is a widely adopted convention, which we shall often use.*]

---

†We should remark, if it is not clear already, that the possibility of cancellations between the numerator and denominator of $H(s)$ shows that we can have realizations of many different orders for a transfer function. The notion of minimal-order realizations is an important one and will be examined in Sec. 2.4.1.

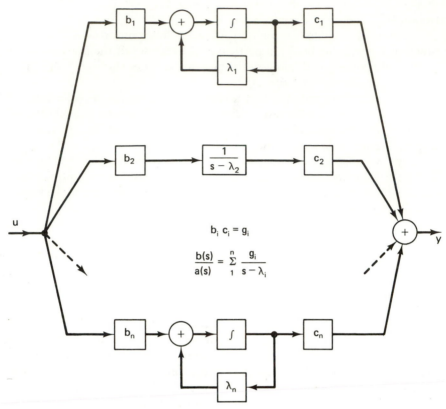

Figure 2.1-11. Diagonal realization for $\sum\limits_{1}^{n} g_i/(s - \lambda_i)$.

It may happen that some roots are *repeated*, e.g., as in

$$\frac{b(s)}{a(s)} = \frac{2s^2 + 6s + 5}{s^3 + 4s^2 + 5s + 2} = \frac{1}{(s + 1)^2} + \frac{1}{s + 1} + \frac{1}{s + 2}$$

This is not hard to handle—see Fig. 2.1-12. Somewhat more thought is required when some of the roots are complex.
　　Suppose that

$$\frac{b(s)}{a(s)} = \frac{2s^2 + 5s + 7}{s^3 + 3s^2 + 7s + 5}$$

$$= \frac{(1 + j/2)/2}{s + 1 + 2j} + \frac{(1 - j/2)/2}{s + 1 - 2j} + \frac{1}{s + 1}$$

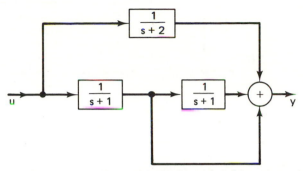

Figure 2.1-12.   Example of sum realization when there are repeated real roots.

This cannot be simulated term by term using real components. The problem can be overcome by working with *second-order* building blocks. Since $b(s)/a(s)$ has real coefficients, the complex roots will always occur in conjugate pairs, so we can regroup $b(s)/a(s)$ as a combination of first-order and second-order terms. In our example we can write

$$\frac{b(s)}{a(s)} = \frac{s+2}{s^2+2s+5} + \frac{1}{s+1}$$

Now the second-order term can be directly simulated in various ways. For example, we could *complete the square* in the denominator polynomial and write, say,

$$\frac{as+b}{cs^2+ds+e} = \frac{\alpha s + \beta}{(s+\gamma)^2 + \delta^2}$$

$$= \frac{\alpha/(s+\gamma)}{1+[\delta^2/(s+\gamma)^2]} + \frac{(\beta-\alpha\gamma)/(s+\gamma)^2}{1+[\delta^2/(s+\gamma)^2]}$$

which can be realized as shown in Fig. 2.1-13.

Product or Cascade Realizations.   By product or cascade realizations we mean realizing, for example,

$$\frac{b(s)}{a(s)} = \frac{s^2+7s+12}{s^3+3s^2+7s+5}$$

as

$$\frac{b(s)}{a(s)} = \frac{s+3}{s^2+2s+5}\frac{s+4}{s+1}$$

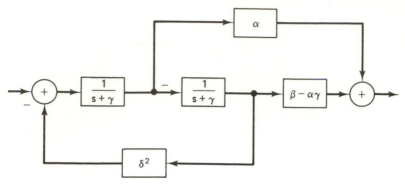

Figure 2.1-13. One realization of the transfer function

$$(\alpha s + \beta)[(s + \gamma)^2 + \delta^2]^{-1}.$$

or as

$$\frac{b(s)}{a(s)} = \frac{s^2 + 7s + 12}{s^2 + 2s + 5} \frac{1}{s + 1}$$

Operational Amplifiers. A basic component in realizing electronic analog computers is the *op amp* or operational amplifier. The ready availability and cheapness of this device has in recent years revolutionized the subject of active circuit synthesis. The op amp is an *ideal* inverting amplifier, with a large voltage gain ($|e_0/e_i|$) of at least 2×10^4 and often over 10^5 and with *infinite* input and "zero" output impedance. Its symbol is shown in Fig. 2.1-14(a). The op amp can be used in combination with resistors and capacitors to synthesize many basic circuits such as the simple amplifier circuit shown in Fig. 2.1-14(b). Referring to the diagram, we note that the voltage e_T at the input to the op amp must be very small, essentially equal to zero, because the op amp has a very large gain and because of the negative feedback. Therefore the current through the resistor R_1 must have value e_i/R_1. This same current, i, must also flow through R_2. Therefore

$$e_0 = -iR_2 = -\frac{R_2}{R_1} e_i$$

The negative sign can be removed by means of an inverting amplifier.

Op amps can also be used to synthesize summers, integrators, gyrators, and negative resistors. A more detailed description of op amp design can be found in [3]–[4] and the references cited therein.

Reprise. We have now introduced several different realizations of a given transfer function,

$$H(s) = \frac{b(s)}{a(s)} = \frac{b_1 s^{n-1} + \cdots + b_n}{s^n + a_1 s^{n-1} + \cdots + a_n}$$

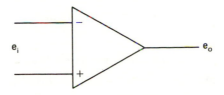

(a) Symbol for an op amp. The gain
$K = |e_o/e_i|$ is taken as infinity.

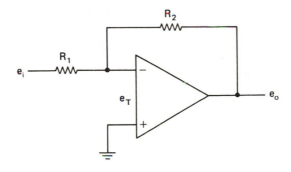

(b) Amplifier with gain $-(R_2/R_1)$.

Figure 2.1-14.   Op-amp circuit.

One application of these procedures is the actual simulation of given transfer functions on an analog computer; for this purpose, the diagonal (when possible) and product (cascade) forms of Sec. 2.1.3 are perhaps the most useful. The main reason for introducing the four other special forms of Sec. 2.1.2 is that, as we shall often see in the rest of this book, they will provide convenient ways of *theoretically* exploring and understanding certain basic aspects of the different systems that can yield a given transfer function. For this purpose, a more compact analytical description is important, and will be provided in the next section.

2.2 STATE EQUATIONS IN THE TIME AND FREQUENCY DOMAINS

The main aim of this section is to introduce certain terminology and notations with which the reader will soon familiarize himself. A first reading should be rather quick; there will be ample opportunity to return to this section from time to time in order to make sure that certain distinctions and usages, which may at first seem pedantic and unnecessary, are well understood.

2.2.1   Matrix Notation and State-Space Equations

Note by comparing Figs. 2.1-4 and 2.1-9 that the *controller* and *observer* forms are duals in the sense that they can be obtained from each other by interchanging y and u, reversing arrows, and replacing nodes by summers and vice versa This duality is not unexpected considering how these forms were deduced. A similar duality holds between the *controllability* form of Fig. 2.1-10 and the *observability* form of Fig. 2.1-7. Some insight into this duality can be obtained from the theory of signal flow graphs. However, for our purposes, more insight can be gathered by considering some algebraic aspects of this duality.

Let us begin with the controller form of Fig. 2.1-4 and label the integrator outputs in that figure, going from left to right, as x_{1c}, x_{2c}, x_{3c}, respectively. Then it is easy to see that the *block diagram* or *wiring diagram* of the analog simulation can equivalently be described by the equations

$$\begin{aligned}
\dot{x}_{1c} &= -a_1 x_{1c} - a_2 x_{2c} - a_3 x_{3c} + u \\
\dot{x}_{2c} &= x_{1c} \\
\dot{x}_{3c} &= x_{2c} \\
y &= b_1 x_{1c} + b_2 x_{2c} + b_3 x_{3c}
\end{aligned} \tag{1}$$

We can write this more compactly in matrix notation as

$$\dot{x}_c = A_c x_c + b_c u, \qquad y = c_c x_c \tag{2}$$

where

$$A_c = \begin{bmatrix} -a_1 & -a_2 & -a_3 \\ 1 & 0 & 0 \\ 0 & 1 & 0 \end{bmatrix}, \qquad b_c = \begin{bmatrix} 1 \\ 0 \\ 0 \end{bmatrix},$$

$$c_c = [b_1 \quad b_2 \quad b_3] \tag{3}$$

Similarly, for the observer form of Fig. 2.1-9 we can label the integrator outputs, now going from *right to left*, as x_{1o}, x_{2o}, x_{3o}, respectively. Then the wiring diagram, Fig. 2.1-9, can be described by the matrix equations

$$\dot{x}_o = A_o x_o + b_o u, \qquad y = c_o x_o \tag{4}$$

where

$$A_o = \begin{bmatrix} -a_1 & 1 & 0 \\ -a_2 & 0 & 1 \\ -a_3 & 0 & 0 \end{bmatrix}, \qquad b_o = \begin{bmatrix} b_1 \\ b_2 \\ b_3 \end{bmatrix},$$

$$c_o = [1 \quad 0 \quad 0] \tag{5}$$

The matrix notation of course provides a much more compact way of conveying the information contained in the block (or wiring) diagrams of the analog-computer simulations.

As a first illustration, the statement about *duality* is reflected into the matrix equations by the relationships

$$A_o = A_c', \qquad b_o = c_c', \qquad c_o = b_c' \tag{6}$$

We can similarly describe the relations between the controllability form of Fig. 2.1-10 and the observability form of Fig. 2.1-7. From Fig. 2.1-10 we can write

$$\dot{x}_{co} = A_{co}x_{co} + b_{co}u, \qquad y = c_{co}x_{co} \tag{7}$$

where

$$A_{co} = \begin{bmatrix} 0 & 0 & -a_3 \\ 1 & 0 & -a_2 \\ 0 & 1 & -a_1 \end{bmatrix}, \qquad b_{co} = \begin{bmatrix} 1 \\ 0 \\ 0 \end{bmatrix}$$

$$c_{co} = [\beta_1 \quad \beta_2 \quad \beta_3] = [b_1 \quad b_2 \quad b_3] \begin{bmatrix} 1 & a_1 & a_2 \\ 0 & 1 & a_1 \\ 0 & 0 & 1 \end{bmatrix}^{-1} \tag{8}$$

The observability form of Fig. 2.1-7 is described by the equations

$$\dot{x}_{ob} = A_{ob}x_{ob} + b_{ob}u, \qquad y = c_{ob}x_{ob}$$

where

$$A_{ob} = \begin{bmatrix} 0 & 1 & 0 \\ 0 & 0 & 1 \\ -a_3 & -a_2 & -a_1 \end{bmatrix},$$

$$b_{ob} = \begin{bmatrix} \beta_1 \\ \beta_2 \\ \beta_3 \end{bmatrix} = \begin{bmatrix} 1 & 0 & 0 \\ a_1 & 1 & 0 \\ a_2 & a_1 & 1 \end{bmatrix}^{-1} \begin{bmatrix} b_1 \\ b_2 \\ b_3 \end{bmatrix}, \qquad c_{ob}' = \begin{bmatrix} 1 \\ 0 \\ 0 \end{bmatrix} \tag{9}$$

We may recall that [cf. (2.1-8d) of Sec. 2.1) in (8) and (9) we have

$$\beta_i = h_i, \qquad \text{where } H(s) = \sum_1^\infty h_i s^{-i} \tag{10}$$

Again, the *duality* between these forms is shown by the relations [cf. (6)]

$$A_{ob} = A_{co}', \qquad b_{ob} = c_{co}', \qquad c_{ob} = b_{co}' \tag{11}$$

The Diagonal and Jordan Forms.   Referring to the diagonal or sum realization shown in Fig. 2.1-11, we can write the matrix equation

$$\dot{x}_d = A_d x_d + b_d u, \qquad y = c_d x_d \qquad (12)$$

where

$$A_d = \text{diag}\{\lambda_1, \ldots, \lambda_n\}$$
$$b'_d = [b_1, \ldots, b_n], \qquad c_d = [c_1, \ldots, c_n]$$

and the output of the ith integrator has been taken as x_{di}.

The modified-diagonal or *Jordan* forms are less simple because there are
several possibilities. For example, the realization of Fig. 2.1-12 can be
described by the equations

$$\begin{bmatrix} \dot{x}_1(t) \\ \dot{x}_2(t) \\ \dot{x}_3(t) \end{bmatrix} = \begin{bmatrix} -1 & 1 & 0 \\ 0 & -1 & 0 \\ 0 & 0 & -2 \end{bmatrix} \begin{bmatrix} x_1(t) \\ x_2(t) \\ x_3(t) \end{bmatrix} + \begin{bmatrix} 0 \\ 1 \\ 1 \end{bmatrix} u(t)$$

$$y(t) = [1 \quad 1 \quad 1][x_1(t) \quad x_2(t) \quad x_3(t)]' \qquad (13)$$

We ask the reader to determine how the $\{x_i(\cdot)\}$ were defined to get the above
choice. The main feature to note is that the A matrix is of the canonical
Jordan type, discussed in Exercise A.52. We can have many different types of
Jordan forms, depending on the actual values of the coefficients in $b(s)$ and
$a(s)$. However, the calculations required to get Jordan forms are numerically
unstable, (see Exercise A.50) and for this and other reasons, we shall call
upon such forms only very occasionally in this book.

Of course, there are also other special realizations and corresponding
matrix descriptions. And by now it should be clear that *to any analog-computer realization there corresponds a set of equations of the form*

$$\dot{x}(t) = Ax(t) + bu(t), \qquad y(t) = cx(t), \qquad (14a)$$

or equivalently a triple $\{A, b, c\}$. The converse is also true. Given equations as
in (14a) it is easy to see how to interconnect n integrators [n is the dimension
of $x(\cdot)$] so as to simulate the equations. *The reader will often find it helpful
to call upon this one-to-one relationship between the matrix equations and
analog-computer simulations in trying to understand or picture various mathematical statements about the equations.*

In this connection, we should perhaps note a slight generalization to the
form

$$\dot{x}(t) = Ax(t) + bu(t)$$
$$y(t) = cx(t) + du(t) \qquad (14b)$$

which defines a matrix quadruple $\{A, b, c, d\}$. However, it is often easier to study the case $d = 0$ and then make the modifications necessary to include the *direct transmission* term $du(\cdot)$, which involves no *dynamics*. Certain generalizations of (14) are sometimes useful, where the output may be allowed to contain derivatives of the input.† There are several results available for such systems as well, some of which we shall discuss in Chapter 8.

We shall say that equations of the form (14) provide *state-space* or *state-variable descriptions* or *realizations* of the given differential equation; the vector $x(\cdot)$ is called the *state vector* and its components the *state variables*, or more simply the *states*, of the realization (14).

The matrix A in the state equations is sometimes called the *state-feedback matrix* because its elements determine the *feedback* connections of the realization: If A is zero, there will be no feedback loops in the realization. For obvious reasons, b is called the *input matrix* and c the *output matrix* of the realization. Moreover, the *duality* that we noted earlier for the four canonical realizations has a more general aspect: For any given realization $\{A, b, c\}$ of a system (transfer function), there is a *dual* realization $\{A', c', b'\}$. We shall find several occasions where the notion of dual realizations will be helpful—in saving work and in suggesting new investigations.

Nonuniqueness of State-Space Realizations.   The connection with analog-computer simulations makes it very clear that a given system differential equation can have many state-space descriptions. Therefore it is meaningless to talk of *the* states of a system; we can only have *states of a realization*. Nevertheless, it is often convenient to use the terms more loosely as, for example, when it is understood which realization one is referring to; no harm is done as long as the proper meaning is clear from the context.

The multiplicity of realizations of a given system is well illustrated by the fact that given a realization (14), we can form another realization by a change of variables (note the order: old variables $= T \cdot$ new variables),

$$x(t) = T\bar{x}(t), \qquad \det T \neq 0 \tag{15a}$$

so that, say,

$$\dot{\bar{x}}(t) = T^{-1}AT \cdot \bar{x}(t) + T^{-1}b \cdot u(t) = \bar{A}\bar{x}(t) + \bar{b}u(t)$$

and   (15b)

$$y(t) = cT \cdot \bar{x}(t) = \bar{c}\bar{x}(t)$$

Since there is a multiple infinity of nonsingular matrices T, there is clearly a

---

†Such systems can arise as mathematical models in studying interconnected subsystems and in certain econometric problems (see, e.g., K. D. Wall, "Rational Expectations and the Control of the Economy," *Proceedings of the 1977 IEEE Conference on Decision and Control*, pp. 1020–1023, New Orleans, Dec. 1977).

multitude of realizations.† In matrix theory, matrices related as

$$\bar{A} = T^{-1}AT$$

are said to be *similar*, and therefore state-variable transformations as in (15) are known as *similarity transformations*. As we shall see in the sequel, such transformations can often be judiciously chosen to give the most convenient state-space description for a given problem, e.g., to give one of the special forms developed in Sec. 2.1.1.

However, we must emphasize that it is not always possible to obtain any desired realization, even of the same order as the original, by a similarity transformation. This can be seen as a purely algebraic fact in the example below (see also Exercise 2.2-11), the physical significance of which will be discussed later, in Sec. 2.4.1. (This example will also provide an opportunity to review some of the matrix theory material in Secs. 10, 11, and 13 in the Appendix.)

Example 2.2-A.   Not All Realizations Can Be Transformed to Diagonal Form

Try to find a *similarity transformation* of a given realization $\{A, b, c\}$ to diagonal form $\{A_d, b_d, c_d\}$.

Solution.   If we can find an invertible matrix T such that, say,

$$T^{-1}AT = A_d = \text{diag}\{\lambda_1, \ldots, \lambda_n\}$$

then b_d and c_d can be found as $T^{-1}b$ and cT, respectively. Now note that if such a T exists, we must have $AT = TA_d$. The columns of T must then satisfy

$$At_i = \lambda_i t_i, \qquad i = 1, \ldots, n$$

from which it is clear that λ_i must be an eigenvalue of A and t_i a corresponding eigenvector. Hence a nonsingular T can be found *if and only if A has n linearly independent eigenvectors*.

It can be shown (Exercise A.42) that the eigenvectors corresponding to *distinct* eigenvalues are independent. Hence if A has n distinct eigenvalues, one can find a T that diagonalizes it. If, however, an eigenvalue is repeated, the situation is more complicated. For example, if λ is an eigenvalue of multiplicity k, then it is possible that there are anywhere from 1 to k independent eigenvectors associated with it; the actual number depends on the particular matrix A (Exercise A.49).

If there are less than n independent eigenvectors, one has to go to generalized eigenvectors (Exercise A.51) to find a complete set of independent vectors to make up T. Now, however, we can at best transform A to Jordan form (Exercise A.52).

As a useful special case, suppose the given realization is actually in controller

---

†We emphasize that there are many other possibilities as well—e.g., the transformation matrix T may be time dependent (Exercise 2.2-9); we may also obtain other realizations by changing the number of states (see Exercise 2.2-10).

canonical form, so that

$$A = A_c = \text{a companion matrix with top row } [-a_1 \quad \cdots \quad -a_n]$$

Then (Exercise A.35) it is easy to verify that for any eigenvalue λ of A_c we have an eigenvector $[\lambda^{n-1} \ \lambda^{n-2} \ \cdots \ \lambda \ 1]'$. Therefore the transformation matrix is

$$T = \begin{bmatrix} \lambda_1^{n-1} & \lambda_2^{n-1} & \cdots & \lambda_n^{n-1} \\ \vdots & \vdots & & \vdots \\ \lambda_1 & \lambda_2 & \cdots & \lambda_n \\ 1 & 1 & & 1 \end{bmatrix} \tag{16}$$

This is the so-called Vandermonde matrix, which will have a nonzero determinant if and only if the $\{\lambda_i\}$ are all distinct (cf. Exercise A.7), and then

$$T^{-1} A_c T = \text{diag}\{\lambda_1, \ldots, \lambda_n\}$$

But what if some of the $\{\lambda_i\}$ are repeated, say $\lambda_1 = \lambda_2$? One eigenvector is $[\lambda_1^{n-1} \ \cdots \ \lambda_1 \ 1]'$, but can we find still another linearly independent eigenvector associated with λ_1? This is possible for some matrices with repeated eigenvalues but not for a matrix in companion form, as can be verified by direct calculation. Companion matrices are an important example of what are called *cyclic* (or *nonderogatory*) matrices, which have only one (normalized) eigenvector associated with each distinct eigenvalue.

In conclusion, a *controller-form* realization can be transformed to *diagonal* form by a *similarity transformation* if and only if all its eigenvalues are distinct. The physical significance of this condition will become clear in Sec. 2.4.1.

We repeat that in principle any realization can be converted to one with the A matrix in modified-diagonal (Jordan) form, but as mentioned before, the actual transformation to such form cannot be made in a numerically stable way. ∎

2.2.2   Obtaining State Equations Directly—Some Examples; Linearization

So far we have only derived state-space equations for systems specified by a differential equation. Often, of course, we are given the system itself, e.g., as an interconnection of resistors, capacitors, and inductors (Rs, Cs and Ls), or mechanical elements, etc. Then we can usually write down state-space equations directly from such specifications, without first computing the governing differential equation. In fact, it often turns out that the state-space equations are easier and more natural to determine than other descriptions of the system, as we shall illustrate with a very familiar example generally studied by non-state-space methods. We shall also illustrate the power of the state-space approach for the less familiar time-variant and nonlinear versions of this same circuit.

Example 2.2-B.   Two-Loop Electric Circuit

The circuit shown in Fig. 2.2-1 is usually analyzed by the so-called loop method. We write equations for each loop current and then eliminate one of the currents in order to get a second-order differential equation describing the other in terms of the excitations u_1 and u_2 and the circuit parameters. After solving this differential equation for one loop current, we can then go back to the loop equations and determine the second one.

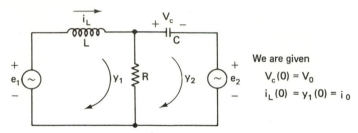

Figure 2.2-1.   Two-loop electric circuit.

For the circuit in Fig. 2.2-1, the loop equations are

$$L\dot{y}_1(t) + R[y_1(t) - y_2(t)] = u_1(t) \tag{17}$$

$$\frac{1}{C}\int_{-\infty}^{t} y_2(\tau)\,d\tau + R[y_2(t) - y_1(t)] = -u_2(t) \tag{18}$$

We can solve for y_1 from (18) and substitute into (17) to get

$$L\dot{y}_2(t) + \frac{L}{RC}y_2(t) + \frac{1}{C}\int_{-\infty}^{t} y_2(\tau)\,d\tau = u_1(t) - \frac{L\dot{u}_2(t)}{R} - u_2(t) \tag{19}$$

which gives the second-order differential equation

$$\ddot{y}_2(t) + \frac{1}{RC}\dot{y}(t) + \frac{1}{LC}y(t) = \frac{\dot{u}_1(t)}{L} - \frac{\dot{u}_2(t)}{L} - \frac{\ddot{u}_2(t)}{R} \tag{20}$$

Now we need to determine the initial conditions $y_2(0)$ and $dy_2(0)/dt$ from the given initial conditions

$$V_c(0) = V_0, \qquad i_L(0) = y_1(0) = i_0$$

The initial value $y_2(0)$ is readily obtained by setting $t = 0$ in (18),

$$y_2(0) = i_0 - \frac{u_2(0) + V_0}{R} \tag{21}$$

For $dy_2(0)/dt$, we have to work a bit harder. From (19) at $t = 0$, we get

$$\frac{dy_2(0)}{dt} = -\frac{1}{RC}y_2(0) - \frac{V_0}{L} + \frac{u_1(0)}{L} - \frac{1}{R}\frac{du_2(0)}{dt} - \frac{1}{L}u_2(0)$$

and now using (21) gives

$$\frac{dy_2(0)}{dt} = -\frac{1}{RC}i_0 + \frac{V_0 + u_2(0)}{R}\left(-\frac{R}{L} + \frac{1}{RC}\right) + \frac{u_1(0)}{L} - \frac{1}{R}\frac{du_2(0)}{dt} \quad (22)$$

Finally we can solve (20) for y_2 and then determine y_1 by solving the first-order equation (17).

This procedure can also be used for multiloop networks, and it is quite straightforward and systematic. *The only complication is really in determining the initial conditions,* which are usually not given in terms of the values at 0 of a loop current and its derivatives but as initial capacitor voltages and initial inductor currents. The complexity of initial condition determination rises sharply with the number of loops. It is therefore *natural to try to describe the system behavior directly in terms of equations for the capacitor voltages and inductor currents.* In our example, we can write

$$\frac{di_L(t)}{dt} = \frac{R}{L}[y_2(t) - i_L(t)] + \frac{u_1(t)}{L}$$

$$= -\frac{R}{L}\frac{1}{R}[u_2(t) + V_c(t)] + \frac{u_1(t)}{L}$$

Also

$$C\frac{dV_c(t)}{dt} = y_2(t) = -\frac{1}{R}[u_2(t) + V_c(t)] + i_L(t)$$

If we set

$$x_1(t) = i_L(t), \qquad \text{the inductor current}$$

$$x_2(t) = V_c(t), \qquad \text{the capacitor voltage}$$

then we have in matrix notation

$$\begin{bmatrix} \dot{x}_1(t) \\ \dot{x}_2(t) \end{bmatrix} = \begin{bmatrix} 0 & -1/L \\ 1/C & -1/RC \end{bmatrix}\begin{bmatrix} x_1(t) \\ x_2(t) \end{bmatrix} + \begin{bmatrix} 1/L & -1/L \\ 0 & -1/RC \end{bmatrix}\begin{bmatrix} u_1(t) \\ u_2(t) \end{bmatrix} \quad (23)$$

Moreover, the initial conditions are immediate,

$$x_1(0) = y_0, \qquad x_2(0) = V_0$$

The loop currents, and in fact any other quantities of interest, can be expressed as algebraic combinations of the state variables and the inputs; e.g.,

$$\begin{bmatrix} y_1(t) \\ y_2(t) \end{bmatrix} = \begin{bmatrix} 1 & 0 \\ 1 & -1/R \end{bmatrix}\begin{bmatrix} x_1(t) \\ x_2(t) \end{bmatrix} + \begin{bmatrix} 0 & 0 \\ 0 & -1/R \end{bmatrix}\begin{bmatrix} u_1(t) \\ u_2(t) \end{bmatrix} \qquad \blacksquare \quad (24)$$

This simple example illustrates some of the advantages of trying to directly describe electrical circuit behavior via state-space equations. The procedure used above of choosing capacitor voltages and inductor currents as state variables and then trying to write down first-order differential equations for these variables can be used for general $\{R, L, C\}$ networks, with a single qualification. Clearly, if there are loops of capacitors in the network, not all capacitor voltages can be chosen as independent state variables, since one voltage can be expressed in terms of the others. A similar situation prevails when only inductors impinge at a (generalized) node—a so-called inductor *cut-set*. Fairly general procedures for obtaining state equations using some simple network topology (as required also for the *loop* and *node* methods) have been described in the literature. It may be of interest that network theorists began to use state equations as early as 1957 (see Bashkow [10]), at about the same time that Bellman [11] and Kalman [12] were reemphasizing their value in control problems. We shall refer to [9] and [13] for general state-description methods. We now discuss some of the special advantages of the state-variable formulation for time-variant and/or nonlinear networks.

Example 2.2-C.   Time-Variant Parameters

Consider the circuit shown in Fig. 2.2-1, but now assume that the elements are time-variant. Then the loop equations (17) and (18) have to be replaced by

$$\frac{d}{dt}[L(t)y_1(t)] + R(t)[y_1(t) - y_2(t)] = u_1(t)$$

$$\frac{1}{C(t)} \int_{-\infty}^{t} y_2(\tau)\, d\tau + R(t)[y_2(t) - y_1(t)] = -u_2(t)$$

If we proceed to eliminate $y_1(\cdot)$ to obtain a differential equation for only $y_2(\cdot)$; high-order derivatives of $L(\cdot)$, $C(\cdot)$, and $R(\cdot)$ will appear.

However, the state equations for $x_1(\cdot) = i_L(\cdot)$ and $x_2(\cdot) = V_c(\cdot)$ will involve only first-order derivatives,

$$\begin{bmatrix} \dot{x}_1(t) \\ \dot{x}_2(t) \end{bmatrix} = -\begin{bmatrix} L^{-1}(t)\dot{L}(t) & L^{-1}(t) \\ -C^{-1}(t) & [\dot{C}(t) + R^{-1}(t)]C^{-1}(t) \end{bmatrix}\begin{bmatrix} x_1(t) \\ x_2(t) \end{bmatrix}$$
$$+ \begin{bmatrix} L^{-1}(t) & -L^{-1}(t) \\ 0 & -R^{-1}(t)C^{-1}(t) \end{bmatrix}\begin{bmatrix} u_1(t) \\ u_2(t) \end{bmatrix} \tag{25}$$

For *time-variant* and/or *nonlinear* systems, it is generally more convenient to use as state variables

$$\Psi(\cdot) = \text{the magnetic flux through the inductor}$$
$$q(\cdot) = \text{the charge on the capacitor}$$

In our example, it is easy to see that

$$\begin{bmatrix} \dot{\Psi}(t) \\ \dot{q}(t) \end{bmatrix} = \begin{bmatrix} 0 & -C^{-1}(t) \\ L^{-1}(t) & -R^{-1}(t)C^{-1}(t) \end{bmatrix}\begin{bmatrix} \Psi(t) \\ q(t) \end{bmatrix} + \begin{bmatrix} 1 & -1 \\ 0 & -R^{-1}(t) \end{bmatrix}\begin{bmatrix} u_1(t) \\ u_2(t) \end{bmatrix} \tag{26}$$

The point is that, with this choice, no derivatives of $L(\cdot)$ and $C(\cdot)$ appear in the state equations. ∎

Example 2.2-D.   Nonlinear Parameters
Suppose that the inductance and capacitance of the circuit in Fig. 2.2-1 are non-linear in that their values depend on the values of the inductor flux and the capacitor charge. This happens when the flux (charge) is a nonlinear function of the inductor current (capacitor voltage). Therefore, let

$$\Psi(t) = \Psi[i_L(t), t], \qquad q(t) = q[V_c(t), t]$$

and assume that we can *invert* these relations, say,

$$i_L(t) = \phi[\Psi(t), t], \qquad V_c(t) = p[q(t), t]$$

Then

$$\frac{d\Psi}{dt} = \frac{\partial \Psi}{\partial t} + \frac{\partial \Psi}{\partial i_L}\frac{di_L}{dt} = \frac{\partial \Psi}{\partial t} + L(i_L, t)\frac{di_L}{dt}, \text{ say,}$$

and

$$\frac{dq}{dt} = \frac{\partial q}{\partial t} + \frac{\partial q}{\partial V_c}\frac{dV_c}{dt} = \frac{\partial q}{\partial t} + C(V_c, t)\frac{dV_c}{dt}, \text{ say}$$

It is easy to see that

$$\dot{\Psi}(t) = u_1(t) - u_2(t) - V_c(t)$$
$$= u_1(t) - u_2(t) - p[q(t), t]$$

and

$$\dot{q}(t) = i_L(t) - \frac{V_c(t) + u_2(t)}{R}$$

$$= \phi[\Psi(t), t] - \frac{p[q(t), t] + u_2(t)}{R}$$

It is perhaps needless to say that loop analysis would be much more complicated for this example.   ∎

Linearization.   The state equations for nonlinear systems will have the general form

$$\dot{x}(t) = f[x(t), u(t), t] \tag{27}$$

It is often helpful to linearize these equations around a nominal solution, say $\{x_0(\cdot), u_0(\cdot)\}$. Thus suppose that

$$x(t) = x_0(t) + \delta x(t), \qquad u(t) = u_0(t) + \delta u(t) \tag{28}$$

where powers $(\delta x)^i$, $(\delta u)^i$, $i > 1$, are "very small" compared to δx, δu. This is usually indicated by the notation

$$(\delta x)^i = o(\delta x, \delta u), \qquad (\delta u)^i = o(\delta x, \delta u), \qquad i > 1 \tag{29}$$

Now suppose that $f(\cdot)$ is smooth enough to have a Taylor series representation

$$f[x(t), u(t), t] = f[x_0(t), u_0(t), t] + f_x(t)\,\delta x(t)$$
$$+ f_u(t)\,\delta u(t) + o(\delta x, \delta u)$$

where

$$f_x(t) = \frac{\partial f}{\partial x}\bigg|_{x_0, u_0}, \qquad f_u(t) = \frac{\partial f}{\partial u}\bigg|_{x_0, u_0} \tag{30}$$

The reader should check that when $f(\cdot)$, $x(\cdot)$, $u(\cdot)$ are vectors, $f_x(\cdot)$ and $f_u(\cdot)$ are matrices of appropriate dimension, which depend on $x_0(\cdot)$, $u_0(\cdot)$. The linearized equations can now be seen to be

$$\delta x = f_x \delta x + f_u \delta u \tag{31}$$

which is often the equation that underlies our standard form

$$\dot{x}(t) = A(t)x(t) + B(t)u(t)$$

or

$$\dot{x}(t) = Ax(t) + bu(t)$$

for a time-invariant, single-input system. It is important to note that the linearization of a *time-invariant nonlinear* system often (when the nominal solution is not constant) gives rise to *time-variant linearized* systems, and this is one of the chief ways time-variant systems are encountered in system analysis. (See Exercise 2.2-21.). In Sec. 2.6.3 we shall show that, under fairly general conditions, linearized models continue to reflect the stability properties of the original system; this is a major reason for the significance of linear systems.

Example 2.2-E.   Linearization: Communications Satellite (Bryson)

We shall consider a multi-input, multi-output problem involving a satellite of mass m in earth orbit specified by its position and velocity in polar coordinates, say

$$x(\cdot) \triangleq [r(\cdot) \quad \dot{r}(\cdot) \quad \theta(\cdot) \quad \dot{\theta}(\cdot) \quad \phi(\cdot) \quad \dot{\phi}(\cdot)]'$$

The input thrusts or forces are written as

$$u(\cdot) \triangleq [u_r(\cdot) \quad u_\theta(\cdot) \quad u_\phi(\cdot)]'$$

and may be applied by using small rocket engines. The equations of motion can then be shown† to be

---

†We cannot enter into the background in mechanics needed to derive these equations. For those with such knowledge, however, we might note that they can be found by using a Lagrangian function $L = T - V$, where $T = m[\dot{r}^2 + (r\dot{\phi})^2 + (r\dot{\theta}\cos\phi)^2]/2$ and $V = -km/r$, $k = $ a known constant.

$$\dot{x} = f(x, u) = \begin{bmatrix} \dot{r} \\ r\dot{\theta}^2 \cos^2 \phi + r\dot{\phi}^2 - k/r^2 + u_r/m \\ \dot{\theta} \\ -2\dot{r}\dot{\theta}/r + 2\dot{\theta}\dot{\phi} \sin \phi/\cos \phi + u_\theta/mr \cos \phi \\ \dot{\phi} \\ -\dot{\theta}^2 \cos \phi \sin \phi - 2\dot{r}\dot{\phi}/r + u_\phi/mr \end{bmatrix}$$

We shall define the output to be the position variables $\{r, \theta, \phi\}$ so that

$$y = Cx = \begin{bmatrix} 1 & 0 & 0 & 0 & 0 & 0 \\ 0 & 0 & 1 & 0 & 0 & 0 \\ 0 & 0 & 0 & 0 & 1 & 0 \end{bmatrix} x$$

A free (undriven) solution of these equations corresponds to the satellite being in a *circular equatorial orbit*,

$$x_0(t) \triangleq [r_0 \quad 0 \quad \omega_0 t \quad \omega_0 \quad 0 \quad 0]', \qquad u_0(t) \equiv 0$$

where the radius r_0 and angular velocity ω_0 are such that $r_0^3 \omega_0^2$ = a constant. However, the satellite will deviate from this orbit due to disturbances, and therefore it is of interest to consider the *linearized equations* about the nominal circular equatorial orbit.

The reader should check that the linearized state-space equations are defined by the matrices

$$A = \begin{bmatrix} 0 & 1 & 0 & 0 & 0 & 0 \\ 3\omega_0^2 & 0 & 0 & 2\omega_0 r_0 & 0 & 0 \\ 0 & 0 & 0 & 1 & 0 & 0 \\ 0 & -2\omega_0/r_0 & 0 & 0 & 0 & 0 \\ \hline 0 & 0 & 0 & 0 & 0 & 1 \\ 0 & 0 & 0 & 0 & -\omega_0^2 & 0 \end{bmatrix}, \quad B = \begin{bmatrix} 0 & 0 & 0 \\ 1/m & 0 & 0 \\ 0 & 0 & 0 \\ 0 & 1/mr_0 & 0 \\ \hline 0 & 0 & 0 \\ 0 & 0 & 1/mr_0 \end{bmatrix}$$

with C as before.

The dashed lines show that this sixth-order set of equations splits into two uncoupled subsets, one involving only the variables describing the motion in the equatorial $\{r, \theta\}$ plane and the other only the (azimuthal) variables $(\phi, \dot{\phi}, u_\phi)$.   ∎

Some Remarks on Solving State Equations.   We have illustrated above that the state-space method of description can be effectively applied even to time-variant and nonlinear circuits. While we can thus get a complete description of the circuits, the question remains of whether the state-space equations can be effectively solved. We make the following remarks:

1. As will be shown often in this book, one can get a lot of information from the state equations without ever explicitly solving them. It is

therefore significant that state-space equations can be formulated for more general networks than are amenable to, say, loop analysis.

2. When different descriptions are available, the work required to actually "solve" the circuit is generally the same, no matter which description is used.

3. However, the first-order differential equations used in the state-space description perhaps lend themselves more easily and accurately to evaluation on a digital computer (by some numerical integration routine).

4. When $n = 2$, there is a useful graphical method called *phase-plane analysis* for examining linear and nonlinear state-space equations (see [14, Chap. 10] and [15]). Exercises 2.2-24 to 2.2-26 provide an introduction to this interesting topic.

2.2.3 A Definition of State

So far we have not given any formal definition of what *states* are. Actually we can use the state equations quite effectively without entering into such questions. However, a deeper examination of the concept of state does have value, and we shall pursue it to some extent in Sec. 5.1. Here it will suffice to make the following brief observations.

Let us, for convenience and definiteness, return to the analog-computer simulations of the linear state-space equations

$$\dot{x}(t) = Ax(t) + bu(t), \qquad y(t) = cx(t)$$

The vector function $x(\cdot)$ describes the evolution of the basic internal variables of the analog-computer realization, viz., the integrator outputs. It is clear (think of how the analog computer works!) that if we know the values of these integrator outputs at *any* given time $t = t_0$, then given knowledge of the input $u(\cdot)$ for $t \geq t_0$ we can calculate *all* present and future values of the output $y(\cdot)$ and of the integrator outputs (in fact, of any signal anywhere in the simulation). Of course, given only the input for all times $(-\infty < t)$, one could also calculate the output values and integrator values for all times. However, the point often is that we may not know what the input was (or at least exactly what it was) up to some time t_0 at which we begin to get more interested in the system; it is then surely comforting to know that the value $x(t_0)$ of the integrator outputs provides a "sufficient statistic," so to say, that enables us to calculate the future $(t \geq t_0)$ response to a new input $\{u(t), t \geq t_0\}$ without worrying about $\{u(t), t < t_0\}$. Note also that more than one past input $\{u(t), t < t_0\}$ can lead to the same value of $x(t_0)$. Therefore $x(t_0)$ is really a *minimal* sufficient statistic; it contains just enough information, no less and no more, to enable us to calculate the future responses without further reference to the old history of inputs and responses. As in more colloquial usage, the knowledge of the state vector at any time specifies the *state* or *condition* of the system at that time. (For example, when faced with a new expenditure, it is perhaps the current state of one's bank balance that is most important, and the many ways it could have reached its present value are not often relevant to the issue at hand.)

It therefore seems reasonable to call, as we have done, the integrator outputs at any time *t* the *state* of the realization. Furthermore, this interpretation is clearly not restricted to analog-computer realizations but applies to the variables of any set of state-space equations, no matter how they are obtained—as realizations of given differential equations or as descriptions of given physical systems.

We emphasize that the state description is not unique and that the same physical system can be described by many different sets of state variables.

Example 2.2-F.   A Simple One-Dimensional System

As a very simple example of the above discussion of the notion of state, we can consider the system represented by Fig. 2.2-2. The input $u(\cdot)$ is a current, and the

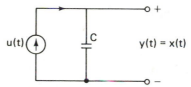

Figure 2.2-2.   Simple example illustrating that the state $x(t)$ can summarize the past up to t.

output $y(\cdot)$ is a voltage, so that the input-output relation for this system can be written as

$$y(t) = \frac{1}{C} \int_{-\infty}^{t} u(\tau)\, d\tau$$

$$= \frac{1}{C} \int_{-\infty}^{t_0} u(\tau)\, d\tau + \frac{1}{C} \int_{t_0}^{t} u(\tau)\, d\tau$$

$$= y(t_0-) + \frac{1}{C} \int_{t_0}^{t} u(\tau)\, d\tau$$

which shows clearly that if we know $y(t_0-)$, it is irrelevant what the values $\{u(\tau),\ -\infty < \tau < t_0\}$ were. Therefore the value $y(t_0-)$ can be regarded as the *state* at time t_0-, and the state equation is

$$\dot{y}(t) = \frac{1}{C} u(t), \qquad t \geq t_0, \qquad y(t_0-) \text{ given } \blacksquare$$

Example 2.2-G. **State Descriptions of a Series RLC Circuit†**

Consider the circuit shown in Fig. 2.2-3, where the input is a voltage $u(\cdot)$ and the output is the capacitor voltage $y(\cdot)$. It is easy to see that

$$u(t) = LC\ddot{y}(t) + RC\dot{y}(t) + y(t) \tag{32}$$

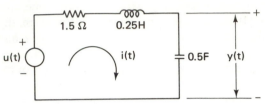

Figure 2.2-3.

so that, with the given values for R, C, L, the transfer function is

$$H(s) = \frac{Y(s)}{U(s)} = \frac{8}{s^2 + 6s + 8}$$

and the impulse response function is

$$h(t) = -4e^{-4t} + 4e^{-2t}, \qquad t \geq 0$$

If we knew the entire input history up to time t and if the system were at rest at $-\infty$, we could calculate the output via the superposition formula,

$$y(t) = \int_{-\infty}^{t} h(t - \tau)u(\tau)\, d\tau$$

But now we may note that by picking any *initial* time t_0 $(t_0 < t)$ we can write

$$y(t) = -4e^{-4(t-t_0)}x_1(t_0) + 4e^{-2(t-t_0)}x_2(t_0) + \int_{t_0}^{t} h(t - \tau)u(\tau)\, d\tau \tag{33a}$$

where

$$x_1(t_0) \triangleq \int_{-\infty}^{t_0} e^{-4(t_0-\tau)}u(\tau)\, d\tau, \qquad x_2(t_0) \triangleq \int_{-\infty}^{t_0} e^{-2(t_0-\tau)}u(\tau)\, d\tau \tag{33b}$$

This shows that, provided we know the values $x_1(t_0)$ and $x_2(t_0)$, the actual input values up to time t_0 are irrelevant. Note also that many different input histories can give rise to the same values $\{x_1(t_0), x_2(t_0)\}$. Therefore the two numbers $\{x_1(t_0), x_2(t_0)\}$ can serve as a very convenient *sufficient statistic* for the entire past up to time t_0. Moreover, t_0 is quite arbitrary, and therefore we can take $\{x_1(t), x_2(t)\}$ as a *state-vector* for the system at any time t.

†This example will also again illustrate the manifold nonuniqueness of state-space descriptions.

By using the rule for differentiation under the integral sign, we obtain the *state equations*

$$\begin{bmatrix} \dot{x}_1(t) \\ \dot{x}_2(t) \end{bmatrix} = \begin{bmatrix} -4 & 0 \\ 0 & -2 \end{bmatrix} \begin{bmatrix} x_1(t) \\ x_2(t) \end{bmatrix} + \begin{bmatrix} 1 \\ 1 \end{bmatrix} u(t) \tag{34}$$

However, it is easy to see that the choice of the state variables is not unique: Any nonsingular linear combination of $x_1(\cdot)$ and $x_2(\cdot)$ will also suffice. In fact, assuming that $u(\cdot)$ is not impulsive at $t = t_0$, we may note from (32) that

$$y(t_0) = -4x_1(t_0) + 4x_2(t_0)$$

Also

$$\dot{y}(t) = 16e^{-4(t-t_0)}x_1(t_0) - 8e^{-2(t-t_0)}x_2(t_0) + h(0)u(t) + \int_{t_0}^{t} \dot{h}(t-\tau)u(\tau)\,d\tau$$

so that at $t = t_0$ we get

$$\dot{y}(t_0) = 16x_1(t_0) - 8x_2(t_0)$$

Therefore

$$\begin{bmatrix} y(t_0) \\ \dot{y}(t_0) \end{bmatrix} = \begin{bmatrix} -4 & 4 \\ 16 & -8 \end{bmatrix} \begin{bmatrix} x_1(t_0) \\ x_2(t_0) \end{bmatrix}$$

and clearly $\{y(\cdot), \dot{y}(\cdot)\}$ can also be used as *state variables*. If we set

$$z_1(\cdot) = y(\cdot), \qquad z_2(\cdot) = \dot{y}(\cdot)$$

then we can get the state equations

$$\begin{bmatrix} \dot{z}_1(t) \\ \dot{z}_2(t) \end{bmatrix} = \begin{bmatrix} 0 & 1 \\ -8 & -6 \end{bmatrix} \begin{bmatrix} z_1(t) \\ z_2(t) \end{bmatrix} + \begin{bmatrix} 0 \\ 8 \end{bmatrix} u(t) \tag{35}$$

Yet another choice of state variables is suggested by our original relation (32), which we can clearly also write as

$$y(t) = -4e^{-4t}\theta_1(t_0) + 4e^{-2t}\theta_2(t_0) + \int_{t_0}^{t} h(t-\tau)u(\tau)\,d\tau$$

where

$$\theta_1(t_0) = \int_{-\infty}^{t_0} e^{4\tau}u(\tau)\,d\tau, \qquad \theta_2(t_0) = \int_{-\infty}^{t_0} e^{2\tau}u(\tau)\,d\tau$$

We can use the $\{\theta_i(\cdot)\}$ as state variables, with state equations

$$\begin{bmatrix} \dot{\theta}_1(t) \\ \dot{\theta}_2(t) \end{bmatrix} = \begin{bmatrix} 0 \end{bmatrix} \begin{bmatrix} \theta_1(t) \\ \theta_2(t) \end{bmatrix} + \begin{bmatrix} e^{4t} \\ e^{2t} \end{bmatrix} u(t) \tag{36}$$

which is now an equation with a zero A matrix and a *time-variant b* matrix. It may seem strange that a time-invariant system has *time-variant* state equations, but reflection will show that this is quite possible since we can admit nonsingular but time-variant transformations of any given set of state variables (see Exercise 2.2-9). We again have here only a slightly more striking example of the manifold nonuniqueness of state-space descriptions of a given system. However, in this book we shall generally confine ourselves to state equations with time-invariant coefficients, except in Chapter 9. ∎

2.2.4 More Names and Definitions†

As stated earlier, the major goal of our study is to try to extract as much qualitative and structural information as we can from various *representations* of the solution rather than from the solution itself. The state equations yield one such representation. For time-invariant systems, an alternative representation is provided by Laplace transformation, which gives

$$X(s) = (sI - A)^{-1}x(0-) + (sI - A)^{-1}bU(s)$$
$$Y(s) = cX(s) \tag{37}$$

where

$$X(s) \triangleq \int_{0-}^{\infty} x(t)e^{-st}\, dt$$

and $U(s)$ and $Y(s)$ are defined likewise. [The matrix notation is shorthand for the fact that the ith element of $X(s)$ is the transform of the ith element of $x(t)$ and so on.]

The matrix $(sI - A)$ is sometimes called the *characteristic matrix* of A. Its determinant,

$$a(s) = \det(sI - A) = s^n + a_1 s^{n-1} + \cdots + a_{n-1}s + a_n$$
$$= (s - \lambda_1)(s - \lambda_2) \cdots (s - \lambda_n)$$

is known as the *characteristic polynomial* of A, and its roots $\{\lambda_i\}$ are called the *eigenvalues* or *characteristic values* of A. The eigenvalues may be complex, provided they occur in complex conjugate pairs (otherwise the coefficients $\{a_i\}$ would not be real, a condition that follows in turn from the reasonable assumption that the matrix A has real-valued elements).

It is clear from (37) that these eigenvalues describe the *natural* or *free* or *unforced* (i.e., $u \equiv 0$) response of the realization $\{A, b, c\}$, and therefore they are often called the *characteristic* or *natural frequencies* (or *modes‡*) of the realization. Note that similar realizations have the same natural frequencies, since

$$a(s) = \det(sI - A) = \det T(sI - A)T^{-1} = \det(sI - T^{-1}AT)$$

†A first reading of this section should be very quick; the definitions will soon become quite familiar or can be looked up when necessary.

‡See also Sec. 2.5.2.

(We should note that the natural frequencies or modes of a realization have to be distinguished from the *poles* of its transfer function; see below.)

A striking property of $a(s)$ is that

$$a(A) = A^n + a_1 A^{n-1} + \cdots + a_n I = 0 \tag{38}$$

This so-called *Cayley-Hamilton theorem*, which will be often used in this course, is discussed in more detail in Sec. 8 in the Appendix.

We also note the important so-called *resolvent formulas*,

$$
\begin{aligned}
\text{Adj}\,(sI - A) &= A^{n-1} + (s + a_1)A^{n-2} + \cdots \\
&\quad + (s^{n-1} + a_1 s^{n-2} + \cdots + a_{n-1})I
\end{aligned} \tag{39a}
$$

$$
\begin{aligned}
&= s^{n-1}I + s^{n-2}(A + a_1 I) + \cdots \\
&\quad + (A^{n-1} + a_1 A^{n-2} + \cdots + a_{n-1}I)
\end{aligned} \tag{39b}
$$

which are discussed further in Exercise A.23. These formulas lead immediately to the expression

$$
\begin{aligned}
b(s) &\triangleq c\,\text{Adj}\,(sI - A)b \\
&= s^{n-1}(cb) + s^{n-2}(cAb + a_1 cb) + \cdots \\
&\quad + (cA^{n-1}b + \cdots + a_{n-1}cb)
\end{aligned} \tag{40}
$$

which we shall use on occasion.

We should also note here that the formal geometric series expansion

$$[sI - A]^{-1} = \frac{1}{s}[I - A/s]^{-1} = \frac{1}{s}[I + A/s + A^2/s^2 + \cdots] \tag{41a}$$

together with the Cayley-Hamilton formula (38) will also lead to (39)–(40). We also remark that taking inverse transforms in (41a) yields

$$
\begin{aligned}
\mathcal{L}^{-1}[sI - A]^{-1} &= I + At + A^2 t^2/2! + \cdots, \qquad t \geq 0 \\
&\triangleq e^{At}, \text{ the matrix exponential}
\end{aligned} \tag{41b}
$$

This function will be discussed in more detail in Sec. 2.5.1.

If common factors are cancelled between $a(s)$ and every element of the numerator matrix in (39a) or (39b), then the resolvent will have the form

$$[sI - A]^{-1} = \Gamma(s)/\mu(s) \tag{41c}$$

where $\mu(s)$ is the *minimal polynomial* of A (cf. Exercise A.29).

Transfer Functions.   The transfer function of the system described by the state-space equations is defined as

$$H(s) = \frac{Y(s)}{U(s)}\bigg|_{x(0-)=0} \tag{42a}$$

and from the Eqs. (37) we see that it can be evaluated as

$$H(s) = c[sI - A]^{-1}b \tag{42b}$$

Although there can be many different realizations of a system (cf. Exercise 2.2-10), they will all necessarily have the same transfer function and the same impulse response,

$$h(t) = \mathcal{L}^{-1}[H(s)] = \mathcal{L}^{-1}[c(sI - A)^{-1}b] \tag{43a}$$

which can be evaluated by using (41b) as

$$h(t) = ce^{At}b \tag{43b}$$

a formula that will be justified in Sec. 2.5 by direct solution of state-space equations.

The expression (42b) for $H(s)$ can be written as, say,

$$c(sI - A)^{-1}b = \frac{c \, \text{Adj} \, (sI - A)b}{\det \, (sI - A)} = \frac{b(s)}{a(s)} \tag{44}$$

Now it may very well happen that $b(s)$ and $a(s)$ have some common factors so that we can write

$$\frac{b(s)}{a(s)} = \frac{b_r(s)}{a_r(s)} \tag{45a}$$

where

$$\{b_r(s), a_r(s)\} \text{ are relatively prime} \tag{45b}$$

i.e., have no common factors (except possibly constants). Relatively prime polynomials are often also called *coprime* polynomials. If we wish to write the transfer function $H(s)$ as a rational fraction, then we must write

$$H(s) = \frac{b_r(s)}{a_r(s)} \tag{46}$$

where $\{b_r(s), a_r(s)\}$ are as in (45). In other words, the transfer function is $c[sI - A]^{-1}b$ expressed in lowest terms.

We note that $b(s)/a(s)$, as defined by (44), is a *representation* of the transfer function, but so of course is

$$\frac{b(s)g(s)}{a(s)g(s)}, \qquad g(s) \text{ arbitrary}$$

The designation *transfer function* is generally reserved for the unique lowest-degree representative, $b_r(s)/a_r(s)$, of the class of rational functions $\{b(s)g(s)/a(s)g(s)\}$. Notice that direct determination of $H(s)$ via (42b) could only give us information about $\{b_r(s), a_r(s)\}$. Nevertheless, the representative $b(s)/a(s)$ of Eq. (44) clearly has a special significance when we talk about a realization $\{A, b, c\}$, and therefore we shall define it as

$$\frac{c \operatorname{Adj} (sI - A)b}{\det (sI - A)} \triangleq \frac{b(s)}{a(s)} \triangleq \text{the } nominal \text{ transfer function}$$

We could also call $b(s)/a(s)$ a polynomial (fraction) description of the realization $\{A, b, c\}$ and should better write it as a couple $\{b(s), a(s)\}$. This is certainly the more mathematically proper way, but it is not so common in engineering discussions. Therefore we shall generally continue to use the nominal transfer function form $H(s) = b(s)/a(s)$, and to avoid being overly pedantic, we shall generally drop the adjective *nominal,* unless the context is insufficient to clarify the distinction between $b(s)/a(s)$ and $b_r(s)/a_r(s)$ when this is significant.

One such occasion† arises in connection with the *poles* and *zeros*‡ of a transfer function, which are the roots of the polynomials $a_r(s)$ and $b_r(s)$, respectively. In particular, the poles are also roots of $a(s) = \det (sI - A)$ and therefore are also eigenvalues of A or characteristic frequencies of a realization $\{A, b, c\}$; however, the converse is not necessarily true, unless $a(s) = a_r(s)$.

External and Internal Descriptions.   The above discussion helps to explain why the transfer function or the impulse response is said to give an *external description* of a system, while the state equations or equivalently the triple $\{A, b, c\}$ or the nominal transfer function $b(s)/a(s)$ give an *internal description.*§ Internal descriptions are highly nonunique—we can modify a triple $\{A, b, c\}$ by similarity transformations, or we can have descriptions with different numbers of states [see Exercise 2.2-10, or think about realizations of $b_r(s)/a_r(s)$ and $b(s)/a(s)$]—but they provide information about the actual realization with which we are dealing, information that may not appear in the external description. This is a relatively obvious statement, but nevertheless

---

†Others will not occur until Secs. 2.4.1 and 3.1.3.

‡We assume these concepts are familiar to the reader from his earlier studies in circuit theory; however, also see Exercises 2.2-18 and 2.2-19.

§An internal description might also be a circuit diagram, for example, or more generally a set of physical components, governed by some interconnection and terminal relations.

lack of attention to it can lead to trouble in designs and analyses based exclusively on external descriptions. The development of the proper concepts and tools for clarifying and resolving such problems is one of the contributions of modern system theory (see, e.g., the discussions in Gilbert [16] and Kalman [17]). We shall begin to encounter some of these new tools and concepts very soon, in Secs. 2.3 and 2.4.

The Markov Parameters.   For convenience of reference, we round out our collection of definitions by introducing the quantities

$$\{cA^{i-1}b, i = 1, 2, \ldots\} \triangleq \text{the } Markov \ parameters \qquad (47)$$

It is easy to see that these quantities are invariant under similarity transformations. In fact, they are uniquely determined by the transfer function,

$$H(s) = c(sI - A)^{-1}b = \sum_{1}^{\infty} h_i s^{-i} \qquad (48a)$$

where, by (41), we can identify

$$h_i = cA^{i-1}b, \qquad i = 1, 2, \ldots \qquad (48b)$$

We may now recall that the $\{h_i\}$ were already introduced in Sec. 2.1 when specifying the parameters of the observability and controllability canonical forms via Eq. (2.1-8d)—in fact, see Exercise 2.2-15.

An important matrix connected with the Markov parameters is

$$M[i, j] = \begin{bmatrix} h_i & h_{i+1} & h_{i+2} & \cdots & h_{i+j} \\ h_{i+1} & h_{i+2} & & & \\ h_{i+2} & & & & \\ \cdot & & \cdot & & \\ \cdot & & \cdot & & \\ \cdot & \cdot & & & \\ h_{i+j} & & & \cdots & h_{i+2j} \end{bmatrix} \qquad (49)$$

The special cases $i = 1, j = n - 1$ and $i = 2, j = n - 1$ will be encountered often. Matrices such as $M[i, j]$ that are constant along the *antidiagonals* are often called *Hankel* matrices.

We may note that if $h(\cdot)$ is the impulse response of the system, defined by

$$h(t) = \mathcal{L}^{-1}H(s) = ce^{At}b$$

then the Markov parameters are the values at the origin of $h(\cdot)$ and its derivatives,

$$cA^i b = h_{i+1} = \frac{d^i}{dt^i}h(t)\Big|_{t=0}, \qquad i = 0, 1, \ldots \qquad (50)$$

The Markov parameters seem somewhat peripheral in the continuous-time case. However, they are important in studying various algebraic properties of state-space realizations, as we already saw in Sec. 2.1 [Eq. (2.1-8d)]. *Moreover, we shall see that they have a very natural and important interpretation in the discrete-time case* [cf. Eqs. (2.3-20)–(2.3-21) of Sec. 2.3.3].

Reprise.   In this section, we have introduced triples $\{A, b, c\}$ as state-space descriptions of a given system. We should appreciate that such descriptions can arise in different ways—as a natural set of equations for a physical system or by mathematical calculations from some other system description, e.g., input-output differential equations or transfer functions.

The one-to-one correspondence with analog-computer simulations gives an intuitive feeling for the notion of *state* and also emphasizes the manifold nonuniqueness of state-space descriptions of a given system. We also made some analysis of transformations between different state-space descriptions; this is an important question, whose further resolution will need the concepts of observability and controllability, to be introduced in the next section.

Finally, at various places and especially in Sec. 2.2.4 we have introduced several names and conventions, which will soon become commonplace through usage.

Exercises

2.2-1.   *Transfer Functions of the Canonical Forms*

Use the properties of companion matrices described in Exercises A.30 and A.31 to verify that $c[sI - A]^{-1}b = b(s)/a(s)$ for the four canonical forms in Sec. 2.1.2.

2.2-2.   *Direct Solution of Simple State-Space Equations*
   a. Solve the state-space equations

$$\dot{x}(t) = \begin{bmatrix} 0 & 1 \\ -1 & 0 \end{bmatrix} x(t), \qquad x(0) = x_0$$

$$y(t) = [1 \quad 0]x(t), \qquad t \geq 0$$

These are the equations for the *simple harmonic motion* of a particle of unit mass.
   b. We have a system described by the state equations

$$\begin{bmatrix} \dot{x}_1(t) \\ \dot{x}_2(t) \end{bmatrix} = \begin{bmatrix} -1 & 0 \\ 1 & 1 \end{bmatrix} \begin{bmatrix} x_1(t) \\ x_2(t) \end{bmatrix} + \begin{bmatrix} -2 \\ 1 \end{bmatrix} u(t), \qquad \begin{bmatrix} x_1(0) \\ x_2(0) \end{bmatrix} = \begin{bmatrix} x_{10} \\ x_{20} \end{bmatrix}$$

$$y(t) = [0 \quad 1]x(t)$$

1. Calculate $x_1(t)$, $x_2(t)$, and $y(t)$ for $t > 0$.
2. What is the transfer function of the system described by these state equations?

3. Does the transfer function give an adequate description of this system? Give reasons for your answer.

2.2-3. *Scaling of Variables*

An important problem in simulation is that of scaling, or choosing the correct units for the variables. Suppose we have a realization given by A, b, c, with the three-dimensional state $x = [x_1 \ x_2 \ x_3]$. Suppose we now change to the variables $[z_1 \ z_2 \ z_3]$, where $z_1 = k_1 x_1$, $z_2 = k_2 x_2$, and $z_3 = k_3 x_3$, and we let $\dot{z} = Fz + gu$, $y = hz$.

a. Write out the matrices F, g, h in terms of the elements of A, b, c and the scale factors k_1, k_2, k_3 (i.e., show the elements f_{ij} of F in terms of the elements a_{ij} of A and the scale factors k_1, k_2, k_3, and similarly for g and h).

b. Suppose we wish to change the time scale and substitute $\tau = a_0 t$ into the equations. Repeat part a, showing how F, g, h depend on the time scale factor a_0 and the elements of A, b, c.

2.2-4. *Realizations of a Nonstrictly Proper H(s)*

Find realizations in controller, observer, controllability, observability, and Jordan canonical forms of the transfer function

$$H(s) = \frac{4s^3 + 25s^2 + 45s + 34}{s^3 + 6s^2 + 10s + 8}$$

Give both state-space equations and block diagrams.

2.2-5. *Ladder Forms*

The reader may know from elementary network synthesis that $H(s)$ can sometimes (e.g., when it is the impedance function of an LC or RC or RL network) be easily realized by first carrying out certain continued-fraction expansions. For example, suppose we can write

$$H(s) = \frac{b_1 s^2 + b_2 s + b_3}{s^3 + a_1 s^2 + a_2 s + a_3} = \cfrac{1}{g_1 s + \cfrac{1}{g_2 + \cfrac{1}{g_3 s + \cfrac{1}{g_4 + \cfrac{1}{g_5 s + \cfrac{1}{g_6}}}}}}$$

and assume also that $g_i > 0$.

a. Show that $H(s)$ can be realized by the triple

$$A = \begin{bmatrix} \dfrac{-1}{g_1 g_2} & \dfrac{-1}{\sqrt{g_3 g_2}\sqrt{g_1}} & 0 \\[3mm] \dfrac{-1}{\sqrt{g_1 g_2}\sqrt{g_3}} & \dfrac{-1}{g_3 g_2} + \dfrac{-1}{g_3 g_4} & \dfrac{-1}{\sqrt{g_3 g_4}\sqrt{g_5}} \\[3mm] 0 & \dfrac{-1}{\sqrt{g_3 g_4}\sqrt{g_5}} & \dfrac{-1}{g_5 g_6} - \dfrac{1}{g_5 g_4} \end{bmatrix}, \quad b = \begin{bmatrix} \dfrac{1}{\sqrt{g_1}} \\[3mm] 0 \\[3mm] 0 \end{bmatrix} = c'$$

Draw a diagram in *ladder form* of this realization.

   b. If the g_i are not all positive, then the above realization cannot be used as it stands, but it can readily be modified so as to work. How?

2.2-6.   Cascade Form

   Draw a block diagram corresponding to the realization

$$
A = \begin{bmatrix}
\lambda_1 & c_2 & c_3 & \cdots & c_{n-1} & 1 \\
 & \lambda_2 & c_3 & & c_{n-1} & 1 \\
 & & \cdot & & c_{n-1} & 1 \\
 & & & \cdot & \cdot & \\
 & \bigcirc & & \ddots & \cdot & \\
 & & & & \lambda_{n-1} & 1 \\
 & & & & & \lambda_n
\end{bmatrix}, \quad
b = \begin{bmatrix}
0 \\ 0 \\ 0 \\ \cdot \\ \cdot \\ 0 \\ b_{n-1}
\end{bmatrix}
$$

$$
c = [c_1 \quad c_2 \quad \cdots \quad c_{n-1} \quad 1]
$$

The block diagram should make clear the reason for the name *cascade form*.

2.2-7.

   Write the wave equation

$$
\frac{\partial^2 u}{\partial t^2} = \Delta u, \qquad \Delta = \sum_1^3 \frac{\partial^2}{\partial x_i^2}, \qquad u = u(x_1, x_2, x_3, t)
$$

in state-variable form, i.e., as a set of *first-order* linear differential equations.

2.2-8.   Interconnections of Subsystems

   Write state equations for two realizations $\{A_i, b_i, c_i\}$ connected in (a) series, (b) parallel, and (c) feedback, with $\{A_1, b_1, c_1\}$ in the forward loop and $\{A_2, b_2, c_2\}$ in the feedback loop.

2.2-9   Time-Dependent Similarity Transformations

   In Eqs. (2.2-15), we discussed the use of similarity transformations

$$
x(t) = Tz(t)
$$

where T is a constant nonsingular matrix. We can also allow time-variant transformations,

$$
x(t) = T(t)z(t)
$$

provided $T(t)$ is nonsingular and differentiable for all t. Find the state equations for $z(\cdot)$ in this case, and show that it is possible to choose $T(\cdot)$ so that the new state-feedback matrix for $z(\cdot)$ is identically zero. Try to find a matrix $T(\cdot)$ to relate the realizations (2.2-34), (2.2-35), and (2.2-36).

2.2-10.   Transfer Functions and Alternative Realizations

   a. If $\{A, b, c\}$ and $\{\bar{A}, \bar{b}, \bar{c}\}$ are related by a constant similarity transformation, show that they have the same transfer function.

b. Realizations can have different numbers of states. Show that the constant realizations

$$\begin{bmatrix} A & A_1 \\ 0 & A_2 \end{bmatrix}, \quad \begin{bmatrix} b \\ 0 \end{bmatrix}, \quad [c \quad q]$$

$$\begin{bmatrix} A & 0 \\ A_1 & A_2 \end{bmatrix}, \quad \begin{bmatrix} b \\ q \end{bmatrix}, \quad [c \quad 0]$$

and $\{A, b, c\}$ all have the same transfer function for all values and (compatible) dimensions of A_1, A_2, q.

2.2-11.

Suppose $\{A_i, b_i, c_i, i = 1, 2\}$ are two realizations of a transfer function $[c_1(sI - A_1)^{-1}b_1 = c_2(sI - A_2)^{-1}b_2]$. In general, even if the A_i have the same order, there may not be a nonsingular linear transformation relating the states of the two realizations. However, if such a transformation does exist, say $x_1(t) = Tx_2(t)$, det $T \neq 0$, prove that we must have

$$A_2 = T^{-1}A_1T, \qquad b_2 = T^{-1}b_1, \qquad c_2 = c_1T$$

Hint: Note that $u(\cdot)$ and $x(0-)$ can be arbitrary.

2.2-12.   *Transformation from Observer to Diagonal Form*

Refer to Example 2.2-A and use duality to determine the transformation between observer form and diagonal form, assuming that A_0 has distinct eigenvalues.

2.2-13.

If $H(s) = b(s)/a(s)$ and $a(s)$ has distinct roots, show that there exists an invertible transformation between the matrices A_c and A_o of the controller and observer realizations of $H(s)$.

2.2-14.   *Two Useful Factorizations of Hankel Matrices*

Refer to Eq. (2.2-49) and show that we can write

$$M[1, j - 1] = \mathcal{O}_j(c, A)\mathcal{C}_j(A, b) \text{ and } M[2, j - 1] = \mathcal{O}_j(c, A)A\mathcal{C}_j(A, b)$$

where

$$\mathcal{O}'_j(c, A) = [c' \quad A'c' \quad \cdots \quad (A')^{j-1}c']', \qquad \mathcal{C}_j(A, b) = [b \quad Ab \quad \cdots \quad A^{j-1}b]$$

2.2-15.

a. Use (2.2-40) to find a relation between the coefficients $\{b_1, \ldots, b_n\}$ of $b(s)$ and the first n Markov parameters $\{h_1, h_2, \ldots, h_n\}$.

b. Is it true that $a_r(s)$, as defined by (2.2-45), is equal to the minimal polynomial of A, as defined by (2.2-41c)? Give reasons for your answer.

2.2-16.   *The Relative Order of a System*

a. Show that the numerator polynomial of the transfer function of a system realization has degree m if and only if

$$cA^ib = 0 \qquad \text{for } i = 0, 1, 2, \ldots, n - m - 2$$

and

$$cA^{n-m-1}b \neq 0$$

b. Conversely, if the difference between the degrees of the numerator and denominator polynomials of a given transfer function is p and if (A, b, c) is a realization of this transfer function, then how many and which of the $\{cA^ib\}_{i=0}^{n-1}$ must be zero? The integer p is known as the *relative order* of the system.

2.2-17. *More Resolvent Identities*

    a. Show, when all inverses exist, that

$$(sI - A)^{-1} - (sI - B)^{-1} = (sI - A)^{-1}(A - B)(sI - B)^{-1}$$

and

$$(sI - A)^{-1} - (vI - A)^{-1} = (sI - A)^{-1}(v - s)(vI - A)^{-1}$$

    b. Use the above results to show that for a realization $\{A, b, c\}$ with

$$u(t) = e^{vt} \cdot 1(t)$$

the output can be written as

$$\mathcal{L}[y(t)] = c(sI - A)^{-1}[x_0 - (vI - A)^{-1}b] + c(vI - A)^{-1}(s - v)^{-1}b$$

    c. Show that a necessary and sufficient condition for an input $u(t) = e^{zt}g$, $t \geq 0$, to yield $y(t) \equiv 0$, $t \geq 0$, is that there exist $\{x_0, g\}$ such that

$$\begin{bmatrix} zI - A & -b \\ c & 0 \end{bmatrix} \begin{bmatrix} x_0 \\ g \end{bmatrix} = \begin{bmatrix} 0 \\ 0 \end{bmatrix}$$

Show that in this case we also have

$$x(t) = e^{zt}x_0, \qquad t \geq 0$$

(Reference: A. MacFarlane and N. Karcanias, *Int. J. Control*, Vol. 24, pp. 33–74, 1976.)

2.2-18. *Dynamical Interpretation of Poles and Zeros*

    a. Let $H(s) = c(sI - A)^{-1}b$, with $a(s) = \det(sI - A)$, $b(s) = c$ Adj $(sI - A)b$ being coprime. Suppose v is not an eigenvalue of A. Show that there exists an initial state x_0 such that the response to $u(t) = e^{vt} \cdot 1(t)$ is $y(t) = H(v)e^{vt} \cdot 1(t)$.

    b. What happens if v is a zero of $H(s)$?

    c. Suppose v is an eigenvalue of A and therefore a pole of $H(s)$. Show that there exists an initial state x_0 such that with no input $[u(t) \equiv 0]$ the response $y(t)$ has the form $\alpha e^{vt} \cdot 1(t)$, $\alpha =$ some constant. Assume that A has distinct eigenvalues.

2.2-19.   *Dynamical Interpretation of Poles*

Suppose $H(s) = b(s)/a(s)$ is irreducible [i.e., $b(s)$ and $a(s)$ are coprime], with deg $b(s) = m \le$ deg $a(s) = n$. Let

$$U(s) = \mathcal{L}[u(t)] = (s - v)^{-1} + p(s)$$

where $p(s)$ is some $(n - 1)$th-degree polynomial.

a. Show that we can write

$$Y(s) = \frac{b(v)}{a(v)} \frac{1}{s - v} + \frac{b(s)p(s) + r(s)}{a(s)}$$

for some polynomial $r(s)$ of degree $n - 1$ or less. [You may wish to use the *remainder theorem*: If a polynomial $q(s)$ is divided by $s - v$, the remainder is $q(v)$; i.e., $q(s) = m(s)(s - v) + q(v)$.]

b. Show that by proper choice of $p(s)$ we can make

$$y(t) = H(v)e^{vt} \cdot 1(t) \qquad \text{for } t > 0$$

Reconcile this with the results in Exercise 2.2-18. *Hint:* $[r(s) + b(s)p(s)]/a(s)$ should be polynomial. Can this be arranged? (See Sec. 2.4.4.)

2.2-20.   *Inverse of a Realization*

a. If $\{A, b, c, d\}$, $d \ne 0$, is a realization with $H(s) = d + c(sI - A)^{-1}b$, show that $\{A - (bc/d), b/d, -c/d, 1/d\}$ is a realization for a system with transfer function $1/H(s)$.

b. If we are given $\{A, b, c, d\}$, $d \ne 0$, show that the zeros of $c(sI - A)^{-1}b + d$ can be computed as the eigenvalues of the matrix $A - bd^{-1}c$.

c. Show also that the zeros can be computed by solving the generalized eigenvalue problem

$$(\lambda E - F)p = 0$$

where

$$E = \begin{bmatrix} I & 0 \\ 0 & 0 \end{bmatrix}, \qquad F = \begin{bmatrix} A & -b \\ c & -d \end{bmatrix}$$

This method is numerically much less sensitive to errors in computation than methods going via initial calculation of the numerator polynomial of $c(sI - A)^{-1}b + d$. (Reference: I. Kaufman, *IEEE Trans. Circuits Syst.*, CT-20, pp. 93–100, March 1973.)

2.2-21.   *Linearization*

a. Linearize the time-invariant nonlinear equations $\dot{h}(t) = -k_2 k_1 \sqrt{h(t)} + k_2 u(t)$, $y(t) = k_1 \sqrt{h(t)}$ about the nominal solution $u_0 \equiv 0$, $2\sqrt{h_0(t)} = 2\sqrt{c} - k_1 k_2 t$. [$c = h_0(0)$.] *Note that the linearized equations are time variant.*

b. The state equations of an inverted pendulum can be shown to be $\dot{x}_1(t)$ $= x_2(t)$, $\dot{x}_2(t) = (g/l) \sin x(t) + u(t)$. Show that the linearized equations about the equilibrium solution $x_1(\cdot) = x_2(\cdot) = u(\cdot) \equiv 0$ have

$$A. = \begin{bmatrix} 0 & 1 \\ g/l & 0 \end{bmatrix}, \quad b = \begin{bmatrix} 0 \\ 1 \end{bmatrix}$$

2.2-22.   A Constant-Resistance Network

a. Show that a realization for the circuit shown in the figure [$u(\cdot)$ is a current and $y(\cdot)$ a voltage] can be written as

$$\dot{x}(t) = \begin{bmatrix} -2R/L & 1/L \\ -1/C & 0 \end{bmatrix} x(t) + \begin{bmatrix} R/L \\ 1/C \end{bmatrix} u(t)$$

$$y(t) = [-R \quad 1]x(t) + Ru(t)$$

if we choose $x_1(t) = i_L(t)$, $x_2(t) = V_c(t)$.

b. Show that the transfer function is given by

$$H(s) = \frac{Y(s)}{U(s)} = \frac{Rs^2 + [(1/C) + (R^2/L)]s + (R/LC)}{s^2 + (2R/L)s + (1/LC)}$$

Note that when $R^2 = L/C$ the transfer function is a constant, $H(s) = R$, for all values of s. This is known as a *constant-resistance network*.

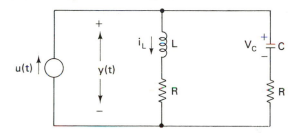

2.2-23.   Another Constant-Resistance Network

Write state equations for the circuit shown in the figure, where $u(\cdot)$ is a current and $y(\cdot)$ a voltage. Determine constraints on $\{L, C, R\}$ that will yield a constant-resistance network.

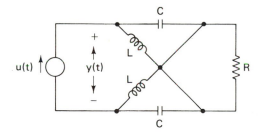

Some Exercises on the Phase Plane.

2.2-24.   *Simple Harmonic Motion*

a. Given the state equations of simple harmonic motion $[x_1(\cdot) = y(\cdot)$, the position, $x_2(\cdot) = \dot{y}(\cdot)$, the velocity],

$$\begin{bmatrix} \dot{x}_1(t) \\ \dot{x}_2(t) \end{bmatrix} = \begin{bmatrix} 0 & 1 \\ -1 & 0 \end{bmatrix} \begin{bmatrix} x_1(t) \\ x_2(t) \end{bmatrix}, \qquad x_1(0) = 1 = x_2(0)$$

make a sketch of the solution as it would appear on the *phase plane*, with the values $x_1(\cdot)$ along the horizontal axis and $x_2(\cdot)$ along the vertical axis.

b. Given any particular point $[x_1(t), x_2(t)]$, sketch the *velocity* vector $[\dot{x}_1(t), \dot{x}_2(t)]$ at this point. Deduce that for all initial conditions $[x_1(0), x_2(0)]$ the phase-plane trajectory will be a circle centered at the origin. Deduce also that the *period* of the motion $x_1(\cdot)$ is independent of the *amplitude* of the motion, a result first realized by Galileo as he idly watched a swinging pendulum in a Florentine church some 400 years ago. Confirm your answers analytically, using the explicit solutions for $x_1(\cdot)$ and $x_2(\cdot)$.

2.2-25.   *Proportional Feedback*

Consider a body of unit mass moving along a line under the influence of a force u. Let $y(t) = $ its displacement at time t, and let $\{y_0, \dot{y}_0\}$ be its initial position and velocity.

a. Let $u = -y$; solve the equations and plot trajectories in the phase plane for

$$\dot{y}_0 = y_0 = 1, \qquad \dot{y}_0 = y_0 = -1, \qquad \dot{y}_0 = y_0 = 4$$

(*Extension:* Let $u = ky$, and discuss the effect of k on the shape and period of the trajectories.) Trajectories behaving as in this part are said to have a *center* at $\dot{y} = y = 0$. Why? Discuss the suitability of $u = -y$ as a control. Find the characteristic equation and its roots. To what root locations does a center correspond?

b. Since part a was not satisfactory, we add velocity feedback and let $u = -3\dot{y} - 2y$ (the numbers are chosen with forethought to give a particular result). Plot the phase-plane portrait and discuss the suitability of this control. [To aid the sketch here, use the *isoclines*: Form $d\dot{y}/dy = \ddot{y}/\dot{y}$, which is the local slope of the path. Set $\ddot{y}/\dot{y} = d\dot{y}/dy = m = $ constant, for example, $m = \infty, 0, 1, -1$. Solve for the curves (lines) where the path slope is known, using the system equations to relate $\ddot{y}$ to y and $\dot{y}$. On these lines, put segments of slope m to guide the sketch. Consider especially the cases $m = -1$ and $m = -2$.] The portrait here is called a *node*. Where are the roots of the characteristic equation?

2.2-26.   *Nonlinear Feedback* ([14] *and* [15])

A proportional feedback can often give excellent results. But a transducer to provide proportional feedback can be expensive. For simplicity and ruggedness, one often uses a relay which is only "on" or "off":

$$u = -\operatorname{sgn} y \triangleq \begin{cases} -1, & y > 0 \\ +1, & y < 0 \end{cases} = -\frac{y}{|y|}$$

(where sgn = *signum* = "sign of").

a. Apply the force $u = -\operatorname{sgn} y$, and sketch the phase-plane paths. Is this control satisfactory?

b. Apply the force $u = -\operatorname{sgn}(y + \dot{y})$, and repeat part a. Include the *switching curve* $\dot{y} = -y$ on your sketch. Show especially that for this *control law* there is a particular solution curve that has *exactly one* reversal of sign of control, u. Show also that for other curves there comes a time when *no solution* is possible. Can you account for this? What do you think happens here? What would a *physical* relay do?

c. Show finally a switching curve that *never* has more than one reversal of sign of control.

Remarks: This nonlinear feedback control problem is a famous one in the theory of automatic control. The fact that the solution curve in part c is *time-optimal* in a certain sense was first proved in the mid-1950s by D. Bushaw and was a spur to the study of optimization techniques. For example, now the result can be obtained by an easy application of the Pontryagin maximum principle (see, e.g., [15, pp. 182–184]).

2.3   INITIAL CONDITIONS FOR ANALOG-COMPUTER SIMULATION; OBSERVABILITY AND CONTROLLABILITY FOR CONTINUOUS- AND DISCRETE-TIME REALIZATIONS

In the first two parts of this section we shall study some simple problems that will introduce two basic concepts—controllability and observability—associated with any realization $\{A, b, c\}$. This discussion will call on some of the matrix theory material in Secs. 2, 4, and 8 of the Appendix.

In Sec. 2.3.3, we further explore and reinforce the ideas and results of all the previous sections by reexamining them for discrete-time systems. We can notice many parallels and also some simplifications in the discrete-time case. In particular, we show how discrete-time simulations can be used to better understand several earlier results and also to naturally indicate some useful new ones.

An important thing to notice in all the discussions is the value of the special realizations of Sec. 2.1 in understanding and illuminating some apparently purely mathematical problems.

As a partial review, Sec. 2.3.4 contains several worked examples to illustrate how various earlier results can be combined to solve some interesting problems.

2.3.1 Determining the Initial Conditions; State Observability

In Sec. 2.1 we found various forms in which a system described by a nth-order input-output differential equation could be realized. However, we had not completed the simulations because of our assumption that the initial values were zero. The problem now facing us is to explain how given nonzero initial conditions $\{y(0-), \dot{y}(0-), \ldots, y^{(n-1)}(0-)\}$ can be translated into appropriate initial values $\{x_i(0-)\}$ for the state variables of any particular realization (or, more concretely, for the integrators in the corresponding analog-computer simulation).

Clearly the first step is to find a convenient expression for $\{y(t), \ldots, y^{(n-1)}(t)\}$. It is easy to see that with

$$\dot{x}(t) = Ax(t) + bu(t), \qquad y(t) = cx(t) \tag{1}$$

we can write

$$y(t) = cx(t)$$
$$\dot{y}(t) = c\dot{x}(t) = cAx(t) + cbu(t)$$
$$\ddot{y}(t) = cA\dot{x}(t) + cb\dot{u}(t)$$
$$\qquad = cA^2x(t) + cAbu(t) + cb\dot{u}(t)$$

and so on, which can be conveniently arranged in matrix form as, say,

$$\mathcal{Y}(t) = \Theta x(t) + \mathbf{T}\mathcal{U}(t) \tag{2}$$

where

$$\mathcal{Y}(t) \triangleq [y(t) \quad \dot{y}(t) \quad \cdots \quad y^{(n-1)}(t)]' \tag{2a}$$
$$\mathcal{U}(t) \triangleq [u(t) \quad \dot{u}(t) \quad \cdots \quad u^{(n-1)}(t)]' \tag{2b}$$
$$\Theta \triangleq \Theta(c, A) \triangleq [c' \quad A'c' \quad \cdots \quad (A')^{n-1}c']' \tag{2c}$$

and

$$\mathbf{T} = \text{a lower triangular Toeplitz matrix with}$$
$$\text{first column } [0 \quad cb \quad \cdots \quad cA^{n-2}b]' \tag{2d}$$

Assuming that $\mathcal{U}(0-) = 0$, we then have

$$\mathcal{Y}(0-) = \Theta x(0-) \tag{3}$$

and the question is whether we can find an *initial-state* vector $x(0-)$ for a given vector $\mathcal{Y}(0-)$. This is a problem in the theory of linear equations (cf. Sec. 4 in the Appendix), and the question is whether $\mathcal{Y}(0-)$ can be formed by a linear combination of the n columns of the matrix Θ.

If these n columns are all linearly independent, i.e., if the $n \times n$ matrix Θ is nonsingular, then we can always find $x(0-)$, for *any* n-vector $\mathcal{Y}(0-)$, as $x(0-) = \Theta^{-1}\mathcal{Y}(0-)$.

On the other hand, if Θ is singular (not of full rank), some columns of Θ will be linearly dependent on others, and therefore we can only find a solution $x(0-)$ for special choices of $\mathcal{Y}(0-)$, namely, those that lie in the (column) range space of Θ.

Now in a differential equation it is generally important to have full freedom of choice in the initial conditions, and this means that of the many possible state-space realizations of the differential equation we should be most interested only in those for which the matrix Θ is such that

$$\Theta(c, A) \text{ has full rank} \qquad (4)\dagger$$

Such realizations are said to be *observable*, and the matrix $\Theta(c, A)$ is called the *observability matrix* of the pair $\{c, A\}$. (These concepts will appear in a more general framework soon.) We first note that, in the present case where Θ is a square matrix, this is equivalent to the condition that

$$\Theta(c, A) \text{ is nonsingular} \qquad (4a)$$

The State-Observability Problem.   The significance of condition (4) stands out even more clearly in a different but closely related problem. Suppose we have a physical system described by the state-space equations

$$\dot{x}(t) = Ax(t) + bu(t), \qquad t \geq 0$$
$$y(t) = cx(t), \qquad x(0) = x_0$$

We assume that we know the matrices $\{A, b, c\}$ and also the input and output functions $\{u(t), t \geq 0\}$, $\{y(t), t \geq 0\}$. The problem is to determine the states $\{x(t), t \geq 0\}$. Such problems can arise in many contexts, e.g., for diagnostic reasons or for purposes of control; several examples will be given in this book.

Now this problem is very close to the initial-condition problem discussed above, because really the only unknown in our problem is the *initial* state x_0: Knowing x_0, $\{A, b, c\}$, and $\{u(t), t > 0\}$, we could set up the equation

$$\dot{x}(t) = Ax(t) + bu(t), \qquad x(0) = x_0$$

and thus obtain $x(t)$ as a function of t.

In fact, if we can determine its value at any time, say t_1, then we can again obtain $x(t)$, $t \geq 0$, by solving the differential equation

$$\dot{x}(t) = Ax(t) + bu(t), \qquad x(t_1) \text{ given}$$

---

†We write $\Theta(c, A)$ from time to time to emphasize the dependence on the given realization.

Now determining $x(t_1)$ brings us back to Eq. (2), which we shall write as

$$\Theta(c, A)x(t_1) = \mathcal{Y}(t_1) - \mathbf{T}\mathcal{U}(t_1)$$

However, when Θ is singular (not of full rank), there is a slight difference from the analog-computer problem. In that problem, we had to assume that the right-hand side [$\mathcal{Y}(0-)$ in that case] was in the (column) range space of the matrix Θ; otherwise the equations had no solution. In the present problem, we start with a realization $\{A, b, c\}$ and an input $u(\cdot)$ and *obtain* $y(\cdot)$ *from it*: Therefore $\mathcal{Y}(t_1) - \mathbf{T}\mathcal{U}(t_1)$ has to lie in the range of Θ.† Thus, whether or not Θ is singular, our equation (2) will always have a solution in the state-determination problem. This would seem to dilute the significance of condition (4), except for a fact that we have not yet taken into account: When Θ is singular, there will be more than one solution of the consistent equations (2). When Θ is singular, it has a nontrivial *null-space*; i.e., there will exist non-zero vectors θ such that

$$\Theta(c, A)\theta = 0$$

Any such θ can be added to a given solution to obtain yet another solution. This degeneracy is of less consequence in the analog-computer problem, where any one of these solutions will suffice to match the initial conditions $\mathcal{Y}(0-)$. However, in the state-determination problem, also called the problem of "observing" the states, this loss of uniqueness is usually fatal. For one thing, we are generally interested in the actual values of the states and not in a whole family of possible values; second, this loss of uniqueness would make meaningless the determination of the state behavior over any time interval.

In other words, the nonsingularity of the matrix $\Theta(c, A)$ is crucial to the problem of observing the states. For these reasons $\Theta(c, A)$ is called the *observability matrix*, and a realization $\{A, b, c\}$ with a full-rank (nonsingular) observability matrix is said to be *observable*.

The question arises of whether matters can be helped when $\Theta(c, A)$ is not of full rank by using further derivatives $\{y^{(n)}(t), y^{(n+1)}(t), \text{etc.}\}$. The answer is no. Taking more derivatives effectively just gives us more equations for $x(t)$, with coefficients cA^n, cA^{n+1}, and so on. But here we recall the important Cayley-Hamilton theorem, which says that for any $n \times n$ matrix A, A^n is a linear combination of the lower powers of A, $\{A^0(= I), A^1, \ldots, A^{n-1}\}$—see Sec. 8 in the Appendix. Therefore all terms $\{cA^{n+i}, i \geq 0\}$ will be linearly dependent on the rows of $\Theta(c, A)$, and no more information about $x(t)$ can be obtained by considering higher-order derivatives of $y(t)$.

---

†Of course we are assuming no errors in the measurement of $y(\cdot)$ and the calculation of $\dot{y}(\cdot)$, $\ddot{y}(\cdot)$, etc. With measurement error we have a different (statistical) problem, which we shall not treat here. See, however, some brief remarks in Secs. 6.2, 9.2 and the final paragraph of this section.

In fact we can say much more: Since $y(\cdot)$ satisfies a differential equation, it is a fairly well-behaved function and can be proved to have a Taylor series expansion around any point. But this means that knowledge of $y(\cdot)$ and all its derivatives at any point t will actually determine the function $y(\cdot)$ at all other points in its region of definition. Therefore if we cannot determine $x(t)$ from $\{y(t), \dot{y}(t), \ldots\}$, i.e., if the realization is not observable, then we cannot determine $x(t)$ in any way even from knowledge of $y(\cdot)$ over a whole (future or past) interval.[†]

It is important to stress that not all realizations need be observable—it depends entirely on the particular pair $\{c, A\}$. However, the two realizations in Sec. 2.1.1 that we called observability [cf. Fig. 2.1-7 and Eq. (2.2-5)] and observer [cf. Fig. 2.1-9 and Eq. (2.2-9)] forms have a particular significance here: They are always guaranteed observable.

By direct calculation we find that

$$\mathcal{O}_{ob} = \mathcal{O}(c_{ob}, A_{ob}) = I \tag{5}$$

so that, for this realization, the determination of $x(0-)$ is trivial, $x(0-) = \mathcal{Y}(0-)$. Examination of the block diagram (Fig. 2.1-7) will show graphically why this is so [recall our assumption that $\mathcal{U}(0-) = 0$].[‡] The simplicity that the *observability form* brings to the state-observation (or state-determination) problem is the reason for its name.

The observer form is not quite so convenient,[§] though it is simple enough: Direct calculation will yield the nice formula

$$\mathcal{O}_o^{-1} = \mathcal{O}^{-1}(c_o, A_o) = \mathcal{Q}_- \tag{6}$$

where

$$\mathcal{Q}_- = \text{a lower triangular Toeplitz matrix}$$

$$\text{with first column } [1 \quad a_1 \quad \cdots \quad a_{n-1}]'$$

Again this fact stands out fairly vividly if the block diagram of the observer form (Fig. 2.1-9) is now examined (starting with the last integrator).

However, scrutiny of the block diagrams for the controller (Fig. 2.1-4) and controllability (Fig. 2.1-10) forms will show that the state-determination problem is no longer so easy, and we really have to solve some simultaneous

---

[†]Note that we have shown this only for constant-parameter continuous-time realizations. In Sec. 2.3.3, we shall see that a more refined analysis can be carried out for discrete-time systems.

[‡]It may be worth noting here explicitly that in the problem of state determination at any time, not just $t = 0-$, one can assume $u(\cdot) \equiv 0$, without loss of generality, since the effect of a nonzero (but known) $u(\cdot)$ is merely to change the right-hand side in the key equation $\mathcal{O}x(t) = \mathcal{Y}(t) - T\mathcal{U}(t)$ to some new known vector. This is often a very convenient assumption.

[§]"The" problem for which the observer form is most natural will be described in Sec. 4.1.

equations to find $x(t)$. In fact we have to invert $\Theta(c, A)$, and for these forms it may or may not happen that $\Theta(c, A)$ has full rank (i.e., is invertible). But if this is the case, why then should we bother with these forms? One reason is that the duality we noted earlier in Sec. 2.2.1 indicates that there must be some problems for which these forms are most "natural"; more generally, several factors enter into the choice of realizations in practical problems.

In this connection, we should note that except perhaps at the point $t = 0-$, where they may be specified, it is clearly difficult to *measure* the values $\{y(t), \dot{y}(t), \ldots, y^{(n-1)}(t), u(t), \ldots, u^{(n-2)}(t)\}$, since differentiation amplifies "noise." Therefore more realistic observation schemes will have to be developed, with inevitable loss in the accuracy of the state "estimates"; a way of doing this will be developed later in Sec. 4.1 (see also Sec. 9.2). Moreover, we should note that the discrete-time analog of the procedure of this section is quite realistic, as will soon be shown in Sec. 2.3.3.

Reprise.   We have defined a realization $\{A, b, c\}$, or just a pair $\{c, A\}$, as being *observable* if its observability matrix $\Theta(c, A)$ has full rank. The presence or absence of this property, together with the fact that the observability and observer realizations are clearly observable, will be helpful in many problems of system theory. In this section we discussed two simple examples of such problems. One was the determination of the initial state of a realization given an arbitrary set of initial conditions $\{y(0-), \ldots, y^{(n-1)}(0-)\}$. The other was the determination of the state at any time given full knowledge of the input and output functions for $t > 0-$. We only gave an idealized (physically unrealistic) solution of the second problem. While more realistic solutions will be developed later (Secs. 4.1 and 9.2), it is worth emphasizing that both problems provide alternative equivalent characterizations of the property of observability. As we shall see on several later occasions, the whole point of having alternative characterizations is that they further illuminate a property, and moreover in new problems one characterization might be more convenient to use than another.

2.3.2 Setting Up Initial Conditions; State Controllability

We have seen that for an observable realization the proper initial conditions (and, in fact, the state at any time) for the realization can be calculated from the input and output functions and their derivatives. It is then natural to ask [prompted also by the duality that we have noticed between the observer (observability) and controller (controllability) forms] how, for a given simulation, we can actually set up any desired initial conditions, i.e., how we can find a suitable input $u(\cdot)$ that will take the system to any desired initial state in a finite (often very "small") time.†

---

†Although we pose this problem in the context of analog-computer simulations, it also applies to any physical system whose state we would like to change at a given time, e.g., as in a *midcourse* correction for a rocket.

To see how this might be done, consider first the simplest case of a scalar (single-integrator) system

$$\dot{x}(t) = ax(t) + bu(t), \qquad t > 0-$$

and observe that if

$$u(t) = g\delta(t)$$

then

$$x(t) = bg \cdot 1(t) + \text{(a continuous function)}, \qquad t > 0-$$

and

$$x(0+) - x(0-) = bg$$

Therefore the system can be taken to any desired initial condition x_0 by suitably choosing g. Moreover, the desired condition can be set up in "zero" time because of the impulsive nature of the input. This is impractical, of course, though it may be noted that in many problems "approximate" impulsive functions can be satisfactorily generated and used—it all depends on the "time constants" of the system under study. State-controllability problems with nonimpulsive inputs will be studied in Sec. 9.2. Here our goal is to pursue an analysis *dual* to that of Sec. 2.3.1.

To generalize the result just obtained, let us (again partly motivated by duality) consider next the controllability canonical form of Fig. 2.1-10. We notice from that figure that if

$$u(t) = g_1\delta(t)$$

then

$$\dot{x}_1(t) = g_1\delta(t) + [-a_3x_3(t)]$$
$$= g_1\delta(t) + \text{(nonimpulsive functions)}$$

and

$$x_1(t) = g_1 \cdot 1(t) + \text{(continuous function)}$$
$$+ [\text{response to } x_1(0-)]$$

Therefore

$$x_1(0+) = g_1 + x_1(0-)$$

and $x_1(0+)$ can be arbitrarily set by properly choosing g_1. It is perhaps not so easy to see how to set $x_2(0+)$ and $x_3(0+)$. However, note that if

$$u(t) = g_1\delta(t) + g_2\delta^{(1)}(t)$$

then

$$\dot{x}_2(t) = g_2\delta(t) + \text{(nonimpulsive functions)}$$

and

$$x_2(t) = g_2 \cdot 1(t) + \text{(continuous functions)}$$
$$+ [\text{response to } x_2(0-)]$$

so that

$$x_2(0+) = g_2 + x_2(0-)$$

which can be set arbitrarily by proper choice of g_2.

But what will happen to $x_1(0+)$ when $u(t) = [g_1\delta(t) + g_2\delta^{(1)}(t)]$ rather than just $g_1\delta(t)$? It is easy to see that $x_3(t)$ is still nonimpulsive so

$$\dot{x}_1(t) = g_1\delta(t) + g_2\delta^{(1)}(t) + \text{(nonimpulsive functions)}$$

Integrating both sides from $0-$ to $0+$ gives

$$x_1(0+) - x_1(0-) = \int_{0-}^{0+} \dot{x}_1(t)\,dt = g_1 \int_{0-}^{0+} \delta(t)\,dt + g_2 \int_{0-}^{0+} \delta^{(1)}(t)\,dt$$
$$+ \int_{0-}^{0+} \text{(nonimpulsive function)}\,dt$$
$$= g_1 + 0 + 0 = g_1$$

as before. This shows how to proceed given a general nth-order *controllability canonical form*: Let

$$u(t) = g_1\delta(t) + g_2\delta^{(1)}(t) + \cdots + g_n\delta^{(n-1)}(t)$$

Then arguments similar to those just used will show that

$$x_i(0+) = g_i + x_i(0-)$$

Thus, for the controllability form, it is easy to *set up* arbitrary initial conditions, just as it was easy to *determine* initial conditions on the dual observability form. But now, as before, we have the question of whether we can find inputs to set up arbitrary initial conditions for a general realization $\{A, b, c\}$. *By duality*, we would expect the answer to be [cf. (3)] yes if and only if the matrix

$$\mathcal{C} = [b \quad Ab \quad \cdots \quad A^{n-1}b] \quad \text{has full rank} \tag{7}$$

i.e., is nonsingular ($\mathcal{C}$ is a square matrix). $\mathcal{C}$ is called the *controllability matrix†* of the particular realization, and a realization for which $\mathcal{C}$ is non-

---

†For reasons that will appear later (see Sec. 2.3.3) this matrix is perhaps better called the *reachability matrix* or perhaps (see Sec. 3.2) the *modal controllability* matrix, but the above terminology is by now well entrenched.

singular is said to be *controllable*. Moreover, as we would expect by duality [cf. (5)], we can show that for the controllability canonical form

$$\mathcal{C}_{co} = I_{nxn} \tag{8}$$

and for the controller form

$$\mathcal{C}_c^{-1} = \mathcal{Q}'_- \tag{9}$$

where [cf. (6)]

$$\mathcal{Q}'_- = \text{an upper triangular Toeplitz matrix}$$
$$\text{with first row } [1 \quad a_1 \quad \cdots \quad a_{n-1}]$$

and the $\{a_i\}$ are the coefficients of the characteristic polynomial of A.

Although the arguments via duality can be made quite completely, it will be useful to have a general proof of our claim that condition (7) is necessary and sufficient to solve our problem. For this, we introduce the following important result.

A General Formula for x(0+) with Impulse Inputs.   If

$$\dot{x}(t) = Ax(t) + bu(t)$$

and

$$u(t) = g_1\delta(t) + \cdots + g_k\delta^{(k-1)}(t), \qquad k \geq 1 \tag{10a}$$

then

$$x(0+) = x(0-) + [b \quad Ab \quad \cdots \quad A^{k-1}b] \cdot$$
$$[g_1 \quad g_2 \quad \cdots \quad g_k]' \tag{10b}$$

This important formula can be proved in several ways.† Here we give a proof that exploits the concept of superposition. Let

$$h(\cdot) = \text{the impulse response of } \dot{x} = Ax + bu$$
$$= \text{the response to } u(t) = \delta(t) \text{ with } x(0-) = 0$$

By linearity it follows that the response to the input

$$u(t) = g_1\delta(t) + g_2\delta^{(1)}(t) + \cdots + g_k\delta^{(k-1)}(t)$$

is

$$x(t) = g_1h(t) + g_2h^{(1)}(t) + \cdots + g_kh^{(k-1)}(t)$$
$$+ \text{(responses to zero input but nonzero initial conditions)}$$

†The simplest uses the fact (easy to prove from first principles) that $x(t) = \int_0^t e^{A(t-\tau)}bu(\tau)d\tau$ if $x(0-) = 0$ (see also Sec. 2.5.1).

Therefore

$$x(0+) = g_1 h(0+) + g_2 h^{(1)}(0+) + \cdots + g_k h^{(k-1)}(0+) + x(0-)$$

Now

$$u(t) = 0, \qquad t \geq 0+$$

so that

$$\dot{h}(t) = Ah(t), \qquad t \geq 0+$$

and

$$h^{(i+1)}(t) = Ah^{(i)}(t), \qquad t \geq 0+$$

In particular, setting $t = 0+$ and recalling (from our initial arguments in this section) that $h(0+) = b$, we shall have

$$h^{(i)}(0+) = A^i h(0+) = A^i b$$

Therefore

$$
\begin{aligned}
x(0+) &= x(0-) + g_1 b + g_2 Ab + \cdots + g_k A^{k-1} b \\
&= x(0-) + [b \quad Ab \quad \cdots \quad A^{k-1} b][g_1 \quad g_2 \quad \cdots \quad g_k]'
\end{aligned}
$$

which is Eq. (10).

State Controllability.   It is easy to see from (10), with $k = n$, that if $\mathcal{C}$ is nonsingular, the coefficients $\{g_i\}$ can be determined so as to (instantaneously) set up *arbitrary* initial conditions. On the other hand, if $\mathcal{C}$ is singular, then it is true that no matter what the $\{g_i\}$, certain vectors $x(0+) - x(0-)$ cannot be obtained from Eq. (10). The columns of $\mathcal{C}$ will be linearly dependent, and we shall not be able to obtain an *arbitrary* n-vector as a linear combination of these columns.† Moreover, as in the case of observability, the Cayley-Hamilton theorem shows that if $\mathcal{C}$ is singular, then higher orders of impulsive inputs will not provide any more linearly independent equations that can serve to determine an appropriate input.

Therefore we have proven that the nonsingularity of $\mathcal{C}$ is necessary and sufficient for the existence of an impulsive input that will change the state from a given value $x(t-)$ (the choice $t = 0$ is just a special case) to an arbitrary desired value at $t+$. This condition is called *state controllability*, and, like observability, it will be seen to be a fundamental property of a realization. As we might expect from the fact that arbitrary inputs can be regarded as an appropriate collection of impulsive inputs, the nonsingularity of $\mathcal{C}$ is also a necessary and sufficient condition for being able to change the

---

†If we have $x(0+) - x(0-)$ in the range of $\mathcal{C}$, then of course a solution always exists by definition (cf. Sec. 4 in the Appendix) whether or not $\mathcal{C}$ is nonsingular.

state arbitrarily in a finite time (i.e., with nonimpulsive inputs). To prove this, we shall need to know how to solve the state equations, and therefore a formal proof is postponed to Example 2.5-1.

A natural question raised by the results of Secs. 2.3.1 and 2.3.2 is whether there are realizations that are both controllable and observable, or, perhaps more fundamentally, why there should be any realizations that are not controllable, or not observable, or even neither? In fact, the reader may wonder why, if the above concepts are so important and if analog-computer realizations are so old (going back about 100 years to Kelvin and about 30 or 40 years to Bush), why it has taken so long for these concepts to be uncovered (basically in the early 1960s).

Insofar as *controllability* goes, one might say that the problem of setting up initial conditions by choice of a suitable system input $u(\cdot)$ is not relevant to analog-computer simulations. In such simulations, all integrators are generally directly accessible, and any desired initial conditions can be individually placed on each integrator (by a simple *charging* circuit). The problem we discussed is relevant for systems or system realizations where such independent access is unfeasible, e.g., because the system is at a remote location. Actually the concept was first encountered as a technical condition for certain optimal control problems and then in a slightly different form in a so-called *finite-settling-time* design problem [12], which we shall discuss in Sec. 2.3.3.

As for *observability*, the problem of determining the initial conditions seems to have been sidestepped by always choosing realizations in which proper initial conditions are easy to calculate, as in the observability or observer forms, or by always implicitly arranging things so that observability was ensured.† Thus it was apparently not necessary to formalize these concepts, though various difficulties, especially with multi-input, multi-output analog-computer simulations, had begun to show the need for a more fundamental analysis.

The present form of the definitions of controllability and observability and the recognition of the simple duality between them were worked out by R. E. Kalman in 1959–1960 (see the historical remarks in [18]). They were very clearly presented in a path-breaking paper [19], which has had a wide influence. It inspired many later researches, including some on the significance of jointly controllable and observable realizations by Gilbert [16], Kalman [17], and Popov [20]; these results will be studied in Sec. 2.4.

Here we shall first pause to show how our discussions so far can be carried over fairly readily to discrete-time [and to some extent also discrete-amplitude (cf. Example 2.3-6)] systems.

---

†For example (cf. Sec. 2.4.1), by working only with transfer functions in reduced form, i.e., without common factors between numerator and denominator.

Reprise.   We have defined a realization $\{A, b, c\}$, or just a pair $\{A, b\}$, as being *controllable* if its controllability matrix $\mathcal{C}(A, b)$ has full rank. We showed that an equivalent characterization was being able to change the state arbitrarily by means of impulsive inputs. Change of state with bounded functions will be discussed in Sec. 9.2. One should note carefully the duality between the characterizations of observability and controllability. For example, make sure you understand completely the statement that a realization $\{A, b, c\}$ is observable (controllable) if and only if the dual realization $\{A', c', b'\}$ is controllable (observable). Note how controllability involves statements about inputs and states, while the dual statements for observability involve outputs and states. With these remarks in mind, reexamine the block diagrams for the four special realizations of Sec. 2.1.2.

One of the important points we wish to make as we proceed through this book is that the different state-space descriptions provide convenient coordinate frames, we might say, for better understanding system behavior. We could always work with a given realization, but different things stand out more clearly in different realizations. We should also repeat here that our use of different realizations is almost purely for conceptual purposes and not for actual numerical implementation. The numerical aspects have to be studied separately, often in each particular problem, though as a general rule, diagonal realizations (when feasible) appear to have the best numerical properties (see the remarks and references on singular-value decompositions in Sec. 15 of the Appendix).

2.3.3   Discrete-Time Systems ; Reachability and Constructibility

In the previous sections, we have studied some aspects of state-space realizations of continuous-time systems. Very similar analyses and results can be obtained for *discrete-time* systems, where it is assumed that the inputs and outputs are known (or are of interest) only at discrete time instants t_0, t_1, t_2, ... It can be arranged that these instants are all integral multiples of some basic unit Δ, say,

$$t_0 = 0, t_1 = \Delta, t_2 = 2\Delta, \ldots$$

in which case Δ is often not explicitly shown and we assume that the time parameter, denoted by k (or sometimes still by t), takes integral values, $k = 0, \pm1, \pm2, \ldots$.

We shall see that there are great similarities between the results for discrete- and continuous-time systems, and therefore most of the discussions in this book are for systems of the latter type. However, there are many problems where the discrete-time language is more intuitive, especially when the Markov parameters enter (cf. Chapter 5). There are also some differences from continuous-time systems (cf. Sec. 3.5.2). Many of these questions can be illuminated by using the discrete-time analogs of the canonical realizations developed in Sec. 2.1.1, and some new results and insights will be developed in this way.

The discrete-time state-space equations are usually written in the form

$$x(k + 1) = Ax(k) + bu(k), \qquad x(0) = x_0$$
$$y(k) = cx(k) \qquad\qquad\qquad k \geq 0 \tag{11}$$

In discrete time, the transfer function is defined via zee transforms [21] rather than Laplace transforms. We write

$$X(z) = \sum_0^\infty x(k)z^{-k} \tag{12}$$

$$\begin{aligned}
Z\{x(k + 1)\} &= \sum_0^\infty x(k + 1)z^{-k} \\
&= x(1) + x(2)z^{-1} + \cdots \\
&= z \sum_0^\infty x(k)z^{-k} - zx(0) \\
&= zX(z) - zx(0)
\end{aligned} \tag{13}$$

Therefore, from the state-space equations we can write

$$Y(z) = c(zI - A)^{-1}b + cz(zI - A)^{-1}x_0 \tag{14}\dagger$$

The transfer function, which is calculated with $x_0 = 0$, is

$$H(z) = \frac{Y(z)}{U(z)} = c(zI - A)^{-1}b \tag{15}$$

Now we notice the important fact that the expression for $H(z)$ is the same as that for the transfer function

$$H(s) = c(sI - A)^{-1}b$$

of the continuous-time system

$$\dot{x}(t) = Ax(t) + bu(t), \qquad y(t) = cx(t)$$

Therefore, modulo certain obvious changes, all results derived by algebraic manipulation of $H(s)$ can be immediately carried over to $H(z)$. As a first example, we note the analogs of the results of Sec. 2.1.

†Note the difference in the last term from its continuous-time analog; this difference could be avoided by taking the sum in (12) from 1 to ∞ rather than 0 to ∞—however, the latter convention is more traditional.

Computer Simulations. An analog of the continuous-time high-order differential equation specification of a system

$$y^{(n)}(t) + a_1 y^{(n-1)}(t) + \cdots + a_n y(t) = b_1 u^{(n-1)}(t) + \cdots + b_n u(t) \qquad (16)$$

is the difference equation

$$\begin{aligned}
y(k+n) + a_1 y(k+n-1) + \cdots + a_n y(k) \\
= b_1 u(k+n-1) + \cdots + b_n u(k), \qquad k \geq 0
\end{aligned} \qquad (17)$$

with initial conditions being given values of $\{y(0), \ldots, y(n-1)\}$. The zee transform, with zero initial conditions, is

$$z^n Y(z) + \cdots + a_n Y(z) = b_1 z^{n-1} U(z) + \cdots + b_n U(z) \qquad (18)$$

so that

$$\frac{Y(z)}{U(z)} = \frac{b_1 z^{n-1} + \cdots + b_n}{z^n + a_1 z^{n-1} + \cdots + a_n} \qquad (19)$$

which is the same, except that z replaces s, as would be obtained by Laplace transformation of the differential equation (16). Therefore, with the obvious change that

$$z^{-1} \text{ can be implemented by a unit delay}$$

(while s^{-1} is implementable by an integrator), all the arguments and results used in Sec. 2.1 can be repeated here. That z^{-1} represents unit delay is evident from the fact that if y is a unit-delayed version of u, i.e., $y(k) = u(k-1)$, then, from (17) and (18), $Y(z) = z^{-1} U(z)$.

We may perhaps also note explicitly that the powers of z^{-1} serve to indicate the position in time at which the associated coefficient occurs. Thus, in particular, polynomials in z correspond to discrete-time functions that start and end at finite "past" times t, $t \leq 0$.

The Markov Parameters Define the Impulse Response. As a matter of fact, several things turn out somewhat more nicely in discrete time. For example, the Markov parameters turn out to be just the values of the discrete-time impulse response. That is, if $\{h_i\}$ are the Markov parameters defined by [cf. (2.2-33)]

$$H(z) = \sum_1^\infty h_i z^{-i} = \sum_1^\infty (cA^{i-1}b) z^{-i} \qquad (20)$$

then the inverse zee transform of $H(z)$ is the discrete-time function

$$h(k) = h_k = cA^{k-1}b, \qquad k \geq 1 \qquad (21)$$

which is also the impulse response of the system; i.e., its response to the discrete-time impulse function

$$\delta(k) = \begin{cases} 1, & k = 0 \\ 0, & k \neq 0 \end{cases} \tag{22}$$

Thus, in discrete time the Markov parameters are much easier to interpret and obtain than they are in continuous time; however, they have the same algebraic properties in both cases, which is certainly a convenience.

Specification of the Controllability and Observability Forms.   As an example, let us consider the controllability canonical form in discrete time, which we show in Fig. 2.3-1 (this is Fig. 2.1-10 drawn with delays in place of the integrators). In this form, the coefficients $\{\beta_1, \beta_2, \beta_3\}$ cannot be written down directly from the transfer function $b(z)/a(z)$. However, by putting an impulse $\delta(k)$ into the realization we see immediately from the realization that

$$\beta_1 = \text{the response } y(\cdot) \text{ at } k = 1$$
$$= h_1, \qquad \text{by definition}$$

and similarly we see that $\beta_2 = h_2$, $\beta_3 = h_3$ (cf. Eq. (2.1-8d) and Exercise 2.2-15). A similar derivation in continuous time is less direct but will be an interesting exercise for the reader.

Examination of the block diagram of the observability form (Fig. 2.3-2) shows similarly that the input vector is given by

$$b_{ob} = [h_1 \quad h_2 \quad h_3]'$$

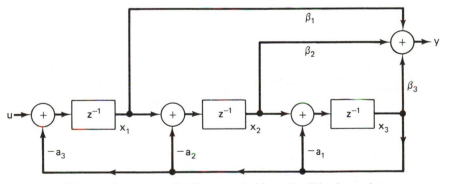

Figure 2.3-1.   Controllability canonical form. It will be shown that this should perhaps better be called the *reachability canonical form* because it is easy to determine an input to take the state from zero at $t = 0$ to any value at $t \geq 3$.

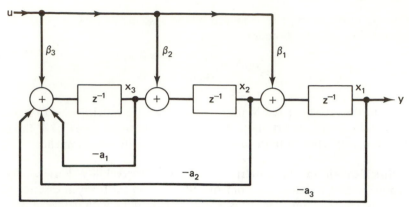

Figure 2.3-2. Observability canonical form. Note how easy it is to "read out" the values of the initial state [especially when $u(\cdot) \equiv 0$].

The several results of Sec. 2.3 on observability and controllability also have simpler discrete-time versions.

Observability.   Thus the analog of the continuous-time equation (2) is

$$
\begin{bmatrix}
y(k) \\
y(k+1) \\
\cdot \\
\cdot \\
\cdot \\
y(k+n-1)
\end{bmatrix}
= \Theta x(k) + \mathbf{T}
\begin{bmatrix}
u(k) \\
u(k+1) \\
\cdot \\
\cdot \\
\cdot \\
u(k+n-1)
\end{bmatrix}
\tag{23}
$$

where

$\Theta' = [c' \quad A'c' \quad \cdots \quad (A')^{n-1}c']$, the (transposed) observability matrix of (c, A)

$\mathbf{T} = $ the impulse response matrix, a lower triangular Toeplitz matrix with first column $[0 \quad h_1 \quad \cdots \quad h_{n-1}]$. $\qquad(24)$

Equation (23) shows that the state at any time k can be uniquely recovered from known inputs and outputs if and only if the observability matrix Θ is nonsingular.

Moreover, the convenience of the observability canonical form is immediately evident from the realization (cf. Fig. 2.3-2): With no input, we can just "read out" the states as (for $n = 3$)

$$y(k) = x_1(k), \qquad y(k+1) = x_2(k), \qquad y(k+2) = x_3(k)$$

We note also that in all cases relation (23) provides a viable calculation method, unlike in continuous time where the analogous method requires differentiation of $y(\cdot)$. (Note, though, that if the discrete-time system is obtained by *sampling* a continuous-time system, then in the limit as the sampling rate gets very high, it can be shown that the discrete-time solution will converge to the continuous-time solution based on derivatives.)

Controllability. Similar simplifications arise when we study the controllability problem in discrete time. Thus we can directly write from the state equation (11) that

$$x(n) = A^n x_0 + [b \quad Ab \quad \cdots \quad A^{n-1}b][u(n-1) \quad \cdots \quad u(0)]' \qquad (25)$$

the continuous analog of which was not so readily obtained [cf. Eq. (10) in Sec. 2.3.2]. We see from (25) that we can transfer *any* initial state to an *arbitrary* state (in not more than n steps) if and only if the matrix $\mathcal{C}(A, b)$ is nonsingular.

Again, note that the discrete problem is somewhat easier and its solution more realistic, at least in that we do not use impulsive inputs. (Note again, though, that in the continuous-time limit the present solution will converge to the impulsive function solution of Sec. 2.3.2—it is in fact an interesting analytical exercise to prove this, as a reader with enough time might wish to show.) In fact, a special case of the above discrete-time problem, viz. the so-called *finite-settling-time problem* [12] in which it is required to find an input to return a perturbed ($x_0 \not\equiv 0$) system to the origin as quickly as possible, was apparently the one in which the notion of state controllability was first explicitly introduced, though the matrix $\mathcal{C}$ had arisen earlier in other control problems (cf. the historical remarks in [18]).

*Controllability to and from the Origin; Reachability. This *controllability-to-the-origin* problem has several interesting aspects, which we shall discuss further in later chapters (e.g., Sec. 3.5.2). Let us consider one special aspect here, because it has been the source of some terminological problems. It arises from the fact that while, as (25) shows, the nonsingularity of $\mathcal{C}$ is *sufficient* to ensure that any arbitrary initial state x_0 can be driven to the origin in a finite time, this condition is not *necessary*. In other words, it may be possible to take any state to the *origin* even if $\mathcal{C}$ is singular. For example, this can happen if the $n \times n$ matrix A has the property that $A^k = 0$ for some k; for such so-called *nilpotent* matrices, any x_0 can be driven to zero with the "zero" input $u(k) \equiv 0$. Another example will be given soon.

Therefore, *controllability p.s.t.o.* (pointwise state to the origin, to use Rosenbrock's terminology [22]) is not always equivalent to the nonsingularity of the matrix

$$\mathcal{C} = [b \quad Ab \quad \cdots \quad A^{n-1}b]$$

On the other hand, *controllability p.s.f.o.* (pointwise state from the origin), or *reachability* as it is often called, is easily seen from (25) to always be equivalent to the nonsingularity of the above matrix, which some people therefore prefer to call the *reachability* matrix.

A necessary and sufficient condition for controllability p.s.t.o. is also clear from (25):

$$A^n x_0 \in \mathfrak{R}[b \quad Ab \quad \cdots \quad A^{n-1}b] \tag{26}$$

where $\mathfrak{R}[\cdots]$ denotes the range space, or the collection of all linear combinations, of the columns $\{b, Ab, \ldots, A^{n-1}b\}$. It is easy to see that controllability p.s.t.o. is in fact completely equivalent to the nonsingularity of $\mathcal{C}$ when the matrix A is nonsingular. For then, from (25), we can write, with $x(n) = 0$,

$$-x_0' = [u(n-1) \quad \cdots \quad u(0)]\mathcal{C}'(A^{-n})' \tag{27}$$

which shows that a suitable input sequence will exist (if and) only if $\mathcal{C}$ is nonsingular. With nonsingular A it seems therefore more appropriate to consider the matrix (rather than $\mathcal{C}$)

$$A^{-n}\mathcal{C} = [A^{-n}b \quad \cdots \quad A^{-1}b] \tag{28}$$

when examining controllability to the origin.

Example 2.3-A.   Controllability Need Not Imply Reachability

Let

$$A = \begin{bmatrix} 1 & 1 \\ 0 & 0 \end{bmatrix}, \qquad b = \begin{bmatrix} 1 \\ 0 \end{bmatrix}$$

The matrix

$$[b \quad Ab] = \begin{bmatrix} 1 & 1 \\ 0 & 0 \end{bmatrix}$$

is singular, so that an arbitrary state, e.g., $[1 \ 1]'$, cannot be reached from the zero state no matter what the input is. On the other hand, any initial state can be returned to zero, for if $x_1(0) = \alpha, x_2(0) = \beta$, let $u(0) = -(\alpha + \beta)$, and then

$$x(1) = \begin{bmatrix} 1 & 1 \\ 0 & 0 \end{bmatrix} \begin{bmatrix} \alpha \\ \beta \end{bmatrix} + \begin{bmatrix} 1 \\ 0 \end{bmatrix} (-\alpha - \beta) \equiv 0$$

(Note that $A^k = A \neq 0$ for all k.)

However, controllability (to the origin) does imply reachability when A is nonsingular.   ∎

Reachability and Controllability Canonical Forms.   It is also useful to explore the difference between reachability (controllability from the origin) and controllability to the origin in terms of (block diagrams of) canonical realizations. Thus if we ask for a realization in which the reachability problem is to be trivial, the realization of Fig. 2.3-1 immediately offers itself; it is clear from the figure that to set up any state $\theta = [\theta_1 \; \theta_2 \; \theta_3]$ we just feed in the inputs

$$u(0) = \theta_3, \qquad u(1) = \theta_2, \qquad u(2) = \theta_1$$

and then we shall have

$$x(3) = \theta$$

Therefore the realization of Fig. 2.3-1 is in fact the *reachability canonical form*.

On the other hand, suppose we wish to take arbitrary states to the origin in the most natural way. If we assume that A is nonsingular,† then formulas (27)–(28) show that we should seek a realization such that

$$[A^{-1}b \quad \cdots \quad A^{-n}b] = I \tag{29}$$

Some analysis‡ shows that the realization (written for $n = 3$)

$$A = \begin{bmatrix} -a_1 & 1 & 0 \\ -a_2 & 0 & 1 \\ -a_3 & 0 & 0 \end{bmatrix}, \qquad b = \begin{bmatrix} -a_1 \\ -a_2 \\ -a_3 \end{bmatrix} \tag{30}$$

has the above property. The output matrix c is not needed here but can be found from the equation ($c = [\gamma_1 \; \gamma_2 \; \gamma_3]$)

$$-\begin{bmatrix} a_1 & a_2 & a_3 \\ a_2 & a_3 & 0 \\ a_3 & 0 & 0 \end{bmatrix} \begin{bmatrix} \gamma_1 \\ \gamma_2 \\ \gamma_3 \end{bmatrix} = \begin{bmatrix} h_1 \\ h_2 \\ h_3 \end{bmatrix} \tag{31}$$

Formula (31) can be obtained by noting that if the input sequence is $\{1, a_1, a_2, a_3\}$, then the output sequence [with $x(0) = 0$] should be $\{0, b_1, b_2, b_3\}$. This calculation is also very evident from the realization, which is shown in Fig. 2.3-3. From the realization, it is also readily evident that with the input

$$u(0) = -x_1(0), \qquad u(1) = -x_2(0), \qquad u(2) = -x_3(0)$$

the state will be zero at $k = 3$:

$$x(0) = [x_1(0) \quad x_2(0) \quad x_3(0)]'$$
$$x(1) = [x_2(0) \quad x_3(0) \quad 0 \;]'$$
$$x(2) = [x_3(0) \quad 0 \quad 0 \;]'$$
$$x(3) = [\; 0 \quad 0 \quad 0 \;]'$$

Therefore the realization in Fig. 2.3-3 should really have the name controllability (p.s.t.o.) canonical form, rather than the one in Fig. 2.3-2.

---

†The case of singular A can also be studied by introducing some notions of *generalized inverses*; we shall not do so here.

‡*Hint*: Use Exercise A.32 on the inverse of a companion matrix.

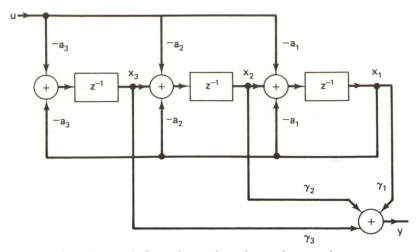

Figure 2.3-3. A form that perhaps better deserves the name controllability (to the origin) canonical form. However, it is necessary here that A be nonsingular. Note how easy it is to determine an input that will take an arbitrary initial state to zero.

However, we shall not insist upon this distinction in this book, chiefly because the two concepts are always equivalent when A is nonsingular and also in continuous time (where the nonsingular matrix exp A takes the place of the matrix A; cf. Sec. 2.5.1). Let us, however, briefly consider the corresponding questions for the observability problem.

Observability and Constructibility.   There must clearly be dual distinctions in the observability problem, and it is useful to examine them. The discrete-time analog of what we called the observability canonical form in Sec. 2.1 (cf. Fig. 2.1-7) was already presented as Fig. 2.3-2. There is no question here that this name is well deserved, because it is clear from the figure that, with the input $u(\cdot) \equiv 0$, we can "read out" the state at any time k as (for $n = 3$)

$$y(k) = x_1(k), \qquad y(k + 1) = x_2(k), \qquad y(k + 2) = x_3(k)$$

Then what corresponds in the observability problem to the distinction between controllability (to the origin) and reachability (from the origin)? Some reflection shows that the dual of reachability of a state (from the origin using past inputs) is observability of a state from future outputs. The dual of controllability of a state (to the origin using future inputs) should then be observability of a state using past outputs. This is called *constructibility* and is equivalent to observability (in the original sense) when A is nonsingular.

Example 2.3-B.   Constructibility Need Not Imply Observability

Let

$$c = [1 \quad 1], \qquad A = \begin{bmatrix} 1 & 1 \\ 1 & 1 \end{bmatrix}$$

Then the observability matrix is

$$\begin{bmatrix} 1 & 1 \\ 2 & 2 \end{bmatrix}$$

which is singular. Therefore the realization is unobservable, as is shown by the equations

$$y(0) = x_1(0) + x_2(0)$$
$$y(1) = 2x_1(0) + 2x_2(0)$$

so that we can only observe the sum $x_1(0) + x_2(0)$.

On the other hand, we see that

$$\begin{bmatrix} x_1(0) \\ x_2(0) \end{bmatrix} = \begin{bmatrix} 1 & 1 \\ 1 & 1 \end{bmatrix} \begin{bmatrix} x_1(-1) \\ x_2(-1) \end{bmatrix} = \begin{bmatrix} x_1(-1) + x_2(-1) \\ x_1(-1) + x_2(-1) \end{bmatrix}$$

so that if we have available

$$y(-1) = x_1(-1) + x_2(-1)$$

then we can construct

$$x_1(0) = y(-1) \quad \text{and} \quad x_2(0) = y(-1)$$

Equivalently, if our observations can only occur for $k \geq 0$, note that we can construct

$$x_1(k + 1) = y(k) = x_2(k + 1) \qquad \blacksquare$$

Clearly, observability implies constructibility but not vice versa, unless A is nonsingular, in which case the two concepts are completely equivalent. In the latter case, we can still ask for a canonical form in which constructibility is trivial. We leave it to the reader to show that such a form is (for $n = 3$)

$$A_{cb} = \begin{bmatrix} -a_1 & -a_2 & -a_3 \\ 1 & 0 & 0 \\ 0 & 1 & 0 \end{bmatrix}, \qquad b_{cb} = \begin{bmatrix} \gamma_1 \\ \gamma_2 \\ \gamma_3 \end{bmatrix} \qquad (32a)$$

$$c_{cb} = [-a_1 \quad -a_2 \quad -a_3] \qquad (32b)$$

which yields

$$\begin{bmatrix} c_{cb} A_{cb}^{-1} \\ c_{cb} A_{cb}^{-2} \\ c_{cb} A_{cb}^{-3} \end{bmatrix} = I \qquad (32c)$$

The $\{\gamma_i\}$ are determined by relation (31), and the constructibility canonical form is shown in Fig. 2.3-4.

Finally, in Fig. 2.3-5 we have summarized the relations among the concepts described above.

Expanded Transfer Relations and Some Useful Identities.†   By definition, feeding $a(z)$ into a system with the transfer function $H(z) = b(z)/a(z)$ (and zero initial condi-

---

†The results of this section will be used only in Sec. 2.4.4.

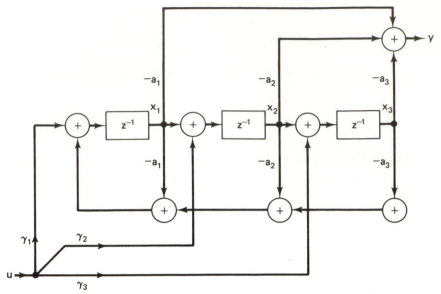

Figure 2.3-4. Constructibility canonical form. Note how easy it is to determine the state at a given time ($t \geq 3$) from knowledge of past $y(\cdot)$.

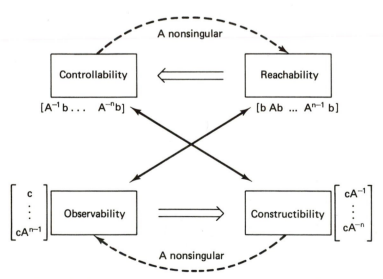

Figure 2.3-5. Relations among four basic concepts in discrete time. For continuous-time constant realizations, the distinction between reachability and controllability and between observability and constructibility disappear (because the nonsingular matrix e^{At} takes the place of A).

tions) will give the output sequence

$$H(z)a(z) = b(z) \tag{33}$$

Comparing powers of z on both sides of (33) will give us the matrix equation

$$
\begin{bmatrix}
0 & & & \\
h_1 & 0 & & \bigcirc \\
h_2 & h_1 & \cdot & \\
\cdot & & \cdot & \\
\cdot & & & \cdot \\
\cdot & & & 0 \\
h_n & \cdots & & h_1 \\
\cdot & & & \cdot \\
\cdot & & & \cdot \\
\cdot & & & \cdot
\end{bmatrix}
\begin{bmatrix}
1 \\ a_1 \\ a_2 \\ \cdot \\ \cdot \\ \cdot \\ a_n
\end{bmatrix}
=
\begin{bmatrix}
0 \\ b_1 \\ b_2 \\ \cdot \\ \cdot \\ \cdot \\ b_n \\ 0 \\ \cdot \\ \cdot
\end{bmatrix}
\tag{34}
$$

Now note that by linearity and time invariance

$$H(z)a(z)z^k = z^k b(z), \qquad k = \text{any integer} \tag{35}$$

In matrix terms, if the input is shifted up or down, the Toeplitz nature of the impulse response matrix in (34) implies a corresponding shift in the output sequence. In particular, therefore, if we consider the n shifts $\{k = 0, -1, \ldots\}$, we can write

$$
\begin{bmatrix}
0 & & & & & \\
h_1 & 0 & & & & \\
\cdot & \cdot & & & \bigcirc & \\
\cdot & & \cdot & & & \\
h_{n-1} & & h_1 & 0 & & \\
h_n & \cdots & & h_1 & & \\
0 & & & & h_1 & \\
\cdot & & & & \cdot & \\
\cdot & & & & \cdot & \cdot \\
h_{2n-1} & \cdots & & h_n & h_{n-1} & \cdots & h_1 & 0
\end{bmatrix}
\begin{bmatrix}
1 & 0 & & & 0 \\
a_1 & 1 & & & \\
\cdot & & & & \\
\cdot & & & & \\
a_{n-1} & a_{n-2} & & & 1 \\
a_n & a_{n-1} & & & a_1 \\
0 & a_n & & & \cdot \\
\cdot & & \cdot & & \cdot \\
\cdot & & & \cdot & \cdot \\
0 & & \cdots & & a_n
\end{bmatrix}
$$

$$
=
\begin{bmatrix}
0 & 0 & \cdots & & 0 \\
b_1 & 0 & & & \\
\cdot & b_1 & \cdot & & \cdot \\
\cdot & & \cdot & & \cdot \\
& & & & 0 \\
b_{n-1} & & b_1 & & \\
b_n & b_{n-1} & b_2 & b_1 & \\
0 & b_n & & b_2 & \\
\cdot & & & \cdot & \\
\cdot & & & \cdot & \\
0 & & \cdots & & b_n
\end{bmatrix}
\tag{36}
$$

or more compactly as [cf. (2.3-24) and (2.2-49)], say,

$$\begin{bmatrix} \mathbf{T} & \bigcirc \\ M[1, n-1]\tilde{I} & \mathbf{T} \end{bmatrix} \begin{bmatrix} \mathfrak{A}_- \\ \mathfrak{A}_+ \end{bmatrix} = \begin{bmatrix} \mathfrak{B}_- \\ \mathfrak{B}_+ \end{bmatrix} \tag{37}$$

where $\tilde{I}$ is the *reversed* identity matrix,

$$\tilde{I} = \begin{bmatrix} & & & 1 \\ & \bigcirc & 1 & \\ & & \cdot & \\ & \cdot & & \bigcirc \\ 1 & & & \end{bmatrix}, \qquad \tilde{I}^2 = I \tag{38}$$

These equations yield some interesting algebraic identities (useful, e.g., in Sec. 2.4.4). Thus, from (37) we immediately obtain a nice factorization of the impulse response matrix,

$$\mathbf{T} = \mathfrak{B}_-\mathfrak{A}_-^{-1} \tag{39a}$$

Moreover, since (lower or upper) triangular Toeplitz matrices commute (cf. Exercise A.6), we can also write

$$\mathbf{T} = \mathfrak{A}_-^{-1}\mathfrak{B}_- \tag{39b}$$

From (38) and (39), we also obtain a formula for the Hankel matrix,

$$M[1, n-1] = [\mathfrak{B}_+ - \mathfrak{B}_-\mathfrak{A}_-^{-1}\mathfrak{A}_+]\mathfrak{A}_-^{-1}\tilde{I} \tag{40}$$

The matrix $\mathfrak{A}_-^{-1}$ has a nice interpretation, which can be obtained by comparing the definition of $\mathfrak{A}_-$ [cf. (36) and (37)] with formula (9) $\mathcal{C}_c^{-1}$, where $\mathcal{C}_c$ is the controllability matrix of the controller form $\{A_c, b_c, c_c\}$. We can identify

$$\mathfrak{A}_-^{-T} = \tilde{I}\mathfrak{A}_-^{-1}\tilde{I} = \mathcal{C}_c, \qquad \mathfrak{A}_- = \tilde{I}\mathcal{C}_c^{-1}\tilde{I} \tag{41}$$

Furthermore, it is easy to verify from the definition (2.2-49) of $M[1, n-1]$ that for any realization $\{A, b, c\}$

$$M[1, n-1] = \mathcal{O}(c, A)\mathcal{C}(A, b) \tag{42}$$

But then by combining (40)–(42), we obtain the nonobvious formula

$$\mathcal{O}_c = \mathcal{O}(c_c, A_c) = [\mathfrak{B}_+ - \mathfrak{B}_-\mathfrak{A}_-^{-1}\mathfrak{A}_+]\tilde{I} \tag{43}$$

For completeness we note here another nonobvious formula for $\mathcal{O}_c$,

$$\mathcal{O}_c = \tilde{I}b(A_c) = \tilde{I}[b_1A_c^{n-1} + \cdots + b_nI] \tag{44}$$

This formula, apparently first given by Wonham and Stuelpnagel [17, p. 178], can be derived by an easy application† of the result of Exercise A.33, part 2. Note that it yields the unexpected matrix identity

$$\tilde{I}b(A_c)\tilde{I} = [\mathcal{B}_+ - \mathcal{B}_-\mathcal{C}_-^{-1}\mathcal{C}_+] \tag{45}$$

Reprise. In this section we have shown that many algebraic properties of a realization $\{A, b, c\}$ are independent of whether it is a continuous- or discrete-time realization. However, there are some differences in the physical characterizations, with the discrete-time problem often leading to more obvious and more elementary proofs (without derivatives or impulses). In this section we also provided some nice illustrations of the power of using the special state-space realizations of Sec. 2.1.2 in various calculations and explorations.

*2.3.4 Some Worked Examples

We shall provide several examples to illustrate some of the concepts introduced in the previous sections. The first two examples basically just show some of the algebra involved in many problems. Example 2.3-2 shows that the algebra can often be replaced, or at least illuminated, by returning to basic definitions; it also points the way to some of the results in the next section. The next three examples illustrate the value of the concepts of controllability and observability in answering some simple questions that arise in just manipulating the many possible state-space realizations of a given system. Finally, in Example 2.3-6, we show how many of the things we have said for systems over the real numbers can be carried over to digital systems operating over the binary field.

Example 2.3-1. Cart with Inverted Pendulums

A cart of mass M has two inverted pendulums on it of lengths l_1 and l_2, both with bobs of mass m. For small $|\theta_1|$ and $|\theta_2|$, the equations of motion can be seen to be

$$M\dot{v} = -mg\theta_1 - mg\theta_2 + u$$
$$m(\dot{v} + l_i\ddot{\theta}_i) = mg\theta_i, \qquad i = 1, 2$$

where v is the velocity of the cart and u is an external force applied to the cart (see the figure).

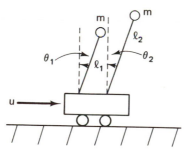

†A detailed derivation is given in Example 2.3-5 in the next section.

1. Is it always possible to "control" both pendulums, i.e., keep them both vertical, by using the input $u(\cdot)$?

2. Is the system observable with output $y = \theta_1$?

Solution. Let

$$x_1 = \theta_1, \qquad x_2 = \theta_2, \qquad x_3 = \dot{\theta}_1, \qquad x_4 = \dot{\theta}_2$$

Then we obtain (after eliminating $\dot{v}$ from the equations of motion) the state equations $\dot{x} = Ax + bu$ with

$$A = \begin{bmatrix} 0 & 0 & 1 & 0 \\ 0 & 0 & 0 & 1 \\ a_1 & a_2 & 0 & 0 \\ a_3 & a_4 & 0 & 0 \end{bmatrix}, \qquad b = \begin{bmatrix} 0 \\ 0 \\ -1/Ml_1 \\ -1/Ml_2 \end{bmatrix}$$

where

$$a_1 = \frac{(M+m)g}{Ml_1}, \qquad a_2 = \frac{mg}{Ml_1}$$

$$a_3 = \frac{mg}{Ml_2} \qquad a_4 = \frac{(M+m)g}{Ml_2}$$

1. To keep the pendulums vertical, i.e., to have $\theta_1 = 0 = \theta_2$, the realization must be controllable from the input. Therefore we have to check det $\mathcal{C}(A, b)$. After some algebra, we can obtain

$$\det \mathcal{C}(A, b) = \frac{2pq}{M^2 l_1 l_2} - \frac{q^2}{M^2 l_1^2} - \frac{p^2}{M^2 l_2^2}$$

where

$$p = \frac{a_1}{Ml_1} + \frac{a_2}{Ml_2}, \qquad q = \frac{a_3}{Ml_1} + \frac{a_4}{Ml_2}$$

Therefore $\mathcal{C}(A, b)$ will be singular if and only if

$$2l_1 l_2 pq - q^2 l_2^2 - p^2 l_1^2 = 0$$

i.e., after some more algebra, if and only if

$$M^2 g^2 l_1^2 l_2^2 (l_1 - l_2)^2 = 0$$

or,

$$l_1 = l_2$$

Therefore, if $l_1 \neq l_2$, the realization is controllable, and we can have $\theta_1 = 0 = \theta_2$ by suitable choice of the input.

2. If $y = \theta_1$, then $c = [1 \; 0 \; 0 \; 0]$, and we can check that

$$\det \Theta(c, A) = -a_2^2 = -\left(\frac{mg}{Ml_1}\right)^2 \neq 0$$

so we shall have observability in all cases (even when $l_1 = l_2$).

Remark: Try to deduce these results by physical arguments. ∎

Example 2.3-2.   Dynamics of a Hot Air Balloon

Approximate equations of motion for a hot air balloon (see the figure) are

$$\dot{\theta} = -\frac{1}{\tau_1}\theta + u$$

$$\dot{v} = -\frac{1}{\tau_2}v + \sigma\theta + \frac{1}{\tau_2}w$$

$$\dot{h} = v$$

where θ = temperature change of air in balloon away from equilibrium temperature, u is proportional to change in heat added to air in balloon (control), v = vertical velocity, h = change in altitude from equilibrium altitude, and w = vertical wind velocity (disturbance).

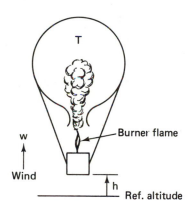

1. Can the temperature change $\theta(\cdot)$ and a constant wind velocity w be observed by a continuous measurement of altitude change h? (Assume, as usual, that u is known.)

2. Determine the transfer function from u to h and from w to h. Is the system completely controllable by u? Is it completely controllable by w?

Solution

1. To the three state equations given, we must add a fourth for w in order to decide whether a constant w can be observed by measuring h. The required equation is evidently

$$\dot{w} = 0$$

Proceeding from fundamentals, to determine the observability of θ and w we must see if we can solve for them from knowledge of h and its derivatives. Now note that

$$\dot{y} = \dot{h} = v$$

which implies that v is observable. We also have

$$\ddot{y} = \ddot{h} = \dot{v} = -\frac{1}{\tau_2}v + \sigma\theta + \frac{1}{\tau_2}w \qquad (*)$$

which implies that $\sigma\theta + (1/\tau_2)w$ is observable. Finally,

$$\dddot{y} = \dddot{h} = \ddot{v} = -\frac{1}{\tau_2}\dot{v} + \sigma\dot{\theta}$$

from which $\sigma\dot{\theta}$ is observable, whence θ is observable (if $\sigma \neq 0$). It then follows from (*) that w also is observable.

To proceed more formally and determine the observability of the whole system, note that

$$\begin{bmatrix} \dot{\theta} \\ \dot{v} \\ \dot{h} \\ \dot{w} \end{bmatrix} = \begin{bmatrix} -1/\tau_1 & 0 & 0 & 0 \\ \sigma & -1/\tau_2 & 0 & 1/\tau_2 \\ 0 & 1 & 0 & 0 \\ 0 & 0 & 0 & 0 \end{bmatrix} \begin{bmatrix} \theta \\ v \\ h \\ w \end{bmatrix} + \begin{bmatrix} 1 \\ 0 \\ 0 \\ 0 \end{bmatrix} u$$

$$y = [0 \quad 0 \quad 1 \quad 0][\theta \quad v \quad h \quad w]'$$

Then it can be checked that $\mathcal{O}$ will be nonsingular when $\sigma \neq 0$.

2. We now consider w as a *second input*. Our state equations are then

$$\begin{bmatrix} \dot{\theta} \\ \dot{v} \\ \dot{h} \end{bmatrix} = \begin{bmatrix} -\tau_1^{-1} & 0 & 0 \\ \sigma & -\tau_2^{-1} & 0 \\ 0 & 1 & 0 \end{bmatrix} \begin{bmatrix} \theta \\ v \\ h \end{bmatrix} + \begin{bmatrix} 1 & 0 \\ 0 & \tau_2^{-1} \\ 0 & 0 \end{bmatrix} \begin{bmatrix} u \\ w \end{bmatrix}; \qquad y = [0 \quad 0 \quad 1] \begin{bmatrix} \theta \\ v \\ h \end{bmatrix}$$

Note that the eigenvalues are $-1/\tau_1$, $-1/\tau_2$, and 0.

The transfer function to the output from one input of a multi-input system is defined with all other inputs zero. Then when all inputs are present we merely use superposition.

Taking transforms with $w = 0$, we obtain, after some algebra,

$$\frac{h(s)}{u(s)}\bigg|_{w=0} = \frac{h(s)}{v(s)}\frac{v(s)}{\theta(s)}\frac{\theta(s)}{u(s)} = \frac{\sigma}{s(s + 1/\tau_2)(s + 1/\tau_1)}$$

Similarly, with $u = 0$, we obtain

$$\frac{h(s)}{w(s)}\bigg|_{u=0} = \frac{h(s)}{v(s)}\frac{v(s)}{w(s)} = \frac{1/\tau_2}{s(s + 1/\tau_2)} = \frac{1}{s(\tau_2 s + 1)}$$

The eigenvalue at $-1/\tau_1$ has evidently been cancelled by the numerator.

To determine controllabilities, we can calculate $\mathcal{C}_u$ and $\mathcal{C}_w$ to find that the system is controllable by u (for $\sigma \neq 0$) but not by w. (The state equations show directly that, with $u = 0$, the equation for θ is undriven; hence θ is uncontrollable by w.) We shall see in Sec. 2.4 that this is related to the fact that all the system eigenvalues appear as poles of the transfer function from u, but the one at $-1/\tau_1$ is absent in the transfer function from w. We show there that no eigenvalues are cancelled out of the nominal transfer function from some input to some output if and only if the system is controllable by that input and observable from that output. Here the system is observable from h, so that any cancellations in a transfer function to h imply uncontrollability by the corresponding input. ■

Example 2.3-3.   Uniqueness of Realizations

We have a realization $\{A, b, c\}$ with known characteristic polynomial $a(s) = \det (sI - A)$ and with $\mathcal{O}(c, A) = I$. Show that this information uniquely determines $\{A, b, c\}$.

Solution.   Since $\mathcal{O} = I$, it is clear that $c = [1 \; 0 \; \cdots \; 0]$. Then cA must equal the first row of A, and, since $\mathcal{O} = I$, this first row must be $[0 \; 1 \; 0 \; \cdots \; 0]$. Proceeding thus, we see that the first $n - 1$ rows of A have the form $[0 \; I_{n-1}]$. Therefore A must be a companion matrix, and it is easy to check that if

$$\det (sI - A) = a(s) = s^n + a_1 s^{n-1} + \cdots + a_n$$

then the last row of A must be $-[a_n \; a_{n-1} \; \cdots \; a_1]$. Therefore the c and A matrices are uniquely determined. To check if there is a unique b we must show that

$$c(sI - A)^{-1}b^{(1)} = c(sI - A)^{-1}b^{(2)} \qquad\qquad (*)$$

implies $b^{(1)} = b^{(2)}$. Now $(*)$ implies that

$$cA^i b^{(1)} = cA^i b^{(2)}, \qquad i = 0, 1, \ldots$$

or

$$\mathcal{O}(b^{(1)} - b^{(2)}) = 0$$

or (since $\mathcal{O} = I$)

$$b^{(1)} - b^{(2)} = 0$$

Note that the fact that $\mathcal{O}(c, A) = I$ is not really necessary to obtain $b_1 = b_2$; it suffices for $\mathcal{O}(c, A)$ to be nonsingular. ■

Example 2.3-4.   Similarity Transformation to Controller Form

Let $\{A, b, c\}$ be a given realization function, and let $\{A_c, b_c, c_c\}$ be another realization in controller from such that $\det (sI - A) = \det (sI - A_c) = s^n + a_1 s^{n-1} + \cdots + a_n = a(s)$. Also let $c(sI - A)^{-1}b = c_c(sI - A_c)^{-1}b_c = b(s)/a(s)$.

Show that we can find a similarity transformation to take the realization $\{A, b, c\}$ to the controller form $\{A_c, b_c, c_c\}$ if and only if $\{A, b\}$ is controllable. *Hint:* Try to determine the transformation explicitly.

Solution.   We have to try to find an invertible matrix T such that

$$A_c = T^{-1}AT, \qquad b_c = T^{-1}b, \qquad c_c = cT$$

An explicit formula for T can be found in various ways, but here let us start from scratch. Let

$$T = [t_1 \quad \cdots \quad t_n]$$

Then the fact that $b'_c = [1 \quad 0 \quad \cdots \quad 0]$ gives

$$b = Tb_c = t_1$$

Now write $A_c = T^{-1}AT$ as

$$TA_c = AT = [At_1 \quad At_2 \quad \cdots \quad At_n]$$

Then, recalling the rule that multiplying a matrix by a column vector gives the weighted sum of the columns of the matrix, we shall have the equations

$$At_1 = -a_1t_1 + t_2 = -a_1b + t_2$$
$$At_2 = -a_2b + t_3, \ldots, At_n = -a_nb$$

which can be rearranged as

$$T = [b \quad Ab \quad \cdots \quad A^{n-1}b]\mathcal{C}'_-$$

where

$$\mathcal{C}'_- = \text{an upper triangular Toeplitz matrix with first row } [1 \quad a_1 \quad \cdots \quad a_{n-1}]$$

This shows immediately that the matrix T will be invertible if and only if $\mathcal{C}(A, b)$ is nonsingular, i.e., if $\{A, b\}$ is controllable.

We note from Eq. (2.3-9) that $\mathcal{C}'_- = \mathcal{C}_c^{-1}$, so that we have the explicit formula (worth remembering)

$$T = \mathcal{C}\mathcal{C}_c^{-1}, \qquad T^{-1} = \mathcal{C}_c\mathcal{C}^{-1} \quad \blacksquare$$

Example 2.3-5.   Observability of Controller Forms

Show that the controller-form realization $\{A_c, b_c, c_c\}$ of a transfer function $b(s)/a(s)$, $a(s) = \det(sI - A_c)$, will be observable if and only if $b(s)$ and $a(s)$ are relatively prime, i.e., have no common roots.

Solution.   We shall use the formula

$$\mathcal{O}_c = \tilde{I}b(A_c)$$

This is easily derived by application of the shifting property of companion matrices (see Exercise A.31):

$$e'_iA_c = e'_{i-1} \qquad \text{for } 2 \le i \le n$$

and

$$e'_1A_c = [-a_1 \quad -a_2 \quad \cdots \quad -a_n]$$

where e_i' is the ith row of the identity matrix I. Now

$$e_n'b(A_c) = b_1 e_n' A_c^{n-1} + \cdots + b_n e_n'$$
$$= b_1 e_1' + b_2 e_2' + \cdots + b_n e_n'$$
$$= [b_1 \quad b_2 \quad \cdots \quad b_n] = c_c$$

Then

$$e_{n-1}'b(A_c) = e_n' A_c b(A_c) = e_n' b(A_c) A_c = c_c A_c$$

since polynomial functions of a given matrix commute. Continuing similarly, we find

$$\mathcal{O}_c = [e_n \quad \cdots \quad e_1]'b(A_c) = \tilde{I}b(A_c)$$

This expression now shows that $\mathcal{O}_c$ will be nonsingular if and only if $\det b(A_c) \neq 0$. Now the determinant of a matrix is equal to the product of its eigenvalues; moreover, if A_c has eigenvalues $\{\lambda_i, i = 1, \ldots, n\}$, then the eigenvalues of the polynomial function $b(A_c)$ will be $\{b(\lambda_i), i = 1, \ldots, n\}$ (cf. Exercise A.36). Therefore

$$\det b(A_c) = \prod_{i=1}^{n} b(\lambda_i)$$

and this will be zero if and only if one or more of the $b(\lambda_i)$ are zero. But, by definition, the $\{\lambda_i\}$ are such that $a(\lambda_i) = \det(\lambda_i I - A_c) = 0$. That is, $\det b(A_c)$ will be nonzero, and hence $\{A_c, b_c, c_c\}$ observable, if and only if $a(s)$ and $b(s)$ have no common roots. ∎

Example 2.3-6.   Systems Over the Binary Field

Consider a discrete-time system where the coefficients come not from the field of real numbers but from the finite field with elements $\{0, 1\}$ with mod-2 addition, and multiplication only by 0 and 1. This is known as $GF(2)$, G for Galois. We can define polynomials in z^{-1} with coefficients in this field. These polynomials may be multiplied, for example,

$$(1 + z^{-1})^2 = 1 + 2z^{-1} + z^{-2} = 1 + z^{-2}$$

or factored, for example,

$$(1 + z^{-15}) = (1 + z^{-1})(1 + z^{-1} + z^{-2} + z^{-3} + z^{-4})(1 + z^{-3} + z^{-4}) \tag{*}$$
$$\cdot (1 + z^{-1} + z^{-4})(1 + z^{-1} + z^{-2})$$

Consider the linear binary *feedback shift register* (FSR) in the figure.

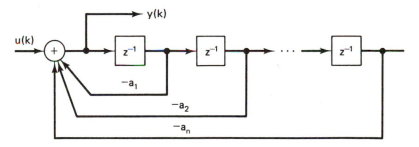

1. Find the transfer function $H(z)$.

2. Let $n = 4$ (i.e., assume only four delay elements), and let $a_1 = a_4 = 1$, $a_2 = a_3 = 0$. Find the impulse response using mod-2 z transforms.

Solution

1. The transfer function is determined from $y(k) = u(k) - a_1 y(k-1) - \cdots - a_n y(k-n)$ to be

$$H(z) = \frac{z^n}{z^n + a_1 z^{n-1} + \cdots + a_n}$$

Note that the numerator degree in this case equals the denominator degree since the present output depends on past *and* present inputs. State-space equations for this system can be written by including a direct feedthrough term in the output equation. Thus

$$H(z) = 1 - \frac{a_1 z^{n-1} + \cdots + a_n}{z^n + a_1 z^{n-1} + \cdots + a_n}$$

yields a realization $\{A, b, c, d\}$ where

$$A = \text{a top-companion matrix with first row } -[a_1 \quad \cdots \quad a_n]$$
$$b = [1 \quad 0 \quad \cdots \quad 0]'$$

2. The impulse response is the inverse z transform of $H(z)$. It is possible to invert the z transform by factoring the denominator polynomial and using partial fractions. However, the algebra satisfied by the roots of this polynomial may not be obvious. Another method (which we employ here) is to use the factorization of $1 + z^{-15}$ given in (∗) above. Thus note that

$$H(z) = \frac{1}{1 + z^{-1} + z^{-4}}$$

$$= \frac{1}{1 + z^{-15}}(1 + z^{-1})(1 + z^{-1} + z^{-2} + z^{-3} + z^{-4})(1 + z^{-3} + z^{-4})$$

$$\cdot (1 + z^{-1} + z^{-2})$$

Expanding the numerator, we get

$$H(z) = \frac{1 + z^{-1} + z^{-2} + z^{-3} + z^{-5} + z^{-7} + z^{-8} + z^{-11}}{1 + z^{-15}}$$

Note now that subtraction in $GF(2)$ is the same as addition $(0 - 0 = 0 + 0, 1 - 0 = 1 + 0, 1 - 1 = 1 + 1, -1 = +1,$ etc.) and therefore

$$\frac{1}{1 + z^{-15}} = 1 + z^{-15} + z^{-30} + \cdots$$

Hence the impulse response is periodic, with period 15. The output during one period is determined by the numerator $1 + z^{-1} + z^{-2} + z^{-3} + z^{-5} + z^{-7} + z^{-8} + z^{-11}$ to be the sequence (1 1 1 1 0 1 0 1 1 0 0 1 0 0 0).

Exercises 2.3-33 and 2.3-34 provide further examples of such $GF(2)$ systems. References [23]–[25] provide detailed discussions, with interesting applications to coding and switching theory. ∎

Exercises

2.3-1.

a. Let $\mathcal{O}_i$ denote the observability matrices of two (nth-order) realizations $\{A_i, b_i, c_i\}$. If

$$\mathcal{O}_1 T = \mathcal{O}_2$$

for some invertible matrix T, by equating the first two rows we obtain

$$c_1 T = c_2, \qquad c_1 A_1 T = c_2 A_2 = c_1 T A_2$$

Can we conclude from this that $A_1 = T A_2 T^{-1}$?

b. We have two controllable realizations $\{A, b, c_1\}$ and $\{A, b, c_2\}$ of a given transfer function. Show that $c_1 = c_2$. Do this problem in as many different ways as you can.

2.3-2.

$\{A, b, c\}$ is an nth-order realization of a given transfer function $b(z)/a(z)$, $a(z) = z^n + a_1 z^{n-1} + \cdots + a_n$. Suppose that $\mathcal{C}(A, b) = I$. Show that this information completely determines $\{A, b\}$.

2.3-3.

Consider the system illustrated in the figure.

a. Give a state-variable representation of this system.

b. Is there any choice of parameters k and/or a for which this system is not controllable *and* not observable?

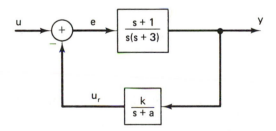

2.3-4.

Choose state variables as shown for the system shown in the figure.

a. Write the state equations.

b. Is this system realization controllable? Observable?

c. What is the transfer function from $u(\cdot)$ to $y(\cdot)$?

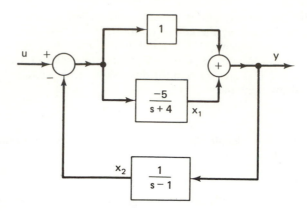

2.3-5.   Initial Conditions for Differential Equations

Suppose that

$$y^{(n)}(t) + a_1 y^{(n-1)}(t) + \cdots + a_n y(t) = b_1 u^{(n-1)}(t) + \cdots + b_n u(t),$$

$$t \geq 0+, \; u(t) = \delta(t)$$

Show that the initial-condition vector

$$Y'(0+) \triangleq [y(0+), \ldots, y^{(n-1)}(0+)], \qquad Y'(0-) \equiv 0$$

can be calculated as $Y(0+) = \mathcal{Q}_-^{-1} b$, where $b' = [b_1 \; \cdots \; b_n]$ and $\mathcal{Q}_-$ is a lower triangular Toeplitz matrix with first column $[1 \; a_1 \; \cdots \; a_{n-1}]'$.

2.3-6.

In Sec. 2.3.1, we noted that when $\mathcal{O}$ was singular, there was an undesirable loss of uniqueness in the solution of the state-determination problem. What is the corresponding phenomenon for the state-controllability problem, and how significant is it?

2.3-7.

Refer to Eq. (2.3-25) and suppose that $\mathcal{C}$ is singular. If $x(n) - A^n x_0$ is in the range of $\mathcal{C}$, then we can always find an input sequence to take x_0 to $x(n)$. In fact, if $\mathcal{C}$ is singular there will be many such sequences. Find the one with minimum energy, $\sum |u(i)|^2$.

2.3-8.   Alternative Proof of (2.3-10)

Use the generalized initial-value theorem of Sec. 1.2 to give an alternative derivation of formula (2.3-10).

2.3-9.   The Moments of a System

a. Given a transfer function $H(s) = b(s)/a(s)$, show that if $a_n \neq 0$ we can always obtain a realization $\{A_{cb}, b_{cb}, c_{cb}\}$, where $b'_{cb} = [\gamma_1, \ldots, \gamma_n]$ and

$$\gamma_k = c_{cb} A_{cb}^{-k} b_{cb}, \qquad k = 1, \ldots, n$$

Compare with formula (2.3-32c).

b. The $\{\gamma_i\}$ are called the *moments* of the system because they can be computed as

$$\int_0^\infty t^{i+1} h(t)\, dt$$

where $h(t) = c e^{At} b$ is the impulse response of the system. Try to prove this formula.

2.3-10.   The Constructibility Problem in Continuous Time

We recall that the Markov parameters $\{cA^{i-1}b, i \geq 1\}$ enter naturally into the observability problem of determining $x(t)$ from $y(t)$ and its derivatives. Show that when A is nonsingular, the moments $\{cA^{-i}b, i < 0\}$ enter naturally into the problem of determining $x^{(n)}(t)$ from $y(t)$ and its derivatives. Compare the most natural canonical forms for these two problems. Compare also with the discrete-time results.

2.3-11.   Another Observable Realization

Given $H(s) = b(s)/a(s)$ as in Sec. 2.1, show that we can always obtain a realization in the form (shown for $n = 3$)

$$A = \begin{bmatrix} -d_1 & 1 & 0 \\ -d_2 & 0 & 1 \\ -d_3 - (b_3 + a_3) & -(b_2 + a_2) & -(b_1 + a_1) \end{bmatrix}, \quad b = \begin{bmatrix} -d_1 \\ -d_2 \\ -d_3 \end{bmatrix}$$

$$c = \begin{bmatrix} 1 & 0 & 0 \end{bmatrix}$$

a. Draw an analog-computer simulation for this realization.
b. Show that this realization is always observable.
c. What statements can you make for the dual form?

2.3-12.

For the constant resistance networks of Exercises 2.2-22 and 2.2-23, determine what relations between R, L, and C are required to make them uncontrollable and/or unobservable.

2.3-13.

Extend the formula for $\mathcal{O}_c$ in Example 2.3-5 to show that for any controllable realization $\{A, b, c\}$ of $H(s) = g(s)/a(s)$ we have

$$g(A) = [p_{n-1}(A')c' \quad \cdots \quad p_0(A')c']'$$

where the $p_i(s)$ are defined by $a(s)(sI - A)^{-1}b = [p_{n-1}(s), \ldots, p_0(s)]'$.

2.3-14.

Show that the pair

$$\left\{ \begin{bmatrix} A & 0 \\ c & 0 \end{bmatrix}, \begin{bmatrix} b \\ 0 \end{bmatrix} \right\}$$

is controllable if and only if $\{A, b\}$ is controllable and

$$\begin{bmatrix} A & b \\ c & 0 \end{bmatrix}$$

has full rank.

2.3-15. Diagonal Forms

Show that a pair $\{A, \beta\}$, where A is diagonal with entries $\{\lambda_i\}$, is controllable if and only if (1) the λ_i are distinct and (2) all components of β are nonzero, with dual results for the observability of a pair $\{\gamma, A\}$. Also try to give simple physical explanations for why, when A is diagonal, repeated eigenvalues cause a loss of controllability and observability. [*Warning:* When A is not diagonal, or cannot be diagonalized, the simple conditions (1) and (2) do not apply; see Exercises 2.3-4 and 2.3-16.]

2.3-16. Controllability and Observability for Jordan Forms

٦. Let

$$J = \begin{bmatrix} \lambda & 1 & 0 \\ 0 & \lambda & 1 \\ 0 & 0 & \lambda \end{bmatrix}, \qquad \beta = \begin{bmatrix} \beta_1 \\ \beta_2 \\ \beta_3 \end{bmatrix}, \qquad \gamma' = \begin{bmatrix} \gamma_1 \\ \gamma_2 \\ \gamma_3 \end{bmatrix}$$

Find necessary and sufficient conditions on the $\{\beta_i\}$ and $\{\gamma_i\}$ for the nonsingularity of $\mathcal{O}(\gamma, J)$ and $\mathcal{C}(J, \beta)$.

b. Repeat for

$$J = \begin{bmatrix} \lambda & 1 & 0 \\ 0 & \lambda & 0 \\ 0 & 0 & \mu \end{bmatrix}$$

c. Generalize to an arbitrary Jordan matrix J.

2.3-17.

Let A be a companion matrix with $-[a_n \; a_{n-1} \; \cdots \; a_1]'$ as the last column. Show that $A' = H^{-1}AH$, where H is an *upper Hankel* matrix with first row $[a_{n-1} \; \cdots \; a_1 \; 1]$.

2.3-18.

Let

$$\mathcal{C}_k = [b \quad Ab \quad \cdots \quad A^{k-1}b]$$

Show that if rank $\mathcal{C}_{k+1} = $ rank $\mathcal{C}_k$ for some k, then rank $\mathcal{C}_{k+i} = $ rank $\mathcal{C}_k$ for all $i \geq 1$.

2.3-19.

Given $\dot{x} = Ax + bu$, $y = cx$, with $\mathcal{C}$ of rank r, show that the transfer function can be written as a ratio of polynomials with denominator of degree r and numerator of degree not greater than $r - 1$. Therefore there must be at least $n - r$ common roots between the numerator and denominator of $c(sI - A)^{-1}b$.

2.3-20.

With

$$(sI - A)^{-1}b = \frac{1}{a(s)}[p_{n-1}(s) \quad \cdots \quad p_1(s) \quad p_0(s)]'$$

show that $\{A, b\}$ is controllable if and only if the matrix P defined by

$$[p_{n-1}(s) \quad \cdots \quad p_1(s) \quad p_0(s)]' = P[s^{n-1} \quad \cdots \quad s \quad 1]'$$

is nonsingular. *Hint:* What is the relation of P to the matrix that transforms $\{A, b\}$ to controller form?

2.3-21.   *Expanded State Equations*

Let

$$\mathcal{Y}_i = [y(0) \quad \cdots \quad y(i-1)]', \qquad \mathcal{U}_i = [u(0) \quad \cdots \quad u(i-1)]'$$
$$\mathcal{C}_i = [b \quad Ab \quad \cdots \quad A^{i-1}b], \qquad \mathcal{O}'_i = [c' \quad A'c' \quad \cdots \quad (A')^{i-1}c']$$

Show that we can write the *expanded* state equations

$$\begin{bmatrix} \mathcal{Y}_i \\ \hline x_i \end{bmatrix} = \begin{bmatrix} \mathcal{O}_i & \vdots & T_i \\ \hline A^i & \vdots & \mathcal{C}_i \end{bmatrix} \begin{bmatrix} x_0 \\ \hline \mathcal{U}_i \end{bmatrix}$$

where T_i is the Toeplitz matrix of the Markov parameters [cf. (2.3-2c)].

2.3-22.   *Unknown-Input Observability*

We know that given the input and output (and their necessary derivatives) for an observable system we can determine its state. Prove that for an observable system we can determine the state *without* knowledge of the *input* if and only if the first $n-1$ Markov parameters, h_1 to h_{n-1}, are zero. Show that this is equivalent to the fact that the transfer function has no (finite) zeros.

2.3-23.   *Zero-Output Reachability*

Show that a state x of a discrete-time system can be reached from the origin in n steps while the output is maintained at zero if and only if

$$x = [A^{n-1}b \quad \cdots \quad Ab \quad b]p$$

for some vector p such that

$$Tp = 0$$

where T is the Toeplitz matrix of Markov parameters defined in Eq. (2.3-24).

2.3-24.   *Second-Order Vector Differential Equations*

To analyze systems with small damping it is often convenient to use sets of coupled second-order equations (e.g., in vibration and circuit analysis):

$$\ddot{x} + D\dot{x} + Kx = Gu$$

where $x = $ an n-vector of generalized coordinates and $u = $ an n-vector of control variables.

a. For a conservative system without gyroscopic coupling, $D = 0$, and K is symmetric. Show that the eigenvalues of such a system occur in pairs $\pm\sigma$ or $\pm j\omega$, where σ, ω are real constants, and that the eigenvectors are orthogonal to each other.

b. For a conservative system with gyroscopic coupling, D is antisymmetric (i.e., $D^T \equiv -D$), and K is symmetric. Show that the eigenvalues of such a system are located symmetrically about both the real and imaginary axes.

c. A frictionless spinning top is an example of a conservative system with gyroscopic coupling. The equations of motion may be normalized to

$$\begin{bmatrix} \ddot{x}_1 \\ \ddot{x}_2 \end{bmatrix} + \begin{bmatrix} 0 & p \\ -p & 0 \end{bmatrix} \begin{bmatrix} \dot{x}_1 \\ \dot{x}_2 \end{bmatrix} - \begin{bmatrix} 1 & 0 \\ 0 & 1 \end{bmatrix} \begin{bmatrix} x_1 \\ x_2 \end{bmatrix} = 0$$

where x_1, x_2 are orthogonal lateral displacements from the vertical position and p is proportional to the spin rate. What is the minimum value of p for which the eigenvalues are pure imaginary?

2.3-25. *Some Zee-Transform Pairs*

Show that (all time functions are zero for $n < 0$)
a. If $x(n) = 1$, $X(z) = z(z - 1)^{-1}$.
b. If $x(n) = e^{-\alpha n}$, $X(z) = z(z - e^{-\alpha})^{-1}$.
c. If $x(n) = n^k$, $X(z) = (-1)^k d^k z(z - 1)^{-1}/dz^k$.
d. If $x(n) = (\alpha^n - \beta^n)/(\alpha - \beta)$, $X(z) = z(z - \alpha)^{-1}(z - \beta)^{-1}$.
e. If $x(n) = \sum_0^n f(k)g(n - k)$, $X(z) = F(z)G(z)$.

2.3-26. *The Fibonacci Sequence*

The Fibonacci sequence $\{0, 1, 1, 2, 3, 5, 8, 13, \ldots\}$ is generated by the equation

$$y_k = y_{k-1} + y_{k-2}, \qquad k \geq 2$$
$$y_0 = 0, \qquad y_1 = 1$$

a. Show that we can write

$$y_n = \frac{1}{\sqrt{5}}(\lambda_+^n - \lambda_-^n), \qquad \lambda_\pm = \frac{1 \pm \sqrt{5}}{2}$$

b. Show that

$$\lim_{n \to \infty} \frac{\ln y_n}{n} = \ln \frac{\sqrt{5} + 1}{2}$$

2.3-27. *Realizations of Moving-Average and Autoregressive Models*
a. For a so-called moving-average (MA) model,

$$Y(z) = [b_0 + b_1 z^{-1} + \cdots + b_m z^{-m}]U(z)$$

give, in matrix ($\{A, b, c, d\}$) and in block diagram form, two different types of realizations (controller and observer types).

b. For a so-called autoregressive (AR) model,

$$Y(z)(1 + a_1 z^{-1} + \cdots + a_m z^{-m}) = U(z)$$

give, in matrix and in block diagram form, controller- and observer-type realizations. Use the results of Exercise 2.2-20 on inverse realizations to deduce realizations of AR models from MA models and vice versa.

c. Given a mixed or ARMA model,

$$\frac{Y(z)}{U(z)} = \frac{b_0 + b_1 z^{-1} + \cdots + b_m z^{-m}}{1 + a_1 z^{-1} + \cdots + a_n z^{-n}}$$

$$= \frac{b_0 z^m + \cdots + b_m}{z^n + a_1 z^{n-1} + \cdots + a_n} \cdot z^{n-m}$$

show how to "merge" the block diagrams of the MA and AR models in (a) and (b) so as to obtain realizations of the ARMA model. Which of these possible mergings will yield realizations with no more than n integrators? Describe the merging procedure in matrix language as well.

2.3-28.   *Time-Variant Models*

Time-variant coefficients can be handled in discrete time much more easily than in continuous time. Thus suppose

$$x(i + 1) = A(i)x(i) + B(i)u(i), \qquad i = 0, 1, \ldots$$

a. Show that

$$x(k) = \Phi(k, j)x(j) + \sum_{i=j}^{k-1} \Phi(k, i + 1)B(i)u(i)$$

where

$$\Phi(k, j) \triangleq A(k - 1), \ldots, A(j), \qquad \Phi(i, i) = I$$

b. Is it always true that $\Phi(i, j)\Phi(j, k) = \Phi(i, k)$, all $i, j, k \geq 0$?

2.3-29.   *An Alternative* ["$u(k + 1)$"] *Model*

In some problems, it turns out to be more natural to consider state-space models of the form

$$\xi(k + 1) = F\xi(k) + gu(k + 1), \qquad k \geq 0$$
$$y(k) = h\xi(k), \qquad\qquad\qquad \xi(0) = \xi_0$$

a. Show that the transfer function is

$$\mathcal{H}(z) = hz(zI - F)^{-1}g$$

Note the extra z in the numerator corresponding to the "advanced" input $u(k + 1)$.

b. Prove that an arbitrary initial state ξ_0 can be transferred to some other arbitrary state in a finite time if and only if the matrix $[g \; Fg \; \cdots \; F^{n-1}g]$ is nonsingular.

c. Consider how to translate results for the "$u(k)$" model to the present model. *Hint:* Define an augmented "state" $[x'(k), u(k)]'$.

2.3-30. *Realization from the Impulse Response*

Suppose we have a discrete-time system with known impulse response $\{h_1, h_2, \ldots\}$. Assume also that somehow we know that the minimal order of any state-space realization of the system is n.

a. Show how to find a minimal realization $\{A, b, c\}$ of this system. *Hint:* Note the special Hankel matrices (cf. Exercise 2.2-14)

$$M[1, n - 1] = \mathcal{O}(c, A)\mathcal{C}(A, b), \qquad M[2, n - 1] = \mathcal{O}A\mathcal{C}$$

and assume that $\{A, b, c\}$ is in controllability canonical form.

b. Show that an irreducible transfer function

$$H(z) = \frac{b(z)}{a(z)} = \frac{b_1 z^{n-1} + \cdots + b_n}{z^n + a_1 z^{n-1} + \cdots + a_n}$$

can be computed via the equations

$$M[1, n - 1][a_n, \ldots, a_1]' = - [h_{n+1}, \ldots, h_{2n}]'$$

and

$$[b_1, \ldots, b_n]' = \mathbf{T}(h)[1 \quad a_1 \quad \cdots \quad a_{n-1}]'$$

where $\mathbf{T}(h)$ is a lower triangular Toeplitz matrix with first column $[h_1, \ldots, h_n]'$. *Remark:* A basic result for such realization questions is the following result of Kronecker (1890): Let $\{h_1, h_2, \ldots\}$ be the impulse response of a discrete-time system. The system admits a finite-dimensional realization of order n (not necessarily minimal) if and only if $\det M[1, n + i] = 0$, $i = 0, 1, 2, \ldots$.

2.3-31. *Polynomial Inputs and the Cayley-Hamilton Theorem*

Note that inputs $u(z)$ that are polynomials in z correspond to inputs occurring at times less than or equal to zero; e.g., $u(z) = z^2 - 1$ corresponds to an input that is 1 at $t = -2$, -1 at $t = 0$, and zero elsewhere. Show that the *response* at $t = 1$ of $x_{k+1} = Ax_k + bu_k$ to the polynomial input $u(z) = g(z) = g_1 + g_2 z + \cdots + g_m z^m$ is $x(1) = g(A)b$ (assuming that the system is at rest before the input is applied). This is the analog of Eq. (2.3-10) in the continuous-time case. *Remark:* The above result can be used to obtain a system-theoretical proof of the Cayley-Hamilton theorem. Thus given a matrix A with characteristic polynomial $a(z) = \det(zI - A)$, choose arbitrary matrices $\{c, b\}$ to form a realization $\{A, b, c\}$. Denote $b(z) = c \, \text{Adj}\,(zI - A)b$. Now apply the input $u(z) = a(z)$ to this realization. By the above result, the response at $t = 1$ is $y(1) = ca(A)b$. But we also have $y(z) = [b(z)/a(z)]a(z) = b(z)$, a polynomial, so that $y(1) = 0$. Therefore $ca(A)b = 0$, which, since $\{c, b\}$ are arbitrary, implies that $a(A) = 0$, the Cayley-Hamilton theorem. Note that this argument can also be carried out in continuous time using impulsive inputs and replacing $t = 1$ by $t = 0+$

2.3-32.   *The Notion of State; Euclidean Division via the Controllability Form*

Continuing with the results of Exercise 2.3-31, recall that by the classical division theorem for polynomials we can write

$$u(z) = q(z)a(z) + r(z), \qquad \deg r(z) < n$$

a. Show that the response at $t = 1$ to the polynomial input $u(z)$ is

$$x(1) = r(A)b$$

This is a striking fact: The entire effect of any past input on the state at $t = 1$ is determined only by at most n numbers—the coefficients of $r(z)$ or any non-singular linear combination of these coefficients (cf. Sec. 2.2.2). Further consideration of this simple result will lead to an abstract definition of state space and state-space realizations; see Sec. 5.1.

b. Suppose the realization $\{A, b\}$ is in controllability form. Show then that $x'[1] = [r_n \; \cdots \; r_2 \; r_1]$, where $r(z) = r_1 z^{n-1} + \cdots + r_n$. Check your calculation by studying the state history $\{x(-2), x(-1), x(0), x(1)\}$ of the response of the controllability-form realization (as in Fig. 2.3-1) to an input $r(z) = r_1 z^2 + r_2 z + r_3$. *Remark:* In other words, we can find the remainder $r(z)$ in dividing a polynomial $u(z)$ by another polynomial $a(z)$ by feeding $u(z)$ into a controllability-form realization of $1/a(z)$ and reading out the components of the state vector at $t = 1$.

2.3-33.   *Maximal Length Shift Register Sequences*

Consider a mod-2 system

$$x(k + 1) = Ax(k), \qquad y(k) = [1 \quad 0 \quad \cdots \quad 0]x(k)$$

If $A^k = I$ for some (smallest) k, then $y(\cdot)$ will clearly form a periodic sequence with period some divisor of k.

a. Show that if

$$A = \begin{bmatrix} 1 & 1 & 0 & 0 \\ 1 & 1 & 1 & 1 \\ 0 & 0 & 0 & 1 \\ 0 & 1 & 1 & 1 \end{bmatrix}$$

then the period is $15 = 2^4 - 1$.

b. Why is this the maximal-length cycle a four-stage shift register can have?

2.3-34.   *Flip-Flops*

Consider a system consisting of a single delay element and a single mod-2 adder, as shown in the figure. With a 0 input, the delay element will retain its state (either 0 or 1). An input 1 will cause the state to change, from 0 to 1 or from 1 to 0. Such a system is called a *trigger flip-flop* or *complementing flip-flop*. The flip-flop is a binary linear system with system function $H_F(z) = 1/(1 + z^{-1})$. Hence, if we have a system with system function $H(z)$ which can be written as $H_1[1/(1 + z^{-1})]$, this system may be constructed from flip-flops instead of delays. Use this fact to design a system using *only* flip-flops (no delays or adders) that is equivalent to the linear binary feedback shift register of Example 2.3-6.

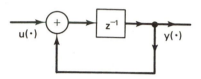

2.4   FURTHER ASPECTS OF CONTROLLABILITY AND OBSERVABILITY

In this section, we shall explore further some of the important properties of the concepts of controllability and observability. In Sec. 2.4.1 we shall bring out their significance for the important problem of deciding when a given realization is minimal, i.e., realized with the smallest possible number of integrators (or delay elements). This is a nice application of some apparently rather special concepts in a general, nondynamical situation (we are not setting up or observing states). Furthermore, we can also now deduce an alternative characterization of minimality in terms of the irreducibility of the associated transfer function. This fact brings out some useful connections with the classical theory of polynomials and their resultants, a topic further explored in Sec. 2.4.4. In Sec. 2.4.2 we shall present some very useful standard forms for noncontrollable and/or nonobservable realizations, which shed further light on the deep relations between controllability and observability and system structure. A nice application of these results is to the development of some very powerful tests for controllability and observability, which we call the PBH tests (Sec. 2.4.3). In our opinion these tests are the most useful in many problems, and we shall apply them often in this book. Finally, in Sec. 2.4.5 we shall present several illustrative examples.

2.4.1   *Joint Observability and Controllability; the Uses of Diagonal Forms*

The question of joint controllability and observability is not trivially resolvable for any of the four canonical forms of Sec. 2.1.2 because, for example, while the observer and observability forms have simple observability matrices, direct calculation of their controllability matrices does not seem to yield anything very transparent. In such situations, the diagonal form corresponding to the sum (parallel) realizations of Sec. 2.1.3 is often helpful in providing some insight into the situation. Thus, consider a realization with

$$A_d = \text{diag}\{\lambda_1, \ldots, \lambda_n\} \tag{1}$$

$$b_d = [\gamma_1 \quad \cdots \quad \gamma_n]', \qquad c_d = [\delta_1 \quad \cdots \quad \delta_n] \tag{2}$$

Then an easy calculation using the formula of Exercise A.7 for the determinant of a Vandermonde matrix shows that

$$\det \mathcal{C} = (\prod \gamma_i) \prod_{i<j} (\lambda_i - \lambda_j) \tag{3}$$

Therefore the realization will be controllable if and only if

$$\lambda_i \neq \lambda_j \tag{4a}$$

and

$$\gamma_i \neq 0, \qquad i, j = 1, \ldots, n \tag{4b}$$

Similarly, it can be proved that the realization will be observable if and only if

$$\lambda_i \neq \lambda_j \tag{5a}$$

and

$$\delta_i \neq 0, \qquad i, j = 1, \ldots, n \tag{5b}$$

It is not hard to see intuitively why controllability and/or observability are lost under the above conditions—we leave that to the reader. But we shall note here the implications for joint observability and controllability: If some of the eigenvalues $\{\lambda_i\}$ are repeated, or if some of the $\{\gamma_i\}$ or $\{\delta_i\}$ are zero, this means that the transfer function

$$H(s) = c_d(sI - A_d)^{-1}b_d = \sum_{1}^{n} \frac{\gamma_i \delta_i}{s - \lambda_i} \tag{6}$$

will in fact have less than n terms in its partial fraction expansion. Therefore the "nominal" representation

$$H(s) = \frac{b(s)}{a(s)}, \qquad a(s) \triangleq \det(sI - A_d) = s^n + a_1 s^{n-1} + \cdots + a_n \tag{7}$$

will in fact have the "reduced" form

$$H(s) = \frac{\bar{b}_1 s^{r-1} + \cdots + \bar{b}_r}{s^r + \bar{a}_1 s^{r-1} + \cdots + \bar{a}_r}, \qquad r < n \tag{8}$$

In other words, we shall have *cancellations* in the transfer function. Reversing this analysis, we see that if we set up an nth-order realization for a transfer function $H(s)$ given as (7) but which is in fact "reducible," then we would *expect* the realization to be either noncontrollable or nonobservable or both.

Why do we say "expect," rather than just assert the above? The reason is that we have only proved this for diagonal realizations. However, we might try to generalize the above arguments as follows: Given an arbitrary realization $\{A, b, c\}$, find an invertible transformation matrix T such that

$$A_d = T^{-1}AT, \qquad b_d = T^{-1}b, \qquad c_d = cT \qquad (9)$$

Because the transformation is invertible, we would expect that all conclusions reached for the transformed realization would also be valid for the original realization. *In particular the reader can easily verify that observability and controllability are preserved under such similarity transformations.*

Unfortunately, while our conjectured result, that an nth-order realization of a transfer function $H(s)$ as in (7) will be both controllable and observable if and only if $b(s)/a(s)$ is *irreducible* [i.e., $b(s)$ and $a(s)$ have no common factors except constants], is indeed correct, the above arguments do not prove this, because it is not true that an arbitrary realization can be transformed to a diagonal realization, as we discussed in some detail in Example 2.2-A. Therefore, more complicated proofs must be used: Either one goes to the rather more complicated Jordan form (Sec. 13 in the Appendix), or one further examines the properties of the earlier canonical forms of Sec. 2.1.1. The latter route will be successfully pursued at the end of this section, though it is probably fair to say that the proofs given there were (as happens so often) stimulated by the fact that the diagonal form had suggested what had to be proved. Similarly, the reader may often find it useful to use the special (restricted) diagonal realization to probe problems for which the right questions or right answers do not seem obvious (see, e.g., the discussion of similarity transformations below).

Potential General Validity of Results True for Diagonalizable Matrices. It may be useful to make some remarks on when results for diagonalizable matrices might hold more generally. This will be the case when the results involve quantities that depend "continuously" on the elements of the matrices involved, i.e., if small changes in the elements produce only small changes in these quantities. For example, the coefficients of the characteristic polynomial, $a(s) = \det(sI - A)$, depend continuously on the values of the elements of A, and so do the roots of the characteristic polynomial (i.e., the eigenvalues of A). However, the rank of A depends discontinuously on the values of A, as do the eigenvectors of A [study, for example, a 2×2 lower triangular matrix with diagonal elements 1 and ϵ and $(2, 1)$-element 1]. Therefore results based on the coefficients of the characteristic polynomial or of the adjugate matrix generally extend to all matrices when they are true for diagonal matrices. Examples are the Cayley-Hamilton theorem and the result noted above on the equivalence of joint controllability and observability to irreducibility of the transfer function. However, the result that a diagonalizable $n \times n$ matrix has n linearly independent eigenvectors should not be expected to hold for arbitrary matrices (and it does not).

A firm mathematical foundation can be laid under these heuristic remarks by using a theorem of H. Weyl called the *principle of the irrelevance of algebraic inequalities*. This theorem states that if a polynomial equation in the entries of a matrix holds for all nonsingular matrices, then the equation holds for singular matrices as well. Several examples of the use of this theorem can be found in [26].

Similarity Transformations Between Realizations.   It is obviously a useful thing to know when two realizations (of course, of a given transfer function) can be connected by a similarity transformation. The realizations must clearly have the same dimension, but is this enough? Using diagonal realizations shows that the answer is no—see Fig. 2.4-1(a). This example shows that

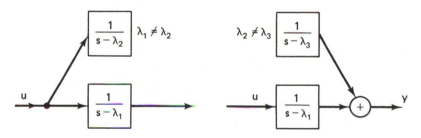

a. The two realizations have the same transfer function, but different eigenvalues and therefore cannot be similar (because similarity transformations do not change the eigenvalues or natural frequencies.

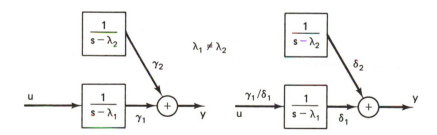

b. These two realizations can always be connected by a similarity transformation if $\lambda_1 \neq \lambda_2$ and $\gamma_i \neq 0$, $\delta_i \neq 0$, $i = 1, 2$. Note that both realizations are observable.

Figure 2.4-1.   Diagonal realizations are useful in studying the problem of similar realizations.

a necessary condition is that the two realizations have the same natural frequencies. But the example in Fig. 2.4-1(a) shows that even this is not enough because, for example, there is no invertible matrix such that

$$1. \quad [1 \quad 0]T = [1 \quad 1] \qquad (cT = \bar{c})$$

and

$$2. \quad T^{-1}\begin{bmatrix} \lambda_1 & 0 \\ 0 & \lambda_2 \end{bmatrix} T = \begin{bmatrix} \lambda_1 & 0 \\ 0 & \lambda_2 \end{bmatrix} \qquad (T^{-1}AT = \bar{A})$$

This is because, to satisfy the second condition, T must be diagonal, and then the first condition can never be met (note that $\lambda_1 \neq \lambda_2$).

Reflection on this fact will lead us to consider the two observable realizations in Fig. 2.4-1(b). Now we can see that we can always achieve

$$[\gamma_1 \quad \gamma_2]T = [\delta_1 \quad \delta_2]$$

with a diagonal matrix, viz., $T = \text{diag}\{\delta_i/\lambda_i\}$. The input matrices are

$$b_1' = [1 \quad 0], \qquad b_2' = [\gamma_1/\delta_1 \quad 0]$$

which are related by $T^{-1}b_1 = b_2$.

Therefore the realizations in Fig. 2.4-1(b) are always similar.

These considerations lead us to conjecture the following general result, for not necessarily diagonal or even diagonalizable realizations.

Lemma 2.4.1.   Similarity of Scalar Realizations

Any two realizations $\{A_i, b_i, c_i\}$ of the same order (and of the same transfer function) can be connected by a similarity transformation if

$$1. \qquad\qquad \det(sI - A_1) = \det(sI - A_2) \qquad\qquad (10)$$

and

2. Both realizations are controllable or both are observable. The appropriate similarity transformation is given by

$$x_1(t) = Tx_2(t), \qquad T = \mathcal{C}(A_1, b_1)\mathcal{C}^{-1}(A_2, b_2) \qquad (10a)$$

if both realizations are controllable, or by

$$x_1(t) = Tx_2(t), \qquad T = \mathcal{O}^{-1}(c_1, A_1)\mathcal{O}(c_2, A_2) \qquad (10b)$$

if both realizations are observable. A general proof of this lemma will be given in Sec. 2.4.5 (Example 2.4-5). (In fact, Example 2.3-5 is already essentially a proof—show this.) ■

The main aim of this discussion is to emphasize the important role of diagonal forms in testing and conjecturing results in system theory. (This is apart from any potential advantages, e.g., in sensitivity, that might accrue from the actual use of diagonal realizations.) However, we must also emphasize the (true) *results* already guessed by this means, and we have summarized them in Fig. 2.4-2. There we have also shown a result that provides an important application of the concepts of controllability and observability— viz., that a controllable and observable realization $\{A, b, c\}$ is also *minimal* (or has *minimal* order). By this we mean that there can be no other realization $\{A_1, b_1, c_1\}$ with A_1 of smaller dimension than A. We recall, of course, that in saying "realization" we mean realization of a transfer function, and therefore what we are saying is that there can be no triple $\{A_1, b_1, c_1\}$ with A_1 "smaller" than A in dimension and such that $c_1(sI - A_1)^{-1}b_1 = c(sI - A)^{-1}b$.

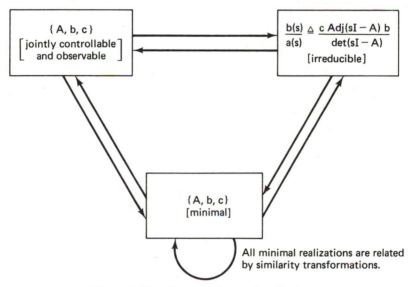

Figure 2.4-2.   Some fundamental realizations.

Note that we now have an *explicit test* for minimality—it would be quite difficult to verify minimality by comparison with all other realizations! Moreover, in view of the relations previously indicated, another explicit test for the minimality of $\{A, b, c\}$ is to check that $a(s) \triangleq \det (sI - A)$ and $b(s) \triangleq a(s)c(sI - A)^{-1}b = c \operatorname{Adj} (sI - A)b$ are *relatively prime*. Both these tests will be much used later—see especially Chapter 6 and also Exercise 2.4-5.

The results just stated on minimality can be proved quite directly from previous results, without assuming diagonal realizations, and we shall do so now. However, it will be valuable to gain as thorough an intuitive understanding as possible of the fundamental relations indicated in Fig. 2.4-2 before proceeding to the formal proofs.

Proofs of Theorems Relating Controllability, Observability, Irreducibility, and Minimality. Earlier in this section, we used the diagonal form to conjecture several important results on the characterization of simultaneously controllable and observable realizations. These results, which are summarized in Fig. 2.4-2, will be proven here. Proofs can be given in many different ways, depending on the order in which the different theorems are considered. The route we choose seems to be the most elementary in that, for example, it avoids the need for any results from linear algebra.

We begin with two preliminary results that are also of value in themselves.

Lemma 2.4-2.

If a transfer function

$$H(s) = \frac{b(s)}{a(s)} = \frac{b_1 s^{n-1} + \cdots + b_n}{s^n + a_1 s^{n-1} + \cdots + a_n}$$

has one controllable and observable nth-order realization, then all nth-order realizations must also be controllable and observable.

Proof. We use the fact that if $\{A, b, c\}$ is a realization, then the Hankel matrix [cf. Eq. (2.2-49)]

$$M[1, n-1] = \begin{bmatrix} cb & cAb & \cdots & cA^{n-1}b \\ cAb & cA^2b & \cdots & cA^nb \\ \vdots & & & \\ cA^{n-1}b & \cdots & & cA^{2n-2}b \end{bmatrix}$$

depends only on the transfer function $[H(s) = \sum_1^\infty (cA^{i-1}b)s^{-i}]$. Furthermore, we can easily check that (cf. Exercise 2.2-14)

$$M[1, n-1] = \mathcal{O}(c, A)\mathcal{C}(A, b)$$

Therefore, if $\{A_1, b_1, c_1\}$ and $\{A_2, b_2, c_2\}$ are two nth-order realizations of $H(s)$,

$$\mathcal{O}(c_1, A_1)\mathcal{C}(A_1, b_1) = \mathcal{O}(c_2, A_2)\mathcal{C}(A_2, b_2)$$

Now since by hypothesis $\mathcal{O}(c_1, A_1)$ and $\mathcal{C}(A_1, b_1)$ are both nonsingular, so is their product, and thus also $\mathcal{O}(c_2, A_2)\mathcal{C}(A_2, b_2)$. But the product of two $n \times n$ matrices is nonsingular if and only if each matrix is nonsingular, so that $\mathcal{O}(c_2, A_2)$ and $\mathcal{C}(A_2, b_2)$ must be nonsingular for any nth-order realization $\{A_2, b_2, c_2\}$. ∎

In view of this result, it suffices to find one case in which joint controllability and observability can be assured. We shall see that the controller form will do.

Lemma 2.4-3

The nth-order controller form of $H(s) = b(s)/a(s)$, $n = \deg a(s)$, will be observable if and only if $b(s)$ and $a(s)$ are coprime, i.e., if $b(s)/a(s)$ is irreducible.

Proof.   This result was proved in Example 2.3-5.   ∎

We can now state the following theorem.

Theorem 2.4-4

A transfer function $H(s) = b(s)/a(s)$ is irreducible if and only if all nth-order realizations, $n = \deg a(s)$, are controllable and observable.

Proof.   This is an immediate consequence of Lemmas 2.4-2 and 2.4-3.   ∎

Next we recall that a realization $\{A, b, c\}$ is *minimal* if it has the smallest order (i.e., the smallest number of state variables) among all realizations having the same transfer function $c(sI - A)^{-1}b$. Clearly, there can be many minimal realizations, but they all share certain important properties.

Theorem 2.4-5

A realization $\{A, b, c\}$ is minimal if and only if $a(s) \triangleq \det(sI - A)$ and $b(s) \triangleq c \operatorname{Adj}(sI - A)b$ are relatively prime.

Proof.   Let

$$H(s) = c(sI - A)^{-1}b = \frac{b(s)}{a(s)}$$

Suppose first that $\{A, b, c\}$ is minimal but that $b(s)/a(s)$ is not irreducible. Then using the *reduced* transfer function, we can obtain a realization with a lower-dimensional state vector. This is a contradiction. Similarly, to prove the converse, assume that $\{A, b, c\}$ is not minimal even though $b(s)/a(s)$ is irreducible. Then any minimal realization of $H(s)$ will have a transfer function with denominator of degree lower than n, the dimension of A. Therefore $b(s)/a(s)$ could not have been irreducible.   ∎

We can combine Theorems 2.4-4 and 2.4-5 to obtain the following important result, which provides a direct test for minimality, instead of having to search over all possible realizations or check the nominal transfer function for irreducibility [17].

Theorem 2.4-6

A realization $\{A, b, c\}$ is minimal if and only if $\{A, b\}$ is controllable and $\{c, A\}$ is observable.

Finally we note that minimal realizations are very tightly related [17].

Theorem 2.4-7

Any two minimal realizations can be connected by a *unique* similarity transformation.

Proof.   Minimal realizations satisfy the conditions of Lemma 2.4-1, and the theorem follows immediately. However, since a proof of Lemma 2.4-1 was deferred, we shall present a direct proof of the present theorem.

Since the two realizations are minimal, they are both observable and controllable. Therefore, if we define

$$T = \mathcal{O}^{-1}(c_1, A_1)\mathcal{O}(c_2, A_2)$$

from the previously noted identity (also easy to verify directly)

$$\mathcal{O}(c_1, A_1)\mathcal{C}(A_1, b_1) = \mathcal{O}(c_2, A_2)\mathcal{C}(A_2, b_2)$$

we can conclude that

$$T = \mathcal{C}(A_1, b_1)\mathcal{C}^{-1}(A_2, b_2)$$

Now we can easily see that

$$T^{-1}b_1 = b_2, \qquad c_1 T = c_2$$

Finally, from the easily verified relation

$$\mathcal{O}(c_1, A_1)A_1\mathcal{C}(A_1, b_1) = \mathcal{O}(c_2, A_2)A_2\mathcal{C}(A_2, b_2)(= M[2, n - 1]),$$

we obtain

$$A_2 = \mathcal{O}^{-1}(c_2, A_2)\mathcal{O}(c_1, A_1)A_1\mathcal{C}(A_1, b_1)\mathcal{C}^{-1}(A_2, b_2) = T^{-1}A_1 T$$

Therefore this T defines a similarity transformation relating the two realizations.

Moreover all matrices $\tilde{T}$ relating $\{A_i, b_i, c_i, i = 1, 2\}$ by similarity must be equal to T as defined above. For if $\tilde{T}$ is any such transformation, we shall have

$$\mathcal{O}(c_1, A_1)(T - \tilde{T}) = 0$$

which implies that $T = \tilde{T}$, since $\mathcal{O}(c_1, A_1)$ is nonsingular. ∎

Example 2.4.6 in Sec. 2.4.5 will provide a nice illustration of the use of this theorem.

Transformations Between the Canonical Realizations.   For various purposes, it will be useful to collect together here the special formulas that enable us to go between the four special realizations of Sec. 2.1.2 for an *irreducible* transfer function $H(s) = b(s)/a(s)$. We do this in Fig. 2.4-3, where the reader should be able to verify all the relations except that relating the observer form to the controller form—this will be treated in Sec. 2.4.4.

2.4.2   Standard Forms for Noncontrollable and/or Nonobservable Systems

In previous sections we have discussed some canonical forms for controllable or observable realizations and have shown their usefulness in special applications. For other applications it will be useful to have standard forms in which noncontrollable and/or nonobservable systems can be represented. By using appropriate similarity transformations, we shall be able to find reali-

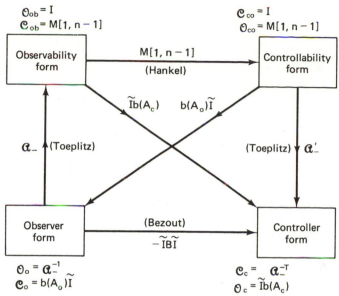

Figure 2.4-3. Transformations among four canonical realizations of an irreducible transfer function. Note that the (invertible) transformation matrix can be calculated as $T = \mathcal{O}_{old}^{-1}\mathcal{O}_{new} = \mathcal{C}_{old}\mathcal{C}_{new}^{-1}$. $\mathcal{Q}_-$ was defined in Eq. (2.3-6), $M[1, n-1]$ in Eq. (2.2-49), and B will be defined in Sec. 2.4-4.

zations in which the noncontrollable and/or nonobservable state variables can be clearly separated out.

Representation of Noncontrollable Realizations. If we have a noncontrollable realization $\dot{x}(t) = Ax(t) + bu(t)$ and A is diagonal with distinct eigenvalues, then we can readily say whether or not a particular state variable is controllable: It will be so if and only if the corresponding element of the input vector b is not zero. However, it is clear that this simple separation will be lost after any change of variables (similarity transformation) that does not again yield a diagonal realization—every new state variable may now have both a controllable and noncontrollable part. In this section we shall show how any given (noncontrollable) realization $\{A, b\}$, with A not necessarily diagonalizable, can be transformed into another realization where the controllable and noncontrollable state variables can be clearly identified. Let $\{A, b, c\}$ be such that

$$\text{rank } \mathcal{C}(A, b) = r < n$$

Then we shall show that a transformation matrix T can always be found such that the realization

$$\{\bar{A} = T^{-1}AT,\ \bar{b} = T^{-1}b,\ \bar{c} = cT\}$$

has the form

$$\bar{A} = \begin{bmatrix} \bar{A}_c & | & \bar{A}_{c\bar{c}} \\ ------ & & \\ 0 & | & \bar{A}_{\bar{c}} \end{bmatrix}\begin{matrix} r \\ \\ n-r \end{matrix}, \qquad \bar{b} = \begin{bmatrix} \bar{b}_c \\ -- \\ 0 \end{bmatrix}, \qquad \bar{c} = [\bar{c}_c \quad \bar{c}_{\bar{c}}] \qquad (11)$$

Any realization in this form has the important properties that

1. The $r \times r$ subsystem $\{\bar{A}_c, \bar{b}_c, \bar{c}_c\}$ is controllable.
2. $\bar{c}(sI - \bar{A})^{-1}\bar{b} = \bar{c}_c(sI - \bar{A}_c)^{-1}\bar{b}_c$; i.e., the subsystem has the same transfer function as the original system.

If the state variables $\bar{x}$ are correspondingly partitioned as $x' = [\bar{x}_c'\ \ \bar{x}_{\bar{c}}']'$, then the variables $\bar{x}_c$ can be said to be controllable and the variables $\bar{x}_{\bar{c}}$ noncontrollable. The realization $\{\bar{A}, \bar{b}, \bar{c}\}$ can be depicted as in Fig. 2.4-4, showing graphically the separation of states.

The above statements can be proved in several different ways. Here let us first check that for a system as in (11)†

$$\bar{c}(sI - \bar{A})^{-1}\bar{b} = \bar{c}\begin{bmatrix} (sI - \bar{A}_c)^{-1} & \text{xx} \\ 0 & (sI - \bar{A}_{\bar{c}})^{-1} \end{bmatrix}\begin{bmatrix} \bar{b}_c \\ 0 \end{bmatrix}$$

$$= [\bar{c}_c \quad \bar{c}_{\bar{c}}]\begin{bmatrix} (sI - \bar{A}_c)^{-1}\bar{b}_c \\ 0 \end{bmatrix} = \bar{c}_c(sI - \bar{A}_c)^{-1}\bar{b}_c$$

[The xx denotes entries whose exact values are unimportant here, though it is easy to show that $\text{xx} = -(sI - \bar{A}_c)^{-1}\bar{A}_{c\bar{c}}(sI - \bar{A}_{\bar{c}})^{-1}$.] Next we note that

$$\mathcal{C}(\bar{A}, \bar{b}) = \begin{bmatrix} \bar{b}_c & \bar{A}_c\bar{b}_c & \cdots & \bar{A}_c^{n-1}\bar{b}_c \\ ------------------ \\ 0 & 0 & & 0 \end{bmatrix}\begin{matrix} r \\ \\ n-r \end{matrix} \qquad (12)$$

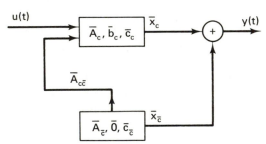

Figure 2.4-4. Decomposition of a noncontrollable realization.

†Another proof is to note that with zero initial conditions, as required when computing transfer functions, the states $x_{\bar{c}}$ will be identically zero.

Now since $\mathcal{C}(\bar{A}, \bar{b}) = T^{-1}\mathcal{C}(A, b)$, $\mathcal{C}(\bar{A}, \bar{b})$ has rank r, and therefore it can have only r linearly independent rows and r linearly independent columns. We shall show that if we start checking the columns from left to right, the first r columns must be linearly independent. For suppose $\bar{A}_c^k \bar{b}_c$ is linearly dependent on $\{\bar{A}_c^i \bar{b}_c\}$, $i < k$. Then clearly $\bar{A}_c^{k+1} \bar{b}_c = \bar{A}_c \bar{A}_c^k \bar{b}_c$ will also be dependent on the set $\{\bar{A}_c^i \bar{b}_c, i < k\}$. In other words, in searching from left to right in (12), once we find a dependent vector, then all subsequent ones must be so too. But since the rank is r, the first r columns $\{\bar{b}_c \cdots \bar{A}_c^{r-1} \bar{b}_c\}$ (and only these) must be linearly independent. This proves statement 1.

The above discussion also suggests a way of finding a suitable transformation matrix T. We require

$$T\mathcal{C}(\bar{A}, \bar{b}) = \mathcal{C}(A, b) \qquad (13)$$

Partitioning these matrices appropriately, we get

$$[T_1 \quad T_2]\begin{bmatrix} \mathcal{C}(\bar{A}_c, \bar{b}_c) & \text{xx} \\ 0 & 0 \end{bmatrix} = [b \quad Ab \quad \cdots \quad A^{r-1}b \quad \text{xx}] \qquad (14)$$

from which

$$T_1 = [b \quad Ab \quad \cdots \quad A^{r-1}b]\mathcal{C}^{-1}(\bar{A}_c, \bar{b}_c) \qquad (15)$$

If, in particular, we wish the new controllable subsystem to be in controllability form, then $\mathcal{C}(\bar{A}_c, \bar{b}_c) = I$, and the first r (mutually independent) columns of T are given by $T_1 = [b \ Ab \ \cdots \ A^{r-1}b]$. The remaining columns, those of T_2, need only to be linearly independent of each other and of the columns of T_1 [and hence of $\mathcal{C}(A, b)$] in order that T be nonsingular. A particular choice for these columns is a set of $(n - r)$ vectors that is *orthogonal* to (and not merely independent of) the columns of $\mathcal{C}(A, b)$.

Examples of some different transformations to standard noncontrollable form are provided in Examples 2.4-1 and 2.4-2—the point is that instead of using a fixed method, the context should be exploited to find the appropriate transformation. For large systems, however, more systematic methods are desirable, and some efficient ones have been proposed by Rosenbrock [22, pp. 80–84] (see also Mayne [27] and Daly [28]). We stress also that in many analytical problems, explicit determination of the transformation is not necessary.

Representation of Nonobservable Realizations.   It should be clear that similar (in fact, dual) statements can be made about nonobservable realizations. Thus if

$$\mathcal{O}(c, A) \text{ has rank } r < n$$

we can find a nonsingular matrix T such that

$$\bar{A} = T^{-1}AT, \qquad \bar{b} = T^{-1}b, \qquad \bar{c} = cT$$

have the form

$$\bar{A} = \begin{bmatrix} \bar{A}_o & \vdots & 0 \\ \hdashline \bar{A}_{\bar{o}o} & \vdots & \bar{A}_{\bar{o}} \end{bmatrix} \begin{matrix} r \\ \\ n-r \end{matrix}, \qquad \bar{c}' = \begin{bmatrix} \bar{c}'_o \\ \hdashline 0 \end{bmatrix}, \qquad \bar{b} = \begin{bmatrix} \bar{b}_o \\ \hdashline \bar{b}_{\bar{o}} \end{bmatrix} \tag{16}$$

and

$$\{\bar{c}_o, \bar{A}_o\} \text{ is observable}$$

$$c(sI - A)^{-1}b = \bar{c}_o(sI - \bar{A}_o)^{-1}\bar{b}_o$$

The separation into observable and nonobservable parts can be depicted as in Figure 2.4-5.

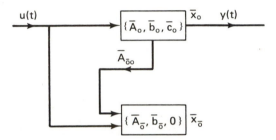

Figure 2.4-5.   Decomposition of nonobservable realizations.

We can combine the above results for noncontrollable or nonobservable systems to obtain a decomposition of arbitrary realizations. Before stating the result, we point out that its content should first be anticipated by considering diagonal realizations, for which the decomposition shown in Fig. 2.4-6 is obvious. For nondiagonal realizations, there can be certain forms of state feedback between the blocks, as shown by the dashed lines in Fig. 2.4-6. The general theorem will be given at the end of this section, but we first interject another useful remark.

Obtaining Minimal Realizations.   The above results suggest one way of obtaining a minimal realization from a given realization or transfer function. Thus, given a transfer function, we can always, for example, write down a controller-form realization by inspection and then separate out any nonobservable part if the realization is not minimal.

General Decomposition Theorem.   We can always find an invertible state transformation that will allow us to rewrite the state equations

$$\dot{x} = Ax + bu, \qquad y = cx$$

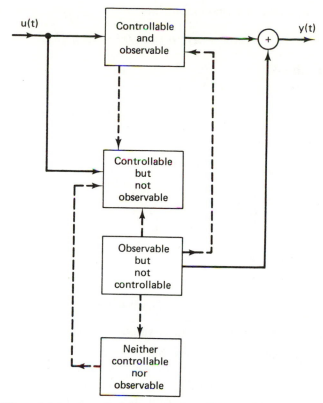

Figure 2.4-6. Canonical decomposition of diagonal realizations.

in the form

$$\dot{\bar{x}} = \bar{A}\bar{x} + \bar{b}u, \qquad y = \bar{c}\bar{x}$$

where

$$\bar{x}' = [\bar{x}'_{c,o} \quad \bar{x}'_{c,\bar{o}} \quad \bar{x}'_{\bar{c},o} \quad \bar{x}'_{\bar{c},\bar{o}}] \tag{17}$$

$$\bar{A} = \begin{bmatrix} \bar{A}_{c,o} & 0 & \bar{A}_{1,3} & 0 \\ \bar{A}_{2,1} & \bar{A}_{c,\bar{o}} & \bar{A}_{2,3} & \bar{A}_{2,4} \\ 0 & 0 & \bar{A}_{\bar{c},o} & 0 \\ 0 & 0 & \bar{A}_{4,3} & \bar{A}_{\bar{c},\bar{o}} \end{bmatrix}, \qquad \bar{b} = \begin{bmatrix} \bar{b}_{c,o} \\ \bar{b}_{c,\bar{o}} \\ 0 \\ 0 \end{bmatrix} \tag{18}$$

$$\bar{c} = [\bar{c}_{c,o} \quad 0 \quad \bar{c}_{\bar{c},o} \quad 0]$$

and

1. The subsystem

$$\{\bar{A}_{c,o}, \bar{b}_{c,o}, \bar{c}_{c,o}\}$$

is controllable and observable, and

$$\bar{c}(sI - \bar{A})^{-1}\bar{b} = \bar{c}_{c,o}(sI - \bar{A}_{c,o})^{-1}\bar{b}_{c,o}$$

2. The subsystem

$$\left\{ \begin{bmatrix} \bar{A}_{c,o} & \vdots & 0 \\ \text{-----} & \vdots & \text{-----} \\ \bar{A}_{2,1} & \vdots & \bar{A}_{c,\bar{o}} \end{bmatrix}, \begin{bmatrix} \bar{b}_{c,o} \\ \text{---} \\ \bar{b}_{c,\bar{o}} \end{bmatrix}, [\bar{c}_{c,o} \quad 0] \right\}$$

is controllable;

3. The subsystem

$$\left\{ \begin{bmatrix} \bar{A}_{c,o} & \vdots & \bar{A}_{1,3} \\ \text{-----} & \vdots & \text{-----} \\ 0 & \vdots & \bar{A}_{\bar{c},o} \end{bmatrix}, \begin{bmatrix} \bar{b}_{c,o} \\ \text{---} \\ 0 \end{bmatrix}, [\bar{c}_{c,o} \quad \bar{c}_{\bar{c},o}] \right\}$$

is observable; and

4. The subsystem

$$\{\bar{A}_{\bar{c},\bar{o}}, [0], [0]\}$$

is neither controllable nor observable.

The actual values of the possible nonzero matrices $\{\bar{A}_{c,o}, \bar{A}_{2,1}, \ldots, \bar{b}_{c,o}, \ldots\}$ are not all unique, because they depend on the particular coordinate transformations used to identify the noncontrollable and nonobservable states, but the dimensions of the various blocks are unique (why?). Note, of course, that the subsystem in 1 has the same transfer function as the original system (prove this also by direct evaluation).

This general decomposition theorem, which was first enunciated by Gilbert [16] and Kalman [17], can be proved by combining the ideas used to analyze noncontrollable (or nonobservable) realizations—we shall omit the detailed construction (but try Exercise 2.4-16 for some physical insight).

Reprise. Given problems involving noncontrollable and/or nonobservable realizations, it is generally a good idea to first assume, as we can do without loss of generality, that they are in the standard forms described in this section—this often simplifies later analyses. It is important to realize that this assumption can be made without having to know a transformation matrix T for taking a given realization to one or the other of these standard forms. Readers should familiarize themselves with the correspondence between Eq. (2.4-11) and Fig. 2.4-4, and Eq. (2.4-16) and Fig. 2.4-5, and should have a good understanding of the reasons for their special names—this again can be achieved without any knowledge of the explicit transformation to these forms.

2.4.3 The Popov-Belevitch-Hautus Tests for Controllability and Observability

The standard forms introduced in the previous section allow us to establish some extremely powerful criteria for testing the controllability and observability of realizations, especially when they arise by combining and arranging subsystems in special ways required by particular problems. These tests are especially useful for theoretical analysis and also in numerical problems whenever determination of matrix eigenvalues and eigenvectors is computationally feasible.

These criteria were introduced independently by several people, especially Popov [29], Belevitch [30, p. 413], Hautus [31], Rosenbrock [22], Hahn (cf. [18, p. 27]), Ford and Johnson ([32] and [33]), and no doubt others. In particular, for the special case where the A matrix is diagonalizable, the test was first given by Gilbert [16]. Since Popov and Belevitch were perhaps the first to give a general statement and Hautus was perhaps the first to note their wide applicability, we shall call them the PBH tests. Several examples of the use of these tests will be given in this book.

Theorem 2.4-8. PBH Eigenvector Tests

1. A pair $\{A, b\}$ will be noncontrollable if and only if there exists a row vector $q \not\equiv 0$ such that

$$qA = \lambda q, \qquad qb = 0 \tag{19}$$

In other words, $\{A, b\}$ will be controllable if and only if there is no row (or left) eigenvector of A that is orthogonal to b.

2. A pair $\{c, A\}$ will be nonobservable if and only if there exists a (column) vector $p \not\equiv 0$ such that

$$Ap = \lambda p, \qquad cp = 0 \tag{20}$$

in other words, if and only if some eigenvector of A is orthogonal to c.

Proof.

1. *The "if" part.* If there exists $q \not\equiv 0$ such that

$$qA = \lambda q, \qquad qb = 0$$

then

$$qAb = \lambda qb = 0$$

and

$$qA^2 b = \lambda qAb = 0$$

and so on, until we see that

$$q\mathcal{C}(A, b) = q[b \quad Ab \quad \cdots \quad A^{n-1}b] = 0$$

which means that the controllability matrix is singular, i.e., $\{A, b\}$ is not controllable.

135

The "only if" part. We have to show that $\{A, b\}$ noncontrollable implies the existence of a vector q as in (19). Now whenever we have to show things about noncontrollable realizations, it is generally a good idea to begin by assuming that the realization has been put into the standard noncontrollable form (11) described in the previous section,

$$
A = \begin{bmatrix} A_c & \vdots & A_{c\bar{c}} \\ \cdots & \vdots & \cdots \\ 0 & \vdots & A_{\bar{c}} \end{bmatrix} \begin{matrix} r \\ \\ n-r \end{matrix} \quad , \qquad b = \begin{bmatrix} b_c \\ \cdots \\ 0 \end{bmatrix}
$$
$$\begin{matrix} r & n-r \end{matrix}$$

where $r = \text{rank } \mathcal{C}(A, b) < n$. Now it is clear that a particular row vector q that is orthogonal to b has the form $q = [0 \mid z]$, and it is perhaps not hard to guess that we should choose z as an eigenvector of $A_{\bar{c}}$,

$$ zA_{\bar{c}} = \lambda z $$

for then

$$ qA = [0 \quad z]A = [0 \quad \lambda z] = \lambda q $$

Therefore we have shown how to find a row vector q satisfying (19), and this completes the proof of part 1.

2. This result is the "dual" of the one in part 1. Let us go through the details of showing this.

We first note that $\{c, A\}$ will be observable if and only if $\{A', c'\}$ is controllable. But, by part 1, this will be true if and only if there exists no row vector p' such that

$$ p'A' = \lambda p' \quad \text{and} \quad p'c' = 0 $$

i.e., such that

$$ Ap = \lambda p \quad \text{and} \quad cp = 0 $$

which is the criterion stated in part 2. ■

Another form of these tests is often useful.

Theorem 2.4-9.   PBH Rank Tests

1. A pair $\{A, b\}$ will be controllable if and only if

$$ \text{rank } [sI - A \quad b] = n \qquad \text{for } all \ s \tag{21} $$

2. A pair $\{c, A\}$ will be observable if and only if

$$ \text{rank } \begin{bmatrix} c \\ sI - A \end{bmatrix} = n \qquad \text{for } all \ s \tag{22} $$

Here, of course, n is the size of A. Note also that these conditions will clearly be met for all s that are not eigenvalues of A, because $\det (sI - A) \neq 0$ for such s; the point of the theorem is that the rank must be n even when s is an eigenvalue of A. We shall see in Sec. 6.3 that conditions (21)–(22) are ways of stating that the matrix polynomials $\{sI - A, b\}$ and $\{c, sI - A\}$ obey certain *relative primeness* (or *co-primeness*) conditions.

Proof.

1. If $[sI - A \quad b]$ has rank n, there cannot be a nonzero row vector q such that

$$q[sI - A \quad b] = 0 \qquad \text{for any } s$$

i.e., such that

$$qb = 0 \quad \text{and} \quad qA = sq$$

But then by Theorem 2.4-8 $\{A, b\}$ must be controllable. The converse follows easily by reversing the above arguments.

2. This can be proved in the same way or via duality. ∎

As noted before, several examples and applications of these tests will appear throughout this book. In fact, when faced with problems of checking for controllability and/or observability, it is a good heuristic rule to first try to apply the PBH tests.

Diagonalizable Realizations (Gilbert [16]).   The special case of diagonalizable realizations provides a good insight into the nature of the PBH tests.

Thus, suppose we have a system

$$\dot{x}(t) = Ax(t) + bu(t)$$

where we can write

$$T^{-1}AT = \Lambda = \text{diag}\,\{\lambda_1, \ldots, \lambda_n\}$$

with

$$T = [p_1 \ldots p_n] \quad \text{and} \quad Ap_i = \lambda_i p_i$$

In other words, $\{\lambda_i\}$ and $\{p_i\}$ are the eigenvalues and eigenvectors of A. Also let

$$q_i = \text{the } i\text{th row of } T^{-1}$$

It is easy to see that

$$q_i p_j = \delta_{ij}$$

and that the $\{q_i\}$ are left eigenvectors of A; i.e., $q_i A = \lambda_i q_i$. Now the realization $\{A, b\}$ can be converted to the diagonal realization

$$\dot{\chi}(t) = \Lambda \chi(t) + \beta u(t), \qquad \beta = T^{-1}b$$

Then as we noted in Sec. 2.4.1, and as is directly evident, this realization will be noncontrollable if any of the components of the input vector is zero. But

$$\beta_i = q_i b, \qquad i = 1, \ldots, n$$

so that we shall have noncontrollability if some left eigenvector of A is orthogonal to b, as claimed by the PBH criterion.

Now we also know that even if all the $\{\beta_i\}$ are nonzero, we can lose controllability if the eigenvalues are repeated, e.g., if $\lambda_1 = \lambda_2$ (cf. Sec. 2.4.1). What happens here? We have, say,

$$q_1 b = \beta_1 \neq 0, \qquad q_2 b = \beta_2 \neq 0$$

with q_1 and q_2 linearly independent.

But because q_1 and q_2 are associated with the same eigenvalue λ, then clearly any linear combination of q_1 and q_2 will also be a left eigenvector associated with λ. Therefore we can readily find an eigenvector orthogonal to b, in fact, $(\beta_2 q_1 - \beta_1 q_2)b = 0$. Therefore for diagonalizable realizations, the PBH eigenvector test is relatively obvious.

For nondiagonalizable realizations, we can prove the test by using the Jordan canonical form, but the proof given above (Theorem 2.4-8) is more self-contained. We should remember, however, from the above argument that whenever more than one independent eigenvector can be associated with a single eigenvalue, we shall lose controllability (and observability).

Example 2.4-A.

Show that any pair $\{A_1, b\}$, where

$$A_1 = \begin{bmatrix} \lambda & 0 & 0 \\ 0 & \lambda & 1 \\ 0 & 0 & \lambda \end{bmatrix}$$

is always noncontrollable.

Solution.   There are three eigenvalues λ. As for eigenvectors, the equations

$$[\alpha_1 \quad \alpha_2 \quad \alpha_3]A = \lambda[\alpha_1 \quad \alpha_2 \quad \alpha_3]$$

yield

$$\alpha_1 \lambda = \lambda \alpha_1, \qquad \alpha_2 \lambda = \lambda \alpha_2, \qquad \alpha_2 + \lambda \alpha_3 = \lambda \alpha_3$$

which can be satisfied by

$$[1 \quad 0 \quad 1] \quad \text{or} \quad [0 \quad 0 \quad 1]$$

and any linear combinations thereof. Clearly, we can always find a combination that will be orthogonal to any 3-vector b.

An alternative solution is provided by the PBH rank test, as we ask the reader to check. ∎

Example 2.4-B.

Find conditions on c so that $\{c, A_2\}$ will always be observable when

$$A_2 = \begin{bmatrix} \lambda & 1 & 0 \\ 0 & \lambda & 1 \\ 0 & 0 & \lambda \end{bmatrix}$$

Solution. By the PBH rank test, we must check the rank of

$$\begin{bmatrix} c_1 & c_2 & c_3 \\ s - \lambda & -1 & 0 \\ 0 & s - \lambda & -1 \\ 0 & 0 & s - \lambda \end{bmatrix}$$

When $s = \lambda$, we see that the rank will be 3 if and only if $c_1 \neq 0$. ∎

We note that one of the PBH tests might be simpler than the other in any particular problem.

Uncontrollable and Controllable Modes and Eigenvalues. The reader should know from earlier experience with linear systems that the eigenvalues of the matrix A determine the so-called *modes* of a realization $\{A, b, c\}$ (cf. Sec. 2.5.2 for a more formal discussion). The PBH tests allow us to say that the modes associated with the eigenvalue λ are *uncontrollable* if and only if some associated left eigenvector is orthogonal to b. (Otherwise the associated modes may be called *controllable*.)

The significance of this definition is very clear for diagonalizable realizations, where there will be no input to the subsystem associated with this mode. In Chapter 3, we shall see also that an uncontrollable mode corresponds to an eigenvalue of A that cannot be changed to some other value by using state feedback. In this sense we can also talk about controllable and uncontrollable eigenvalues (or natural frequencies).

It is worth noting that every mode and every eigenvalue of a realization can be classed as completely controllable or completely uncontrollable, while this is not possible for any *state variable* in a realization, unless the realization happens to be in the standard form of Sec. 2.4.1.

It also should go without saying that all the statements we have made above about controllability properties go over, with obvious changes, to observability.

Reprise. The PBH eigenvector and rank tests for controllability and observability are very useful. Their intuitive meaning can be brought out by recalling how (diagonalizable) realizations are transformed to diagonal form. The tests also allow us to identify controllable and/or observable modes and natural frequencies (eigenvalues).

*2.4.4 Some Tests for Relatively Prime Polynomials

In Sec. 2.4.1 we noted that the minimality of a realization $\{A, b, c\}$ could be determined either by checking the nonsingularity of the observability and controllability matrices $\mathcal{O}(c, A)$ and $\mathcal{C}(A, b)$ or by checking the relative primeness of $a(z) = \det(zI - A)$ and $b(z) = c \, \text{Adj} \, (zI - A)b$. Now while controllability and observability may be relatively new concepts, that of relatively prime polynomials is a very old one, and there exist many tests for it. We shall give three of the better known ones, due, respectively, to Sylvester [34], Bezout [35], and MacDuffee [36]. There should of course be intimate relations between these criteria and the notions of controllability and observability because both provide tests for minimality; we shall in fact display some of these.

The question of checking the relative primeness of two polynomials $a(z)$ and $b(z)$ can be regarded as a special case of the problem of finding the greatest common divisor (gcd) of two polynomials. This can be done by using the celebrated Euclidean algorithm, which Knuth [37, p. 294 ff.] has remarked is the oldest nontrivial algorithm that has survived to the present day.

The Euclidean algorithm is based on the fact that given two polynomials

$$a(z) = a_0 z^n + a_1 z^{n-1} + \cdots + a_n, \qquad a_0 \neq 0 \tag{23a}$$

$$b(z) = b_0 z^m + \cdots + b_m, \qquad m \leq n \tag{23b}$$

there exists a unique quotient polynomial $q(z)$ and a unique remainder polynomial $r(z)$ such that

$$a(z) = q(z)b(z) + r(z), \qquad \deg r(z) < \deg b(z)$$

Suppose that

$$\deg b(z) \leq \deg a(z)$$

Then by successive use of the above polynomial division formula we can write

$$a(z) = q_1(z)b(z) + r_1(z), \qquad \deg r_1 < \deg b$$
$$b(z) = q_2(z)r_1(z) + r_2(z), \qquad \deg r_2 < \deg r_1$$
$$\vdots$$
$$r_{p-3}(z) = q_{p-1}(z)r_{p-2}(z) + r_{p-1}(z), \qquad \deg r_{p-1} < \deg r_{p-2}$$
$$r_{p-2}(z) = q_p(z)r_{p-1}(z) + 0$$

The algorithm stops when the remainder $r_p(z) = 0$ and then the gcd of $a(z)$ and $b(z)$ is $r_{p-1}(z)$.

The reader should check this for himself with simple numerical examples. A formal proof is not very difficult. Note first that the last equation shows that $r_{p-1}(z)$ divides $r_{p-2}(z)$ (written $r_{p-1} | r_{p-2}$). The last-but-one equation can be written

$$r_{p-3} = q_{p-1}(q_p r_{p-1}) + r_{p-1}$$

so we see that $r_{p-1} | r_{p-3}$. Proceeding in this way gives

$$r_{p-1}(z) | b(z), \qquad r_{p-1}(z) | a(z)$$

To show that this is the greatest common divisor, we have to show that any other divisor of $a(z)$ and $b(z)$ also divides $r_{p-1}(z)$. For this, we first note that successive substitution in the equations

$$r_{p-1} = r_{p-3} - q_{p-1} r_{p-2}, \qquad r_{p-2} = r_{p-4} - q_{p-2} r_{p-3}, \qquad \cdots$$

shows that there exist two polynomials $x(z)$ and $y(z)$ such that

$$r_{p-1}(z) = x(z)a(z) + y(z)b(z) \tag{24}$$

Now it is clear that any common factor of $a(z)$ and $b(z)$ will also be a factor of $r_{p-1}(z)$, which is thereby proved to be the gcd. The gcd is only unique up to a constant, but it can be made unique by requiring it to be monic, i.e., to have highest-order coefficient equal to unity.

The algebra involved in computing the gcd can be organized in several different ways, which we shall not pursue here. (One of the most efficient algorithms can be found in [38, Sec. 8.4]). Our main interest is in the case where the polynomials are relatively prime (or coprime, as we shall often say). For this case, we can state the following slight strengthening of the result (23).

Lemma 2.4-10.
The polynomials $a(z)$ and $b(z)$ in (24) will be coprime if and only if there exist two polynomials $\tilde{x}(z)$ and $\tilde{y}(z)$ such that

$$\tilde{x}(z)a(z) + \tilde{y}(z)b(z) = 1 \tag{25a}$$

and

$$\deg \tilde{x}(z) < m, \qquad \deg \tilde{y}(z) < n \tag{25b}$$

Moreover, if $a(z)$ and $b(z)$ are coprime, then $\tilde{x}(z)$ and $\tilde{y}(z)$ will be unique.

Proof. Everything has been proved already except for the possibility of the degree constraints on $x(z)$ and $y(z)$. That is, by (24) we know that $a(z)$ and $b(z)$ are coprime if and only if there exist $\{x(z), y(z)\}$ such that $x(z)a(z) + y(z)b(z) = 1$. But now let $y(z) = q(z)a(z) + r(z)$, $\deg r < n$, and define

$$\tilde{y}(z) = r(z) = y(z) - q(z)a(z), \qquad \tilde{x}(z) = x(z) + q(z)b(z)$$

Then

$$\tilde{x}(z)a(z) + \tilde{y}(z)b(z) = x(z)a(z) + y(z)b(z) + \{q(z)b(z)a(z) - q(z)a(z)b(z)\}$$
$$= 1 + 0 = 1$$

Also note that

$$\deg [\tilde{x}(z)a(z)] = \deg [1 - \tilde{y}(z)b(z)] < m + n$$

so that

$$\deg \tilde{x}(z) < m + n - \deg a(z) = m$$

To prove uniqueness, suppose that $\{a(z), b(z)\}$ are coprime but that there exist two sets $\{\tilde{x}_i(z), \tilde{y}_i(z), i = 1, 2\}$ satisfying (25). Then

$$[\tilde{x}_1(z) - \tilde{x}_2(z)]a(z) + [\tilde{y}_1(z) - \tilde{y}_2(z)]b(z) = 0$$

which implies that

$$\frac{a(z)}{b(z)} = \frac{\tilde{y}_2(z) - \tilde{y}_1(z)}{\tilde{x}_1(z) - \tilde{x}_2(z)}$$

But since $\deg[\tilde{y}_2(z) - \tilde{y}_1(z)] < m$ and $\deg[\tilde{x}_1(z) - \tilde{x}_2(z)] < n$, this means that $b(z)$ and $a(z)$ must have a common factor, contradicting our initial assumption. Hence. . . . ∎

The result of Lemma 2.4-10 will be used in Sec. 4.5.3 on compensator design. In that connection, we should note that there are several other criteria and tests for coprimeness, and here we shall note three of the best known ones.

Sylvester's Resultant (1840).   Given polynomials $\{a(z), b(z)\}$ as in (25), define a so-called *Sylvester matrix* (shown for $n = 3, m = 2$)

$$\tilde{S}(a, b) = \begin{bmatrix} a_0 & a_1 & a_2 & a_3 & 0 \\ 0 & a_0 & a_1 & a_2 & a_3 \\ 0 & 0 & b_0 & b_1 & b_2 \\ 0 & b_0 & b_1 & b_2 & 0 \\ b_0 & b_1 & b_2 & 0 & 0 \end{bmatrix} \tag{26}$$

Sylvester showed that

$$\{a(z), b(z)\} \text{ are coprime if and only if } \det \tilde{S}(a, b) \neq 0 \tag{27}$$

The determinant of $\tilde{S}(a, b)$ is known as the *Sylvester resultant*. We note that since elementary row and column operations leave the determinant unchanged, we could rearrange $\tilde{S}(a, b)$ in many ways. In (26), we have shown a form favored by Jury ([39] and [40]), which displays a "left triangle" of zeros. Another useful form (especially when $m = n$, as shown below for $m = 3 = n$) is

$$S(a, b) = \left[\begin{array}{ccc:ccc} a_0 & 0 & 0 & b_0 & 0 & 0 \\ a_1 & a_0 & 0 & b_1 & b_0 & 0 \\ a_2 & a_1 & a_0 & b_2 & b_1 & b_0 \\ \hdashline a_3 & a_2 & a_1 & b_3 & b_2 & b_1 \\ 0 & a_3 & a_2 & 0 & b_3 & b_2 \\ 0 & 0 & a_3 & 0 & 0 & b_3 \end{array} \right]$$

$$= \left[\begin{array}{c:c} \mathcal{Q}_- & \mathcal{B}_- \\ \hdashline \mathcal{Q}_+ & \mathcal{B}_+ \end{array} \right] \tag{28}$$

where the notation $\mathcal{A}_-$, $\mathcal{B}_-$ agrees with definitions introduced earlier in Eqs. (2.3-36 and 2.3-37). The Sylvester test (27) can be established in many ways, one of which is described in Exercise 2.4-18.

MacDuffee's Resultant (1950).   MacDuffee [36] showed that polynomials $b(z)$ and $a(z)$ will be relatively prime if and only if

$$\det [b(A_c)] \neq 0 \tag{29a}$$

or equivalently

$$\det [a(B_c)] \neq 0 \tag{29b}$$

where A_c and B_c are companion matrices of $a(z)$ and $b(z)$, with the coefficients of $a(z)$ and $b(z)$ in the top rows. The determinants in (29) are often known as *MacDuffee resultants;* whether (29a) or (29b) is used will depend on whether m is less than n or greater than n.

A proof follows easily from the fact that if $\{\lambda_1, \ldots, \lambda_n\}$ are the eigenvalues of A_c, then the eigenvalues of the matrix polynomial $b(A_c)$ are (cf. Exercise A.36) $\{b(\lambda_1), \ldots, b(\lambda_n)\}$. Therefore

$$\det b(A_c) = (-1)^n \det [zI - b(A_c)]\bigg|_{z=0} = \prod_1^n b(\lambda_i)$$

and clearly $\det b(A_c)$ will be zero if and only if at least one of the $\{b(\lambda_i)\}$ is zero, i.e., if and only if $b(z)$ and $a(z)$ have at least one common factor, and similarly for the test involving $a(B_c)$.

Bezout's Resultant (1764).   It will be convenient here to assume that the polynomials $a(z)$ and $b(z)$ have equal degrees $(m = n)$, which can of course always be arranged by using zero coefficients. With this assumption, we return to the Sylvester matrix in the form (28). The Bezout test becomes evident in trying to simplify $S(a, b)$ by reducing it to triangular form. Thus note that

$$\begin{bmatrix} \mathcal{A}_- & \mathcal{B}_- \\ \mathcal{A}_+ & \mathcal{B}_+ \end{bmatrix} \begin{bmatrix} I & \mathcal{B}_- \\ 0 & -\mathcal{A}_- \end{bmatrix} = \begin{bmatrix} \mathcal{A}_- & 0 \\ \mathcal{A}_+ & \tilde{B} \end{bmatrix} \tag{30}$$

where we have used the fact that lower triangular Toeplitz matrices commute (cf. Exercise A.6) and have defined

$$\tilde{B} = \mathcal{A}_+ \mathcal{B}_- - \mathcal{B}_+ \mathcal{A}_- \tag{31}$$

Since $\mathcal{A}_-$ is always nonsingular $(a_0 \neq 0$ by assumption), it is clear that

$$\det S(a, b) = 0 \Longleftrightarrow \det \tilde{B} = 0 \tag{32}$$

and therefore $\det \tilde{B}$ can also be used as a resultant. It turns out to be somewhat more convenient to introduce the *Bezout matrix* or *Bezoutian*

$$B = \tilde{I}\tilde{B} = \tilde{I}[\mathcal{A}_+ \mathcal{B}_- - \mathcal{B}_+ \mathcal{A}_-] \tag{33}$$

because it will follow (as we ask the reader to show—cf. Exercise 2.4-21) that B will be symmetric, unlike $\tilde{B}$. We shall define

$$\text{Bezout's resultant} \triangleq \det B$$

We may note that the Bezoutian can also be introduced via the bilinear form

$$B(s, \sigma) = \frac{a(s)b(\sigma) - b(s)a(\sigma)}{\sigma - s} \tag{34a}$$

$$= \sum_{i,j=1}^{n} B_{ij} s^{i-1} \sigma^{j-1} \tag{34b}$$

The matrix $[B_{ij}]$ in (34) must clearly be symmetric, and it can be verified that it equals the expression in (33).

System-Theoretic Interpretations.   As stated in the introduction, because of the results of Sec. 2.4-1, we would expect some relationships between the above tests for relative primeness and the concepts of controllability and observability. In fact, for example, MacDuffee's test was rediscovered by Kalman ([17] and [41]) on the basis of the result [cf. (2.3-44)] that for strictly proper systems ($b_0 = 0$, $a_0 = 1$)

$$\tilde{I}b(A_c) = \mathcal{O}_c, \quad \text{the observability matrix of the controller-form}$$
$$\text{realization } \{A_c, b_c, c_c\} \text{ of } b(z)/a(z) \tag{35}$$

Now we know (cf. Fig. 2.4-2) that $\{A_c, b_c, c_c\}$ will be jointly controllable and observable if and only if $b(z)$ and $a(z)$ are coprime. Since the controller form is always controllable, this means that $b(z)$ and $a(z)$ will be coprime if and only if $\mathcal{O}_c$ is nonsingular, or, using (35), if and only if $b(A_c)$ is nonsingular, which is Mac-Duffee's test (29).

Next recall that in Sec. 2.3.3 we had obtained another expression for $\mathcal{O}_c$, namely [cf. Eq. (2.3-43)]

$$\mathcal{O}_c = [\mathcal{B}_+ - \mathcal{B}_-\mathcal{C}_-^{-1}\mathcal{C}_+]\tilde{I} = [\mathcal{B}_+ - \mathcal{C}_-^{-1}\mathcal{B}_-\mathcal{C}_+]\tilde{I}$$

Then some algebra yields [use (33)]

$$-\tilde{I}\mathcal{C}_-\mathcal{O}_c\tilde{I} = \tilde{I}[\mathcal{B}_-\mathcal{C}_+ - \mathcal{C}_-\mathcal{B}_+] = B' = B \tag{36}$$

which shows that, like MacDuffee's test, the Bezout test is just another way of saying that the controller form of $b(s)/a(s)$ will be observable if and only if $b(s)$ and $a(s)$ are coprime.

Furthermore, we now recall that [cf. (2.3-6)] $\mathcal{O}_o^{-1} = \mathcal{C}_-$, so (36) can be rewritten as (cf. [42, Theorem 4])

$$-\tilde{I}B\tilde{I} = \mathcal{O}_o^{-1}\mathcal{O}_c \tag{37}$$

which we can recognize [cf. (2.4-10)] as the transformation matrix from the observer-form realization of $b(s)/a(s)$ to its controller-form realization, a transformation that will only exist when $b(s)$ and $a(s)$ are coprime (why?).

There are numerous other variants and identities for the resultants, especially concerning transformations from a (minimal) realization to its dual (see Fig. 2.4-3) and they are closely connected to classical results on root location, Padé approximation, etc. (see, e.g., [42]–[48]). However, we do not have space to pursue such questions here.

*2.4.5 Some Worked Examples

We present several examples to reinforce some of the results and tests given in Secs. 2.3 and 2.4.

The first two examples illustrate some computational as well as conceptual features. As with all worked examples, the reader will profit most by first trying to work out the problems for himself. In any case, however, it will be useful for the reader to make for himself a list of important facts and ideas gained by a close study of these examples. Here we draw attention only to the methods described for computing the standard noncontrollable form and to the facts that controllability and observability depend on the variables chosen for the state-space description of a given physical system. The remaining examples are more theoretical but also bring out several things worth remembering. Example 2.4-3 provides another example of a useful result that can be easily conjectured by using the diagonal form (cf. the discussion in Sec. 2.4.1). The results of Example 2.4-4 (and of the closely related Exercises 2.4-6 and 2.4-7) are very helpful in avoiding some perhaps tedious algebra in many problems. In Example 2.4-5 we give a previously promised proof of Lemma 2.4-1. Finally in Example 2.4-6, we show an application of the fact that a unique similarity transformation relates any two minimal realizations (Theorem 2.4-7)—this problem arose in the study of reciprocal networks and was in fact one of the motivations for the development of the theorem by Youla [49], independently of Kalman [17].

Example 2.4-1.   Radial and Tangential Control of a Satellite

In Example 2.2-E, we noted that the linearized equations for a satellite in a circular equatorial orbit are given by

$$\dot{x}(t) = Ax(t) + Bu(t), \qquad x' = [r \quad \dot{r} \quad \theta \quad \dot{\theta}]$$

where (with $m = 1 = r_0$, $\omega_0 = \omega$, as compared to Example 2.2-E)

$$A = \begin{bmatrix} 0 & 1 & 0 & 0 \\ 3\omega^2 & 0 & 0 & 2\omega \\ 0 & 0 & 0 & 1 \\ 0 & -2\omega & 0 & 0 \end{bmatrix}, \qquad B = \begin{bmatrix} 0 & 0 \\ 1 & 0 \\ 0 & 0 \\ 0 & 1 \end{bmatrix}, \qquad u = \begin{bmatrix} u_1 \\ u_2 \end{bmatrix}$$

and the control $u_1(\cdot)$ represents radial thrust and $u_2(\cdot)$ represents tangential thrust.

Determine if the system is controllable when

1. There is only radial thrust $[u_2(\cdot) \equiv 0]$.
2. There is only tangential thrust $[u_1(\cdot) \equiv 0]$.

Also transform the realization into standard noncontrollable form when appropriate.

Solution.

1. When $u_2(\cdot) \equiv 0$ (only radial thrust), $b'_1 = [0\ 1\ 0\ 0]$, B becomes a column matrix, and we can calculate

$$
\mathcal{C}(A, b_1) =
\begin{bmatrix}
0 & 1 & 0 & -\omega^2 \\
1 & 0 & -\omega^2 & 0 \\
0 & 0 & -2\omega & 0 \\
0 & -2\omega & 0 & 2\omega^3
\end{bmatrix}
$$

We may note that the last column is ω^2 times the second one, so that $\mathcal{C}(A, b_1)$ is singular and the system is not controllable using only a radial force. The PBH tests could also be used—thus one can check that $[sI - A\ b_1]$ loses rank at $s = 0$ or equivalently that $[2\omega\ 0\ 0\ 1]$ is a left eigenvector of A (associated with the eigenvalue 0), which is orthogonal to b_1.

Although the state equations in this problem are fairly simple, it is not quite obvious from them which variables (or, rather, combinations of variables) are noncontrollable. To determine these, we should transform the given equations to a standard noncontrollable form, as described in the discussion following Eq. (13). *One* way to choose the transformation matrix T is

$$
T = [b_1\quad Ab_1\quad A^2 b_1\quad t]
$$

where t is independent of the preceding vectors (and may be chosen to be orthogonal to them). For example, we may choose

$$
T =
\begin{bmatrix}
0 & 1 & 0 & 2\omega \\
1 & 0 & -\omega^2 & 0 \\
0 & 0 & -2\omega & 0 \\
0 & -2\omega & 0 & 1
\end{bmatrix}
$$

Then some algebra yields†

$$
T^{-1}AT = \bar{A} =
\begin{bmatrix}
0 & 0 & 0 & 6\omega^3 + \dfrac{3\omega}{2} \\
1 & 0 & -\omega^2 & 0 \\
0 & 1 & 0 & -(1/2\omega) \\
\hline
0 & 0 & 0 & 0
\end{bmatrix}
$$

Also, since $T\bar{b} = b$, we must have $\bar{b}' = [1\ 0\ 0\ \vdots\ 0]$. $\bar{A}$ and $\bar{b}$ are now in the standard form for displaying noncontrollability. The uncontrollable part has characteristic polynomial s, as already indicated by the PBH tests. Note that from $\bar{A}$ it is easy to

---

†It is worth remarking here that the algebraic labor involved in the above calculations can be conceptually illuminated by using some basic linear algebra. It would therefore be useful to read Secs. 5.2.1 and 5.2.2 at this point.

determine the overall characteristic polynomial as $s \cdot (s^2 + \omega^2)s$, so that the eigenvalues are $\{0, 0, \pm j\omega\}$.

Another route to obtaining a suitable T for transforming to the standard form is to note that we require

$$T^{-1}A = \left[\begin{array}{c|c} A_c & A_{12} \\ \hline 0 & \lambda \end{array}\right] T^{-1}, \qquad T^{-1}b = \left[\begin{array}{c} b_c \\ \hline 0 \end{array}\right]$$

where λ is an uncontrollable eigenvalue.

If the last row of T^{-1} is denoted by t_n, we require

$$t_n A = \lambda t_n, \qquad t_n b = 0$$

It is clear from the PBH test that a suitable t_n is then

$$t_n = [2\omega \quad 0 \quad 0 \quad 1]$$

The remaining rows of T^{-1} may be arbitrarily chosen to form an independent set (so that the transformation matrix is invertible). Let us take

$$T^{-1} = \begin{bmatrix} 1 & 0 & 0 & 0 \\ 0 & 1 & 0 & 0 \\ 0 & 0 & 1 & 0 \\ 2\omega & 0 & 0 & 1 \end{bmatrix}$$

We then obtain,

$$\bar{A}_1 = T^{-1}AT = \left[\begin{array}{ccc|c} 0 & 1 & 0 & 0 \\ -\omega^2 & 0 & 0 & 2\omega \\ -2\omega & 0 & 0 & 1 \\ \hline 0 & 0 & 0 & 0 \end{array}\right], \qquad \bar{b}_1 = T^{-1}b = \left[\begin{array}{c} 0 \\ 1 \\ 0 \\ \hline 0 \end{array}\right]$$

which again immediately displays the noncontrollability of the eigenvalue 0. We should also note the nonuniqueness of the standard forms: $\{\bar{A}, \bar{b}\}$ is not the same as $\{\bar{A}_1, \bar{b}_1\}$. But in both forms we can see that the uncontrollable variable is $2\omega x_1 + x_4 \triangleq 2\omega r + \dot{\theta}$ (a fact also evident from the PBH eigenvector test). There is actually a nice physical interpretation for this: Radial thrust cannot change the angular momentum, and the interested reader can check (referring back to Example 2.2-E) that if the angular momentum is to be the same after a small perturbation from the nominal orbit, then we must have $2\omega r + \dot{\theta} = 0$.

2. When $u_1(\cdot) \equiv 0$ (tangential thrust only), $b_2' = [0 \ 0 \ 0 \ 1]$, and we can see readily that $\mathcal{C}(A, b_2)$ is nonsingular [e.g., its determinant is $-2\omega(-8\omega^3 + 2\omega^3) = 12\omega^4$]. Therefore tangential thrust is enough to provide controllability.

The PBH tests are not quite so simple now, since we must show that there is *no*

s for which rank $[sI - A\ b_2]$ is less than n or that there is *no* left eigenvector of A orthogonal to b_2. We shall not go through the labor here.

It is worth mentioning here that in a multi-input system, as in this example, we do not usually require the system to be controllable by each input acting *separately*; the system is controllable if the inputs acting together in combination can set up arbitrary states. It will be shown in Sec. 6.2 that the test for controllability is then that the matrix $\mathcal{C} = [B\ AB\ \cdots\ A^{n-1}B]$ have full rank, and *not* that the $\mathcal{C}_i = [b_i\ Ab_i\ \cdots\ A^{n-1}b_i]$ all have full rank. ∎

Example 2.4-2. An Inverted Pendulum on a Cart

An important idealized control problem is that of a pendulum mounted on a moving carriage. After linearization, the equations of motion can be written as

$$\dot{x}(t) = Ax(t) + bu(t)$$

where

$$A = \begin{bmatrix} 0 & 1 & 0 & 0 \\ 0 & -F/M & 0 & 0 \\ 0 & 0 & 0 & 1 \\ -g/L & 0 & g/L & 0 \end{bmatrix}, \qquad b = \begin{bmatrix} 0 \\ 1/M \\ 0 \\ 0 \end{bmatrix}$$

and M is the mass of the carriage (assumed much greater than the mass m of the pendulum), F is the friction coefficient for the motion of the carriage, g is the gravitational constant, and $L = (J + ml^2)/ml$, where l is the distance of the center of gravity from the pivot point and J is the moment of inertia about this point. The input $u(\cdot)$ is the force on the cart. The state variables are

$$x'(t) = [d(t)\quad \dot{d}(t)\quad d(t) + L\phi(t)\quad \dot{d}(t) + L\dot{\phi}(t)]$$

where $d(\cdot)$ is the displacement of the cart from the origin and ϕ is the deviation of the pendulum from the vertical (see the figure).

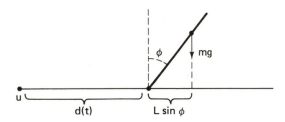

1. Find the eigenvalues and left and right eigenvectors of A.
2. Determine the controllability properties of the system.
3. Assume that the output equation is $y(\cdot) = \phi(\cdot)$. Show that the system is unobservable and determine the observable and unobservable variables. With this knowledge, what observations would be needed to observe all the states?

Solution.

1. Because A is *block triangular*, we can see immediately that the eigenvalues are $\{0,\ -F/M,\ \pm\sqrt{g/L}\}$.

We note that the corresponding right eigenvectors are

$$p_1' = [1 \quad 0 \quad 1 \quad 0], \qquad\qquad p_2' = [1 \quad -F/M \quad \alpha \quad -\alpha F/M]$$
$$p_3' = [0 \quad 0 \quad 1 \quad \sqrt{g/L}], \qquad p_4' = [0 \quad 0 \quad 1 \quad -\sqrt{g/L}]$$

where $\alpha = (g/L)/(gL^{-1} - F^2 M^{-2})$. The left eigenvectors can be found as

$$q_1 = [F/M \quad 1 \quad 0 \quad 0], \qquad\qquad q_2 = [0 \quad 1 \quad 0 \quad 0]$$
$$q_3 = [-\sqrt{g/L} \quad \beta_+ \quad \sqrt{g/L} \quad 1], \qquad q_4 = [\sqrt{g/L} \quad \beta_- \quad -\sqrt{g/L} \quad 1]$$

where $\beta_\pm = -\sqrt{g/L}/[(F/M) \pm \sqrt{g/L}]$. As a check, notice that $q_i p_j = 0, i \neq j$, while $q_i p_i \neq 0$.

2. We see that $q_i b \neq 0$, and so by the **PBH** test the realization is controllable. (This can also be seen quite easily from the **PBH** rank test or by direct calculation of the controllability matrix.)

3. Since $\phi(t) = [x_3(t) - x_1(t)]/L$, the output equation is

$$y(t) = [-1/L \quad 0 \quad 1/L \quad 0]x(t)$$

By the **PBH** eigenvector test we see that

$$cp_1 = 0, \qquad cp_i \neq 0, \qquad i = 2, 3, 4$$

Therefore the unstable mode corresponding to $s = 0$ is unobservable.

This can also be seen by computing the transfer function from $u(\cdot)$ to $y(\cdot)$, which turns out to be (e.g., use Exercise A.13, part 2) $H(s) = -s/[LM(s - F/M)(s^2 - g/L)]$. The pole at $s = 0$ evidently has been cancelled out of the transfer function. Since the system is controllable, the cancellation must correspond to an unobservable natural frequency or unobservable mode of oscillation.

The unobservability of the system using observations of ϕ alone may have been suspected on purely physical grounds. It would seem that knowledge of ϕ can never tell us what is d, the displacement of the cart; a different initial d could still give us the same ϕ history.

We attempt therefore to rewrite the state equations in such a way as to make the conjectured unobservability of d obvious. Choosing a new set of state variables as, for example,

$$\bar{x} = [d \quad L\phi \quad L\dot\phi \quad d]'$$

we get [cf. (2.4-16)]

$$\bar{A} = \begin{bmatrix} -F/M & 0 & 0 & 0 \\ 0 & 0 & 1 & 0 \\ -1 & g/L & 0 & 0 \\ \hline 1 & 0 & 0 & 0 \end{bmatrix} = \begin{bmatrix} A_o & 0 \\ A_{21} & A_{\bar{o}} \end{bmatrix}$$

and

$$\bar{c} = [0 \quad 1/L \quad 0 \;\vdots\; 0] = [c_o \quad 0]$$

It is immediately seen that d (and its associated eigenvalue of zero) is unobservable. The zero eigenvalue is a consequence of the fact that, with no input and no initial condition on any of the other state variables, any initial condition on d remains constant.

To make the original realization observable, we could measure $d(\cdot)$ as well, giving a two-output system, say,

$$\begin{bmatrix} y_1(t) \\ y_2(t) \end{bmatrix} = \begin{bmatrix} -1/L & 0 & 1/L & 0 \\ 1 & 0 & 0 & 0 \end{bmatrix} x(t)$$
$$= Cx(t)$$

This system is now clearly observable, and though we have not studied multiple-output systems as yet, the reader can verify that the appropriate algebraic criteria for observability are that either (1) rank $[C' \ A'C' \ \cdots \ (A')^{n-1}C'] = n$, or (2) no right eigenvector of A is orthogonal to all the rows of C, or that (3) $[sI - A' \ C']$ has full rank for all s.

We can, however, make the system observable without converting it to a two-output system. The reader can verify that with the single measurement $y = d + L\phi = [0 \ 0 \ 1 \ 0]x$, the system is observable. ∎

Example 2.4-3.

Let

$$(sI - A)^{-1}b = \frac{[p_1(s), p_2(s), \ldots, p_n(s)]'}{a(s)}, \qquad a(s) = \det(sI - A)$$

Use the diagonal form to guess the relationship between the controllability of $\{A, b\}$ and the fact that the $n + 1$ polynomials $\{p_1(s), \ldots, p_n(s), a(s)\}$ have no nontrivial common factors.

Then give general proofs (in both directions) of your conjectured relation.

Solution. Let $n = 3$ and $A = \text{diag}\{\lambda_1, \lambda_2, \lambda_3\}$, $b' = [b_1 \ b_2 \ b_3]$. Then

$$(sI - A)^{-1}b = [b_1(s - \lambda_1)^{-1} \quad b_2(s - \lambda_2)^{-1} \quad b_3(s - \lambda_3)^{-1}]'$$

and $p_1(s) = b_1(s - \lambda_2)(s - \lambda_3)$, $p_2(s) = b_2(s - \lambda_1)(s - \lambda_3)$, and $p_3(s) = b_3(s - \lambda_1)(s - \lambda_2)$, while $a(s) = (s - \lambda_1)(s - \lambda_2)(s - \lambda_3)$. The controllability of $\{A, b\}$ is equivalent to the conditions $b_i \neq 0$, $\lambda_1 \neq \lambda_2 \neq \lambda_3$. In this case the $\{p_i(s), a(s)\}$ will have no common roots, whereas they clearly will if one of the $\{b_i\}$ is zero or if the $\{\lambda_i\}$ are not all distinct. Thus we may make the conjecture that

$$\{A, b\} \text{ controllable} \iff \{p_i(s), a(s)\} \text{ have no common roots}$$

We can in fact strengthen the conjecture to

$$\{A, b\} \text{ controllable} \iff \{p_i(s)\} \text{ have no common roots}$$

We now can try to prove the first conjecture directly; the second we leave to the reader.

First, if $\{A, b\}$ is controllable, then without loss of generality we can assume that we are in controller form $\{A_c, b_c\}$. In this case, it is easy to see (cf. Exercise A.33) that

$$p_{ic}(s) = s^{n-i}, \qquad i = 1, \ldots, n$$

so that $\{p_{ic}(s), a(s)\}$ will have no common factors (except the trivial factor unity). This result must also hold for the original pair $\{A, b\}$, because we have made an invertible transformation to get to $\{A_c, b_c\}$. But to get more confidence with such arguments, let us confirm this by a direct calculation. Let

$$A_c = T^{-1}AT, \qquad b_c = T^{-1}b$$

Then

$$(sI - A)^{-1}b = T(sI - A_c)^{-1}b_c$$

which shows that

$$[p_1(s), \ldots, p_n(s)]' = T[p_{1c}(s), \ldots, p_{nc}(s)]'$$

Let λ be a root of $a(s) = 0$. Then it is clear that $[p_{1c}(\lambda), \ldots, p_{nc}(\lambda)] \neq 0$ (because $p_{nc} = 1$). But since T is invertible, it must also hold that $[p_1(\lambda), \ldots, p_n(\lambda)] \neq 0$, which is the desired result. However, we should emphasize that such confirmation is not necessary—it is enough to know that the transformation involved is invertible. Note also that we do not need to know the explicit form of the transformation.

We shall use similar arguments to prove the converse. Thus, suppose now that $\{A, b\}$ is not controllable. We can without loss of generality assume that we have the standard noncontrollable form (2.4-11), so that $[(sI - \bar{A})^{-1}\bar{b}]' = [((sI - \bar{A}_c)^{-1}b_c)' \ 0]$, while $a(s) = \det(sI - \bar{A}_c) \det(sI - \bar{A}_{\bar{c}}) = a_c(s)a_{\bar{c}}(s)$, say. Therefore

$$[\bar{p}_1(s), \ldots, \bar{p}_n(s)] = [p_{1c}(s)a_{\bar{c}}(s), \ldots, p_{rc}(s)a_{\bar{c}}(s), 0, \ldots, 0]$$

so that $\{\bar{p}_i(s), a(s)\}$ have the common factor $a_{\bar{c}}(s)$.   ∎

Example 2.4-4.   Observability of Series Realizations

Let $\{A_i, b_i, c_i\}$ be observable realizations of order n_i of the transfer functions $H_i(s) = g_i(s)/a_i(s)$, $i = 1, 2$, where $\deg a_i(s) = n_i$.

1. Write a natural set of state equations for the series (or cascade) connection of $H_1(s)$ followed by $H_2(s)$.

2. Show that these equations will be observable if and only if $a_1(s)$ and $g_2(s)$ are coprime.

Solution.

1. From the equations

$$\dot{x}_1 = A_1 x_1 + b_1 u, \qquad \dot{x}_2 = A_2 x_2 + b_2 y_1$$

$$y_1 = c_1 x_1, \qquad\qquad y_2 = c_2 x_2$$

we can write

$$\begin{bmatrix} \dot{x}_1 \\ \dot{x}_2 \end{bmatrix} = \begin{bmatrix} A_1 & 0 \\ b_2 c_1 & A_2 \end{bmatrix} \begin{bmatrix} x_1 \\ x_2 \end{bmatrix} + \begin{bmatrix} b_1 \\ 0 \end{bmatrix} u$$

$$y = [0 \quad c_2][x_1' \quad x_2']'$$

2. This problem can be approached in many ways, but here we shall use the PBH rank test. A preliminary result, which is useful in its own right, is first stated.

The matrix $sI - A$, where A is a left-companion matrix with $-[a_1 \ \cdots \ a_n]'$ as the first column, can be reduced to the form

$$\begin{bmatrix} & -1 & & \bigcirc \\ \bigcirc & & \ddots & \\ & \bigcirc & & -1 \\ \hline a(s) & & \bigcirc & \end{bmatrix}, \qquad a(s) = s^n + a_1 s^{n-1} + \cdots + a_n$$

by elementary row and column operations corresponding to pre- and postmultiplication by certain well-structured matrices that we ask the reader to determine.

Now assume without loss of generality that the two given observable realizations are in observer form. Noting that $c_1 = [1 \ 0 \ \cdots \ 0] = c_2$, we see that

$$\begin{bmatrix} sI - A \\ c \end{bmatrix} = \left[\begin{array}{ccccc|ccc} & sI - A_1 & & & 0 & & & \\ -b_2 & 0 & \cdots & 0 & & sI - A_2 & & \\ \hline & 0 & \cdots & 0 & & 1 & 0 & \cdots & 0 \end{array} \right]$$

Using the elementary row and column operations mentioned above, this matrix can be reduced to the form

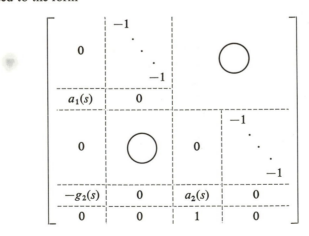

The composite system is observable if and only if this $(n_1 + n_2 + 1) \times (n_1 + n_2)$

matrix has rank $n_1 + n_2$ for all values of s, i.e., if it has all its columns independent for all s. This will clearly be so if and only if the first column is never zero, i.e., if and only if there is *no* λ for which $a_1(\lambda) = g_2(\lambda) = 0$.

In Exercise 2.4-9, we suggest that another simple proof can be obtained by using the (dual of the) criterion of Example 2.4-3. ∎

Example 2.4-5.   Transformation Between Controllable Realizations
(Lemma 2.4-1)

Show that any two controllable realizations (of the same transfer function) with the same characteristic polynomial can be related by a similarity transformation. (This will essentially prove Lemma 2.4-1.)

Solution.   One method is to note that there exists an invertible transformation from any controllable realization to a controller-form realization with the same characteristic polynomial (see Example 2.3-4). Here we shall give a somewhat more direct proof.

If $x_1(t) = Tx_2(t)$, we know $T^{-1}A_1T = A_2, T^{-1}b_1 = b_2$. Therefore $\mathcal{C}_2 = T^{-1}\mathcal{C}_1$ or $T = \mathcal{C}_1\mathcal{C}_2^{-1}$ if both realizations are controllable. Now let us try to reverse this argument. Since we are given $\{A_i, b_i, c_i, i = 1, 2\}$ both controllable, we can certainly define $T = \mathcal{C}_1\mathcal{C}_2^{-1}$, but we have to see if this works, i.e., if with this T, $T^{-1}A_1T = A_2, T^{-1}b_1 = b_2, c_1T = c_2$. Now

$$T^{-1}b_1 = \mathcal{C}_2\mathcal{C}_1^{-1}b_1 = \mathcal{C}_2[b_1 \quad Ab_1 \quad \cdots \quad A^{n-1}b_1]^{-1}b_1 = b_2$$

Next we examine

$$T^{-1}A_1T = \mathcal{C}_2\mathcal{C}_1^{-1}A_1\mathcal{C}_1\mathcal{C}_2^{-1}$$

Consider for simplicity $n = 3$, and

$$\mathcal{C}_1^{-1}A_1\mathcal{C}_1 = [b_1 \quad A_1b_1 \quad A_1^2b_1]^{-1}[A_1b_1 \quad A_1^2b_1 \quad A_1^3b_1]$$

Now, by the Cayley-Hamilton theorem,

$$(A_1^3 + a_1A_1^2 + a_2A_1 + a_3I)b = 0$$

where the $\{a_i\}$ have no subscript because by assumption they are the same for both realizations. Therefore

$$\mathcal{C}_1^{-1}A_1\mathcal{C}_1 = \begin{bmatrix} 0 & 0 & -a_1 \\ 1 & 0 & -a_2 \\ 0 & 1 & -a_3 \end{bmatrix}$$

which is the *same* for both realizations. Therefore

$$\mathcal{C}_1^{-1}A_1\mathcal{C}_1 = \mathcal{C}_2^{-1}A_2\mathcal{C}_2$$

and

$$T^{-1}A_1T = \mathcal{C}_2\mathcal{C}_1^{-1}A_1\mathcal{C}_1\mathcal{C}_2^{-1} = \mathcal{C}_2\mathcal{C}_2^{-1}A_2\mathcal{C}_2\mathcal{C}_2^{-1} = A_2$$

Finally, to bring in the $\{c_i\}$, we have to refer to the transfer function or to the Markov parameters, which yield

$$c_1 b_1 = c_2 b_2, \qquad c_1 A_1 b_1 = c_2 A_2 b_2, \qquad \cdots$$

or

$$c_1 \mathcal{C}_1 = c_2 \mathcal{C}_2$$

or

$$c_2 = c_1 \mathcal{C}_1 \mathcal{C}_2^{-1} = c_1 T, \qquad \text{as desired}$$

Note that the assumption that the $\{c_i, A_i\}$ are observable is not necessary. ∎

Example 2.4-6.   Transformation Between Dual Realizations

Let $\{A, b, c\}$ be a minimal realization of a scalar transfer function $H(s)$. Show that there exists a unique *symmetric* matrix T satisfying the relations

$$TA' = AT \quad \text{and} \quad cT = b'$$

Solution.   Since $H(s)$ is scalar, we have $H(s) = H'(s)$, so that

$$c(sI - A)^{-1}b = b'(sI - A')^{-1}c'$$

Therefore if $\{A, b, c\}$ is a minimal realization, so is $\{A', c', b'\}$ (by Theorem 2.4-6 and Lemma 2.4-2). Therefore by Theorem 2.4-7 there must be a *unique* invertible matrix such that

$$TA' = AT, \qquad Tc' = b, \qquad b' = cT$$

Taking transposes, we obtain the equalities

$$T'A = A'T', \qquad cT' = b', \qquad T'c' = b$$

which show that T' is also a similarity transformation between the two realizations. But by Theorem 2.4-7 there can be only one such transformation. Therefore $T = T'$, which completes the proof.

This result was discovered by Youla in his studies of the properties of *reciprocal networks*—see [49]; in fact, it appears that the above problem led Youla to an independent proof of Theorem 2.4-7. ∎

Exercises

At this stage the reader will profit from another look at the exercises of the previous sections, many of which will be simpler to do now.

2.4-1.

Consider a realization with

$$A = \text{block diag} \left\{ \begin{bmatrix} -2 & 1 \\ 0 & -2 \end{bmatrix}, \begin{bmatrix} 1 & 1 \\ 0 & 1 \end{bmatrix} \right\}$$

$$b' = [1 \quad 0 \quad 1 \quad 1], \qquad c = [1 \quad 0 \quad 0 \quad 0]$$

Draw a block diagram of the realization and check its controllability in as many ways as you can.

2.4-2.
Consider a realization $\{A, b, c\}$ with

$$A = \begin{bmatrix} -0.5 & 1 & 0 \\ -1 & -0.5 & 0 \\ 0 & 1 & 0 \end{bmatrix}, \quad b = \begin{bmatrix} 1 \\ 2 \\ 0 \end{bmatrix}, \quad c' = \begin{bmatrix} 0 \\ 0 \\ 1 \end{bmatrix}$$

a. Is the system observable? If not completely observable, what quantities are unobservable?

b. Is the system controllable? If not completely controllable, what quantities are uncontrollable?

2.4-3.   (*DeBra*)
An inverted pendulum, of mass m, is hinged at A. A gyro with spin angular momentum, h, is attached to the pendulum but is free to rotate about the pendulum axis (angle ϕ) as shown in the figure. A control torque, Q, can be applied to the gyro from the pendulum. The equations of motion are $I\ddot{\theta} = mgl\theta - h\dot{\phi}$ and $J\ddot{\phi} = h\dot{\theta} + Q$, where $I =$ the moment of inertia of the pendulum plus gyro about A, $J =$ the moment of inertia of the gyro about axis AC, and $C =$ the mass center of the pendulum plus gyro.

a. Compute the transfer functions from $u(\cdot)$ to $\phi(\cdot)$ and $u(\cdot)$ to $\theta(\cdot)$.

b. Show that the system is controllable by Q, observable with ϕ, and unobservable with θ.

c. Show that the system is always unstable.

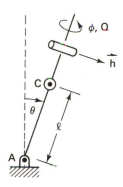

2.4-4.   *A Model Common in System Identification Problems*
Consider the equation

$$y(t) + a_1 y(t - 1) + \cdots + a_n y(t - n) = b_1 u(t - 1) + \cdots + b_n u(t - n)$$

Assume that the associated polynomials $a(z)$ and $b(z)$ are relatively prime, and introduce as a "state" vector

$$\theta'(t) = [y(t) \quad y(t-1) \quad \cdots \quad y(t-n+1) \quad u(t-1) \quad \cdots \quad u(t-n+1)]$$

a. Write the associated state equations for the system.

b. Show that this state-space realization is not minimal and determine the eigenvalues corresponding to the "hidden" modes.

2.4-5.

a. Consider the cascade connections of minimal realizations of $H_1(s)$ and $H_2(s)$ as $H_1(s)H_2(s)$ and $H_2(s)H_1(s)$, where $H_1(s) = 1/(s+1)$ and $H_2(s) = (s+1)/(s+2)(s+3)$. For each connection, determine the uncontrollable and unobservable modes, if any.

b. Repeat for the realizations connected in feedback form, first with $H_1(s)$ in the feedforward path and $H_2(s)$ in the feedback path, and then vice versa.

c. In Sec. 2.0, we noted that the behavior of the cascade connection of systems with $H_1(s) = 1/(s-1)$ and $H_2(s) = (s-1)/(s+1)$ depended on the order in which they were connected. Can you give a simple explanation of the differences?

2.4-6. *Controllability and Observability of Interconnected Subsystems*

Let $\{A_i, b_i, c_i, i = 1, 2\}$ be realizations of order n_i of the transfer functions $H_i(s) = g_i(s)/a_i(s)$, also of order n_i.

a. Show that if the realizations are controllable, then the *series* combination of system 1 followed by 2 is controllable if and only if $g_1(s)$ and $a_2(s)$ are coprime.

b. Show that if the realizations are observable (controllable), then the *parallel* combination is observable (controllable) if and only if $a_1(s)$ and $a_2(s)$ are coprime.

c. Show that if the realizations are observable (controllable), then the *feedback* configuration with system 1 in the forward path and 2 in the feedback path is observable (controllable) if and only if $g_1(s)$ and $a_2(s)$ are coprime. *Note:* One way to prove part c is to show, using general arguments, that the feedback system is equivalent, insofar as observability (controllability) goes, to the series combination of system 2 followed by system 1 (system 1 followed by system 2).

d. Extend the results to the case of systems with direct feedthrough from input to output.

2.4-7. *Characteristic Polynomial of Interconnected Systems*

Let $\{A_i, b_i, c_i, i = 1, 2\}$ be two minimal realizations with characteristic polynomials $a_i(s) = \det (sI - A_i)$.

a. Show that the characteristic polynomial of the

    (1) Series connection (of the two systems) is $a_1(s)a_2(s)$.

    (2) Parallel connection is $a_1(s)a_2(s)$.

    (3) Feedback connection, with $\{A_1, b_1, c_1\}$ in the forward path and $\{A_2, b_2, c_2\}$ in the feedback path, is $a_1(s)a_2(s) + b_1(s)b_2(s)$.

b. Let $H_i(s) = b_i(s)/a_i(s)$, $i = 1, 2$. Show that the denominators of the nominal (i.e., without cancellation of common factors) transfer functions of the series and parallel combinations are just the characteristic polynomials found in parts a(1) and a(2). For the feedback connection, we can write

$$H_f(s) = \frac{H_1(s)}{1 + H_1(s)H_2(s)}$$

$$= \frac{b_1(s)}{a_1(s)[a_1(s)a_2(s) + b_1(s)b_2(s)]} a_1(s)a_2(s)$$

$$= \frac{b_1(s)a_2(s)}{a_1(s)a_2(s) + b_1(s)b_2(s)}$$

where the denominator is again the characteristic polynomial of the feedback combination [cf. part a(3)]. Yet there was a cancellation of $a_1(s)$ in forming the overall transfer function $H_f(s)$. How do you reconcile these two facts?

2.4-8.

a. Use the PBH tests to show that the controller-form realization of $b(s)/a(s)$ will be observable if and only if $\{b(s), a(s)\}$ are coprime. (Another proof was given in Example 2.3-5.) (*Hint:* See Exercise A.35.)

b. The differential equation $\dot{x}(t) = Ax(t) + bu(t)$ can be approximated by the equations $x_{k+1} = (I + A\Delta)x_k + \Delta bu_k$. If $\{A, b\}$ is controllable, what can you say about the controllability of $\{I + A\Delta, \Delta b\}$?

2.4-9.   *Alternative Tests for Observability*

a. Show that $\{c, A\}$ is observable if and only if the $n + 1$ polynomials $\{q_i(s), a(s)\}$ defined by $c(sI - A)^{-1} = [q_1(s), \ldots, q_n(s)]/a(s)$ have no non-trivial common factors.

b. Use this result to obtain an alternative proof of the condition of Example 2.4-4 for observability of a series combination of two observable subsystems.

2.4-10.   *Alternative Characterizations of Controllability*

a. Prove that $\{A, b\}$ is controllable if and only if $\{A - bk, b\}$ is controllable for all k.

b. Show that $\{A, b\}$ is controllable if and only if the only $n \times n$ matrix X such that $AX = XA$ and $Xb = 0$ is the matrix $X \equiv 0$.

2.4-11.   *Simple Multivariable Systems*

a. Let

$$Y(s) = \begin{bmatrix} Y_1(s) \\ Y_2(s) \end{bmatrix} = \begin{bmatrix} \dfrac{1}{s+1} & \dfrac{2}{s+1} \\ \dfrac{-1}{(s+1)(s+2)} & \dfrac{1}{s+2} \end{bmatrix} \begin{bmatrix} U_1(s) \\ U_2(s) \end{bmatrix} = H(s)U(s)$$

be the transfer function of a two-input, two-output linear system. Find a realization $\{A, B, C\}$ of order not greater than 4 and another realization of order 3. Prove that 3 is the minimal order by using the theorem that minimality is equivalent to simultaneous controllability and observability (cf. Example 2.4-1).

b. A two-input, two-output system is described by the equations

$$\dot{y}_1(t) + y_2(t) = u_1(t) + u_2(t)$$
$$\dot{y}_2(t) + y_1(t) = u_2(t)$$

Calculate the transfer function matrix. Try to give at least two different analog-computer simulations, say one using three integrators and one using two integrators.

2.4-12.   *A Nice Form for System Identification*

Let $\{y(\cdot), u(\cdot)\}$ be the input and output of a system with a strictly proper irreducible transfer function, with denominator of degree n. For certain *adaptive identification* schemes, it is useful to reparametrize the system as shown in the figure. Here we choose

$$N(s) = \begin{bmatrix} (sI - F)^{-1}g \\ 0 \end{bmatrix}, \qquad M(s) = \begin{bmatrix} 0 \\ (sI - F)^{-1}g \end{bmatrix}$$

where $\{F, g\}$ is an *arbitrary* controllable pair, F being $n \times n$ and g $n \times 1$, where n is the order (assumed known) of the unknown system.

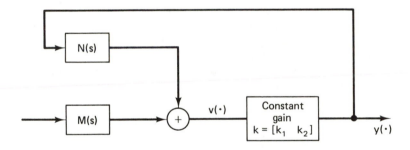

a. Show that the transfer function of the system shown can be written as

$$W(s) = \frac{Y(s)}{U(s)} = [1 - k_1(sI - F)^{-1}g]^{-1}k_2(sI - F)^{-1}g$$
$$= k_2[sI - (F + gk_1)]^{-1}g$$

b. Show that if $\{F, g\}$ is controllable, then by proper choice of $[k_1 \ k_2]$ we can make $W(s)$ have an arbitrary nth-degree denominator polynomial and an arbitrary numerator polynomial of degree less than or equal to $n - 1$; i.e., we have a "model" of the given system.

c. What will happen if n happens to be larger than the order of the given system? *Remark:* For the application of this model to the so-called *model reference adaptive identification* techniques, see a recent account by B.D.O. Anderson (*Automatica,* Vol. 13, pp. 401–408, 1977).

2.4-13.

Suppose $\{A, b, c\}$ is minimal and $a(s) = \det(sI - A)$ has a repeated root. Prove that A cannot be diagonalized by a similarity transformation.

2.4-14. (*Brockett*)

Suppose that $\{A, b, c\}$ is minimal. Prove that A and bc cannot commute if $n \geq 2$ (A is $n \times n$).

2.4-15.

a. If $\{A, b\}$ is given and *not* controllable, is it always possible to choose c so that $[c, A]$ is observable? A full explanation or a counterexample will suffice.

b. If $\{A, b\}$ is given and is controllable, can we always choose c so that $\{c, A\}$ is observable?

2.4-16.

Use the state equations to give a physical explanation of why the $\bar{A}_{1,3}$ block in Eq. (2.4-18) can be nonzero but not the $\bar{A}_{1,4}$ block. (All the other blocks have more obvious explanations.)

2.4-17.   *A Classical Resultant Formula*

Suppose $a(s) = a_0 \prod_1^n (s - \alpha_i), a_0 \neq 0,$ and $b(s) = b_0 \prod_1^m (s - \beta_i),$ $b_0 \neq 0$. Show that

$$\det \tilde{S}(a, b) = a_0^m b_0^n \prod (\alpha_i - \beta_j)$$

where $\tilde{S}(a, b)$ is the Sylvester matrix of $\{a(s), b(s)\}$.

2.4-18.   *Alternative Derivation of the Sylvester Test*

Let $\tilde{S}(a, b)$ be the Sylvester matrix for the polynomials $a(z)$ and $b(z)$ of Eq. (2.4-23) (with $a_n \neq 0 \neq b_m$). Define a matrix R with first column $[z^{n+m-1} \cdots z \; 1]$, ones on the diagonal [except in the $(1, 1)$ location], and zeros everywhere else. Show by evaluating the determinant in two different ways that

$$\det [\tilde{S}(a, b)R] = [\det \tilde{S}(a, b)]z^{n+m-1} = a(z)f(z) + b(z)g(z)$$

where $\{f(z), g(z)\}$ are polynomials of degree *at most* $n - 1$ and $m - 1$, respectively. Show from this relation that $\{a(z), b(z)\}$ have a nontrivial common factor if and only if $\det \tilde{S}(a, b) = 0$. (Reference: H. Skala, *Am. Math. Mon.*, Vol. 78, pp. 889–890, 1971.)

2.4-19.   *Another Test for Coprimeness*

a. From a systems point of view another test for relative primeness is that the controllability-form realization of $b(z)/a(z)$ be observable. Show that this fact leads to a Hankel matrix test for coprimeness.

b. Deduce this test in another way by proving the identity $\beta_j = -\mathcal{C}'_- M[1, n-1]\mathcal{C}_-$. (*Hint:* See Fig. 2.4-3.)

2.4-20.   *The Sylvester Matrix and an Observability Matrix*

Let $a(z) = z^n + a_1 z^{n-1} + \cdots + a_n$ and $b(z) = b_1 z^{n-1} + \cdots + b_n$. Show that by elementary row *and* column operations we can reduce the Sylvester matrix to the form block diag $\{I_{n-1}, \Theta(b, A_c)\}$, where $b = [b_1 \; \cdots \; b_n]$ and A_c is a companion matrix with $-[a_1 \; \cdots \; a_n]$ as the first row. Use this to give another proof of the Sylvester test for coprimeness.

2.4-21. *Symmetry of the Bezoutian*

Prove that the matrix B defined by (2.4-33) is symmetric. *Hint:* Use the matrix identities corresponding to the trivial polynomial identity $a(z)b(z) = b(z)a(z)$.

*2.5 SOLUTIONS OF STATE EQUATIONS AND MODAL DECOMPOSITIONS

So far in this chapter we have deliberately said very little about actually solving the state equation, because, as we have seen already and shall see even more as we proceed, one does not need explicit solutions in order to gain useful information about the system. This is fortunate, because often the solutions may be impossible to get in any convenient analytical form, especially for systems with time-variant coefficients. And furthermore, in many studies it is more important to have a nice "representation" of the solution rather than an actual "solution." Long usage often tends to blur this distinction, especially with scalar equations. For example, the solution of the scalar equation

$$\dot{x}(t) = ax(t), \qquad x(0) = 1$$

is

$$x(t) = e^{at} \quad \text{or} \quad \exp at$$

Now we can readily plot the solution using a table of exponentials or by some direct method of evaluating exponentials. However, often we do not need any explicit values of exp at but only the "defining" property

$$\frac{de^{at}}{dt} = ae^{at}$$

In this case exp at is more a useful "representation" of a solution than a solution itself. The significance of this statement will become clearer as the reader studies matrix exponentials in Sec. 2.5.1 and their time-variant generalizations in Chapter 9.

In Sec. 2.5.2, we shall give a brief-discussion of the concept of modes and modal decompositions of a realization, a topic that has both conceptual and numerical value.

---

*This section may be omitted without loss of continuity, especially since, to get an exposure to some more significant applications, the reader could at this point proceed directly to Chapter 3. Section 2.5.1 can be taken up just before the study of time-variant systems in Chapter 9.

2.5.1 Time-Invariant Equations and Matrix Exponentials

As the reader may recall from his studies of elementary circuit theory, the solution of the homogeneous part of a differential equation essentially also determines the solution of the inhomogeneous equation. Therefore we shall start with the equations

$$\dot{x}(t) = Ax(t), \qquad t \geq 0-, \qquad x(0-) = x_0 \tag{1}$$

We have already studied such equations via their Laplace transforms,

$$sX(s) - x_0 = AX(s)$$

which gives

$$X(s) = [sI - A]^{-1}x_0 \tag{2}$$

This is one representation of the solution. Our goal here is to see how the inverse Laplace transform

$$x(t) = \mathcal{L}^{-1}[sI - A]^{-1}x_0 \tag{3}$$

can be expressed in the time domain. We can proceed in various ways.

A direct time-domain method is to *discretize* Eq. (1), a necessary step if (1) is to be solved on a digital computer. Now the discretization can be done in many ways, with the choice between them depending on many factors: the nature and size of the matrix A, the time interval over which the solution is required, the accuracy that is important, speed of solution, etc., etc. Such factors have to be considered in any actual application, but for "theoretical" purposes, keeping in mind that by a limiting procedure we would like to return to the continuous-time case, the following *Euler discretization* method is often the simplest and most suggestive.

We consider only discrete time instants,

$$t = kh, \qquad k = 0, 1, 2, \ldots$$

where h is the *discretization interval*. Then we approximate the derivative $\dot{x}(t)$ as

$$\dot{x}(t) \doteq \frac{x(t+h) - x(t)}{h} \tag{4}$$

so that (1) can be approximated by

$$x(t+h) = (I + Ah)x(t), \qquad x(0-) = x_0 \tag{5}$$

This is a vector difference equation which is easily solved to give

$$x(kh) = (I + Ah)^k x_0, \qquad k = 1, 2, \ldots \tag{6}$$

By choosing h sufficiently small, we could in principle get a good approximation to the function $x(\cdot)$, but, as stated earlier, this is not, except in very simple problems, a recommended numerical procedure. But this solution serves well in returning to the continuous limit, which we can do by fixing a value t and letting

$$k \longrightarrow \infty, \; h \longrightarrow 0 \text{ such that } t = kh$$

Then

$$x(t) = \lim_{k \to \infty} \left(I + \frac{At}{k}\right)^k x_0 \tag{7}$$

But analogy with the definition of the number e in elementary calculus will suggest defining

$$\lim_{k \to \infty} \left(I + \frac{At}{k}\right)^k = e^{At} \quad \text{or} \quad \exp At \tag{8}$$

This is the so-called *matrix exponential*, and it turns out that we have the properties expected from the scalar analogy, especially

1. $\dfrac{d}{dt} \exp At = A \exp At = (\exp At)A$ (9)

2. $\exp A(t_1 + t_2) = \exp At_1 \exp At_2$ (10)

3. $\exp At$ is nonsingular   and $[\exp At]^{-1} = \exp(-At)$ (11)

4. $\exp At = \lim_{k \to \infty} \sum_0^k \dfrac{A^j t^j}{j!} = I + At + \dfrac{A^2 t^2}{2!} + \cdots$ (12)†

Accepting the matrix exponential, the solution of the differential equation (1) is of course

$$x(t) = e^{At} x_0 \tag{13}$$

But now that we have brought up the scalar analogy, note that we could also have guessed from the Laplace transform formula that we should write

$$\mathcal{L}^{-1}[sI - A]^{-1} = e^{At} \tag{14}$$

Moreover, we can readily "see" the truth of (12) via

$$\mathcal{L}^{-1}[sI - A]^{-1} = \mathcal{L}^{-1}[Is^{-1} + As^{-2} + A^2 s^{-3} + \cdots] = I + At + \frac{A^2 t^2}{2!} + \cdots \tag{15}$$

†The expression in (12) is usually the way the matrix exponential is first defined; it is interesting that the definition via (8) is not only closer to the approach in elementary calculus but also turns out to be more general (applying, for example, also to unbounded operators A as arise in quantum mechanics).

These facts should be persuasive evidence for the correctness of the solution (13). It is also clear that what is needed for a definitive proof is assurance that the limits in (8) or (12) converge. In fact they do, and sufficiently well that term-by-term differentiation of (12) is permissible; however, the actual proof is not particularly significant for us, nor is it particularly difficult. Therefore we shall merely refer the reader to any textbook on differential equations. Here we shall just assume as true the result that $[\exp At]x_0$ is a solution of (1). However, is it the only solution? The answer is yes, because if (1) had two solutions, x_1 and x_2, their difference $z = x_1 - x_2$ would obey the equation

$$\dot{z}(t) = Az(t), \qquad z(0) = 0 \tag{16}$$

Taking Laplace transforms gives

$$Z(s) = 0$$

so that by the uniqueness theorem for Laplace transforms

$$x_1(t) - x_2(t) = z(t) = 0, \qquad t \geq 0$$

Therefore the vector differential equation (1) has a *unique* solution $[\exp At]x_0$.

Another (purely time-domain) proof is the following. Suppose $\mathcal{E}(t)$ is *any* solution of $\dot{x}(t) = Ax(t)$, $x(0) = x_0$, $t \geq 0$. Then

$$\frac{d}{dt}[e^{-At}\mathcal{E}(t)] = -Ae^{-At}\mathcal{E}(t) + e^{-At}\dot{\mathcal{E}}(t) = (-A + A)e^{-At}\mathcal{E}(t) = 0, \qquad t \geq 0$$

Therefore

$$e^{-At}\mathcal{E}(t) = c, \qquad \text{a constant for all } t \tag{17}$$

so that, with $t = 0$, we see that

$$c = \mathcal{E}(0) = x_0$$

Therefore $\mathcal{E}(\cdot)$ has to have the (unique) form $\mathcal{E}(t) = e^{At}x_0$.

Explicit Calculations. Of course, so far we have not explicitly computed $\exp At$ for any A, but, as stated earlier, such explicit solutions are rarely necessary in system analysis or design. In fact, $\exp At$ is basically a collection of scalar exponentials whose particular form is rarely as suggestive as the representations $\exp At$ or $\mathcal{L}^{-1}[sI - A]^{-1}$. There is continuing activity on the best method of actually calculating $\exp At$ on a digital computer. One might just, as in Exercises 2.5-1 and 2.5-10, compute partial sums of the series $I + At + A^2t^2/2! + \ldots$ until no significant change can be seen in the values of the sums. This method is reasonable whenever there is not too large a

spread between the largest and smallest eigenvalues of A. More sophisticated techniques are discussed in the literature, and if actual computation is necessary, one should first do a search for the currently best computer routine.†
We repeat that for us the main significance of exp At is its representational value, which will be the major theme of the following discussions.

The State-Transition Matrix.   From the fact that

$$x(t) = (\exp At)x_0$$

we see that exp At can be regarded as the *state-transition matrix*, which (in the absence of any input) takes an initial state x_0 to a state $x(t)$ by time t. Note that choosing $t = 0$ as the initial time is purely a convenience, and that because of stationarity (time invariance), which arises from the constancy of the parameters, the differential equation

$$\dot{x}(t) = Ax(t)$$

is satisfied by

$$x(t) = [\exp A(t - \tau)]x(\tau), \qquad \text{all } t, \tau \tag{18}$$

This fact underlies the solution of the inhomogeneous equation.
    Let us also note that the formula

$$x(t) = (\exp At)x_0$$

shows that

> the ith column of exp At = solution of $\dot{x}(t) = Ax(t)$,
> for the particular initial
> condition $x_0 = e_i$, the
> ith unit vector   (19)

In other words, exp At is composed of n solutions of the differential equation $\dot{x}(t) = Ax(t)$, corresponding to the n (linearly independent) initial-condition vectors $\{e_i, i = 1, \ldots, n\}$. This interpretation can also be used to define the state-transition matrix for time-variant $A(\cdot)$, as we shall discuss further in Sec. 9.1.

The Inhomogeneous Equation.   By direct analogy with the corresponding scalar equation, we can write down a solution to

$$\dot{x}(t) = Ax(t) + bu(t), \qquad x(0-) = x_0 \tag{20}$$

---

†See, for example, a recent paper: C. B. Moler and C. F. Van Loan, "Nineteen Dubious Ways to Compute the Exponential of a Matrix," *SIAM Rev.*, 20. pp. 801–836, Oct. 1978.

as

$$x(t) = (\exp At)x_0 + \int_0^t e^{A(t-\tau)}bu(\tau)\, d\tau \tag{21}$$

The validity of this solution can be proved by verifying that it does in fact satisfy the differential equation, which can be shown to have a unique solution whenever the homogeneous equation does. We may note that in carrying out the above prescription, and in other proofs, it may be conceptually clearer to use the notation

$$\phi(t; 0, x_0) = \text{solution at } t \text{ of the differential equation}$$
$$\text{with initial condition } x_0 \text{ at } t = 0 \tag{22}$$

rather than to use the same symbol $x(\cdot)$ for the "variable" in the differential equation and for the solution. However, while the explicit notation is mathematically and logically more meaningful, it is more cumbersome. With a little attention, one can learn to work just as clearly with the notation $x(\cdot)$, despite its dual meaning. Moreover, we often do make the distinction, when necessary, without explicitly using the notation (22).

Thus let us consider obtaining (21) by using the impulse response and transfer function of the realization (20). Therefore let

$$u(t) = \delta(t) \quad \text{and} \quad x(0-) = 0 \tag{23}$$

Then from the differential equation we can conclude that $x(\cdot)$ must have a discontinuity between $0-$ and $0+$ in order to accommodate the driving impulse $bu(\cdot)$. Moreover, the magnitude of the discontinuity must be such that $x(0+) = b$. Therefore, for $t \geq 0+$, we have the equation

$$\dot{x}(t) = Ax(t), \qquad x(0+) = b, \qquad t \geq 0+$$

for which the solution is

$$x(t) = [\exp At]b \tag{24}$$

This is the impulse response vector of our system, and normally we would call it $h(t)$ for clarity rather than $x(t)$. That is, after (23) we would have said "Let $h(\cdot)$ be the solution of

$$\dot{x}(t) = Ax(t) + b\delta(t), \qquad x(0-) = 0$$

and then we would have used $h(\cdot)$ in place of $x(\cdot)$ in all the arguments leading to (24). So this is an example of how we often really do what the notation (22) requires, but less formally. In any case, since (24) gives the impulse response, by using superposition (Sec. 1.3), the response of the system to an arbitrary input $u(\cdot)$ can be written as

$$\int_0^t [\exp A(t - \tau)]bu(\tau)\, d\tau \tag{25}$$

Adding to this the contribution of any nonzero initial condition $x(0-) = x_0$, we again have the formula

$$x(t) = (\exp At)x_0 + \int_0^t e^{A(t-\tau)}bu(\tau)\, d\tau$$

Alternatively, the transfer function of the realization (20) can be calculated by taking Laplace transforms, with $x(0-) = 0$, to get

$$X(s) = (sI - A)^{-1}bU(s)$$

Inverse transformation yields (25), which when augmented by the initial-condition term, $(\exp At)x_0$, again leads to formula (21). (Notice here that we are more willing to tolerate the notational ambiguity, perhaps because the chain of argument is shorter.)

Example 2.5-1.   Singular $\mathcal{C}$ Implies Lack of State Controllability

We can now establish a statement claimed on heuristic grounds in Sec. 2.3.2, namely that a realization whose controllability matrix $\mathcal{C}$ is singular cannot be point-wise controllable. For if $\mathcal{C}$ is singular there exists a vector q' such that

$$q'\mathcal{C} = 0, \qquad q \neq 0 \tag{26}$$

But then we cannot find an input $u(\cdot)$ to take $x_0 = 0$ to $x(T) = q$ in any finite or infinite time T, for any such $u(\cdot)$ would have to satisfy

$$q = \int_0^T e^{A(T-\tau)}bu(\tau)\, d\tau$$

or

$$q'q = \int_0^T q' e^{A(T-\tau)}bu(\tau)\, d\tau$$

But

$$q' e^{A(T-\tau)}b = q'b + q'Ab(T - \tau) + \cdots + q'A^ib\frac{(T - \tau)^i}{i!} + \cdots$$

and by hypothesis (26)

$$q'b = 0, \qquad q'Ab = 0, \ldots, q'A^{n-1}b = 0$$

and therefore by the Cayley-Hamilton theorem $q'q \equiv 0$, which contradicts the fact that $q \not\equiv 0$. ∎

Note: The converse result is that if $\mathcal{C}$ is nonsingular, there always exists an input $u(\cdot)$ that will take an arbitrary initial state $x(0)$ to an arbitrary state q in some specified time T. This result is true, but its proof requires some properties of (Gramians of) linearly independent functions, which we shall not really need until we study time-variant systems. Therefore this proof is postponed to Sec. 9.2, but it may of course be explored as an exercise.

Example 2.5-2. A Finite Form for exp At

By the Cayley-Hamilton theorem, the matrices $\{A^{n+k}, k \geq 0\}$ can be expressed in terms of the first n powers $\{A^0 = I, A, \ldots, A^{n-1}\}$. Therefore we see that the matrix exponential can be written as

$$\exp At = I + At + \frac{A^2 t^2}{2!} + \cdots$$

$$= \xi_0(t)I + \xi_1(t)A + \cdots + \xi_{n-1}(t)A^{n-1} \tag{27}$$

for some functions $\{\xi_i(\cdot)\}$. The functions may not be unique, but the point is that one can always find a finite set of such scalar functions, and moreover we can always specify a set with the property that the $\{\xi_i(\cdot)\}$ are linearly independent (over any interval). To see this, let us first substitute the desired expression (27) into the formula (9), $d(\exp At)/dt = A \exp At$, to obtain

$$I\dot{\xi}_0 + A\dot{\xi}_1 + \cdots + A^{n-1}\dot{\xi}_{n-1} = A\xi_0 + A^2\xi_1 + \cdots + A^n\xi_{n-1}$$

$$= A\xi_0 + \cdots + A^{n-1}\xi_{n-2}$$

$$+ (-a_n I - a_{n-1}A - \cdots - a_1 A^{n-1})\xi_{n-1} \tag{28}$$

where, in the second step, we have again used the Cayley-Hamilton theorem. Since at $t = 0$, $\exp At$ is equal to I, we can choose $\xi_0(0) = 1$, $\xi_i(0) = 0$, $i \geq 1$. Furthermore, *one way* of choosing the $\{\xi_i(\cdot)\}$ is to define them via the differential equations

$$\dot{\xi}_0 = -a_n \xi_{n-1}$$

$$\dot{\xi}_1 = -a_{n-1}\xi_{n-1} + \xi_0, \qquad \xi_0(0) = 1, \xi_i(0) = 0, i \geq 1$$

$$\vdots \tag{29}$$

$$\dot{\xi}_{n-1} = -a_1 \xi_{n-1} + \xi_{n-2}$$

Clearly $\{\xi_i(\cdot)\}$ defined in this way will give us equality in (28). However, note that it is not the only choice, and that in particular (29) need not be intepreted as obtained by comparing coefficients of powers of A in (28). The main advantage of defining the $\{\xi_i(\cdot)\}$ as in (29) is that they are guaranteed linearly independent, the proof of which fact we shall leave to the reader. [*Hint:* If, $\sum_1^n c_i \xi_{i-1}(t) = c'\xi(t) = 0$, show by considering (enough) derivatives at $t = 0$ that $c \equiv 0$.] ∎

2.5.2 Modes of Oscillation and Modal Decompositions

We have used the term modes (of oscillation) of a realization on several occasions, relying on the familiarity and intuitive feeling most readers will have for this concept. However, a more explicit discussion can be useful, especially if numerical work is being considered, and is therefore presented here.

The modes of a realization $\{A, b, c\}$ are a useful description of its "natural" or "unforced" behavior. The modes are determined by the eigenvalues $\{\lambda_i\}$ of A, i.e., the roots of the equation $a(s) = \det (sI - A) = 0$. For simplicity of present discussion, we shall assume that

$$\text{the } \{\lambda_i\} \text{ are distinct}$$

Now consider an unforced system

$$\dot{x}(t) = Ax(t), \qquad x(0-) = x_0$$

Taking transforms and making a partial fraction expansion, we can write

$$X(s) = (sI - A)^{-1}x_0 = \sum_{i=1}^{n} \frac{R_i}{s - \lambda_i} x_0 \tag{30}$$

where the *residue* matrices R_i are given by (cf. Exercises A.44–A.46, especially A.46, part 3)

$$R_i = p_i q_i' \tag{31}$$

and p_i and q_i' are the right (or column) and left (or row) eigenvectors of A associated with the eigenvalue λ_i,

$$Ap_i = \lambda_i p_i, \qquad q_i'A = \lambda_i q_i', \qquad and \qquad q_i'p_i = 1 \tag{32}$$

(The reader may wish to spend some time on the above-cited exercises before proceeding further.)

Therefore we have the decomposition

$$X(s) = \sum_{1}^{n} \frac{(q_i' x_0)}{s - \lambda_i} p_i \tag{33}$$

or, by taking (inverse) Laplace transforms,

$$x(t) = \sum_{1}^{n} \alpha_i e^{\lambda_i t} p_i, \qquad \alpha_i = q_i' x_0 \tag{34}$$

Now we shall say that only one mode of oscillation, or briefly one *mode*, in particular the ith mode, of the realization is excited if

$$x(t) = e^{\lambda_i t} c$$

where c is a constant vector proportional to the eigenvector p_i. The significant thing is that *every component of $x(\cdot)$ varies with time in the same way, namely as exp $\lambda_i t$.* The modes have several important properties, noted here under the assumption of distinct eigenvalues:

1. There are n modes, one for each eigenvalue λ_i.
2. An arbitrary initial condition will in general excite all the modes, but the amount, $\alpha_i = q_i' x_0$, of excitation of any mode is clearly independent of that of any other mode.
3. To excite only a single mode, say the ith one, we should make x_0 proportional to the ith eigenvector p_i (cf. Exercise A.44).
4. The decomposition (34) is unique; i.e., if

$$x(t) = \sum e^{\lambda_i t} \alpha_i p_i = \sum e^{\lambda_i t} \beta_i p_i$$

then $\alpha_i = \beta_i$. This follows from the linear independence of the $\{p_i\}$, which in turn follows from the assumption that the $\{\lambda_i\}$ are distinct (cf. Exercise A.42).
5. If the matrix A is symmetric, the $\{p_i\}$ will be mutually orthogonal (cf. Exercise A.43). In this case, the modes are often called *normal* modes. At one time, the definition of mode was restricted to such (normal) modes, and as a consequence it was believed that modal decompositions could only be found for restricted classes of systems, e.g., two-element-kind (RC or LC or RL) networks. The state-space description showed immediately that modes could be defined for any realization, though of course they would not always be normal. The adjective *natural* is often used for these general (non-normal) modes.

We should note also that though we have assumed distinct eigenvalues, or actually diagonalizable realizations, the basic concepts go through to the general case as well. For example, if the eigenvalue λ_1 is repeated twice, then the modes corresponding to this eigenvalue will in general be (exp $\lambda_1 t$)c and (t exp $\lambda_1 t$)d, where c and d are constant vectors. For further details (not needed for this course), see Zadeh and Desoer [50, Chap. V]; we should also note that numerical determination of the modal decomposition in the repeated roots Jordan form case is generally difficult.

Example 2.5-3.

Let

$$A = \begin{bmatrix} 0 & 1 \\ -4 & -5 \end{bmatrix}$$

so that $\lambda_1 = -1$, $\lambda_2 = -4$, and we can take (as is readily verified),

$$p_1 = \begin{bmatrix} 1 \\ -1 \end{bmatrix}, \quad p_2 = \begin{bmatrix} -1 \\ 4 \end{bmatrix}$$

$$q_1 = \frac{1}{3}\begin{bmatrix} 4 \\ 1 \end{bmatrix}, \quad q_2 = \frac{1}{3}\begin{bmatrix} 1 \\ 1 \end{bmatrix}$$

We also have

$$x_1(t) = \tfrac{1}{3}(4x_{10} + x_{20})e^{-t} - \tfrac{1}{3}(x_{10} + x_{20})e^{-4t}$$
$$x_2(t) = -\tfrac{1}{3}(4x_{10} + x_{20})e^{-t} + \tfrac{4}{3}(x_{10} + x_{20})e^{-4t}$$

Note that if x_0 is proportional to p_1 only the first mode, e^{-t}, will be excited, while the reverse is true if x_0 is proportional to p_2.

For this second-order system, it is possible to get a nice geometric interpretation of the modal decomposition by plotting $x(\cdot)$ on a *phase plane*, as shown in Fig. 2.5-1.

Let $\vec{\chi}(t)$ be the vector that denotes the state (position in the phase plane) at time t. Let $\vec{i}_1$ and $\vec{i}_2$ be unit vectors as shown. Then we can write

$$\vec{\chi}(t) = x_1(t)\vec{i}_1 + x_2(t)\vec{i}_2$$

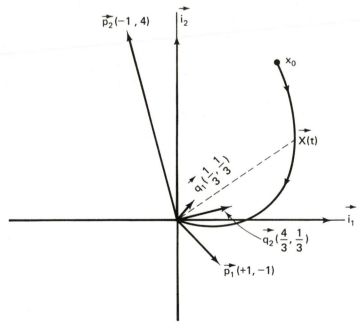

Figure 2.5-1.   State evolution in terms of modes.

where the vectors $\vec{i}_1$, $\vec{i}_2$ have the representations

$$\begin{bmatrix} 1 \\ 0 \end{bmatrix}, \quad \begin{bmatrix} 0 \\ 1 \end{bmatrix}$$

That is, we have expressed $\vec{\chi}$ in terms of its components along the coordinate axes $\vec{i}_1$, $\vec{i}_2$. However, there is no need to choose the particular axes $\vec{i}_1$, $\vec{i}_2$. Let us choose (oblique) axes $\vec{p}_1$, $\vec{p}_2$, as shown. Then we can write $\vec{\chi}(t) = \xi_1(t)\vec{p}_1 + \xi_2(t)\vec{p}_2$. How do we find the $\{\xi_i(t)\}$? Because the axes $\vec{p}_1$, $\vec{p}_2$ are not "rectangular," we cannot find the $\{\xi_i(t)\}$ by "projecting" $\chi(t)$ on the $\{\vec{p}_i\}$. Instead we proceed as follows: Let q_1 be a vector such that $\vec{q}_1 \perp \vec{p}_2$ and $\vec{q}_1\vec{p}_1 = 1$. [It is easy to see that $(\frac{4}{3}, \frac{1}{3})$ provides such a vector.] Then $\vec{\chi}(t)\vec{q}_1 = \xi_1(t)$. In this way we find that $\xi_1(t) = e^{-t}(4x_{10} + x_{20})/3$. Similarly, we can calculate $\xi_2(t) = e^{-4t}(-x_{10} - x_{20})/3$.

Comparison with our earlier discussions shows that describing the motion along the vectors $\vec{p}_1$, $\vec{p}_2$ (which are the eigenvectors of A) gives us the decomposition into natural modes. Note that if x_0 lies anywhere on the line from the origin to $\vec{p}_1$, then $x(t)$ remains on this line for all t. So also if x_0 lies on the line from the origin to $\vec{p}_2$, then $x(t)$ remains proportional to p_2 for all t. In this sense the basis $\{\vec{p}_1, \vec{p}_2\}$ is more convenient for describing the motion $\vec{\chi}(t)$ than the basis $\{\vec{i}_1, \vec{i}_2\}$. (However, for ease of computations, the basis $\{\vec{i}_1, \vec{i}_2\}$ is usually preferred.) The vectors $\{\vec{q}_1, \vec{q}_2\}$ are said to form a basis *dual* or *reciprocal* to the basis $\{\vec{p}_1, \vec{p}_2\}$. ∎

Applications of the Modal Decomposition.   The point of the modal decomposition is that large interconnected systems can be decomposed into a parallel combination of simple first-order systems. Many systems problems become easier (and numerically better conditioned) in this framework—recall, for example, the analysis of controllability and observability in Sec. 2.4.1. Similarly, we can obtain useful insight into the PBH tests of Sec. 2.4.3. Thus, consider the simple example studied above and let us attempt, assuming $x(0-) = 0$ and starting from $x(0-) = 0$, to set up an arbitrary state $x(0+)$ using an input $g_1\delta(t) + g_2\delta^{(1)}(t)$. Then $x(0+) = g_1b + g_2Ab$. Suppose now that the column vector b lies along p_1 (and hence $q_2'b = 0$). Then $Ab = \lambda_1 b$, and $x(0+) = (g_1 + g_2\lambda_1)b$. Clearly, the only initial states we can set up are those that lie along b (and hence p_1). These form the set of controllable states. Note also that the mode corresponding to λ_2 can never be excited through the input since all attainable $x(0+)$ excite only the other mode: $e^{-\lambda_2 t}$ is then an uncontrollable mode. Although such an interpretation is harder for the case of, say, repeated eigenvalues, the essential idea is the same.

The modal decomposition is also important when it is desired to focus attention on some important subset of the modes, e.g., the fastest ones, or the low-frequency ones, etc. In such cases we can reduce the "dimension" of the problems treated by ignoring all the other modes.

Exercises

2.5-1.

The series (12) for exp At is especially easy to evaluate for certain matrices. Do so for

a. $A = \begin{bmatrix} 0 & 1 \\ 0 & 0 \end{bmatrix}$ b. $A = \begin{bmatrix} 0 & 1 \\ -1 & 0 \end{bmatrix}$.

c. A matrix with ones on the first superdiagonal and zeros everywhere else.

d. A matrix with λs on the diagonal, ones on the first superdiagonal, and zeros everywhere else, i.e., a Jordan block.

e. A matrix A such that $A^k = 0$ for some $k \geq 2$ (called nilpotent matrices).

f. $A = \text{block diag}\{A_1, \ldots, A_N\}$.

2.5-2.

If

$$A = \begin{bmatrix} \sigma & \omega \\ -\omega & \sigma \end{bmatrix}$$

show that

$$\exp At = \begin{bmatrix} (\exp \sigma t)\cos \omega t & (\exp \sigma t)\sin \omega t \\ -(\exp \sigma t)\sin \omega t & (\exp \sigma t)\cos \omega t \end{bmatrix}$$

Note: When

$$F = \begin{bmatrix} \sigma & \omega \\ \omega & -\sigma \end{bmatrix}$$

it turns out that

$$e^F =$$

$$\begin{bmatrix} \cosh(\omega^2 + \sigma^2)^{1/2} + \dfrac{\sigma \sinh(\omega^2 + \sigma^2)^{1/2}}{(\omega^2 + \sigma^2)^{1/2}} & \dfrac{\omega \sinh(\omega^2 + \sigma^2)^{1/2}}{(\omega^2 + \sigma^2)^{1/2}} \\[2ex] \dfrac{\omega \sinh(\omega^2 + \sigma^2)^{1/2}}{(\omega^2 + \sigma^2)^{1/2}} & \cosh(\omega^2 + \sigma^2)^{1/2} - \dfrac{\sigma \sinh(\omega^2 + \sigma^2)^{1/2}}{(\omega^2 + \sigma^2)^{1/2}} \end{bmatrix}$$

This expression was found by using the symbol manipulation computer language MACSYMA.

2.5-3.

If $\bar{A} = T^{-1}AT$, show that $\bar{A}^k = T^{-1}A^kT$ and that $\exp At = T(\exp \bar{A}t)T^{-1}$.

2.5-4.

Since $\exp At = \mathcal{L}^{-1}[sI - A]^{-1}$, we might argue that $[\exp At]^{-1} = \mathcal{L}^{-1}[sI - A]$. Is this result true or false? Why?

2.5-5.   *Matrix Convolution*

We define the convolution of matrix $F = [f_{ij}]$ with matrix $G = [g_{ij}]$ by $F * G = [\sum_k f_{ik} * g_{kj}]$.

a. Is matrix convolution commutative; i.e., is $F * G = G * F$? If not, under what conditions is it commutative?

b. Show that $e^{At} * e^{-At} = \mathcal{L}^{-1}(s^2I - A^2)^{-1} = e^{-At} * e^{At}$.

c. Write $(s^2I - A^2)^{-1}$ in a power series in A/s to express exp At * exp $-At$ as a power series in t.

2.5-6.   Series Method of Solution

Another approach to the solution of the vector equation (1) is to first assume a solution in the form of a Taylor series $x(t) = \Gamma_0 + \Gamma_1 t + \Gamma_2 t^2 + \cdots$. Show that $\Gamma_k = A^k x_0 / k!$ and hence that

$$x(t) = \left[I + At + A^2 \frac{t^2}{2!} + \cdots \right] x_0$$

2.5-7.

a. Establish formula (10) by using the infinite series representation (12). Show how to obtain (11) from (10).

b. Show that exp At is nonsingular by proving that det exp $At =$ exp (tr A)t, where tr A (the trace of A) is the sum of the diagonal elements of A.

2.5-8.

Establish the properties (9)–(12) of the matrix exponential as defined by (8).

2.5-9.

a. Write state equations for the network shown in the figure.

b. Assume that the input $u(\cdot)$ is turned off at $t = 0$. Show how to choose $i_1(0)$, $i_2(0)$, subject to the constraint $i_1^2(0) + i_2^2(0) = 1$, so that the system returns to rest at the fastest possible rate.

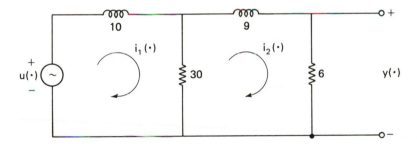

2.5-10.   Another Proof of the Cayley-Hamilton Theorem

Let $\{A, b, c\}$ be a state-space realization of a discrete-time system and assume that

$$a(z) = \det (zI - A) = \prod_1^n (z - \lambda_i)$$

a. Show that the response at $k = 1$ to the input
  (1) $U(z) = z - \lambda_1$ is $c (A - \lambda_1 I) b$.
  (2) $U(z) = (z - \lambda_1)(z - \lambda_2)$ is $c (A - \lambda_1 I)(A - \lambda_2 I) b$.

b. Use these facts to obtain a system-theoretical proof of the Cayley-Hamilton theorem. (Compare with Exercise 2.3-31.)

c. Show that the response for $k > 0$ to $U(z) = z - \lambda_1$ will have no component along the eigenvector corresponding to λ_1. *Hint:* Use the modal decomposition for $X(z)$.

2.5-11.

Given a realization

$$\dot{x}(t) = Ax(t), \qquad y = cx(t) = ce^{At}x(0)$$

suppose we try to calculate $x(0)$ from values of $y(t)$, $t = 0, \Delta, \ldots, (n-1)\Delta$.

a. Show that we can do this if and only if $\{c, e^{A\Delta}\}$ is observable.

b. Give proofs and/or counterexamples to check if

(1) $\{c, e^{A\Delta}\}$ observable implies $\{c, A\}$ observable.

(2) $\{c, A\}$ observable implies $\{c, e^{A\Delta}\}$ observable.

2.5-12.   *Sampled-Data Systems*

To interface with a digital computer used to compute a control input, we may assume that the input is piecewise constant, say,

$$u(t) = u_k, \qquad k\Delta < t \leq (k+1)\Delta$$

and that the output $y(\cdot)$ is observed only at the instants $k\Delta$, $k = 0, 1, \ldots,$ where Δ is some fixed sampling interval. Show that for this purpose an underlying continuous-time realization

$$\dot{x}(t) = Ax(t) + bu(t), \qquad y(t) = cx(t)$$

may be replaced by the discrete-time system

$$x_{k+1} = \Phi x_k + \Gamma u_k, \qquad y_k = cx_k$$

where

$$y_k = y(k\Delta), \qquad x_k = x(k\Delta)$$

$$\Phi = \exp A\Delta, \qquad \Gamma = \int_0^\Delta (\exp A\tau)\, d\tau \cdot b$$

Note that in such so-called *sampled-data* systems, the state-feedback matrix Φ is always nonsingular. *Hint:* Start with

$$x(t) = \exp A(t - s)x(s) + \int_s^t \exp A(t - \tau)bu(\tau)\, d\tau, \quad \text{and   let }\; t = (k+1)\Delta,$$
$$s = k\Delta.$$

2.5-13.   *Preservation of Controllability and Observability*
**   *Under Sampling***

If $\{A, b, c\}$ is a controllable (observable) continuous-time realization, show that the associated sampled data realization $\{\Phi, \Gamma, c\}$ (cf. Exercise 2.5.12) will be controllable (observable) if and only if the sampling interval Δ is such that

Im $[\lambda_i - \lambda_j] \neq 2\pi n/\Delta$, whenever Re $[\lambda_i - \lambda_j] = 0$ [the $\{\lambda_i\}$ are the eigenvalues of A.] (*Remark:* The original proof involved the Jordan canonical form; a proof based on the PBH tests can be found in [31].)

2.6 A GLIMPSE OF STABILITY THEORY

We assume that the reader has had some exposure to stability results for linear time-invariant systems, at least to the extent of knowing that stability is connected with the roots of the characteristic polynomial of a realization having negative real parts. This knowledge will suffice for most of our references to stability in the rest of the book. However, for convenience and some measure of completeness, we shall briefly present some aspects of the stability problem in this section.

We first define what we shall call external, or bounded-input, bounded-output, stability for a linear time-invariant system described by its impulse response. Then the internal stability of a realization is defined by the requirement that all the roots of the characteristic polynomial have negative real parts. We note that external stability may not be equivalent to internal stability, except for minimal realizations. Next we briefly present the celebrated method of Lyapunov for investigating stability. We use it to obtain a useful analytical criterion for the internal stability of linear time-invariant systems and also to show the important (and reassuring) result of Poincaré and Lyapunov that, under reasonable conditions, the stability of a nonlinear system can be inferred from the stability of a linear approximation.

In conclusion, we should emphasize that our presentation is quite sketchy and that many things are omitted, e.g., the Nyquist criterion. However, there are several surveys and textbooks devoted largely to stability theory; we list, for example [39]–[48] and [51]–[60].

2.6.1 *External and Internal Stability*

We shall say that a causal system is *externally stable* if a bounded input, $u(t) < M_1$, $-\infty < -T \leq t < \infty$, produces a bounded output $y(t) < M_2$, $-T \leq t < \infty$. A well-known necessary and sufficient condition for such *bibo* (bounded-input, bounded-output) *stability* is that the impulse response $h(\cdot)$ be such that

$$\int_0^\infty |h(t)| \, dt < M < \infty \tag{1}$$

This condition was apparently first presented in [57]. In discussing external stability, we shall assume zero initial conditions. Nonzero initial conditions can be set up by prior bounded inputs (starting from systems at rest), and therefore there is no loss of generality here. However, nonzero initial conditions may also arise from stray voltages and parasitic excitations, but these cannot be incorporated into the input-output description, which is only the specification of the relationship between the output $y(\cdot)$ and specified input $u(\cdot)$, not between the output and any other (stray) inputs. The state-space or internal description does permit us to incorporate the effects of such stray phenomena, as we have seen before and shall study further below in discussing internal stability. First we establish the above-stated criterion for bibo stability.

To prove the sufficiency (only if) part, we note that

$$y(t) = \int_0^\infty h(\tau)u(t - \tau)\, d\tau$$

so that under the assumptions on $h(\cdot)$ and $u(\cdot)$,

$$|y(t)| \le \int_0^\infty |h(\tau)||u(t - \tau)|\, d\tau \le M_1 \int_0^\infty |h(\tau)|\, d\tau \le M, \qquad M < \infty$$

To show necessity (the "if" part), assume that

$$\int_0^\infty |h(\tau)|\, d\tau = \infty$$

but that all bounded inputs give bounded outputs. We shall establish a contradiction here, thus proving the necessity of condition (1). For this, consider a bounded input defined via

$$u(t_1 - \cdot) = \operatorname{sgn} h(\cdot) \triangleq \begin{cases} +1, & \text{if } h(\cdot) > 0 \\ 0, & \text{if } h(\cdot) = 0 \\ -1, & \text{if } h(\cdot) < 0 \end{cases}$$

where t_1 is some fixed time instant. Then

$$y(t_1) = \int_0^\infty h(\tau)u(t_1 - \tau)\, d\tau$$

$$= \int_0^\infty |h(\tau)|\, d\tau = \infty, \qquad \text{by assumption}$$

so that $y(\cdot)$ is not bounded. Hence. . . .

Internal Stability.   Internal stability refers to the stability of a realization of a system. We shall say that a realization

$$\dot{x}(t) = Ax(t) + bu(t), \qquad y(t) = cx(t)$$

is *internally stable* or *stable in the sense of Lyapunov* if the solution of

$$\dot{x}(t) = Ax(t), \qquad x(t_0) = x_0, \qquad t \ge t_0$$

tends toward zero as $t \longrightarrow \infty$ for arbitrary x_0.

By examining the solution in the Laplace transform domain

$$X(s) = (sI - A)^{-1}x_0$$

and recalling that

$$t^k e^{-at} \longleftrightarrow (s + a)^{-k}$$

we can see that the realization will be stable if and only if

$$\text{Re}\{\lambda_i(A)\} < 0 \tag{2}$$

where $\{\lambda_i(A)\}$ are the eigenvalues of A.

The above argument also shows that an internally stable realization will always have an impulse response that satisfies condition (1); in other words, it will also be externally stable. However, the converse is not true, as can be seen by considering diagonal realizations that are not controllable (or not observable). Therefore, by using the general theorem decomposing a realization into its co, $c\bar{o}$, $\bar{c}o$, $\overline{co}$ parts (cf. Sec. 2.4.2), it can be seen that external stability will be equivalent to internal stability if the realization is minimal, i.e., both controllable and observable. Recalling that the interconnection of minimal realizations can often lead to a nonminimal overall realization, we should realize the necessity for caution in assuming internal stability of the overall realization either from the stability of its subsystems or the stability of its impulse response.†

The above arguments can all be repeated for discrete-time systems with obvious changes. A necessary and sufficient condition for bibo stability is that

$$\sum_{0}^{\infty} |h(k)| < M < \infty \tag{3}$$

A realization will be internally stable if and only if

$$|\lambda_i(A)| < 1 \tag{4}$$

and this will be equivalent to condition (3) if the realization is minimal.

It may be remarked that what we have defined above as internal stability is also often called *asymptotic stability*, whereas a realization in which $x(t)$ remains bounded as $t \longrightarrow \infty$ (without necessarily going to zero) is just called *stable*. The point is that nonrepeated purely imaginary eigenvalues of A (roots on the unit circle in discrete-time systems) are consistent with stability in this sense (but not with asymptotic stability). Lossless (conservative) systems are examples of stable but not asymptotically stable systems.

We turn now to certain equivalent methods of stating (2) [and (4)] based on a very general method introduced by Lyapunov in the 1890s to study the stability of linear and nonlinear systems.

2.6.2   The Lyapunov Criterion

We shall use the matrix exponential studied in Sec. 2.5.1 to establish the following important result.

Theorem 2.6-1.   (Lyapunov)

A matrix A is a *stability* matrix, i.e., $\text{Re}\{\lambda(A)\} < 0$ for all eigenvalues of A, if and only if for any given positive-definite symmetric matrix Q there exists a *positive-*

†See Sec. 3.1.

definite (symmetric) matrix P that satisfies

$$A'P + PA = -Q \tag{5}$$

This theorem is rarely used for direct numerical verification of stability. Instead certain special tests are used, for the derivation of which the following variant of this theorem, essentially due to Kalman [58], is often useful.

Corollary 2.6-2.   (Kalman)

In the preceding theorem we can take Q to be *positive semidefinite*, provided $x'(t)Qx(t)$ is not identically zero along any nonzero solution of $\dot{x}(t) = Ax(t)$.

The physical basis of these results is the following. The quantity

$$V[x(t)] = x'(t)Px(t) \tag{6}$$

can be regarded as a *generalized energy* associated with the realization. In a stable system the energy should decay with time, consistent with the calculation

$$\begin{aligned} \frac{d}{dt} V[x(t)] &= \dot{x}'(t)Px(t) + x'(t)P\dot{x}(t) \\ &= x'(t)[PA + A'P]x(t) \\ &= -x'(t)Qx(t) \end{aligned} \tag{7}$$

From this we can conclude that $x(t) \longrightarrow 0$ as $t \longrightarrow \infty$ provided Q is positive definite (Theorem 2.6-1) or at least provided $x'(t)Qx(t)$ is not identically zero along any trajectory of $\dot{x}(t) = Ax(t)$ (Corollary 2.6-2).

By finding appropriate functionals $V[x(t)]$, useful stability criteria can also be established for several classes of nonlinear systems (see, e.g., [51]–[60]).

We should also remark that the physical argument shows that the additional hypothesis in Corollary 2.6-2 is really an observability requirement for a multi-output system with output $y(t) = Cx(t)$, where $Q = C'C$.

Proof of Theorem 2.6-1.   *Necessity:* If A is such that Re $\lambda_i(A) < 0$, then we shall prove that a suitable matrix P is given by

$$P = \int_0^\infty e^{A't}Qe^{At}\, dt \tag{8}$$

The integrand is a sum of terms of the form $t^k e^{\lambda t}$, where λ is an eigenvalue of A, and since Re $\lambda(A) < 0$, the integral will exist. Next, by substitution,

$$\begin{aligned} A'P + PA &= \int_0^\infty A'e^{A't}Qe^{At}\, dt + \int_0^\infty e^{A't}Qe^{At}A\, dt \\ &= \int_0^\infty \frac{d}{dt} e^{A't}Qe^{At}\, dt \\ &= e^{A't}Qe^{At}\Big|_0^\infty = -Q \end{aligned}$$

Sufficiency: The heuristic argument given earlier using the generalized energy functional $V[x(t)]$ can be formalized to prove sufficiency, see [53].   ∎

The following result is often useful.

Corollary 2.6-3.

If A is a stability matrix, then the Lyapunov equation

$$A'P + PA + Q = 0$$

has a *unique* solution for every Q.

Proof.   Suppose there are two solutions P_1 and P_2 so that

$$A'(P_1 - P_2) + (P_1 - P_2)A = 0$$

Therefore

$$0 = e^{A't}[A'(P_1 - P_2) + (P_1 - P_2)A]e^{At}$$

$$= \frac{d}{dt} e^{A't}(P_1 - P_2)e^{At}$$

and

$$e^{A't}(P_1 - P_2)e^{At} = \text{a constant for all } t$$

In particular, taking $t = 0$ and $t = T$, we must have

$$P_1 - P_2 = e^{A'T}(P_1 - P_2)e^{AT}$$

Now letting $T \longrightarrow \infty$ and using the stability of A, we obtain

$$P_1 - P_2 = 0 \quad \text{or} \quad P_1 = P_2$$

Actually uniqueness can be proved in many ways. One is to use a matrix theorem (Exercise A.40) that the matrix equation $AX + XB = 0$ has a unique solution (namely $X = 0$) if $\lambda_i(A) + \lambda_j(B) \neq 0$, any i, j. In our case $B = A'$, and this condition will be met since Re $\lambda_i(A) < 0$ for all i.

Perhaps the most elegant proof of uniqueness was the one given by Lyapunov himself in 1893: For any Q, the integral (8) defines a solution of the n^2 linear equations (5) in the n^2 unknowns P_{ij}. Since *every* system (5) has *some* solution, this solution must be *unique* by the theory of linear algebraic equations (the *range* is full, and therefore the *null-space* must be empty—see Sec. 4 in the Appendix).   ∎

Discrete-Time Systems.   By similar arguments we can also study discrete-time systems

$$x(k + 1) = Ax(k)$$

This will give the analogs of Theorem 2.6-1 and Corollaries 2.6-2 and 2.6-3 with the only change that the continuous-time Lyapunov equation (5) is replaced by

$$P - A'PA = Q \tag{9}$$

It is useful to recall here that many discrete-time results can also be obtained from
the continuous-time results by use of the well-known bilinear transformation

$$s = \frac{z-1}{z+1}, \qquad z = \frac{1+s}{1-s} \tag{10}$$

which maps the left half plane into the unit circle and vice versa. The appropriate
transformations for systems in state-space form are [59, p. 688]

$$A \longrightarrow (A_d - I)(A_d + I)^{-1}, \qquad Q \longrightarrow \tfrac{1}{2}(A_d' + I)^{-1} Q_d (A_d + I)^{-1}, \qquad P \longrightarrow P_d \tag{11}$$

where the subscript d is temporarily used to denote discrete-time quantities. Now
we can verify that with these substitutions

$$A'P + PA = -Q \text{ goes into } P_d - A_d' P_d A_d = Q_d$$

2.6.3   A Stability Result for Linearized Systems

Many linear systems arise from nonlinear systems by linearization about an
equilibrium point. The relationship between the stability of the linearized model
and that of the original system is therefore of great interest. One of the most notable
early achievements of the Lyapunov theory was a contribution to this question.
Let

$$\dot{x}(t) = f[x(t)] \quad \text{and} \quad f(x_e) = 0 \tag{12}$$

Then if

$$x(t) = z(t) + x_e$$

we can write

$$f[z(t) + x_e] = f(x_e) + \frac{\partial f(x, t)}{\partial x}\bigg|_{x=x_e} z(t) + g[z(t)]$$

where

$$\frac{\|g[z(t)]\|}{\|z\|} \longrightarrow 0 \qquad \text{as } \|z\| \longrightarrow 0 \tag{13}$$

and $\|z\|^2 = \sum_1^n z_i^2$. This gives the perturbation equation

$$\dot{z}(t) = Az(t) + g[z(t)]$$

where

$$A \triangleq \left[\frac{\partial f_i[x(t)]}{\partial x_j}\bigg|_{x_j = x_{je}} \right]$$

and the linearized equation is

$$\dot{z}(t) = Az(t) \tag{14}$$

Now Lyapunov and Poincaré proved that if A is a stability matrix, then the original
nonlinear equation (12) is also asymptotically stable under "small" perturbations
$z(\cdot)$ from the equilibrium position x_e.

To justify this we can examine the generalized energy function (or Lyapunov function)

$$V[z(t)] = z'(t)Pz(t)$$

where P is the unique symmetric positive-definite matrix associated with the stability matrix A through the Lyapunov equation $A'P + PA = -I$. Now

$$\frac{d}{dt} V[z(t)] = \dot{z}'(t)Pz(t) + z'(t)P\dot{z}(t)$$

$$= z'(t)(A'P + PA')z(t)$$

$$+ 2g'[z(t)]Pz(t)$$

$$= -z'(t)z(t)\left(1 - \frac{2g'[z(t)]Pz(t)}{z'(t)z(t)}\right)$$

For sufficiently small $\|z(t)\|$, we see by using (13) that

$$\frac{d}{dt} V[z(t)] < 0 \qquad \text{for } \|z(t)\| \longrightarrow 0$$

so that $z'(t)Pz(t) \longrightarrow 0$ as $t \longrightarrow \infty$, and because P is positive definite, $z(t) \longrightarrow 0$ as $t \longrightarrow \infty$. More formal proofs of this important result can be found, for example, in [60]. ∎

This concludes our brief look at some aspects of stability theory. The problems of stability are crucial in control system design and in the study of large-scale systems and will have to be pursued further by those going into those fields.

Exercises

2.6-1.

If P satisfies the linear matrix (Lyapunov) equation

$$AP + PA' + Q = 0$$

show that we can also write it as

$$P = e^{At}Pe^{A't} + \int_0^t e^{As}Qe^{A's}\, ds$$

2.6-2.

Show that the discrete-time Lyapunov equation can be rewritten as

$$[I - A'\otimes A']p = q$$

where $\otimes$ denotes the Kronecker product and p and q denote column vectors obtained in a one-one manner from the matrices P and Q. Hence, show that the Lyapunov equation has a unique solution if and only if $1 - \lambda_i\lambda_j \neq 0$, all i, j, where $\{\lambda_i\}$ are the eigenvalues of A. Note that $\{|\lambda_i| < 1\}$ is a sufficient condition.

2.6-3.

For the realization

$$\dot{x}_1 = x_2, \qquad \dot{x}_2 = -6x_1 - 5x_2$$

what can you say about its stability and asymptotic stability from the candidate Lyapunov functions

a. $V_1 = 6x_1^2 + x_2^2$.
b. $V_2 = x_1^2 + x_2^2 - x_1x_2$.
c. $V_3 = x_1^2 + x_2^2 + 2x_1x_2$.

Solution.   a. Note that $V_1 = 6x_1^2 + x_2^2 > 0$. Then $\dot{V}_1 = -10x_2^2 \le 0$ $\Rightarrow$ stable. Note that $\dot{V}_1 \equiv 0 \Rightarrow x_2 \equiv 0 \Rightarrow \dot{x}_1 \equiv 0 \equiv \dot{x}_2 \Rightarrow$ system is asymptotically stable.

b. $V_2 > 0$, but $\dot{V}_2 = 6x_1^2 - 5x_1x_2 - 11x_2^2 \not< 0$, so we can conclude nothing about the stability of the system.

c. $V_3 \ge 0$, $\dot{V}_3 = -4[3x_1^2 + 5x_1x_2 + 2x_2^2] < 0$, so we cannot conclude anything about the stability.

2.6-4.

A self-tracking antenna has a dominating nonlinearity because of the variation of returned power as a function of the angle between the beam centerline and the target. An approximation to the situation is given in the figure for one possibility.

a. Define states so that the linear system is in observer canonical form, and write the complete (nonlinear) equations of motion for the system.

b. Let $V = x_1^2 + x_2^2$, and show that V is a Lyapunov function for this system.

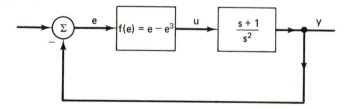

2.6-5.   *A Schur-Cohn Test for Stability Within the Unit Circle*
Let

$$a(z) = a_0 z^n + a_1 z^{n-1} + \cdots + a_n$$

and form the matrix

$$S = [s_{ij}], \qquad s_{ij} = \sum_{k=0}^{\min(i,\,j)} (a_{i-k}a_{j-k} - a_{n+k-i}a_{n+k-j}), \qquad 0 \le i, j \le n-1$$

or, more compactly,

$$S = LL' - \tilde{L}\tilde{L}'$$

where L is a lower triangular Toeplitz matrix with first column $[a_0, \ldots, a_{n-1}]'$ and $\tilde{L}$ is similar but with first column $[a_n, \ldots, a_1]'$. Take as a discrete-time Lyapunov function the quadratic form

$$V[x(k)] = x'(k)Sx(k)$$

Show that the matrix S will be positive definite if and only if

$$V[x(k+1)] - V[x(k)] = -[(1 - a_n^2)x_1(k) + (a_1 - a_{n-1}a_n)x_2(k) + \cdots$$
$$+ (a_{n-1} - a_1 a_n)x_n(k)]^2$$

is negative and not identically zero except when $x(\cdot) \equiv 0$. Deduce a (so-called Schur-Cohn) test for checking if the roots of $a(z)$ all lie within the unit circle. *Remark:* This test can be described in other forms, closely related to the theory of orthogonal polynomials (see, e.g., A. Vieira and T. Kailath, *IEEE Trans. Circuits Systems*, Vol. CT-24, pp. 218–220, April 1977).

REFERENCES

1. THOMSON, W. (Lord Kelvin), "Mechanical Integration of the General Linear Differential Equation of Any Order with Variable Coefficients," *Proc. Roy. Soc.*, vol. 24, pp. 271–275, 1876.

2. LEVINE, L., *Methods for Solving Engineering Problems Using Analog Computers*, McGraw-Hill, New York, 1964.

3. J. E. SOLOMON, "The Monolithic Op Amp: A Tutorial Study," *IEEE J. Solid-State Circuits*, vol. SC-9, pp. 314–332, Dec. 1974.

4. R. MEYER, ed., *Integrated-Circuit Operational Amplifiers*, IEEE Press, New York, 1978.

5. A. V. OPPENHEIM and R. W. SCHAFER, *Digital Signal Processing*, Prentice-Hall, Englewood Cliffs, N.J., 1975.

6. A. PELED and B. LIU, *Digital Signal Processing-Theory, Design, and Implementation*, Wiley, New York, 1976.

7. J. F. KAISER, "Some Practical Considerations in the Realization of Linear Digital Filters," in *Proceedings of the Third Allerton Conference on Circuit and System Theory*, pp. 621–633, 1965; reprinted in *Digital Signal Processing* (L. R. Rabiner and C. M. Rader, eds.), IEEE Press, New York, 1972.

8. S. Y. Hwang, "Minimum Uncorrelated Unit Noise in State-Space Digital Filtering", *IEEE Trans. Acoustics, Speech and Signal Processing*, vol. ASSP-25, pp. 273–281, Aug. 1977.

9. B. D. O. ANDERSON and S. VONGPANITLERD, *Network Analysis and Synthesis, A Modern Systems Theory Approach*, Prentice-Hall, Englewood Cliffs, N.J., 1973.

10. T. R. BASHKOW, "The A Matrix: A New Network Description," *IRE Trans. Circuit Theory*, vol. CT-4, pp. 117–120, September 1957.

11. R. E. BELLMAN, *Dynamic Programming*, Princeton University Press, Princeton, N.J., 1957.

12. R. E. KALMAN, "Optimal Nonlinear Control of Saturating Systems by Intermittent Control," *IRE WESCON Rec.*, Sec. IV, pp. 130–135, 1957.

13. R. A. ROHRER, *Circuit Theory: An Introduction to the State-Variable Approach*, McGraw-Hill, New York, 1970.

14. H. S. TSIEN, *Engineering Cybernetics*, McGraw-Hill, New York, 1954.

15. I. FLUGGE-LOTZ, *Discontinuous and Optimal Control*, McGraw-Hill, New York, 1968.

16. E. GILBERT, "Controllability and Observability in Multivariable Control Systems," *SIAM J. Control*, vol. 1, pp. 128–151, 1963.

17. R. E. KALMAN, "Mathematical Description of Linear Systems," *SIAM J. Control*, vol. 1, pp. 152–192, 1963.

18. R. E. KALMAN, *Lectures on Controllability and Observability*, C.I.M.E., Bologna, 1968.

19. R. E. KALMAN, "On the General Theory of Control Systems," in *Proceedings of the First International Congress on Automatic Control*, Butterworth's, London, 1960, pp. 481–493.

20. V. M. POPOV, "On a New Problem of Stability for Control Systems," *Autom. Remote Control*, vol. 24, no. 1, pp. 1–23, 1963 (English trans.).

21. E. I. JURY, *Theory and Application of the Zee-Transform Method*, Wiley, New York, 1964.

22. H. H. ROSENBROCK, *Multivariable and State-Space Theory*, Wiley, New York, 1970.

23. W. W. PETERSON and E. J. WELDON, *Error-Correcting Codes*, M.I.T. Press, Cambridge, Mass., 1971.

24. J. L. MASSEY and M. K. SAIN, "Codes, Automata and Continuous Systems: Explicit Interconnections," *IEEE Trans. Autom. Control*, vol. AC-12, pp. 644–650, 1970.

25. C. M. RADER and J. H. MCCLELLAN, *Number Theory in Digital Signal Processing*, Prentice-Hall, Englewood Cliffs, N.J., 1979.

26. W. WATKINS, "Polynomial Identities for Matrices," *Am. Math. Mon.*, vol. 82, pp. 364–368, April 1975.

27. D. Q. MAYNE, "An Elementary Derivation of Rosenbrock's Minimal Realization Algorithm," *IEEE Trans. Autom. Control*, vol. AC-18, pp. 306–307, 1973.

28. K. C. DALY, "The Computation of Luenberger Canonical Forms Using Elementary Similarity Transformations", *Int. J. Syst. Sci.*, vol. 7, pp. 1–15, January 1976.

29. V. M. POPOV, *Hyperstability of Control Systems*, Springer, Berlin, 1973, (trans. of Romanian ed., 1966).

30. V. BELEVITCH, *Classical Network Theory*, Holden-Day, San Francisco, 1968.

31. M. L. J. HAUTUS, "Controllability and Observability Conditions of Linear Autonomous Systems," *Ned. Akad. Wetenschappen, Proc. Ser. A*, vol. 72, pp. 443–448, 1969.

32. C. D. JOHNSON, "Invariant Hyperplanes for Linear Dynamical Systems," *IEEE Trans. Autom. Control*, vol. AC-11, pp. 113–116, January 1966.

33. D. A. FORD and C. D. JOHNSON, "Invariant Subspaces and the Controllability and Observability of Linear Dynamical Systems," *SIAM J. Control*, vol. 6, no. 4, pp. 553–558, 1968.

34. J. J. SYLVESTER, "A Method of Determining by Mere Inspection the Derivatives from Two Equations of Any Degree," *Philos. Mag.*, vol. XVI, pp. 132–135, 1840; also *Collected Math. Pap.*, vol. 1, pp. 54–57.

35. E. BEZOUT, "Recherches sur le degree des équations résultantes de l'evanouissement des inconnues, et sur les moyens qu'il convient d'employer pour trouver ces équations," *Hist. l'Acad. Roy. Sci., Paris*, pp. 288–338, 1764.

36. C. C. MACDUFFEE, "Some Applications of Matrices in the Theory of Equations," *Am. Math. Mon.*, vol. 57, pp. 154–161, 1930.

37. D. KNUTH, *The Art of Computer Programming, Vol. II: Seminumerical Algorithms*, Addison-Wesley, Reading, Mass., 1969.

38. A. V. AHO, J. E. HOPCROFT and J. D. ULLMAN, *The Design and Analysis of Computer Algorithms*, Addison-Wesley, Reading, Mass., 1974.

39. E. I. JURY, *Inners and Stability of Dynamic Systems*, Wiley, New York, 1974.

40. E. I. JURY, "The Theory and Application of the Inners," *Proc. IEEE*, vol. 63, pp. 1044–1068, 1975.

41. R. E. KALMAN, "Some Computational Problems and Methods Related to Invarient Factors and Control Theory," in *Computational Problems of Abstract Algebra* (J. Leech, ed.), Pergamon, Elmsford, N.Y., 1970, pp. 393–398.

42. B. D. O. ANDERSON, "On the Computation of the Cauchy Index," *Quarterly Applied Math.*, vol. 29, pp. 577–582, January 1972.

43. A. S. HOUSEHOLDER, "Bézoutians, Elimination and Localization," *SIAM Rev.*, vol. 12, pp. 73–78, 1970.

44. S. BARNETT, "Matrices, Polynomials and Linear Time-Invariant Systems," *IEEE Trans. Autom. Control*, vol. AC-18, pp. 1–10, Feb. 1973.

45. S. BARNETT, "Some Topics in Algebraic Systems Theory: A Survey," in *Recent Mathematical Developments in Control* (D. A. Bell, ed.), Academic Press, New York, 1973, pp. 323–344.

46. V. M. POPOV, "La Bézoutienne dans la theorie de la stabilité," *Rev. d'électrotechnique d'énergétique*, vol. 3, pp. 95–110, 1958.

47. S. BARNETT, and D. D. SILJAK, "Routh's Algorithm, A Centennial Survey," *SIAM Review*, vol. 19, pp. 472–489, April 1977.

48. E. I. JURY, "Stability of Multidimensional Scalar and Matrix Polynomials," *Proc. IEEE*, vol. 66, no. 9, pp. 1018–1047, September 1978.

49. D. C. YOULA and P. TISSI, "N-Port Synthesis Via Reactance Extraction—Pt. I," *IEEE Int. Conv. Rec.*, vol. 14, pt. 7, pp. 183–208, 1966.

50. L. A. ZADEH and C. A. DESOER, *Linear System Theory—A State-Space Approach*, McGraw-Hill, New York, 1963.

51. P. C. PARKS, "Analytical Methods for Investigating Stability—Linear and Nonlinear Systems: A Survey," *Proc. Inst. Mech. Eng.*, vol. 178, pt. 3M, pp. 23–35, 1964.

52. R. E. KALMAN and J. E. BERTRAM, "Control System Analysis and Design via the Second Method of Lyapunov," *Trans. ASME J. Basic Eng.*, vol. 82, pp. 371–392, 1960.

53. J. P. LASALLE and S. LEFSCHETZ, *Stability by Lyapunov's Direct Method with Applications*, Academic Press, New York, 1961.

54. S. LEHNIGK, *Stability Theorems for Linear Motions*, Prentice-Hall, Englewood Cliffs, N.J., 1966.

55. C. A. DESOER and M. VIDYASAGAR, *Feedback Systems: Input-Output Properties*, Academic Press, New York, 1975.

56. A. G. J. MACFARLANE, ed., *Frequency-Response Methods in Automatic Control*, IEEE Press, New York, 1979.

57. H. M. JAMES, N. B. NICHOLS, and R. S. PHILLIPS, *Theory of Servomechanisms*, M.I.T. Radiation Lab. Ser., Vol. 25, McGraw-Hill, New York, 1974.

58. R. E. KALMAN, "Lyapunov Functions for the Problem of Lur'e in Automatic Control," *Proc. Nat. Acad. Sci.*, vol. 49, pp. 201–205, February 1963.

59. V. M. POPOV, "Hyperstability and Optimality of Automatic Systems with Several Control Functions," *Rev. Roum. Sci. Tech.*, vol. 9, pp. 629–690, 1964.

60. R. E. BELLMAN, *Stability Theory of Differential Equations*, Dover, New York, 1953.

LINEAR
STATE-VARIABLE FEEDBACK

3

3.0 INTRODUCTION

One of the first applications of state-space methods to linear systems was that of using feedback of the state variables to relocate the eigenvalues of a given system. J. Bertram in 1959 was perhaps the first to realize (according to Kalman et al. [1, p. 49]) that if a given system realization was state controllable, then any desired characteristic polynomial could be obtained by state-variable feedback. In 1962 Rosenbrock [2] discussed, among other things, the use of state feedback to relocate the eigenvalues of a plant so as to achieve better response characteristics, but a complete analysis was not pursued. Bertram apparently obtained his result by the use of root-locus techniques. A direct analytical attack on the problem is quite feasible, and Rissanen was the first to publish, in a little-known 1960 paper [3], a statement and complete proof of this result. The result was independently deduced in almost the same way by Popov in 1964 [4], who in fact treated the multiple-input problem. Both these works were stimulated by Kalman's fundamental paper [5].

In Sec. 3.1, we shall give some motivation for the eigenvalue-shifting problem (often also called the pole-shifting problem, because for minimal systems the poles and eigenvalues coincide) and its relation to earlier attempts at system stabilization. In Sec. 3.2, we shall show how to calculate the gain vector for the appropriate state feedback in several different ways. We shall also briefly explore some other aspects of state-feedback solutions, including the fact that they cannot relocate the zeros, and also some applications to regulator problems with nonzero set points and constant input disturbances. In Sec. 3.3, we shall present several worked examples.

A question that is always raised is how one can translate the many, partly subjective, criteria of "desirable" system response and behavior into the specification of a desired set of closed-loop poles. In Sec. 3.4 we shall study a problem where a definite and tractable *quadratic performance criterion* can be imposed, so that some insight can be gained into a possibly good (*optimal* under the above criterion) set of poles. We shall emphasize the important role of controllability *and* observability in various aspects of this problem.

In Sec. 3.5, we shall examine pole shifting for discrete-time systems and present an interesting connection between pole shifting and the problem of taking an arbitrary state to the origin. This connection brings up the so-called Principle of Optimality and the technique of dynamic programming, which we shall use later to derive the standard Riccati equation solution to the discrete-time quadratic regulator problem. In Sec. 3.5.4, we shall present alternative standard and fast square-root algorithms, which promise several computational benefits in this and several other linear system calculations.

3.1   ANALYSIS OF STABILIZATION BY OUTPUT FEEDBACK

We shall begin with an analysis of some feedback compensation schemes for modifying the transfer function of a given system and shall try to give the reader an idea of the situation in the late 1950s that set the stage for state-space methods.

In Sec. 2.0 we discussed the stabilization of an unstable system with the transfer function $1/(s-1)$ by the use of a series compensator with the transfer function $(s-1)/(s+1)$.

However, we saw that this was not satisfactory because the cancellation in the transfer function did not mean that the unstable natural frequency had disappeared from the overall realization. This emphasized the distinction between internal and external descriptions, which we pursued at some length in Chapter 2.

When we look more closely at the problem in Sec. 2.0, we can see that some kind of feedback compensation will be necessary. For, from Fig. 2.0-2, we may see that a small nonzero value of $x_2(0)$ will give rise to an exponentially growing or exponentially decreasing value of $x_2(\cdot)$ according to whether $x_2(0)$ is positive or negative. However, without feedback, there is no way of knowing at the input whether $x_2(\cdot)$ is growing or falling, and therefore there is no way of introducing a control $v(\cdot)$ to compensate for this. Therefore the use of feedback is almost unavoidable in most control problems.

In the present problem, it is reasonably obvious that simple output feedback, as shown in Fig. 3.1-1, will be satisfactory. The state equations for the system of Fig. 3.1-1 are

$$\dot{x} = x - kx + v, \qquad y = x$$

so that the new transfer function is

$$H(s) = \frac{1/(s-1)}{1 + k/(s-1)} = \frac{1}{s+k-1}$$

Clearly, by choosing the feedback gain suitably, we can easily stabilize the system, *and there will be no unstable hidden modes,* since the realization of the overall transfer function is obviously minimal.

This is nice, but of course we must ask how the method works for more complicated systems. Thus, consider a system with transfer function $1/(s^2 - s)$, which has poles at 1 and 0 and is therefore unstable. With output feedback, it is easy to see that the characteristic polynomial will be

$$a_k(s) = s^2 - s + k$$

whose roots are at

$$s = \frac{1 \pm \sqrt{1 - 4k}}{2}$$

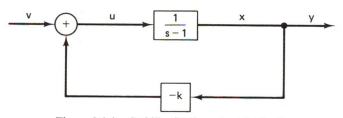

Figure 3.1-1.   Stabilization by output feedback.

It is clear that no choice of k will make the real parts of the roots negative so that output feedback will not stabilize this system.

This is a good place to mention the so-called *root-locus technique,* which shows the dependence on a parameter (k in our case) of the roots of a given polynomial [$a_k(s)$ in our case]. The root loci for our problem can be easily determined and are as in Fig. 3.1-2. The figure shows clearly that output feedback is not sufficient for stabilization. (Here, for simplicity, the loci are drawn only for positive k. The reader should add the loci for $k < 0$.) The result that arbitrary pole relocation is impossible by varying k is also clear analytically, and any reader unfamiliar with root-locus techniques can use direct methods. [We may note that the goal of this section is to provide some

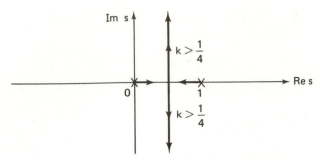

Figure 3.1-2.   Root loci for $s^2 - s + k$.

motivation and therefore it is not necessary to completely appreciate every claim or statement. The root-locus display was invented by Walter Evans in 1948, who also presented a set of simple rules for rapidly sketching root loci in various situations. Good descriptions of the technique can be found in almost all books on (classical) control theory—we mention here the treatments by Cannon [6] and Ogata [7]. Root loci are a valuable aid in system design, especially in showing stability *margins* and in suggesting how poles and zeros (viz., a compensator) can be added to improve the performance of a given system. Readers interested in control theory will probably already have some familiarity with root-locus techniques here. However, such knowledge is not important for understanding the material of this chapter, and all discussions referring to them may be omitted.]

Returning to our specific problem, if output feedback fails, we might try feeding back more information—say, not just $y(\cdot)$ (the position) but also $\dot{y}(\cdot)$ (the velocity). That is, we measure $\dot{y}(\cdot)$ suitably (e.g., by a velocity sensor) and feed back a linear combination, $-[k_1 y(\cdot) + k_2 \dot{y}(\cdot)]$, as shown in Fig. 3.1-3. The new characteristic polynomial is

$$a_k(s) = s^2 - s + k_1 + k_2 s$$

$$= s^2 - s + k_2(s + \delta), \qquad \delta = \frac{k_1}{k_2}$$

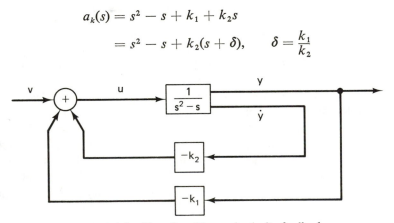

Figure 3.1-3.   Use of position and velocity feedback.

We now have an extra degree of freedom with the parameter δ, and the root-locus plot of Fig. 3.1-4 shows clearly that the system can now readily be stabilized. Furthermore, arbitrary relocation of poles can be achieved by suitable choice of k_1 and k_2. This is also easy to see analytically because $a_k(s) = s^2 + (k_2 - 1)s + k_1$ can be made into an *arbitrary* second-order polynomial by suitable choice of k_1 and k_2. Moreover, it is easy to see that the same statements hold for any second-order system of the form $H(s) = 1/(s^2 + a_1 s + a_2)$.

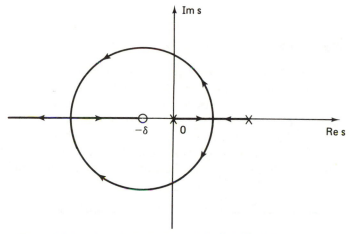

Figure 3.1-4. Root loci of $s^2 - s + k_2(s + \delta)$, δ fixed and k_2 variable.

This is a nice result, but how general is it? Thus, now consider a system with the transfer function

$$H(s) = \frac{(s - 1)(s - 3)}{s(s - 2)(s - 4)}$$

With output and derivative feedback, the new characteristic polynomial is

$$a_k(s) = s(s - 2)(s - 4) + k_2(s - 1)(s - 3)(s + \delta), \qquad \delta = \frac{k_1}{k_2}$$

The root loci for a particular choice of δ are as in Fig. 3.1-5, and we see that stabilization, not to speak of arbitrary pole location, cannot be achieved. It is easy to see that the same is true for *any* choice of δ and k_2 (i.e., k_1 and k_2). This is not unexpected, since we have a third-order system and we are using only two parameters. We might try to introduce more parameters by also using acceleration feedback, but such differentiation of the output y (to obtain $\ddot{y}$) is usually not realistic. Also, in this case, the denominator degree

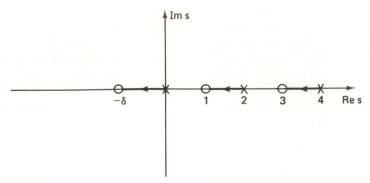

Figure 3.1-5. Root loci of $s(s-2)(s-4) + k_2(s-1)(s-3)(s+\delta)$.

will increase from 3 to 4. Therefore we have to try other methods. Classical ways of introducing more parameters are to use the configurations shown in Figs. 3.1-6 and 3.1-7 (see, e.g., Truxal [8] and Schultz and Melsa [9]). Notice that in these problems the new transfer function will have denominator of degree 5 rather than 3. Clearly, more complicated techniques have to be sought to achieve stabilization while keeping the overall denominator polynomial of degree 3. But it is not clear how to proceed, though with hindsight

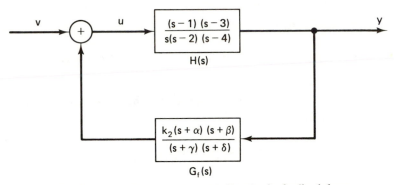

Figure 3.1-6. Compensation with filter in the feedback loop.

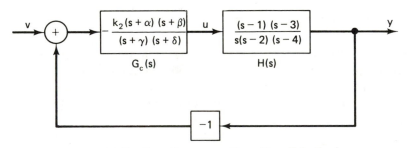

Figure 3.1-7. Cascade compensation with unit feedback.

(as we shall discuss in Sec. 4.5.3) a direct solution is not hard, at least for single-input, single-output systems. We should stress that part of the difficulty is one of formulation. What are the proper requirements? Why should we need to have the new transfer function of the same degree as the old one? Different configurations may have different numbers of internal states: Should one prefer one to the other? And so on. The following examples show some of the difficulties.

Example 3.1-1.

Given a transfer function $H(s) = 1/s(s - 2)$, we wish to choose $G_c(s)$ in a unity-feedback configuration (as in Fig. 3.1-7) so that the overall transfer function has the form

$$H_o(s) = \frac{\beta(s)}{(s+1)^2}, \qquad \deg \beta(s) \le 2$$

In other words, we want the new poles to be at $s = -1$, but no requirement is put on the zeros.

Solution.   Let $G_c(s) = c(s)/d(s)$, with $\deg c(s) \le \deg d(s)$. Then the overall transfer function is readily seen to be

$$H_o(s) = \frac{c(s)}{s(s - 2)d(s) + c(s)} = \frac{\beta(s)}{(s + 1)^2}$$

where the second equality is to be achieved by proper choice of $c(s)$ and $d(s)$.

Now some algebra shows that

$$G_c(s) = \frac{c(s)}{d(s)} = \frac{s(s - 2)\beta(s)}{s^2(1 - \beta_0) + s(2 - \beta_1) + 1 - \beta_2}$$

where $\beta(s) = \beta_0 s^2 + \beta_1 s + \beta_2$. We see that to keep $G_c(s)$ proper, we must have $\beta_0 = 0 = \beta_1$, and then

$$G_c(s) = \frac{\beta_2 s(s - 2)}{s^2 + 2s + 1 - \beta_2}$$

Several different compensators $G_c(s)$ can be obtained by taking different values of β_2. For example,

$$\beta_2 = 9 \Longrightarrow G_c(s) = \frac{9s}{s + 4}$$

while

$$\beta_2 = 4 \Longrightarrow G_c(s) = \frac{4s(s - 2)}{s^2 + 2s - 3}$$

which is not stable. However, as noted earlier, the possible stability or instability of $G_c(s)$ is not the issue, because all the $G_c(s)$ will be unsatisfactory. The point is that in all cases we shall be cancelling the *unstable* natural frequencies s or $s - 2$ or both, depending on which $G_c(s)$ we choose. This is clear from the *cancellations* in

$$H_o(s) = \frac{G_c(s)H(s)}{1 + G_c(s)H(s)} = \frac{\beta_2 s(s-2)}{s(s-2)(s^2 + 2s + 1 - \beta_2) + \beta_2 s(s-2)}$$

Therefore the overall realization will always be internally unstable, and this seems to be a consequence of our requirement that the new denominator $(s+1)^2$ have the same degree as the old one, $s(s-2)$. At this point it is not quite clear what we might do, and our doubts might be compounded by the next example.

Example 3.1-2.

1. Given a system with the transfer function $H(s) = (s-1)/(s^2 + 0.5s)$, find a compensator $G_c(s) = c_0$ in a unity-feedback configuration as in Example 3.1-1 so that the overall system has one pole at $-\frac{2}{3}$, with the other pole being free. Show that the solution is not satisfactory.

2. Try to achieve a better result by using a compensator $G_c(s) = (c_0 s + c_1)/(s + d_1)$ to ensure that all the modes of the overall system (including the ones that might be *hidden*) are stable. However, still keep one pole at $-\frac{2}{3}$.

Solution.   1. With $G_c(s) = c_0$,

$$H_o(s) = \frac{c_0(s-1)}{s(s+0.5) + c_0(s-1)}$$

so that to have a pole at $s = -\frac{2}{3}$, we must have

$$0 = -\frac{2}{3}(-\frac{2}{3} + \frac{1}{2}) + c_0(-\frac{2}{3} - 1)$$

or $c_0 = +\frac{1}{15}$. Noting that $-c_0$ is also the product of the roots, we see that the other root is at $s = \frac{1}{10}$, and therefore $H_o(s)$ is unstable.

2. With the more complicated compensator

$$H_o(s) = \frac{(c_0 s + c_1)(s-1)}{s(s+0.5)(s+d_1) + (c_0 s + c_1)(s-1)}$$

and to have a second-order transfer function, clearly $c_0 s + c_1$ must divide $s + 0.5$. It cannot divide s since this is an unstable mode. If it divided $s + d_1$, we would be back to the solution in part 1. Note also that $s + d_1$ cannot be used to cancel $s - 1$ since this would again give us a cancelled or hidden mode that was unstable. Therefore with $c_0 s + c_1 = c(s + 0.5)$,

$$H_o(s) = \frac{c(s-1)}{s(s+d_1) + c(s-1)}$$

We must now have c negative in order for the poles to be stable. For example, $c = -\frac{4}{3}$, $d_1 = 4$ gives us the internally and externally stable system

$$H_o(s) = \frac{-\frac{4}{3}(s-1)}{(s + \frac{2}{3})(s+2)}$$

with one pole at $s = -\frac{2}{3}$. However, the inversion of the input (due to negative c) may be undesirable.

At this stage it is not quite clear what exactly can and cannot be done (see also Exercise 3.1-4).

It was at some such point of vague confusion that, motivated by Kalman's work [5] on state-variable descriptions of linear systems, Rissanen [3] argued that instead of feeding back $y(\cdot)$ and its derivatives [or feeding back $y(\cdot)$ through a linear *compensating* network], the proper thing to feed back was the *state* $x(\cdot)$ of a realization of the system, since, after all, the *state* summarizes all the "current information" about the system.† Therefore, anything we can do with $y(\cdot)$ or $\dot{y}(\cdot)$, etc., we must also be able to do with the states, and, more important, anything we cannot do with the states probably cannot be done in any other general way. Rissanen showed that *state feedback* could be used to modify at will the natural frequencies of the system and, in particular, to make them all stable, provided only that the realization used to define the states of the system is state controllable in the sense defined in Sec. 2.3.2., i.e., the matrix $\mathcal{C} = [b \; Ab \; \cdots \; A^{n-1} \; b]$ of the realization $\{A, b, c\}$ is nonsingular.

This is a striking result and a good justification of the importance of the concept of *state*. However, as stressed by Rissanen, the usefulness of this result is dependent on our ability to obtain the states, and a full consideration of this point will return us to the earlier situation (cf. Sec. 4.2 and 4.5). However, partly for clarity of discussion and also for historical and pedagogical reasons, we shall treat the two problems—of determining the states and feeding back the states—separately for a while. For the moment we assume that by some means the states can be made available. [Notice that in order to obtain the states, we need to know (cf. Sec. 2.3.1) not only the output $y(\cdot)$ and its derivatives but also the input $u(\cdot)$ (and, in general, we need $\{y(\cdot), \dot{y}(\cdot), \ldots, y^{(n-1)}(\cdot), u(\cdot), \ldots, u^{(n-2)}(\cdot)\}$). Therefore, we would expect *that a suitable feedback configuration should involve both the output and some of its derivatives and the input and some of its derivatives*. This is a crucial insight provided by the state-variable solution, and with this insight we can directly obtain compensator configurations that will allow arbitrary pole relocation while still guaranteeing internal stability of the overall configuration and keeping the overall transfer function of the same (denominator) degree as the original. If we are only interested in arbitrary poles and internal stability, without the degree restriction on the new transfer function, then it turns out that direct use of the input is not necessary. These facts will be established in Sec. 4.5.2, after we have built up some insights via the state-variable approach.]

---

†It should also be noted that the work of R. Bellman had shown in the early 1950s [10], [11] that state-variable feedback (not necessarily linear) was an *optimal* solution for many problems.

Exercises

3.1-1.

Given the configuration of Fig. 3.1-7, with

$$H(s) = \frac{s - 0.5}{s(s - 1)}, \qquad G_c(s) = -\frac{k_2(s + \delta)}{s + \gamma}$$

find k_2, δ, γ so that the overall system will have all its poles at $s = -1$.

3.1-2.

Given the configuration of Fig. 3.1-7, with a general second-order $H(s)$, $H(s) = (s + b_1)/(s^2 + a_1 s + a_2)$, we wish to find a series compensator $G_c(s) = (c_1 s + c_2)/(s + d_1)$ so that the overall system will have a specified denominator, $\alpha(s) = s^3 + \alpha_1 s^2 + \alpha_2 s + \alpha_3$. Show that (c_1, c_2, d_1) can be found by solving the equation

$$\begin{bmatrix} \alpha_1 \\ \alpha_2 \\ \alpha_3 \end{bmatrix} = \begin{bmatrix} 1 & 0 & a_1 & 1 \\ b_1 & 1 & a_2 & a_1 \\ 0 & b_1 & 0 & a_2 \end{bmatrix} \begin{bmatrix} c_1 \\ c_2 \\ 1 \\ d_1 \end{bmatrix}$$

When will this equation have a solution for all $\{\alpha_i\}$? Write down state equations for the overall system, and find the characteristic polynomial, $\det (sI - A)$.

3.1-3.

The linearized equations of a free pendulum are $\ddot{\theta} + \omega_0^2 \theta = u$. Show that output feedback $-k\theta$ will not stabilize the system but that this can be done by using a system $(s + \alpha)/(s + \beta)$ in the feedback path, provided $-\beta\omega_0^2 < \alpha < \beta, \beta > 0$.

3.1-4.

a. Refer to Example 3.1-1, and suppose that in some other design calculation we have computed a block $G_e(s) = (13s + 1)/(s + 1)$ and that we now wish to find $G_c(s)$ as in the figure so that the overall transfer function is

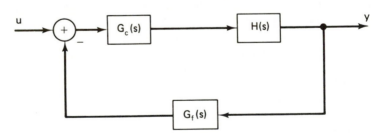

$\beta(s)/(s + 1)^2$ and the overall system is stable. Show how to do this, and compare with the results of Example 3.1-1.

b. Refer to Example 3.1-2, and repeat the problem for $H(s) = (s - 1)/(s^2 - 0.5s)$. Compare with the results of Examples 3.1-1 and 3.1-2.

3.2 STATE-VARIABLE FEEDBACK AND MODAL CONTROLLABILITY

Motivated by the discussions in Sec. 3.1.1, we shall consider the following problem. We are given a realization

$$\dot{x}(t) = Ax(t) + bu(t), \qquad y(t) = cx(t) \tag{1}$$

with characteristic polynomial

$$a(s) = \det(sI - A) = s^n + a_1 s^{n-1} + \cdots + a_n \tag{2}$$

If a transfer function is specified, say,

$$H(s) = \frac{b(s)}{a(s)} = \frac{b_1 s^{n-1} + \cdots + b_n}{s^n + a_1 s^{n-1} + \cdots + a_n} \tag{3}$$

we shall assume that $\{A, b, c\}$ is some n-state realization of this transfer function.

We wish to modify the given system by the use of state-variable feedback so as to obtain a new system with specified eigenvalues or, equivalently, a specified characteristic polynomial, say,

$$\alpha(s) = s^n + \alpha_1 s^{n-1} + \cdots + \alpha_n \tag{4}$$

Now state-variable feedback is obtained (cf. Fig. 3.2-1) by the substitution

$$u(\cdot) \longrightarrow v(\cdot) - kx(\cdot), \qquad k = [k_1 \quad \cdots \quad k_n] \tag{5}$$

where

$$v(\cdot) = \text{the new external input}$$

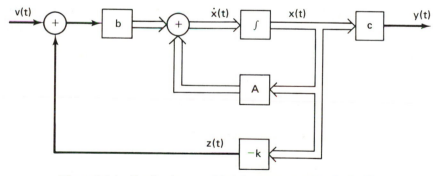

Figure 3.2-1. Realization modified by state-variable feedback. The reader may profit by drawing out a complete diagram for a two- or three-state realization $\{A, b, c\}$. The double lines are used for vector quantities in this very symbolic representation.

197

The use of $-kx(\cdot)$ rather than $kx(\cdot)$ is purely conventional—feedback is usually *negative*. In any case, after feedback we have the realization

$$\dot{x}(t) = (A - bk)x(t) + bv(t), \qquad y(t) = cx(t) \tag{6}$$

which has the characteristic polynomial $a_k(s) = \det(sI - A + bk)$. Our task is to choose k so that $a_k(s) = \alpha(s)$. There are many ways of determining a suitable k, and we shall present some of the most insightful ones in Sec. 3.2.1. In Sec. 3.2.2, we shall describe how, using hindsight, these *state-space* formulas can also be obtained by an appropriate transfer function analysis.

3.2.1   Some Formulas for the Feedback Gain

Perhaps the most direct method is that of Bass and Gura [12], which uses the determinant identity of Exercise A-12 to rewrite $a_k(s)$ as

$$\begin{aligned}
a_k(s) &= \det(sI - A + bk) \\
&= \det\{(sI - A)[I + (sI - A)^{-1}bk]\} \\
&= \det(sI - A)\det[I + (sI - A)^{-1}bk] \\
&= a(s)[1 + k(sI - A)^{-1}b]
\end{aligned} \tag{7}$$

Therefore

$$a_k(s) - a(s) = a(s)k(sI - A)^{-1}b \tag{8}$$

Now, both sides are polynomials in s, and hence k can be found by equating the coefficients of the powers of s on both sides. To do this we recall the resolvent formula (Exercise A-23),

$$(sI - A)^{-1} = \frac{1}{a(s)}[s^{n-1}I + s^{n-2}(A + a_1I) + s^{n-3}(A^2 + a_1A + a_2I) + \cdots] \tag{9}$$

which in (8) yields the equations

$$\alpha_1 - a_1 = kb, \qquad \alpha_2 - a_2 = kAb + a_1kb,$$
$$\alpha_3 - a_3 = kA^2b + a_1kAb + a_2kb$$

and so on, which can be collected in matrix form as

$$\alpha - a = k\mathcal{C}\alpha'_- \tag{10}$$

where we have written

$$\alpha = [\alpha_1 \quad \alpha_2 \quad \cdots \quad \alpha_n], \qquad a = [a_1 \quad a_2 \quad \cdots \quad a_n]$$
$$\mathcal{C} = [b \quad Ab \quad \cdots \quad A^{n-1}b]$$

and

$$\mathcal{C}_- = \text{a lower triangular Toeplitz matrix with}$$
$$\text{first column } [1 \quad a_1 \quad \cdots \quad a_{n-1}]' \tag{11}$$

Since the matrix $\mathcal{C}_-$ is always nonsingular, we can solve (10) for k for arbitrary α and a if and only if $\mathcal{C}$ is nonsingular. We have thus proved that

state feedback can provide arbitrary relocation of
eigenvalues of a realization $\{A, b, c\}$ if and only
if $\mathcal{C} = [b \quad Ab \quad \cdots \quad A^{n-1}b]$ is nonsingular $\tag{12}$

Moreover, the required feedback gain can be calculated as

$$k = (\alpha - a)\mathcal{C}_-^{-T}\mathcal{C}^{-1}, \text{ where } \mathcal{C}_-^{-T} \triangleq [\mathcal{C}_-^{-1}]'. \tag{13}$$

We shall call this the *Bass-Gura formula* for k.

The ability to arbitrarily relocate eigenvalues by state-variable feedback has been called *modal controllability* (first apparently by Simon and Mitter [13]). It is rather striking that the condition for modal controllability is the same as the condition for *pointwise state controllability*, which is the ability to move the state from a given value to an arbitrary value by suitable choice of input (Sec. 2.3.2). A connection between the two concepts will be explored in Sec. 3.5.2.

We see from (13) that large changes in the coefficients of $a(s)$ will generally need larger values of the feedback gain vector k, with consequent difficulties in implementation and operation (we might move away from the operating point into a nonlinear region, have transient distortion, cause noise enhancement, etc.). We see also that large k can also arise if $\mathcal{C}$ is nearly singular. These factors have to be kept in mind in applications, but in the rest of this section we shall concentrate on some further analytical exploration of formula (13) and its significance.

Significance of the Controller Canonical Form.   Given the explicit formula for k, it is natural to ask about realizations in which k is most easily obtained. For state controllability we found in Sec. 2.3.2 that the controllability canonical form (for which $\mathcal{C} = I$) was the most convenient. Here, too, having $\mathcal{C} = I$ would certainly be useful, but it is possible to find a simpler form. Namely, recall (or verify) that the special matrix $\mathcal{C}_-'$ can be regarded as the inverse of the controllability matrix $\mathcal{C}_c$ of the controller canonical form $\{A_c, b_c, c_c\}$, where A_c is a companion matrix with the $\{-a_i\}$ in the top row and b_c has unity for the first entry and zeros elsewhere. Clearly, if we start in controller form,

$$k_c = (\alpha - a)\mathcal{C}_-^{-T}\mathcal{C}_c^{-1} = \alpha - a \tag{14}$$

This simple result can also be directly obtained by going back to formula (8) and using the fact that for the controller realization (cf. Exercise A-33)

$$(sI - A_c)^{-1}b_c = \frac{[s^{n-1} \quad s^{n-2} \quad \cdots \quad s \quad 1]'}{a(s)} \tag{15}$$

Another derivation, which serves at the same time as a mnemonic for recalling the form of the controller realization, is to note that because of the special form of b_c,

$$A_c - b_c k_c = \begin{bmatrix} -(a_1 + k_{c1}) & \cdots & -(a_n + k_{cn}) \\ 1 & & 0 \\ \cdot & \cdot & \cdot \\ \cdot & \cdot & \cdot \\ \cdot & \cdot & \cdot \\ 0 & 1 & 0 \end{bmatrix} \tag{16}$$

This is still in controller form, and therefore

$$\det (sI - A_c + b_c k_c) = s^n + (a_1 + k_{c1})s^{n-1} + \cdots + (a_n + k_{cn}) \tag{17}$$

which immediately yields (14). The reader should check that this result follows even more graphically if one draws out the feedback loops for the controller-form block diagram of Fig. 2.1-4.

This simple solution, of course, suggests that a way of solving the problem for a general realization $\{A, b, c\}$ is [3] to first convert it to controller form by means of a suitable change of variables

$$x(t) = Tx_c(t), \qquad \det T \neq 0 \tag{18}$$

Then since

$$A = TA_c T^{-1}, \qquad b = Tb_c \tag{19}$$

we have

$$a_k(s) = \det (sI - A + bk) = \det (sI - A_c + b_c kT) \tag{20}$$

and

$$kT = k_c \tag{21}$$

Now $k_c = \alpha - a$, and we know (cf. Example 2.3-4) that T is given by

$$T = \mathscr{C}\mathscr{C}_c^{-1} = \mathscr{C}\alpha'_- \tag{22}$$

so that we again recover (13).

Other Formulas for the Feedback Gain.   Equation (13) gives a formula for k in terms of the coefficients of the old and the new characteristic polynomials. This can be written in several other ways, each of which has its own usefulness.

Ackermann's Formula.   Ackermann [14] noted that

$$k = q'_n \alpha(A) \tag{23a}$$

where $\alpha(s)$ is the desired characteristic polynomial and

$$q'_n = [0 \quad \cdots \quad 0 \quad 1] \mathcal{C}^{-1} = \text{the last row of } \mathcal{C}^{-1} \tag{23b}$$

Therefore explicit knowledge of the original polynomial $a(s) = \det(sI - A)$ is not required; this is sometimes convenient for theoretical analyses (see, e.g., Sec. 3.5.2), though from a numerical point of view the work required to determine $\alpha(A)$ is not much different from that required to compute $a(s)$ by the Leverrier-Souriau-Faddeev method (cf. Exercise A-23).

To show that the above formula is correct, we can derive it from the previous formula (13). Now, as we have already seen, it should be simpler to first assume that we are in controller form, so that what we have to show is that

$$k_c = \alpha - a = q'_{c,n} \alpha(A_c), \qquad \text{the last row of } \alpha(A_c) \tag{24}$$

Now

$$\alpha(A_c) = A_c^n + \alpha_1 A_c^{n-1} + \cdots + \alpha_n I$$

and, by the Cayley-Hamilton theorem,

$$A_c^n = -\sum_1^n a_i A_c^{n-i}$$

where the $\{a_i\}$ are the coefficients of the characteristic polynomial of A_c. Therefore

$$\alpha(A_c) = \sum_1^n (\alpha_i - a_i) A_c^{n-i} \tag{25}$$

Finally, by the shifting property of companion matrices (Exercise A-33), as reflected, for example, in the fact that

$$q'_{c,n} A_c = [0 \quad \cdots \quad 1 \quad 0]$$

it is easy to see that (24) holds. We may note that Ackermann's original proof was somewhat different [14].

Feedback Gains in Terms of Eigenvalues—the Mayne-Murdoch Formula.
We have given our formulas for the feedback gain in terms of the characteristic polynomials, though in many problems it is actually the eigenvalues that
may be specified. Of course, the two specifications are theoretically equivalent,
but for numerical purposes it will generally be preferable to work directly
from the specified eigenvalues, at least when they are distinct.

Suppose, then, that A is diagonal with eigenvalues $\{\lambda_1, \ldots, \lambda_n\}$ and that
the desired roots are at $\{\mu_1, \ldots, \mu_n\}$. To find the feedback gain, it is perhaps
simplest to return to the early formula (8),

$$a_k(s) - a(s) = a(s)k(sI - A)^{-1}b$$

and rewrite it as

$$\frac{a_k(s)}{a(s)} = 1 + \sum_1^n \frac{k_i b_i}{s - \lambda_i} \tag{26}$$

This shows that one rule is to make a partial fraction expansion of
$a_k(s)/a(s)$ and divide the coefficient of the term $(s - \lambda_i)^{-1}$ by b_i. More explicitly, multiply both sides by $s - \lambda_i$ and then set $s = \lambda_i$ to get

$$k_i b_i = \frac{\prod_j (\lambda_i - \mu_j)}{\prod_{j \neq i} (\lambda_i - \lambda_j)} \tag{27}$$

This formula was apparently first given by Mayne and Murdoch [15]. Its
most interesting consequence is that it shows clearly that the feedback gain
increases as we increase the separation $|\lambda_i - \mu_j|$ between the open- and
closed-loop poles. Large shifts in the eigenvalues will be reflected in large
shifts in the coefficients of the characteristic polynomial and thus also show
up in our earlier formula (13). Explicit formulas for the case where A is not
diagonalizable can also be obtained [16], though they are of less significance.

3.2.2   A Transfer Function Approach

Our use of (26) in deriving the Mayne-Murdoch formula suggests that
it might be useful to reexamine all the formulas for k from a more classical
transfer function perspective once the basic state-feedback configuration of
Fig. 3.2-1 has been assumed.

Thus let us define a *fictitious* output

$$z(t) = kx(t)$$

and referring to Fig. 3.2-1, note that

$$\frac{Z(s)}{U(s)} = k(sI - A)^{-1}b = \frac{g(s)}{a(s)}, \text{ say}$$

As a result of simple unity feedback, this is changed to

$$\frac{Z(s)}{V(s)} = \frac{g(s)}{a(s) + g(s)}$$

Then the new transfer function is

$$\begin{aligned}
\frac{Y(s)}{V(s)} &= \frac{Y(s)}{U(s)} \frac{U(s)}{Z(s)} \frac{Z(s)}{V(s)} \\
&= \frac{b(s)}{a(s)} \frac{a(s)}{g(s)} \frac{g(s)}{a(s) + g(s)} \\
&= \frac{b(s)}{a(s) + g(s)}
\end{aligned} \tag{28}$$

Therefore the new characteristic polynomial after state-variable feedback is

$$a(s) + g(s) = \alpha(s)$$

But then the reader can recognize that this is just formula (8), which we obtained in the state-space context by using a determinant identity.

Moreover, the Markov parameters of $g(s)/a(s)$ are determined from

$$\frac{g(s)}{a(s)} = \sum_{1}^{\infty} \frac{kA^{i-1}b}{s^i}$$

so that by comparing coefficients we can readily obtain the Bass-Gura formula (13).

Furthermore, (28) implies that for any realization $\{A, b, c\}$

$$\alpha(A) = a(A) + g(A) = g(A)$$

since $a(A) = 0$ by the Cayley-Hamilton theorem. Now suppose we actually have a controller-form realization. Then (cf. Exercise A-33)

$$\frac{g(s)}{a(s)} = k_c(sI - A_c)^{-1}b_c = \sum_{1}^{n} k_{ci} \frac{s^{n-i}}{a(s)}$$

Also we recall from Example 2.3-5 that

$$g(A_c) = \check{I}\Theta(g, A_c) = \check{I}\Theta(k_c, A_c)$$

Therefore we are led to the Ackermann formula (24), namely

$$\begin{aligned}
k_c &\triangleq [0 \quad \cdots \quad 0 \quad 1]\check{I}\Theta(k_c, A_c) \\
&= [0 \quad \cdots \quad 0 \quad 1]g(A_c) = [0 \quad \cdots \quad 0 \quad 1]\alpha(A_c)
\end{aligned}$$

Finally, for the Mayne-Murdoch formulas (26) and (27), we note that

$$\frac{V(s)}{U(s)} = \frac{\alpha(s)}{a(s)} = 1 + \frac{g(s)}{a(s)} = 1 + \sum_{1}^{n} k_{di} b_{di}(s - \lambda_i)^{-1}$$

for a diagonal realization $\{A_d, b_d\}$ with corresponding feedback gain k_d.

The transfer function analysis is somewhat simpler than the preceding state-space analysis and also leads naturally to the various formulas, especially that of Ackermann, which was introduced without much motivation in Sec. 3.2.1. Such simplicity is often characteristic of transfer function analysis, but we should note that the basic configuration of Fig. 3.2-1 was first introduced by state-space arguments and intuitions. The point is that a combination of state-space and transfer function analysis is very powerful, and we shall see several examples of this later (e.g., Secs. 3.4 and 4.5).

3.2.3   Some Aspects of State-Variable Feedback

The introduction of the notion of state feedback has allowed us to pose and answer a clear question: when and how can we shift the modes of a (given or assumed) realization of a system? However, we cannot make a fair comparison of this result with the apparently inconclusive analyses of Sec. 3.1, first because we have made a new and strong assumption that all the state variables are available for feedback and second because we have not examined all the properties of the state-feedback solution. We shall explore some of these properties in this section, but a fuller comparison with the problems of Sec. 3.1 must be deferred to Secs. 4.2 and 4.5. It should be mentioned that the state-feedback solution has been the object of a great deal of study, as indicated by a glance at the control literature of the mid-1960s to early 1970s. We shall not make a detailed study here, but shall confine ourselves to some simple aspects, accessible without much elaboration or introduction. More comprehensive treatments, which get more into control theory than is the intention in this book, can be found in [17] and [18] and in the current technical literature.

State Feedback and the Zeros of the Transfer Function.   We have shown that by using state feedback we can change the denominator of the transfer function from $a(s)$ to any (monic) polynomial $\alpha(s)$ of the same degree. In Sec. 3.1.1, we proposed to change both numerator and denominator (zeros and poles) of the transfer function, and therefore we should study the effect of state feedback on the zeros.

This can be done in several ways, but one of the simplest† is to assume a controller canonical form $\{A_c, b_c, c_c\}$, where A_c is a companion matrix with top row $-[a_1 \cdots a_n]$, $b'_c = [1 \ 0 \ \cdots \ 0]$, and c_c contains the coefficients of $b(s)$,

---

†The result also follows directly from the transfer function analysis of Sec. 2.3.2—see Eq. (28).

$$c_c = [b_1 \quad b_2 \quad \cdots \quad b_n]$$

Now we see that after state feedback, the realization is $\{A_c - b_c k_c, b_c, c_c\}$, which is still in controller form, so that the transfer function is

$$\frac{Y(s)}{V(s)} = \frac{b(s)}{\alpha(s)}$$

In other words, state feedback does not affect the zeros of the transfer function, unless, of course, they are cancelled by appropriate choice of the new denominator polynomial $\alpha(s)$. This result, which was first noted by Morgan [19], indicates that one of the difficulties with the efforts in Sec. 3.1.1 was the attempt to change both poles and zeros. If we had constrained ourselves to only changing the poles, then things might have been clearer in Sec. 3.1.1, as we shall see when we reconsider such problems in Sec. 4.5.

On the other hand, the gain in definiteness of the state-feedback solution still leaves us with the problem of what to do about zeros in undesirable locations, with consequent limitations on the delay and phase characteristics of the system; this has to be regarded as one of the limitations of state feedback.

Noncontrollable Realizations and Stabilizability; Controllable and Uncontrollable Modes.   It is natural to ask what can be done about noncontrollable systems. Even though the eigenvalues cannot be arbitrarily shifted, can some of them perhaps be shifted arbitrarily, or can they perhaps all be moved with certain limits? Some insight into the problem can be obtained by studying systems with *distinct eigenvalues*. We can assume that such a system has already been put into the diagonal canonical form

$$\dot{x} = \Lambda x + bu, \qquad \Lambda = \text{diag}\,\{\lambda_1, \ldots, \lambda_n\}, \lambda_i \neq \lambda_j$$

We have seen earlier (Sec. 2.4.1) that such a realization is controllable if and only if none of the elements of b is zero. But if, for instance, $b_1 = b_3 = 0$, then the input is decoupled from the corresponding modes λ_1 and λ_3, and clearly no feedback can affect these modes! Therefore, in a diagonal noncontrollable realization certain eigenvalues cannot be shifted, or equivalently, certain modes cannot be affected by state-variable feedback. If these modes are stable, then, in many problems, it may not be important that they cannot be affected. However, if these modes are unstable, then we shall usually have trouble, and our mathematical model of the physical problem must be reexamined. We shall say that a realization is *stabilizable* if all the unstable eigenvalues (i.e., those with nonnegative real parts) can be arbitrarily relocated by state-variable feedback. This is clearly a weaker requirement than complete modal controllability.

A diagonal realization is stabilizable if and only if the $\{b_i\}$ corresponding to the unstable $\{\lambda_i\}$ are nonzero, i.e., if the unstable modes are all controllable. In general, we use the result of Sec. 2.4.2 that a realization $\{A, b\}$ with controllability matrix of rank r can be transformed into a pair $\{\bar{A}, \bar{b}\}$, where

$$\bar{A} = \begin{bmatrix} A_c & A_{12} \\ 0 & A_{\bar{c}} \end{bmatrix}, \qquad \bar{b} = \begin{bmatrix} b_c \\ 0 \end{bmatrix}_{n-r}^{r}$$
$$_{r} _{n-r}$$

Then $\{\bar{A}, \bar{b}\}$ and hence $\{A, b\}$ *will be stabilizable if and only if all the eigenvalues of $A_{\bar{c}}$ have negative real parts.* To see this, note that with state-variable feedback

$$a_k(s) \triangleq \det(sI - \bar{A} + \bar{b}\bar{k})$$
$$= \det(sI - A_c + b_c k_c) \det(sI - A_{\bar{c}})$$

where $\bar{k} = [k_c \ k_{\bar{c}}]$. On the other hand,

$$a(s) = \det(sI - A) = \det(sI - \bar{A})$$
$$= \det(sI - A_c) \det(sI - A_{\bar{c}})$$

Therefore the eigenvalues of $A_{\bar{c}}$ cannot be shifted by state-variable feedback, and they specify the *uncontrollable modes* of a realization $\{A, b, c\}$ (and all realizations similar to it). Clearly, a realization $\{A, b, c\}$ is *stabilizable* if and only if the uncontrollable modes are stable.

We leave it to the reader to check that the uncontrollable eigenvalues are precisely those that have left eigenvectors orthogonal to b; at these eigenvalues $[sI - A \ b]$ will have rank less than n (cf. the PBH tests of Sec. 2.4.3).

Regulator Problems, Nonzero Set Points, and Tracking [18, p. 270]. Given a controllable realization $\{A, b, c\}$, we have shown that with state feedback, $u(\cdot) = -kx(\cdot) + v(\cdot)$, we can obtain a realization $\{A - bk, b, c\}$ with $\det(sI - A + bk)$ arbitrary. This result is usually used in *regulation* problems, where the nonzero initial condition, $x(0) = x_0 \neq 0$, arises from some *disturbance* and feedback, $-kx(\cdot)$, is used to *restore* the state to zero at a rate determined by the eigenvalues of $A - bk$. Very rapid decay can be obtained by moving the eigenvalues far off into the left half plane, but a price has to be paid in terms of large control energy (large k), increased bandwidth of the closed-loop system and hence increased sensitivity to noise, etc. There should be a trade-off between these conflicting features, and one specific way of doing this will be discussed at some length in Sec. 3.4. Here, however, we shall note some closely related problems that can be treated by small modifications of the solution to the regulator problem.

Nonzero Set Points.   In the regulator problem, the goal is to return the state $x(\cdot)$ to zero, and therefore the *external* input $v(\cdot)$ is taken to be zero.

Suppose, however, that we wish to have

$$y(t) = cx(t) \longrightarrow y_d$$

where y_d is some desired (or *commanded*) nonzero steady-state value of the combination $cx(t)$. (Other combinations can be similarly handled.)

This can be achieved by using a constant command input v_d such that

$$\dot{x} = 0 = (A - bk)x_d + bv_d$$
$$y_d = cx_d = -c(A - bk)^{-1}bv_d$$
$$= H_k(0)v_d$$

where $H_k(s)$ is the closed-loop transfer function

$$H_k(s) = c(sI - A + bk)^{-1}b$$

Note that $(A - bk)^{-1}$ exists because k was presumably chosen to make the eigenvalues of $A - bk$ have sufficiently negative real parts (and in particular no eigenvalue of $A - bk$ is zero). Now we can uniquely determine v_d as

$$v_d = H_k^{-1}(0)y_d$$

provided $H_k(0) \neq 0$, i.e., provided the closed-loop transfer function has no zeros at the origin. But, as noted earlier in this section, the zeros of $H_k(s)$ are the same as the zeros of the original transfer function $H(s)$. Therefore, the command input v_d can be determined if and only if

$$H(0) = -cA^{-1}b \neq 0$$

Once v_d is found, the response of the system can be found, using super-position, as the sum of the solutions for the cases

1. $v(\cdot) \equiv v_d$, $x(0) = x_d$ and
2. $v(\cdot) \equiv 0$, $x(0) = x_0 - x_d$.

Therefore the response $y(\cdot)$ can be written

$$y(t) = y_d + c\tilde{x}(t)$$

where

$$\dot{\tilde{x}}(t) = (A - bk)\tilde{x}(t), \qquad \tilde{x}(0) = x_0 - x_d$$

We see that $\tilde{x}(t) \longrightarrow 0$ at a rate determined by the eigenvalues of $A - bk$ and therefore that $y(t) \longrightarrow y_d$ at this rate.

Note that if the command value y_d is changed "slowly enough," the above scheme can do a reasonable job of *tracking*, i.e., making $y(\cdot)$ follow $y_c(\cdot)$. Example 3.3-4 will illustrate the calculations of this section.

Constant Input Disturbances and Integral Feedback [18, p. 277].   Suppose that in our model we have a constant, but unknown, disturbance vector w,

$$\dot{x}(t) = Ax(t) + bu(t) + w, \qquad x(0) = x_0$$
$$y(t) = cx(t)$$

If we use state feedback, $u(t) = -kx(t)$, to stabilize the original system, then the presence of w will yield a nonzero steady-state value. This can be reduced by increasing k, but this has limits, because of saturation and noise effects.

A reasonable approach might be to attempt to at least estimate the unknown w in some fashion and use this estimate to cancel out the disturbance. This approach will be explored in Chapter 4 (Example 4.2-2). Here we may note that the effects of constant disturbance vectors can often be eliminated by using the so-called *integral-error* feedback. Thus, introduce an additional state variable

$$\dot{q}(t) = y(t)$$

and use the feedback

$$u(t) = -kx(t) - k_q q(t)$$

The augmented closed-loop system is

$$\begin{bmatrix} \dot{x}(t) \\ \dot{q}(t) \end{bmatrix} = \begin{bmatrix} A - bk & -bk_q \\ c & 0 \end{bmatrix} \begin{bmatrix} x(t) \\ q(t) \end{bmatrix} + \begin{bmatrix} w \\ 0 \end{bmatrix}$$

and if $\{k, k_q\}$ are chosen to make this system stable, then the steady-state value of $y(\cdot)$ will be zero, since the second equation gives $0 = cx(\infty) = y(\infty)$.

It is worth noting that the steady-state error (or bias) has been brought to zero without any knowledge of the disturbance w. We should note that by using a command input v_d in addition to integral feedback we can obtain a desired nonzero set point [i.e., a desired value of $y(\infty)$].

For more detailed studies of the uses of state feedback, we refer to textbooks devoted chiefly to control theory, e.g., [9, Chap. 9], [17], [18] and [20] and the technical literature, especially the *IEEE Transactions on Automatic Control*, the *International Journal of Control*, and *Automatica*.

Exercises

3.2-1.

Let $(sI - A)^{-1}b = [p_{n-1}(s) \cdots p_0(s)]'/a(s)$. Show that the common roots of the $n + 1$ polynomials $\{p_{n-1}(s), \ldots, p_0(s), a(s)\}$ specify exactly the uncontrollable natural frequencies of $\{A, b\}$.

3.2-2.

Suppose $\{A, b\}$ is controllable and we make a change of state variables such that $T^{-1}b = \bar{b} = [b_1 \ 0 \ \cdots \ 0]'$, say and $T^{-1}AT = \bar{A}$, where

$$\bar{A} = \begin{bmatrix} A_{11} & A_{12} \\ A_{21} & A_{22} \end{bmatrix}$$

and A_{11} is a scalar. Show that $\{A_{22}, A_{21}\}$ is controllable. Try to think of an application of this result.

3.2-3.   *Relative Order Under State Feedback*

Show that the relative order of a linear system, which is defined as the difference between the degrees of the denominator and numerator polynomials of its transfer function, is not affected by state-variable feedback.

3.2-4.   *The Partial State*

Let $b(s)/a(s)$ be an irreducible transfer function, and write $a(s)\xi(s) = u(s)$, $y(s) = b(s)\xi(s)$.

a. Show that knowledge of $\xi(\cdot)$ and its derivatives completely determines the state variables of any minimal realization of $b(s)/a(s)$. (*Hint:* Start with Fig. 2.1-4 on the controller-form realization.) Therefore $\xi(s)$ is often called the *partial state* of the system.

b. Show that constant state-feedback corresponds to polynomial feedback of the partial state $\xi(\cdot)$: $v(s) = u(s) - g(s)\xi(s)$ for some polynomial $g(s)$ of degree less than or equal to $n - 1$.

c. Show that with such feedback the new transfer function is $b(s)/[a(s) + g(s)]$, as in Eq. (3.2-28).

*3.3   SOME WORKED EXAMPLES

We shall present several examples to illustrate some of the calculations and considerations that arise in state-feedback problems. The results of Examples 3.3-5 and 3.3-6 on the effect of state feedback on controllability and observability are also useful in other contexts.

Example 3.3-1.   Balancing a Pointer

Consider the act of balancing, say, a pointer on your fingertip. Let us idealize the situation as shown in the figure. The bottom end of the pointer is moving along the x axis, with your input $u(t)$ being the acceleration of this point: $u(t) = \ddot{\xi}(t)$. The length of the stick is L; assume that its mass m is concentrated at the top end. Suppose φ is small (so that $\sin \varphi \approx \varphi$ and $\cos \varphi \approx 1$). A force from your fingertip can be applied only in the direction of the stick. Therefore, by equating the forces acting vertically (no acceleration along the vertical axis), we get $mg = F$, and the force component acting in the x direction is

$$mg\,\varphi(t) = F_x(t) = m\ddot{x}(t)$$

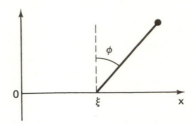

while the center of gravity has the x coordinate given by

$$x(t) = \xi(t) + L\varphi(t)$$

Consider the balancing as a dynamical system with input $u(\cdot)$ (the acceleration of your fingertip in the x direction) and output $y(\cdot) = \varphi(\cdot)$ (the stick's angle to the vertical).

1. Introduce φ and $\dot{\varphi}$ as state variables, and give a state-space description of the system. Determine the transfer function of this system.

2. Is this system controllable? Observable? Determine its eigenvalues. Is it stable?

3. Can the system be stabilized by proportional feedback, $u(t) = -ky(t)$?

4. Determine a state-feedback law, $u(t) = -k_1\varphi(t) - k_2\dot{\varphi}(t)$, so that the closed-loop system has both poles at -1.

5. What do parts 3 and 4 tell us about hand-eye coordination?

Solution. 1. Let

$$z_1 = \varphi, \qquad z_2 = \dot{\varphi}$$

From the previous equations we see that $g\varphi = \ddot{\xi} + L\ddot{\varphi}$, which yields the state-space equations

$$\dot{z} = \begin{bmatrix} 0 & 1 \\ g/L & 0 \end{bmatrix} z + \begin{bmatrix} 0 \\ -1 \end{bmatrix} \frac{\ddot{\xi}}{L}$$

$$y = [1 \quad 0]z$$

For convenience let us work with the normalized input $\ddot{\xi}/L$, which we shall denote by u. Then the characteristic polynomial is

$$a(s) = s^2 - g/L$$

and

$$H(s) = -1/(s^2 - g/L)$$

2.

$$\mathcal{C} = \begin{bmatrix} 0 & -1 \\ -1 & 0 \end{bmatrix} \quad \text{and} \quad \mathcal{O} = \begin{bmatrix} 1 & 0 \\ 0 & 1 \end{bmatrix}$$

so the system is controllable and observable. Alternatively, note there are no cancellations in the nominal transfer function.

The system eigenvalues are at $\pm\sqrt{g/L}$ so the system is unstable.

3. With

$$u = -ky = [-k \quad 0]z$$

we obtain a new system that has eigenvalues at $\pm\sqrt{g/L+1}$. We have thus not been able to get rid of the positive eigenvalue, and the system remains unstable.

4. Perhaps the most straightforward way to find $k = [k_1 \; k_2]$ in this simple example is to set $\det(sI - A + bk) = \alpha(s)$ and *equate coefficients*. This leads easily to $k_1 = g/L - 1$, $k_2 = -2$. The reader should also try to find k via the Bass-Gura, Ackermann, and Mayne-Murdoch formulas (take $g/L = 9$, for simplicity).

Example 3.3-2.   Station Keeping Control for a Libration-Point Satellite (Bryson)

On the line connecting the center of the earth to the center of the moon, there is a so-called libration point L_1 where the pull of the earth on a satellite (in an orbit about the earth with the same period as the moon's orbit) exactly equals the pull of the moon plus the centrifugal force. However, we shall see that this point is an unstable equilibrium point. Then we shall show that by using state feedback (via a small reaction engine) a satellite at that point can be stabilized.

The dynamic equations for small deviations in position away from the libration point can be shown to be (see the figure)

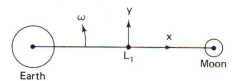

$$\ddot{x} - 2\omega\dot{y} - 9\omega^2 x = 0$$
$$\ddot{y} + 2\omega\dot{x} + 4\omega^2 y = u$$

where x = radial position perturbation, y = azimuthal position perturbation, $u = F/m\omega^2$, F = engine thrust in the y direction, m = satellite mass, and $\omega = 2\pi/29$ rad day^{-1}.

1. With $u = 0$, show that the equilibrium point $x = y = 0$ is unstable.
2. To stabilize the position, use state-variable feedback

$$u = -k_1 x - k_2\dot{x} - k_3 y - k_4\dot{y}$$

Determine the $\{k_i\}$ so that the closed-loop system has poles at $s = -3\omega$, $s = -4\omega$, and $s = (-3 \pm 3j)\omega$.

Solution. 1. We can use $x, \dot{x}, y, \dot{y}$ as state variables, which yields $\{A, b\}$ as

$$A = \begin{bmatrix} 0 & 1 & 0 & 0 \\ 9\omega^2 & 0 & 0 & 2\omega \\ 0 & 0 & 0 & 1 \\ 0 & -2\omega & -4\omega^2 & 0 \end{bmatrix}, \qquad b = \begin{bmatrix} 0 \\ 0 \\ 0 \\ 1 \end{bmatrix}.$$

The characteristic polynomial is

$$a(s) = \det(sI - A) = s^4 - \omega^2 s^2 - 36\omega^4$$

whose roots are determined by

$$s^2 = \frac{\omega^2 \pm \sqrt{\omega^4 + 144\omega^4}}{2} = \frac{\omega^2(1 \pm \sqrt{145})}{2}$$

Therefore the eigenvalues are

$$\left\{ \pm \frac{\omega}{\sqrt{2}} (1 + \sqrt{145})^{1/2}, \pm \frac{j\omega}{\sqrt{2}} (\sqrt{145} - 1)^{1/2} \right\} = \{\pm\omega 2.35, \pm j\omega 2.55\}$$

and the system is clearly unstable.

2. However, the equations are controllable, since $\mathcal{C}(A, b)$ is readily seen to be nonsingular (check that $\det \mathcal{C} = -36\omega^4$). Therefore we can (arbitrarily) relocate the eigenvalues by state feedback. The desired characteristic polynomial is

$$\alpha(s) = (s + 3\omega)(s + 4\omega)(s + 3\omega + 3j\omega)(s + 3\omega - 3j\omega)$$
$$= s^4 + 13\omega s^3 + 72\omega^2 s^2 + 198\omega^3 s + 216\omega^4$$

The feedback gains can now be found by the Bass-Gura formula, but here again it is easier to use comparison of coefficients, which yields

$$k_4 = 13\omega, \qquad k_3 = -28\omega^2, \qquad k_2 = 50.5\omega, \qquad k_1 = 157.5\omega^2$$

Note: Some simplification can be achieved in the above calculations by measuring time in units of $1/\omega$, which reduces the characteristic equation to $s^4 - s^2 - 36 = 0$. In physical problems, one should always look for convenient normalizations.

Example 3.3-3. (De Bra)

The motions of a nearly vertical rod with an attached pendulum (see the figure) can be described by the equations

$$\ddot{\psi}(t) - \sigma^2\psi(t) - n^2[\psi(t) + \theta(t)] = u(t)$$
$$\ddot{\theta}(t) + \omega_0^2\theta(t) + \epsilon\ddot{\psi}(t) = 0$$

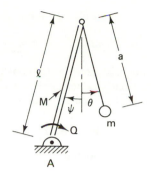

where

$$\sigma^2 = \frac{3g}{2l}, \qquad \omega^2 = \frac{3mg}{Ml}, \qquad \epsilon = \frac{l}{a}$$

$$\omega_0^2 = \frac{g}{a}, \qquad u = \frac{Q}{Ml}$$

Q = torque applied at A

g = gravitational force per unit mass

Assume that $a = l$ and $m = M/3$, and design a state-feedback controller so that the mass m will stop swinging with a time constant of about $T/2$, where $T = 2\pi/\sqrt{a/g}$ = the period of oscillation of the hanging mass.

Solution.   If we divide the given equations by ω_0^2 and use $a = l$, $m = M/3$, we shall have

$$\frac{\sigma^2}{\omega_0^2} = \frac{3}{2}, \qquad \frac{\omega^2}{\omega_0^2} = 1$$

Then if we measure time in units of $1/\omega_0$ and $u(\cdot)$ in units of ω_0^2, the equations can be written as

$$\ddot{\psi}(t) - \tfrac{3}{2}\psi(t) - [\psi(t) + \theta(t)] = u(t)$$
$$\ddot{\theta}(t) + \theta(t) + \ddot{\psi}(t) = 0$$

Let us introduce the state variables

$$x_1 = \psi, \qquad x_2 = \dot{\psi}, \qquad x_3 = \theta, \qquad x_4 = \dot{\theta}$$

Then

$$A = \begin{bmatrix} 0 & 1 & 0 & 0 \\ 2.5 & 0 & 1 & 0 \\ 0 & 0 & 0 & 1 \\ -2.5 & 0 & -2 & 0 \end{bmatrix}, \qquad b = \begin{bmatrix} 0 \\ 1 \\ 0 \\ -1 \end{bmatrix}$$

and

$$a(s) = \det{(sI - A)} = s^4 - 0.5s^2 - 2.5$$

so that the open-loop eigenvalues are at $\pm j1.162$ and ± 1.361.

To obtain time constants of about $\omega_0 T/2$ (in our normalized units) we require

$$\text{Re}[s] \sim -2/\omega_0 T = -1/\pi \approx -1/3$$

We could choose four poles consistent with this constraint, but this still leaves a lot of choice. By means of arguments to be described in the next section (cf. Example 3.4-1), we can take as a "reasonable" choice, eigenvalues at

$$-\tfrac{1}{3} \pm \tfrac{3}{2}j, \qquad -\tfrac{3}{2} \pm \tfrac{1}{2}j$$

yielding a desired characteristic polynomial

$$\alpha(s) = s^4 + 3.667s^3 + 6.861s^2 + 8.75s + 5.903$$

We can now use various methods to compute a suitable feedback gain vector as

$$k = [8.75 \quad 5.083 \quad 8.403 \quad 1.042]$$

Example 3.3-4.   Nonzero Set Points

1. Using the solution to Example 3.3-2 (the station-keeping satellite), design a y-position command controller for the station-keeping satellite at the L_1 libration point.

2. Explain why an x-position controller cannot be designed for the satellite.

Solution.   1. Referring to the discussion in Sec. 3.2.3, we first calculate $H_k(0) = -c(A - bk)^{-1}b$, where in our problem $c = [0\ 0\ 1\ 0]$. Some algebra gives

$$H_k(0) = -\tfrac{1}{24}\omega^2$$

Therefore

$$v_d = H_k^{-1}(0)y_d = -24\omega^2 y_d$$

and the feedback solution is

$$u(t) = -kx(t) - 24\omega^2 y_d$$

where, as found in Example 3.3-2,

$$k = [157.5\omega^2 \quad 50.5\omega \quad -28\omega^2 \quad 13\omega]$$

2. To command the x variable, we should take $c = [1\ 0\ 0\ 0]$. But now it turns out that

$$-c(A - bk)^{-1}b = 0 \qquad \text{for all } k$$

so that it is impossible to find an input v_d to set up any desired value of x.

Example 3.3-5.   Effect of State Feedback on Controllability [21]

Discuss the effect of state feedback on the controllability of a controllable realization $\{A, b, c\}$.

Solution. If a realization $\{A, b\}$ is controllable, then without loss of generality we can assume that it is in controller form $\{A_c, b_c\}$, for which it is obvious by inspection that $\{A_c - b_c k_c, b_c\}$ is still in controller form and therefore still controllable.

This result can be proved in many other ways. For example, one could use the PBH tests. Another method is to note the following algebraic identity, which might also be useful in other contexts: Let $F = A - BK$, where F, A, B, K are matrices of comparable dimensions. Then some simple algebra will show that

$$[B \quad AB \quad \cdots \quad A^{i-1}B] = [B \quad FB \quad \cdots \quad F^{i-1}B]D$$

where

$$D = \begin{bmatrix} I & KB & KAB & \cdots & KA^{i-2}B \\ & I & KB & \ddots & \vdots \\ & & I & \ddots & \vdots \\ & & & \ddots & KAB \\ & \bigcirc & & \ddots & KB \\ & & & & I \end{bmatrix}$$

Note that when $i = n$ the first two matrices can be regarded as the controllability matrices of the *open-loop* realization $\{A, b\}$ and the *closed-loop* realization $\{A - bk, b\}$, respectively. Since the triangular matrix has full rank, this identity again shows that state feedback does not affect controllability (or the lack of it).

Further insight into this property, and in fact a significant generalization, comes from recalling that nonsingularity of $\mathcal{C}(A, b)$ is also equivalent to being able to go from any state to the zero state by means of a suitable input. Suppose we are in a given state, say x_0, and that a particular input $u_*(\cdot)$ [e.g., some string of impulsive functions as in Eq. (2.3-10)] can be found that will take the state to zero. Now suppose the given realization is modified by state feedback, $-kx(\cdot)$. Then can we still go from x_0 to zero with a suitable input? The answer is yes—just use the input $v(\cdot) = u_*(\cdot) + kx(\cdot)$. In fact this is clearly true even for *nonlinear* state feedback, $k[x(\cdot)]$, or even for arbitrary state and output feedback, $k[x(\cdot), y(\cdot)]$.

Example 3.3-6. Observability Under Feedback [21]

Can the observability of a minimal realization $\{A, b, c\}$ be affected by

1. State feedback, $u(t) \longrightarrow v(t) - kx(t)$?
2. Output feedback, $u(t) \longrightarrow v(t) - ly(t)$?

Solution. Since $\{A, b, c\}$ is minimal $\{A, b\}$ is controllable and $\{c, A\}$ is observable. Now after state feedback we have a realization $\{A - bk, b, c\}$ in which $\{A - bk, b\}$ is still controllable for all k (cf. Example 3.3-5). So $\{c, A - bk\}$ will be observable if and only if $\{A - bk, b, c\}$ is minimal, or equivalently if and only if

$$c(sI - A + bk)^{-1}b = \frac{b(s)}{a_k(s)}$$

is irreducible [where $c(sI - A)^{-1}b = b(s)/a(s)$]. But this will be true if and only if $a_k(s)$ and $b(s)$ have no common factors, i.e., provided we do not shift eigenvalues of A to coincide with zeros of $b(s)$. This answers part 1 of the question.

Output feedback is a special case of state feedback, but it is not hard to see that output feedback is insufficient to do what general state feedback can, namely create a nonobservable mode by pole-zero cancellation. For, given the input and output of the new system with output feedback, we can uniquely find the input and output of the original system without feedback and hence, by observability of the original system, can deduce its state. Thus knowing the input and output of the new system suffices to uniquely determine its state, and the new system is hence observable.

More specifically, with output feedback we have

$$\dot{x} = Ax + b(v - ly), \qquad y = cx$$

where l is a scalar gain. Hence, given y (and its first $n - 1$ derivatives) and the external input v (and its derivatives), we can uniquely determine $u = v - ly$ (and its derivatives). It follows that we can uniquely determine the state x, since the system

$$\dot{x} = Ax + bu, \qquad y = cx$$

is given to be observable. Note in fact that this result remains true for arbitrary (linear or nonlinear) feedback of y (and its first $n - 1$ derivatives).

Therefore the answer to part 2 is that output feedback cannot affect the observability of a minimal system. It follows also that output feedback cannot move an eigenvalue of A to a zero of $b(s)$ [if $b(s)/a(s)$ is irreducible]. (Try to deduce the last result by other means as well.)

The above discussions can be straightforwardly extended to the case of nonminimal systems by using the canonical forms of Sec. 2.4.2.

Exercises

3.3-1.   Inverted Pendulum on a Cart

In appropriate dimensionless units the equations of motion for a cart of mass M with a uniform stick of mass m pivoted on top may be written as (see the figure) $\ddot{\theta} = \theta + u$, $\ddot{x} = -\beta\theta - u$, where $\beta = \frac{3}{4}[m/(M + m)] =$ a parameter of the system and $u =$ torque applied to the wheels of the cart by an electric motor. We wish to find a feedback control that will balance the stick (i.e., keep $\theta \approx 0$) *and* keep the cart near $x = 0$. To do this, find gains $k_1, k_2,$ $k_3,$ and k_4 in the state-variable feedback $u = -k_1\theta - k_2\dot{\theta} - k_3x - k_4\dot{x}$ such that the closed-loop system has a double pole at $s = -1$ and a pair of complex poles at $s = -1 \pm 1j$.

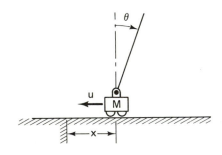

3.3-2.

A helicopter near hover can be described by the equations

$$\begin{bmatrix} \dot{x}_1 \\ \dot{x}_2 \\ \dot{x}_3 \end{bmatrix} = \begin{bmatrix} -0.02 & -1.4 & 9.8 \\ -0.01 & -0.4 & 0 \\ 0 & 1.0 & 0 \end{bmatrix} \begin{bmatrix} x_1 \\ x_2 \\ x_3 \end{bmatrix} + \begin{bmatrix} 9.8 \\ 6.3 \\ 0 \end{bmatrix} u$$

where

$$x_1 = \text{horizontal velocity}, \; x_2 = \text{pitch rate},$$
$$x_3 = \text{pitch angle}, \; u = \text{rotor tilt angle}.$$

a. Find the open-loop poles.

b. Show that a state-feedback law to move the poles to $s = -2$, $s = -1 \pm j$ is $k = [0.0628 \; 0.4706 \; 0.9949]$.

c. Design a velocity-command controller for this system.

3.3-3.

Consider the helicopter of Exercise 3.3-2 with a constant (but unknown) headwind w, so that the equations of motion have an added term on the right-hand side, $[-0.02 \; -0.01 \; 0]'w$. Design a u-velocity-command controller that incorporates integral-error feedback. Use state feedback to shift the integrator pole at $s = 0$ to $s = -1$, leaving the other poles at $s = -2$, $-1 \pm j$.

3.3-4.

a. Using the solution to Exercise 3.3-1, design an x-position command controller for the stick-balancing cart.

b. If a command is given to move the cart from $\dot{x} = x = \theta = \dot{\theta} = 0$ to $x = x_t > 0$, show that the cart starts in the opposite direction; i.e., $\ddot{x}(0) < 0$. Explain this anomalous behavior.

3.3-5. *Altitude-Hold Autopilot for an Airplane (Bryson)*

The approximate equations of motion in pitch and plunge for an airplane in nearly steady, horizontal flight can be shown to be (see the figure) $\tau\dot{\gamma} = \alpha$, $\ddot{\theta} = -\omega_0^2(\alpha - Q\delta)$, $\dot{h} = V\gamma$, where $\gamma \equiv \theta - \alpha = $ flight path angle relative to horizontal, $\theta = $ pitch angle perturbation, $\alpha = $ angle-of-attack perturbation, $V = $ magnitude of velocity with respect to ground (assumed constant), $u = $ elevator deflection (control), $Q = $ elevator effectiveness (constant), $\tau = $ lift time constant, and $\omega_0 = $ undamped natural frequency in pitch-plunge

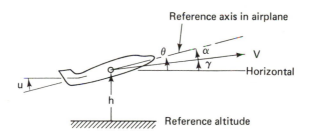

Reference axis in airplane

(constant). An autopilot is to be designed to keep $h \approx 0$ in the presence of the vertical wind disturbances.

a. Using an altimeter to measure h, show that the proportional error feedback, $u = -kh$, does *not* produce a stable closed-loop system for any value of k.

b. Using a gyro to also measure θ, show that a stable closed-loop system can be built with the feedback $u = -k_1 h - k_2 \theta$.

Hint: Many tests can be used to study stability, e.g., Routh's test or root-locus plots ([6] and [7]).

3.4 QUADRATIC REGULATOR THEORY FOR CONTINUOUS-TIME SYSTEMS

We shall consider a system

$$\dot{x}(t) = Ax(t) + bu(t)$$

that is normally at rest, $x(\cdot) \equiv 0$, but due to some (impulsive) disturbance the state at some time t_0 is displaced to $x(t_0) = x_0 \neq 0$. Therefore $x(t) \neq 0$, $t \geq t_0$, and the regulator problem is to apply a control $u(\cdot)$ such that the state $x(\cdot)$ is returned to zero as quickly as possible. Based on our previous results, there are at least two ways of attacking this problem.

If $\{A, b\}$ is controllable, we saw in Sec. 2.3.2 that we can use an impulsive input of the form $\sum_0^{n-1} g_i \delta^{(i)}(t)$ to *instantaneously* restore the state to zero. The price we pay for this solution is of course very high (infinite) energy inputs.

The result of Sec. 3.2.1 suggests on the other hand that if $\{A, b\}$ is controllable, we can obtain a finite energy input as $u(\cdot) = -kx(\cdot)$, so that

$$\dot{x}(t) = (A - bk)x(t)$$

and by choosing k suitably we can make $x(\cdot)$ decay to zero as fast as we wish. The rate of decay depends on how negative the real parts of the eigenvalues of $A - bk$ are. Now from Eq. (3.2-27), we can see that the more negative these are, the larger will be the value of k and therefore the higher the required signal energy. In fact it is an interesting algebraic exercise to show that as the eigenvalues of $A - bk$ are forced to $-\infty$, the input $u(\cdot) = -kx(\cdot)$ goes exactly to the string of impulsive functions of Sec. 2.3.2 that will restore $x(\cdot)$ instantaneously to zero.

These facts suggest that we should try to make a trade-off between the rate of decay of $x(\cdot)$ and the energy of the input. In the *quadratic regulator* problem, this is done by choosing $u(\cdot)$ to minimize

$$J = \int_{t_0}^{t_f} [x'(t)Qx(t) + ru^2(t)] \, dt + x'(t_f)Q_f x(t_f)$$

subject to

$$\dot{x}(t) = Ax(t) + bu(t), \qquad x(t_0) = x_0$$

By choice of r, Q, t_f, and Q_f we can give different weightings to the cost of control and the cost of deviations from the desired state (which is 0 for all t). The choice of these quantities is again more of an art than a science but one that is being investigated in various ways (see, e.g., [22]).

Our interest here lies in the fact that under certain observability and controllability conditions, the optimal control when $t_f = \infty$ turns out to be a linear feedback control,

$$u(t) = -\bar{k}x(t), \qquad t_0 \leq t \leq \infty$$

where $\bar{k}$ depends on the parameters $\{A, b, Q, r\}$.

This is of course exactly the kind of constant state feedback studied in Sec. 3.2, and clearly the *optimum* feedback gain must be associated with an optimum set of pole locations. In Sec. 3.4.1 we shall present the main results of the *steady-state* ($t_f = \infty$) quadratic regulator theory and note the insights it provides into the question of optimum pole location. We also stress that perhaps the main use of the theory is to enable a convenient parametrization of "good" trial solutions that can then be improved by an interactive procedure based on modifying Q and r. A partial justification of these results will be given in Sec. 3.4.2, where we shall emphasize the fundamental way in which the concepts of controllability and observability enter into the solution. In Sec. 3.4.3 we shall briefly introduce one of the most widely studied equations in "modern" control theory, the so-called algebraic Riccati equation (ARE).

We should note here that a comprehensive study of the general quadratic regulator problem, in which new results and new methods continue to be found, is beyond our scope; however, we feel it is important to provide an introduction to some of the key facts.

3.4.1   Optimum Steady-State Solutions

We shall study the problem of choosing $u(\cdot)$ in

$$\dot{x}(t) = Ax(t) + bu(t), \qquad x(t_0) = x_0 \qquad (1)$$
$$y(t) = cx(t), \qquad\qquad t \geq t_0 \qquad (2)$$

so as to minimize the *cost functional*

$$J_\infty = \int_0^\infty [y^2(t) + ru^2(t)]\, dt, \qquad r > 0 \qquad (3)$$

Let us assume also (for reasons that will only appear later, in Sec. 3.4.2) that

$$\{A, b, c\} \text{ is minimal} \tag{4a}$$

or equivalently that

$$\{A, b\} \text{ is controllable and } \{c, A\} \text{ is observable} \tag{4b}$$

or equivalently that

$$c(sI - A)^{-1}b = \frac{b(s)}{a(s)} \text{ is irreducible} \tag{4c}$$

where

$$a(s) = \det (sI - A) \tag{5}$$

Then it will be shown that the optimal solution is to use linear state feedback

$$u(t) = -\bar{k}x(t) \tag{6}$$

where $\bar{k}$ is such that

$$\det (sI - A + b\bar{k}) = \prod_{1}^{n} (s - z_i) \tag{7a}$$

and

$$\{z_i\} = \{\text{the left-half-plane roots of } \Delta(s)\} \tag{7b}$$

with $\Delta(s)$ being the $2n$-degree polynomial

$$\Delta(s) = a(s)a(-s) + r^{-1}b(s)b(-s) \tag{8}$$

It is clear that if λ is a root of $\Delta(s)$, then so is $-\lambda$, so that $\Delta(s)$ will have n left-half-plane roots with n mirror images in the right half plane. We can see that $\Delta(j\omega) > 0$, so that there are no roots on the $j\omega$ axis.† Note also that because of the symmetric location of the roots of $\Delta(s)$ we can write

$$\Delta(s) = a_{\bar{k}}(s)a_{\bar{k}}(-s) = a(s)a(-s) + r^{-1}b(s)b(-s) \tag{9}$$

The rule (7)–(8), due to Letov [23] and Chang [24], can be used to obtain the optimal pole configuration (and the associated optimal feedback gain) for any specified weighting parameter r. It should be noted here that there are other ways of determining the optimal $\bar{k}$ and thereby the optimal pole configuration (cf. Sec. 3.4.3). However, the present method has the advantage that it allows us to more easily study the *structure* of the optimal configuration

---

†We assume of course that $a(s)$ has real coefficients, so that $a(j\omega)a(-j\omega) = |a(j\omega)|^2$ and so also for $b(s)$. Now use (4c).

as a function of r, e.g., by using root-locus techniques, and thereby to gain more insight into the solution. Furthermore, quite explicit results can be obtained in two limiting cases ([24] and [25]).

High Cost of Control $(r \longrightarrow \infty)$. Thus suppose that $r \longrightarrow \infty$, so that the control energy is weighted very heavily in the cost functional (3). Then we see from (8) and (9) that

$$a_{\hat{k}}(s)a_{\hat{k}}(-s) \longrightarrow a(s)a(-s) \qquad \text{as } r \longrightarrow \infty \qquad (10)$$

Since by (7) all the roots of $a_{\hat{k}}(s)$ must be in the left half plane, this means that (cf. Fig. 3.4-1)

1. We should not move the stable roots of the original characteristic polynomial $a(s)$, and
2. We should reflect the unstable roots about the $j\omega$ axis.

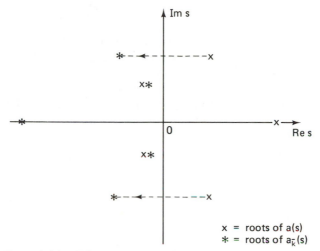

Figure 3.4-1. When control cost is high $(r \longrightarrow \infty)$, the unstable roots of $a(s)$ are reflected about the $j\omega$ axis; the other roots are not disturbed.

It is reasonable that when control is expensive we should not expend any effort in moving the stable poles around, but one might have thought that the unstable poles should have been moved as little as possible, namely just into the left half plane rather than being reflected about the $j\omega$ axis. However, the need to *trade off* feedback gain vs. rate of decay of $x(\cdot)$ dictates the latter requirement (see Example 3.4-1).

Low Cost of Control $(r \longrightarrow 0)$. On the other hand, as $r \longrightarrow 0$, so that control is "cheap," we should move the roots far into the left half plane. We see from (8) and (9) that

$$a_{\hat{k}}(s)a_{\hat{k}}(-s) \longrightarrow r^{-1}b(s)b(-s) \qquad \text{as } r \longrightarrow 0 \qquad (11)$$

This shows that any zero of $a(s)a(-s) + r^{-1}b(s)b(-s)$ that remains finite as $r \rightarrow 0$ must tend toward a zero of $b(s)b(-s)$. But from (1) and (2) we know that the degree of $b(s)b(-s)$, say $2m$, is less than the degree, $2n$, of $a_k(s)a_k(-s)$. Therefore, $(n - m)$ *zeros of* $a_k(s)$ *must tend toward infinity and the remaining* m *to the left-half-plane roots of* $b(s)b(-s)$. For very large s, we can ignore all but the highest powers of s in (8), so that for the roots that tend toward infinity we shall have the approximate relation

$$0 = (-1)^n s^{2n} + r^{-1}(-1)^m b_0^2 s^{2m} \qquad (12)$$

where

$$b(s) = b_0 s^m + \cdots + b_m$$

Therefore, we can write

$$s^{2(n-m)} = (-1)^{n-m+1} b_0^2 r^{-1} \qquad (13)$$

The $2(n - m)$ solutions of (13) lie on a circle of radius $(b_0^2/r)^{1/2(n-m)}$ in a particular pattern known in network theory as a Butterworth configuration. Some examples are sketched in Fig. 3.4-2.

As argued in the introduction, it is not surprising that if no attention is paid to the cost of control the poles should be moved out to infinity (so that

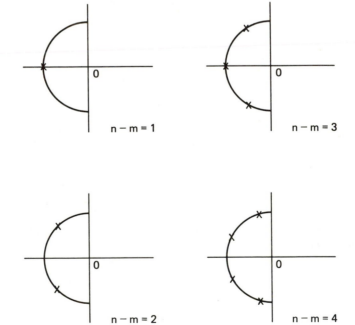

Figure 3.4-2.   Some Butterworth configurations.

the disturbance is brought to zero more quickly), but the particular Butterworth pattern may not have been expected. However, the Butterworth pattern is a familiar one in networks and control, and it is one that classical designers often begin with. It has some advantages and some disadvantages (e.g., high overshoot), which of course only shows that a quadratically optimal solution is not necessarily satisfactory in all respects (see also Exercise 3.4-17).

Example 3.4-1.   A First-Order System

Explicit formulas are difficult to get except for first- and second-order systems. Here we shall consider a first-order system

$$\dot{x}(t) = ax(t) + u(t), \qquad x(0) = x_0$$
$$y(t) = x(t)$$

with

$$J = \int_0^\infty [y^2(t) + ru^2(t)] \, dt$$

Let us assume a feedback control of the form

$$u(t) = -kx(t)$$

and try to find the optimum k by direct minimization of the cost functional. We note first that

$$x(t) = x(0) \exp (a - k)t$$

so that after some algebra we can find that

$$J(k) = \begin{cases} \dfrac{-(1 + rk^2)(a - k)^{-1}}{2}, & a - k < 0 \\[2mm] \infty, & a - k > 0 \end{cases}$$

If $a < 0$, i.e., the original plant is stable, then we can choose $k = 0$ and have a finite cost $-(2a)^{-1}$. If $a > 0$ (unstable plant), then we must choose $k > a$ for a finite cost; i.e., we must move the unstable pole at $s = a$ at least into the left half plane (LHP), $s = a - k < 0$. If we just move ϵ into the LHP, however, $x(t) = x(0) \exp (a - k)t$, and the cost $J(k = \epsilon + a)$ may be quite high. There will be an optimum value of k that will minimize the cost. To find this, we differentiate $J(k)$ with respect to k and set the result equal to zero to get

$$\bar{k}^2 r - \bar{k} 2ar - 1 = 0$$

so that

$$\bar{k} = a \pm \sqrt{a^2 + r^{-1}}$$

As $r \longrightarrow \infty$ (expensive control), we see that

$$\bar{k} \longrightarrow 0 \quad \text{or} \quad 2a$$

The value 0 will give an infinite cost if $a > 0$, and so in this case we must choose $\bar{k} = 2a$, so that the closed-loop pole is at $s = a - \bar{k} = -a$, the mirror image of the unstable pole at $s = a$. On the other hand, if $a < 0$ (stable system), then the optimal choice is $\bar{k} = 0$, and the closed-loop pole is left at the location of the open-loop pole.

If control is cheap ($r \longrightarrow 0$), then $\bar{k} \longrightarrow \pm\sqrt{1/r}$, but only the $+$ sign will give a finite cost (we need $\bar{k} > a$). The cost can be made as small as we wish by choosing $\bar{k}$ large enough. We see that the results are in conformity with the general rules described earlier.

For nonextreme values of r, the results are often most transparently presented in terms of a root-locus plot, as we shall explain now.

＊General Control Cost—The Symmetric Root Locus.†   Returning to our general solution, we note that Eq. (8) is in a form that suggests that root-locus techniques (which indicate the variation of the zeros of a polynomial with respect to a scalar parameter; see [6] and [7]) can be useful. We are interested in the solutions of $a(s)a(-s) + r^{-1}b(s)b(-s) = 0$ or [in standard (Evans) root-locus form] the solution of

$$(-1)^{n-m}r^{-1}\frac{\prod(s + z_i)(s - z_i)}{\prod(s + p_i)(s - p_i)} = -1 \tag{14}$$

where the z_i and p_i are the zeros and poles of $b(s)/a(s)$. The variation of the solutions of (14) with variation in r is given by the usual root-locus plot. We plot the *180° locus* (using the associated rules) if $n - m$ is even and the *0° locus* (with its own rules) if $n - m$ is odd. (Note that $r^{-1} > 0$.)

Thus, given $H(s) = b(s)/a(s)$, we simply reflect its poles and zeros across the imaginary axis to get those of $b(s)b(-s)/a(s)a(-s)$. We plot the usual root locus, which, because of the symmetric pole-zero configuration, is itself symmetric with respect to the real *and* imaginary axes and hence is called the *symmetric root locus*. The left-half-plane locus gives the optimum root locations, and the previous discussions for the limiting values $r \longrightarrow 0$ and $r \longrightarrow \infty$ now apply to the extreme points of the root-locus plots.

Example 3.4-2.   Optimum Pole Locations for the Pointer-Balancing Problem

In Example 3.3-3, we considered stabilizing a pendulum attached to a nearly vertical rod. We had, as governing equations,

$$\ddot{\psi} - \tfrac{3}{2}\psi - (\psi + \theta) = u$$
$$\ddot{\theta} + \theta + \ddot{\psi} = 0$$

where ψ and θ were the angular deviations of the rod and pendulum, respectively, from the vertical position.

---

†The rest of this section may be omitted by those unfamiliar with root-locus techniques, though some efforts to learn this simple technique will not be unrewarding.

Determine the optimum root loci for regulation of the quantity $y = \psi + \theta$, and select an optimum pole configuration such that the dominant time constant is $\approx -\frac{1}{3}$.

Solution. Taking transforms of the governing equations, we obtain a transfer function

$$\frac{1}{s^4 - 0.5s^2 - 2.5}$$

the poles of which are at

$$s = \pm 1.162j, \qquad s = \pm 1.361$$

There are no finite zeros. Following the rules described in this section, we obtain the loci of optimum roots shown in the figure.

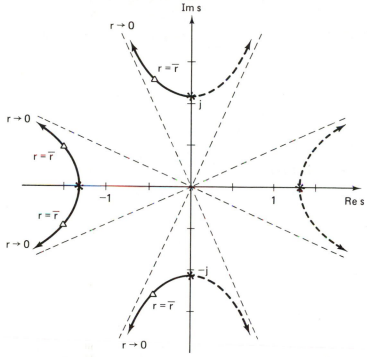

Note in particular that for $r \longrightarrow \infty$ the optimum poles are at $\pm 1.162j$, -1.361, -1.361 and that for $r \longrightarrow 0$ the poles move to infinity along asymptotes at $112.5°$, $157.5°$, $202.5°$, and $247.5°$. The triangles $\triangle$ indicate a set of poles, corresponding to some value of r, say $r = \bar{r}$, for which the dominant time constant is $\approx -\frac{1}{3}$. These poles are approximately given by

$$s = -\tfrac{1}{3} \pm \tfrac{3}{2}j \quad \text{and} \quad s = -\tfrac{3}{2} \pm \tfrac{1}{2}j$$

and are the poles chosen in Example 3.3-4.

*3.4.2 *Plausibility of the Selection Rule*
for the Optimal Poles †

A standard way of solving problems like (1)–(3), of minimization under constraints, is to use Lagrange multipliers and the calculus of variations [28]. We shall quickly outline one way of approaching a general nonlinear variational problem and then apply it to our special case.

Consider a system governed by the (nonlinear) state equations

$$\dot{x} = f(x, u, t), \qquad x(0) = x_0 \tag{15}$$

We wish to choose $u(\cdot)$ so as to minimize the cost criterion:

$$J = \phi[x(t_f)] + \int_0^{t_f} L(x, u, t)\, dt \tag{16}$$

where the final time, $t_f < \infty$, is specified and $\phi[x(t_f)]$ represents a terminal cost. We adjoin the constraint $f - \dot{x} = 0$ to the cost, using a (vector) Lagrange multiplier $\lambda(\cdot)$, to get

$$J = \phi + \int_0^{t_f} [L + \lambda'(f - \dot{x})]\, dt$$

$$= \phi + \int_0^{t_f} [H - \lambda'\dot{x}]\, dt \tag{17a}$$

where the *Hamiltonian H* is defined by

$$H = L + \lambda'f \tag{17b}$$

Integrating the last term in (17a) by parts, to eliminate $\dot{x}$, we get

$$J = \phi + \lambda'x \Big|_{t_f}^{0} + \int_0^{t_f} [H + \dot{\lambda}'x]\, dt \tag{18}$$

Now if a minimizing u has been found, then small but arbitrary variations δu in u should, to first order, produce no change in J; i.e., we should have $\delta J = 0$. From (18) we get

$$\delta J = \left(\frac{\partial \phi}{\partial x} - \lambda'\right) \delta x \Big|_{t_f} + \lambda'\, \delta x \Big|_{0} + \int_0^{t_f} \left[\left(\frac{\partial H}{\partial x} + \dot{\lambda}'\right)\delta x + \frac{\partial H}{\partial u}\delta u\right] dt \tag{19}$$

†There are many ways of establishing the validity of the rules of Sec. 3.4.1. A detailed proof is quite long (see, e.g., [18] or [26]), and here we shall not give a complete argument. Our approach is chosen to bring out the significance of the Hamiltonian matrix (27) and Eqs. (24). Further study of these equations leads to an efficient computational algorithm (see Sec. 3.4.3) and to a unifying and powerful scattering-theoretical framework for such problems (cf. [27]). Another approach, based on the so-called algebraic Riccati equation, is described in Exercises 3.4-14 to 3.4-16.

Here $\partial H/\partial x$ denotes a row vector whose ith component is $\partial H/\partial x_i$. Note that $\delta x|_0 = 0$ since $x(0)$ is specified. A convenient choice for λ is now evident. Letting

$$\dot{\lambda}' = -\frac{\partial H}{\partial x}, \qquad \lambda'(t_f) = \frac{\partial \phi}{\partial x}\bigg|_{t_f} \tag{20}$$

simplifies expression (19) for δJ to

$$\delta J = \int_0^{t_f} \frac{\partial H}{\partial u} \delta u \, dt$$

A necessary condition that $\delta J = 0$ for arbitrary δu is then that

$$\frac{\partial H}{\partial u} = 0 \tag{21}$$

Equations (20) and (21) are the famous *Euler-Lagrange equations*, giving necessary conditions for $u(\cdot)$ to minimize J.

In our problem we have

$$f = Ax + bu; \qquad \phi = 0$$
$$L = \tfrac{1}{2}(y^2 + ru^2) = \tfrac{1}{2}(x'c'cx + ru^2) \tag{22a}$$
$$H = \tfrac{1}{2}(x'c'cx + ru^2) + \lambda'(Ax + bu) \tag{22b}$$

The Euler-Lagrange equations now become

$$\dot{\lambda}(t) = -A'\lambda(t) - c'cx(t), \qquad \lambda(t_f) = 0 \tag{23a}$$

and

$$u(t) = -r^{-1}b'\lambda(t) \tag{23b}$$

Recalling that

$$\dot{x}(t) = Ax(t) + bu(t), \qquad x(0) = x_0$$

we see that we actually have $2n$ linear differential equations

$$\begin{bmatrix} \dot{x}(t) \\ \dot{\lambda}(t) \end{bmatrix} = \begin{bmatrix} A & -br^{-1}b' \\ -c'c & -A' \end{bmatrix} \begin{bmatrix} x(t) \\ \lambda(t) \end{bmatrix} \tag{24a}$$

but with *mixed* or *two-point* boundary conditions

$$x(0) = x_0, \qquad \lambda(t_f) = 0 \tag{24b}$$

The fact that the boundary conditions are not both given at the same point (0 or t_f) makes this equation harder to solve than the *one-point* equations of Sec. 2.5.1.

There are general methods for solving such equations (see [18, Sec. 3.3.2] and Exercise 9.1-11) and for studying their limiting behavior as $t_f \longrightarrow \infty$ (see [18, Sec. 3.4] and [29]). It will be too big a digression for us to pursue this analysis here, and we shall be content just to state, without proof, that these methods show that as $t_f \longrightarrow \infty$, and under the assumptions that $\{A, b\}$ is controllable (or just stabilizable) and $\{c, A\}$ is observable (or just detectable), $\lambda(\cdot)$ and $x(\cdot)$ can be related as

$$\lambda(t) = \bar{P}x(t) \tag{25}$$

where $\bar{P}$ is a certain $n \times n$ matrix. More will be said about this matrix in Sec. 3.4.3, but here our major point is that from (23) and (25) we can see that the optimal control has a feedback form

$$u(t) = -\bar{k}x(t) \tag{26a}$$

where $\bar{k}$ is a certain row vector. Moreover, it is shown in [18] and [29] that this $\bar{k}$ is such that the closed-loop matrix

$$A - b\bar{k} \text{ is stable} \tag{26b}$$

Clearly, $\bar{k}$ can be calculated as

$$\bar{k} = b'r^{-1}\bar{P}$$

but there are other ways of obtaining it without explicitly computing $\bar{P}$. In fact, we want to show that $\bar{k}$ can be computed by the procedure of Sec. 3.4.1. Once the fact that $u(\cdot)$ has the form (26) is granted, then we can say that the n characteristic frequencies of the optimal closed-loop system

$$\dot{x}(t) = (A - b\bar{k})x(t), \qquad x(0) = x_0$$

must be among the $2n$ characteristic frequencies of the differential equation system (24). These are determined by the roots of the characteristic equation of (24), namely by the roots of

$$\Delta(s) = \det(sI - \mathbf{M}) = 0, \qquad \mathbf{M} \triangleq \begin{bmatrix} A & -br^{-1}b' \\ -c'c & -A' \end{bmatrix} \tag{27}$$

By using standard matrix and determinant manipulations (cf. Exercise A-11), we can write

$$\Delta(s) = \det(sI - A)\det[sI + A' - c'c(sI - A)^{-1}br^{-1}b']$$
$$= \det(sI - A)\det[I - c'c(sI - A)^{-1}br^{-1}b'(sI + A')^{-1}]\det(sI + A')$$

Then using Exercise A-12 and the fact that $\det(sI + A') = (-1)^n \det(-sI - A)$, we can see that

$$(-1)^n \Delta(s) = a(s)a(-s)\left[1 + r^{-1}\frac{b(-s)b(s)}{a(-s)a(s)}\right]$$
$$= a(s)a(-s) + r^{-1}b(s)b(-s) \tag{28}$$

which is the basic quantity (8) introduced earlier in specifying the optimum solution. Now the n roots of the characteristic polynomial of the optimum closed-loop system must be among the $2n$ roots of $\Delta(s) = 0$. And because the roots of $\Delta(s)$ are symmetrically distributed about the $j\omega$ axis,† the optimum solution is determined by the n stable roots of $\Delta(s) = 0$. But this is just rule (7) described in Sec. 3.4.1.

Of course in our justification of rule (7), we did not show how to obtain the key results (26a) and (26b). This was done deliberately, since we did not wish to enter into a long discussion of the basic equation (24), which has value beyond the particular problem of justifying rule (7). However, we should repeat that other approaches are possible which do not bring in (24) but focus directly on the matrix $\bar{P}$ of (25). One such approach is outlined in Exercise 3.4.16, which an interested reader could attempt after reading Sec. 3.4.3.

To close the present discussion, however, it will be useful to briefly note some aspects of the way controllability and observability enter into the convergence proofs that we omitted here.

To establish the convergence of the solutions of (24) as $t_f \rightarrow \infty$, it is necessary to know that the minimum cost is finite for all $t_f \leq \infty$ ([18] and [29]).

Controllability Ensures $J_{\min} < \infty$.   The proof of this statement will be in two stages.

1. Suppose A is *stable*; i.e., all its eigenvalues have negative real parts.

Then if we choose $u(\cdot) \equiv 0$, the cost, say J_0, will obviously be greater than or equal to $J_{\min}$ [because $u(\cdot) \equiv 0$ is not the optimal solution].

We shall see that $J_0 < \infty$. For when $u(\cdot) \equiv 0$, $x(t)$ tends toward zero *exponentially* as $t \rightarrow \infty$. Therefore

$$J_0 = \int_0^\infty x'(t)Qx(t)\, dt < \infty, \qquad Q = c'c$$

Controllability is not needed so far. But suppose the following.

2. A is *unstable*, but $\{A, b\}$ is controllable. Then by a preliminary feedback, say $fx(\cdot)$, we can make a new system

$$\dot{x}_f(t) = (A - bf)x_f(t) + bu(t)$$

such that $A - bf$ is stable. Now we use $u(\cdot) \equiv 0$ as before. Then the cost will be

$$J_{\min} \leq \int_0^\infty \{x_f'(t)Qx_f(t) + r[fx(t)]^2\}\, dt < \infty \tag{29}$$

Thus controllability ensures boundedness of $J_{\min}$ whether or not A is stable. In fact, we see that all we really need to assume is the stabilizability of $\{A, b\}$.

An alternative argument will also be instructive. In Sec. 2.3.2 we stated that a bounded input $u(\cdot)$ could be used to drive the state of a controllable realization to zero in a finite time (this fact is proved in Sec. 9.2.1). If we then applied no input, the state would continue at zero, and clearly the cost with this strategy would be finite. Clearly, then, the optimum strategy will have a cost no higher than this, and therefore again we must have $J_{\min} < \infty$.

†Note that $\Delta(s) = 0$ if and only if $\Delta(-s) = 0$.

Observability Ensures Stability.   It might appear that the finiteness of $J_{\min}$ would imply that the optimal system is stable, as claimed in (26b). However, this is not necessarily so, unless $\{c, A\}$ is observable or at least detectable. The point is that there could be an undetectable (i.e., unstable and unobservable) subsystem whose output would not contribute to $J_{\min}$.

More precisely, since $J_{\min} < \infty$, we must have $y \longrightarrow 0$ and $u \longrightarrow 0$. But then the basic relation (2.3-6) (applied to the optimal system)

$$
\Theta x = \begin{bmatrix} y(t) \\ \cdot \\ \cdot \\ \cdot \\ y^{(n-1)}(t) \end{bmatrix} + \begin{bmatrix} h_0 & & & & \\ h_1 & \cdot & & \bigcirc & \\ \cdot & \cdot & \cdot & & \\ \cdot & \cdot & \cdot & \cdot & \\ h_{n-1} & \cdot & \cdot & h_1 & h_0 \end{bmatrix} \begin{bmatrix} u(t) \\ \cdot \\ \cdot \\ \cdot \\ u^{(n-1)}(t) \end{bmatrix} \tag{30}
$$

shows that $\Theta x \longrightarrow 0$. Then the nonsingularity of Θ implies that $x \longrightarrow 0$, so that the system must be stable. It should be clear how to argue if we only had detectability of $\{c, A\}$.

It is at least of theoretical interest to ask if optimal solutions might not exist even if $\{A, b, c\}$ is not minimal. This question is beyond our scope here, but the answer is a qualified yes: Provided $\{A, b\}$ is controllable, there can be optimal but unstable solutions. For further discussion of such solutions, their significance, calculation, etc., refer to the work of Mårtensson [30].

*3.4.3   The Algebraic Riccati Equation

In Sec. 3.4.2, we remarked that the optimal solution being in fact a feedback solution depended on the possibility that the solutions $\lambda(\cdot)$ and $x(\cdot)$ of the two-point boundary-value equations (24) could be related as in (25), viz., as

$$
\lambda(t) = \bar{P}x(t) \tag{31}
$$

for some $n \times n$ matrix $\bar{P}$. To discover more about the properties that such a matrix must have, we substitute the above relation into (23)–(24) to find

$$
\begin{aligned}
\bar{P}\dot{x}(t) &= \bar{P}Ax(t) + \bar{P}bu(t) \\
&= (\bar{P}A - \bar{P}br^{-1}b'\bar{P})x(t) \\
&= \dot{\lambda}(t) = (-A'\bar{P} - c'c)x(t)
\end{aligned} \tag{32}
$$

Now we see that Eq. (32) [and (24)] can be satisfied if we choose $\bar{P}$ as a solution of the equation

$$
\bar{P}A - \bar{P}br^{-1}b'\bar{P} = -A'\bar{P} - c'c
$$

i.e.,

$$
A'\bar{P} + \bar{P}A - \bar{P}br^{-1}b'\bar{P} + c'c = 0 \tag{33}
$$

By taking transposes, we see that if $\bar{P}$ satisfies (33), so does $\bar{P}'$, so that we can assume that

$$\bar{P} \text{ is symmetric} \qquad (34)$$

This is not a fact that might have been evident without (33). This nonlinear (quadratic) algebraic equation, which is a celebrated one in state-space control theory, is known as the *algebraic Riccati equation* (ARE) because it is the steady-state form of a famous differential equation called the Riccati equation—see Exercise 3.4-11.

Once $\bar{P}$ is known, the optimal control can also be found as [cf. (26)]

$$u(t) = -\bar{k}x(t), \qquad \bar{k} = r^{-1}b'\bar{P}$$

and for this and other reasons (see, e.g., Exercise 3.4-8) it is of interest to further explore the ARE.

We must first face the fact that the (quadratic) ARE may or may not have a solution or that it may have more than one solution, and it is not clear which one should be chosen to give $\bar{k}$. The question of existence can be settled in several ways —Exercise 3.4-10 gives a proof under the assumption that the Hamiltonian matrix (27) has distinct eigenvalues. Another proof is outlined in Exercise 3.4-16. The convergence results of [18, Sec. 3.4.2] and [29] (noted without proof in Sec. 3.4.2) also establish the existence of a solution under the assumptions of stabilizability and detectability.

However, we also have the problem that the ARE can have many solutions. It would have been nice if $\bar{k}$ were the same no matter which solution $\bar{P}$ was chosen, but unfortunately that is not true. For example, take

$$n = 1, \qquad b = 1, \qquad r = 1, \qquad a = 4, \qquad c = 3$$

Then the ARE is

$$\bar{P}^2 - 8\bar{P} - 9 = (\bar{P} - 9)(\bar{P} + 1) = 0$$

so that

$$\bar{k} = \bar{P} = +9 \quad \text{or} \quad -1$$

To help choose between these solutions, we may note that when $\bar{k} = 9$

$$sI - A + bk = s + 5, \qquad \text{stable}$$

while when $\bar{k} = -1$

$$sI - A + bk = s - 5, \qquad \text{unstable}$$

We can conclude that $\bar{k} = 9$ is the optimal solution, since the obvious controllability and observability of $\{A, b, c\}$ imply that the closed-loop characteristic polynomial must be stable (cf. Sec. 3.4.2).

In the general case as well, it turns out that the assumption of controllability and observability of $\{A, b, c\}$ resolves the problem of uniqueness. For controllability and observability require, as proven in Sec. 3.4.2, that $A - b\bar{k}$ be stable, and the following lemma shows that there is only one solution of the ARE yielding a $\bar{k}$ with this property.

Lemma 3.4-1.

There is only one solution, $\bar{P}_s$ say, of the ARE that yields a stable closed-loop characteristic polynomial

$$a_{\bar{k}}(s) = \det(sI - A + b\bar{k}), \qquad \bar{k} = r^{-1}b'\bar{P}_s$$

Proof.   Suppose there exist two solutions $\bar{P}_1$ and $\bar{P}_2$ such that

$$\bar{A}_i = A - br^{-1}b'\bar{P}_i \text{ is stable,} \qquad i = 1, 2$$

Now it is easy to check that

$$(\bar{P}_1 - \bar{P}_2)\bar{A}_1 + \bar{A}'_2(\bar{P}_1 - \bar{P}_2) = 0 \qquad (35)$$

By Exercise A-40, this matrix equation for $\bar{P}_1 - \bar{P}_2$ will have a unique solution if no eigenvalue of $\bar{A}'_2$ is the negative of any eigenvalue of $\bar{A}_1$. This condition is clearly met since both $\bar{A}_1$ and $\bar{A}_2$ are stable by hypothesis. Therefore the solution is

$$\bar{P}_1 - \bar{P}_2 = 0$$

i.e., there is only one *stabilizing* solution of the ARE.

The question now is how we can find this unique solution $\bar{P}_s$ (without first finding all solutions and checking the stability of the corresponding closed-loop filters). A direct characterization of $\bar{P}_s$, namely that $\bar{P}_s$ is the only solution of the ARE (under the assumptions of $\{A, b, c\}$ minimal) that is positive definite, is noted in Exercise 3.4-7, but again this is not useful for computation. A popular computational method is the following.

The MacFarlane-Potter-Fath Methods ([31]–[33]).   Assume that the Hamiltonian matrix

$$\mathbf{M} = \begin{bmatrix} A & -br^{-1}b' \\ -c'c & -A' \end{bmatrix}$$

has *distinct* eigenvalues, and form the eigenvalues and eigenvectors of $\mathbf{M}$, say,

$$\mathbf{M}\begin{bmatrix} f_i \\ g_i \end{bmatrix} = \lambda_i \begin{bmatrix} f_i \\ g_i \end{bmatrix}, \qquad i = 1, 2, \ldots, 2n \qquad (36)$$

where the $\{f_i\}$ and $\{g_i\}$ are sets of n vectors.

Now choose the n eigenvalues, say $\{\bar{\lambda}_i, i = 1, \ldots, n\}$, with negative real parts, and let $\{\bar{f}_i, \bar{g}_i\}$ be the corresponding eigenvectors. Then $\bar{P}_s$ can be calculated as

$$\bar{P}_s = [\bar{g}_1 \quad \cdots \quad \bar{g}_n][\bar{f}_1 \quad \cdots \quad \bar{f}_n]^{-1} \qquad (37)$$

The order of the $\{\bar{f}_i, \bar{g}_i\}$ is irrelevant.

This procedure can be justified by showing that (Exercises 3.4-9 and 3.4-10) every solution $\bar{P}$ of the ARE can be written in the form (37) for some choice of n

eigenvalues $\{\lambda_i\}$ and their corresponding eigenvectors and moreover that these eigenvalues will also be the eigenvalues of the corresponding closed-loop matrix $\bar{A} = A - br^{-1}b'\bar{P}$. Now we need only recall that there is only one solution $\bar{P}$ that will make $\bar{A}$ stable and also, because of the symmetry of the roots of det $(sI - \mathbf{M})$, that there is only one choice of n (*stable*) eigenvalues with negative real parts.

This construction for $\bar{P}_s$ was discovered independently by MacFarlane [31] and Potter [32]; Fath [33] used the QR algorithm to compute the eigenvectors and eigenvalues of **M**. A computer package, called OPTSYS, based on these ideas has been developed by Bryson and Hall [34]. The case of nondistinct eigenvalues is less direct—the computation of the eigenvalues and eigenvectors can now be quite difficult, but one can compute other bases for the subspace spanned by the eigenvectors associated with stable eigenvalues of the Hamiltonian matrix (see, e.g., [35]). Laub [36] and Holley and Wei [37] have taken some steps in this direction.

There are several other approaches to the computation of $\bar{P}_s$ (see, e.g., [17, Section 15.3] and [38]–[42]); though no definitive comparisons between these methods are presently available, we may refer to [42] and [42a] for some useful results and for some information on the care necessary to obtain meaningful comparisons.

As stated before, there is a vast literature on the properties of the ARE and its significance for quadratic minimization problems. We shall stop here because our aim has been only to introduce some of the basic facts.

Exercises

3.4-1.   *Guaranteeing a Stability Margin* [17]

Suppose we wish to minimize $J_\alpha = \int_0^\infty [ru^2(t) + x'(t)Qx(t)]e^{2\alpha t}\, dt$. Show that this problem can be reduced to the standard one by introducing $x_\alpha(t) = x(t)e^{\alpha t}$, $u_\alpha(t) = u(t)e^{\alpha t}$. Also show that the closed-loop poles of the system minimizing J_α will have poles with real parts all less than α.

3.4-2.   *Optimum Root Loci*

Determine the loci of optimum closed-loop poles for the problem of Example 3.4-1.

3.4-3.   *Optimum Root Loci*

For the harmonic oscillator $\dot{x}_1 = x_2$, $\dot{x}_2 = -\omega_0^2 x_1 + u$, determine the loci of optimum closed-loop poles for regulating (a) $y = x_1$, (b) $y = x_2$, and (c) $y = x_1 + x_2$.

3.4-4.   *More on the Case $r \rightarrow 0$*

In Sec. 3.4.1 we discussed the fact that for low cost of control ($r \rightarrow 0$) some of the poles moved off to infinity in a Butterworth pattern with radius given by (13). Show that the radius can also be identified as the frequency ω_0 such that $|H(j\omega_0)| = |c(j\omega_0 I - A)^{-1}b| = \sqrt{r}$. (Reference: R. J. Leake, *IEEE Trans. Autom. Control*, **AC-10**, pp. 342–344, 1965.)

3.4-5.   *A Tracking or Servo Problem Reducible to a Regulator Problem*

Consider a realization $\{A, b, c\}$ in *observability* canonical form and a cost functional $J = \int_0^\infty \{[x_1(t) - p(t)]^2 + ru^2(t)\}\, dt$, where $p(\cdot)$ satisfies the equation $a(D)p(t) = 0$, $a(s) = \det(sI - A)$, $D = d/dt$. By defining new state variables, $\theta_i(t) = x_i(t) - p^{(i-1)}(t)$, show that the above optimal tracking problem can be reduced to an optimal regulator problem. (Reference: K. Ramar and B. Ramaswamy, *IEEE Trans. Autom. Control*, AC-15, pp. 500–501, 1970.)

3.4-6.   Multiple Solutions of the ARE [25]

Show that the ARE can have many solutions, even nonnegative-definite ones, by checking that

$$P_1 = \begin{bmatrix} \sqrt{2} - 1 & -\sqrt{2} + 1 \\ -\sqrt{2} + 1 & \sqrt{2} - 1 \end{bmatrix}, \qquad P_2 = \begin{bmatrix} 3 + \sqrt{2} & 1 + \sqrt{2} \\ 1 + \sqrt{2} & 1 + \sqrt{2} \end{bmatrix}$$

both satisfy the ARE with

$$A = \begin{bmatrix} 0 & 1 \\ 1 & 0 \end{bmatrix}, \qquad b = \begin{bmatrix} 0 \\ 1 \end{bmatrix}, \qquad c = \begin{bmatrix} 1 & -1 \end{bmatrix}, \qquad r = 1$$

3.4-7.   An Alternative Form of the ARE

a. Show that the ARE can be rewritten as

$$\bar{A}'\bar{P}_s + \bar{P}_s\bar{A} = -\begin{bmatrix} c' & k' \end{bmatrix} \begin{bmatrix} 1 & 0 \\ 0 & r \end{bmatrix} \begin{bmatrix} c \\ k \end{bmatrix}$$

where $\bar{A} = A - bk$, $k = r^{-1}b'\bar{P}_s$.

b. Now apply Corollary 2.6-2 (check the implications of $\dot{V} \equiv 0$) to show that if $\{c, A\}$ is observable, then $\bar{P}_s$ is the only stabilizing and *positive-definite* (as defined in Sec. A.14) solution of the ARE. *Remark:* For part b, one can also use the fact that $\{c, A\}$ is observable if and only if

$$\left\{ \begin{bmatrix} c \\ k \end{bmatrix}, A - bk \right\}$$

is observable, which can be proved by using the **PBH** rank criterion for observability.

3.4-8.   An Expression for the Minimal Cost

Show that $J_{\min} = x_0'\bar{P}_s x_0$. *Hint:* By direct substitution, show that $J_{\min} = x_0'Px_0$, where $P \triangleq \int_0^\infty e^{\bar{A}'t}(c'c + \bar{P}_sbr^{-1}b'\bar{P}_s)e^{\bar{A}t}\,dt$ and $\bar{A} = A - b\bar{k}$. Now show by direct calculation (using the stability of $\bar{A}$) that $\bar{A}'P + P\bar{A} = -(c'c + \bar{P}_sbr^{-1}b'\bar{P}_s)$, and compare with the ARE as written in Exercise 3.4-7. (See also Exercise 3.4-12.)

3.4-9.   The Hamiltonian Matrix M

The so-called Hamiltonian matrix of (3.4-27), written for the multivariable case as

$$\mathbf{M} = \begin{bmatrix} A & -BR^{-1}B' \\ -Q & -A' \end{bmatrix}, \qquad R = \text{a symmetric matrix}$$

has many interesting properties. A few of these are explored here, especially those that would supplement the discussions in Sec. 3.4.3. Let P be any solution of the ARE $0 = A'P + PA - PBR^{-1}B'P + Q$, and let

$$T = \begin{bmatrix} I & 0 \\ P & I \end{bmatrix}$$

Show that

a. $T^{-1}MT = \begin{bmatrix} A - BK & -BR^{-1}B' \\ 0 & -(A - BK)' \end{bmatrix}$, $K \triangleq R^{-1}B'P$, and

b. $\det(sI - M) = \tilde{a}(s)\tilde{a}(-s)(-1)^n$, where $\tilde{a}(s) = \det(sI - A + BK)$.

3.4-10. *Eigenvalues and Eigenvectors of* **M**

Suppose the Hamiltonian matrix **M** of Exercise 3.4-9 has $2n$ distinct eigenvalues. Let

$$\mathbf{M}\begin{bmatrix} f_i \\ g_i \end{bmatrix} = \lambda_i \begin{bmatrix} f_i \\ g_i \end{bmatrix}, \qquad i = 1, \ldots, 2n$$

and $\mathcal{F} \triangleq [f_1, \ldots, f_n]$, $\mathcal{G} \triangleq [g_1, \ldots, g_n]$, where we choose n eigenvectors such that $\mathcal{F}^{-1}$ exists.

a. If λ_i is an eigenvalue of **M**, show that $-\lambda_i$ is also.

b. Show that $P = \mathcal{G}\mathcal{F}^{-1}$ is a solution of the ARE.

c. Show that with P as in part b the matrix $A - BK$ has eigenvalues $\{\lambda_i\}$ associated with eigenvectors $\{f_i, i = 1, \ldots, n\}$, $K \triangleq R^{-1}B'P$.

d. Show that all solutions of the ARE have the form $\mathcal{G}\mathcal{F}^{-1}$.

e. Hence show again that there can be at most one solution $\bar{P}$ of the ARE such that $A - BK$ is stable.

Hint for part d: Note first that

$$\mathbf{M}\begin{bmatrix} I \\ P \end{bmatrix} = \begin{bmatrix} I \\ P \end{bmatrix}A_K, \qquad A_K \triangleq A - BK$$

and $\mathcal{F}^{-1}A_K\mathcal{F} = \Lambda = \mathrm{diag}\{\lambda_1, \ldots, \lambda_n\}$, which shows that the columns of $[\mathcal{F}' \ \mathcal{F}'P']'$ are eigenvectors of **M**. Hence $\mathcal{F}'P' = \mathcal{G}'$.

3.4-11. *The Riccati Equation*

Show that to be consistent with the Euler-Lagrange equations (23) the matrix $P(\cdot)$ defined by $\lambda(t) = P(t)x(t)$ [cf. (3.4-25)] must obey the so-called Riccati differential equation $-\dot{P}(t) = P(t)A + A'P(t) + c'c - P(t)br^{-1}b'P(t)$, $t \leq t_f$, with $P(t_f) = 0$. *Remarks:* The Riccati equation was studied ca. 1700 by James Bernoulli and by Jacopo Francesco, Count Riccati, and the name was bestowed by D'Alembert in 1763. Legendre introduced the Riccati equation into the calculus of variations in 1786 (cf. Exercise 3.4-12). Its first appearance in control theory was in a 1957 paper by R. Bellman (*Q. Appl. Math.*, **14**, pp. 353–359). In 1960 R. E. Kalman (*Bol. Soc. Mat. Mex.*, **5**, pp. 102–119) introduced the matrix version and showed the importance of the notions of controllability and observability in studying the properties of the

Riccati equation in the quadratic regulator problem, and since then there has been an intensive study of this equation in the literature of "modern" control theory (see, e.g., [22]). In the author's opinion, this emphasis on the *nonlinear* Riccati equation has diverted attention from the basically linear underlying aspects of the quadratic regulator problem and has led to a neglect of other not less promising approaches. However, this question is beyond the scope of this book, see, e.g., [41].

3.4-12. *Legendre's Method for the General Quadratic Regulator Problem*

Let $\dot{x}(t) = A(t)x(t) + B(t)u(t)$, $x(t_0) = x_0$. Show that the following identity holds for any differentiable symmetric matrix $P(\cdot)$: $x'(t)P(t)x(t)\Big|_{t_0}^{t_f} = \int_{t_0}^{t_f} \phi[x(t), u(t)]\, dt$, where $\phi(\cdot, \cdot)$ is a quadratic form:

$$\phi[x(t), u(t)] = [x' \quad u'] \begin{bmatrix} \dot{P} + A'P + PA & PB \\ B'P & 0 \end{bmatrix} \begin{bmatrix} x \\ u \end{bmatrix}$$

Use this identity to show that $u(\cdot) = -R^{-1}B'P(\cdot)x(\cdot)$ will minimize $J = x'(t_f)Q_f x(t_f) + \int_{t_0}^{t_f} [x'(t)Q(t)x(t) + u'(t)R(t)u(t)]\, dt$ *if there exists* $P(\cdot) = P'(\cdot)$ satisfying the Riccati equation $-\dot{P}(t) = A'(t)P(t) + P(t)A(t) + Q(t) - P(t)B(t)R^{-1}(t)B'(t)P(t)$, $P(t_f) = Q_f$. Also show that $J_{\min} = x_0'P(t_0)x_0$. *Hint:* Write $J - x_0'P(t_0)x_0$ as the integral of a perfect square. [Legendre (1783) treated the scalar version of this problem, with $A(\cdot) \equiv 0$, $B(\cdot) \equiv 1$.]

3.4-13. *An Example*

Show that a control $u(\cdot)$ that will minimize $J = q_f x^2(t_f) + \int_0^{t_f} u^2(t)\, dt$ subject to $\dot{x} = u$, $x(0) = x_0$ is $u(t) = -(t_f - t + q_f^{-1})^{-1}$, $0 \le t \le t_f$. What happens in the limit as $q_f \to \infty$? $t_f \to \infty$? Try to explain the limiting results.

3.4-14. *Chang-Letov Formula from the ARE*

Show how to deduce the Chang-Letov formula (3.4-9) from the expression $\bar{k} = r^{-1}b'\bar{P}$, where $\bar{P}$ is any solution of the ARE. *Hint:* Write the ARE as $\bar{P}(sI - A)r^{-1} + (-sI - A')\bar{P}r^{-1} + \bar{P}br^{-2}b'\bar{P} = r^{-1}c'c$; pre- and postmultiply both sides by $b'(-sI - A')^{-1}$ and $(sI - A)^{-1}b$, respectively; and use the result (3.4-7) that $a_k(s) = a(s)[1 + k(sI - A)^{-1}b]$.

3.4-15. *Quadratic Minimization via the ARE*

Use the method of Exercise 3.4-12 to show that if there exists a solution $\bar{P}$ of the ARE (3.4-33) such that $A - br^{-1}b'\bar{P}$ is stable, then the cost J of Eqs. (3.4-1)–(3.4-3) is minimized by choosing $u(\cdot) = -r^{-1}b'\bar{P}x(\cdot)$. One method of proving the existence of such a $\bar{P}$ (when $\{A, b, c\}$ is minimal) is described in the Exercise 3.4-16. Others can be found in [18] and [29].

3.4-16. *Existence of a Positive-Definite Solution of the ARE*

Suppose that $\{A, b\}$ is controllable. Then we can find a feedback law $u = -k_1 x$ such that $A - bk_1$ is (asymptotically) stable. Assume also that $\{c, A\}$

is observable, and define a matrix $P_1 = P_1'$ as the unique positive-definite solution of the linear Lyapunov matrix equation (cf. Corollary 2.6-2) $P_1(A - bk_1) + (A - bk_1)'P_1 = -c'c - k_1'k_1$.

a. Show that with $u = -k_1x$, $\int_0^\infty [u^2 + (cx)^2]\, dt = x_0'P_1x_0$. (Note that the integral must always be strictly > 0, unless $x_0 = 0$, which proves again that P_1 is positive definite.)

b. Define $k_2 = b'P_1$, and show that $P_1(A - bk_2) + (A - bk_2)'P_1 = -c'c - k_2'k_2 - (k_2 - k_1)'(k_2 - k_1)$. Conclude that $A - bk_2$ is stable.

c. Define $P_2 = P_2'$ as the unique positive-definite solution of the Lyapunov equation $P_2(A - bk_2) + (A_2 - bk_2)'P_2 = -c'c - k_2'k_2$, and show that with

$$u = -k_2x, \quad x_0'P_2x_0 = \int_0^\infty [u^2 + (cx)^2]\, dt \leq x_0'P_1x_0.$$

d. Continue in this way to obtain a sequence of gains $\{k_i = b'P_{i-1}\}$ and a decreasing sequence of costs $\{x_0'P_ix_0\}$. Since the costs are bounded below by zero, this decreasing sequence has a limit $x_0'\bar{P}x_0$, and since x_0 is arbitrary, we can say also that $\lim P_i = \bar{P}$. It follows also that $\lim k_i = \bar{k} = b'\bar{P}$ and $\bar{P}(A - b\bar{k}) + (A - b\bar{k})\bar{P} = -c'c - \bar{k}'\bar{k}$.

e. Conclude that $\bar{P}$ obeys the ARE (for $r = 1$) and that $A - b\bar{k}$ is stable. *Remarks:* This scheme is in fact the so-called Newton-Raphson method for solving the ARE (cf. [38] and [18, Sec. 3.5.4]).

3.4-17.   Sensitivity Properties

a. For the optimal quadratic regulator use Eq. (3.2-7) to show that

$$|1 + \bar{k}(j\omega I - A)^{-1}b| \geq 1$$

b. Show that a so-called Nyquist plot in the complex plane of $\bar{k}(j\omega - A)^{-1}b$, as ω varies from $-\infty$ to ∞, always remains outside a circle of center $-1 + j0$ and radius 1.

c. Those familiar with classical control concepts should try to show that part b implies that the optimal regulator has a gain margin of ∞ (i.e., we can increase the loop gain arbitrarily without affecting stability) and a phase margin of at least 60°.

[Reference [17] provides further discussion of these topics and of the good tolerance of the optimal regulator to nonlinearities in the feedback loop. It is worth noting however that these good properties may be lost if the states are not directly available but have first to be estimated, a problem discussed in the next chapter.]

3.5   DISCRETE-TIME SYSTEMS

Several of the previous results carry over easily to discrete-time systems, as we shall note in Sec. 3.5.1. In Sec. 3.5.2, we shall exhibit a nice relationship between the state-controllability problem of taking an arbitary state to the origin and the feedback solution to the modal controllability problem. In the course of this analysis, we see that the celebrated *Principle of Optimality* (useful in *Dynamic Programming*) arises very naturally. An early application of dynamic programming was to the discrete-time quadratic regulator problem, which we examine briefly in Sec. 3.5.3, also noting the discrete-time analog of the symmetric root-locus method of Sec. 3.4.1. The finite-time regulator problem is generally solved via a Riccati-type difference equation. But to provide a glimpse of more recent developments, in Sec. 3.5.4 we shall briefly present alternative *standard* and *fast* square-root control algorithms, which are also important in several other linear systems calculations.

3.5.1 Modal Controllability

Suppose we have a realization

$$x(i + 1) = Ax(i) + bu(i)$$

with

$$a(z) = \det (zI - A)$$

and we use state-variable feedback $u \rightarrow v - kx$ to get a new realization

$$x(i + 1) = (A - bk)x(i) + bv(i)$$

The question is whether the new characteristic polynomial

$$\alpha(z) = \det (zI - A + bk)$$

can be specified arbitrarily by proper choice of k. It is clear that if z is replaced by s, we shall have exactly the problem discussed in Sec. 3.2, and exactly the same algebraic arguments will show that the realization will be modally controllable [i.e., $\alpha(z)$ can be an arbitrary nth-degree polynomial in z] if and only if $\{A, b\}$ is controllable; i.e., $\mathcal{C}(A, b) = [b \ Ab \ \cdots \ A^{n-1}b]$ is nonsingular. All the other results and comments of Secs. 3.2 and 3.3 carry over as well. For example, the required gain can be computed as [cf. (3.2-14)]

$$k = (\alpha - a)\mathfrak{Q}_-^{-T}\mathcal{C}^{-1}$$

or, via Ackermann's formula (3.2-23),

$$k = [0 \ \cdots \ 1]\mathcal{C}^{-1}\alpha(A)$$

Incidentally, one distinction should perhaps be explicitly noted. In continuous time, a realization is stable if its eigenvalues (the roots of its characteristic polynomial) have negative real parts; in discrete time, the condition is that the roots must lie within the unit circle (cf. Sec. 2.6).

We recall now that in continuous time the conditions for modal controllability and state controllability coincide. This is essentially also true in discrete time, but we have to be a bit careful because, as noted in Sec. 2.3.3, in discrete time we have to distinguish between controllability *to* the origin (controllability) and controllability *from* the origin (reachability) unless the matrix A is nonsingular. Also, there is an interesting connection between the controllability to the origin problem and the use of state-variable feedback, which we shall now explore.

3.5.2 Controllability to the Origin, State-Variable Feedback, and the Principle of Optimality

Suppose we have a controllable realization

$$x(i + 1) = Ax(i) + bu(i), \qquad x(0) = x_0 \tag{1}$$

Then it is clear (see also Sec. 2.3.3) that the unique input sequence that will take an arbitrary x_0 to zero in (not more than) n steps can be obtained ([5] and [43]) by solving the equation

$$0 = x(n) = A^n x_0 + A^{n-1} bu_*(0) + \cdots + bu_*(n - 1) \tag{2}$$

For simplicity of argument,† let us assume also that

$$A \text{ is nonsingular} \tag{3}$$

so·that we can write

$$\begin{bmatrix} u_*(0) \\ \cdot \\ \cdot \\ \cdot \\ u_*(n - 1) \end{bmatrix} = -[A^{-1}b \quad \cdots \quad A^{-n}b]^{-1} x_0 \tag{4}$$

Let us further write this as

$$u_*(0) = -q_1 x_0, \qquad u_*(1) = -q_2 x_0, \ldots, u_*(n - 1) = -q_n x_0 \tag{5}$$

where

$$q_i = i\text{th row of } [A^{-1}b \quad \cdots \quad A^{-n}b]^{-1}$$

The above-determined sequence of controls will take an arbitrary (initial) state to the origin in n steps. Of course, there are states that can be driven to the origin in less than n steps. For example, if

$$x_0 = \gamma_0 A^{-1} b$$

---

†The more general case can also be studied by using *pseudo-inverses* or, more basically, by making a change of variables so that $T^{-1}AT =$ block diag $\{\bar{A}, N\}$, $\bar{A}$ nonsingular, N nilpotent.

then we see from our explicit solution that

$$u_*(0) = -q_1 x_0 = -\gamma_0, \qquad u_*(i) = 0, \qquad i = 1, \ldots, n-1$$

so that any such state can be driven to the origin in *one step*. Similarly, any state of the form

$$x_0 = \gamma_0 A^{-1} b + \gamma_1 A^{-2} b$$

can be taken to the origin in two steps by the input

$$u_*(0) = -\gamma_0, \qquad u_*(1) = -\gamma_1, \qquad u_*(i) = 0, \qquad i = 2, \ldots, n-1$$

It is interesting to note the details of the transition to zero. We first have

$$\begin{aligned}
x_*(1) &= A x_0 + b u_*(0) \\
&= A(\gamma_0 A^{-1} b + \gamma_1 A^{-2} b) + b(-\gamma_0) \\
&= \gamma_1 A^{-1} b
\end{aligned}$$

But now $x_*(1)$ is a state that can be taken to zero in one step,

$$\begin{aligned}
x_*(2) &= A(\gamma_1 A^{-1} b) + b u_*(1) \\
&= \gamma_1 b + b(-\gamma_1) = 0
\end{aligned}$$

Similarly, a state that requires three steps to be forced to the origin will in the first step be reduced to a state that requires only two steps to go to the origin, and so on.† The interesting point shown by this argument is that *the control computed via (5) will take any initial state to the origin in the shortest possible time.*

The above discussion illustrates a nice property of an *optimal* trajectory, a general result that has been called the *Principle of Optimality* [44]: "An optimal trajectory has the property that at an intermediate point, no matter how it was reached, the rest of the trajectory must coincide with an optimal trajectory as computed from this intermediate point as the initial point." This is, of course, almost a truism, and a proof by contradiction is immediate. Nevertheless, it is a very powerful idea, and we shall get a very nice result by using it here. Other applications abound in the literature.

[It may be of interest that the principle apparently first arose in connection with the famous Brachystochrone problem, which had been published as a challenge in 1696 to all the mathematicians of his day by Johann Bernoulli (*Problema Novum, Ad cujus Solutionem Mathematici invitantur*). Newton

---

†The reader might wish to develop the analogies between these facts and the results of Sec. 2.3.2 on using impulses to change the state of a continuous-time system.

presented a solution in January 1697, anonymously, but Bernoulli recognized the style (*ex ungue leonem*—"by the claws of the lion"). Johann, and his brother Jakob, each published solutions in May 1697. Jakob used the Principle of Optimality, which was explicitly restated by Johann in 1706 [45]: "parceque toute Courbe, qui doit donner un *Maximum*, conserve aussi dans toutes ses parties les loix de ce même Maximum" ("because all curves, which should give a maximum, preserve also in all their sections the laws of the same maximum"). Euler stated this principle explicitly in a 1744 textbook [46, Sec. 38], but it was largely ignored until its use was revived in recent times by Carathéodory in the 1930s [47] and by Bellman in the 1950s [44]. Our historical remarks are drawn from an article by Wishart [48] and a little book by Woodhouse [49] in which James Bernoulli's original solution is reproduced [49, pp. 4–5]. English translations of relevant sections of the references to Euler and Bernoulli can be found in [49a, pp. 391–401].]

Feedback Form.   To return to our problem, an important implication of the principle of optimality is the following: When we start at 0 with an initial state x_0, the optimal control that has to be applied at 0 is [cf. (5)]

$$u_*(0) = -q_1 x_0 \qquad (6)$$

Now suppose we are at time i and the state along the optimal trajectory is $x_*(i)$. The $*$ is used to denote states on the optimal trajectory, the one that would be followed if the optimal inputs were used. But considering this intermediate time i, when we are in state $x_*(i)$, as a new initial time, the principle of optimality says that the optimal control input should be

$$u_*(i) = -q_1 x_*(i) \qquad (7)$$

which should be compared with the value given by (5), $u_*(i) = -q_{i+1} x_0$. Therefore we have the identity

$$-q_1 x_*(i) = -q_{i+1} x_0, \qquad i = 1, \ldots, n-1 \qquad (8)$$

whose truth can also readily be established by direct calculation ([43] and [5]). We leave this to the exercises, and go on here to explore the significance† of the closed-loop solution (7).

Formula (7) means that the control at any step can be computed as a fixed linear combination of the state variables at that time; in contrast, our original equations (5) determined all the controls (at the time $t = 0$) as different linear combinations of the initial state. The original formulas give us what is called an *open-loop control*; the present formulas yield a *feedback*

---

†The reader should explore the continuous-time analog of this result. (*Hint:* Reread the initial discussion in Sec. 3.4.)

control, which, as expected, is better in terms of sensitivity and in reduction of the effects of external disturbances. We shall explain this presently, but first we note that such feedback control with a constant gain vector was already employed in Sec. 3.2 in connection with the pole-shifting problem. It is therefore natural to ask what effect use of the feedback control law

$$u(i) = -q_1 x(i), \qquad i = 0, 1, \ldots$$

has on the eigenvalues of the system

$$x(k + 1) = Ax(k) + bu(k) = (A - bq_1)x(k) \tag{9}$$

We could calculate $\det (zI - A + bq_1)$ directly, but we can obtain the answer without any calculation. For this, note that with the specified control we shall have

$$0 = x(n) = (A - bq_1)^n x_0$$

so that, since x_0 is arbitrary, $A - bq_1$ is a matrix such that

$$(A - bq_1)^n = 0 \tag{10}$$

A matrix is said to be *nilpotent* if some power of the matrix is zero, and it is easy to see that all the eigenvalues of such a matrix must be zero. [For suppose that $A^k = 0$ for some $k < \infty$ and that λ is an eigenvalue of A with associated eigenvector p, $Ap = \lambda p$. Then $0 = A^k p = \lambda^k p$, so that, since eigenvectors are assumed to have nonzero (unit) length, we must have $\lambda = 0$.] Therefore we must have

$$\det (zI - A + bq_1) = z^n \tag{11}$$

We also see that the effect of the feedback control is to shift all the eigenvalues of A to the origin. It is easy to see in retrospect that doing this would solve the controllability (p.s.t.o.) problem, and, in fact, this was the solution used in the early non-state-variable analyses of the finite-time-settling problem! According to Kalman [5], however, in these earlier solutions unnecessary conditions, e.g., stability of A, were imposed and a clear understanding was gained only with the state-variable solution.

To return to our problem, it is of interest to determine, by the methods of Sec. 3.2, the gain that is necessary to force all the eigenvalues to the origin and to check that this gain is equal to the row vector q_1. For this, it will be easiest to use the special formula due to Ackermann, Eq. (3.2-23), which gives the gain vector directly in terms of the desired characteristic polynomial,

$$\alpha(z) = a_k(z) = z^n \tag{12}$$

as

$$k = [0 \quad \cdots \quad 0 \quad 1] \mathcal{C}^{-1} A^n$$

$$= [0 \quad \cdots \quad 0 \quad 1][A^{-n}b \quad \cdots \quad A^{-1}b]^{-1} = q_1 \tag{13}$$

The Advantage of Feedback Control. We started with an *open-loop* solution (5) of the controllability p.s.t.o. problem and showed that it could be rewritten in a *closed-loop* form (7). Let us now explain why the closed-loop solution is preferred in practice, though for the mathematical problem we posed it is completely equivalent to the open-loop control.

For simplicity of discussion, suppose that we have a 10-state system at rest but that at some time, say $t = 0$, there is a state disturbance, say x_0. Suppose also that x_0 is such that control will have to be applied over 10 time instants before the system can be restored to equilibrium, i.e., to the zero state. We now calculate and apply the open-loop control that will do this. However, suppose that there is another disturbance to the state at $t = 4$. Then the state at time 10 will not be zero as we had expected when we computed the control at $t = 0$. We shall now have to recompute another open-loop control that will restore equilibrium in not more than another 10 instants (i.e., by $t \leq 20$), unless there is still another disturbance between $t = 11$ and $t = 20$. However, let us see what happens with the closed-loop control. If there were no disturbance at $t = 4$, then equilibrium would be restored by $t = 10$, just as with the open-loop control. However, if there is a disturbance at $t = 4$, then the closed-loop control will drive this to zero in not more than another 10 instants, i.e., by $t \leq 14$, provided, as before, that there is no other disturbance. Therefore, we see that with feedback control the system will respond to disturbances more quickly, a fact well known to control engineers.

Dynamic Programming. In the above we used the principle of optimality to show the equality of the optimal open-loop control to a feedback or closed-loop control. In fact, the principle can be used to directly obtain the closed-loop control, as we ask active readers to try to show. This will be an application of the *dynamic programming* method. An early application of dynamic programming in linear control theory was to the discrete-time quadratic regulator problem (see esp. [50]), with results that we shall describe next.

3.5.3   The Discrete-Time Quadratic Regulator Problem

For the linear system

$$x(k + 1) = Ax(k) + bu(k), \qquad x(0) = x_0 \tag{14}$$

the problem is to choose $\{u(\cdot)\}$ so as to minimize

$$J_N = x_N' P_f x_N + \sum_0^{N-1} x'(i) Q x(i) + r \sum_0^{N-1} u^2(i) \tag{15}$$

The optimum solution will be shown to be (using subscripts for notational compactness)

$$u_{*i} = -k_i \tilde{R}_i^{-1} x_{*i} \tag{16}†$$

where

$$k_{i-1} \triangleq b' P_i A, \qquad \tilde{R}_{i-1} \triangleq r + b' P_i b \tag{17}$$

and $P(\cdot)$ obeys a Riccati-type difference equation

$$P_{i-1} = A' P_i A + Q - k'_{i-1} \tilde{R}_{i-1}^{-1} k_{i-1}, \qquad i = N, \dots, 1, 0 \tag{18}$$

with terminal condition

$$P_N = P_f \tag{19}$$

The optimal cost can be calculated as

$$J_{*N} = x'_0 P_0 x_0 \tag{20}$$

This result can be derived in several ways, e.g., by the analog of the method of Sec. 3.4.1. We shall describe another (so-called *square-root*) approach at the end of Sec. 3.5.4.

The Steady-State Solution.   If the final time is very large so that we can assume that $N = \infty$ and $P_f = 0$, then it can be shown that if $\{A, b\}$ is controllable (or just stabilizable) and $\{c, A\}$ is observable (or just detectable), where $Q = c'c$, then P_i and k_i tend toward constancy. The optimal steady-state gain $\bar{k}$ can then be characterized (cf. Sec. 3.4.1) as being such that

$$\det (zI - A + b\bar{k}) = \prod_1^n (z - z_i)$$

where the $\{z_i\}$ are the n roots lying within the unit circle of the $2n$th-degree polynomial equation $\Delta(z) = 0$, where

$$\Delta(z) = 1 + H(z^{-1}) r^{-1} H(z), \qquad H(z) \triangleq c(zI - A)^{-1} b$$

Note that in discrete time z^{-1} takes the place of $-s$, and therefore instead of a *symmetric* root locus as in Sec. 3.4.1, we now have what we may call a *reciprocal* root locus.

For a simple example, we shall consider the root-locus plot necessary to solve part c of Exercise 3.5-6. It will turn out there that

$$H(z) = \frac{X_1(z)}{U(z)} = \frac{\Delta^2(z + 1)}{2(z - 1)^2}$$

†The $*$ is used to designate the optimum trajectory; i.e., $x_*(i + 1) = A x_*(i) + b u_*(i)$.

so that the reciprocal root characteristic equation is

$$\frac{(z^{-1} + 1)(z + 1)}{(z^{-1} - 1)^2(z - 1)^2} = -\frac{4r}{\Delta^4}$$

and a sketch of the upper half of the reciprocal root locus is given in the figure. Exercises 3.5-7–3.5-9 provide further illustrations.

As in the continuous-time case, the discrete-time quadratic regulator problem has been and is being extensively studied in the current technical literature. Such study is beyond our scope here, though we shall briefly present some interesting recent developments in Sec. 3.5.4, including a derivation of Eqs. (16)–(20).

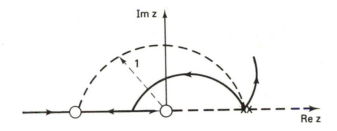

*3.5.4   Square-Root and Related Algorithms

For reasons that will be explained at greater length below, it is of interest to consider how to propagate not the Riccati matrix P_i but its "square root." Thus, let

$$P_i^{1/2} \triangleq \text{a square root of } P_i,$$
$$= \text{any } n \times n \text{ matrix such that } P_i^{T/2}P_i^{1/2} = P_i$$

Here the superscript $T/2$ is used to denote the transpose of $P_i^{1/2}$. We note that square roots are not unique because if S is any orthogonal matrix, i.e., if S is such that

$$S'S = I = SS'$$

then clearly

$$P_i = P_i^{T/2}P_i^{1/2} = P_i^{T/2}S'SP_i^{1/2}$$

We could define a unique square root by taking $P^{1/2}$ to be symmetric or to be lower triangular with positive diagonal elements, etc., but it is not necessary to specify a unique choice if our ultimate interest is in the squared quantity. In any case, let us now consider the following algorithm ([51] and [52]). Assume that we have computed $P_i^{1/2}$. Then $P_{i-1}^{1/2}$ can be determined as follows: Apply to the left of the $(n + 2) \times (n + 1)$-dimensional

$$\text{prearray} \triangleq \begin{matrix} & \overset{n}{} & \overset{1}{} & \\ \begin{bmatrix} c & 0 \\ 0 & r^{1/2} \\ P_i^{1/2}A & P_i^{1/2}b \end{bmatrix} & \begin{matrix} 1 \\ 1 \\ n \end{matrix} \end{matrix} \qquad (21)$$

any $(n + 2) \times (n + 2)$-orthogonal transformation S that will force n zeros in the positions shown, and label the entries in the remaining positions by $\{X, g, f\}$ as shown:

$$\text{postarray} \triangleq \begin{array}{c} 1 \\ n \\ 1 \end{array} \overset{\begin{array}{cc} n & 1 \end{array}}{\begin{bmatrix} 0 & 0 \\ X & 0 \\ g & f \end{bmatrix}} \tag{22}$$

Then by *squaring up* the two arrays and comparing terms, we shall find that

$$f^2 = r + b'P_i b \triangleq \tilde{R}_{i-1}$$

$$fg = b'P_i A = k_{i-1}$$

$$X'X + g'g = Q + A'P_i A, \qquad Q = c'c$$

so that we can identify

$$g = \tilde{R}_{i-1}^{-1/2} k_{i-1} \tag{23a}$$

$$X'X = Q + A'P_i A - k'_{i-1}\tilde{R}_{i-1}^{-1} k_{i-1} = P_{i-1} \tag{23b}$$

and

$$X = P_{i-1}^{1/2}, \qquad \text{a square root of } P_{i-1} \tag{23c}$$

This is an intriguing algorithm, for several reasons First, the transformation from (21) to (22) can be effected without explicitly determining the (nonunique) orthogonal transformation S by using one of several numerically stable algorithms, e.g., the so-called Householder or Givens or modified Gram-Schmidt transformations (see, e.g., [53], [54] and [55] for recent discussions). A second point is that the $\{P_i\}$ obtained in this way will be guaranteed nonnegative definite (because they are obtained by *squaring* and so will always be nonnegative definite), which theoretical requirement may not be achieved due to numerical errors in propagating P_i via the Riccati equation (18). Furthermore in certain applications we can use lower-precision arithmetic when working with square-roots, because they have a lower dynamic range of values than the squared quantities (e.g., $(-10, 10)$ as compared to $(-100, 100)$).

These possible numerical advantages often make it feasible to compute the steady-state value of the Riccati variable by propagating the above algorithm for many steps. But in fact if only the steady-state value is desired, we can find this with substantially less computation.

Steady-State Value—A Square-Root Doubling Algorithm.   For notational convenience, assume that we reverse the direction of time and instead of starting at $t = N$, where N is allowed to be very large, we start at $t = 0$ with $P_0 = 0$ and wish to compute P_i for very large i. Then we can use a *square-root doubling* algorithm to successively compute

$$P_1^{1/2}, P_2^{1/2}, P_4^{1/2}, \ldots$$

so that, for example, in 20 steps we obtain what would have taken more than 10^6 steps of direct iteration of the Riccati equation. The algorithm can be described as follows ([56]):

At time i, we assume we have found $P_i^{1/2}$ and two auxiliary quantities $\{W_i^{1/2}, \Phi_i\}$. These auxiliary quantities have a certain physical significance as well, though we shall not pursue this here. With the given quantities, we form prearray I and apply to it *any* orthogonal transformation to give an array with zeros in the positions indicated in postarray I:

$$
\begin{bmatrix} P_i^{1/2} & 0 \\ 0 & I \\ P_i^{1/2}\Phi_i & P_i^{1/2}W_i^{T/2} \end{bmatrix} \sim \begin{bmatrix} 0 & 0 \\ P_{2i}^{1/2} & 0 \\ \hat{K}_i & \hat{R}_i^{1/2} \end{bmatrix} \tag{24a}
$$
$$
\quad\;\; \text{\textit{prearray I}} \qquad\qquad \text{\textit{postarray I}}
$$

Then in the remaining positions we can read off the square root at time $2i$ and also certain other quantities $\{\hat{K}_i, \hat{R}_i\}$. To proceed to the next step we need to know $W_{2i}^{1/2}$ and Φ_{2i}. The latter can be computed as

$$
\Phi_{2i} = \Phi_i[\Phi_i - W_i^{T/2}(\hat{R}_i^{1/2})^{-1}\hat{K}_i] \tag{24b}
$$

For $W_{2i}^{1/2}$, we form prearray II and transform it to postarray II with any orthogonal transformation forcing zeros as shown:

$$
\begin{bmatrix} (\hat{R}_i^{T/2})^{-1}W_i^{1/2}\Phi_i' \\ W_i^{1/2} \end{bmatrix} \sim \begin{bmatrix} 0 \\ W_{2i}^{1/2} \end{bmatrix} \tag{24c}
$$
$$
\;\;\text{\textit{prearray II}} \quad\; \text{\textit{postarray II}}
$$

For the initial values we should take

$$
\Phi_1 = A, \qquad P_1^{1/2} = [c \;\; 0], \qquad W_1^{1/2} = r^{-1/2}\begin{bmatrix} b' \\ 0 \end{bmatrix} \tag{24d}
$$

Some numerical examples of the successful use of this algorithm and comparisons with other methods of computing the steady-state value of P are given in [42]. The basis for this algorithm lies in a certain *scattering-theoretical* interpretation [27], [42] of the Hamiltonian equations underlying the solution of the discrete-time quadratic regulator problem—space limitations forbid our discussing this here, though at the end of this section we shall show how the Hamiltonian equations arise in the problem.

Before that, however, let us briefly indicate another important variation of the basic square-root algorithm (21)–(23).

Fast Algorithms for Constant-Parameter Systems. Although we have written the basic algorithm (21)–(23) for constant-parameter systems, the algorithm actually applies also to *time-variant* systems—we merely have to explicitly introduce the time variation by replacing A, b, c, r by $\{A_i, b_i, c_i, r_i\}$. However, this generality is at the same time a weakness when we have *time-invariant* or (*constant-parameter*) systems, because we should be able to exploit the constancy. In fact it has been shown (cf. [51] and [57]) that this can be done by working with the differences $\delta P(i) = P(i-1) - P(i)$ rather than with $P(\cdot)$ itself.

As a simple illustration, let us consider the special case of no terminal cost

$$P(N) = P_f = 0 \tag{25a}$$

Then it can be shown [after some algebra (see [57])] that

$$\delta P(i) = P(i-1) - P(i) \geq 0, \qquad i = N, \ldots, 1 \tag{25b}$$

i.e., $\delta P(i)$ is nonnegative definite for all i. Therefore we can write

$$\delta P(i) = L'(i-1)L(i-1), \qquad \delta P(N) = Q \tag{26}$$

where the $L(\cdot)$ are $p \times n$ matrices and p is the rank of Q (which is 1 if as assumed above $Q = c'c$, c a row vector). Now we can present the following algorithm for propagating $k(\cdot)$ and $L(\cdot)$ ([51] and [52]).

Apply to the rows of the prearray

$$\begin{bmatrix} \tilde{R}_i^{-1/2} k_i & \tilde{R}_i^{1/2} \\ L_i A & L_i b \end{bmatrix} \tag{27}$$

any orthogonal transformation that will take it to the form

$$\begin{bmatrix} Y & Z \\ X & 0 \end{bmatrix} \tag{28}$$

Then we can identify

$$X = L_{i-1}, \qquad Y = \tilde{R}_{i-1}^{-1/2} k_{i-1}, \qquad Z = \tilde{R}_{i-1} \tag{29}$$

The simplification available here is that the arrays will have $(n+1)(1+p)$ elements as opposed to the $(n+1)(1+p+n)$ of arrays (21) and (22). If, as may often happen, $p \ll n$ (e.g., $p = 1$ when $Q = c'c$), this means a substantial reduction in storage and computational requirements.

To show how the fast algorithm (27)–(29) follows from the general algorithm we note that by using the decomposition [cf. (26)]

$$P_i = P_{i+1} + L_i' L_i$$

(where we have used subscripts for notational convenience) the array in (21) can be expanded as

$$\begin{bmatrix} c & 0 \\ 0 & r^{1/2} \\ P_i^{1/2} A & P_i^{1/2} b \end{bmatrix} \sim \begin{bmatrix} c & 0 \\ 0 & r^{1/2} \\ P_{i+1}^{1/2} A & P_{i+1}^{1/2} b \\ L_i A & L_i b \end{bmatrix} \tag{30}$$

where $\sim$ signifies that the two arrays are the same after *squaring*; i.e., the two arrays differ only by an orthogonal transformation.

Next, just as (21) could be transformed to (22) by an orthogonal transformation, we can replace the larger array in (30) by

$$\begin{bmatrix} c & 0 \\ 0 & r^{1/2} \\ P_{i+1}^{1/2}A & P_{i+1}^{1/2}b \\ \hdashline L_iA & L_ib \end{bmatrix} \sim \begin{bmatrix} 0 & 0 \\ 0 & 0 \\ \tilde{R}_i^{-1/2}k_i & \tilde{R}_i^{1/2} \\ \hdashline L_iA & L_ib \end{bmatrix} \tag{31}$$

At the same time the postarray in (22) can itself be rewritten as

$$\begin{bmatrix} 0 & 0 \\ P_{i-1}^{1/2} & 0 \\ \tilde{R}_{i-1}^{-1/2}k_{i-1} & \tilde{R}_{i-1}^{1/2} \end{bmatrix} \sim \begin{bmatrix} 0 & 0 \\ P^{1/2} & 0 \\ L_{i-1} & 0 \\ \tilde{R}_{i-1}^{-1/2}k_{i-1} & \tilde{R}_{i-1}^{1/2} \end{bmatrix} \tag{32}$$

Comparing appropriate columns of the two right-hand arrays in (31) and (32) gives us the fast algorithm (27)–(29).

For further results on these so-called discrete-time Chandrasekhar-type algorithms, refer to [51] and [57]. We may note that these algorithms also play an important role in elucidating certain structural aspects (connected with the poles and zeros) of constant multivariable systems (cf. Sec. 7.6).

At this point, we should say something about the origin of these square-root array methods. Several explanations can be given (see, e.g., [51] and [52]), but here we shall give one based on a particular kind of dynamic programming approach to the quadratic regulator problem.

Square-Root Approach to the Quadratic Regulator Problem. Here we shall briefly describe, following a streamlined version of [52] as worked out by G. Verghese, a simple but comprehensive (including the square-root, Riccati, and Hamiltonian equations) approach to the discrete-time quadratic regulator problem.

Consider a dynamical system

$$x_{i+1} = A_i x_i + B_i u_i, \qquad x_0 \text{ given} \tag{33}$$

where we are to choose $\{u_i, i \geq 0\}$ so as to minimize

$$J_N = x'_N P_N x_N + \sum_0^{N-1} [x'_i \; u'_i] \begin{bmatrix} Q_i & H_i \\ H'_i & R_i \end{bmatrix} \begin{bmatrix} x_i \\ u_i \end{bmatrix} \tag{34}$$

which is a generalization (to time-variant and multivariable-input systems)† of the

†We treat this general case here to indicate that the extension of our discussions to the time-invariant multivariable case is sometimes largely a matter of notation.

problems (14)–(15) introduced in Sec. 3.5.3. Let us assume that the weighting matrix in (34) is nonnegative definite and that it is factored as

$$\begin{bmatrix} Q_i & H_i \\ H_i' & R_i \end{bmatrix} = \begin{bmatrix} C_i' \\ D_i' \end{bmatrix} [C_i \quad D_i] \tag{35a}$$

Assume also that $P_N \geq 0$ and that it is factored as

$$P_N = M_N' M_N \tag{35b}$$

Then we can rewrite J_N as

$$J_N = x_N' M_N' M_N x_N + \sum_0^{N-1} y_i' y_i \tag{36}$$

where

$$y_i = C_i x_i + D_i u_i \tag{37}$$

To obtain a recursive solution, let us rewrite this further as

$$J_N = \| Y_{N-2} \|^2 + \left\| \begin{bmatrix} y_{N-1} \\ M_N x_N \end{bmatrix} \right\|^2 \tag{38}$$

where $\| \ \ \|$ denotes the norm of a vector ($\| a \|^2 = a'a$) and

$$Y_{N-2}' \triangleq [y_0' \quad \cdots \quad y_{N-2}']$$

Now note that because

$$\begin{bmatrix} y_{N-1} \\ M_N x_N \end{bmatrix} = \begin{bmatrix} C_{N-1} & D_{N-1} \\ M_N A_{N-1} & M_N B_{N-1} \end{bmatrix} \begin{bmatrix} x_{N-1} \\ u_{N-1} \end{bmatrix} \tag{39}$$

the $\{ y_{N-1}, x_N \}$ are determined entirely by x_{N-1} and u_{N-1}; furthermore, u_{N-1} has no effect on Y_{N-2}. We can thus find the value of u_{N-1} that minimizes J_N as a function of x_{N-1}. This value is such that it minimizes the second term in (38), and it can be found as follows.

Note first that if S_N is any orthogonal matrix, i.e., $S_N S_N' = I = S_N' S_N$, then

$$\left\| S_N \begin{bmatrix} y_{N-1} \\ M_N x_N \end{bmatrix} \right\| = \left\| \begin{bmatrix} y_{N-1} \\ M_N x_N \end{bmatrix} \right\|$$

We can use this freedom to display precisely the extent to which u_{N-1} can affect this norm. Thus, choose S_N so that

$$S_N \begin{bmatrix} y_{N-1} \\ M_N x_N \end{bmatrix} \triangleq S_N \begin{bmatrix} C_{N-1} & D_{N-1} \\ M_N A_{N-1} & M_N B_{N-1} \end{bmatrix} \begin{bmatrix} x_{N-1} \\ u_{N-1} \end{bmatrix} = \begin{bmatrix} M_{N-1} & 0 \\ G_{N-1} & F_{N-1} \end{bmatrix} \begin{bmatrix} x_{N-1} \\ u_{N-1} \end{bmatrix} \tag{40}$$

where

$$F_{N-1} \text{ has } \textit{full row rank}$$

Then

$$J_N = \| Y_{N-2} \|^2 + \| M_{N-1} x_{N-1} \|^2 + \| G_{N-1} x_{N-1} + F_{N-1} u_{N-1} \|^2 \qquad (41)$$

which is minimized by choosing u_{N-1} such that

$$F_{N-1} u_{N-1} = -G_{N-1} x_{N-1}$$

(recall that x_{N-1} does not depend on u_{N-1}). The fact that F_{N-1} has full row rank ensures that this equation always has a solution. The problem is now reduced to minimizing

$$J_{N-1} = \| Y_{N-2} \|^2 + x'_{N-1} M'_{N-1} M_{N-1} x_{N-1} \triangleq \| Y_{N-3} \|^2 + \left\| \begin{bmatrix} y_{N-2} \\ M_{N-1} x_{N-1} \end{bmatrix} \right\|^2 \qquad (42)$$

which is of the same form as the problem just solved but over a reduced interval.

We thus have a terminating recursive algorithm: Choose orthogonal matrices $\{S_i\}$ so that

$$S_i \begin{bmatrix} C_{i-1} & D_{i-1} \\ M_i A_{i-1} & M_i B_{i-1} \end{bmatrix} = \begin{bmatrix} M_{i-1} & 0 \\ G_{i-1} & F_{i-1} \end{bmatrix} \qquad (43)$$

where F_{i-1} has full row rank. Then the optimal $\{u_i\}$ are solutions of

$$F_i u_i = -G_i x_i, \qquad i = N - 1, \ldots, 0 \qquad (44)$$

Note also that

$$J_{\min} = x'_0 M'_0 M_0 x_0. \qquad (45)$$

Results for $H_i \equiv 0$.   To slightly reduce the notational burden and to make for easier comparisons with earlier more special results, *we shall assume from now on that the cross-terms* H_i *of (34) are identically zero*, i.e.,

$$H_i = C'_i D_i = 0, \qquad (46a)$$

and that the control costs are strictly positive (the so-called nonsingular case), i.e.,

$$R_i = D'_i D_i > 0, \qquad (46b)$$

The latter assumption implies that the D_i have full column rank, a property that is inherited by the F_i, as (43) shows. In this case, then, the F_i are nonsingular (since they have full row and column rank) and the u_{*i} are uniquely given by

$$u_{*i} = -F_i^{-1} G_i x_i \qquad (44a)$$

Riccati Equation.   Comparing this with the previously derived Eq. (20) leads us to suspect that if we define

$$P_i = M_i'M_i, \qquad i = 0, \ldots, N - 1 \tag{47a}$$

then the P_i should obey a Riccati equation. In fact by squaring both sides of (43), we can see that

$$P_{i-1} = A_{i-1}'P_iA_{i-1} + C_{i-1}'C_{i-1} - G_{i-1}'G_{i-1} \tag{47b}$$

$$F_{i-1}'G_{i-1} = B_{i-1}'P_iA_{i-1} \tag{47c}$$

$$F_{i-1}'F_{i-1} = R_{i-1} + B_{i-1}'P_iB_{i-1} \tag{47d}$$

which on elimination gives a Riccati equation:

$$P_{i-1} = A_{i-1}'P_iA_{i-1} + Q_{i-1} - A_{i-1}P_iB_{i-1}(R_i + B_{i-1}'P_iB_{i-1})^{-1}B_{i-1}'P_iA_{i-1}' \tag{48}$$

It can be checked that (48) is equivalent to the form given earlier in (18) and (19) (for the scalar case with $B_i = b$, $R_i = r$, $A_i = A$).

Square-Root Algorithms.   We see now that we can identify M_i as a square root of P_i and that (43) is an algorithm for updating the $\{M_i\}$ (starting with $i = N$ and going to $i = 0$). When specialized to the time-invariant and scalar case, we shall have exactly the square-root algorithm (21)–(23). To see this, we only have to check that choosing

$$C_i = \begin{bmatrix} c \\ 0 \end{bmatrix}, \qquad D_i = \begin{bmatrix} 0 \\ r^{1/2} \end{bmatrix}$$

will reduce the general problem of (33)–(37) to the special time-invariant scalar problem studied earlier in this section.

The Discrete Hamiltonian Equations.   From (40), we can write

$$\begin{bmatrix} y_{i-1} \\ M_ix_i \end{bmatrix} = \begin{bmatrix} C_{i-1} & D_{i-1} \\ M_iA_{i-1} & M_iB_{i-1} \end{bmatrix} \begin{bmatrix} x_{i-1} \\ u_{i-1} \end{bmatrix} = S_i' \begin{bmatrix} M_{i-1} & 0 \\ G_{i-1} & F_{i-1} \end{bmatrix} \begin{bmatrix} x_{i-1} \\ u_{i-1} \end{bmatrix}$$

$$= S_i' \begin{bmatrix} M_{i-1}x_{i-1} \\ 0 \end{bmatrix} \tag{49}$$

where the last equality follows from using (44) for the optimal u_{i-1}. Now using the identity in (43), we can write

$$\begin{bmatrix} C_{i-1}' & A_{i-1}'M_i' \\ D_{i-1}' & B_{i-1}'M_i' \end{bmatrix} \begin{bmatrix} y_{i-1} \\ M_ix_i \end{bmatrix} = \begin{bmatrix} P_{i-1}x_{i-1} \\ 0 \end{bmatrix} \tag{50}$$

Then defining

$$\lambda_{i-1} = P_{i-1}x_{i-1} \tag{51}$$

we see from (50) that

$$\lambda_{i-1} = A_{i-1}'\lambda_i + C_{i-1}'y_{i-1} \tag{52}$$

and also that

$$0 = D'_{i-1}y_{i-1} + B'_{i-1}\lambda_i = D'_{i-1}(C_{i-1}x_{i-1} + D_{i-1}u_{i-1}) + B'_{i-1}\lambda_i$$
$$= R_{i-1}u_{i-1} + B'_{i-1}\lambda_i$$

so we shall have

$$\begin{bmatrix} x_{i+1} \\ \lambda_i \end{bmatrix} = \begin{bmatrix} A_i & -B_iR_i^{-1}B'_i \\ Q_i & A'_i \end{bmatrix}\begin{bmatrix} x_i \\ \lambda_{i+1} \end{bmatrix} \tag{53}$$

These are the discrete-time Hamiltonian equations [cf. Eqs. (3.4-23) and (3.4-24) in the continuous-time problem]. Representation of these equations in a flow graph leads quickly to a *scattering-theory* framework (see e.g., [27]) and thence, among other things, to the square-root doubling algorithm (24) (see [42]). However, we shall terminate our discussions here.

Exercises

3.5-1.

If $\{A, b\}$ is controllable, A is nonsingular, and $A - bk$ is nilpotent, show that the only eigenvector of $A - bk$ is $A^{-1}b$.

3.5-2.

Show that if $\mathcal{C}(A, b)$ has rank less than n, a state θ_0 at $k = 0$ can be taken to a state θ_1 in not more than n steps if θ_1 and θ_0 lie in the space spanned by the columns of $\mathcal{C}(A, b)$, i.e., if θ_1 and θ_0 can be written as linear combinations of these columns.

3.5-3.

a. Find a scalar control law u which will take the output y of the system

$$x_{n+1} = \begin{bmatrix} 0 & 1 \\ a_0 & a_1 \end{bmatrix}x_n + \begin{bmatrix} 0 \\ 1 \end{bmatrix}u_n, \qquad x_0 = \begin{bmatrix} 0 \\ 0 \end{bmatrix}$$
$$y_n = [1 \quad 0]x_n$$

from 0 to 1 in a minimum number of steps and will maintain this state thereafter.

b. If $a_0 = a_1 = 0$, can you find a way to construct this control law in the feedback configuration as shown in the figure? *Hint:* The solution to part a is $u_0 = 1$, $u_1 = 1 - a_1$, $u_k = 1 - a_0 - a_1$, $k \geq 2$. The solution to part b is $K(z) = [1 - z^{-2}]^{-1}$.

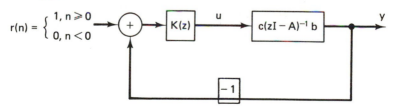

3.5-4.

Give an algebraic proof of the identity (3.5-8).

3.5-5.

Use the principle of optimality to find a closed-loop solution to the controllability-to-the-origin problem.

3.5-6. *Discrete-Time Version of Double-Integrator System*

We wish to control the plant $\dot{x}_1 = x_2$, $\dot{x}_2 = u$, $y = x_1$ by a digital computer that updates u at discrete time intervals, $0, \Delta, \ldots$.

a. Find A and b in

$$\begin{bmatrix} x_1(n+1) \\ x_2(n+1) \end{bmatrix} = A \begin{bmatrix} x_1(n) \\ x_2(n) \end{bmatrix} + bu(n)$$

so that $x(t)|_{t=n\Delta} = x(n)$.

b. Find k in $u(n) = -kx(n)$ so that $x(2) = 0$ for arbitrary $x(0)$. Verify that $(A - bk)^2 = 0$.

c. Find the reciprocal root characteristic equation and sketch the reciprocal root locus for this system with performance index $J = \sum_{n=0}^{\infty} \{[y(n+1)]^2 + r[u(n)]^2\}$.

3.5-7.

Draw the reciprocal root locus for the scalar system $x(i+1) = ax(i) + bu(i)$, $y(i) = x(i)$, and determine the optimal feedback gain k for zero control cost.

3.5-8.

a. Consider a system

$$x(i+1) = \begin{bmatrix} \sqrt{2} & 1 \\ 1 & \sqrt{2} \end{bmatrix} x(i) + \begin{bmatrix} \sqrt{2}-1 \\ 1 \end{bmatrix} u(i)$$

Let $y(i) = x_1(i) = [1 \ 0]x(i)$, and draw the reciprocal root locus for optimum quadratic regulation of $y(\cdot)$.

b. Determine the feedback gain vector k required to obtain the optimum pole positions when control cost is negligible.

3.5-9. *Asymptotic Behavior of the Optimal Quadratic Regulator*

These results should be compared with those in Sec. 3.4.1. We suppose that the *open-loop* transfer function in (3.5-22) has the form $H(z) = b(z)/a(z)$, $a(z) = \det(sI - A)$, $a(z) = z^{n-q} \prod_1^q (z - v_i)$, $b(z) = \alpha z^{s-p} \prod_1^p (z - \epsilon_i)$, with $p \le q$, $\{v_i \neq 0 \neq \epsilon_i\}$.

a. Show that $n - q$ closed-loop poles always remain at the origin.

b. Show that as $r \to \infty$ (high cost of control) the q nonzero closed-loop poles approach the values

$$\hat{v}_i = \begin{cases} v_i & \text{if } |v_i| \le 1 \\ \dfrac{1}{v_i} & \text{if } |v_i| > 1 \end{cases}$$

c. Show that as $r \longrightarrow 0$ (cheap control) $q - p$ poles go to zero, while the remaining $n - (n - q) - (q - p) = p$ poles go to the values

$$\hat{\epsilon}_i = \begin{cases} \epsilon_i & \text{if } |\epsilon_i| \leq 1 \\ \dfrac{1}{\epsilon_i} & \text{if } |\epsilon_i| > 1 \end{cases}$$

d. What are the corresponding results for $p > q$?

3.5-10. *Continuous Limit of Discrete Regulator Solution*

Consider the discrete-time system $x(k + 1) = (I + A\Delta)x(k) + b\Delta u(k)$, $x(0) = x_0$, where the $\{x(k), u(k)\}$ stand for $\{x(k\Delta), u(k\Delta)\}$ and Δ is the sample spacing used to discretize the equation $\dot{x}(t) = Ax(t) + bu(t)$. Determine the optimal quadratic regulator [minimizing (3.5-15)], and find the limit as $\Delta \longrightarrow 0$ of the formulas (3.5-16)–(3.5-20) for the optimal feedback control.

REFERENCES

1. R. E. KALMAN, P. FALB, and M. A. ARBIB, *Topics in Mathematical System Theory*, McGraw-Hill, New York, 1969.

2. H. H. ROSENBROCK, "Distinctive Problems of Process Control," *Chem. Eng. Prog.*, **58**, pp. 43–50, Sept. 1962.

3. J. RISSANEN, "Control System Synthesis by Analogue Computer Based on the 'Generalized Linear Feedback' Concept," in *Proceedings of the Symposium on Analog Computation Applied to the Study of Chemical Processes*, International Seminar, Brussels, pp. 1–13, November 21–23, 1960. Presses Académiques Européennes, Bruxelles, 1961.

4. V. M. POPOV, "Hyperstability and Optimality of Automatic Systems with Several Control Functions," *Rev. Roum. Sci. Tech. Ser. Electrotech. Energ.*, **9**, pp. 629–690, 1964. See also V. M. Popov, *Hyperstability of Control Systems*, Springer-Verlag New York, Inc., New York, 1973 (Rumanian ed., Bucharest, 1966).

5. R. E. KALMAN, "On the General Theory of Control Systems," in *Proceedings of the First IFAC Congress*, Vol. 1, Butterworth's, London, 1960, pp. 481–491.

6. R. H. CANNON, JR., *Dynamics of Physical Systems*, McGraw-Hill, New York, 1967.

7. K. OGATA, *Modern Control Engineering*, Prentice-Hall, Englewood Cliffs, N.J., 1970.

8. J. G. TRUXAL, *Control System Synthesis*, McGraw-Hill, New York, 1955.

9. D. G. SCHULTZ and J. L. MELSA, *State Functions and Linear Control Systems*, McGraw-Hill, New York, 1967.

10. R. BELLMAN, "The Theory of Dynamic Programming," *Proc. Nat. Acad. Sci. USA*, 38, pp. 716–719, 1952; also *Bull. Amer. Math. Socy.*, **60**, pp. 503–516, 1954.

11. R. BELLMAN, "On the Application of the Theory of Dynamic Programming to

the Study of Control Processes," *Proc. Symp. on Nonlinear Circuit Analysis*, pp. 199–213. Polytechnic Institute of Brooklyn Press, New York, 1956.

12. R. W. Bass and I. Gura, "High Order Design via State-Space Considerations," *Proceedings of the 1965 Joint Automatic Control Conference, Troy, N.Y.*, 1965, pp. 311–318. See also R. W. Bass, "Control Synthesis and Optimization," NASA Langley Research Center, Aug. 1961 (multilithed lecture notes).

13. J. D. Simon and S. K. Mitter, "A Theory of Modal Control," *Inf. Control*, 6, pp. 659–680, 1968.

14. J. Ackermann, "Der Entwurf Linearer Regelungssysteme im Zustandsraum," *Regulungestechnik und Prozessedatenverarbeitung*, 7, pp. 297–300, 1972. See also J. Ackermann, *Abtastregulung*, Springer, Berlin, 1972.

15. P. Murdoch, "Explicit Results in Modal Control," M.Sc. Rept., Imperial College, London, Sept. 1968. See also D. Q. Mayne and P. Murdoch, "Modal Control of Linear Time-Invariant Systems," *Int. J. Control*, 11, pp. 223–227, 1970, and the discussion of this paper by B. Porter and P. Murdoch in *IEEE Trans. Autom. Control*, AC-20, p. 582, Aug. 1975.

16. L. A. Gould, A. T. Murphy, and E. F. Berkman, "On the Simon-Mitter Pole Allocation Algorithm—Explicit Gains for Repeated Eigenvalues," *IEEE Trans. Autom. Control*, AC-15, pp. 259–260, 1970.

17. B. D. O. Anderson and J. B. Moore, *Linear Optimal Control*, Prentice-Hall, Englewood Cliffs, N.J., 1971.

18. H. Kwakernaak and R. Sivan, *Linear Optimal Control Systems*, Wiley, New York, 1972.

19. B. S. Morgan, "The Synthesis of Linear Multivariable Systems by State-Variable Feedback," *Proceedings 1964 Joint Automatic Control Conference*, pp. 468–472, Stanford University, Stanford, Ca., June 1964.

20. F. Fallside, ed., *Control System Design by Pole-Zero Assignment*, New York: Academic Press, 1977.

21. R. W. Brockett, "Poles, Zeros and Feedback: State Space Interpretation," *IEEE Trans. Autom. Control*, AC-10, pp. 129–135, April 1965.

22. M. Athans, "The Role and Use of the Stochastic Linear-Quadratic-Gaussian Problem in Control Systems Design," *IEEE Trans. Autom. Control*, AC-16, pp. 529–552, Dec. 1971. This special issue contains several other papers on the quadratic regulator problem.

23. A. M. Letov, "Analytical Controller Design, I, II," *Autom. Remote Control*, 21, pp. 303–306, 1960.

24. S. S. L. Chang, *Synthesis of Optimum Control Systems*, McGraw-Hill, New York, 1961.

25. R. E. Kalman, "When Is a Linear Control System Optimal?," *Trans. ASME Ser. D. J. Basic Eng.*, 86, pp. 51–60, 1964.

26. R. W. Brockett, *Finite-Dimensional Linear Systems*, Wiley, New York, 1970.

27. T. Kailath, "Redheffer Scattering Theory and Linear State-Space Estimation and Control Problems," *Ricerche di Automatica,* Special Issue on System Theory and Physics, Jan. 1979.

28. A. E. BRYSON and Y. C. HO, *Applied Optimal Control*, Halsted Press, New York, 1968.

29. T. KAILATH and L. LJUNG, "The Asymptotic Behavior of Constant-Coefficient Riccati Differential Equations," *IEEE Trans. on Auto. Control*, AC-21, pp. 385–388, June 1976.

30. K. MÅRTENSSON, "On the Matrix Riccati Equation," *Inf. Sci.*, 3, pp. 17–49, 1971.

31. A. G. J. MACFARLANE, "An Eigenvector Solution of the Optimal Linear Regulator," *J. Electron. Control.*, 14, pp. 643–654, June 1963.

32. J. E. POTTER, "Matrix Quadratic Solutions," *SIAM J. Appl. Math.*, 14, pp. 496–501, May 1966.

33. A. F. FATH, "Computational Aspects of the Linear Optimal Regulator Problem," *IEEE Trans. Autom. Control*, AC-14, pp. 547–550, Oct. 1969.

34. A. E. BRYSON and W. E. HALL, "Optimal Control and Filter Synthesis by Eigenvector Decomposition," *Rept. No. 436*, Dept. of Aeronautics & Astronautics, Stanford University, Stanford, Calif., Dec. 1971.

35. G. W. STEWART, "Error and Perturbation Bounds for Subspaces Associated with Certain Eigenvalue Problems," *SIAM Review* 15, pp. 727–764, Oct. 1973.

36. A. J. LAUB, "A Schur Method for Solving Algebraic Riccati Equations," *Proc. 1978 IEEE Trans. Autom. Control*, AC-24, pp. 913–921, Dec. 1979.

37. W. E. HOLLEY and S. Y. WEI, "An Improvement in the MacFarlane-Potter Method for Solving the Algebraic Riccati Equation," *Proc. 1979 Joint Automatic Control Conference*, pp. 921–923, Denver, Colorado, June 1979.

38. D. L. KLEINMAN, "On an Iterative Technique for Riccati Equation Computation," *IEEE Trans. Autom. Control*, AC-13, pp. 114–115, 1968.

39. J. D. ROBERTS, "Linear Model Reduction and Solution of Algebraic Riccati Equations by Use of the Sign Function," Rept. CUED/B-Control/TR13, Dept. of Eng., University of Cambridge, Cambridge, England, 1971.

40. M. WOMBLE and J. POTTER, "A Prefiltering Version of the Kalman Filter and New Numerical Integration Formulas for Riccati Equations," *IEEE Trans. Autom. Control*, AC-20, pp. 378–380, 1975.

41. T. KAILATH, "Some New Algorithms for Recursive Estimation in Constant Linear Systems," *IEEE Trans. Inform. Thy.*, IT-19, pp. 750–760, November 1973.

42. J. NEWKIRK, "Some Computational Issues in Kalman Filtering," Ph.D. Dissertation, Dept. of Electrical Engineering, Stanford University, Stanford, California, June 1979.

42a. F. A. FARRAR and R. C. DI PIETRO, "Comparative Evaluation of Numerical Methods for Solving the Algebraic Matrix Riccati Equation," *United Technologies Research Center Rept. No. R'76-140268-1*, East Hartford, Ct., Dec. 1976.

43. R. E. KALMAN and J. BERTRAM, "General Synthesis Procedure for Computer Control of Single and Multi-Loop Linear Systems," *Trans. AIEE*, 77, pp. 602–609, 1958.

44. R. BELLMAN, *Dynamic Programming*, Princeton University Press, Princeton, N.J., 1957.

45. J. Bernoulli, "Solution du Problème Proposé par M. Jacques Bernoulli, etc. (1706)," in *Collected Works*, G. Cramer, ed., Vol. 1, pp. 424–435, Lausanne, 1742.

46. L. EULER, "Methodus Inveniendi Linear Curvas Maximi Minimive Proprietate Gaundentes Sive Solutio Problematis Isoperimetrici Latissimo Sensu Accepti, Lausanne, 1744," in *Collected Works*, Ser. 1, Vol. 24, Teubner, Leipzig, 1938.

47. C. CARATHÉODORY, *Calculus of Variations and Partial Differential Equations of the First Order*, Teubner, Leipzig, 1935 (English trans., Holden-Day, San Francisco, 1967).

48. D. M. G. WISHART, "The Continuity of Ideas in the Calculus of Variations," *Control Theory Centre Rept. No. 21*, University of Warwick, Warwick, England, 1972.

49. R. WOODHOUSE, *A History of the Calculus of Variations in the Eighteenth Century*, Cambridge University Press, London, 1810 (reprint, Chelsea Publishing Co., New York, 1966).

49a. D. J. STRUIK, ed., *A Source Book in Mathematics, 1200–1800*, Harvard University Press, Cambridge, Mass., 1969.

50. R. E. KALMAN and R. W. KOEPCKE, "Optimal Synthesis of Linear Sampling Control Systems Using Generalized Performance Indexes," *Trans. ASME*, 80, pp. 1800–1826, 1958.

51. M. MORF and T. KAILATH, "Square-Root Algorithms for Least-Squares Estimation," *IEEE Trans. Autom. Control*, AC-20, pp. 487–497, Aug. 1975.

52. L. SILVERMAN, "Discrete Riccati Equations: Alternative Algorithms, Asymptotic Properties, and System Theory Interpretations," in *Control and Dynamic Systems*, Vol. 12 (C. T. Leondes, ed.), 1976, pp. 313–386, Academic Press, New York.

53. G. W. STEWART, *Introduction to Matrix Computations*, Academic Press, New York, 1973.

54. G. BIERMAN, *Factorization Methods for Discrete Estimation*, Academic Press, New York, 1977.

55. G. SEBER, *Linear Regression Analysis*, Wiley, New York, 1977.

56. M. MORF, J. DOBBINS, B. FRIEDLANDER, and T. KAILATH, "Square-Root Algorithms for Parallel Processing in Optimal Estimation," *Automatica*, 15, pp. 299–306, May 1979.

57. M. MORF, G. S. SIDHU, and T. KAILATH, "Some New Algorithms for Recursive Estimation in Constant, Linear, Discrete-Time Systems," *IEEE Trans. Autom. Control*, AC-19, pp. 315–323, Aug. 1974.

ASYMPTOTIC OBSERVERS AND COMPENSATOR DESIGN

4

4.0 INTRODUCTION

In Chapter 3, we showed that if a realization $\{A, b, c\}$ is controllable, then state-variable feedback can modify the eigenvalues of A at will. We shall now discuss the problem of actually obtaining the states of the realization from knowledge only of the system input $u(\cdot)$ and system output $y(\cdot)$. We have already seen in Sec. 2.3 that if the realization $\{A, b, c\}$ is also observable, then a differentiation technique can be used to calculate the states. However, this technique is clearly impractical, and therefore in Sec. 4.1 we shall develop a more realistic state estimator, which is usually known as an asymptotic observer. The name arises from the fact that the states can only be obtained with an error, but one that can be made to go to zero at any specified exponential rate.

In Sec. 4.2 we shall discuss the use of such state estimates in place of the possibly unavailable true states. We shall find that the design equations for the controller are not affected by the fact that approximate states are being used instead of the true states. More crucially, however, we shall find the important and a priori nonobvious fact that the overall observer-controller configuration is *internally stable*, which was an issue not completely faced by classical design methods. However, use of the estimated instead of the true states, for feedback, may lead in general to a deterioration of the transient response (see Example 4.2-1).

The observer studied in Sec. 4.1 was apparently first introduced in unpublished work by Bertram in 1961 (see the comments in Astrom [1, p. 158]) and by Bass in 1963 (see the comment of Kalman et al. [2, p. 55]). An independent and rather different approach was published by Luenberger in 1964 [3] (see also his Ph.D. thesis, Stanford University, Stanford, Calif., 1963). Luenberger treats a somewhat more general estimation problem which, when specialized to our situation, actually yields an observer with one less state variable than the observer of Sec. 4.1. This so-called reduced-order or Luenberger observer is discussed in Sec. 4.3 by a method due to Gopinath [4] and Cumming [5] (see also the survey papers [6] and [7]).

In Sec. 4.4, we shall briefly discuss the question of optimum observer pole locations, which will lead us to some results *dual* to those developed in Sec. 3.4.1 for the optimal quadratic regulator. Analogs of the results in the other sections of Secs. 3.4 and 3.5 can be developed, though we shall not do so here. Instead, in our concluding Sec. 4.5, we shall explore at some length the fact that it is possible to obtain the combined observer-controller configuration directly by transfer function analysis, thus obviating the need for the notions of state, controllability, and observability. This was shown by Chen ([8]) although, as we shall explain in Sec. 4.5.1, without the insights provided by the state-variable design, certain judicious choices and assumptions in the arguments of Reference [8] might not have been apparent. However, once the transfer function method has been introduced, we shall find that this suggests certain extended results that are not immediately obvious in the state-space approach; moreover, this method will extend nicely to the multivariable case (Sec. 7.5).

4.1 ASYMPTOTIC OBSERVERS FOR STATE MEASUREMENT

We shall now begin to explore the question of methods of actually determining the states of a realization

$$\begin{aligned}
\dot{x}(t) &= Ax(t) + bu(t), & x(0-) &= x_0 \\
y(t) &= cx(t) & t &> 0-
\end{aligned} \tag{1}$$

given knowledge only of $y(\cdot)$ and $u(\cdot)$.

In Sec. 2.3. we have already described one method of determining the states at any time t. But this method involves differentiation and is therefore impractical.

When we reflect on the fact that we know $A, b, c, y(\cdot)$, and $u(\cdot)$, which is really quite a lot, we wonder why $x(\cdot)$ cannot be reconstructed by forming a *dummy* system $\{A, b\}$ and driving it with $u(\cdot)$. The problem,† of course, is

---

†It is important to note that the dummy system need not use the same physical components as the original system—it can be a miniaturized electronic analog computer simulation or a special-purpose digital computation package. For the present purpose, quite large systems—chemical plants, satellites, etc.—may be quite cheaply simulated in this way.

that we do not know the initial condition $x(0-) = x_0$. This again has to be found by the differentiation technique, but we have a slight simplification since it is reasonable to assume that $u(t) = 0$, $t \leq 0-$. Then some calculation shows that $x(0-)$ can be obtained from the equation

$$\Theta x(0-) = [y(0-), \ldots, y^{(n-1)}(0-)]' \tag{2}$$

and in many problems the values $\{y(0-), \ldots, y^{(n-1)}(0-)\}$ are given to us as part of the problem statement.

An Open-Loop Observer.   Thus, by setting up a dummy system $\{A, b, c\}$ with the proper initial conditions and driving it with the known input $\{u(t), t > 0-\}$, we can obtain $\{x(t), t > 0-\}$. Note that the values of the output $\{y(t), t > 0-\}$ are no longer required. Unfortunately, however, this last fact is really more of a disadvantage than a help. The point is that strategies that are the same no matter what the output is (called *open-loop* strategies as opposed to *feedback* or *closed-loop* methods, which adjust parameters according to the current value of the system output or system state) are obviously susceptible to disturbances that may arise during system operation, and, furthermore, they allow no means of compensating for any errors, possibly small but still almost inevitable, in the predetermined strategy.

Thus, suppose that the initial condition that we use for our dummy system is slightly in error—say we have not x_0 but

$$\hat{x}_0 \triangleq x_0 + \epsilon, \qquad \|\epsilon\| \ll \|x_0\| \tag{3}$$

where we shall use, both here and elsewhere, the caret ($\wedge$) to denote *estimate of*. What is the effect of this small error on the states calculated from our dummy system? The states will no longer be $x(t)$ but will be a different function, say $\hat{x}(t)$, which satisfies the equation

$$\dot{\hat{x}}(t) = A\hat{x}(t) + bu(t), \qquad \hat{x}(0-) = \hat{x}_0 = x_0 + \epsilon \tag{4}$$

Clearly, there will be an error

$$\tilde{x}(t) \triangleq x(t) - \hat{x}(t)$$

which will satisfy the differential equation

$$\dot{\tilde{x}}(t) = A\tilde{x}(t), \qquad \tilde{x}(0-) = \epsilon \tag{5}$$

That is, the error ϵ in the initial condition will produce an error, $\tilde{x}(\cdot)$, at all later times. Now, as can be seen from a partial fraction expansion of

$$\tilde{X}(s) = (sI - A)^{-1}\epsilon$$

$\tilde{x}(t)$ will be a sum of terms of the form $\{e^{\lambda_i t}, t^j e^{\lambda_i t}, \ldots\}$, where the $\{\lambda_i\}$ are the eigenvalues of A. Clearly, if the system is unstable (recall that we are interested in determining states to be fed back to achieve stabilization), then the error $\tilde{x}(t)$ will become *arbitrarily large* as $t \to \infty$, no matter *how small* the initial error is. Less dramatically, even if the system is stable but some eigenvalues have real parts that are very small, the effects of errors in the initial estimates will take a long time to die out.

A Closed-Loop Observer. There may be situations in which open-loop estimators are fairly satisfactory, especially if periodic *resetting* is feasible so as to eliminate, or at least suitably control, the effects of initial errors and later disturbances.

However, the classical way of overcoming the potential difficulties of open-loop systems is to use *feedback*, viz., to try to *zero the error* by driving the system with a term proportional to the error in the estimate. But how can we determine this "error?" In classical feedback problems, this is provided by the *reference* signal.† In our problem, the error is $x(\cdot) - \hat{x}(\cdot) = \tilde{x}(\cdot)$, but of course $x(\cdot)$ is not available! Therefore, we have to obtain a reference signal in some other way. The system output $y(\cdot)$, which was not used in the open-loop solution, comes in now because it is related to the quantity $x(\cdot)$ we are interested in: $y(\cdot) = cx(\cdot)$, and $y(\cdot)$ is clearly available. Therefore, an *error* signal can be generated as

$$y(t) - \hat{y}(t) = y(t) - c\hat{x}(t) = c[x(t) - \hat{x}(t)] = c\tilde{x}(t)$$

and it can be used to drive the estimator equation. These considerations lead us‡ to consider an estimator for $x(\cdot)$ of the form (see Fig. 4.1-1)

$$\dot{\hat{x}}(t) = A\hat{x}(t) + bu(t) + l[y(t) - c\hat{x}(t)], \qquad \hat{x}(0) = \hat{x}_0 \qquad (6)$$

where

$$\hat{x}_0 = \text{an estimated initial state vector}$$

and

$$l = \text{a } \textit{feedback gain vector}, \text{ to be suitably chosen}$$

The system yielding $\hat{x}(\cdot)$ is called an observer, or, actually, for reasons that will become clearer presently, an *asymptotic observer*.

We should, of course, choose l so as to properly control the error $\tilde{x}(\cdot)$. Now the error $\tilde{x}(\cdot)$ obeys the differential equation

$$\dot{\tilde{x}}(t) = \dot{x}(t) - \dot{\hat{x}}(t)$$
$$= (A - lc)\tilde{x}(t), \qquad \tilde{x}(0) = x_0 - \hat{x}_0 \qquad (7)$$

---

†For example, in a thermostat, the reference signal is the desired ambient temperature.
‡See also Exercise 4.1-2.

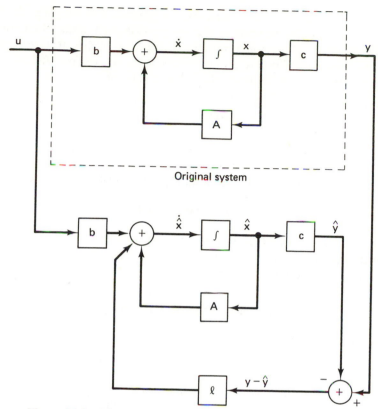

Figure 4.1-1. Block diagram of asymptotic observer. Access to the original system is assumed to be possible at the input and output terminals, while this restriction is not necessary for the observer.

Notice that when $l = 0$ this equation reduces to the open-loop error equation (5). The effect of the *error feedback* $l[y(\cdot) - \hat{y}(\cdot)] = lc[x(\cdot) - \hat{x}(\cdot)]$ is to give us some control over the behavior of the error $\tilde{x}(\cdot)$. In fact, the natural frequencies will be the eigenvalues of $A - lc$, and what we can try to do is to choose l so that the error will die out as rapidly as we deem suitable. Notice that the actual value of the initial estimate $\hat{x}_0$ is unimportant —if we have no special information, we often take $\hat{x}_0 = 0$. The reason for the name *asymptotic observer* should now be clear.

But can we always find a suitable l? We shall prove that if $\{c, A\}$ is observable, then we can choose l so that $A - lc$ has arbitrary eigenvalues, or equivalently

$$\det (sI - A + lc) = \alpha(s) = s^n + \alpha_1 s^{n-1} + \cdots + \alpha_n \qquad (8)$$

where the $\{\alpha_i\}$ are completely arbitrary. The reader will recognize that this problem is essentially the same as the one we discussed in Sec. 3.2 in determining the feedback vector k required to give a controllable realization arbitrary dynamics. There is in fact a very close connection, and we shall actually solve our problem by *dualizing* the results of Sec. 3.2.

Formulas for the Observer Gain.   In Sec. 3.2 we showed that given any realization $\{A, b, c\}$ with $\{A, b\}$ controllable we could in several ways find k so that

$$\det (sI - A + bk) = s^n + \alpha_1 s^{n-1} + \cdots + \alpha_n \tag{9}$$

for any $\{\alpha_i\}$. For example, one formula is [cf. (3.2-13)]

$$k = (\alpha - a)\mathfrak{a}_-^{-1}\mathfrak{C}^{-T}(A, b) \tag{10}$$

where a is the vector of coefficients of $a(s) = \det (sI - A)$ and $\mathfrak{a}_-^T$ is an upper triangular Toeplitz matrix with $[1 \ a_1 \ \cdots \ a_{n-1}]$ as the first row.

To recast our observer problem into this form we merely have to note that

$$\det (sI - A + lc) = \det (sI - A' + c'l')$$

so that if we let

$$A \longrightarrow A', \qquad b \longrightarrow c', \qquad k \longrightarrow l'$$

in the solution of the controller problem, we deduce that we shall be able to find l if and only if $\mathfrak{C}(A', c')$ is nonsingular. Then

$$l' = (\alpha - a)\mathfrak{a}_-^{-T}\mathfrak{C}^{-1}(A', c')$$

But

$$\mathfrak{C}(A', c') = [c' \quad A'c' \quad \cdots \quad A'^{(n-1)}c']$$
$$= \mathfrak{O}'(A, c), \qquad \text{the transpose of the observability matrix of } \{A, b, c\} \tag{11}$$

Hence the observer gain vector l can be calculated as

$$l = \mathfrak{O}^{-1}(A, c)\mathfrak{a}_-^{-1}(\alpha - a)' \tag{12}$$

It is perhaps somewhat unexpected that asymptotic observability requires the same condition—the nonsingularity of $\mathfrak{O}(A, c)$—as exact observability does. But some reflection on what *unobservable states* are (cf. Sec. 2.4.2) will provide an explanation.

The duality between the asymptotic observer problem and the modal controller problem is quite striking and useful and can be used as above to translate almost all the results of Chapter 3 to the present context. We shall leave these extensions to the reader (and partly to the exercises).

Example 4.1-1.   The Pointer-Balancing Problem

For the pointer-balancing problem of Example 3.3-1, design an observer with modes at $-10 \pm j10$.

For $\varphi(0) = 0.1$, $\dot{\varphi}(0) = 0$, calculate and plot the observer errors $\tilde{\varphi}(\cdot)$ and $\tilde{\dot{\varphi}}(\cdot)$.

Solution.   Recall that with $g/L = 9$ and $z = [\varphi \ \dot{\varphi}]'$ we have

$$\dot{z} = \begin{bmatrix} 0 & 1 \\ 9 & 0 \end{bmatrix} z + \begin{bmatrix} 0 \\ -1 \end{bmatrix}$$

$$y = [1 \ \ 0]z$$

We want $\det (sI - A + lc) = (s + 10)^2 + 10^2 = s^2 + 20s + 200 = \alpha_0(s)$, and a simple calculation yields $l' = [20 \ \ 209]$.
Let $\hat{z}(0) = 0$; then $\tilde{z}(0) = z(0) = [0.1 \ 0]'$ and

$$\tilde{Z}(s) = (sI - A + lc)^{-1}\tilde{z}(0)$$

$$= \frac{1}{\alpha_0(s)} \begin{bmatrix} s & 1 \\ -200 & s + 20 \end{bmatrix} \begin{bmatrix} 0.1 \\ 0 \end{bmatrix}$$

Therefore

$$\tilde{z}_1(t) = \tilde{\varphi}(t) = \mathcal{L}^{-1}\left[\frac{0.1s}{\alpha_0(s)}\right]$$

$$= 0.1\mathcal{L}^{-1}\left[\frac{(s + 10)}{(s + 10)^2 + 10^2} - \frac{10}{(s + 10)^2 + 10^2}\right]$$

$$= 0.1e^{-10t}(\cos 10t - \sin 10t)$$

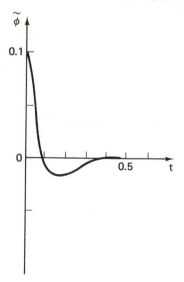

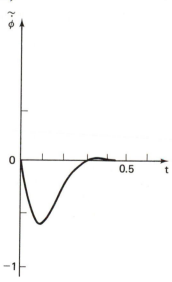

Similarly, $\tilde{\phi}(t) = -2e^{-10t} \sin 10t$. These are sketched in the figure. Note that in about three time constants $(3 \times \frac{1}{10})$ the observer error is down to a small fraction of its maximum value.

Example 4.1-2.   The Need for Proper Modeling

Discuss the design, as $\epsilon \longrightarrow 0$, of observers for a realization

$$\begin{bmatrix} \dot{x}_1(t) \\ \dot{x}_2(t) \end{bmatrix} = \begin{bmatrix} -1 & 0 \\ 2 & -1 \end{bmatrix} \begin{bmatrix} x_1(t) \\ x_2(t) \end{bmatrix} + \begin{bmatrix} 1 \\ 0 \end{bmatrix} u(t)$$

$$y(t) = x_1(t) + \epsilon x_2(t), \qquad \epsilon \ll 1$$

Note that $x_1(\cdot)$ does not depend on $x_2(\cdot)$, and therefore as $\epsilon \longrightarrow 0$, $x_2(\cdot)$ tends to disappear completely from the observation $y(\cdot)$. Therefore $x_2(\cdot)$ becomes unobservable as $\epsilon \longrightarrow 0$, and we should expect some difficulties in trying to estimate it.

Solution.   Since this is an observer problem, we can, without loss of generality, assume that $u(\cdot) \equiv 0$. We can now readily calculate the observer gain required to obtain any set of observer poles, but we shall see that the parameter ϵ will give us indications of potential difficulties. For example,

$$\mathcal{O} = \begin{bmatrix} 1 & \epsilon \\ -1 + 2\epsilon & -\epsilon \end{bmatrix}, \qquad \det \mathcal{O} = -2\epsilon^2$$

so that $\mathcal{O}$ is almost singular for small ϵ. Therefore we would expect trouble even if perfect differentiation were feasible. Thus note that

$$\mathcal{O}^{-1} = -\frac{1}{2\epsilon^2} \begin{bmatrix} -\epsilon & -\epsilon \\ 1 - 2\epsilon & 1 \end{bmatrix}$$

Therefore, if, for example, $\epsilon = 0.01$, we see that the ideal observer will give

$$\hat{x}_1(t) = 50[y(t) + \dot{y}(t)]$$
$$\hat{x}_2(t) = -4999.99y(t) - 5000\dot{y}(t)$$

while if $\epsilon = 0.02$, the estimates change quite a bit to

$$\hat{x}_1(t) = 25[y(t) + \dot{y}(t)]$$
$$\hat{x}_2(t) = -2499.98y(t) - 2500\dot{y}(t)$$

Such difficulties will persist no matter what method of state estimation is used. In this problem, when ϵ is small we see that $y(\cdot)$ depends almost completely on $x_1(\cdot)$, and it is better to represent the model by the equations

$$\dot{x}_1(t) = -x_1(t) + u(t)$$
$$y(t) = x_1(t)$$

for which no observer is needed. Of course this is because the $x_2(\cdot)$ equation is stable so that $x_2(\cdot)$ will tend to die out with time. If $x_2(\cdot)$ is unstable, then we shall have trouble, and we should reexamine the original model to see if it can be modified.

The point is that one should not just blindly plug into a mathematical formula but should also examine the solution and the model for sensitivity to parameter changes, *singular* phenomena, etc.

In this problem, some other clues to the potential difficulties might have been recognized from the transfer function,

$$H(s) = \frac{s + 1 + 2\epsilon}{(s + 1)^2}$$

which will show that, for small ϵ, we really have a first-order system. The pole-zero cancellation for $\epsilon = 0$ shows that we shall lose observability or controllability of the realization. A block diagram of the realization (see the figure) shows that as $\epsilon \longrightarrow 0$, a pole of the first subsystem cancels a zero of the second subsystem, showing (according to Example 2.4-5) that we have unobservability of the overall realization.

The active reader should try to develop the *duals* for controllability of the above problem and remarks.

Exercises

4.1-1.   *Ackermann's Formula*

Obtain a formula for l in terms of $\{A, c\}$ and the coefficients $\{\alpha_i\}$ of the desired characteristic polynomial, $\alpha(s)$.

4.1-2.   *Why Not Feedback to the Input?*

In Chapter 3, we used feedback to the input according to $u(t) = v(t) - kx(t)$. We could do this for the observer $u(t) = v(t) + l[y(t) - \hat{y}(t)]$, where, of course, l is now a scalar. What can be achieved with such feedback? Why is it that we use feedback to the states in the observer problem but not in the controller problem?

4.1-3.

Is the asymptotic observer of an observable system itself observable for all possible l? Give a proof.

4.1-4.   *Feedback and Observability*

Show that the observability of a realization $\{A, b, c\}$ is not invariant under general state feedback $(u \longrightarrow v - kx)$ but is invariant under linear or nonlinear output feedback $\{u(t) \longrightarrow v(t) - f[y(s)], s \leq t\}$.

4.1-5.   *Another Approach to Observers* [3]

If $\dot{x}(t) = Ax(t) + bu(t)$, $y(t) = cx(t)$, $x(t_0) = x_0$, let $\hat{x}(\cdot)$ obey $\dot{\hat{x}}(t) = F\hat{x}(t) + gu(t) + hy(t)$, $\hat{x}(t_0) = \hat{x}_0$. The second equation can be said to define an *observer* for the first if $x_0 = \hat{x}_0 \Rightarrow x(t) = \hat{x}(t)$, $t \geq t_0$. Show that a necessary and sufficient condition for this is that $F = A - kc$, $h = k$, $g = b$, where k is an arbitrary $n \times 1$ vector.

4.1-6.   *Effects of Mismatched Models*

Given a realization $\dot{x}(t) = Ax(t) + bu(t), y(t) = cx(t)$, consider an *observer* $\dot{\chi}(t) = \hat{A}\chi(t) + \hat{b}u(t) + v(t)$, $\hat{y}(t) = \hat{c}\chi(t)$, $v(t) = \hat{b}l(y - \hat{y})$, where the $\{\hat{A}, \hat{b}, \hat{c}\}$ are *estimates* (approximations to) of $\{A, b, c\}$. Show that $\dot{\epsilon}(t) = [\hat{A} - \hat{b}l\hat{c}]\epsilon(t) + [\delta A - \hat{b}l(\delta c)]x(t) + (\delta b)u(t)$, $\epsilon(t) = x(t) - \chi(t)$, where $\delta A = A - \hat{A}, \delta b = b - \hat{b}$, and $\delta c = c - \hat{c}$. *Note:* Further analysis allows us to relate these results to questions of system sensitivity (as originally conceived by Bode (1945): See W. A. Porter, *IEEE Trans. Autom. Control*, **AC-22**, pp. 144–146, Feb. 1977.

4.1-7.

Design an observer for the oscillatory system $\dot{x}(t) = v(t), \dot{v}(t) = -\omega_0^2 x(t)$ using measurements of the velocity $v(\cdot)$. Place both observer poles at $s = -\omega_0$.

4.1-8.   *Station-Keeping Satellite*

For the station-keeping satellite of Example 3.3-2, design an observer using measurements $y(\cdot)$ of azimuthal position perturbation. Place the observer poles at $s = -2\omega$, $s = -3\omega$, $s = -3\omega \pm j3\omega$, which means that the estimate errors will decay in about $2\frac{1}{2}$ days ($\doteq \frac{1}{2}\omega$, where $\omega = 2\pi/29.3$ rad/day).

4.1-9.   *Deadbeat Observers in Discrete-Time*

Develop observer designs for discrete-time systems and show how to design an observer for which the error will go to zero in no more than n steps, where n is the number of states.

4.2 COMBINED OBSERVER-CONTROLLER COMPENSATORS

We were led to the observer problem by the need to obtain the states for use in the controller. Now we only have asymptotically correct estimates of the states rather than the states themselves, and a natural question is whether our previous result on arbitrary pole placement via state-variable feedback will continue to hold when only such estimates of the actual states are available. We have no option but to see what happens with the estimators that we have available.

In steady state there should clearly be no loss in using the asymptotic observer, since the error in the estimates will be zero. Therefore, as we shall soon confirm, the transfer function of the combined observer-controller will be just that of a pure controller (with perfect state feedback). However, this is not our major worry since, as noted in Sec. 3.1.1, there are many ways of arranging for a desired transfer function. The real question for the present scheme is whether the incorporation of the observer dynamical system into the feedback loop will affect the *stability* of the overall system—the point being that interconnections of stable subsystems may lead to unstable overall systems (cf. Sec. 3.1). However, we shall now prove the nice result that *incorporation of a stable asymptotic observer does not impair stability*.

We begin our analysis by setting up a joint observer-controller system. This is provided by Fig. 4.2-1, which the reader should be able by now to set up on his own. However, some brief words of explanation might be helpful. The original system may be described by its transfer function $H(s)$ or by a realization $\{A, b, c\}$. If we have $H(s)$, we must set up a realization in any form convenient for us (e.g., controller form) so as to be able to use state-variable feedback. If the states are directly measurable, we can close the loop through the feedback gain vector k. Otherwise we have to use state estimates $\hat{x}(\cdot)$, which we obtain by setting up a dummy system driven by the *error* term $l[y(\cdot) - c\hat{x}(\cdot)]$ and *also* by the same input $[u(\cdot) = v(\cdot) - k\hat{x}(\cdot)]$ as enters the state equations of the original system. This last fact should be well understood: If this input is not fed back to the observer, its states $\hat{x}(\cdot)$ will certainly not follow the states $x(\cdot)$. We can see this mathematically as well. Thus, for the scheme shown in Fig. 4.2-1, the equations for the over-all system can be written down by inspection as

$$\dot{x}(t) = Ax(t) - bk\hat{x}(t) + bv(t), \qquad x(0-) = x_0$$
$$\dot{\hat{x}}(t) = lcx(t) + (A - lc - bk)\hat{x}(t) + bv(t), \qquad \hat{x}(0-) = \hat{x}_0$$

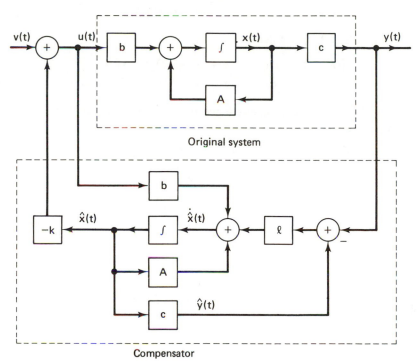

Figure 4.2-1. Combined observer-controller. Note that the observer is driven by both the output $y(\cdot)$ and the input $u(\cdot)$ $[= v(\cdot) - kx(\cdot)]$ of the original system.

The error in the estimates obeys the equation

$$\tilde{x}(t) = \dot{x}(t) - \dot{\hat{x}}(t)$$

$$= Ax(t) - A\hat{x}(t) - lcx(t) + lc\hat{x}(t)$$

$$= (A - lc)\tilde{x}(t), \qquad \tilde{x}(0-) = x_0 - \hat{x}_0 \tag{1}$$

and the effect of the input, $u(\cdot) = v(\cdot) - k\hat{x}(\cdot)$, is cancelled out. To study the overall system we arrange the equations for $x(t)$ and $\hat{x}(t)$ in matrix form as

$$\begin{bmatrix} \dot{x}(t) \\ \dot{\hat{x}}(t) \end{bmatrix} = \begin{bmatrix} A & -bk \\ lc & A - lc - bk \end{bmatrix} \begin{bmatrix} x(t) \\ \hat{x}(t) \end{bmatrix} + \begin{bmatrix} b \\ b \end{bmatrix} v(t), \qquad \begin{bmatrix} x(0-) \\ \hat{x}(0-) \end{bmatrix} = \begin{bmatrix} x_0 \\ \hat{x}_0 \end{bmatrix} \tag{2a}$$

$$y(t) = cx(t) \tag{2b}$$

Calculation of the Transfer Function. When computing transfer functions we have to assume zero initial conditions. Doing this and then taking Laplace transforms, we obtain

$$sX(s) = AX(s) - bk\hat{X}(s) + bV(s) \tag{3a}$$

$$s\hat{X}(s) = lcX(s) + (A - lc - bk)\hat{X}(s) + bV(s) \tag{3b}$$

Then eliminating $\hat{X}(s)$, we get

$$[sI - A + bk(sI - A + bk + lc)^{-1}lc]X(s)$$
$$= [I - bk(sI - A + bk + lc)^{-1}]bV(s) \tag{4}$$

It seems difficult to proceed, but if we use the important matrix-inversion identity (Exercise A-20)

$$[I + C(sI - A)^{-1}B]^{-1} = I - C(sI - A + BC)^{-1}B$$

we can rewrite (4) as

$$bV(s) = [I + bk(sI - A + lc)^{-1}]\{sI - A + lc$$
$$\qquad - [I - bk(sI - A + bk + lc)^{-1}]lc\}X(s)$$
$$= (sI - A + lc + bk - lc)X(s)$$
$$= (sI - A + bk)X(s)$$

Therefore

$$X(s) = (sI - A + bk)^{-1}bV(s)$$

so that the transfer function of the overall system is

$$H_{o\text{-}c}(s) = c(sI - A + bk)^{-1}b \tag{5}$$

(where the subscript *o-c* stands for observer-controller). These are the same equations we would have had with perfect state feedback; i.e., the overall transfer function is just that of the controlled system and does not depend on the dynamics of the observer. In other words, the transfer function does not depend on how quickly the error in the state estimates goes to zero.

The explanation is that when the initial conditions for $x(\cdot)$ and $\hat{x}(\cdot)$ are both zero, and therefore the *same*, then of course the observer, when driven with the same inputs as the original system, will have the same outputs as the original system (see our earlier discussions in Sec. 4.1). That is, $x(t) = \hat{x}(t)$ when $x(0) = 0 = \hat{x}(0)$. Therefore, when making transfer function calculations, the asymptotic observer is the same as the perfect observer!

This is a nice result, but our major concern of course is with the modes of the overall realization. These will be the roots of the characteristic equation of realization (2), namely

$$a_{o\text{-}c}(s) = \det \begin{bmatrix} sI - A & bk \\ -lc & sI - A + lc + bk \end{bmatrix} \qquad (6)$$

We can simplify this determinant by obvious row and column transformations to obtain,

$$a_{o\text{-}c}(s) = \det (sI - A + bk) \det (sI - A + lc) \qquad (7)$$

$$= a_{\text{cont}}(s)a_{\text{obs}}(s), \text{ say.} \qquad (8)$$

That is, the characteristic polynomial of the overall system is just the product of the characteristic polynomial of the observer and the characteristic polynomial of the controlled system assuming perfect knowledge of the states. This is nice, because it means that the natural frequencies or modes of the overall system can always be arranged to be stable. In fact, they can be chosen completely arbitrarily (if the original realization is controllable and observable). For by choosing k as in Sec. 3.2, we can make $a_{\text{cont}}(s) = \det (sI - A + bk)$ arbitrary. Therefore, if $a_{\text{cont}}(s)$ is stable (i.e., has no roots with zero or positive real parts) and $a_{\text{obs}}(s)$ is stable, then $a_{o\text{-}c}(s)$ is stable. This is a fundamental result and one that is not a priori obvious because, as noted earlier, there are many situations where the interconnection of stable systems leads to an unstable system.

Another useful consequence of (7) and (8) is that the *controller and observer can be designed independently of each other.* Whether the true states are available, or only asymptotically correct estimates of the states, is immaterial to the calculation of the feedback gains k; similarly, the dynamics of the asymptotic observer can be calculated from knowledge of A and c without caring if the observer is to be combined with a feedback controller or not. This is the so-called *separation property* of the observer-controller design procedure. However there will generally be a deterioration in the transient response of the combined system—see Example 4.2-1 below.

Implementation of the Observer. The fact that the observer has as many states as the original system might be disturbing—does this mean that we essentially have to replicate the original system in order to achieve the above nice results? The original system might be a rather complicated power plant, chemical processor, etc. Fortunately, we recognize that except under very special circumstances the observer need not be constructed in the same way as the original system—in particular, we can build the observer with (miniaturized) electronic circuitry, with great advantages of size, ruggedness, cost, etc. In fact, with modern developments in integrated electronics, we can readily envisage the use of very high-order observers for help in the control of quite complicated physical systems. Thus, in several applications, it will be feasible to implement the observer via a microprocessor specially designed to integrate the observer state equations.

Summary. We have now obtained some clear and definite answers to several of the questions we raised in Sec. 3.1.1. Briefly, by using feedback of the states of a completely controllable and completely observable realization of the original transfer function, we can obtain a new internally stable realization whose natural frequencies are completely under our control. While we have obtained this result only for scalar (single-input, single-output) systems, it also has a fairly natural extension to the multivariable case (Sec. 7.4) and in this sense represents perhaps the major triumph in the state-space approach to linear systems. These points are worth further discussion, which we shall begin to pursue further in Example 4.2-3.

Example 4.2-1.

In this example we shall illustrate the sort of response obtained by using an observer to generate state estimates for feedback. The results were obtained by simulation on an analog computer.

The system we shall consider corresponds to the pointer-balancing problem of Example 3.3-1, except that we now take $g/L = 1$ for convenience of simulation. We then have

$$\dot{z} = \begin{bmatrix} 0 & 1 \\ 1 & 0 \end{bmatrix} z + \begin{bmatrix} 0 \\ -1 \end{bmatrix} u$$

$$y = [1 \quad 0]z$$

and the system modes are at $\lambda = \pm 1$. An analog simulation of the system is shown in Fig. a.

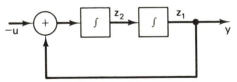

Figure a. Simulation of system.

The system is clearly unstable, as indicated in Fig. b by the response to a nonzero initial condition.

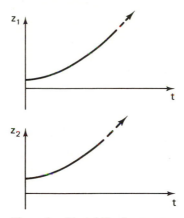

Figure b.   Unstabilized response.

We use state feedback $u = -kz$ to move the system poles to $-0.5 \pm j0.5$. The required k is easily obtained as $k = [-1.5 \quad -1]$. The decay of the system from an initial state to the origin, using this direct state feedback, is shown in Fig. c.

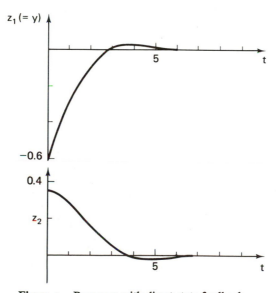

Figure c.   Response with direct state feedback.

We now construct an observer for the system. The observer modes are chosen to lie at $-1 \pm j1$. (They are faster than the desired closed-loop poles but slow enough for us to see clearly their effect on the system response. A brief discussion of the factors involved in selecting reasonable observer modes is presented in Sec. 4.4.)

The observer gain is readily found as $l = [2 \quad 3]'$, and the observer is then easily constructed.

We now use the observer states to generate the required control, $u = -k\hat{z}$, and the resulting compensator is shown in Fig. d.

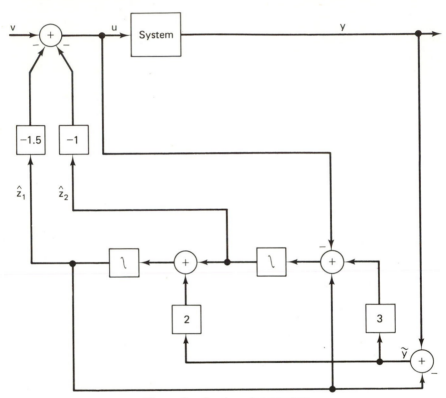

Figure d.   Compensator structure.

The response of this closed-loop system is indicated in Fig. e. Note the manner in which the observer error decays; it is clear that the observer error goes to zero at about twice the rate at which the system settles to zero, as is to be expected from the fact that the error poles are twice the closed-loop poles. Comparison of Figs. c and d indicates the deterioration in response due to the use of estimated rather than actual states for feedback. In Fig. f we compare the controls resulting from direct state feedback and from use of the observer-controller compensator structure; the initial large-amplitude oscillation of the control in the latter case is again a result of the large initial uncertainty as to the true state of the system.

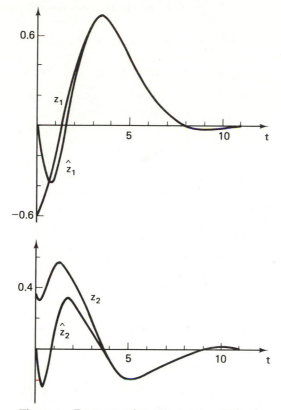

Figure e. Response using compensator of Fig. d.

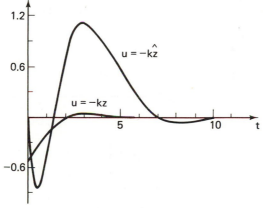

Figure f. Comparison of controls generated by direct state feedback and using the compensator.

275

Example 4.2-2.   Constant Disturbances and Integral Feedback

For a system driven by a *constant* unknown disturbance w, design an observer to estimate w, and use this to compensate for the disturbance.

Solution.   We have

$$\dot{x} = Ax + bu + bw$$
$$\dot{w} = 0$$
$$y = cx$$

where u is the control input and y the observed output. The constant disturbance w is modeled as the output of an undriven integrator. We then have the *augmented* system shown in Fig. a.

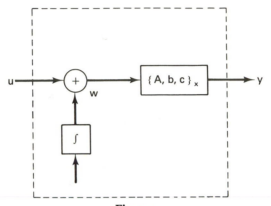

Figure a.

If now we had an estimate $\hat{w}$ of w, we could set $u = -\hat{w}$ to attempt to cancel out the disturbance. This motivates us to set up an observer to estimate w.

An observer for the augmented system is given by

$$\begin{bmatrix} \dot{\hat{x}} \\ \dot{\hat{w}} \end{bmatrix} = \begin{bmatrix} A & b \\ 0 & 0 \end{bmatrix} \begin{bmatrix} \hat{x} \\ \hat{w} \end{bmatrix} + \begin{bmatrix} b \\ 0 \end{bmatrix} u + l(y - c\hat{x})$$
$$\hat{x}(0) = 0, \qquad \hat{w}(0) = 0$$

where l is an $(n + 1) \times 1$ vector. Partitioning l as $[l_1'\ \ l_2']'$, with l_2 a scalar, we get

$$\begin{bmatrix} \dot{\hat{x}} \\ \dot{\hat{w}} \end{bmatrix} = \begin{bmatrix} A - l_1 c & b \\ -l_2 c & 0 \end{bmatrix} \begin{bmatrix} \hat{x} \\ \hat{w} \end{bmatrix} + \begin{bmatrix} b \\ 0 \end{bmatrix} u + \begin{bmatrix} l_1 \\ l_2 \end{bmatrix} y$$

The observer structure is then as shown in Fig. b. Now *if* the augmented system is observable, we can choose l so as to obtain arbitrary error decay modes and thus ensure that $\hat{w}$ approaches w asymptotically. Let us temporarily ignore the question

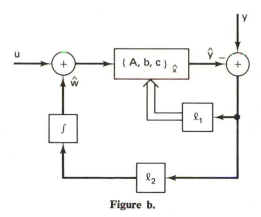

Figure b.

of observability. We also make a particular choice of l as $[0 \; l_2]$, which will simplify our observer considerably; for the moment we shall not worry about whether this can still ensure that the error-decay modes are stable. Now, on setting $u = -\hat{w}$, our observer equation reduces to

$$\begin{bmatrix} \dot{\hat{x}} \\ \dot{\hat{w}} \end{bmatrix} = \begin{bmatrix} A & 0 \\ -l_2 c & 0 \end{bmatrix} \begin{bmatrix} \hat{x} \\ \hat{w} \end{bmatrix} + \begin{bmatrix} 0 \\ l_2 \end{bmatrix} y, \qquad \hat{x}(0) = 0, \qquad \hat{w}(0) = 0$$

Since the equation for $\hat{x}$ is undriven and the initial condition is zero, we have $\hat{x} \equiv 0$, and our observer is simply

$$\dot{\hat{w}} = l_2 y, \qquad \hat{w}(0) = 0$$

The resulting overall compensation scheme is shown in Fig. c (where the dashed lines indicate parts that drop out of the compensator).

The result of the above procedure is thus precisely the technique that was presented in Sec. 3.2.3 for compensation of constant unknown disturbances, namely *integral feedback*. It arises here in a more natural and motivated manner.

The question we have so far avoided is whether proper choice of l_2 can ensure that $\hat{w}$ approaches w.

Our earlier observer equation shows that the observer error behavior is determined by the roots of (see Exercise A-11)

$$\alpha(s) = \det \begin{bmatrix} sI - A & -b \\ l_2 c & s \end{bmatrix} = \det(sI - A) \det[s + l_2 c(sI - A)^{-1}b]$$

$$= sa(s) + l_2 b(s) = 0$$

We assume now that the original system $\{A, b, c\}$ was stable (or stabilized) and hence that $a(s)$ is stable, i.e., has roots with strictly negative real parts. It can then be shown that proper choice of l_2 can give stable $\alpha(s)$ if and only if $b(s)$ has no root

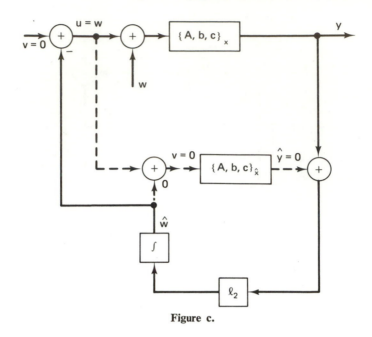

Figure c.

at the origin. [Necessity is obvious, since if $b(s)$ contained s as a factor, then $\alpha(s)$ would also contain this unstable factor; initial observer error would then not decay to zero. Sufficiency is easily proved using root-locus-type arguments, but we shall omit these here.]

The case where $b(s)$ has a root at the origin (i.e., where the original system has a zero at the origin) is, however, rather trivial, since the presence of a zero at the origin makes the output of the system insensitive to a constant input anyway (see Sec. 3.2.3). Another way of understanding the above fact is to note (see Example 2.4-4) that if $b(s)$ contained s as a factor then the augmented system (from the disturbance initial conditions to $y(\cdot)$) would be unobservable due to the cancellation of the term s.

This is a simple example of the so-called "internal model" principle for the rejection of disturbances.

Example 4.2-3. A Review of the Overall Observer-Controller Design Problem

We are given a system with transfer function $H(s)$. The *system* may be a process, power plant, machine, market, bacterial colony, etc. We have control over one variable, the *input*, and are interested in controlling the behavior (evolution, response) of some other system variable, the *output*. The system given to us is characterized by a certain input-output behavior, say,

$$H(s) = \frac{(s-2)(s-3)}{s(s-1)(s-4)} = \frac{s^2 - 5s + 6}{s^3 - 5s^2 + 4s}$$

The system may be unstable (as above), or its response to a given input may not be what we wish it to be. Suppose we wish to relocate the poles at $-1 \pm j$ and -5 (guided perhaps by the optimum root locus—draw this!).

Cascade compensation, which cancels the unstable poles, is unsatisfactory.

Output feedback, $u \longrightarrow v - ky$, will not suffice; we have one free parameter and three quantities to be independently adjusted.

Position, velocity, and acceleration feedback, $(k_2 + k_1 s + k_0 s^2)y$, gives us enough parameters, but the system order increases to 4,

$$H(s) = \frac{(s-2)(s-3)}{s(s-1)(s-4) + (k_2 + k_1 s + k_0 s^2)(s-2)(s-3)}$$

Other strategies can be tried, as in Sec. 3.1.1, without leading to any obvious clear answer, and therefore we shall try to approach the problem with state-space ideas. We know that to modify the system arbitrarily we must know what the system is doing. And this (by definition, or by construction, . . .) is told us by the state variables. Suppose then that someone gives us these state variables (or states, for short). This statement by itself is meaningless. One cannot talk of states apart from the *realization* they correspond to. To make sense, we have to be given measurements (terminals, . . .) that define the state variables $x(\cdot)$ *of a realization* $\{A, b, c\}$. How was the realization obtained by the person who gives us the states? Again, it is meaningless to talk of *the* realization. What the person had was a set of equations describing the system; these were transformed (and perhaps linearized) to *a* set of first-order differential equations in the variables $x(\cdot)$, which are now made available. If the realization is controllable, we can use linear state feedback, $kx(\cdot)$, with k chosen to obtain any desired free response (natural frequencies).

Assume that we have available the states of a controller-form realization of $H(s)$,

$$A_c = \begin{bmatrix} +5 & -4 & 0 \\ 1 & 0 & 0 \\ 0 & 1 & 0 \end{bmatrix}, \qquad b_c = \begin{bmatrix} 1 \\ 0 \\ 0 \end{bmatrix}$$

$$c_c = \begin{bmatrix} 1 & -5 & 6 \end{bmatrix}$$

We would like a new characteristic polynomial

$$\alpha(s) = (s+5)(s+1+j)(s+1-j)$$
$$= s^3 + 7s^2 + 12s + 10$$

Now with state feedback,

$$A_c - b_c k_c = \begin{bmatrix} -7 & -12 & -10 \\ 1 & 0 & 0 \\ 0 & 1 & 0 \end{bmatrix}$$

which shows that

$$k_c = (\alpha - a) = \begin{bmatrix} 12 & 8 & 10 \end{bmatrix}$$

Now how realistic is the assumption that the states are available? What is involved in making the states available? Note that by making the states available we are in effect defining n new outputs. It is then perhaps not so surprising that we can achieve so much more (namely arbitrary pole location) with state feedback than with feedback of the single original output. The obvious cost involved in making the states available is the cost of the sensors required to measure the states and the transducers needed to convert these measurements to useful quantities for control.

(There is a whole *dual* range of possibilities for system modification that arises from the definition of additional *inputs*, or, equivalently, from postulating access to the *inputs* of the *integrators* of the realization, but we shall not go into this here.)

So what do we do if the states are not available (or the cost of making them available is prohibitive)? If the system is observable, we can determine its state at any time by differentiation of the output and input. This is unrealistic because of noise considerations. If, however, we know the state of the system at some instant, we can construct a dummy system (realization) with this given initial state and drive it with the same input as the original system from that time on. (Note that this dummy system is only mathematically equivalent to the original—its actual implementation may be quite different, perhaps an analog-computer simulation, or often a *digital-computer algorithm* for integrating the state equations.) The (simulated) states are now available.

It is desirable to use an additional input to the dummy, namely $y - \hat{y} = \tilde{y}$ (the difference between the actual and dummy outputs). This allows asymptotic correction of errors in initial-state determination and in fact obviates the need for initial-state determination altogether. The system is still required to be observable, however, in order that the error dynamics may be arbitrarily chosen.

Returning to our example, we construct a dummy system $\{A_c, b_c, c_c\}$ with states $\hat{x}_c$, driven by an additional input $l_c(y - c_c\hat{x}_c)$. Suppose we want the observer poles at $-6 \pm j6$ and -6 (a "reasonable" choice for our given closed-loop pole locations). Then l_c must be chosen such that the characteristic polynomial of $A_c - l_c c_c$ is

$$\alpha_l(s) = (s + 6)(s^2 + 12s + 72) = s^3 + 18s^2 + 144s + 432$$

Some algebra then shows that

$$l_c = [807.5 \quad 127.5 \quad -24.5]'$$

The combined observer-controller can now be implemented as in Figure 4.2-1. In practice, however, there are still problems to be overcome in the actual realization, one of the most important being *scaling*: We may need to transform the realization to a form where the required gains are of reasonable magnitude—see Exercise 2.2-19.

Exercises

4.2-1. *Alternative State Equations*

The fact that $\tilde{X}(s) = X(s) - \hat{X}(s) = 0$ in (3a) and (3b) suggests that perhaps a more convenient set of state variables for the observer-controller system is $[x, \tilde{x}]$. Write state equations for these variables, and use them to calculate $H_{o\text{-}c}(s)$ and $a_{o\text{-}c}(s)$.

4.2-2. *Controllability and Observability of the Observer-Controller Realization*

The realization in Fig. 4.2-1 is clearly not minimal.

a. Show that this realization is not controllable. What are the noncontrollable state variables?

b. Show that the realization will be nonobservable if and only if at least one of the following holds:

(1) $\{k, A - lc\}$ is not observable,

(2) $\{c, A - bk\}$ is not observable, or

(3) A pole of the observer cancels a zero of the original transfer function $H(s)$.

It is, of course, assumed that $\{A, b, c\}$ is minimal.

4.2-3.

In the combined controller-observer design we select k and l so that $a_c(s)$ and $a_o(s)$ both have poles in the left half plane. Is it true that the resulting design must be stable even if the loop is broken open at y, for instance? Explain your answer briefly.

4.2-4.

Consider the undamped harmonic oscillator $\dot{x}_1(t) = x_2(t)$, $\dot{x}_2(t) = -\omega_0^2 x_1(t) + u(t)$. Using an observation of velocity, $y(\cdot) = x_2(\cdot)$, design an observer/state-feedback compensator to control the position $x_1(\cdot)$. Place the state-feedback controller poles at $s = -\omega_0 \pm j\omega_0$ and both observer poles at $s = -\omega_0$.

*4.3 REDUCED-ORDER OBSERVERS

The observer design method described in Sec. 4.1 is due, as noted earlier, to Bertram (1961) and Bass (1963). In his Ph.D. thesis (Stanford, 1963) Luenberger took a different approach to the observer problem. We shall not really discuss this approach here but shall focus instead on its most significant contribution. The observer obtained in Sec. 4.1 had n states, where n was the order of the realization whose states were being observed. Luenberger [3] pointed out that the order of the observer can actually be less than this because the observed output provides a linear relationship $y(t) = cx(t)$ between the state variables. Therefore it suffices to observe $n - 1$ of the states and then to calculate the final one from this linear relationship.

The reduction by one in observer dimension is not particularly significant, especially when the observer is implemented with (integrated-circuit) electronic logic. However, the question has some interesting theoretical aspects, which we shall explore here. Moreover, for multioutput systems, somewhat more substantial reductions can be obtained (cf. Sec. 7.3). In this section, we shall essentially follow the historical approach to the topic, which starts with the unreduced observer of Sec. 4.1. The reduced-order observer will arise in a more direct way in the frequency-domain design procedure to be presented in Sec. 4.5.1.

We shall first explain why reduction is possible and describe Luenberger's idea [3] for exploiting it. However for the detailed calculations we shall use a different technique ([4] and [5]).

Let us begin with a realization $\{A, b, c\}$ and make an easily found (and non-unique) similarity transformation such that the output vector has the form†

$$c = [0 \quad \cdots \quad 1] \tag{1}$$

and

$$y = [0 \quad \cdots \quad 1]x(t) = x_n(t) \tag{2}$$

Therefore we have clearly displayed the fact that one state is directly observable,

$$\hat{x}_n(t) = y(t) = x_n(t)$$

and we only need to estimate the states $x_r = [x_1 \quad \cdots \quad x_{n-1}]$, where the subscript r is used to denote the states that remain to be estimated. It should be possible to estimate these $n - 1$ states by an $(n - 1)$-dimensional observer. This is true, but the design has to be approached with a little care. Thus, consider the partitioned state equations

$$\begin{bmatrix} \dot{x}_r(t) \\ \dot{x}_n(t) \end{bmatrix} = \begin{bmatrix} A_r & b_r \\ c_r & a_{nn} \end{bmatrix} \begin{bmatrix} x_r(t) \\ x_n(t) \end{bmatrix} + \begin{bmatrix} g_r \\ g_n \end{bmatrix} u(t), \qquad y(t) = x_n(t) \tag{3}$$

where the reasons for our notation will become clear later. Now we set up a dummy system for $\hat{x}_r(t)$,

$$\dot{\hat{x}}_r(t) = A_r\hat{x}_r(t) + b_ry(t) + g_ru(t) + \text{(feedback term)} \tag{4}$$

The feedback term should be derived from the error $x_r(t) - \hat{x}_r(t)$, but not only is this term inaccessible, just as $x(t) - \hat{x}(t)$ was in the nth-order observer of Sec. 4.1, but the accessible quantity $[y(t) - c\hat{x}(t)]$ used in Sec. 4.1 is identically zero [with our choice (2) for c] if we insist that $\hat{x}_n(t) = y(t)$. Therefore, it seems difficult to get a *feedback* estimator for $x_r(t)$. However, it is worthwhile to persist a bit.

Note that without feedback, the *open-loop* equation (4) for the error $\tilde{x}_r(t) = \hat{x}_r - x_r(t)$ is

$$\dot{\tilde{x}}_r(t) = \dot{x}_r(t) - \dot{\hat{x}}_r(t) = A_r\tilde{x}_r(t), \qquad \tilde{x}_r(0) = x_r(0) - \hat{x}_r(0) \tag{5}$$

[Note again that we might as well have assumed the *known* inputs $b_ry(\cdot)$ and $g_ru(\cdot)$ to be zero.] We had a similar equation in Sec. 4.1 for the n-dimensional open-loop observer,

$$\dot{\tilde{x}}(t) = A\tilde{x}(t), \qquad \tilde{x}(0) = x(0) - \hat{x}(0) \tag{6}$$

†One such realization is obtained from our usual observer form $\{A_0, b_0, c_0\}$ by merely labeling the states in the reverse order to our conventioual order. In fact, the reader will find it helpful to interpret various results stated for observable realizations by assuming the realizations to be in observer (or relabeled observer) form. For actual computation, however, it may be generally simpler to avoid transformation all the way to observer form and to merely find some convenient transformation that gives $c = [0 \quad \cdots \quad 0 \quad 1]$.

and we concluded that the error dynamics were beyond our control because the eigenvalues of A could not be changed by similarity transformations taking A to $T^{-1}AT$. However, it was at this point that Luenberger made the key observation that even though the eigenvalues of a matrix A could not be changed by similarity transformations, this is *not* true of eigenvalues *of submatrices of A.* (Note that if $\bar{A} = T^{-1}AT$, it is unlikely that [cf. (3) for the notation] $\bar{A}_r$ is similar to A_r.) Luenberger showed in fact that if the original equations are observable, then state transformations can be found that will yield a *submatrix* with arbitrary eigenvalues (see Exercise 4.3-1).

Luenberger's technique is ingenious and makes interesting use of different state realizations. However, it is somewhat confusing (at least initially) to keep changing realizations, since after all we are basically interested in only one realization, viz., the one that we assumed for the original system that was to be controlled by state feedback.

Another Approach (Gopinath [4] and Cumming [5]).   For another approach we can return to the original equations (3) and note that so far we have only really examined the subset of equations describing the evolution of $x_r(\cdot)$. A direct approach using this set did not succeed—see (4)—because we could not obtain a suitable *feedback term.* However, we did not use the remaining equation

$$\dot{x}_n(t) = a_{nn}x_n(t) + c_r x_r(t) + g_n u(t)$$

which certainly does provide some more information about $x_r(\cdot)$. Thus, note that if we define

$$y_r(t) = \dot{y}(t) - a_{nn}y(t) - g_n u(t) \tag{7}$$

then we can write

$$\dot{x}_r(t) = A_r x_r(t) + b_r y(t) + g_r u(t) \tag{8}$$

$$y_r(t) = c_r x_r(t) \tag{9}$$

The function $y_r(\cdot)$ is completely determined by $y(\cdot)$ and $u(\cdot)$ and can therefore be regarded as an observation process. The difficulty is that $y_r(\cdot)$ contains the derivative $\dot{y}(\cdot)$ and cannot therefore be realistically obtained from $y(\cdot)$. However, if we temporarily ignore this difficulty, we see that Eqs. (8) and (9) display the $n-1$ states $x_r(\cdot)$ in a form to which the feedback design of Sec. 4.1 can be applied.

That is, we can set up an observer of the form

$$\dot{\hat{x}}_r(t) = A_r \hat{x}_r(t) + b_r y(t) + g_r u(t) + l_r[y_r(t) - c_r \hat{x}_r(t)] \tag{10}$$

where l_r is an $(n-1) \times 1$ matrix. Now

$$\dot{\tilde{x}}_r(t) = (A_r - l_r c_r)\tilde{x}_r(t), \qquad \tilde{x}_r(0) = x_r(0) - \hat{x}_r(0) \tag{11}$$

and we see that if

$$\{c_r, A_r\} \text{ is observable} \tag{12}$$

then $\tilde{x}_r(\cdot)$ can be made to go to zero arbitrarily fast (but, of course, not faster than *exponentially*).

The observability of $\{c_r, A_r\}$ is to be expected from that of $\{c, A\}$ because (3) shows that the only way $y(\cdot) = x_n(\cdot)$ gives information about the remaining states $x_r(\cdot)$ is through $y_r(\cdot)$. In any case, a simple proof of this fact can be obtained by using the PBH tests.

Therefore, we now have a reduced-order observer, except for the fact that $y_r(\cdot)$ contains a derivative of the observation $y(\cdot)$, which is generally unacceptable. However, our experience with analog-computer simulations in Sec. 2.1 shows how to avoid this difficulty. The method is clear from Figs. 4.3-1 and 4.3-2. [Note that removing the differentiator (cf. Fig. 4.3-2) changes the input to the integrator from $\hat{x}_r(\cdot)$ to $\hat{x}_r(\cdot) - l_r\dot{y}(\cdot)$ and changes the integrator output to $\hat{x}_r(\cdot) - l_r y(\cdot)$; therefore, $\hat{x}_r(\cdot)$ can be recovered by adding $l_r y(\cdot)$ to the output of the integrator.]

State equations can be written down from Fig. 4.3-2 to yield

$$\dot{\theta}(t) = (A_r - l_r c_r)\theta(t) + (b_r - l_r a_{nn} + A_r l_r - l_r c_r l_r)y(t)$$
$$+ (g_r - l_r g_n)u(t) \tag{13}$$
$$\hat{x}_r(t) = \theta(t) + l_r y(t) \tag{14a}$$
$$\hat{x}_n(t) = x_n(t) = y(t) \tag{14b}$$

as the equations for the observer.

Of course, there are other methods of implementing the equations we first obtained for the observer. Thus we could take Laplace transforms of (7)–(10) to

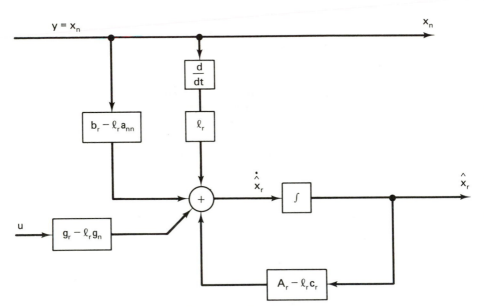

Figure 4.3-1. Implementation using differentiators.

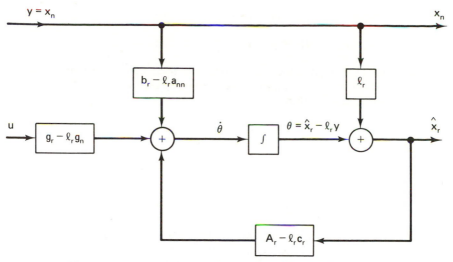

Figure 4.3-2.   Equivalent implementation without differentiators.

obtain the transfer function

$$\hat{X}(s) = (sI - A_r + l_r c_r)^{-1}[(sl_r + b_r - l_r a_{nn})Y(s) + (g_r - l_r g_n)U(s)] \quad (15)$$

which can be simplified and then realized in several different ways. The advantage of the first technique is that the realization (Fig. 4.3-2) it yields contains the parameters $\{A_r, b_r, c_r, a_{nn}, l_r\}$ in a direct way, while if (15) is used, these various parameters may be recombined in a way that destroys their identity.

From Fig. 4.3-2 and (13)–(15) we see that the reduced-order observer can be straightforwardly incorporated into our compensator scheme.

Example 4.3-1.   Pointer-Balancing Problem

We shall present here the results obtained by analog-computer simulation of the system of Example 4.2-1, now compensated with a reduced-order observer.

Recall that the system was described by

$$\dot{z} = \begin{bmatrix} 0 & 1 \\ 1 & 0 \end{bmatrix} z + \begin{bmatrix} 0 \\ -1 \end{bmatrix} u, \qquad y = [1 \quad 0]z$$

and we had determined a state-feedback gain vector to move the poles from ± 1 to $-0.5 \pm j0.5$, namely $k = [-1.5 \ -1]$.

We design from first principles a reduced-order observer with pole at -1.5. Since z_1 is directly observed, we construct an observer for z_2. We thus have

$$\dot{\hat{z}}_2 = z_1 - u + l \times (\text{error signal})$$

Noting that $\dot{z}_1 = \dot{y} = z_2$, we see that a convenient error signal can be generated, using only available measurements, as $\dot{y} - \hat{z}_2$, to give

$$\dot{\hat{z}}_2 = -l\hat{z}_2 + y - u + l\dot{y} \qquad \text{when } \dot{\hat{z}}_2 = -l\hat{z}_2$$

Choosing $l = 1.5$ gives the desired observer-error decay. To implement the observer without differentiation of y, we work with a new variable $\theta = \hat{z}_2 - ly$. Our observer equation is then

$$\dot{\theta} = -l\theta - l^2 y + y - u$$
$$= -1.5\theta - 1.25y - u$$

and

$$\hat{z}_1 = y, \qquad \hat{z}_2 = \theta + 1.5y$$

Figure 4.3-3(a) shows the resulting compensator structure. In Fig. 4.3-3(b) we illustrate the closed-loop response using this compensator; the results shown are for $\theta(0) = +0.5$ (a value chosen for convenience of simulation), and, as before, $z(0) = [-0.6 \ 0.35]$. This response may be compared with that obtained by direct state feedback and by using a full-order observer in the compensator—see Example 4.2-1. Figure 4.3-3(c) shows the control generated by the reduced-order compensator.

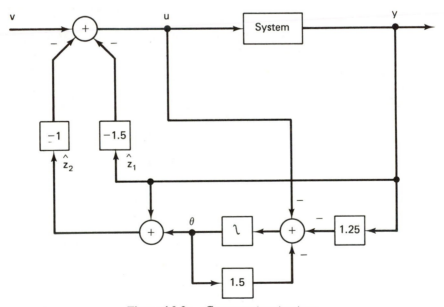

Figure 4.3-3a. Compensator structure.

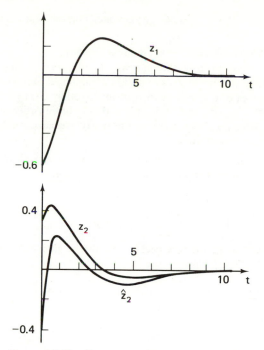

Figure 4.3-3b. Response using compensator of (a).

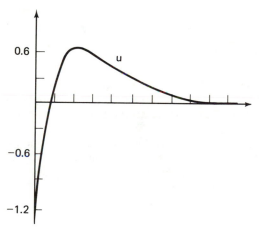

Figure 4.3-3c. Control generated by compensator of (a).

Example 4.3-2.   Observers and Some Classical Designs

This example is solved in a different way (using several canonical forms) in Chen [9, p. 296–298]. The problem is to design a combined observer-controller that will relocate the poles of a system with transfer function $H(s) = 1/s(s + 1)$ at the new locations $\{s = -1 \pm j\}$, and to do this with a reduced-order observer that has a pole at $s = -2$. Because of the separation property, we can consider the controller and observer problems separately.

Design of Controller.   We shall work with a realization of $H(s)$ in controller canonical form. This form can be obtained by inspection,

$$A_c = \begin{bmatrix} -1 & 0 \\ 1 & 0 \end{bmatrix}, \qquad b_c = \begin{bmatrix} 1 \\ 0 \end{bmatrix}, \qquad c_c = \begin{bmatrix} 0 & 1 \end{bmatrix}$$

Using state feedback $k_c x_c$, the row vector k_c must be such that the characteristic polynomial of $[A_c - b_c k_c]$ will be

$$a_{\text{cont}}(s) = (s + 1 + j)(s + 1 - j) = s^2 + 2s + 2$$

It follows easily that

$$k_{c2} = 2$$

Design of Reduced-Order Observer.   Without loss of generality we can assume that $u(\cdot) \equiv 0$. Then $x_{c2}(\cdot) = y(\cdot)$, and we only need to estimate $x_{c1}(\cdot)$. We have

$$\dot{x}_{c1}(t) = -x_{c1}(t), \qquad \dot{y}(t) = \dot{x}_{c2}(t) = x_{c1}(t)$$

and we define

$$y_r(t) = \dot{y}(t) = x_{c1}(t)$$

Therefore the reduced-order observer is

$$\dot{\hat{x}}_{c1}(t) = -\hat{x}_{c1}(t) + l[y_r(t) - \hat{x}_{c1}(t)]$$

where l is to be such that $a_{\text{obs}}(s) = s + 2 = \det(sI - A_r + l c_r) = s + 1 + l$, which gives $l = 1$.

Combined Observer-Controller.   We can combine the above results as shown in Fig. 4.3-4. Note that the input $u(\cdot) = v(\cdot) - k\hat{x}(\cdot)$ has to be fed to the observer. We have also moved $y(\cdot)$ across the integrator in the observer system so as to obviate the need for differentiation.

It will be of interest to consider various alternative configurations obtainable by transfer function manipulation. We begin by computing the transfer functions from $y(\cdot)$ and $v(\cdot)$ to $u(\cdot)$, the actual input to the original system. From the block diagram (Fig. 4.3-4) we can write the equations

$$\dot{\theta}(t) = -2\theta(t) - 2y(t) + u(t)$$
$$u(t) = v(t) - 2y(t) - [y(t) + \theta(t)]$$

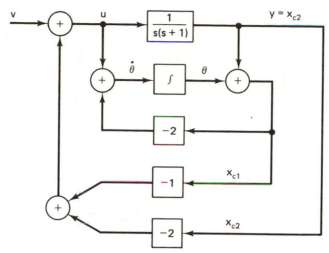

Figure 4.3-4. Combined observer-controller for Example 4.3-2.

Taking Laplace transforms and eliminating θ, we obtain

$$U(s) = \left(-3 + \frac{2}{s+2}\right)Y(s) + V(s) - U(s)(s+2)^{-1} \tag{16}$$

Therefore

$$U(s) = -\frac{3s+4}{s+3}Y(s) + \frac{s+2}{s+3}V(s) \tag{17}$$

Relation (16) can be realized as shown in Fig. 4.3-5(a), which may be rearranged as in Fig. 4.3-5(b), a classical configuration. By the methods of Sec. 3.1.2 we can check that this realization has two hidden natural frequencies at $s = -2$, which are cancelled out of the transfer function from $v(\cdot)$ to $y(\cdot)$. The configuration is stable but uses one integrator more than necessary.

Relation (17) suggests the realization of Fig. 4.3-5(c), which can be shown to have hidden natural frequencies at $s = -2$ and $s = -3$. The frequency $s = -3$ is fixed by our original choice of transfer function poles and observer poles; in this case it *happens* to be a stable natural frequency, so the configuration is usable, though again it has one more integrator than necessary.

Another classical configuration, with *unity* feedback, is shown in Fig. 4.3-6, where $\mathcal{Y}(s)$ is determined by choosing it to make the overall transfer function equal to its desired value of $(s^2 + 2s + 2)^{-1}$. We have

$$\frac{\mathcal{Y}(s)H(s)}{1 + \mathcal{Y}(s)H(s)} = \frac{1}{s^2 + 2s + 2}$$

which yields

$$\mathcal{Y}(s) = s(s+1)^{-1}$$

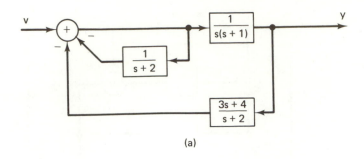

(a)

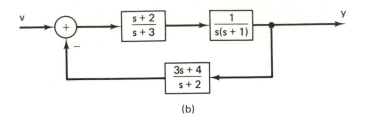

(b)

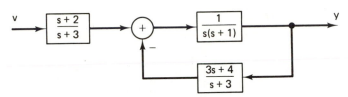

Figure 4.3-5. Various transfer function arrangements of the realization in Fig. 4.3-4. (a) Direct implementation of (16). (b) A classical configuration. (c) Direct implementation of (17).

This is a very simple solution which seems to avoid all the fuss about separate controller design, the design of reduced-order observers, and the combination of controller and observer. Why don't we just use it directly? The reason is (cf. the discussion in Sec. 3.1) that this particular procedure cannot give us full control over the hidden modes; in fact, we can see (by writing down a state-space realization or, more simply, by using the transfer function method of Exer. 2.4-7) that the configuration of Fig. 4.3-6 will have an unstable hidden mode at $s = 0$. However by a

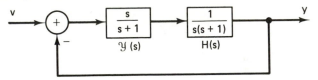

Figure 4.3-6. Classical unity-feedback configuration.

slightly more complicated design procedure, we can in fact achieve fairly satisfactory results with this configuration, as we shall show in Sec. 4.5.2.

Exercises

4.3-1.   *Eigenvalues of Submatrices* [3]

Suppose we have a realization $\{c, A\}$ in the (column-transposed observer) form with $c = [0 \cdots 0\ 1]$ and A a *right-companion* matrix with $-[a_n \cdots a_1]$ in the last column. Suppose also that we wish to find a nonsingular matrix T so that $\bar{A}_r$, defined as the $(n - 1) \times (n - 1)$ left-hand submatrix of $T^{-1}AT$, has characteristic polynomial $\det(sI - \bar{A}_r) = \alpha(s) = s^{n-1} + \alpha_1 s^{n-2} + \cdots + \alpha_{n-1}$. Show that choosing T to have last column $[\alpha_{n-1} \cdots \alpha_1\ 1]'$ with 1s for the remaining diagonal elements and 0s elsewhere will yield $\bar{c} = c$ and $\bar{A}_r$ as a right-companion matrix with last column $-[\alpha_{n-1} \cdots \alpha_1]'$.

4.3-2.   *Another Approach to Reduced-Order Observers*

Show how to use the result of Exercise 4.3-1 to determine a reduced-order observer for an arbitrary but observable realization (A, b, c). Do Examples 4.3-1 and 4.3-2 by this method.

4.3-3.

Show that the subsystem $\{c_r, A_r\}$ defined by Eqs. (4.3-3) will be observable if $\{c, A\}$ is observable. Do this
   a. By using the result of Exercise 4.3-1,
   b. By direct algebraic manipulation of $\Theta(c_r, A_r)$, and
   c. By using the PBH tests of Sec. 2.4.3.

4.3-4.   *Updating an Inertial Navigator with a Velocity Measurement.*

An error model of the east-velocity channel of an inertial navigator is (in normalized variables)

$$\begin{bmatrix} \dot{v} \\ \dot{\varphi} \\ \dot{\epsilon} \end{bmatrix} = \begin{bmatrix} 0 & -1 & 0 \\ 1 & 0 & 1 \\ 0 & 0 & 0 \end{bmatrix} \begin{bmatrix} v \\ \varphi \\ \epsilon \end{bmatrix} + \begin{bmatrix} 0 \\ 0 \\ w \end{bmatrix}$$

where $v =$ east-velocity error, $\varphi =$ platform tilt about north axis, $\epsilon =$ north-gyro drift, and $w =$ gyro drift rate of change.
   a. Show that the open-loop eigenvalues are $\lambda_1 = 0$, $\lambda_{2,3} = \pm j$.
   b. Construct an observer using $z = v$ as the observation [made by a Doppler radar (airplane) or an EM log (ship)], placing the estimate-error poles at $\{-10, -0.1, -0.1\}$.
   c. Construct a *second-order* observer using $z = v$ as the observation, placing the estimate-error poles at $\{-0.1, -0.1\}$.

4.3-5.

We are given a system with transfer function $H(s) = 1/s(s - 2)$ and wish to use the unity-feedback configuration of Fig. a in order to design a compensator $\mathcal{Y}(s)$ that will shift both poles of $H(s)$ to -1. No requirement is put on the zeros. You can choose any polynomial $\beta(s)$ (of degree not greater than 2) that you wish.

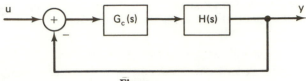

Figure a.

a. Find such a compensator. Is G_c stable?

b. Do you think that the compensator procedure is satisfactory? Give reasons.

c. If we set up some minimal realizations for G_c and H and connect them as in Fig. a, then the overall system will be a realization of $\beta(s)/(s + 1)^2$. Is it a minimal realization? A controllable realization? An observable realization?

d. Suppose that in some other design calculation we have computed a block $G_f(s) = (13s + 1)/(s + 1)$ and that we now want to find $G_c(s)$ as in Fig. b so that the overall transfer function is $\beta(s)/(s + 1)^2$ and the overall system is stable. Find such a $G_c(s)$.

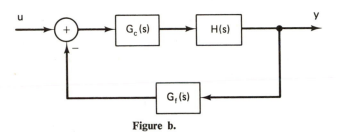

Figure b.

Compare the results of parts b and d.

4.3-6.

A system is described by the transfer function $H(s) = (s + 1)/s^2 = Y(s)/U(s)$.

a. Find a first order dynamic feedback compensator so that the transfer function from reference input V to output Y is $Y(s)/V(s) = [(s^2 + 2s + 4)(s + 2)]^{-1}(s + 8)(s + 1)$.

b. Draw a block diagram of the resulting design.

c. Find a first order dynamic feedback compensator so that the transfer function from reference input to output is $Y(s)/V(s) = (s^2 + 2s + 4)^{-1}(s + 1)$.

d. Draw a block diagram of the design of part c.

e. Is the system of part a controllable from V? Observable from Y?

4.3-7.

Carry out the steps required to obtain reduced-order observers for discrete-time systems.

4.4 AN OPTIMALITY CRITERION
FOR CHOOSING OBSERVER POLES

In Sec. 3.4.1, we discussed the relation between the modal controller and the so-called *optimal quadratic regulator*. Since the asymptotic observer is *dual* to the modal controller, it is natural to ask for its relationship to the dual of the quadratic regulator, and we shall briefly discuss this here.

We begin by noting that we could make the observer error decay as rapidly as we wished by putting the observer poles sufficiently far into the left half plane. In fact, one can show that in the limit the asymptotic observer reduces exactly to the ideal differentiating observer of Sec. 2.3.1. Of course, this fact brings up one reason we should not try to get too high a speed of reconstruction for the state estimates: The resulting observer will have (a high bandwidth and) a high susceptibility to the almost-inevitable measurement noise.

The conflict between speed of reconstruction and protection against measurement noise can be introduced into a quadratic optimality criterion as follows. Suppose that our actual state equations are

$$\dot{x}(t) = Ax(t) + bu(t) + gv_1(t) \tag{1a}$$

$$y(t) = cx(t) + v_2(t) \tag{1b}$$

where $v_1(\cdot)$ and $v_2(\cdot)$ are uncorrelated zero-mean wide-band (white) noise processes with spectral intensities unity and r, respectively; i.e., the noise power in a frequency band $(-B/2, B/2)$ is B or rB, respectively. We assume that the unknown initial condition $x(0)$ is also random but that it is not correlated with the noises $v_1(\cdot)$ and $v_2(\cdot)$. We have no direct knowledge of $v_1(\cdot)$ and $v_2(\cdot)$ and therefore can only set up an observer as before,

$$\dot{\hat{x}}(t) = A\hat{x}(t) + bu(t) + l[y(t) - c\hat{x}(t)], \qquad \hat{x}(0) = 0 \tag{2}$$

But now the error will depend not only on $x(0) - \hat{x}(0)$ but also on the noises $v_1(\cdot)$ and $v_2(\cdot)$,

$$\dot{\tilde{x}}(t) = (A - lc)\tilde{x}(t) + gv_1(t) - lv_2(t), \qquad \tilde{x}(0) = x(0) \tag{3}$$

Again, by choosing l sufficiently large we can make the expected or mean value of $\tilde{x}$ tend toward zero, $E[\tilde{x}] \to 0$, as fast as desired. But we still have to worry about the fluctuations of $\tilde{x}(\cdot)$—choosing large values for l will accentuate the driving noise term $lv_2(\cdot)$. One criterion is to choose l so as to minimize the mean-square error, i.e., to minimize

$$J = E\|\tilde{x}(t)\|^2 \tag{4}$$

This problem can be attacked in many different ways and in fact has an extensive literature (see, e.g., [1] and [10]–[12] among many others).

We shall not pursue the calculations here—they need some knowledge of statistics and of the solution of stochastic differential equations—but shall just quote the most relevant result. Suppose that

$$\{A, g\} \text{ is controllable,} \qquad \{c, A\} \text{ is observable} \tag{5}$$

Then the roots of the characteristic polynomial $\det(sI - A + \bar{l}c)$ of the optimal observer can be found as the left-half-plane roots of the $2n$-degree polynomial

$$g(s)g(-s) + ra(s)a(-s) = 0 \tag{6}$$

where

$$a(s) = \det(sI - A), \qquad g(s) = c \, Adj \, (sI - A)g$$

This should be compared with the rule (3.4-7)–(3.4-8) for the optimal regulator of Sec. 3.4.1—the two solutions will be identical if we make the dual interchanges†

$$br^{-1}b' \longleftrightarrow gg', \qquad c'c \longleftrightarrow c'r^{-1}c \tag{7}$$

The implications of (6) for the noisy observer can therefore be studied for various values of the noise parameter r by the same arguments as in Sec. 3.4.1. For example, if $r \to \infty$ (high observation noise), the optimal observer poles are obtained as the stable roots of the original characteristic polynomial $a(s)$ along with the unstable roots reflected across the $j\omega$ axis. On the other hand, when $r \to 0$, we can have $n - m$ poles going off to infinity in a Butterworth configuration, with the other m poles going to the zeros of $g(s) = c \, Adj \, (sI - A)g$. For intermediate values of r, the symmetric root-locus plot can be useful, as we shall now illustrate.

Example 4.4-1. Harmonic Oscillator

For the undamped harmonic oscillator driven by a disturbance input w of unit intensity,

$$\begin{bmatrix} \dot{x}_1 \\ \dot{x}_2 \end{bmatrix} = \begin{bmatrix} 0 & 1 \\ -\omega_0^2 & 0 \end{bmatrix} \begin{bmatrix} x_1 \\ x_2 \end{bmatrix} + \begin{bmatrix} 0 \\ 1 \end{bmatrix} w$$

determine the locus of optimum modes for an observer that uses a noisy measurement $y = x_2 + v$, where v is observation noise of intensity r.

---

†It should be noted that more than one (variational) dual problem can be associated with the quadratic regulator problem and that they can be useful in obtaining approximate solutions and bounding solutions—see, for example, References [13] and [26].

Solution

$$\frac{y(s)}{w(s)} = [0 \quad 1] \begin{bmatrix} s & -1 \\ \omega_0^2 & s \end{bmatrix}^{-1} \begin{bmatrix} 0 \\ 1 \end{bmatrix} = \frac{s}{s^2 + \omega_0^2}$$

so the modes of the optimum observer are the stable solutions of

$$-s^2/(s^2 + \omega_0^2)^{-2} + r = 0$$

The corresponding root loci are shown in the figure.

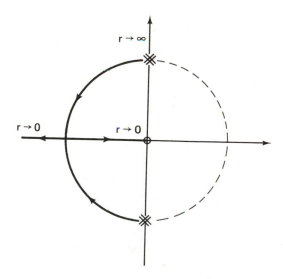

Note that as the observation noise tends toward zero one observer mode approaches the origin and hence is shielded from the driving noise $w(\cdot)$ by the system zero at the origin; the other mode tends toward $-\infty$ and results effectively in differentiation of the system output to determine the state.

Other Factors in Selecting Observer Poles.   It should be noted that non-stochastic factors are also important in the selection of observer dynamics. References [14]–[15] discuss how one can select them to compensate for the effect of deterministic modeling errors, component tolerances, and amplifier biases (see Exercise 4.1-6). Here we may remark that the reduced-order methods of Sec. 4.3 cannot be used when the observed process contains white (broadband) noise, because differentiation would greatly accentuate the noise. Observer designs for colored additive noise have also been studied (see, e.g., [16] and [17]).

A Stochastic Separation Property.   As in the purely deterministic case studied in Section 4.2, the problem of optimal quadratic regulation when the states and the measurements are corrupted by noise as in (1) also reveals a so-called *separation property*. That is, the expected cost

$$EJ = E \int_0^\infty \{|c\, x(t)|^2 + r|u(t)|^2\}\, dt$$

is minimized by a feedback law

$$u(t) = -\bar{k}\hat{x}(t)$$

where $\bar{k}$ is determined as in the case of complete state availability, and $\hat{x}(\cdot)$ is the optimal least-squares estimate of $x(\cdot)$ as determined by (6). We shall not enter further into this result and the special conditions under which it holds, except to refer to the derivations in, for example, [1, pp. 256–292]. We may remark also that, as mentioned in Sec. 3.4.1, the use of state-estimates may cause a deterioration in some of the special robustness (e.g., guaranteed gain and phase margins) properties of the optimal quadratic regulator with completely accessible states (see [18]).

Exercises

4.4-1.   *Variational Approach to the Estimator*

It can be shown that the solution of the following optimization problem also leads to the optimal noisy observer. Find $x(\cdot)$ and $v_1(\cdot)$ to minimize

$J = \int_0^\infty \{[y(t) - cx(t)]'r^{-1}[y(t) - cx(t)] + v_1'(t)v_1(t)\}\, dt$ subject to the equations $\dot{x}(t) = Ax(t) + gv_1(t), y(t) = cx(t) + v_2(t)$. It might be argued that this is a reasonable criterion for the observer problem, whether or not its minimization is equivalent to minimizing the mean-square error $E[x(t) - \hat{x}(t)]'[x(t) - \hat{x}(t)]$; this equivalence is hard to show a priori, but it can be justified *ex post facto* by verifying that its solution yields the same result as the minimum mean-square-error problem. Therefore, solve this problem by using the (Lagrange multiplier, calculus of variations) method of Sec. 3.4.1 and obtain the solution rule described in Sec. 4.4. *Note:* This method yields the linear two-point equations [$x(t_0)$ and $\lambda(t_f)$ are given]

$$\begin{bmatrix} \dot{x}(t) \\ \dot{\lambda}(t) \end{bmatrix} = \begin{bmatrix} A & -gg' \\ -c'r^{-1}c & -A' \end{bmatrix} \begin{bmatrix} x(t) \\ \lambda(t) \end{bmatrix} + \begin{bmatrix} 0 \\ c'r^{-1}y(t) \end{bmatrix}$$

which can be very effectively studied by scattering-theory methods. (See Reference 27 of Chapter 3.)

4.4-2.   *Optimum Root Loci*

Consider the station-keeping satellite of Example 3.3-2, with input disturbances w_x and w_y, and $\ddot{x} = 9x + 2\dot{y} + w_x$, $\ddot{y} = -2\dot{x} - 4y + w_y$. Determine the loci of optimum observer modes when we have measurements of y corrupted by white noise of intensity r and
  a. $w_x =$ unit intensity white noise and $w_y = 0$ and
  b. $w_x = 0$ and $w_y =$ unit intensity white noise.

c. Extend (6) to the case of systems with multiple uncorrelated white noise inputs. Now find the loci when both w_x and w_y are (uncorrelated) unit intensity white noise disturbances. *Hint:* If $H_x(s)$ and $H_y(s)$ are the transfer functions from w_x and w_y, respectively, to the output y, show that the observer modes are the left-half-plane roots of $H_x(s)H_x(-s) + H_y(s)H_y(-s) + r = 0$.

4.5   DIRECT TRANSFER FUNCTION DESIGN PROCEDURES

The combined observer-controller gives a complete solution, at least in theory, to the problem we raised in Sec. 3.1 of using feedback to modify the dynamic behavior of a given system. Let us briefly review the situation. We pointed out in Sec. 3.1 that simple output feedback or even feedback of the output and some derivatives did not give us enough information to carry out our desired objectives in all situations. Moreover, it was not very clear exactly what was to be done and exactly what could be achieved. The fact that the states gave a complete description of the system then led us to examine the possibility of feeding back not the output and its derivatives but the states of a realization of the given system. We showed in Sec. 3.2 that for a *controllable* realization such feedback could, in fact, allow us to re-locate the poles (or rather, the natural frequencies) arbitrarily. But, of course, in general the states may not be directly available, and this led us in Sec. 4.1 to the use of asymptotic observers, which can provide asymptotically good estimates of the states of any *observable* realization. The difference between using output feedback (by which term we shall henceforth also imply possible feedback of some derivatives of the output) and using an observer is that in the latter both input and output are fed back (see Fig. 4.2-1, redrawn more schematically as Fig. 4.5-1). However, by some simple block diagram manipulations, Fig. 4.5-1 can be redrawn in the form of several classical output-feedback configurations, which is perhaps why it was never explicitly considered. Of course, the fact that we are starting from the special observer-controller configuration imposes certain constraints on the derived classical configurations. Moreover, we now know that the point is that such block diagram manipulations may impair stability by introducing additional *hidden modes*, which may or may not be stable. [We should stress that block diagram manipulations are not in general equivalent to similarity transformations between different realizations—in particular, block diagram manipulations can change not only the natural frequencies but also the number of states.]

However, we should not abandon the possibilities of transfer function analysis so easily. We know that by watching for cancellations in the *nominal* transfer function we can keep track of hidden modes (cf. Exercise 2.4-7) and therefore have full knowledge of the internal behavior of the system.

This thought suggests that we carefully examine the special observer-controller configuration from a transfer function point of view to see whether it may be possible to achieve a direct transfer function design without using internal state-space descriptions. This can indeed be done, as we shall show in Sec. 4.5.1, and with definite advantages in simplicity and directness. Moreover, the analysis in Sec. 4.5.1 also suggests certain flexibilities in the design procedure and objectives that are by no means as evident in the state-space approach (Sec. 4.5.2). A more purely transfer function analysis, in which we work exclusively with polynomial equations, is developed in Sec. 4.5.3.

4.5.1 A Transfer Function Reformulation of the Observer-Controller Design

As we might expect from the block diagram configurations shown in Figs. 4.2-1 and 4.3-3, the transfer function approach will be essentially the same whether we use an observer or a reduced-order observer. For simplicity of explanation, however, we shall start our analysis with the full-order observer and then show how easily the reduced-order observer design can be incorporated into the transfer function approach.

The observer-controller configuration of Fig. 4.2-1 can be redrawn more schematically as shown in Fig. 4.5-1. The transfer functions of the blocks denoted $H_u(s)$ and $H_y(s)$ can be calculated from Eq. (4.2-2) for $\hat{x}(\cdot)$,

$$\dot{\hat{x}}(t) = A\hat{x}(t) + l[y(t) - c\hat{x}(t)] + bu(t)$$

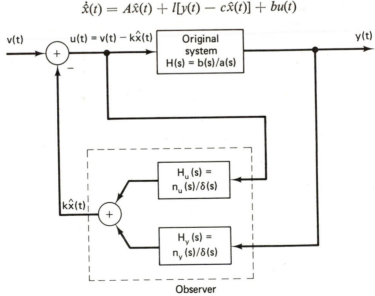

Figure 4.5-1. Block diagram of the combined observer-controller of Fig. 4.2-1.

so that

$$k\hat{X}(s) = k(sI - A + lc)^{-1}[lY(s) + bU(s)]$$
$$= H_y(s)Y(s) + H_u(s)U(s) \tag{1}$$

where

$$H_y(s) = k(sI - A + lc)^{-1}l, \qquad H_u(s) = k(sI - A + lc)^{-1}b \tag{2}$$

As we expect (see the more detailed diagram, Fig. 4.2-1), the natural frequencies of $H_u(s)$ and $H_y(s)$ are the same, viz., the roots of the characteristic polynomial of the observer, det $(sI - A + lc)$. We know also that the overall transfer function must be

$$\frac{Y(s)}{V(s)} = \frac{b(s)}{\det (sI - A + bk)} = \frac{b(s)}{\alpha(s)} \tag{3}$$

as compared to the transfer function of the original system:

$$H(s) = \frac{b_1 s^{n-1} + \cdots + b_n}{s^n + a_1 s^{n-1} + \cdots + a_n} = \frac{b(s)}{a(s)} \tag{4a}$$

Now once the basic structure of Fig. 4.5-1 has been given (or has been "guessed"), it seems reasonable to try to directly calculate transfer functions $\{H_y(s), H_u(s)\}$ that can yield the desired overall transfer function (3).

Therefore let us assume that these transfer functions have the form

$$H_y(s) = \frac{\beta_1 s^{n-1} + \cdots + \beta_n}{s^n + \delta_1 s^{n-1} + \cdots + \delta_n} = \frac{n_y(s)}{\delta(s)} \tag{4b}$$

and

$$H_u(s) = \frac{\gamma_1 s^{n-1} + \cdots + \gamma_n}{s^n + \delta_1 s^{n-1} + \cdots + \delta_n} = \frac{n_u(s)}{\delta(s)} \tag{4c}$$

Now $n_y(s)$, $n_u(s)$, and $\delta(s)$ are to be chosen so that the transfer function of the overall configuration of Fig. 4.5-1 has a desired form (3). The overall transfer function can be computed as follows: Note first that the loop through $H_u(s)$ can be replaced by a transfer function $\delta(s)/[n_u(s) + \delta(s)]$ in series with $H(s) = b(s)/a(s)$. Then we can write

$$\frac{Y(s)}{V(s)} = \frac{\delta(s)b(s)/[n_u(s) + \delta(s)]a(s)}{1 + ([\delta(s)b(s)n_y(s)]/\{[n_u(s) + \delta(s)]a(s)\delta(s)\})}$$
$$= \frac{\delta^2(s)b(s)}{\delta(s)\{[n_u(s) + \delta(s)]a(s) + b(s)n_y(s)\}}$$

Therefore we have a characteristic polynomial of degree $3n$. However, it is obvious from Fig. 4.5-1 that we can use the same n integrators to implement $H_u(s)$ and $H_y(s)$. If we do this, we can "cancel" $\delta(s)$ to have a realization with the transfer function

$$\frac{Y(s)}{V(s)} = \frac{b(s)\delta(s)}{a(s)\delta(s) + a(s)n_u(s) + b(s)n_y(s)} \qquad (5)$$

and a characteristic polynomial of degree $2n$, i.e., with only $2n$ modes. Now by proper choice of $[n_y(s), n_u(s), \delta(s)]$ we can hope to obtain an arbitrary $2n$-th-degree characteristic polynomial, say, $p(s)$, where

$$p(s) = s^{2n} + \cdots + p_{2n}$$

While this degree of generality may be useful, the results of Sec. 4.2 suggest that one choice is to make

$$p(s) = \alpha(s)\delta(s) \qquad (6)$$

where $\alpha(s) = $ the desired nth-degree characteristic polynomial and $\delta(s) = $ an arbitrary nth-degree polynomial, for then the overall transfer function will be

$$\frac{Y(s)}{V(s)} = \frac{b(s)\delta(s)}{\alpha(s)\delta(s)} = \frac{b(s)}{\alpha(s)} \qquad (7)$$

as obtained with the observer-controller configuration of Sec. 4.2. Of course $\delta(s)$ should not be left completely free but should be chosen to have its roots sufficiently stable that the *hidden modes* we shall now have due to the cancelled $\delta(s)$ in (7) will not seriously perturb the overall system.

However, we have not yet said, in the present framework, what is needed to be able to achieve the choice $p(s) = \alpha(s)\delta(s)$. We shall prove that this choice of $p(s)$, and in fact *any other choice of $p(s)$ as an arbitrary 2nth-degree polynomial*, can be achieved if and only if

$$a(s) \text{ and } b(s) \text{ are relatively prime} \qquad (8)$$

that is, whenever there are no cancellations in the original transfer function, $H(s) = b(s)/a(s)$. Of course, this result is consistent with the result of Sec. 4.2 because (cf. Sec. 2.4.1) in such a case any nth-order controllable realization (and such a realization can *always* be obtained) of $H(s)$ will also be observable, or (vice versa) any nth-order observable realization will also be controllable. Nevertheless, the present approach shows that no explicit knowledge of controllability or observability, or even of the concept of state, is necessary to solve the problem of modifying the dynamics of a given system by feedback. Of course this neglects the vital fact that it is the state-variable concept—and the analysis of modal controllability and asymptotic observability—that led to consideration of the structure of Fig. 4.5-1 in the first place. The reader will recall that in Sec. 3.1.1, when we first introduced this problem, we did not really have much clue as to this particular structure. We tried some simple structures that did not work, and perhaps by trial and error we may have been led to the structure of Fig. 4.5-1. But as it happened, the development along state-variable lines took place first, and apparently it was only in 1968 that a transfer function interpretation was sought (cf. Chen [8] and [9]); related results had been obtained earlier by Mortensen [19] and Shipley [20] (see Exercises 3.1-2 and 3.1-3), but at that time there was not the full awareness of the importance of explicit attention to hidden modes.

We drop these speculations and return to the analysis of the basic relation

$$p(s) = a(s)[\delta(s) + n_u(s)] + b(s)n_y(s) \tag{9}$$

where $a(s)$, $b(s)$, and $\delta(s)$ are specified nth-degree polynomials, and we have to try to find the nth-degree polynomials $n_y(s)$ and $n_u(s)$ such that the right-hand side of (9) is some specified [e.g., by (6)] $2n$th-degree polynomial. We can explore this question in several ways. Perhaps the simplest way to begin is to see if we can obtain from (9) enough equations to uniquely specify the coefficients of $n_y(s)$ and $n_u(s)$. Therefore, we expand both sides of (9) and equate the coefficients of corresponding powers of s to get (cf. also Exercise A-6)

$$S(a, b)z = w \tag{10}$$

where

$$z' = [\delta_1 + \gamma_1 \cdots \delta_n + \gamma_n \quad \beta_1 \cdots \beta_n]$$
$$w' = [p_1 - a_1 \cdots p_n - a_n \quad p_{n+1} \cdots p_{2n}]$$

and $S(a, b)$ is the $2n \times 2n$ matrix (shown for $n = 3$)

$$S(a, b) = \begin{bmatrix} 1 & 0 & 0 & 0 & 0 & 0 \\ a_1 & 1 & 0 & b_1 & 0 & 0 \\ a_2 & a_1 & 1 & b_2 & b_1 & 0 \\ a_3 & a_2 & a_1 & b_3 & b_2 & b_1 \\ 0 & a_3 & a_2 & 0 & b_3 & b_2 \\ 0 & 0 & a_3 & 0 & 0 & b_3 \end{bmatrix} \tag{11}$$

We thus have $2n$ equations for the $2n$ unknowns $\{\beta_1, \ldots, \beta_n, \gamma_1, \ldots, \gamma_n\}$, and these equations will have a solution for an *arbitrary* right-hand side w if and only if det $S(a, b) \neq 0$. But $S(a, b)$ is the well-known Sylvester matrix, and it is known (see Sec. 2.4.4) that

$$\det S(a, b) \neq 0 \iff \{a(s), b(s)\} \text{ are coprime} \qquad (12)$$

which is the condition (8) that we quoted earlier. However, to keep this section self-contained, let us give a direct proof of result (12). Thus note that $S(a, b)$ will be nonsingular if and only if the only solution of the equation

$$S(a, b)[p_0 \quad p_1 \quad \cdots \quad p_{n-1} \quad q_0 \quad \cdots \quad q_{n-1}]' = [0 \quad \cdots \quad 0]' \qquad (13)$$

is the trivial solution

$$p_0 = p_1 = \cdots = q_0 = \cdots = q_{n-1} = 0$$

Now (13) can be written in polynomial form as (cf. Exercise A-6)

$$a(s)p(s) + b(s)q(s) = 0 \qquad (14)$$

where $p(s) = p_0 s^{n-1} + \cdots + p_{n-1}$ and similarly for $q(s)$, while $a(s)$ has the higher degree n. But if (14) is true for $p(s) \neq 0$ and $q(s) \neq 0$, we shall have

$$b(s)/a(s) = -p(s)/q(s)$$

which is impossible if and only if $b(s)$ and $a(s)$ are coprime. Therefore $p(s) \equiv 0 \equiv q(s)$, i.e., $S(a, b)$ will be nonsingular, if and only if $b(s)$ and $a(s)$ are coprime.

This completes the proof of (12) and justifies our new design procedure: Choose $\delta(s)$ arbitrarily, and solve for the coefficients of $n_y(s)$ and $n_u(s)$ from Eq. (10); now set up the configuration of Fig. 4.5-1 using the same n integrators to implement the common denominator $\delta(s)$ of $H_y(s)$ and $H_u(s)$.

Reduced-Order Compensators. Actually, a little thought shows that only $n - 1$ integrators are really necessary, corresponding to the use of reduced-order observers in the state-space method.

Thus in (4) we used an nth-degree denominator polynomial $\delta(s)$ for $H_y(s)$ and $H_u(s)$, but clearly an $(n - 1)$th-degree polynomial will suffice [less than $n - 1$ would make $H_y(s)$ and $H_u(s)$ improper, since Eq. (10) shows that the degrees of $n_u(s)$ and $n_y(s)$ can be as high as $n - 1$]. Therefore, let us replace $\delta(s)$ by, say,

$$\delta_r(s) = s^{n-1} + \cdots + \delta_{n-1} \qquad (15)$$

and notice that this will make the overall characteristic polynomial have degree $2n - 1$ rather than $2n$, say,

$$p_r(s) = p_1 s^{2n-1} + \cdots + p_{2n} \qquad (16)$$

The new design equation is [cf. (9)]

$$p_r(s) = [\delta_r(s) + n_u(s)]a(s) + n_y(s)b(s) \tag{17}$$

which is equivalent to a somewhat simpler matrix equation than (10), viz.,

$$S(a, b)z_r = w_r \tag{18}$$

where

$$z_r' = [1 + \gamma_1 \cdots \delta_{n-1} + \gamma_n \quad \beta_1 \cdots \beta_n]$$

and

$$w_r' = [p_1 \cdots p_n \quad p_{n+1} \cdots p_{2n}]$$

Of course these changes do not affect the solvability of the equation, which will still yield a solution for arbitrary w_r if and only if $\{a(s), b(s)\}$ are coprime.

Example 4.5-1

To illustrate the transfer function method, let us rework Example 4.3-2 by the present method. The problem was, given $H(s) = 1/s(s + 1)$, to shift the poles to $\{-1 \pm j\}$ using a reduced-order observer with pole at -2.

In the present method, this means that we choose

$$\delta_r(s) = s + 2$$

and

$$p_r(s) = (s + 1 + j)(s + 1 - j)(s + 2)$$

Then we have to find $n_y(s)$ and $n_u(s)$ such that

$$a(s)[n_u(s) + \delta_r(s)] + b(s)n_y(s) = p_r(s)$$

In such simple low-degree cases, it is best to assume the expected forms for $n_y(s)$ and $n_u(s)$ and compare coefficients. This leads to

$$(s^2 + s)(\gamma_1 s + \gamma_2 + s + 2) + (\beta_1 s + \beta_2) = s^3 + 4s^2 + 6s + 4$$

and

$$\gamma_1 = 0, \qquad \gamma_2 = 1, \qquad \beta_1 = 3, \qquad \beta_2 = 4$$

Therefore the compensator is defined by connecting

$$H_y(s) = \frac{3s + 4}{s + 2}, \qquad H_u(s) = \frac{1}{s + 2}$$

as in Fig. 4.5-1. We can verify that this coincides with the results obtained with considerably more effort in Example 4.3-2.

The transfer function method will always be simpler when the original system is specified by its transfer function $b(s)/a(s)$, and, depending on the circumstances, the transfer function method might be simpler even otherwise. Note that the highly structured (Toeplitz) form of $S(a, b)$ will lend itself to efficient solution methods (see, e.g., F. Gustavsson and D. Yun, *IEEE Transactions on Circuits and Systems*, **CAS-26**, Sept. 1979).

4.5.2 Some Variants of the Observer-Controller Design

The transfer function analysis also suggests some interesting variants that are not quite so obvious in a state-space analysis.

We return to the block diagram of Fig. 4.5-1 and recall that [cf. (17)]

$$\frac{Y(s)}{V(s)} = \frac{b(s)\delta_r(s)}{p_r(s)} \qquad p_r(s) = [\delta_r(s) + n_u(s)]a(s) + n_y(s)b(s) \qquad (19)$$

where $\delta_r(s)$ and $p_r(s)$ are arbitrary (of degrees $n-1$ and $2n-1$, respectively) and $n_u(s)$ and $n_y(s)$ (of degrees not greater than $n-1$) are to be determined. We showed that if $\{a(s), b(s)\}$ were relatively prime we could choose $p_r(s)$ as an arbitrary polynomial of degree $2n-1$. In particular, by choosing it to have the special form

$$p_r(s) = \alpha(s)\delta_r(s) \qquad (20a)$$

we obtained the observer-controller design,

$$\frac{Y(s)}{V(s)} = \frac{b(s)}{\alpha(s)} \qquad (20b)$$

where $\alpha(s)$ is an arbitrary nth-degree polynomial and the cancelled (or hidden) modes corresponding to $\delta_r(s)$ are also completely under our control.

However, it should be emphasized that unlike what may appear from the state-space analysis, we have choices other than always choosing $p_r(s)$ to cancel out $\delta_r(s)$ from the overall transfer function; in particular, we could leave in all or some of the factors of $\delta_r(s)$. We still always have $2n-1$ natural frequencies, but the number of poles could vary from 1 to $2n-1$ and we could add up to $n-1$ additional zeros [beyond those left uncancelled in $b(s)$]. It is not quite clear exactly how this additional freedom should be exploited, but in Exercise 4.5-2 we shall indicate one possibility of interest in classical control system theory.

If we are not at all interested in the zero locations but only wish to have $2n-1$ arbitrary natural frequencies, then the configuration of Fig. 4.5-1 can be simplified somewhat to the one shown in Fig. 4.5-2. Namely, from

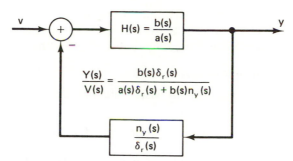

Figure 4.5-2.   Configuration that can provide an arbitrary set of natural frequencies but with a fixed added set of zeros.

(17) we see that we can set

$$n_u(s) \equiv 0 \qquad (21)$$

and still find $\{\delta_r(s), n_y(s)\}$ to set up an arbitrary $(2n-1)$th-degree characteristic polynomial $p_r(s)$. Therefore, feedback of the input $u(\cdot)$ is not essential in setting up an internally stable realization. What is lost by making assumption (21) is that now $\delta_r(s)$ is fixed by our choice of $p_r(s)$, and therefore we cannot in general obtain an overall transfer function with denominator polynomial of degree n [the degree of the denominator $a(s)$ of the original system $H(s)$]. Moreover, the design procedure introduces some fixed zeros [the roots of $\delta_r(s)$] into the overall transfer function.

We leave it to the reader to show that results similar to these can be obtained for the classical unity-feedback compensation scheme of Fig. 4.5-3. The results associated with Figs. 4.5-2 and 4.5-3 were first obtained by Pearson ([21] and [22]), who used state-space proofs.

We note that the role of the explicit feedback of $u(\cdot)$ is to give us some extra degrees of freedom—namely, if $n_u(s) \not\equiv 0$, then we can choose $\delta_r(s)$ as an arbitrary $(n-1)$th-order polynomial and still solve Eq. (17) for any $p_r(s)$, e.g., in the observer-controller form (20) or otherwise.

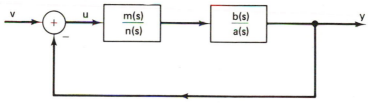

Figure 4.5-3.   The classical unity-freedback scheme can provide $2n-1$ arbitrary modes, but there is no freedom in choosing the zeros.

However, we should note here that in the schemes presented so far the so-called *relative order* of the (nominal or actual) transfer function is the same; i.e., the difference between the number of poles and zeros is constant.

4.5.3 Design Via Polynomial Equations

In Sec. 4.5.1, we saw that the basic equations to be solved were of the form [cf. (9) and (17)]

$$p(s)a(s) + q(s)b(s) = c(s) \qquad (22)$$

where $\{a(s), b(s), c(s)\}$ were given polynomials and $\{p(s), q(s)\}$ are polynomials to be determined under certain degree constraints. In Sec. 4.5.1 we studied these equations by converting them to equivalent matrix form [cf. (10) or (18)]. However, we can also give a more direct solution of these so-called Diophantine† equations. This solution will suggest another approach to the compensation problem, which will lead us more directly (without appealing to state-space insights) to the basic design equations (9) and (17); this approach will also extend nicely to the multivariable case (cf. Sec. 7.5.1).

Direct Analysis of the Diophantine Equation (22). Let us start with Eq. (22). We see first of all that any common factor of $\{a(s), b(s)\}$ must also be present in $c(s)$, and therefore if $c(s)$ is to be arbitrary, a *necessary* condition for the success of our design method is that $\{a(s), b(s)\}$ are coprime.

To show the sufficiency of this condition is not hard either if we recall that coprimeness is equivalent to the fact (cf. Sec. 2.4.4) that there exist polynomials $\{z(s), w(s)\}$ such that

$$z(s)a(s) + w(s)b(s) = 1 \qquad (23)$$

But then clearly

$$f(s) \triangleq c(s)z(s), \qquad g(s) \triangleq c(s)w(s) \qquad (24)$$

will satisfy the original equation (22). However, in our special problem there have to be degree constraints on any solution pair $\{\bar{f}(s), \bar{g}(s)\}$ because they will be used as numerators and denominators of certain transfer functions, which must be proper if they are to be physically realizable. Such constraints are not hard to achieve.

We shall show that when $c(s)$ is of degree $2n - 1$, $a(s)$ of degree n, and $b(s)$ of degree $n - 1$, then we can replace $\{\bar{f}(s), \bar{g}(s)\}$ by polynomials $\{f_0(s), g_0(s)\}$ of degree not greater than $n - 1$. This will be a satisfactory solution

---

†Because of the restriction that $\{p(s), q(s)\}$ be polynomials. Diophantus of Alexandria wrote a book in the third century A.D. about the (mathematically identical) problem of finding integer solutions to $pa + qb = c$ when $\{a, b, c\}$ are given integers. More recent discussions can be found in books on number theory or continued fractions (see, e.g., the very elementary books [23, Chaps. 1, 2] and [24, pp. 1–8] and also [25]).

for the observer and reduced-order observer problems. We do this by using the Euclidean division algorithm (in a way already illustrated in Sec. 2.4.4). Thus we start by writing

$$\begin{bmatrix} \bar{f}(s) & \bar{g}(s) \\ b(s) & -a(s) \end{bmatrix} \begin{bmatrix} a(s) \\ b(s) \end{bmatrix} = \begin{bmatrix} c(s) \\ 0 \end{bmatrix} \tag{25}$$

By polynomial division, we can write

$$\bar{g}(s) = l(s)a(s) + g_0(s), \qquad \deg g_0 < n \tag{26a}$$

Then the (elementary row) operation of adding $l(s)$ times the second equation in (25) to the first equation will give

$$\begin{bmatrix} f_0(s) & g_0(s) \\ b(s) & -a(s) \end{bmatrix} \begin{bmatrix} a(s) \\ b(s) \end{bmatrix} = \begin{bmatrix} c(s) \\ 0 \end{bmatrix} \tag{26b}$$

where

$$f_0(s) = \bar{f}(s) - l(s)b(s) \tag{26c}$$

As for the degree of $f_0(s)$, we note that [cf. (26b)]

$$\deg f_0(s) + \deg a(s) = \deg [c(s) - g_0(s)b(s)] \leq 2n - 1$$

so that

$$\deg f_0(s) \leq n - 1$$

Therefore $\{f_0(s), g_0(s)\}$ have the claimed properties.

Physical Interpretations and an Alternative Approach.   Reviewing the above, we see that the polynomial criterion (23) for the coprimeness of $\{a(s), b(s)\}$ was at the heart of our solution.

To better understand the physical significance of this condition, we note that $\{b(s), a(s)\}$ are not arbitrary polynomials but are related to the system input and output by

$$y(s) = \frac{b(s)}{a(s)} u(s) \tag{27}†$$

Now we can rewrite (27) as

$$y(s) = b(s)\xi(s), \qquad a(s)\xi(s) = u(s) \tag{28}$$

where we should recall that the *partial state* $\xi(\cdot)$ and its derivatives completely

---

†For notational convenience, we do not use capital letters for the transforms of $\{y(\cdot), x(\cdot), u(\cdot)\}$.

determine the state of any realization.† The interesting thing about (23) and (28) is that they show immediately how to recover the partial state if it is not directly accessible, namely

$$z(s)u(s) + w(s)y(s) = z(s)a(s)\xi(s) + w(s)b(s)\xi(s) = \xi(s) \qquad (29)$$

This suggests the configuration of Fig. 4.5-4, where we have shown the feedback through an additional operation—multiplication by a possibly rational function, $m(s)$. The point is that the resulting transfer function is

$$\frac{y(s)}{v(s)} = \frac{b(s)}{a(s) + m(s)} \qquad (30)$$

and therefore can be made to have an arbitrary denominator by proper choice of $m(s)$; e.g., $m(s) = \alpha(s) - a(s)$ (cf. Sec. 3.2.2).

Of course, the fact that $\xi(s)$ can be reconstructed is not unexpected, since we are effectively using the *ideal observer*, i.e., differentiation, to obtain the (partial) state. However, it is not hard to obtain a physically realizable (asymptotic observer) version of the above scheme. Thus, if $m(s)\xi(s)$ is the quantity desired for feedback to the input, let us introduce a *denominator polynomial* $\delta(s)$ as shown in Fig. 4.5-5(a). Now $w(s)m(s)\delta(s)$ and $z(s)m(s)\delta(s)$ can be *reduced* as in (26) to have degree less than or equal to that of $\delta(s)$, yielding physically realizable implementation as shown in Fig. 4.5-5(b), which is identical to Fig. 4.5-1. In more detail, note from (23) that

$$[z(s)m(s)\delta(s)]a(s) + [w(s)m(s)\delta(s)]b(s) = m(s)\delta(s) \qquad (31a)$$

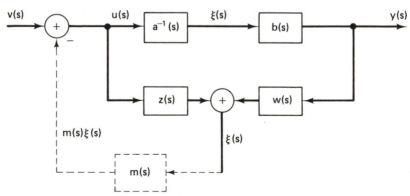

Figure 4.5-4. Reconstruction and feedback of the state to get a transfer function $b(s)[a(s) + m(s)]^{-1}$.

†See Exercise 3.2-5. Note first that $\xi(s)$ determines the state vector $x_c(\cdot)$ of a controller-form realization; then the states of any other minimal realization can be obtained by a nonsingular (similarity) transformation.

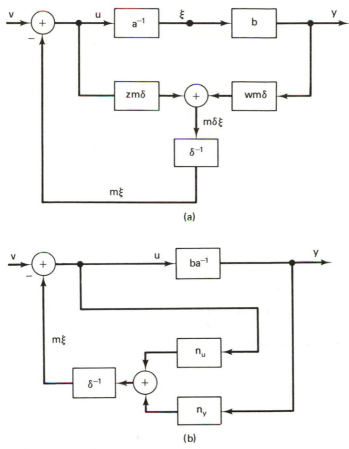

Figure 4.5-5. Realizable compensator. (a) Toward a realizable compensator. (b) Reducing the order in (a)

which we can reduce as in (26) to

$$n_u(s)a(s) + n_y(s)b(s) = m(s)\delta(s) \tag{31b}$$

Let us assume that the *feedback* function $m(s)$ is a polynomial, and let

$$\deg m(s) \leq n - 1 \tag{32a}$$

to correspond to the fact that no more than $n - 1$ derivatives of the partial state ξ are needed to yield the states of the realization.

Now by construction, $\deg n_y(s) < n = \deg a(s)$, and from (31)–(32) we can see that

$$\deg n_u(s) < \deg \delta(s) \qquad (32b)$$

Therefore, choosing

$$\deg \delta(s) \geq n - 1 \qquad (33)$$

will suffice to make $n_u(s)/\delta(s)$ and $n_y(s)/\delta(s)$ realizable, i.e., proper rational functions. Now (31b) directly yields Fig. 4.5-5(b). To complete the identification with the results of Sec. 4.5.1, we just have to choose

$$m(s) = \alpha(s) - a(s) \qquad (34)$$

With this substitution in (31b) we see that

$$a(s)[n_u(s) + \delta(s)] + b(s)n_y(s) = \alpha(s)\delta(s) \qquad (35)$$

which is precisely Eq. (9) of Sec. 4.5.1, thus yielding the promised rederivation of the results of Sec. 4.5.1.

The insight gained in this calculation as to the significance of (23) can be applied to obtain the results of Sec. 4.5.2 and in fact some useful extensions thereof, especially to changing the relative order. However we shall not pursue these extensions here.

Exercises

4.5-1.

Given $H(s) = (s - 1)/s^2$, find a compensator that will yield a new third-order realization with characteristic polynomial $(s + 5)(s^2 + s + 1)$. Is there a unique solution of this problem? If not, give some criteria for choosing among the different possible solutions.

4.5-2. (*Franklin*)

Consider a system with the transfer function $1/s(s + 1)$, which we wish to modify using feedback so as to have a new characteristic polynomial $s^2 + 4s + 8$. Show that the steady-state error in the response to a unit ramp is $1/K_v$, $K_v = 2$. It can be shown that for a system with closed-loop poles at $\{-p_i\}$ and closed-loop zeros at $\{-z_i\}$,

$$\frac{1}{K_v} = \Sigma \frac{1}{p_i} - \Sigma \frac{1}{z_i}$$

Show that by choosing $p_3 = 0.1$ and $-z_3 = -1/10.4$, we can achieve $K_v = 10$. This illustrates one way in which flexibility in zero assignment might be used.

4.5-3.

Given a system with the transfer function $b(s)/a(s)$ and a desired stable transfer function $b(s)\delta(s)/p(s)$, where $\deg p(s) - \deg \delta(s) \geq n$, show that we can obtain an internally stable realization as follows: Use state feedback (or an observer-controller compensator) to change $b(s)/a(s)$ to $b(s)/\alpha(s)$, where $\alpha(s)$ contains n roots of $p(s)$; then cascade this with a realization of $\delta(s)/\epsilon(s)$, where $\alpha(s)\epsilon(s) = p(s)$.

4.5-4. *Idealized Compensators*

In Eqs. (4.5-2) and (4.5-3), suppose we assume that the denominator polynomial $\delta(s)$ is absent; i.e., assume $\delta(s) = 1$. Show that we can achieve a desired transfer function $H(s) = b(s)/\alpha(s)$, where $\alpha(s)$ is a specified nth-order polynomial, by choosing the coefficients of $\beta(s)$ and $\gamma(s)$ so that $[\beta' \; \gamma']S' = [0 \; \alpha' - a']$, where S is the Sylvester matrix of $\{a(s), b(s)\}$, β is the column vector of the coefficients of $\beta(s)$, and similarly for a, α, and γ. Compare with the results of Sec. 4.5.3.

4.5-5. *Reconciliation to State-Space Designs*

Suppose the original transfer function is realized in controller form. Calculate the feedback gain k_c required to give a new characteristic polynomial $\alpha(s)$, and assume that the states $x(t)$ are provided by an ideal observer. Calculate $H_u(s)$ and $H_y(s)$ for this compensator (cf. Fig. 4.3-1), and show that they coincide with $\beta(s)$ and $\gamma(s)$ as found in Exercise 4.5-4.

4.5-6.

Consider the compensator scheme depicted in Fig. 4.3-6. Show that we can find a stable transfer function $\mathcal{Y}(s)$ that will make all the modes of the closed-loop system stable if and only if $H(s)$ has the following *parity interlacing property*: for every RHP zero of $H(s)$, there must be an even number (counted according to multiplicity) of poles of $H(s)$ to the right of it. *Reference:* D. C. Youla, J. J. Bongiorno, and C. N. Lu, *Automatica*, **10**, pp. 159–174, 1974; see also *Ibid.*, **12**, pp. 387–388, 1976.

REFERENCES

1. K. J. Åstrom, *Introduction to Stochastic Control Theory*, Academic Press, New York, 1970.

2. R. E. Kalman, P. Falb, and M. A. Arbib, *Topics in Mathematical System Theory*, McGraw-Hill, New York, 1969.

3. D. G. Luenberger, "Observing the State of a Linear System," *IEEE Trans. Mil. Electron.*, **MIL-8**, pp. 74–80, 1964. (Also see Ph.D. thesis, Stanford University, Stanford, Calif., 1963.)

4. B. Gopinath, "On the Control of Linear Multiple Input-Output Systems," *Bell Syst. Tech. J.*, **50**, pp. 1063–1081, March 1971. (Also see Ph.D. thesis, Stanford University, Stanford, Calif., 1968.)

5. D. G. Cumming, "Design of Observers of Reduced Dynamics," *Electron. Lett.*, **5**, no. 10, pp. 213–214, May 15, 1969.

6. D. G. Luenberger, "An Introduction to Observers," *IEEE Trans. Autom. Control*, **AC-16**, pp. 596–603, Dec. 1971.

7. A. E. BRYSON and D. G. LUENBERGER, "The Synthesis of Regulator Logic Using State-Variable Concepts," *Proc. IEEE*, **58**, pp. 1803–1811, Nov. 1970.

8. C. T. CHEN, "A New Look at Transfer Function Design," *Proc. IEEE*, **59**, pp. 1580–1585, Nov. 1971. See also Proc. Natl. Electronics Conf., **25**, pp. 46–51, 1969.

9. C. T. CHEN, *Introduction to Linear System Theory*, Holt, Rinehart and Winston, New York, 1970.

10. R. E. KALMAN and R. S. BUCY, "New Results in Linear Filtering and Prediction Theory," *Trans. ASME Ser. D. J. Basic Eng.*, **83**, pp. 95–107, Dec. 1961.

11. H. KWAKERNAAK and R. SIVAN, *Linear Optimal Control Systems*, Wiley, New York, 1972.

12. T. KAILATH, *Lectures on Linear Least-Squares Estimation*, CISM Courses and Lectures No. 140, Springer-Verlag, New York, 1978.

13. W. L. CHAN, "Variational Dualities in the Linear Regulator and Estimation Problems," *J. Inst. Math. Appl.*, **18**, pp. 237–248, Oct. 1976.

14. J. D. POWELL, "Mass Center Estimation in Spinning Drag-Free Satellites," *J. Spacecr. Rockets*, **9**, pp. 399–405, 1972. (Also see Ph.D. thesis, Dept. of Aeronautics and Astronautics, Stanford University, Stanford, Calif., May 1970.)

15. F. E. THAU and A. KESTENBAUM, "The Effect of Modeling Errors on Linear State Reconstructors and Regulators," *J. Dyn. Syst. Meas. Control*, pp. 454–459, Dec. 1974.

16. T. KAILATH and R. GEESEY, "An Innovations Approach to Least-Squares Estimation, Part V: Innovations Representations and Recursive Estimation in Colored Noise," *IEEE Trans. Autom. Control*, **AC-18**, no. 5, pp. 435–453, Oct. 1973.

17. L. M. NOVAK, "Discrete-Time Optimal Stochastic Observers," in *Control and Dynamic Systems*, Vol. 12 (C. T. Leondes, ed.), Academic Press, New York, 1976, pp. 259–311.

18a. J. C. DOYLE, "Guaranteed Margins for LQG Regulators," *IEEE Trans. Autom. Control*, **AC-23**, pp. 756–757, Aug. 1978.

18b. J. C. DOYLE and G. STEIN, "Robustness with Observers," *IEEE Trans. Automat. Contr.*, **AC-24**, Oct. 1979.

19. R. E. MORTENSEN, "The Determination of Compensation Functions for Linear Feedback Systems To Produce Specified Closed-Loop Poles," *Internal Technical Rept. TR-59-0000-00781*, Space Tech. Laboratories, Los Angeles, Aug. 1959. (See also *IEEE Trans. Autom. Control*, **AC-8**, p. 386, Oct. 1963.)

20. P. P. SHIPLEY, "A Unified Approach to Synthesis of Linear Systems," *IEEE Trans. Autom. Control*, **AC-8**, pp. 114–120, April 1963.

21. J. B. PEARSON, "Compensator Design for Dynamic Optimization," *Int. J. Control*, **9**, pp. 413–482, 1969.

22. F. M. BRASCH and J. B. PEARSON, "Pole Placement Using Dynamic Compensators," *IEEE Trans. Autom. Control*, **AC-15**, pp. 34–43, 1970.

23. C. D. OLDS, *Continued Fractions*, Random House, New York, 1963.

24. A. YA. KHINCHIN, *Continued Fractions*, Chicago University Press, Chicago, 1964 (Russian ed., 1935).

25. V. KUČERA, *Discrete Linear Control: The Polynomial Equation Approach*, J. Wiley, London, 1979.

26. U. B. DESAI and H. L. WEINERT, "Generalized Control-Estimation Duality and Inverse Projections," *Proceedings 1979 Conference on Information Sciences and Systems*, Johns Hopkins University, Baltimore, Maryland, March 1979.

SOME ALGEBRAIC COMPLEMENTS 5

5.0 INTRODUCTION

In this chapter we shall attempt to deepen our understanding of some of the basic results of Chapter 2, partly to provide a better basis for their extension to the multivariable case in the next chapter.

We begin by showing how a simple idea of Nerode's [1] for defining the *state space* associated with an input-output map can be used to lead us abstractly to the state-space equations we had obtained in Secs. 2.1 and 2.2 by calling upon results for analog-computer simulation of transfer functions. We are now able to develop those realizations in a more basic way. The insights from this approach are then used in Sec. 5.1.2 to introduce the problem of obtaining finite-dimensional realizations from impulse-response data (Markov parameters in continuous time). This special realization problem has a large literature, especially in a subject sometimes called *algebraic system theory*, which attempts to develop system theory in a highly abstract and algebraic way. Our approach here is much more concrete, and can be used as a guide for further exploration by interested readers (cf. Sec. 5.1.3).

In Sec. 5.2, we shall show how to set up a geometric structure in which different state-space realizations correspond to different choices of bases in n-space. This will provide a nice way of interpreting similarity transformations, which is useful both in appreciating earlier results and also in later chapters.

314

5.1 ABSTRACT APPROACH TO STATE-SPACE
REALIZATION METHODS; NERODE EQUIVALENCE

In Sec. 2.2.3, we gave an intuitive definition of the *state* of a system as the *smallest* entity that summarized the past history of the system. Here, following the work of Nerode [1] (which originated in automata theory), we shall show how to formalize this notion. That this is more than just a pretty mathematical exercise will be shown by using it to deduce in a natural way the state-space realization procedures of Sec. 2.1, which, the reader may recall, were introduced in a somewhat ad hoc and certainly nonmathematical way. Then in Sec. 5.1.2 we shall consider the closely related problem of trying to find minimal realizations of a given impulse response sequence.

5.1.1 Realization from Scalar Transfer Functions

It will be conceptually and notationally simpler to confine ourselves here to discrete-time systems and to use z transforms to describe the input-output relations. If we have an input $\{u(k), -\infty \leq k \leq \infty\}$, we shall write

$$u(z) = \sum_{-\infty}^{\infty} u(k)z^{-k}$$

where z can be regarded merely as a *placeholder*—positive powers of z correspond to time instants before zero and negative powers to instants after zero. We shall denote output sequences (or strings) by

$$y(z) = \sum_{-\infty}^{\infty} y(k)z^{-k}$$

Now given a (causal time-invariant) system that associates with each input string $u(z)$ an output string $y(z)$, not necessarily in a linear fashion, Nerode proposed the following method for introducing the notion of state.

Pick some convenient reference time, *say $t = 0$*, and consider *all inputs that terminate at zero*. Also make the reasonable assumption that the inputs begin at some *finite* time in the past. In other words, these two assumptions mean that all the inputs we shall consider are *polynomials* in z; i.e.,

$$u(z) = \sum_{-N}^{0} u(i)z^{-i} \qquad \text{for some } N < \infty \tag{1}$$

By causality, the *response* of the system to such a (polynomial) input string will have the form

$$y(z) = \sum_{-N}^{\infty} y(i)z^{-i} \tag{2}†$$

†To avoid some degeneracies, we shall also assume that the zero input gives a zero output.

315

where we note that even in simple systems the response can continue indefinitely into the future. [Therefore the $y(z)$ are in general *power series* rather than polynomials.]

With these conventions, we can now declare the following: Input strings $u_1(z), u_2(z), \ldots$ can be said to leave the system in the *same state* at $t = 1$ (after the inputs end) if the *responses* $y_1(z), y_2(z), \ldots$ are all the *same* for $t \geq 1$. Moreover, we can identify this particular state (at $t = 1$) with the *class* of all input strings with the above property. We can now break up the set of all possible input strings (ending at zero) into classes such that for all inputs in any class the response for $t \geq 1$ is the same. Then *each class* can be associated with a *state*. Of course, this was all with reference to the states at $t = 1$, but because of time invariance we can replace $t = 1$ by any other time.

To get a more concrete feeling for this abstract and general construction, which is known as the *Nerode equivalence* construction, let us see how it works for *linear* systems.

For linear systems, any class of inputs with the above property can be further characterized as follows: Any two inputs, say $u_1(z)$ and $u_2(z)$, in a class differ by an input

$$u_0(z) \triangleq u_1(z) - u_2(z)$$

such that the response to $u_0(z)$ is zero for $t \geq 1$; i.e.,

$$y_0(z) \triangleq y_1(z) - y_2(z)$$

$$= \sum_{-N}^{0} y_0(i)z^{-i} \qquad \text{for some } N < \infty$$

In other words, the response $y_0(z)$ to such an input $u_0(z)$ can at most be a polynomial in z (no terms in z^{-1}).

Let us be even more specific. Suppose the linear system has a transfer function

$$H(z) = \frac{b(z)}{a(z)} = \frac{b_1 z^{n-1} + \cdots + b_n}{z^n + a_1 z^{n-1} + \cdots + a_n} \tag{3}$$

[where $b(z)$ and $a(z)$ are *relatively prime*]. Then it is easy to find an input $u_0(z)$ that will give a polynomial output $y_0(z)$:

$$\text{if } u_0(z) = a(z), \text{ then } y_0(z) = b(z) \tag{4}$$

Therefore we have at least one such input. But clearly all inputs of the form

$$u_0(z) = p(z)a(z), \qquad p(z) = \text{an arbitrary polynomial}$$

will also give

$$y_0(z) = H(z)u_0(z) = p(z)b(z) = \text{a polynomial in } z$$

Furthermore, these are the only such inputs, for an arbitrary input $u(z)$ can be decomposed (Euclidean algorithm) as

$$u(z) = q(z)a(z) + r(z) \tag{5}$$

where

$$\deg r(z) < \deg a(z) = n$$

The response to this $u(z)$ will be

$$y(z) = b(z)q(z) + \frac{b(z)r(z)}{a(z)} \tag{6}$$

$$= (\text{polynomial in } z) + \left[\text{the strictly proper part of } \frac{b(z)r(z)}{a(z)}\right] \tag{7}$$

which will be a polynomial in z only if $r(z) \equiv 0$. [Note that the strictly proper part has to be *expanded* in powers of z^{-1}; expansion in powers of z would mean that we have a noncausal system. *Example*: $(z + 1)^{-1} \neq 1 - z + z^2 + \cdots$, but $(z + 1)^{-1} = z^{-1}(1 + z^{-1})^{-1} = z^{-1}(1 - z^{-1} + z^{-2} + \cdots)$.] Therefore the class $\mathfrak{M}$ of all inputs $u_0(z)$ for which $y_0(z)$ is a polynomial in z is of the form

$$\mathfrak{M} = \{p(z)a(z)\}, \qquad p(z) = \text{an arbitrary polynomial} \tag{8}$$

Now the set of all inputs can be broken up as follows into classes that each define a unique *state*: Take an arbitrary input $u_1(z)$ and add to it all elements in $\mathfrak{M}$. (This clearly defines a class with the desired property that the difference of any two inputs in the class yields a polynomial response.) Now take an input $u_2(z)$ that is not in the first class $\{u_1(z) + \mathfrak{M}\}$, and form $\{u_2(z) + \mathfrak{M}\}$. Now take $u_3(z)$ not in either of these classes and form $\{u_3(z) + \mathfrak{M}\}$. Continue in this way until all inputs are in some class.

In our specific case, this collection of classes will be n-dimensional, because we know already that $H(z) = b(z)/a(z)$ has an n-dimensional state space realization. How can we see this here? We note first that any class $\{u_i(z) + \mathfrak{M}\}$ can be represented by the polynomial of *least* degree in that class. Moreover, this least degree must be less than n, because if not, the remainder after division by $a(z)$ would also clearly be in the class and would have lower degree. Therefore the collection of classes, or equivalently the collection of states, can be represented by a collection of polynomials of the form

$$r(z) = x_1 + x_2 z + \cdots + x_n z^{n-1} \tag{9}$$

Furthermore, each such polynomial represents a distinct class. Clearly it takes n numbers (e.g., the coefficients $\{x_1, \ldots, x_n\}$) to specify any of these polynomials and hence any of the states. Therefore we can say that the state space is *n-dimensional*, or that the state (at $t = 1$) can be represented by an *n*-vector, say $x'(1) = [x_1, \ldots, x_n]$. [Clearly, instead of the coefficients $\{x_i\}$ of the polynomial, we could use (suitable) linear combinations of these coefficients to also uniquely identify the polynomial $r(z)$.]

State Equations. It is now natural to ask how the state evolves with time as further inputs (past $t = 0$) are applied. This is not hard to do in the present framework either. Thus, suppose we are in some initial state $x(1)$ at $t = 1$ and we put in an input $u(1)$ at $t = 1$. What is $x(2)$?

Note first that $x(1)$ is a representation of some polynomial of degree $< n$, say in fact the polynomial

$$r_0(z) = x_1(1) + x_2(1)z + \cdots + x_n(1)z^{n-1} \tag{10}$$

[where the $x_i(1)$ are the components of $x(1)$]. Moreover, recall also that $r_0(z)$ is the lowest degree (shortest) input that will leave the system in state $x(1)$ at $t = 1$. Now the polynomial corresponding to $x(2)$ can be calculated as follows: After $u(1)$ is applied, let us redefine $t = 1$ as the new *origin* (we can clearly do this because of our assumption of time invariance) so that we can associate state $x(2)$ with the input

$$r_1(z) \triangleq u(1) + zr_0(z) \tag{11}$$

Note that $zr_0(z)$ is the input $r_0(z)$ shifted backwards by one unit. A simple sketch of the input defined by $r_0(z)$ followed by the input $u(1)$ will resolve any confusion about the preceding statement. However, we cannot identify state $x(2)$ with $r_1(z)$ because $r_1(z)$ may not have degree *less than n*. But it is easy to see what to do: We take the remainder after division by $a(z)$. For this, we write

$$
\begin{aligned}
r_1(z) &= zr_0(z) + u(1) \\
&= z[x_1(1) + x_2(1)z + \cdots + x_n(1)z^{n-1}] + u(1) \\
&= u(1) + x_1(1)z + \cdots + x_{n-1}(1)z^{n-1} + x_n(1)z^n \tag{12}
\end{aligned}
$$

But

$$z^n = a(z) - a_1 z^{n-1} - \cdots - a_n \tag{13}$$

so that

$$
\begin{aligned}
r_1(z) &= x_n(1)a(z) + [u(1) - a_n x_n(1)] + [x_1(1) - a_{n-1}x_n(1)]z \\
&\quad + \cdots + [x_{n-1}(1) - a_1 x_n(1)]z^{n-1} \tag{14}
\end{aligned}
$$

Therefore we see that we can identify the components of $x(2)$ as the coefficients of the terms $\{z^i, 0 \leq i \leq n - 1\}$ in the *remainder* after dividing $r_1(z)$

by $a(z)$. In matrix notation we have

$$
\begin{bmatrix} x_1(2) \\ x_2(2) \\ \vdots \\ x_n(2) \end{bmatrix} = \begin{bmatrix} 1 \\ 0 \\ \vdots \\ 0 \end{bmatrix} u(1) + \begin{bmatrix} 0 & \cdots & 0 & -a_n \\ 1 & \cdots & 0 & -a_{n-1} \\ \vdots & & & \vdots \\ 0 & & 1 & -a_1 \end{bmatrix} \begin{bmatrix} x_1(1) \\ x_2(1) \\ \vdots \\ x_n(1) \end{bmatrix}
\tag{15}
$$

or, say,

$$
x(2) = A_{co}x(1) + b_{co}u(1)
\tag{16a}
$$

The output vector c_{co} can be obtained from the impulse response or Markov parameters: The response to the unit (impulse) input $u(k) = \delta_{k0}$ is

$$
y(1) = h_1 = c_{co}b_{co}
$$
$$
y(2) = h_2 = c_{co}A_{co}b_{co}
$$

which identifies c_{co} as

$$
c_{co} = [h_1 \quad h_2 \quad \cdots \quad h_n]
\tag{16b}
$$

Therefore we have obtained a state-space realization in the *controllability canonical form* of Secs. 2.1 and 2.2 [cf. Eq. (2.2-8)].

The reader will see that we could obtain other forms by choosing different combinations of the coefficients of the *state* polynomials to define the *state* vector.

Controller Canonical Form.   To illustrate this, let us define the states as follows, taking $n = 3$ for convenience: If $r(z)$ is as in (9), let us define the state at $t = 1$ by the following combination of coefficients of $r_0(z)$:

$$
\xi_1(1) = x_1(1) - a_1x_2(1) + (a_1^2 - a_2)x_3(1)
$$
$$
\xi_2(1) = x_2(1) - a_1x_3(1)
\tag{17}
$$
$$
\xi_3(1) = x_3(1)
$$

There is no "loss" in doing this, since the calculation can easily be reversed. Now the next state $\xi(2)$ can be constructed from the $r_1(z)$ of (14) as, in matrix notation,

$$
\begin{bmatrix} \xi_1(2) \\ \xi_2(2) \\ \xi_3(2) \end{bmatrix} = \begin{bmatrix} -a_1 & -a_2 & -a_3 \\ 1 & 0 & 0 \\ 0 & 1 & 0 \end{bmatrix} \begin{bmatrix} \xi_1(1) \\ \xi_2(1) \\ \xi_3(1) \end{bmatrix} + \begin{bmatrix} 1 \\ 0 \\ 0 \end{bmatrix} u(1)
\tag{18}
$$

which the reader will recognize as the controller canonical form.

Clearly, one way of getting different realizations is to use different methods of *parametrizing* or *coordinatizing* the state space. From this point of view, the (similarity) transformations between different realizations essentially correspond to different choices of *bases* in n space. This reasonably obvious geometric interpretation will be elaborated further in Sec. 5.2. However, all the realizations we shall get in this way will be controllable, and we cannot go from them to observable realizations by similarity transformations unless the nominal transfer function $b(z)/a(z)$ is irreducible, i.e., unless the polynomial $a(z)$ is really the lowest degree polynomial such that $u(z) = a(z)$ will give a polynomial response $y(z)$. We did in fact assume this when we began, and therefore there is no need to worry separately about observable realizations.

Nevertheless it should be of interest to investigate a direct way of getting observable realizations, and in fact this is quite easy.

Observability Canonical Form. If we go back to our initial discussion of "Nerode equivalence" [see, especially, the paragraph below (2)], we shall find that another way of describing states is not by different equivalence classes of *inputs* but by the different outputs for $t = 1, 2, \ldots$, provided of course that there are no further inputs (i.e., $u(t) = 0$, $t \geq 1$). Then each state at $t = 1$ will produce a *unique* output sequence for $t \geq 1$. Let us denote this sequence by $[y(z)]_+ = y_1 z^{-1} + y_2 z^{-2} + \cdots$, where the $[\ ]_+$ means that we only consider the response for $t > 0$ (i.e., $t \geq 1$). This description will lead us naturally to the observability canonical realization.

We begin by noting that every permissible output sequence must have the form

$$y(z) = H(z)u(z) = \frac{b(z)}{a(z)}u(z) \tag{19}$$

for some *polynomial* input $u(z)$, i.e., an input that stops at $t = 0$. Therefore all such output sequences $y(z)$ obey

$$a(z)y(z) = b(z)u(z) = \text{a polynomial}$$

i.e.,

$$[a(z)y(z)]_+ = 0 \tag{20}$$

How can we characterize such output sequences? Clearly there must be n constraints on the family of such sequences, where

$$n = \deg a(z)$$

And some reflection suggests that knowledge of the n coefficients $\{y_1, y_2,$

$\ldots, y_n\}$ suffices to characterize such sequences, for if, in an obvious notation,

$$y(z) = [y(z)]_- + [y(z)]_+$$

then

$$[a(z)y(z)]_+ = [a(z)[y(z)]_-]_+ + [a(z)[y(z)]_+]_+$$

The first term on the right-hand side is zero, and therefore

$$[a(z)y(z)]_+ = [(z^n + a_1 z^{n-1} + \cdots + a_n)(y_1 z^{-1} + y_2 z^{-2} + \cdots)]_+$$

Now by (20) the coefficient of z^{-1} must be zero,

$$a_n y_1 + a_{n-1} y_2 + \cdots + a_1 y_n + y_{n+1} = 0$$

i.e.,

$$y_{n+1} = -a_1 y_n - \cdots - a_n y_1 \tag{21}$$

Similarly, we can see that $\{y_{n+2}, y_{n+3}, \ldots\}$ will also be determined by the first n coefficients $\{y_1, \ldots, y_n\}$. Therefore we may choose

$$x'_{ob} = [y_1 \quad y_2 \quad \cdots \quad y_n] \tag{22}$$

as a *state vector*. Then the output can be written

$$y_1 = [1 \quad 0 \quad \cdots \quad 0]x_{ob} \tag{23a}$$

so that

$$c_{ob} = [1 \quad 0 \quad \cdots \quad 0] \tag{23b}$$

The *free* [i.e., with $u(k) = 0, k \geq 1$] evolution of the state is determined by checking the representation of

$$[zy(z)]_+ = y_2 z^{-1} + \cdots + y_{n+1} z^{-n} + \cdots$$

But

$$
\begin{bmatrix} y_2 \\ y_3 \\ \cdot \\ \cdot \\ \cdot \\ y_{n+1} \end{bmatrix}
=
\underbrace{\begin{bmatrix} 0 & 1 & 0 & \cdots & 0 \\ 0 & 0 & 1 & \cdots & 0 \\ & & & & \cdot \\ & & & & \cdot \\ & & & & 1 \\ -a_n & & \cdots & & -a_1 \end{bmatrix}}_{A_{ob}}
\begin{bmatrix} y_1 \\ y_2 \\ \cdot \\ \cdot \\ \cdot \\ y_n \end{bmatrix}
\tag{24a}
$$

so that the state-transition map is

$$x_{ob}(2) = A_{ob}x_{ob}(1) \tag{24b}$$

Finally, if we assume $x_{ob}(0) = 0$, the contribution to $x_{ob}(1)$ of an input $u(z) = u(0) = 1$ is (by definition of the impulse response) equal to

$$[y_1 \quad y_2 \quad \cdots \quad y_n] = [h_1 \quad \cdots \quad h_n] \qquad (25a)$$

which means that

$$b'_{ob} = [h_1 \quad \cdots \quad h_n] \qquad (25b)$$

We have thus obtained the *observability canonical form* of Secs. 2.1 and 2.2 [cf. Eq. (2.2-9)]. We leave it to the reader to determine the special parametrization of states in terms of $\{y_1, \ldots, y_n\}$ that would lead to yet other realizations. Moreover, combinations of the above two basic procedures can be made in various ways. For example, we can write

$$y(z) = b(z)\frac{1}{a(z)}u(z)$$

$$y(z) = b(z)r(z), \qquad a(z)r(z) = u(z) \qquad (26)$$

realize $1/a(z)$ by the parametrization used for the observability form, and then notice that $y(z) = b(z)r(z)$ can be easily read out. This will give the controller canonical form. And so on.

We thus hope to have shown that the realization procedures of Sec. 2.1.2 need not just have been "pulled out of the air." We shall see a further confirmation of this in the multivariable case (cf. Sec. 6.6). The abstract approach we have described is usually cast in a much more algebraic way ([2, Chap. 10] and [3, Chap. 8]); the interested reader may find the more concrete approach used above to be helpful in penetrating the algebraic theory (see also Sec. 5.1.3).

On the other hand, our discussions (e.g., of the different parametrizations used to obtain the controllability and controller forms) do indicate that it would be helpful to seek a *geometric* or *coordinate-free* interpretation of some of our methods and results, and this will be studied in Sec. 5.2. At that time, we shall also show how that language can lead to a more compact description of the methods used in this section (see Example 5.2-5).

5.1.2   Realization from the Markov Parameters

Here we shall show how the basic Nerode approach to the definition of state can be applied when our basic data are the *impulse response values* of a finite-dimensional (discrete-time) system or, equivalently, the *Markov parameters* of the system. (These could as well be the Markov parameters of a continuous-time system, though the corresponding realization problem is of less relevance now.)

We recall from the first arguments in Sec. 5.1.1 that we can identify the state space at time $t = 1$ with a set of *equivalence* classes of (polynomial) inputs terminating at $t = 0$. Two such inputs belong to the same equivalence class if they produce the same system response for $t \geq 1$. If the system is linear, we can say that two inputs are equivalent if the response to their difference is zero for $t \geq 1$. Now the system is finite dimensional if the dimension of this set of equivalence classes is n, $n < \infty$. What conclusions can we draw from this hypothesis? Clearly that there can be no more than n "different" inputs; or, to put it another way, if we take any $n + 1$ inputs, it must be possible to express one of the inputs in terms of the others.

Let us see how to reflect this fact into a statement about the quantities that are available to us, namely the impulse response or the Markov parameters. Consider the table of unit-impulse inputs and their responses (shifted versions of the impulse response) in Fig. 5.1-1. If $n = 3$, there must be a

...	-3	-2	-1	$t = 0$	1	2	3	4	...	→ t
...	0	0	0	1	h_1	h_2	h_3	h_4	...	
...	0	0	1	h_1	h_2	h_3	h_4	h_5	...	
...	0	1	h_1	h_2	h_3	h_4	h_5	h_6	...	
...0	1	h_1	h_2	h_3	h_4	h_5	h_6	h_7	...	
Inputs←	→ Outputs				→ Output for $t > 0$					

Figure 5.1-1.   How the Hankel matrix arises; the inputs are zero to the right of the diagonal line.

linear combination of three or more inputs such that the corresponding response measured for $t > 0$ (i.e., $t = 1, 2, \ldots$) is identically zero. Therefore for some $\{a_1, a_2, a_3\}$ it must be true that

$$-(\text{4th row of Hankel matrix}) = a_1 \cdot \text{3rd row} + a_2 \cdot \text{2nd row}$$
$$+ a_3 \cdot \text{1st row}$$

where the Hankel matrix is the infinite matrix set off by dashed lines in the array of Fig. 5.1-1. Equivalently, we see that the input

$$a(z) = z^3 + a_1 z^2 + a_2 z + a_3$$

gives a zero response for $t \geq 1$ and that it is the shortest-length input that does this. Therefore $a(z)$ must be equal to the denominator polynomial of the transfer function of the system. Moreover, once we have $a(z)$ and we know the Markov parameters, then we can calculate the numerator polynomial $b(z)$ by comparing coefficients in the equality

$$b(z) = H(z)a(z) = \left(\sum_1^\infty h_i z^{-i} \right) a(z) \tag{27}$$

More explicitly, this yields the formula

$$[b_1, \ldots, b_n]' = T(h)[1 \quad a_1 \quad \cdots \quad a_{n-1}]' \tag{28}$$

where

$$T(h) = \text{a lower triangular Toeplitz matrix}$$
$$\text{with first column } [h_1, \ldots, h_n]'$$

We should note also that to determine the $\{a_i\}$ it is not necessary to work with the rows of infinite length shown in Fig. 5.1-1. Since the column rank of a matrix is equal to its row rank, it suffices to take only $n = 3$ columns of the infinite Hankel matrix, so that the equation for the $\{a_i\}$ can be written

$$[a_3 \quad a_2 \quad a_1 \quad 1]\begin{bmatrix} h_1 & h_2 & h_3 \\ h_2 & h_3 & h_4 \\ h_3 & h_4 & h_5 \\ h_4 & h_5 & h_6 \end{bmatrix} = [0 \quad 0 \quad 0]$$

In general, therefore, we have the following method for determining a realization, known to have minimal order n, from knowledge of the Markov parameters $\{h_1, h_2, \ldots\}$. First find the coefficients of the denominator polynomial by solving the equation

$$[a_n \quad \cdots \quad a_1 \quad 1]M_0(n+1, n) = [0 \ldots 0] \tag{29}$$

where

$$M_0(n+1, n) \triangleq \text{an } (n+1) \times n \text{ Hankel matrix with first}$$
$$\text{row } [h_1, \ldots, h_n] \text{ and last column}$$
$$[h_n, \ldots, h_{2n}]' \tag{30}$$

(This is a modification of the notation of Sec. 2.2.4 [cf. Eq. (2.2-49)], which applied only to square matrices. For square matrices we can use the old notation; e.g., $M_0(n, n) = M[1, n]$). Once $a(z)$ is determined, $b(z)$ can be found via (28). Of course, from $\{b(z), a(z)\}$ many state-space realizations $\{A, b, c\}$ can immediately be obtained.

The major question of course is how the minimal order n can be determined from the $\{h_i\}$. In principle n could be determined as

$$n = \text{rank } \mathbf{M}$$

where **M** is the infinite Hankel matrix

$$\mathbf{M} = \begin{bmatrix} h_1 & h_2 & h_3 & \cdots \\ h_2 & h_3 & h_4 & \\ h_3 & h_4 & & \\ \cdot & & & \\ \cdot & & & \\ \cdot & & & \end{bmatrix}$$

This follows from the readily verified factorization

$$M_0(i, i) = [c' \quad A'c' \quad \cdots \quad (A')^{i-1}c']'[b \quad Ab \quad \cdots \quad A^{i-1}b], \quad i = 1, 2, \ldots$$

and the fact that a minimal realization must be observable and controllable. In practice of course we must somehow have an upper bound on the rank of **M** in order to be able to stop checking the ranks of the $\{M_0(i, i), i = 1, \ldots\}$.

In fact in many problems one will only have available a finite sequence of Markov parameters, $\{h_1, \ldots, h_N\}$ say, and here one might wish to find the smallest-order realization whose first N Markov coefficients equal the given sequence $\{h_1, \ldots, h_N\}$. This so called *partial realization* problem is in fact a version of the classical Padé approximation problem, which arises in many contexts in physics, network theory, and numerical analysis and has an extensive literature, at least in the scalar case (see, e.g., [4] and [5] for recent surveys).

From a system-theoretical point of view, the problem of realization given the Markov parameters is often regarded as being somewhat academic, at least when working over the real (rather than some finite) field. One reason is that matching an ever-larger number of terms in the sequence of Markov parameters means placing increasing emphasis on the (more noise-susceptible) high-frequency behavior of the system. This is not very realistic, and it may be better to make some good reduced-order "average" fit to a set of Markov parameters rather than to try to exactly match them. We might also try to match some other system parameters. A further difficulty is that often we do not directly have the Markov parameters but only noisy records of input and output measurements from an operating system. The many problems of fitting realizations to noisy data are considered in the literature of *system identification* and *model reduction*. This is far beyond our scope here, especially since the subject is a vast one and needs to draw upon both system theory and statistics; the interested reader could start with the references [6], [7] and [8].

Nevertheless, the realization problem studied in this section has been the main object of attention in a subject often called *algebraic system theory*

[2]. This is a proposed general approach to linear systems, one of whose objectives is the study of systems operating not on real numbers but, for example, also on finite fields and on the integers. Such systems arise in coding theory and in certain approaches to eliminating round-off error in digital filtering (see, e.g., [9]–[11]).

We should mention here that the original system-theoretical interest in the minimal realization problem grew out of the studies of the multivariable version made independently by Youla [12], Ho [13], and Silverman [14]. The multivariable partial realization problem was first explored by Kalman [15] and Tether [16]. Later, Rissanen [17] gave a nice recursive solution of the scalar version of this problem. His results turned out to be closely related to a fast algorithm of Berlekamp and Massey for decoding BCH codes (see [9] and [10]). Kung and Kailath showed (see [18, Chap. 4]) that these recursive algorithms were natural extensions of a method of Lanczos for solving linear equations with Hankel coefficient matrices. In the case where all the principal minors of this matrix are nonsingular, Lanczos [19] showed that the solution could be compactly described in terms of certain (Lanczos) orthogonal polynomial recursions that are well known in numerical analysis. In the partial realization problem, several minors will often be zero and in [18] it is shown how the Lanczos algorithm can be extended (nonuniquely) to handle this—a particularly natural choice leads to the Berlekamp-Massey-Rissanen method.

A somewhat different nonrecursive approach was developed by Sugiyama, et al. [20], who used a construction based on the Euclidean division algorithm. This algorithm can be executed in a fast way (cf. [21, Sec. 8.9]), which then leads to a solution with $0(n \log^2 n)$ computations, as compared to $0(n^2)$ for the recursive methods (see, e.g., [22]–[23]). At least when working over the reals, however, these algorithms have poor numerical properties and stabler approximation methods may be more appropriate (see, e.g., [8]).

*5.1.3 Some Algebraic Language

Some diligent readers (or those with a mathematics background) might wish to pursue the algebraic aspects of the above discussions in more detail. To guide their further reading, we provide the following "glossary" of terms that might have been used in our discussions in Secs. 5.1.1 and 5.1.2. Of course, for more details it will be necessary to refer to one of the standard treatments, e.g., [24] or [25].

The collection of all inputs (starting at some finite time and terminating at $t = 0$) forms a *ring of polynomials*, denoted $K[z]$, which is closed under addition and multiplication. The set of input polynomials that gives zero response for $t \geq 1$ is the *kernel* of the *input-output map*, which takes polynomial inputs into power series formed by the system response for $t \geq 1$. It is easy to see that the kernel $\mathcal{I}$ is an *ideal* in the ring of polynomials $K[z]$; i.e., $\mathcal{I}$ is closed under addition and under multiplication by elements of $K[z]$. Moreover, the kernel is a *principal ideal* in the sense that all elements in the ideal can be written as (polynomial) multiples of a certain lowest-degree polynomial [in our case this will be $a(z)$ if $H(z) = b(z)/a(z)$ is irreducible]. This polynomial is said to be a *generator* of the ideal. It can be shown by using the Euclidean algorithm that all ideals in the ring $K[z]$ are principal, i.e., that $K[z]$ is a *principal ideal ring* and in fact a *principal ideal domain,* namely one in which the product of two elements cannot be zero unless one of the elements is zero.

Given an ideal $\mathcal{I}$ in a ring $K[z]$, we can form a *quotient ring* $K[z]/\mathcal{I}$ whose elements (or *residue classes*) are collections of polynomials that differ by an element of $\mathcal{I}$. It is this quotient ring that defines the state space for a (controllable) realization of a time-invariant system. In Sec. 5.1.1 we saw that the quotient ring was isomorphic to a collection of polynomials of degree n. This collection is a *vector space*, closed under addition and under multiplication by scalars that are elements of the field K (in our case, real or complex numbers, though any *field* can be used). This polynomial vector space is also isomorphic to a vector space of $n \times 1$ vectors with elements from the field. In this latter space, we saw that state evolution with time corresponds to multiplication by a matrix $A : x(1) = Ax(0)$. In the associated polynomial vector space, this corresponds to multiplication by z, *mod* $a(z)$ (i.e., if $x(0)$ is associated with $r_0(z)$, then $x(1)$ is associated with the polynomial $zr_0(z) \bmod a(z)$.). It is easy to check that the collection of polynomials defining the state-space is closed under "multiplication" by polynomials provided the multiplication, say $\otimes$, is defined as

$$\alpha(z) \otimes r(z) = \alpha(z)r(z) \bmod a(z)$$

This fact allows us to regard the polynomial state-space as a *module* over $K[z]$, i.e., as a collection of objects (e.g., polynomials or vectors of polynomials) closed under addition and under "scalar multiplication" by elements of a ring ($K[z]$ in the present case). We note that the space of polynomial inputs is also a module over $K[z]$, as is the space of outputs (which are power series in z^{-1}), and in fact the kernel of the input-output map (which is also a principal ideal domain in the scalar case, though not in the matrix case—see Section 6.6).

The approach used in the abstract theory to define the state space is to note that the input-output map

$$u(z) \longmapsto [H(z)u(z)]_+$$

where $[\ \ ]_+$ denotes the response for $t \geq 1$, is a *module homomorphism;* i.e., it is a mapping from the input module (polynomials over $K[z]$) to the output module (power series in z^{-1} over $K[z]$) that preserves the structure of the input and output spaces. Now it is a standard theorem that any module homomorphism has a unique factorization into the sequence of an *onto* map (or epimorphism) followed by a 1-1 map (or monomorphism)—see Fig. 5.1-2. The range of the onto map then qualifies as the state space of an observable realization of the system. Then standard module-decomposition theorems can be used to get different state-space realizations, but we shall not pursue these things here. We may note in closing only that this brief list of definitions will be of help only to the somewhat knowledgeable or exceptionally diligent reader who would like to get a brief glimpse at what lies ahead in algebraic system theory. Suitable further reading might include References [26] to [30], which give an account of applications to other than finite constant

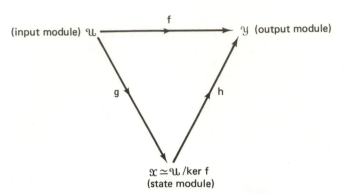

Figure 5.1-2. Canonical factorization of the input-output map $u(z) \mapsto [H(z)u(z)]_+$.

systems. The concrete approach of Secs. 5.1-1 and 5.1-2 (see also Sec. 6.6) might be a useful guiding light in these ventures.

Exercises

5.1-1. *Toeplitz Equations*

We noted in the text that once we found the denominator coefficients by solving (5.1-29) the numerator coefficients could then be found via Eq. (5.1-28). Show that we can combine the above steps into one by rearranging them as shown below for $n = 2$ (for simplicity):

$$\begin{bmatrix} h_1 & 0 & 0 \\ h_2 & h_1 & 0 \\ \hline h_3 & h_2 & h_1 \\ h_4 & h_3 & h_2 \\ h_5 & h_4 & h_3 \end{bmatrix} \begin{bmatrix} 1 \\ a_1 \\ a_2 \end{bmatrix} = \begin{bmatrix} b_1 \\ b_2 \\ 0 \\ 0 \\ 0 \end{bmatrix}$$

Then show that the Toeplitz equations can be directly obtained by comparing powers of s on both sides of the relation $b_1 s^{n-1} + \cdots + b_n = (\sum_1^\infty h_i s^{-i})(s^n + a_1 s^{n-1} + \cdots + a_n)$.

5.1-2.

a. If the input to a discrete-time system $\mathbb{S}$ is the sequence $\{u_0 = 1, u_1 = 0, \ldots, u_k = 0, \ldots\}$, we obtain the output sequence $\{y_0 = 0, y_1 = 0, y_2 = 1, y_3 = 0, \ldots, y_k = 0, \ldots\}$. Determine a minimal realization for $\mathbb{S}$.

b. If the input to $\mathbb{S}$ is the sequence $\{u_0 = 1, u_1 = 0, \ldots, u_k = 0, \ldots\}$ and the response is the sequence $\{y_0 = 0, y_1 = 1, y_2 = 0, y_3 = -1, y_4 = 1, y_5 = 0, y_6 = -1, \ldots\}$, determine a minimal realization for $\mathbb{S}$.

5.1-3. *An Identification Problem*

Let $\{A, b, c\}$ be a realization with $\mathcal{O}(c, A) = I$. Also define $\mathcal{Y}'(k) \triangleq [y(k), \ldots, y(k + n - 1)]$ and similarly $\mathcal{U}(k)$.

a. Show that we can write $\mathcal{Y}(k+1) = A\mathcal{Y}(k) + R\mathcal{U}(k)$, where

$$R = \begin{bmatrix} & & \bigcirc \\ \beta_{n-1} & \cdots & \beta_0 \end{bmatrix}, \qquad \beta_{n-1} = b_1, \ \beta_{n-2} = b_2 + a_1 b_1, \ \ldots$$

and $\det(sI - A) = s^n + a_1 s^{n-1} + \cdots + a_n$, $b' = [b_1 \ \cdots \ b_n]$.

b. Show that we can write $b = \beta_{n-1} A^{n-1} + \cdots + \beta_1 A + \beta_0 I$.
Remark: By assumption $c = [1 \ 0 \ \cdots \ 0]$. Therefore if we know A and R, we can calculate b. But A and R can be found from the equation

$$\sum_{k=1}^{n} \mathcal{Y}(k+1)[\mathcal{Y}'(k) \ \ \mathcal{U}'(k)] = [A \ \ R]\Gamma$$

provided

$$\Gamma \triangleq \sum_{k=1}^{n} \begin{bmatrix} \mathcal{Y}(k) \\ \mathcal{U}(k) \end{bmatrix} [\mathcal{Y}'(k) \ \ \mathcal{U}'(k)] \ \text{is nonsingular}$$

B. Gopinath (*Bell Syst. Tech. J.*, **48**, pp. 1101–1113, 1969) has shown Γ^{-1} exists for almost all input sequences $\{u(0), \ldots, u(n-1)\}$. *Hint:* Use Eq. (2.3-2).

5.2 GEOMETRIC INTERPRETATION
OF SIMILARITY TRANSFORMATIONS;
LINEAR VECTOR SPACES

In Sec. 5.1 we noted that different state-space realizations of a transfer function corresponded to different ways of *coordinatizing* an abstract (n-dimensional) state space. This fact should of course have a geometrical interpretation, which we shall develop in this section. In particular, we shall present a geometrical interpretation of similarity transformations as corresponding to different choices of bases for the n-dimensional state space. We shall also present the concept of a *dual basis* as an aid to simplifying (at least conceptually) the algebraic burden of similarity transformations in certain important cases.

These concepts are part of the subject of linear algebra. Our exposition is not meant to compete with the standard presentations of this subject. On the other hand, our main goal is to use a combination of explanations, intuition, and deduction to try to give the reader without a linear-algebra background a compact way, based on geometric intuition, of thinking about similarity transformations of state-space equations. For more details, and a complete logical development, we refer to standard books on the subject; two books close in spirit to our presentation are those of Noble and Daniel [31] and Strang [32].

5.2.1 Vectors in n-Space: Linear Independence

An $n \times 1$ (column) matrix can be thought of as a point in n-space or as a *vector* (from the origin to this point) in that space. We assume that the reader is familiar with the geometry of n-space, which (at least for all the properties of interest to us) is just a natural generalization of 3-space.

Vectors in n-space will be written as column matrices (column vectors), e.g., $\mathbf{p}$, $\mathbf{p}_1$, $\mathbf{p}_2$, . . . , $\mathbf{x}$, etc. The elements of the matrices are real numbers.† *In this section, for emphasis, we shall use boldface to distinguish vectors and matrices from scalars.*

A set of vectors $\{\mathbf{p}_1, \mathbf{p}_2, \ldots, \mathbf{p}_m\}$ will be said to be *linearly independent* if the only scalars (complex numbers) $\{\alpha_1, \ldots, \alpha_m\}$ for which

$$\alpha_1 \mathbf{p}_1 + \alpha_2 \mathbf{p}_2 + \cdots + \alpha_m \mathbf{p}_m = \mathbf{0}$$

are $\alpha_1 = \alpha_2 = \cdots = \alpha_m = 0$. ($\mathbf{0}$ is of course the column matrix with all elements zero.) Otherwise, the $\{\mathbf{p}_i\}$ will be said to be *linearly dependent*.

To determine if a collection of vectors is independent, we must solve n equations with m unknowns $\{\alpha_1, \alpha_2, \ldots, \alpha_m\}$ and see if the *only* solution is $\alpha_1 = \cdots = \alpha_m = 0$. (Note that $\alpha_1 = \cdots = \alpha_m = 0$ is always a solution.) We don't actually need to *solve* the equations but only to see if there is a nonzero solution or not. For example, suppose that

$$\mathbf{p}_1 = \begin{bmatrix} 1 \\ 2 \\ 3 \end{bmatrix}, \qquad \mathbf{p}_2 = \begin{bmatrix} 3 \\ -1 \\ 9 \end{bmatrix}$$

Then

$$\alpha_1 \mathbf{p}_1 + \alpha_2 \mathbf{p}_2 = \begin{bmatrix} \alpha_1 + 3\alpha_2 \\ 2\alpha_1 - \alpha_2 \\ 3\alpha_1 + 9\alpha_2 \end{bmatrix}$$

so that the equations are $\alpha_1 + 3\alpha_2 = 0$, $2\alpha_1 - \alpha_2 = 0$, $3\alpha_1 + 9\alpha_2 = 0$, which have the unique solution $\alpha_1 = 0 = \alpha_2$. Therefore $\mathbf{p}_1$ and $\mathbf{p}_2$ are linearly independent.

We shall say that a vector $\mathbf{p}$ is a *linear combination* of $\{\mathbf{p}_1, \mathbf{p}_2, \ldots, \mathbf{p}_k\}$ if there exist scalars $\{\alpha_1, \ldots, \alpha_k\}$ such that

$$\mathbf{p} = \alpha_1 \mathbf{p}_1 + \cdots + \alpha_k \mathbf{p}_k$$

---

†It is sometimes useful to allow complex entries, as, for example, when dealing with eigenvalues and eigenvectors. The necessary modifications in our discussion are straightforward and essentially consist of replacing transposes by conjugate transposes (called *Hermitian transposes*). For more general entries, see the remarks in Sec. 5.2.4.

It is easy to see that the vectors $\{\mathbf{p}_1, \mathbf{p}_2 \ldots, \mathbf{p}_m\}$ are linearly dependent if and only if at least one of them is a linear combination of the others.

Basis Vectors and Change of Bases. We "know" that n-space is n dimensional, but what does this mean? It means that there are no more than n linearly independent vectors in n-space and that any vector in the space can be expressed (uniquely) as a linear combination of the elements of any set, called a *basis*, of n linearly independent vectors. Bases are of course non-unique, and it should be obvious that the columns of any nonsingular $n \times n$ matrix form a basis.

The *standard basis* for n-space is the collection of vectors

$$e_1 = [1 \quad 0 \quad \cdots \quad 0]', \qquad e_2 = [0 \quad 1 \quad 0 \quad \cdots \quad 0]', \ldots,$$

$$e_n = [0 \quad \cdots \quad 0 \quad 1]'$$

Unless a basis is otherwise specified, we may assume that any given column matrix is a vector expressed with respect to the standard basis; i.e., we can think of it as

$$\mathbf{x} = [x_1, \ldots, x_n]' = x_1 e_1 + \cdots + x_n e_n$$

However, sometimes other bases are helpful and we would like to see how a vector given in one basis (say the standard basis) can be represented in terms of some other basis. Let $\{\mathbf{p}_1, \ldots, \mathbf{p}_n\}$ be a collection of n independent vectors in n-space and hence a basis of n-space. How can an arbitrary n-vector $\mathbf{x}$ be written in this basis? That is, what are the coefficients, say $\bar{x}_i$, such that

$$\mathbf{x} = \sum_1^n x_i e_i = \sum_1^n \bar{x}_i \mathbf{p}_i$$

Example: Suppose x is a column matrix; $\mathbf{x}' = [1 \ 2 \ 4 \ -1]$. Then with respect to the standard basis (e_1, e_2, e_3, e_4) the representation is just $\mathbf{x}' = [1 \ 2 \ 4 \ -1]$, whose elements are the same as the elements of $\mathbf{x}$. However, suppose we take as a basis the vectors

$$\mathbf{p}_1 = [0 \quad 0 \quad 0 \quad 1]', \qquad \mathbf{p}_2 = [0 \quad 0 \quad 1 \quad 1]'$$
$$\mathbf{p}_3 = [0 \quad 1 \quad 1 \quad 1]', \qquad \mathbf{p}_4 = [1 \quad 1 \quad 1 \quad 1]'$$

We have

$$\mathbf{x} = (-5)\mathbf{p}_1 + 2\mathbf{p}_2 + 1\mathbf{p}_3 + 1\mathbf{p}_4$$

so the representation now is $\bar{\mathbf{x}}' = [-5 \ 2 \ 1 \ 1]$.

Bases and Dual Bases; Geometric Interpretation. If we arrange the basis vectors in a matrix as

$$\mathbf{P} = [\mathbf{p}_1 \ldots \mathbf{p}_n]$$

then the coefficients in the two representations are connected by

$$Ix = P\bar{x} \quad \text{or} \quad \bar{x} = P^{-1}x$$

Let

$$q_i' = \text{the } i\text{th } row \text{ of } P^{-1}$$

so that

$$P^{-1} = [q_1, \dots, q_n]'$$

Then note that the ith component of $\bar{x} = P^{-1}x$ is

$$\bar{x}_i = q_i'x, \qquad i = 1, \dots, n$$

so that we can write

$$\bar{x} = \sum_1^n \bar{x}_i p_i = \sum (q_i'x)p_i$$

We shall now give a *geometric interpretation* to this calculation. Note first that if the $\{p_i\}$ are *orthogonal* to one another, then we can clearly find the coefficients $\{\bar{x}_i\}$ quite directly as

$$\bar{x}_i = \text{the } dot \text{ } product \text{ of } x \text{ and } p_i = x'p_i = p_i'x$$

The inner (or dot) product of two n-vectors x and y is defined as

$$\langle x, y \rangle = x'y = y'x = \sum_1^n x_i y_i$$

x and y are orthogonal if their inner product is zero. (If the vectors have complex coefficients, we combine conjugation with the transpose.)

If the $\{p_i\}$ are not orthogonal, then things are more complicated and, as we have just seen, we have to solve simultaneous equations (or, equivalently, invert matrices) to find the $\{\bar{x}_i\}$. This procedure can be made conceptually more transparent by introducing a set of vectors $\{q_1, \dots, q_n\}$ such that

$$q_i'p_j = \delta_{ij}$$

That is, q_1 is orthogonal to $\{p_2, \dots, p_m\}$ and has unit projection on p_1, and similarly for $q_2, q_3, \dots$. This geometric interpretation of the $\{q_i\}$ shows how to visualize them in simple examples. Since the $\{p_i\}$ are linearly independent, it should be clear that the $\{q_i\}$ will be also. (Show this.) Therefore, the $\{q_i\}$ can also be used as a basis for n-space, and we shall call it the *dual basis* of the basis $\{p_i\}$.

The significance of the $\{q_i\}$ is that if

$$\bar{x} = \sum_1^n \bar{x}_i p_i$$

then clearly we can find the coefficients as

$$\bar{x}_i = \text{dot product of } \mathbf{x} \text{ and } \mathbf{q}_i = \mathbf{x}'\mathbf{q}_i = \mathbf{q}_i'\mathbf{x}$$

which is the result we obtained in a more algebraic way before.

5.2.2 Matrices and Transformations

We have said that any $n \times 1$ matrix can be regarded as a vector in n-space. What about an $n \times n$ matrix $\mathbf{A} = [a_{ij}]_1^n$? One point of view is just to regard $\mathbf{A}$ as a particular set of column matrices and therefore as a particular set of n-vectors expressed in the standard basis. This interpretation is a very limited one, which, for example, does not allow us to answer the question of what happens if we choose a basis other than the standard basis.

It turns out that a useful way to interpret a matrix is as a representation of a particular transformation of n-space into itself. That is, a matrix $\mathbf{A}$ takes every n-vector $\mathbf{x}$ into another n-vector $\mathbf{y}$ according to the rule $\mathbf{Ax} = \mathbf{y}$. Now to understand how a transformation operates on an arbitrary vector it suffices to know how it operates on a set of basis vectors, e.g., the vectors of the standard basis. But

$$\begin{bmatrix} a_{11} & \cdots & a_{1i} & \cdots & a_{1n} \\ & & & & \\ & & & & \\ \vdots & & & & \\ & & & & \\ a_{n1} & & a_{ni} & & a_{nn} \end{bmatrix} \begin{bmatrix} 0 \\ \vdots \\ 1 \\ \vdots \\ 0 \end{bmatrix} = \begin{bmatrix} a_{1i} \\ \vdots \\ \vdots \\ a_{ni} \end{bmatrix}, \qquad i = 1, \cdots, n$$

or, more compactly,

$$\mathbf{Ae}_i = \mathbf{a}_i, \qquad i = 1, \ldots, n$$

where $\mathbf{a}_i$ is the ith column of the matrix $\mathbf{A}$.

Therefore, just given an $n \times n$ matrix $\mathbf{A}$ without any specification of vector spaces or bases, it is clear that we can interpret $\mathbf{A}$ as being a transformation that takes the ith vector in the standard basis into the vector represented in the standard basis by the ith column of $\mathbf{A}$.

But suppose we wish to consider not the standard basis but some other basis for n-space, say $\{\mathbf{p}_1, \ldots, \mathbf{p}_n\}$. What will a transformation, described by $\mathbf{A}$ in the standard basis, look like with respect to the new basis? To work this out, recall that the equation

$$\mathbf{Ax} = \mathbf{y}$$

can be regarded as representing *in the standard basis* the action of a transformation **A** on a vector **x**. Now with respect to the basis $\{\mathbf{p}_1, \ldots, \mathbf{p}_n\}$, we have seen earlier that

$$\mathbf{x} = \mathbf{P}\bar{\mathbf{x}} \text{ and } \mathbf{y} = \mathbf{P}\bar{\mathbf{y}}$$

where

$$\mathbf{P} = [\mathbf{p}_1 \, \mathbf{p}_2 \cdots \mathbf{p}_n]$$

Therefore we can rewrite $\mathbf{Ax} = \mathbf{y}$ as $\mathbf{AP}\bar{\mathbf{x}} = \mathbf{Py}$, or as

$$\bar{\mathbf{A}}\bar{\mathbf{x}} = \bar{\mathbf{y}}$$

where

$$\bar{\mathbf{A}} = \mathbf{P}^{-1}\mathbf{AP}$$

This means that with respect to the basis $\{\mathbf{p}_1, \ldots, \mathbf{p}_n\}$ a transformation represented by the matrix **A** in the standard basis will be represented by the matrix $\bar{\mathbf{A}} = \mathbf{P}^{-1}\mathbf{AP}$. We have noted before that two matrices connected in this way by a nonsingular matrix **P** are said in matrix theory to be *similar* and that the transformation from **A** to $\bar{\mathbf{A}}$ is called a *similarity transformation*. What we have now gained is the geometric insight that what a similarity transformation does is describe the effect of a *change of basis*. The point is that with respect to certain bases certain facts may be displayed more clearly than with respect to other bases. We have seen this before.

We shall use the preceding concepts to describe a simple way of calculating $\bar{\mathbf{A}}$ that in many cases circumvents the need for explicit calculation of the matrix $\mathbf{P}^{-1}$. For this, note that we can write

$$\bar{\mathbf{A}} = \mathbf{P}^{-1}\mathbf{AP} = \mathbf{Q}(\mathbf{AP}) = \begin{bmatrix} \mathbf{q}_1' \\ \cdot \\ \cdot \\ \cdot \\ \mathbf{q}_n' \end{bmatrix} [\mathbf{Ap}_1 \quad \cdots \quad \mathbf{Ap}_n]$$

$$= \begin{bmatrix} \mathbf{q}_1'(\mathbf{Ap}_1) & \mathbf{q}_1'(\mathbf{Ap}_2) & \cdots & \mathbf{q}_1'(\mathbf{Ap}_n) \\ \cdot & \cdot & & \cdot \\ \cdot & \cdot & & \cdot \\ \cdot & \cdot & & \cdot \\ \mathbf{q}_n'(\mathbf{Ap}_1) & \mathbf{q}_n'(\mathbf{Ap}_2) & & \mathbf{q}_n'(\mathbf{Ap}_n) \end{bmatrix}$$

This shows that the

> ith column of $\bar{\mathbf{A}}$ = the coefficients of the representation of the vector $\mathbf{Ap}_i$ in the basis $\{\mathbf{p}_i\}$

Note also that if **b** is any vector and

$$\bar{\mathbf{b}} = \mathbf{P}^{-1}\mathbf{b} = [\mathbf{q}_1'\mathbf{b} \quad \cdots \quad \mathbf{q}_n'\mathbf{b}]'$$

then the

ith component of $\bar{\mathbf{b}}$ = ith coefficient of the representation of
b in the basis $\{\mathbf{p}_i\}$

We turn now to some examples that arise often in this book.

Example 5.2-1.   Diagonal Form (Eigenvector Basis)

Let A be a matrix with distinct eigenvalues $\{\lambda_i\}$ and corresponding eigenvectors $\{\mathbf{u}_i\}$. These eigenvectors are known to be linearly independent (Exercise A-42) and hence can be taken as a basis.

We have

$$\mathbf{A}\mathbf{u}_i = \lambda_i\mathbf{u}_i = 0 \cdot u_1 + \cdots + \lambda_i \cdot u_i + \cdots + 0 \cdot u_n$$

which shows that $\bar{\mathbf{A}} = \text{diag}\{\lambda_1, \ldots, \lambda_n\}$.

Example 5.2-2.   Controllability Form

Suppose we have a controllable realization $\{\mathbf{A}, \mathbf{b}\}$, so that $\mathcal{C} = [\mathbf{b}\ \mathbf{Ab}\ \cdots\ \mathbf{A}^{n-1}\mathbf{b}]$ is nonsingular and its columns can be used as a basis for n-space. The representation of **b** in this basis is $\bar{\mathbf{b}} = \mathcal{C}^{-1}\mathbf{b}$, which can be evaluated directly. However, in terms of the method described above, we can find $\bar{\mathbf{b}}$ by seeking a representation of **b** with respect to the basis $\{\mathbf{b}, \ldots, \mathbf{A}^{n-1}\mathbf{b}\}$. Now

$$\mathbf{b} = 1 \cdot \mathbf{b} + 0 \cdot \mathbf{Ab} + \cdots + 0 \cdot \mathbf{A}^{n-1}\mathbf{b}$$

so that

$$\bar{\mathbf{b}} = [1 \quad 0 \quad \cdots \quad 0]'$$

The representation of **A** with respect to the new basis is $\bar{\mathbf{A}} = \mathcal{C}^{-1}\mathbf{A}\mathcal{C}$. To calculate it we note that

$$\mathbf{Ab} = 0 \cdot \mathbf{b} + 1 \cdot \mathbf{Ab} + \cdots + 0 \cdot \mathbf{A}^{n-1}\mathbf{b}$$

so that the first column of $\bar{\mathbf{A}}$ is the transpose of $[0\ 1\ 0\ \cdots\ 0]$. Next

$$\mathbf{A} \cdot \mathbf{Ab} = \mathbf{A}^2\mathbf{b} = 0 \cdot \mathbf{b} + 0 \cdot \mathbf{Ab} + 1 \cdot \mathbf{A}^2\mathbf{b} + \cdots + 0 \cdot \mathbf{A}^{n-1}\mathbf{b}$$

so that the second column of $\bar{\mathbf{A}}$ is the transpose of $[0\ 0\ 1\ \cdots\ 0]$. Proceeding, for the last column we have

$$\mathbf{A} \cdot \mathbf{A}^{n-1}\mathbf{b} = \mathbf{A}^n\mathbf{b}$$

which can be represented in terms of the new basis by using the Cayley-Hamilton theorem. This yields

$$\mathbf{A}^n\mathbf{b} = -a_1\mathbf{A}^{n-1}\mathbf{b} - \cdots - a_n\mathbf{Ib}$$

so that the last column of $\bar{\mathbf{A}}$ is the transpose of $[-a_n \cdots -a_1]$. We see that $\{\bar{\mathbf{A}}, \bar{\mathbf{b}}\}$ is in controllability form.

Example 5.2-3.   Controller Form

If $\{\mathbf{b}, \mathbf{Ab}, \ldots, \mathbf{A}^{n-1}\mathbf{b}\}$ are linearly independent, so will be various linear combinations of them. In particular, let us develop a basis as follows: We can rewrite the Cayley-Hamilton equation $a(\mathbf{A}) = 0$ as

$$\mathbf{Ap}_n = -a_n\mathbf{b}$$

where

$$\mathbf{p}_n \triangleq \mathbf{A}^{n-1}\mathbf{b} + a_1\mathbf{A}^{n-2}\mathbf{b} + \cdots + a_{n-1}\mathbf{b}$$

Again we can rewrite the last expression as

$$\mathbf{p}_n - a_{n-1}\mathbf{b} = \mathbf{Ap}_{n-1}$$

where

$$\mathbf{p}_{n-1} \triangleq \mathbf{A}^{n-2}\mathbf{b} + \cdots + a_{n-2}\mathbf{b}$$

Proceeding in this fashion, we shall get

$$\mathbf{p}_{n-2} \triangleq \mathbf{A}^{n-3}\mathbf{b} + \cdots + a_{n-3}\mathbf{b}, \ldots, \mathbf{p}_2 = \mathbf{Ab} + a_1\mathbf{b}, \mathbf{p}_1 = \mathbf{b}$$

The $\{\mathbf{p}_i\}$ so defined are linearly independent because each $\mathbf{p}_i$ includes one more term of the linearly independent sequence $\{\mathbf{A}^i\mathbf{b}\}$. Therefore the $\{\mathbf{p}_i\}$ form a basis and in fact a quite useful one. With respect to this basis, we shall obtain the realization $\{\bar{A}, \bar{b}\}$ in controller form.

Example 5.2-4.   Observability Form

For an observable pair $\{\mathbf{c}, \mathbf{A}\}$ the *rows* $\{\mathbf{c}, \mathbf{cA}, \ldots, \mathbf{cA}^{n-1}\}$ of $\mathbf{O}(\mathbf{c}, \mathbf{A})$ are linearly independent.

Now, although our discussions so far have been restricted to column matrices and column vectors, it is clear that with small and obvious changes everything can be carried out with row matrices and row vectors. Therefore we can ask what the above system will look like not in the standard row basis but in the new basis represented by the rows of $\mathbf{O}(\mathbf{c}, \mathbf{A})$. We ask the reader to confirm for himself that the answer is a pair $\{\bar{\mathbf{c}}, \bar{\mathbf{A}}\}$, where

$$i\text{th row of } \bar{\mathbf{A}} = \text{the coefficients of the representation}$$
$$\text{of the row vector } \mathbf{cA}^{i-1} \cdot \mathbf{A} \text{ in the new basis}$$

This will show that $\bar{\mathbf{A}}$ will be a lower-companion matrix with last row $-[a_n \cdots a_1]$. Clearly also $\bar{\mathbf{c}} = [1 \ 0 \ \cdots \ 0]$, so that $\{\bar{\mathbf{c}}, \bar{\mathbf{A}}\}$ is in observability canonical form. Notice, of course, that the same results can be obtained by working with the new column vector basis $\{\mathbf{c}', \mathbf{A}'\mathbf{c}', \ldots, \mathbf{A}'^{n-1}\mathbf{c}'\}$ and calculating $\bar{\mathbf{A}}'$ and $\bar{\mathbf{c}}'$ with respect to this new basis. Nevertheless, it is worthwhile to develop some facility with row arguments as well, and therefore the earlier method is useful.

For further practice, we suggest that the reader work out for himself the discussions for the observer canonical form (see, of course, Example 5.2-3 on the controller form).

Example 5.2-5.   Polynomial Bases

We shall consider a slightly different example, where we have an n-dimensional space of column vectors and a uniquely related (isomorphic) linear space of polynomials $\{r(z), \deg r(z) < n\}$. The relation is that such a polynomial

$$r(z) = r_1 z^{n-1} + \cdots + r_n$$

can be naturally associated with the vector $[r_1 \ \cdots \ r_n]'$. We see also that the basis

$$\{[1 \quad 0 \quad \cdots \quad 0]', [0 \quad 1 \quad \cdots \quad 0]', \ldots, [0 \quad \cdots \quad 0 \quad 1]'\}$$

is naturally associated with the basis

$$\{1, z, \ldots, z^{n-1}\}$$

for the space of polynomials of degree less than n.

Now this space of polynomials arose in Sec. 5.1, where they were used to define the state space of a system. Moreover, we saw [cf. Eqs. (5.1-10)–(5.1-15)] that the evolution of the system was determined by evaluating

$$[zr(z) + u(k)] \bmod a(z)$$

which is the *remainder* obtained when $zr(z) + u(k)$ is divided by $a(z)$. Now since $u(k)$ is a constant,

$$[zr(z) + u(k)] \bmod a(z) = zr(z) \bmod a(z) + u(k)$$

Now we note that the effect of multiplication by z is easy to determine in the basis $\{1, z, \ldots, z^{n-1}\}$. In fact, under multiplication by z,

$$1 \longrightarrow z, z \longrightarrow z^2, z^{n-1} \longrightarrow z^n = -a_1 z^{n-1} \cdots -a_n$$

Translated into the space of vectors this says that the evolution of the system can be described by the relation between coefficients of the corresponding state polynomials $r(z)$ and $[zr(z) \bmod a(z) + u(k)]$, say

$$x(k+1) = Ax(k) + bu(k), \qquad b = [1 \quad 0 \quad \cdots \quad 0]'$$

where A should be

$$A = \begin{bmatrix} 0 & 0 & & & -a_n \\ 1 & 0 & & & \\ & 1 & & & \\ & & \cdot & & \\ & & & \cdot & \\ 0 & & & 1 & -a_1 \end{bmatrix}$$

This derivation of the state equations is not much different from that given in Sec. 5.1, but it is no doubt a little bit simpler. The simplicity will become much more evident when we consider the analogous multivariable problem in Sec. 6.6. It is worth repeating a fact shown by the above that the **A** matrix of a state-space realization is the matrix corresponding to the application of the shift operation (multiplication by z) to the polynomial representing the present state.

Another point made clear by Example 5.2-5 is that we need not restrict ourselves, as we have basically done so far, to only considering linear vector spaces of $n \times 1$ column (or row) vectors. The elements of a linear vector space can be quite varied, and we have just seen a simple example of this where the elements are polynomials of a certain maximum degree. This space has a natural relation to n-space, and so we do not need to worry much about its properties. But we are going to encounter more complicated spaces (of polynomial vectors) in our study of multivariable systems. Hence it will be useful to have a general definition of a linear vector space, and this will be done at the end of this section. Let us, however, first complete our discussion of n-space results with a few remarks on vector subspaces.

5.2.3   Vector Subspaces

In the space of all n-vectors, let us consider for a moment the set of all vectors *with 0 in the first entry*. This set has the property that for *any* two vectors $\{x_1, x_2\}$ in the set and *any* two scalars $\{\alpha_1, \alpha_2\}$, the vector $\alpha_1 x_1 + \alpha_2 x_1$ also belongs to the set. But this property is exactly the property that characterizes vector spaces, so that it is natural to call these sets *vector subspaces*.

Some Examples of Vector Subspaces in 4-Space

1. All vectors with first and third entries equal to zero.
2. All linear combinations of the vectors $\{[1 \ 2 \ 3 \ -1]', [0 \ 1 \ 1 \ 1]'\}$.
3. All linear combinations of the vectors $\{[1 \ 1 \ 3 \ -1]', [0 \ 1 \ 1 \ 1]', [0 \ 0 \ 1 \ 1]', [0 \ 1 \ 2 \ 2]'\}$. Note that the fourth vector is a linear combination of the previous ones, so that this subspace is the same as the subspace of all linear combinations of the vectors $\{[1 \ 2 \ 3 \ -1]', [0 \ 1 \ 1 \ 1]', [0 \ 0 \ 1 \ 1]'\}$.
4. All linear combinations of $\{[1 \ 2 \ 3 \ -1]', [0 \ 1 \ 1 \ 1]', [0 \ 0 \ 1 \ 1]', [0 \ 0 \ 0 \ 1]'\}$. This subspace is actually the whole space, because the four vectors are linearly independent so that they form a basis of 4-space.
5. Let **A** be any $n \times n$ matrix. The set of n-vectors **Au**, $u = $ any $n \times 1$ matrix, is a subspace that is called the *range space* of **A**. The set of all vectors **u** such that $Au = 0$ is also a subspace called the *null-space* of **A** (cf. Sec. A.4 for the important role of these subspaces in solving linear equations).

Given m vectors $\{x_1, x_2, \ldots, x_m\}$, the collection of all their linear combinations is called the *subspace spanned* by the $\{x_1, \ldots, x_m\}$.

Exercise: Is the set of all vectors with the first entry equal to 1 a subspace? Why?

Bases for Subspaces.   Since a subspace has the same properties as vector spaces, we can define a *basis* in it as a set of linearly independent vectors with the property that every vector in the subspace can be written as a linear combination of vectors of this set.

Examples: For the subspaces defined above we have the bases

1. $\{[0 \ 1 \ 0 \ 0]', [0 \ 0 \ 0 \ 1]'\}$,
2. $\{[1 \ 2 \ 3 \ -1]', [0 \ 1 \ 1 \ 1]'\}$,
3. $\{[1 \ 2 \ 3 \ -1]', [0 \ 1 \ 1 \ 1]', [0 \ 0 \ 1 \ 1]'\}$, and
4. The standard basis.

The number of elements in the basis is called the *dimension* of the subspace. Clearly the dimension of the whole space is greater than or equal to the dimension of *any* of its subspaces.

Rank and Nullity.   The dimension of the range space, $\mathcal{R}(A)$, of A is the rank of A. The dimension of the null space, $\mathcal{N}(A)$, of A is called the *nullity* of A. It is an important result that if A is an $m \times n$ matrix, then

$$\text{rank } A + \text{nullity } A = n$$

One proof of this result starts with a basis $\{f_1, \ldots, f_v\}$ for $\mathcal{N}(A)$, and augments it to a basis for n-space, say $\{f_1, \ldots, f_v, g_1, g_2 \ldots g_\mu\}$, where $v = \dim \mathcal{N}(A)$ and $\mu = n - v$. Now for any n-vector $x = \sum_1^v a_i f_i + \sum_1^\mu b_i g_i$, $Ax = \sum b_i A g_i$, since the $\{Af_i\}$ are all zero. Therefore $\{Ag_1, \ldots, Ag_\mu\}$ spans $\mathcal{R}(A)$. We shall prove that these vectors are linearly independent, so that in fact they form a basis for $\mathcal{R}(A)$. But then $\mu = \dim \mathcal{R}(A) = \text{rank of } A$, and the desired result that $\mu + v = n$ follows. To check the linear independence, suppose there exist $\{c_i\}$ such that $\sum_i c_i A g_i = 0$. Then $\sum c_i g_i$ belongs to $\mathcal{N}(A)$ and therefore we can write $\sum c_i g_i = \sum d_i f_i$ for some $\{d_i\}$. But this implies a linear dependence between the $\{f_i, g_i\}$, which is impossible unless all the $\{c_i\}$ and $\{d_i\}$ are zero. That is, $\sum c_i A g_i = 0$ implies the $\{c_i\}$ are all zero, which means that the $\{Ag_i\}$ are indeed linearly independent.

Invariant Subspaces.   Given an $n \times n$ matrix A in n-space, a subspace S is said to be *invariant under* A if

$$As \in S \qquad \text{for all } s \in S$$

Example: Let u_1 and u_2 be two eigenvectors of A corresponding to distinct eigenvalues λ_1 and λ_2. Then the set of all linear combinations $\{\alpha_1 u_1 + \alpha_2 u_2\}$ is a subspace that is invariant under A.

Example: Suppose the controllability matrix

$$\mathfrak{C}(A, b) = [b \;\vdots\; Ab \;\vdots\; \ldots \quad A^{n-1}b]$$

has rank $r \le n$. Then by the Cayley-Hamilton theorem, it is easy to see that the range space of $\mathfrak{C}$ (i.e., the set of all n-vectors of the form $\mathfrak{C}u$, u an arbitrary n-vector) is an r-dimensional subspace of n-space that is invariant under A.

Bases Determined by Invariant Subspaces.   Suppose we have an r-dimensional subspace S of an n-dimensional space of column vectors and suppose that S is invariant under A, an $n \times n$ matrix. Let

$$\{s_1, \ldots, s_r\} = \text{a basis for } S$$

We can obtain a basis for the whole space by augmenting the basis for S by $n - r$ linearly independent vectors, say $\{p_{r+1}, \ldots, p_n\}$. Therefore

$$\{s_1, \ldots, s_r, p_{r+1}, \ldots, p_n\} \text{ is a basis for } n\text{-space}$$

We can now ask how the original matrix A will look in this new basis. By the same arguments as used previously, we readily obtain the representation

$$\bar{A} = \begin{bmatrix} A_{11} & \vdots & A_{12} \\ \hdashline 0 & \vdots & A_{22} \end{bmatrix} \begin{matrix} r \\ \\ n\text{-}r \end{matrix}$$
$$\phantom{\bar{A} = \begin{bmatrix} \end{bmatrix}} {\scriptstyle r \qquad n\text{-}r}$$

where the A_{ij} denote block matrices (of appropriate dimensions) that are in general nonzero. The significant point is that the $(n - r) \times r$ block of $\bar{A}$ in the lower left corner is identically zero. (The reason for this is hopefully quite clear by now, but, to repeat, it is that the representations of $\{As_1, As_2, \ldots, As_r\}$ with respect to the new basis will be combinations only of $\{s_1, \ldots, s_r\}$ because the $\{As_i\}_1^r$ also belong to S.)

Orthogonal Complements.   If S is a subspace, we shall write

$$S^\perp = \{\text{the set of all vectors orthogonal to every vector in } S\}$$

and call it the *orthogonal complement* of S in n-space. It is easy to check that $S^\perp$ is also a subspace.

If $S = \mathfrak{R}(A)$, then it turns out that $S^\perp = \mathfrak{N}(A')$. (See Exercise A.15.)

Exercise.   Let

$$\mathfrak{C}(A, b) = [b \quad Ab \quad \ldots \quad A^{n-1}b]$$

have rank r, and let $S =$ the range of $\mathfrak{C}$. Then show that S is an r-dimensional invariant subspace of A for which $\{b, Ab, \ldots, A^{r-1}b\}$ can be taken as a basis.

If this basis is augmented in an arbitrary way to get n linearly independent vectors, find the representations of **A** and **b** with respect to this new basis (see Sec. 2.4.2).

5.2.4   Abstract Linear Vector Spaces

We have treated the space of all n-vectors with real or complex entries and (in Example 5.1-5) the space of all polynomials of degree less than n as *linear vector spaces*. We allowed linear combinations with real or complex coefficients. In applications it often arises that we need to consider more general situations. For example, in multivariable systems we need to consider vectors whose entries are polynomials (with real or complex coefficients), and it will be useful to form linear combinations of such vectors with coefficients that are rational functions (see Sec. 6.3). Therefore it will be useful to have a general definition of linear vector spaces, so that we may know whether the properties and results discussed above for the special vector spaces can be carried over to more complicated situations. The definition can be found in the opening pages of most books in linear algebra, but for convenience of reference we shall present the basic axiomatic definition here.

An *abstract linear vector space V* over a *field F* is a collection of *elements* (also called *vectors*), denoted $\{p, q, r, \ldots\}$, equipped with

1. An operation $+$, called *addition*, such that for any two vectors $\{p, q\}$ in V the *sum* $p + q$ is also a vector (i.e., an element of V), and
2. An operation $\cdot$, called *scalar multiplication*, such that for $\alpha \in F$ (the elements of F are called *scalars*) and $p \in V$, $\alpha \cdot p$ (often written αp) is also a vector in V.

Moreover, the addition rule has the following properties (for $\{p, q, r \in V\}$):

1. $p + q = q + p$.
2. $(p + q) + r = p + (q + r)$.
3. There exists a *zero element* θ such that $p + \theta = p$.
4. For every $p \in V$, there exists an *inverse* $q \in V$ such that $p + q = \theta$.

The scalar multiplication has the following properties (for $\alpha, \beta \in F$, $p, q \in V$):

5. $\alpha(\beta p) = (\alpha\beta)p$.
6. $(\alpha + \beta)p = \alpha p + \beta p$.
7. $\alpha(p + q) = \alpha p + \alpha q$.
8. $1 \cdot p = p$, where 1 is the *identity* element of the field F.

We note that a *field F* is a collection of elements along with an operation $+$ obeying properties 1–4 (with $\{p, q, r\}$ now elements of F) and an operation $\cdot$ obeying properties 5–7, but in addition there exists an element $1 \neq 0$ such that $1 \cdot \alpha = \alpha$, for all $\alpha \in F$, and there exists a multiplicative inverse α^{-1} such that $\alpha^{-1}\alpha = 1$ for all $\alpha \neq 0$.

Note the difference between the zero element 0 in F and the zero element θ in V—often things are clear enough from the context that we can use the same symbol 0 for both quantities.

Example A: The reader should verify that the collection of rational functions is a field. Also verify that the collection of column matrices (or column vectors) of the form $[p_1(s), \ldots, p_n(s)]'$, where the $\{p_i(s)\}$ are polynomials, is a linear vector space over the field of rational functions.

Example B: The reader should verify that the space of solutions to the differential equation $y^{(n)}(t) + a_1 y^{(n-1)}(t) + \cdots + a_n y(t) = u(t)$, $0 \leq t \leq 1$, is a linear vector space over the field of complex numbers. (Also consider the special case where all the $\{a_i\}$ are zero.)

The advantage of the abstract definition is that the basic properties of linear combinations, linear dependence, linear independence, rank, and subspaces that we discussed for real n-vectors can be carried over to every linear vector space. However, things that are trivial for real n-space may require a longer proof in the general case—for example, show from the formal definition that $0 \cdot p = \theta$, and compare this with the trivial proof for the case that p is an n-vector of real numbers.

For the purposes of this book, this is all that we shall need to say about linear vector spaces and linear algebra—we hope that the applications already described and those to be made in the following chapters will provide some motivation and some guidance for further mathematical studies of this useful subject.

Exercise: Show that for $p \in V$, $\alpha \in F$, (a) $-\alpha p = (-\alpha) \cdot p$, (b) $\alpha p = 0$ and $\alpha \neq 0$ imply $p = 0$, and (c) $\alpha p = 0$ and $p \neq 0$ imply $\alpha = 0$.

Exercise: Would the discussion in Sec. 5.2.1 of inner products and dual bases apply to linear vector spaces over the field of rational functions?

REFERENCES

1. A. NERODE, "Linear Automaton Transformations," Proc. Amer. Math. Soc., **9**, pp. 541–544, 1958.

2. R. E. KALMAN, P. A. FALB, and M. A. ARBIB, *Topics in Mathematical System Theory*, McGraw-Hill, New York, 1969.

3. L. PADULO and M. A. ARBIB, *System Theory*, Saunders, Philadelphia, 1974.

4. W. B. GRAGG, "The Padé Table and its Relation to Certain Algorithms of Numerical Analysis," *SIAM Rev.*, **14**, pp. 1–62, 1972.

5. J. GILEWICZ, *Approximants de Padé*, Lecture Notes in Mathematics, vol. 667, Springer-Verlag, N.Y., 1978.

6. K. J. ASTROM and P. EYKHOFF, "System Identification—A Survey," *Automatica*, **7**, pp. 123–162, March 1971.

7. G. C. GOODWIN and R. L. PAYNE, *Dynamic System Identification*, Academic Press, New York, 1977.

8. S. Kung, "A New Low-Order Approximation Algorithm via Singular Value Decompositions," *Proc. 1979 IEEE Conference on Decision & Control*, Clearwater, Fla., December 1979. Also *IEEE Trans. Automat. Contr.*, 1980.

9. E. Berlekamp, *Algebraic Coding Theory*, McGraw-Hill, New York, 1968.

10. J. L. Massey, "Shift-Register Synthesis and BCH Decoding," *IEEE Trans. Inf. Theory*, **IT-15**, pp. 122–127, 1969.

11. R. P. Kurshan and B. Gopinath, "Digital Single-Tone Generator-Detectors," *Bell Syst. Tech. J.*, **55**, pp. 469–496, April 1976.

12. D. C. Youla, "The Synthesis of Linear Dynamical Systems from Prescribed Weighting Patterns," *SIAM J. Appl. Math.*, **14**, pp. 527–549, 1966.

13. B. L. Ho, "An Effective Construction of Realizations from Input/Output Descriptions," Ph.D. dissertation, Stanford University, Stanford, Calif., 1966.

14. L. M. Silverman, "Representation and Realization of Time-Variable Linear Systems," Ph.D. Dissertation, Dept. of Electrical Engineering, Columbia University, New York, June 1966.

15. R. E. Kalman, "On Partial Realizations of a Linear Input/Output Map," in *Aspects of Network and System Theory* (R. E. Kalman and N. DeClaris, eds.,) Holt, Rinehart and Winston, New York, 1970.

16. A. J. Tether, "Construction of Minimal Linear State-Variable Models from Finite Input-Output Data," *IEEE Trans. Autom. Control*, **AC-17**, pp. 427–436, 1970.

17. J. Rissanen, "Recursive Identification of Linear Systems," *SIAM J. Control*, **9**, pp. 420–430, 1971.

18. S. Kung, "Multivariable and Multidimensional Systems: Analysis and Design," Ph.D. Dissertation, Dept. of Elec. Engr., Stanford University, Stanford, Calif., June 1977.

19. C. Lanczos, "Solution of Systems of Linear Equations by Minimized Iterations," *J. Res. Nat. Bur. Stand.*, **49**, pp. 35–53, 1952.

20. Y. Sugiyama, M. Kasahara, S. Hirasawa and T. Namekawa, "A Method for Solving Key Equation for Decoding Goppa Codes," *Inform. Contr.*, **27**, pp. 87–99, January 1975.

21. A. V. Aho, J. E. Hopcroft and J. D. Ullman, *The Design and Analysis of Computer Algorithms*, Addison-Wesley, Reading, Mass., 1974.

22. J. Justesen, "On the Complexity of Decoding of Reed-Solomon Codes," *IEEE Trans. on Inform. Theory*, **IT-22**, pp. 237–238, March 1976.

23. F. G. Gustavsson and D. Y. Y. Yun, "Fast Computation of Rational Interpolants and Toeplitz Systems of Equations via the Fast Extended Euclidean Algorithm," *IEEE Trans. on Circuits and Systems*, **CAS-26**, September 1979.

24. S. MacLane and G. Birkhoff, *Algebra*, Macmillan, New York, 1967.

25. B. Hartley and T. O. Hawkes, *Rings, Modules and Linear Algebra*, Chapman & Hall, London, 1970.

26. E. Sontag, "Linear Systems over Commutative Rings: A Survey," *Ric. Autom.*, **7**, pp. 1–34, 1976.

27. E. SONTAG, "On Linear Systems and Noncommutative Rings," *Math. Syst. Theory*, **9**, pp. 327–344, 1976.

28. E. KAMEN, Lectures on Algebraic System Theory: Linear Systems over Rings, *NASA Contractor Rept. 3016*, July 1978.

29. M. SAIN, "The Growing Algebraic Presence in Systems Engineering: An Introduction," *Proc. IEEE*, **64**, pp. 96–111, Jan. 1976.

30. P. A. FUHRMANN, *Linear Operators and Systems in Hilbert Space*, McGraw-Hill, New York, 1979.

31. B. NOBLE and J. DANIEL, *Applied Linear Algebra*, Prentice-Hall, Englewood Cliffs, N.J., 1969; 2nd rev. ed., 1977.

32. G. STRANG, *Linear Algebra and Its Applications*, Academic Press, New York, 1976.

STATE-SPACE AND
MATRIX-FRACTION
DESCRIPTIONS OF MULTIVARIABLE
SYSTEMS

6

6.0 INTRODUCTION

This long chapter is the counterpart for multi-input, multi-output (or multi-variable for short) systems of the basic Chapter II for single-input, single-output (scalar) systems. As with scalar systems, we shall begin with a study of realization techniques for multivariable transfer functions and show how this problem leads naturally to the observability and controllability concepts. Now, however, all the problems are more involved because, unlike the scalar case, there does not seem to be a single unique (or "canonical") choice of realizations, and observation and control schemes. Moreover, the close connections with irreducible transfer functions that were developed in Sec. 2.4 for scalar systems do not seem obvious now. However, a much closer analogy with the scalar results can be achieved by using the so-called *matrix-fraction descriptions* (MFDs) of rational matrices as the "ratio" of two relatively-prime polynomial matrices (Sec. 6.2.3). To pursue this further, it is necessary to develop the relatively unfamiliar theory of polynomial matrices. This is done in some detail in Sec. 6.3, which contains some classical and some very recent results. Sec. 6.5 similarly presents some old and new mathematical results on rational matrices. No similar comprehensive sections appear in any single source in the literature. The use of MFDs allows us to develop state-space realizations (Sec. 6.4) that are much better coupled to the transfer functions than are those of Sec. 6.1. It is an interesting fact that the development in Sec. 6.4 is closely related to that in the scalar case

345

of Sec. 2.1, and, in fact, the matrix procedures tends to be somewhat better motivated and more illuminating than the scalar ones. The constructions in Sec. 6.4 provide a set of simple and useful models for multivariable systems, and thereby provide a convenient and concrete coordinate system for the discussion of several general properties (see, e.g., Secs. 7.2, 7.3, 8.2, and 8.3). We draw particular attention to the concept of "partial state" and to the algebraic properties noted in Sec. 6.4.2. The brief Sec. 6.6 shows how the idea of Nerode equivalence discussed in Sec. 5.1 can also be used in the multivariable case to provide a natural way of proceeding from external matrix-fraction descriptions to the internal state-space descriptions of Sec. 6.4. Sec. 6.7 further pursues the relations between state-space models and MFDs by introducing two important (canonical) MFDs, the triangular Hermite form and the polynomial echelon Popov form, and their associated state-space descriptions. Some important applications are also discussed.

6.1   SOME DIRECT REALIZATIONS OF MULTIVARIABLE TRANSFER FUNCTIONS

In the (scalar) single-input, single-output case, given a transfer function

$$H(s) = \frac{b(s)}{a(s)}$$

we could readily write down a variety of nice state-space realizations of $H(s)$. This was done in Sec. 2.1, and among the "nice" properties were the fact that the order of the state-space realization could readily be arranged to be no more than n, the degree of $a(s)$, and that controllable or observable realizations could be easily specified.

It is not hard to write down state-space realizations in the (matrix) multi-input, multi-output case, but it is less easy to decide what "nice" properties are.

It is best to proceed by example. Thus, consider different ways of realizing a two-input, two-output system with the transfer function (matrix)

$$H(s) = \begin{bmatrix} \dfrac{1}{(s-1)^2} & \dfrac{1}{(s-1)(s+3)} \\ \dfrac{-6}{(s-1)(s+3)^2} & \dfrac{s-2}{(s+3)^2} \end{bmatrix}$$

Perhaps the most straightforward way is to make a realization of each entry (by any of the methods of Sec. 2.1) and then connect them appropriately. For example, if we use controller canonical realizations for each entry, it can be seen that a realization of $H(s)$ is

$$\dot{x}(t) = Ax(t) + Bu(t), \qquad y(t) = Cx(t)$$

where

$$
A = \begin{bmatrix}
2 & -1 & & & & & & & \\
1 & 0 & & & & & & & \\
& & -2 & 3 & & & & & \\
& & 1 & 0 & & & & & \\
& & & & -5 & -3 & 9 & & \\
& & & & 1 & 0 & 0 & & \\
& & & & 0 & 1 & 0 & & \\
& & & & & & & -6 & -9 \\
& & & & & & & 1 & 0
\end{bmatrix}
$$

$$
B = \begin{bmatrix}
1 & 0 & 0 & 0 & 1 & 0 & 0 & 0 & 0 \\
0 & 0 & 1 & 0 & 0 & 0 & 0 & 1 & 0
\end{bmatrix}'
$$

$$
C = \begin{bmatrix}
0 & 1 & 0 & 1 & 0 & 0 & 0 & 0 & 0 \\
0 & 0 & 0 & 0 & 0 & 0 & -6 & 1 & -2
\end{bmatrix}
$$

This realization has nine states (the sum of the degrees of the denominators of the different entries) and has a certain structure, but only by extension from the scalar case. To try to get a better analog of some of the canonical forms of Sec. 2.1, let us rewrite $H(s)$ as

$$
H(s) = \frac{N(s)}{d(s)}
$$

where

$d(s) = $ the least common multiple of the denominators
       of the entries of $H(s)$

$\qquad = s^r + d_1 s^{r-1} + \cdots + d_r$

In our example,

$$
d(s) = (s-1)^2(s+3)^2 = s^4 + 4s^3 - 2s^2 - 12s + 1
$$

and

$$
N(s) = N_1 s^3 + N_2 s^2 + N_3 s + N_4
$$

where

$$
N_1 = \begin{bmatrix} 0 & 0 \\ 0 & 1 \end{bmatrix}, \qquad N_2 = \begin{bmatrix} 1 & 1 \\ 0 & -4 \end{bmatrix}
$$

$$
N_3 = \begin{bmatrix} 6 & 2 \\ -6 & 5 \end{bmatrix}, \qquad N_4 = \begin{bmatrix} 9 & -3 \\ 6 & -2 \end{bmatrix}
$$

Then it is easy to verify that the following is a realization of $H(s)$:

$$
A = \begin{bmatrix} -d_1 I_m & -d_2 I_m & \cdots & -d_r I_m \\ I_m & 0 & \cdots & 0 \\ & \cdot & & \\ & & \cdot & \\ & & & \cdot \\ 0 & & I_m & 0 \end{bmatrix}, \quad B = \begin{bmatrix} I_m \\ 0 \\ 0 \\ \cdot \\ \cdot \\ \cdot \\ 0 \end{bmatrix} \quad (1)
$$

$$
C = [\; N_1 \quad \cdots \quad N_r \;]
$$

where m is the number of inputs and I_m is the $m \times m$ identity matrix. This form has a structure that is very similar to that of the scalar controller canonical form [Eq. (2.2-3)], and therefore we shall call it a *block controller form*. (*Warning:* The name is not standard.) However, unlike the scalar case, the order of this realization (the size of the matrix A) is $r \times m = 4 \times 2 = 8$, which is larger than the degree 4 of the denominator polynomial $d(s)$. Moreover, it is not clear as yet whether the form we have has any special *controllability* properties that would justify the name. We shall explore this point further in Sec. 6.2.

But we might note that the order 8 of the present realization is less than that of the first form of realization, which had order 9. This raises the question of what the minimal order is. In the scalar case, if there were no cancellations between the numerator and denominator polynomials, i.e., if the nominal scalar transfer function was irreducible, then we saw (Sec. 2.4) that the minimal order was the degree of the denominator polynomial. In the matrix case, we might by analogy think that if there is no value of s for which $N(s_0) = 0 = d(s_0) = 0$, then the minimal order should be the degree of $d(s)$. In our example, therefore, we might expect $n_{\min} = 4$, but there is no obvious way of getting a four-state realization and in fact we shall see later that $n_{\min}$ will generally be greater than the degree of $d(s)$. Therefore we shall have to look deeper for the proper analog of the scalar property. This will be obtained in Sec. 6.3, but for the moment let us proceed (as in Sec. 2.1) with some other "obvious" realizations of $H(s)$.

Thus, by analogy with the scalar case, we may guess and can verify that another realization of $H(s)$ is the *block observer form*

$$
A = \begin{bmatrix} -d_1 I_p & I_p & 0 & \cdots & 0 \\ -d_2 I_p & 0 & I_p & \cdots & 0 \\ \cdot & \cdot & \cdot & \cdot & \\ \cdot & \cdot & \cdot & & \cdot \\ \cdot & \cdot & \cdot & & I_p \\ -d_r I_p & 0 & 0 & \cdots & 0 \end{bmatrix}, \quad B = \begin{bmatrix} N_1 \\ N_2 \\ \cdot \\ \cdot \\ \cdot \\ N_r \end{bmatrix} \quad (2)
$$

$$
C = [\; I_p \quad 0 \quad 0 \quad \cdots \quad 0 \;]
$$

where p is the number of outputs. Here the number of states is rp, which is still 8 in our two-input, two-output example but could obviously be different from the number rm of states in the block controller form whenever $m \neq p$. We can also obtain *block controllability* and *block observability* forms with orders rm and rp, respectively.

There is almost a dual relationship between realizations (1) and (2)—the A matrix in (2) is almost the transpose of the one in (1), *except for dimension*, and similarly for the B and C matrices. The difference in dimension is unavoidable unless $m = p$. Modulo this difference, however, we can again use duality to help in organizing our results and guessing new results. This will be the spirit in which the word *dual* will often be used for multivariable systems.

The point we wish to make from the preceding discussions is that the natural analogs of scalar realizations do not seem to give us much guidance as to the minimal possible order of realizations of a matrix transfer function. This is a fairly serious difficulty. The lack of knowledge of the true (minimal) order of a matrix transfer function and, more importantly, the lack of awareness that inattention to this point could give rise to differences between the behavior of the *internal* and *external* descriptions of a system apparently led to several errors and misjudgments in early treatments of multi-input, multi-output systems (see the references in [1]). This situation persisted in fact until the early 1960s, when Gilbert [1] and Kalman [2] independently made contributions that clarified the problem by using the concepts of controllability and observability.

Gilbert's Diagonal Realizations.   The first result was due to Gilbert, who obtained it for (perhaps not surprisingly, cf. Sec. 2.4.1) *diagonalizable* systems. Thus, consider systems in which the denominator polynomial has distinct roots,

$$d(s) = \prod_{1}^{r} (s - \lambda_i), \qquad \lambda_i \neq \lambda_j \tag{3a}$$

so that we can expand $H(s)$ into partial fractions:

$$H(s) = \frac{N(s)}{d(s)} = \sum_{1}^{r} \frac{R_i}{s - \lambda_i} \tag{3b}$$

where the residue matrices $\{R_i\}$ can be obtained as

$$R_i = \lim_{s \to \lambda_i} (s - \lambda_i)H(s) \tag{4a}$$

Now let

$$\rho_i = \text{the rank of } R_i$$

so that (Exercise A.17) we can write

$$R_i = C_i B_i, \qquad C_i \text{ is } p \times \rho_i, \qquad B_i \text{ is } \rho_i \times m$$

Then it is easy to verify that

$$A = \text{block diag } \{\lambda_i I_{\rho_i}, i = 1, \ldots, r\} \tag{4b}$$

$$B' = [B_1', \ldots, B_r'], \qquad C = [C_1, \ldots, C_r] \tag{4c}$$

is a realization of $H(s)$. Moreover, the realization has order

$$n = \sum_1^r \rho_i \tag{4d}$$

Example 6.1-1.

Let

$$H(s) = \begin{bmatrix} \dfrac{1}{s+1} & \dfrac{2}{s+1} \\ \dfrac{-1}{(s+1)(s+2)} & \dfrac{1}{s+2} \end{bmatrix}$$

$$= \frac{\begin{bmatrix} 1 & 2 \\ -1 & 0 \end{bmatrix}}{s+1} + \frac{\begin{bmatrix} 0 & 0 \\ 1 & 1 \end{bmatrix}}{s+2}$$

The ranks of R_1 and R_2 are $\rho_1 = 2$, $\rho_2 = 1$, so that $n = 3$. This is less than the order 4 of the block controller or observer realizations but is greater than the degree 2 of the denominator polynomial $d(s)$. However, it can be proven that 3 is the smallest possible order of any state-space realization of $H(s)$. ∎

In general, *whenever $d(s)$ has distinct roots*, it is true that

$$n_{\min} = \sum_1^r \rho_i \tag{4e}$$

As stated before, this result is due to Gilbert [1] and was proved using a matrix version of the result of Sec. 2.4.1 that a (scalar) realization is minimal if and only if it is both controllable and observable. Therefore the next step in our study will be to examine the notions of controllability and observability for MIMO (multiple-input, multiple-output) systems. After that we shall return to this fairly nontrivial question of specifying the minimal order for general (not necessarily diagonalizable) realizations.

However, before going on we should draw attention to a special class of *multivariable* (or *MIMO*) systems that for many purposes has essentially the same properties as *scalar* (or *SISO*, single-input, single-output) systems. These are multivariable systems with either

1. One input ($m = 1$) and many outputs ($p > 1$) (SIMO systems) or
2. Many inputs ($m > 1$) but only one output ($p = 1$) (MISO systems).

Single-Input, Multiple-Output Systems.   When $m = 1$, the transfer function matrix is a column,

$$H(s) = \left[\frac{b_1(s)}{a_1(s)} \quad \cdots \quad \frac{b_p(s)}{a_p(s)}\right]'$$ (5a)

which we can rewrite as, say,

$$H(s) = \frac{[n_1(s) \quad \cdots \quad n_p(s)]'}{d(s)}$$ (5b)

where

$$d(s) = \text{the least common multiple of the denominators } a_i(s), i = 1, \ldots, p$$ (5c)

and

$$n_i(s) = \frac{b_i(s)d(s)}{a_i(s)}$$

Note that if the $\{b_i(s)/a_i(s)\}$ are strictly proper, as we shall assume, then so will be the $\{n_i(s)/d(s)\}$. Now we can readily obtain a controller-form realization for $H(s)$. We first realize $1/d(s)$ by Kelvin's method (Fig. 2.1-2) and then "read-out" the numerators $n_1(s), n_2(s), \ldots, n_p(s)$—see Fig. 6.1-1. The state-

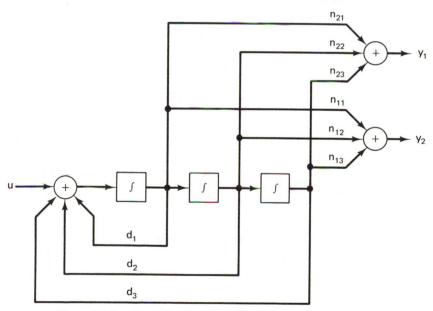

Figure 6.1-1.   Controller canonical form for a one-input, two-output system.

space equations will be defined by

$$A = \begin{bmatrix} -d_1 & \cdots & & -d_n \\ 1 & & & \\ & \cdot & & \\ & & \cdot & \\ & & & \cdot \\ & & 1 & 0 \end{bmatrix}, \quad b = \begin{bmatrix} 1 \\ 0 \\ \cdot \\ \cdot \\ \cdot \\ 0 \end{bmatrix}$$

and

$$C = \begin{bmatrix} n_{11} & n_{12} & \cdots & n_{1n} \\ \cdot & \cdot & & \cdot \\ \cdot & \cdot & & \cdot \\ \cdot & \cdot & & \cdot \\ n_{p1} & n_{p2} & \cdots & n_{pn} \end{bmatrix}$$

as in (1). This realization is obviously controllable. As for observability, for convenience we shall postpone a discussion to Sec. 6.2, but the interested reader can try to prove for himself that the above realization will be observable if and only if the polynomials $\{d(s), n_1(s), \ldots, n_p(s)\}$ have no common factors. Pole shifting is easy to do in this form—there is no difference from the scalar problem. Observer design is a multivariable problem and will be discussed in Chapter 7.

It will be a useful exercise for the reader to try to obtain the controllability canonical realizations for the above $H(s)$ and to go through the dual arguments for multiple-input, single-output systems. Here we shall go on to further study of the truly multivariable problems, with multiple inputs and multiple outputs.

Exercises

6.1-1.

　　a. Show how to obtain block controllability (and block observability) realizations of the 2×2 matrix transfer function in Example 6.1-1.

　　b. Show how to do this for any $p \times 1$ ($1 \times m$) transfer function.

6.1-2.

　　Verify that $C(sI - A)^{-1}B$ for the block controller form (6.1-1) does in fact give the matrix transfer function $H(s)$ *Hint:* Use Kronecker products for a general proof (see Exercises A.1 and A.41).

6.2 STATE OBSERVABILITY AND CONTROLLABILITY; MATRIX-FRACTION DESCRIPTIONS

In this section, we shall derive the multivariable analogs of the scalar results of Secs. 2.3 and 2.4. In Sec. 6.2.1, we shall use certain idealized problems of determining and setting up initial conditions to show how certain observability and controllability matrices arise naturally. Sec. 6.2.2 will be devoted to the behavior of realizations under certain similarity transformations and to the use of these results in relating minimality and the property of simultaneous controllability and observability. We shall also present here the very useful PBH tests for controllability and observability, which will lead us to consider certain coprime polynomial matrices. This should bring to mind Sec. 2.4, where we explored the transfer function significance of the controllability, observability, and irreducibility of scalar transfer functions. In Sec. 6.2.3, we shall indicate how these results may be extended to the matrix (multivariable) case by using what we shall call matrix-fraction descriptions (MFDs).

6.2.1   The Observability and Controllability Matrices

The basic ideas and arguments of this section will follow the pattern of the scalar case treated in Secs. 2.3 and 2.4, which may profitably be reviewed at this point. In addition, we shall use some simple results from the theory of linear equations.

An Observability Problem.   A realization $\{A, B, C\}$ will be said to be (state) *observable* if we can *uniquely* determine the state $x(t)$ of the realization

$$\dot{x}(t) = Ax(t) + Bu(t), \qquad x(0) = x_0$$
$$y(t) = Cx(t) \tag{1}$$

given knowledge of $\{A, B, C\}$ and $\{y(t), u(t), t \geq 0\}$. As in the scalar case (Sec. 2.3.1), a solution can be obtained in principle (since differentiation is unrealistic in practice) from the equations

$$\begin{bmatrix} y(t) \\ \dot{y}(t) \\ \cdot \\ \cdot \\ \cdot \\ y^{(n-1)}(t) \end{bmatrix} = \Theta x(t) + \mathbf{T} \begin{bmatrix} u(t) \\ \cdot \\ \cdot \\ \cdot \\ u^{(n-1)}(t) \end{bmatrix} \tag{2}$$

where

$$\Theta' = [C' \quad A'C' \quad \cdots \quad (A')^{n-1}C'] \tag{3}$$

and $\mathbf{T}$ is the Toeplitz *transmission* matrix,

$$\mathbf{T} = \begin{bmatrix} h_0 & & & \bigcirc \\ h_1 & \cdot & & \\ \cdot & \cdot & \cdot & \\ \cdot & & \cdot & \cdot \\ \cdot & & & \cdot & \cdot \\ h_{n-1} & & h_1 & h_0 \end{bmatrix} \tag{4}$$

with

$$h_0 = 0, \qquad h_i = CA^{i-1}B, \qquad \text{the } Markov \ parameters \qquad (5)$$

The Markov parameters are uniquely defined by the transfer function via the expansion

$$H(s) = \sum_{1}^{\infty} h_i s^{-i} \tag{6}$$

For convenience and without loss of generality, note that we could assume that the known input $u(\cdot)$ is identically zero. If it is not, its only effect is to modify the observations $y(\cdot)$ in a known way. Therefore this assumption will often be made.

Now if the $np \times n$ matrix $\mathcal{O}$ has rank less than n, then some linear combination of the n columns must add to zero, and therefore there are states that will not appear in $y(\cdot)$ or any of its derivatives (use the Cayley-Hamilton theorem). Therefore the condition that $\mathcal{O}$ has full rank is *necessary for observability*. To prove that it is also *sufficient* we first multiply (2) on both sides by $\mathcal{O}'$ to obtain

$$\mathcal{O}'\mathcal{Y}(t) = \mathcal{O}'\mathcal{O}x(t) + \mathcal{O}'\mathbf{T}\mathcal{U}(t) \tag{7}$$

where we define

$$\mathcal{Y}'(t) = [y'(t) \quad \dot{y}'(t) \quad \cdots \quad y'^{(n-1)}(t)] \tag{8}$$

$$\mathcal{U}'(t) = [u'(t) \quad \dot{u}'(t) \quad \cdots \quad u'^{(n-1)}(t)] \tag{9}$$

Now if $\mathcal{O}$ has rank n, it is not hard to show by contradiction that

$$\mathcal{O} \text{ full rank} \iff \mathcal{O}'\mathcal{O} \text{ nonsingular} \tag{10}$$

which means that $x(t)$ can be directly obtained from (7) as

$$x(t) = (\mathcal{O}'\mathcal{O})^{-1}\mathcal{O}'[\mathcal{Y}(t) - \mathbf{T}\mathcal{U}(t)] \tag{11}$$

This is the only solution of (7), but we might ask whether it is the only solution of (2). The answer is yes, for if $x_1(t)$ and $x_2(t)$ are two different solutions of (2), then we shall have

$$\mathcal{O}[x_1(t) - x_2(t)] = 0$$

which means that some combination of the columns of $\mathcal{O}$ is zero, which contradicts the assumption that $\mathcal{O}$ has full rank. Therefore we have proved that

A realization $\{A, B, C\}$ is state observable if
and only if the $np \times n$ *observability* $\tag{12}$
matrix $\mathcal{O}(C, A)$ has full rank n

The sufficiency part has been proven using differentiability, but as in the scalar case more realistic solutions can also be achieved—cf. Secs. 7.3 and 9.2. As in the scalar case, there are special forms in which observability is obvious and the determination of $x(t)$ particularly easy. Thus, for the block observer canonical form [Eq. (6.1-2)], it can be seen that Θ will be an $np \times n$ lower triangular matrix with ones along the main diagonal; such a matrix will clearly have full rank. In fact, as in the scalar case, for the state-observability problem it is somewhat simpler to use the block observability form, for which the reader should check that the first n rows of Θ will form an $n \times n$ identity matrix.

The block forms are thus reasonable generalizations of the scalar forms— the only problem is that, as stated in Sec. 6.1, the block forms will rarely be minimal. Moreover, it will not in general be possible to always transform an arbitrary observable pair $\{C, A\}$ to one of these block forms because n may not be a multiple of p. We shall pursue this point further in Sec. 6.2.2 but shall continue here with a further study of the observability and controllability matrices.

A State-Controllability Problem.   Proceeding by analogy with Sec. 2.3, let us study the following problem as a means of introducing the concept of state controllability. Given

$$\dot{x}(t) = Ax(t) + Bu(t), \qquad x(0-) = \text{arbitrary}$$

suppose we wish to set up some desired state vector at $t = 0+$, say $x(0+)$, by using an impulsive input. If this can always be done, we shall say that the realization $\{A, B, C\}$, or often just the pair $\{A, B\}$, is state controllable. Now since B is a matrix, and not a column as it was in Sec. 2.3.2, let us denote

$$B = [b_1 \quad \cdots \quad b_m]$$

so that we can write

$$\dot{x}(t) = Ax(t) + b_1 u_1(t) + \cdots + b_m u_m(t)$$

Then by using superposition we can [when $x(0-) = 0$] reduce this problem to a set of single-input problems. By this, or other means, we can see that if

$$u(t) = g_1 \delta(t) + \cdots + g_n \delta^{(n-1)}(t) \tag{13}$$

where each $\{g_i\}$ is an m-vector, then we can write

$$x(0+) = x(0-) + \mathcal{C}[g_1' \quad \cdots \quad g_n']' \tag{14}$$

where $\mathcal{C}$ is the $n \times nm$ matrix

$$\mathcal{C} = [B \quad AB \quad \cdots \quad A^{n-1}B] \tag{15}$$

Now if $\mathcal{C}$ has full rank, so that it contains n linearly independent columns, then we can find a combination of columns that will give any arbitrary n-vector $x(0+) - x(0-)$. This combination will determine the vector $[g_1' \cdots g_n']'$ and thereby the appropriate input vector $u(\cdot)$. On the other hand, if $\mathcal{C}$ does not have full rank, then no combination of its columns will yield an arbitrary n-vector $x(0+) - x(0-)$. Therefore we have the following result, which should be compared with (12):

> A necessary and sufficient condition for state controllability is that the $n \times nm$ *controllability* matrix $\mathcal{C}(A, B)$ has full rank n (16)

Of course, we have only proved this result for impulsive inputs, but, as in the scalar case, this is inessential and it can be shown that the above condition is necessary and sufficient even when finite energy inputs are used (try to extend the argument in Example 2.5-1).

An Explicit Solution.   It is of interest to actually determine a set of $\{g_i\}$ that will determine a satisfactory input when $\mathcal{C}$ has rank n. We have to solve Eq. (14). Since $\mathcal{C}$ has rank n, it is easy to show by contradiction (see also Sec. A.5) that

$$\mathcal{C} \text{ has rank } n \Longleftrightarrow \mathcal{C}\mathcal{C}' \text{ nonsingular} \qquad (17)$$

and then one can verify that a solution is given by

$$[g_1' \cdots g_n']' = \mathcal{C}'[\mathcal{C}\mathcal{C}']^{-1}[x(0+) - x(0-)] = \gamma, \text{ say.} \qquad (18)$$

By now it should not be unexpected that if the realization is in block controllability form, then $\mathcal{C}$ has the form $[I_n \quad X]$, and the determination of the $\{g_i\}$ is quite simple.

The Controllability Index and the Observability Index.   The fact that the controllability matrix

$$\mathcal{C} = [B \quad AB \quad \cdots \quad A^{n-1}B]$$

has rank n means that there are n linearly independent columns in $\mathcal{C}$. It may be that we can find n such columns in the *partial controllability matrix*

$$\mathcal{C}_q = [B \quad AB \quad \cdots \quad A^{q-1}B], \qquad 1 \le q \le n$$

and the smallest q, say μ, such that $\mathcal{C}_q$ has rank n will be called the *controllability* index of $\{A, B\}$.

The *observability index* of $\{C, A\}$ can be defined in a similar way, as the smallest integer q, say ν, such that

$$\mathcal{O}_q' = [C' \quad A'C' \quad \cdots \quad (A')^{q-1}C'], \qquad 1 \le q \le n$$

has full rank n.

The significance of these quantities will be seen later (e.g., in Sec. 7.2); here the reader may wish to show that we need at most ν derivatives of $y(\cdot)$ to determine x_0 and at most a μth-order delta function input to change the state of a controllable system.

It should be obvious by now that a *duality* can be established in which

> State controllability (observability) of a realization $\{A, B, C\}$ is the same as state observability (controllability) of a realization $\{A', C', B'\}$

A duality can also be built up between solutions (11) and (18) of the observability and controllability problems. We shall not do this here, though we shall study a property of solution (18), consideration of the dual of which will lead us to reexamine (11). However, this discussion may be omitted without loss of continuity.

*Nonuniqueness of Solutions to (18); Minimum-Length Solutions.   We can see from (14) that the solution γ given by (18) is not unique because we can clearly add to it any vector θ such that

$$\mathcal{C}\theta = 0 \tag{19}$$

There will always be such vectors because only n of the np columns of $\mathcal{C}$ are linearly independent. Therefore there are many solutions, but it turns out that the solution γ has a *minimum-length* property

$$\|\gamma + \theta\| \ge \|\gamma\| \tag{20}$$

for all θ satisfying (19). The proof follows by noting that

$$\|\gamma + \theta\|^2 \triangleq (\gamma + \theta)'(\gamma + \theta) = \gamma'\gamma + \theta'\theta + 2\gamma'\theta$$

But by (18) and (19)

$$\gamma'\theta = [x'(0+) - x'(0-)][\mathcal{C}\mathcal{C}']^{-1}\mathcal{C}\theta = 0 \tag{21}$$

so that, as claimed in (20),

$$\|\gamma + \theta\|^2 = \|\gamma\|^2 + \|\theta\|^2 \ge \|\gamma\|^2$$

A geometric interpretation of this result is instructive. We say that

$$\mathcal{C}\theta = 0 \implies \theta \in \mathfrak{N}(\mathcal{C}), \qquad \text{the } \textit{null space of } \mathcal{C}$$

Now $\mathfrak{N}(\mathcal{C})$ is a linear space [i.e., it is closed under addition and under multiplication by (real or complex) scalars], and therefore we can decompose the whole space into two *orthogonal subspaces* as schematically indicated in Fig. 6.2-1. The vectors θ lie in $\mathfrak{N}(\mathcal{C})$. Also we have just shown that $\gamma'\theta = 0$ for all θ in $\mathfrak{N}(\mathcal{C})$ so that γ is orthogonal to $\mathfrak{N}(\mathcal{C})$ and we have shown it in the subspace $\mathfrak{N}^{\perp}(\mathcal{C})$ of all such vectors. The family of all solutions to (14) lies along the hyperplane parallel to $\mathfrak{N}(\mathcal{C})$, and it is easy to see that γ is the unique minimum-length solution.

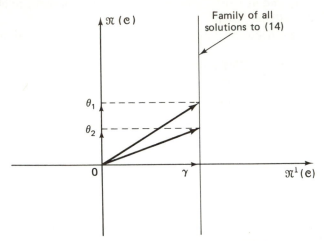

Figure 6.2-1.   Geometry of solutions of (14).

Nonexistence of Solutions to (2); Least-Squares "Solutions".   The special property we have found for the solution of Eq. (14) for the controllability problem should lead us, because of duality if nothing else, to wonder if Eq. (2) for the observability problem also has any special properties. For simplicity let us take $u(\cdot) \equiv 0$ in (2) and examine the equation

$$\mathcal{O}x(t) = \mathcal{Y}(t) \tag{22}$$

Now we proved earlier that a solution of this equation must be unique, so that we do not have a family of solutions as in the controllability problem. However, there is still something special about Eq. (22), namely that there may not be *any* solution for an arbitrary right-hand side. This is because in general we cannot construct an arbitrary $np \times 1$ vector as a linear combination of at most n such $np \times 1$ vectors (the columns of $\mathcal{O}$)! There can only be a solution (unique whenever it exists) if the right-hand side of (22) is in fact some linear combination of the columns of $\mathcal{O}$, i.e., only if the right-hand side belongs to $\mathfrak{R}(\mathcal{O})$, the *range space* of $\mathcal{O}$, in which case the equations are *consistent*.

Least-Squares "Solutions" of Inconsistent Equations.   The reason we did not bring up existence when we first discussed (2) was that we knew the $y(\cdot)$ arose from $x(\cdot)$ and $u(\cdot)$ and therefore the equation had to be consistent.

However, there often are *measurement* errors in recording $y(\cdot)$ or working with $y(\cdot)$, and therefore in the observability problem we shall often have to deal with inconsistent equations. What do we do then? Now (22) will not have any solution, but what is often done is to try to find a vector $\hat{x}(t)$ such that $\Theta\hat{x}(t)$ matches $\mathcal{Y}(t)$ as closely as possible in the *least-squares* sense, viz., to find $\hat{x}(t)$ so that

$$\| \Theta\hat{x}(t) - \mathcal{Y}(t)\|^2 = \text{minimum} \tag{23}$$

It can be shown in several ways (see Exercises 6.2-1 and 6.2-2) that there is a unique $\hat{x}(t)$ with this property if Θ has rank n, namely

$$\hat{x}(t) = (\Theta'\Theta)^{-1}\Theta'\mathcal{Y}(t) \tag{24}$$

Since $\hat{x}(t)$ coincides with the solution when one exists [cf. (11)] and otherwise gives a best least-squares fit, we call

$$\hat{x}(t) \triangleq \text{the least-squares solution of (22)}$$

However, it is important to remember that $\hat{x}(t)$ is not a solution in the ordinary sense; i.e., it does not necessarily satisfy $\Theta x(t) = \mathcal{Y}(t)$.

At this point an alert reader may raise the following question. Why is not $\hat{x}(t)$ always a true solution of (22), since it can always be obtained by the following permissible operations? Multiply both sides of

$$\Theta x(t) = \mathcal{Y}(t) \tag{25}$$

by Θ' to obtain

$$\Theta'\Theta x(t) = \Theta'\mathcal{Y}(t) \tag{26}$$

Now since Θ has full rank, $\Theta'\Theta$ will be nonsingular, and we can obtain $x(t)$ as

$$(\Theta'\Theta)^{-1}\Theta'\mathcal{Y}(t) \triangleq \hat{x}(t)$$

The answer of course is that $\hat{x}(t)$ always satisfies the (consistent) equations (26), but $\hat{x}(t)$ satisfies (25) only when (25) is consistent. A geometric picture will give further insight into the significance of the least-squares solution $\hat{x}(t)$ (cf. Fig. 6.2-2). If the equations (25) are not consistent, $\mathcal{Y}(t)$ will not belong to $\mathcal{R}(\Theta)$, but we can project it onto $\mathcal{R}(\Theta)$ to get a new vector $\hat{\mathcal{Y}}(t)$ in $\mathcal{R}(\Theta)$. Then we can solve the equations (always consistent)

$$\Theta\hat{x}(t) = \hat{\mathcal{Y}}(t)$$

to get the solution

$$\hat{x}(t) = (\Theta'\Theta)^{-1}\Theta'\hat{\mathcal{Y}}(t)$$

Because $\mathcal{Y}(t) - \hat{\mathcal{Y}}(t) \perp \mathcal{R}(\Theta)$, this can also be written as in (24), namely

$$\hat{x}(t) = (\Theta'\Theta)^{-1}\Theta'\mathcal{Y}(t)$$

Actually we could replace $\hat{\mathcal{Y}}(t)$ by any vector, say $\hat{\hat{\mathcal{Y}}}(t)$, in $\mathcal{R}(\Theta)$ and obtain a corresponding "solution" $\hat{\hat{x}}(t)$. In this family of solutions $\hat{x}(t)$ clearly has the *least-*

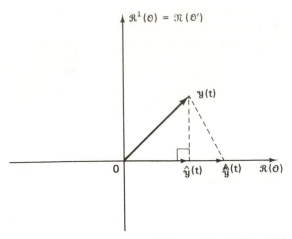

Figure 6.2-2.   The geometry of the equation $\Theta x(t) = \mathcal{Y}(t)$.

squares property that

$$\| \mathcal{Y}(t) - \Theta \hat{x}(t) \|^2 \leq \| \mathcal{Y}(t) - \Theta \overset{\ast}{x}(t) \|^2 \tag{27}$$

Least-squares solutions are widely used in statistics and numerical analysis. Although we shall not pursue these matters here, comparisons (which we leave to the interested reader) between Figs. 6.2-1 and 6.2-2 (and the associated results) may make it seem a little less mysterious that certain quadratic regulator (minimum-length) control problems are *dual* to certain statistical (least-squares) estimation problems (cf. Sec. 4.4). For more detailed discussions of the linear equation theory that we have briefly developed above, refer to the books [3]–[5].

6.2.2   Standard Forms for Noncontrollable/Nonobservable Realizations; Minimal Realizations

We shall say that two realizations are similar to each other (or related by a similarity transformation) if there exists a nonsingular matrix T such that

$$\bar{A} = T^{-1}AT, \qquad \bar{B} = T^{-1}B, \qquad \bar{C} = CT \tag{28}$$

It can be seen that the states of the two realizations can then be related as

$$x(t) = T\bar{x}(t)$$

but that the transfer functions are unaffected,

$$H(s) = C(sI - A)^{-1}B = \bar{C}(sI - \bar{A})^{-1}\bar{B}$$

Furthermore, we have

$$\bar{\Theta} = \Theta T, \qquad \bar{\mathcal{C}} = T^{-1}\mathcal{C} \tag{29}$$

so that the ranks of the observability and controllability matrices are preserved under similarity transformation.

It is therefore natural to look for similarity transformations that will give realizations in special (canonical) forms that may be especially suited to particular applications. For example, suppose we have an arbitrary controllable realization $\{A, B, C\}$ for which we wish to set up an arbitrary state $x(0+)$ at time $t = 0+$ by using a suitable input signal. Then from the discussions of the last section it would seem that we might try first to transform the realization to block controllability form. Unfortunately, this can be done (by a similarity transformation) only if the number of inputs divides the number of states. Therefore the block controllability form, and for similar reasons all the other block companion forms, are not very useful in general, and we shall have to work harder to determine "nice" general forms for controllable and/or observable realizations. Therefore this will be postponed to Sec. 6.4. However, it turns out to be easy to obtain certain general forms for *non-*controllable and/or *non*observable realizations. We shall give these here, along with a very important application (Theorem 6.2-3).

Theorem 6.2-1.   Displaying the Controllable and Noncontrollable States

Suppose $\{A, B, C\}$ is a realization such that

$$\mathcal{C}\{A, B\} \text{ has rank } r \leq n$$

Then we can find a similarity transformation (matrix) T such that

$$\bar{A} = T^{-1}AT, \quad \bar{B} = T^{-1}B, \quad \text{and} \quad \bar{C} = CT$$

have the forms

$$\bar{A} = \begin{bmatrix} \bar{A}_c & \vdots & \bar{A}_{12} \\ \text{---} & \text{---} & \text{---} \\ 0 & \vdots & \bar{A}_{\bar{c}} \end{bmatrix} \begin{smallmatrix} r \\ \\ n-r \end{smallmatrix}, \quad \bar{B} = \begin{bmatrix} \bar{B}_c \\ \text{---} \\ 0 \end{bmatrix} \begin{smallmatrix} r \\ \\ n-r \end{smallmatrix}, \quad \bar{C}' = \begin{bmatrix} \bar{C}'_c \\ \text{---} \\ \bar{C}'_{\bar{c}} \end{bmatrix} \quad (30)$$

where the important feature is the location of the identically zero matrices in $\bar{A}$ and $\bar{B}$. As a consequence of this it also follows that

1. $\{\bar{A}_c, \bar{B}_c\}$ is controllable,
2. The states of $\{\bar{A}, \bar{B}, \bar{C}\}$ can be clearly separated into controllable and non-controllable states, and
3. $\bar{C}_c(sI - \bar{A}_c)^{-1}\bar{B}_c = C(sI - A)^{-1}B$.

Proof.   Since $\mathcal{C}\{A, B\}$ has rank r, we can find r linearly independent columns, $\{f_1, \ldots, f_r\}$ say, that span the nm columns of $\mathcal{C}\{A, B\}$. Let $\{\theta_{r+1}, \ldots, \theta_n\}$ be $n - r$ columns such that

$$T \triangleq [f_1 \quad \cdots \quad f_r \quad \theta_{r+1} \quad \cdots \quad \theta_n]$$

is nonsingular. Then direct calculation (see the discussion in Secs. 2.4.2 and 5.2) will show that $\bar{A}, \bar{B}, \bar{C}$ will have the stated forms. To prove result 1, we first note that

$$\mathcal{C}(\bar{A}, \bar{B}) = \begin{matrix} r \\ n-r \end{matrix}\begin{bmatrix} \bar{B}_c & \bar{A}_c\bar{B}_c & \cdots & \bar{A}_c^{r-1}\bar{B}_c & \cdots & \bar{A}_c^{n-1}\bar{B}_c \\ 0 & 0 & & 0 & & 0 \end{bmatrix}$$

Since $\mathcal{C}(A, B) = \mathcal{C}(\bar{A}, \bar{B})T^{-1}$ has rank r, the first r rows of $\mathcal{C}(\bar{A}, \bar{B})$ must be linearly independent. Also there must be r linearly independent columns of $\mathcal{C}(\bar{A}, \bar{B})$. But, by the Cayley-Hamilton theorem applied to the $r \times r$ matrix $\bar{A}_c$, we see that these must occur in the first rm columns of $\mathcal{C}(\bar{A}, \bar{B})$, which of course proves that $\mathcal{C}(\bar{A}_c, \bar{B}_c)$ will have rank r.

For result 2, we note from the state equations

$$\dot{\bar{x}}_c(t) = \bar{A}_{12}\bar{X}_{\bar{c}}(t) + \bar{B}_c u(t)$$
$$\dot{\bar{x}}_{\bar{c}}(t) = \bar{A}_{\bar{c}}\bar{x}_{\bar{c}}(t) + 0 \cdot u(t)$$

that the states $\bar{x}_{\bar{c}}(\cdot)$ cannot be affected by the input and can therefore be said to be noncontrollable. On the other hand, by result 1, the states $\bar{x}_c(t)$ can be said to be controllable. Clearly also the eigenvalues of $\bar{A}_{\bar{c}}$ may be said to specify the uncontrollable modes of the realization $\{\bar{A}, \bar{B}\}$.

Finally, we can obtain result 3 by taking transforms of the above state equations for $\bar{x}(0) = 0$. Then $\bar{x}_{\bar{c}}(\cdot) \equiv 0$, and result 3 follows easily. Alternatively, result 3 could be obtained by direct algebraic manipulation of $\bar{C}(sI - \bar{A})^{-1}\bar{B}$. ∎

Similarly, we can always find a choice of state variables in which observable and nonobservable states can be clearly identified.

Theorem 6.2-2.   Displaying the Observable and Nonobservable States

Suppose $\{A, B, C\}$ is a realization such that

$$\mathcal{O}\{C, A\} \text{ has rank } r \leq n$$

Then we can find a similarity transformation matrix T such that

$$\bar{A} = T^{-1}AT, \quad \bar{B} = T^{-1}B, \quad \text{and} \quad \bar{C} = CT$$

have the forms

$$\bar{A} = \begin{bmatrix} \bar{A}_o & 0 \\ \bar{A}_{21} & \bar{A}_{\bar{o}} \end{bmatrix}\begin{matrix} r \\ n-r \end{matrix}, \quad \bar{B} = \begin{bmatrix} \bar{B}_o \\ \bar{B}_{\bar{o}} \end{bmatrix}, \quad \bar{C}' = \begin{bmatrix} \bar{C}'_o \\ 0 \end{bmatrix} \tag{31}$$

where

1. $\{\bar{C}_o, \bar{A}_o\}$ is observable,
2. The states of $\{\bar{A}, \bar{B}, \bar{C}\}$ can be clearly separated into observable and nonobservable states, and
3. $\bar{C}_o(sI - \bar{A}_o)^{-1}\bar{B}_o = C(sI - A)^{-1}B$.

Proof.   This can be done by direct specification of a suitable T (using rows instead of columns this time) or by duality. We leave the proof to the reader as a useful review exercise. ∎

The above results can be put together to obtain a decomposition of any realization into four parts that clearly have the properties of joint controllability and observability, controllability but not observability, observability but not controllability, and neither controllability nor observability. Since the form is exactly the same as in the scalar case [Eqs. 2.4-18], we shall not restate the results here.

One of the most important applications of Theorems 6.2-1 and 6.2-2 is the following theorem, which gives us a useful characterization of minimal realizations.

A *minimal realization* is one that has the smallest-size A matrix for all triples $\{A, B, C\}$ satisfying

$$C(sI - A)^{-1}B = H(s), \qquad \text{a given transfer function}$$

Theorem 6.2-3.

A realization $\{A, B, C\}$ is minimal if and only if it is controllable and observable.

Proof.   We shall first establish the *only if* part by showing that a noncontrollable or nonobservable realization cannot be minimal. If $\{A, B\}$ were noncontrollable, say, then by Theorem 6.2-1 we could obtain another realization $\{\bar{A}_c, \bar{B}_c, \bar{C}_c\}$ with the same transfer function but a smaller number of states. Therefore $\{A, B, C\}$ could not have been minimal.

For the *if* part, we have to prove that a controllable and observable realization $\{A, B, C\}$ must be minimal. If it is not, let $\{\bar{A}, \bar{B}, \bar{C}\}$ be another controllable and observable realization with a lower state dimension, $\bar{n} < n$. Now since

$$CA^iB = \bar{C}\bar{A}^i\bar{B}, \qquad \text{all } i$$

note that

$$\mathcal{O}\mathcal{C} = \bar{\mathcal{O}}_{n-1}\bar{\mathcal{C}}_{n-1} \tag{32}$$

where (note that the last power of $\bar{A}$ is $n - 1$, not $\bar{n} - 1$)

$$\bar{\mathcal{C}}_{n-1} \triangleq [\bar{B} \quad \bar{A}\bar{B} \quad \cdots \quad \bar{A}^{n-1}\bar{B}]$$

and similarly for $\bar{\mathcal{O}}_{n-1}$. Next we note that from Sylvester's inequality (Exercise A.18) we have for the rank of $\mathcal{O}\mathcal{C}$ the bounds

$$\rho(\mathcal{O}) + \rho(\mathcal{C}) - n \leq \rho(\mathcal{O}\mathcal{C}) \leq \min\{\rho(\mathcal{O}), \rho(\mathcal{C})\} \tag{33}$$

Since $\mathcal{O}$ and $\mathcal{C}$ have rank n by hypothesis, we see that

$$\rho(\mathcal{O}\mathcal{C}) = n$$

But if $\{\bar{A}, \bar{B}, \bar{C}\}$ is also controllable and observable, it must follow similarly that

$$\rho(\bar{\mathcal{O}}_{n-1}\bar{\mathcal{C}}_{n-1}) = \bar{n}$$

But by (32) this would imply that $\bar{n} = n$, which is a contradiction. Therefore, if $\{A, B, C\}$ is controllable and observable, there can be no lower-dimensional realization that is also controllable and observable. This completes the proof of the theorem. ∎

An important property of minimal realizations is that they can all be related to one another by a unique similarity transformation.

Theorem 6.2-4.

If $\{A_i, B_i, C_i, i = 1, 2\}$ are two minimal realizations of a transfer function, there exists a unique invertible matrix T such that

$$A_2 = T^{-1}A_1T, \qquad B_2 = T^{-1}B_1, \qquad C_2 = C_1T \tag{34a}$$

Furthermore, T can be specified as

$$T = \mathcal{C}_1\mathcal{C}_2'(\mathcal{C}_2\mathcal{C}_2')^{-1} \tag{34b}$$

or

$$T^{-1} = (\mathcal{O}_2'\mathcal{O}_2)^{-1}\mathcal{O}_2'\mathcal{O}_1 \tag{34c}$$

Proof. Since the realizations are minimal, they are controllable and observable (by the previous theorem), and therefore

$$\rho(\mathcal{O}_i) = n = \rho(\mathcal{C}_i), \qquad i = 1, 2$$

and

$$\mathcal{O}_i'\mathcal{O}_i \text{ and } \mathcal{C}_i\mathcal{C}_i' \text{ are nonsingular}$$

so that from the equality

$$\mathcal{O}_1\mathcal{C}_1 = \mathcal{O}_2\mathcal{C}_2 \tag{35}$$

we can conclude that, say,

$$\mathcal{C}_2 = (\mathcal{O}_2'\mathcal{O}_2)^{-1}\mathcal{O}_2'\mathcal{O}_1\mathcal{C}_1 = T_1\mathcal{C}_1$$

and

$$\mathcal{O}_2 = \mathcal{O}_1\mathcal{C}_1\mathcal{C}_2'(\mathcal{C}_2\mathcal{C}_2')^{-1} = \mathcal{O}_1T$$

Since $\mathcal{C}_1$ and $\mathcal{C}_2$ have full rank, T_1 must be nonsingular and similarly for T. In fact,

$$T_1 = T^{-1}$$

since

$$T_1T = (\mathcal{O}_2'\mathcal{O}_2)^{-1}\mathcal{O}_2'\underbrace{\mathcal{O}_1\mathcal{C}_1}\mathcal{C}_2'(\mathcal{C}_2\mathcal{C}_2')^{-1}$$

$$= (\mathcal{O}_2'\mathcal{O}_2)^{-1}\underbrace{\mathcal{O}_2'\mathcal{O}_2}\underbrace{\mathcal{C}_2\mathcal{C}_2'}(\mathcal{C}_2\mathcal{C}_2')^{-1} = I \cdot I$$

Therefore,

$$\mathcal{C}_2 = T^{-1}\mathcal{C}_1, \qquad \mathcal{O}_2 = \mathcal{O}_1T$$

from which, by equating the first columns and rows, we obtain

$$B_2 = T^{-1}B_1, \qquad C_2 = C_1 T$$

To find the relation between A_1 and A_2, we start from the easily verified equality

$$\mathcal{O}_1 A_1 \mathcal{C}_1 = \mathcal{O}_2 A_2 \mathcal{C}_2 \qquad (36)$$

to obtain

$$A_2 = (\mathcal{O}_2'\mathcal{O}_2)^{-1}\mathcal{O}_2'\mathcal{O}_1 A_1 \mathcal{C}_1 \mathcal{C}_2'(\mathcal{C}_2\mathcal{C}_2')^{-1} = T^{-1}A_1 T$$

Finally for uniqueness, we note that if $\tilde{T}$ were another matrix relating the $\{A_i, B_i, C_i\}$, then

$$\mathcal{O}_1(T - \tilde{T}) = 0.$$

But since $\mathcal{O}_1$ has full rank this can only hold if $T \equiv \tilde{T}$. ∎

Theorems 6.2-3 and 6.2-4 are the multivariable forms of results obtained in Sec. 2.4-1 (cf. Fig. 2.4-2). In the scalar case we also connected the concept of minimality and simultaneous controllability and observability to the notion of irreducible transfer functions. The multivariable version of that connection will be treated in Sec. 6.2.3. Before going on to that, we note a previously promised application to the Gilbert realization procedure.

Minimality of the Gilbert Realization.   We shall verify controllability and observability of the Gilbert realization. The controllability matrix for [cf. (6.1-3)–(6.1-4)]

$$A = \text{diag}\,\{\lambda_i I_{\rho_i}\}, \qquad B' = [B_1', \ldots, B_r'], \qquad \lambda_i \text{ distinct}$$

can be written as

$$\mathcal{C} = \begin{bmatrix} B_1 & & \\ & B_2 & \bigcirc \\ & & \ddots \\ \bigcirc & & B_r \end{bmatrix} \begin{bmatrix} I_m & \lambda_1 I_m & \cdots & \lambda_1^{n-1}I_m \\ \vdots & & & \vdots \\ I_m & \lambda_r I_m & \cdots & \lambda_r^{n-1}I_m \end{bmatrix} = \mathcal{B}\mathcal{U} \qquad (37)$$

Now $\mathcal{U}$ is a *block Vandermonde* matrix,[†] which is of full rank as long as $\lambda_i \neq \lambda_j$ (cf. Exercise A.7), as we assume in the Gilbert procedure. Therefore $\mathcal{C}$ will be full rank if and only if $\mathcal{B}$ has full rank. But clearly, by the Gilbert

---

†The matrix is actually $V \otimes I_m$, the Kronecker product of the ordinary Vandermonde matrix V and I_m, for which $\det V \otimes I_m = (\det V)^m$—see Exercise A.41.

construction,

$$\rho(\mathcal{B}) = \sum_1^r \rho(B_i) = \sum_1^r \rho_i \triangleq n$$

so that the realization is controllable. The observability can be similarly proved.

The PBH Tests for Controllability and Observability. In Sec. 6.2.1, we showed that controllability and observability could be checked by determining the rank of certain rectangular controllability and observability matrices. In the scalar case, we found that the PBH tests of Sec. 2.4.3 were often more useful for analytical purposes, and this turns out to be true in the matrix case as well.

The reader can profitably review Sec. 2.4.3 at this point, since our results will be fairly obvious generalizations of the scalar results. For the same reason, we shall omit the proofs, leaving them as a useful review for the reader.

Theorem 6.2-5.   The PBH Eigenvector Tests

1. A pair $\{A, B\}$ will be controllable if and only if there exists no left eigenvector of A that is orthogonal to all the columns of B, i.e., if and only if

$$p'A = \lambda p' \qquad p'B = 0 \Longrightarrow p \equiv 0 \tag{38}$$

2. A pair $\{C, A\}$ will be observable if and only if there exists no eigenvector of A that is orthogonal to all the rows of C, i.e., if and only if

$$Ap = \lambda p, \qquad Cp = 0 \Longrightarrow p \equiv 0 \tag{39}$$

Theorem 6.2-6.   The PBH Rank Tests

1. A pair $\{A, B\}$ will be controllable if and only if the matrix

$$[sI - A \quad B] \text{ has rank } n \text{ for } all \text{ } s \tag{40}$$

2. A pair $\{C, A\}$ will be observable if and only if the matrix

$$\begin{bmatrix} C \\ sI - A \end{bmatrix} \text{ has rank } n \text{ for } all \text{ } s \tag{41}$$

In the theory of matrix polynomials, matrices $sI - A$ and B obeying condition (40) are said to be *relatively left prime* (or *left coprime*), while matrices C and $sI - A$ obeying (41) are said to be *relatively right prime* (or *right coprime*). (See Sec. 6.3.1.)

We shall next indicate how the coprimeness notions are useful in connecting the concepts of controllability, observability, and minimality to that of *irreducible* matrix transfer function descriptions.

6.2.3 Matrix-Fraction Descriptions

In Sec. 6.1, we wrote

$$H(s) = \frac{N(s)}{d(s)}$$

but found that, except when $N(s)$ was scalar, the order of any obvious realization was greater than the degree, r, of $d(s)$. For block controller-controllability forms, the order was $n = rm$, while for block observer-observability forms, the order was $n = rp$.

However, now notice that if we write $H(s)$ as a *matrix fraction*,

$$H(s) = N_R(s)D_R^{-1}(s), \qquad D_R(s) = d(s)I_m, \qquad N_R(s) = N(s) \qquad (42)$$

and define the *degree* of the *denominator* matrix as

$$\deg D_R(s) \triangleq \deg \det D_R(s) = rm$$

then the order of the block controller realization of Sec. 6.1 would be equal to rm, the degree of $D_R(s)$.

Similarly, if we write

$$H(s) = D_L^{-1}(s)N_L(s), \qquad D_L(s) = d(s)I_p, \qquad N_L(s) = N(s) \qquad (43)$$

then the degree of the *denominator* matrix is

$$\deg D_L(s) \triangleq \deg \det D_L(s) = rp$$

and we can easily associate with $H(s)$ in this form a block observer realization of order equal to this degree. This rewriting of $H(s)$ already brings us closer to the scalar case, except that because matrices are involved we have to pay attention to the order in which they are written, so that we have *right* and *left* denominators.

There can be (see Example 6.2-1) many right and left matrix-fraction descriptions (MFDs) of $H(s)$. For any of these descriptions, we shall be able to obtain similar results. Thus in Sec. 6.4 we shall establish the following.

Claim A: Given *any* right MFD of $H(s)$,

$$H(s) = N_R(s)D_R^{-1}(s)$$

we can always obtain a controllable state-space realization $\{A, B, C\}$ of order

$$n = \deg \det D_R(s) \triangleq \text{the degree of the MFD}$$

Claim B: Given any left MFD of $H(s)$,

$$H(s) = D_L^{-1}(s)N_L(s)$$

we can always obtain an observable state-space realization $\{A, B, C\}$ of order

$$n = \deg \det D_L(s) \triangleq \text{the degree of the MFD}$$

Moreover, we shall show in Sec. 6.5 that the minimal determinantal degree of the denominators of all (right or left) MFDs of $H(s)$ is also the minimal order of any state-space realization of $H(s)$.

The above facts, and the other results to be described later, show that matrix-fraction descriptions provide a natural generalization of the scalar rational function representation of $H(s)$, though in the multivariable case we have to distinguish between right and left descriptions. However, from Claims A and B, we may note that there is a certain *duality* in these descriptions. Therefore, in the sequel we shall for the most part give detailed derivations only for right MFDs and controllable state-space realizations. *Furthermore, in view of this, we shall often omit the subscript R, restoring it only when necessary for emphasis or clarity.*

Example 6.2-1.   Alternative MFDs for a Transfer Function

Let us consider the transfer function

$$H(s) = \begin{bmatrix} \dfrac{s}{(s+1)^2(s+2)^2} & \dfrac{s}{(s+2)^2} \\ \dfrac{-s}{(s+2)^2} & \dfrac{-s}{(s+2)^2} \end{bmatrix} \tag{44}$$

for which one MFD is

$$H(s) = \begin{bmatrix} s & s(s+1)^2 \\ -s(s+1)^2 & -s(s+1)^2 \end{bmatrix} \begin{bmatrix} (s+1)^2(s+2)^2 & 0 \\ 0 & (s+1)^2(s+2)^2 \end{bmatrix}^{-1}$$

$$= N_1(s) D_1^{-1}(s) \tag{45}$$

where

$$\deg \det D_1(s) = 8$$

We can obtain another MFD as follows. We multiply the first column of $H(s)$ by the least common multiple (lcm) of the denominators of the entries in this first column, and similarly for the second column. Doing this, we obtain

$$H(s) \underbrace{\begin{bmatrix} (s+1)^2(s+2)^2 & 0 \\ 0 & (s+2)^2 \end{bmatrix}}_{D_2(s)} = \underbrace{\begin{bmatrix} s & s \\ -s(s+1)^2 & -s \end{bmatrix}}_{N_2(s)} \tag{46}$$

or

$$H(s) = N_2(s) D_2^{-1}(s)$$

where now

$$\deg \det D_2(s) = 6 < 8 = \deg \det D_1(s)$$

However, it is possible that we can obtain an MFD with a denominator matrix of even lower degree. Thus we can verify that

$$H(s) = \begin{bmatrix} s & 0 \\ -s(s+1)^2 & s^2 \end{bmatrix} \begin{bmatrix} (s+1)^2(s+2)^2 & -(s+1)^2(s+2) \\ 0 & s+2 \end{bmatrix}^{-1}$$

$$= N_3(s)D_3^{-1}(s) \tag{47}$$

is another MFD with

$$\deg \det D_3(s) = 5$$

And yet another is

$$H(s) = \begin{bmatrix} s & 0 \\ -s & s^2 \end{bmatrix} \begin{bmatrix} 0 & -(s+1)^2(s+2) \\ (s+2)^2 & s+2 \end{bmatrix}^{-1}$$

$$= N_4(s)D_4^{-1}(s) \tag{48}$$

with

$$\deg \det D_4(s) = 5$$

We shall prove later that 5 is the minimal degree of any matrix denominator of $H(s)$ and also the minimal order of any state-space realization of $H(s)$. ∎

Right Divisors and Irreducible MFDs.   To further pursue the analysis of Example 6.2-1, we note that given an MFD, an infinity of others can be obtained by choosing any nonsingular† polynomial matrix $W(s)$ such that

$$\bar{N}(s) = N(s)W^{-1}(s), \qquad \bar{D}(s) = D(s)W^{-1}(s)$$

are polynomial matrices, for then

$$H(s) = N(s)D^{-1}(s) = \bar{N}(s)\bar{D}^{-1}(s)$$

Since

$$N(s) = \bar{N}(s)W(s), \qquad D(s) = \bar{D}(s)W(s)$$

we can call $W(s)$ a *right divisor* of $N(s)$ and $D(s)$. Moreover, since

$$\deg \det D(s) = \deg \det \bar{D}(s) + \deg \det W(s) \tag{49}$$

we shall have

$$\deg \det D(s) \geq \deg \det \bar{D}(s) \tag{50}$$

In other words, the degree of the MFD (i.e., the degree of the determinant of

---

†We repeat a fact already implicit in (42) and (43) that a polynomial matrix is nonsingular if det $D(s)$ is not identically zero; this means that $D(s)$ will be nonsingular for almost all values of s [all except those that make det $D(s) = 0$].

the denominator matrix) can be reduced by removing right divisors of the numerator and denominator matrices. Therefore, we should expect that we would get a minimum-degree MFD by extracting a *greatest common right divisor* (gcrd) of $N(s)$ and $D(s)$. The concept of a gcrd can be defined in several ways,† but we can see from (49) that the concept must be related to the question of equality in (50). That is, we shall have extracted a gcrd from $N(s)$ and $D(s)$ if and only if

$$\deg \det D(s) = \deg \det \bar{D}(s)$$

for *all* nonsingular right divisors $W(s)$ of $N(s)$ and $D(s)$. This equality will clearly hold if and only if all $W(s)$ have the property

$$\det W(s) = \text{a nonzero constant, independent of } s \qquad (51)$$

Such matrices are said to be *unimodular* and will be studied in more detail below.

With the above definitions, several useful statements follow naturally:

1. Two polynomial matrices $N(s)$ and $D(s)$ with the same number of columns will be said to be *relatively right prime* (or *right coprime*) if they only have unimodular common right divisors.
2. An MFD $H(s) = N(s)D^{-1}(s)$ will be said to be *irreducible* if $N(s)$ and $D(s)$ are right coprime.
3. Irreducible MFDs are not unique, because if $N(s)D^{-1}(s)$ is irreducible, so is $N(s)W(s)[D(s)W(s)]^{-1}$ for any unimodular $W(s)$.
   (Therefore gcrds are also not unique, which is why we said *a* gcrd rather than *the* gcrd on several occasions above.)

Similar statements, with obvious changes, can be made for *left* MFDs, and we shall not give them here.

From the above results, one would expect that the minimal order of any state-space realization of a given transfer function will be equal to the degree of any irreducible right or left MFD of the transfer function. A formal proof of this important result will be given in Sec. 6.5.1.

It should be clear that we should set out toward that goal by first developing some properties of polynomial matrices, and this will be the subject of Sec. 6.3.

Exercises

6.2-1.

Consider the equation $\Theta x(t) = \mathcal{Y}(t)$, where Θ is $np \times n$ and of full rank but $\mathcal{Y}(t)$ is not in the range of Θ. Use ordinary calculus (differentiation) to find $\hat{x}(t)$ such that $\| \Theta \hat{x}(t) - \mathcal{Y}(t) \|^2 = \text{minimum}$.

†See Sec. 6.3.1.

6.2-2.

Use the same notations as in Exercise 6.2-1, and let $x_0(t) = (\Theta'\Theta)^{-1}\Theta'\mathcal{Y}(t)$, $\Delta = x(t) - x_0(t)$. Evaluate $(\Theta\Delta)'[\Theta x_0(t) - \mathcal{Y}(t)]$, and use the result to show that $\|\Theta x(t) - \mathcal{Y}(t)\|^2 \geq \|\Theta x_0(t) - \mathcal{Y}(t)\|^2$ and hence that $x_0(t) = \hat{x}(t)$. Give a geometric interpretation of this argument.

6.2-3.

With the same notations as in Exercise 6.2-1, let T be any $np \times np$ matrix such that $\Theta'T\Theta$ is nonsingular.

a. Show that if the equations $\Theta x(t) = \mathcal{Y}(t)$ are consistent, then $(\Theta'T\Theta)^{-1}\Theta'T\mathcal{Y}(t)$ is independent of T.

b. Is this true when the equations are not consistent?

6.2-4.

Prove Theorems 6.2-5 and 6.2-6.

6.2-5.

a. Given any polynomial n-vector $X_0(s)$ with elements of degree $n - 1$ or less, show that we can find polynomial vectors $X(s)$ and $U(s)$ such that $(sI - A)X(s) - BU(s) = X_0(s)$ if and only if $\{A, B\}$ is controllable. Show that the elements of $X(s)$ need have degree no higher than $n - 2$, while those of $U(s)$ need have degree no higher than $n - 1$. (*Remark:* Actually we can replace n by q in these restrictions on degree, where q is any integer not less than the degree of the minimal polynomial of A.)

b. Use the result of part a to show that if and only if $\{A, B\}$ is controllable will there exist an input $u(\cdot)$ that will take an arbitrary initial state $x(0-) = x_0$ to the origin.

6.2-6.

Suppose A_c is a companion matrix with the coefficients of the characteristic polynomial $a(s)$ in the top row. Show that $\{C, A_c\}$ will be observable if and only if the polynomials $a(s) = \det(sI - A_c)$ and $\{c_i(s) = \sum_{j=1}^n c_{ij}s^{n-j}, i = 1, \ldots, p\}$ have no common zero.

6.2-7.

Prove that $\{A, B\}$ is controllable if and only if $\{A - BK, B\}$ is controllable for all $m \times n$ matrices K.

6.2-8.

Refer to (6.1-5), and let $n_i(s) = n_{i1}s^{n-1} + \cdots + n_{in}$, $i = 1, \ldots, p$, and $d(s) = s^n + d_1s^{n-1} + \cdots + d_n$.

a. Show that $\{d(s), n_1(s), \ldots, n_p(s)\}$ will be relatively prime, i.e., have no common roots, if and only if the matrix Q has rank n, where A_c is a companion matrix with first row $-[d_1 \cdots d_n]$ and Q is a block matrix with ith row $n_i(A_c)$, $i = 1, \ldots, p$.

b. Rewrite Q as an observability matrix of a certain pair of matrices.

6.2-9.

Let $N(s) = N_0s^r + N_1s^{r-1} + \cdots + N_r$. Show that the block controller realization of $H(s) = N(s)/s^r$ will be observable if N_r has full rank.

6.2-10.

Suppose

$$A = \begin{bmatrix} 1 & 0 & 0 \\ 0 & -1 & 0 \\ 0 & 0 & -3 \end{bmatrix}, \quad B = \begin{bmatrix} 0 \\ 1 \\ 1 \end{bmatrix}, \quad C' = \begin{bmatrix} 1 & 0 \\ -1 & 2 \\ 0 & 1 \end{bmatrix}$$

a. Is the realization controllable? Observable?

b. Try to write $C(sI - A)^{-1}B$ as $D_L^{-1}(s)N_L(s)$ and $N_R(s)D_R^{-1}(s)$, where $D_L(s)$, $D_R(s)$ have determinants of degree 3.

c. Repeat part b if the determinants are to be of degree 2.

6.3   SOME PROPERTIES OF POLYNOMIAL MATRICES†

The basic mathematical theory of polynomial matrices is drawn from Mac-Duffee [6], Gantmakher [7], Wedderburn [8], and Newman [9]. The application to systems problems seems to have first been made by Belevitch [10], Popov [11], Rosenbrock (cf. [12] and the references therein), Forney [13], and Wolovich (cf. [14] and the references therein); in so doing, some new mathematical results have also been added by these authors and others, see especially Forney [15].

Many of the results to be described here were first obtained [e.g., the Smith form (1861)] for matrices with integer entries; it was only realized much later that they also held for entries that are polynomials (see also the discussion in Sec. 4.5.1) and more generally for entries drawn from any principal ideal domain (cf. [9]). While this generality can be useful in systems problems (see, e.g., [13], [16], and [17]), in our presentation we shall only consider matrices with entries that are polynomials with coefficients that are real or complex numbers. (Readers wishing to go further will find that generally only small changes in language are necessary to accommodate other classes of coefficients, though certain results just do not extend, e.g., Lemma 6.3-6. We shall say no more about such matters here, except to mention [6] and [9] as suitable references.)

In Sec. 6.3.1, we shall introduce some basic facts about polynomial matrices, especially the Hermite form and several results on greatest common divisors and coprimeness (cf. Sec. 6.2.3). In Sec. 6.3.2 we shall introduce the notion of row- and column-reduced matrices, which, as we shall see in Sec. 6.4, will be important in realization problems. Then in Sec. 6.3.3 we shall discuss the Smith canonical form and some related facts. Finally, in Sec. 6.3.4 we shall show how to transform polynomial matrices to linear polyno-

---

†A first reading of this section should be a quick one, with a closer look at relevant topics as they arise in later sections.

mial form, for which certain important results of Kronecker are available. This will then suggest the need for further examination of the behavior at $s = \infty$, which will in turn point to the study of rational matrices, which we shall take up in Sec. 6.5.

6.3.1   Unimodular Matrices; the Hermite Form and Coprime Polynomial Matrices

A polynomial matrix $P(s)$ is said to be nonsingular if $\det P(s) \neq 0$. For example,

$$Q(s) = \begin{bmatrix} s + 1 & s + 3 \\ s^2 + 3s + 2 & s^2 + 5s + 4 \end{bmatrix}$$

is nonsingular because

$$\det Q(s) = (s + 1)(s^2 + 5s + 4) - (s + 3)(s^2 + 3s + 2) = -2s - 2$$

is not identically zero (it is zero only when $s = -1$). However, the matrix

$$P(s) = \begin{bmatrix} s + 1 & s + 3 \\ s^2 + 3s + 2 & s^2 + 5s + 6 \end{bmatrix}$$

is *singular* because

$$\det P(s) = (s + 1)(s^2 + 5s + 6) - (s + 3)(s^2 + 3s + 2) \equiv 0$$

We would expect a singular matrix to have dependent rows and columns, and let us examine this for the matrix $P(s)$. The columns of $P(s)$, say $p_1(s)$ and $p_2(s)$, appear to be linearly independent because it does not seem that they are proportional to each other. But this is so only if we restrict ourselves to using real or complex numbers as the proportionality constants. However, the linear algebraic notions of dependence are more general, and, as noted at the end of Sec. 5.2, they allow us to consider situations in which the *scalars* can be members of any *field*. Now rational fractions form a field, and we can see that with coefficients drawn from this field we can write

$$p_2(s) = \frac{s + 3}{s + 1} p_1(s)$$

so that $p_2(s)$ is proportional to $p_1(s)$, consistent with the singularity of $P(s)$. We can say that $P(s)$ has rank 1. Note that when $s = -1$, $Q(-1)$ is singular and has rank 1. But this is the only value of s for which the rank is not 1, and therefore we may say that the *normal* rank of $Q(s)$ is 2. However, the adjective *normal* will often not be explicitly used. In general a polynomial matrix has rank r if r is the largest of the orders of the minors that are not identically zero.

Therefore in talking about linear-dependence properties of polynomial matrices (e.g., their column ranks or row ranks or their range spaces or null-spaces), we can use the field of rational functions in carrying over the basic facts familiar to us for Euclidean 3-space or n-space. However, a slight simplification can be obtained here. Note that

$$p_1(s) - \frac{s+3}{s+1}p_2(s) = 0$$

can be rewritten as

$$(s+1)p_1(s) - (s+3)p_2(s) = 0$$

This leads to the useful fact that

> Polynomial vectors are dependent over the field of rational functions if and only if they can be made dependent using only polynomial coefficients

for if

$$\sum_1^n r_i(s)p_i(s) = 0$$

multiplication by the least common multiple of the denominators of the $\{r_i(s)\}$ will show that the $\{p_i(s)\}$ are linearly dependent with polynomial coefficients.

With this fact in mind, we see that the *elementary row and column operations* for polynomial matrices will have the form

1. Interchange of any two columns (or rows),
2. Addition to any column (row) of a polynomial multiple of any other column (row), and
3. Scaling any column (row) by any nonzero real or complex number.

These elementary operations can be represented by *elementary matrices*, postmultiplication by which correspond to elementary column operations, while premultiplication yields elementary row operations. Some examples are

$$\begin{bmatrix} 1 & 0 & 0 \\ 0 & 0 & 1 \\ 0 & 1 & 0 \end{bmatrix}, \quad \begin{bmatrix} 1 & 0 & 0 \\ \alpha(s) & 1 & 0 \\ 0 & 0 & 1 \end{bmatrix}, \quad \begin{bmatrix} 1 & 0 & 0 \\ 0 & 3 & 0 \\ 0 & 0 & 1 \end{bmatrix}$$

The reader should check that, as expected, the inverses of elementary matrices are also elementary matrices. Notice also that the elementary matrices are strongly nonsingular in that their determinants are independent of s (and thus nonzero for *all* values of *s*, not just for almost all *s* as with just "normally" nonsingular matrices). This is a useful property, which can also apply to nonelementary matrices, e.g., to

$$P(s) = \begin{bmatrix} s+1 & s+3 \\ s+2 & s+4 \end{bmatrix}$$

A nonsingular polynomial matrix whose determinant is not a function of *s* is called *unimodular*.

Lemma 6.3-1.   Characterization of Unimodular Matrices

A polynomial matrix $W(s)$ is unimodular if and only if its inverse $W^{-1}(s)$ is also polynomial.

Proof. If $W(s)$ is unimodular, det $W(s)$ is a scalar (constant), and $W^{-1}(s) =$ [Adj $W(s)$]/det $W(s)$ is clearly polynomial. Conversely, if $W(s)$ and $W^{-1}(s)$ are both polynomial matrices, let det $W(s) = a_1(s)$, det $W^{-1}(s) = a_2(s)$. Clearly, $a_1(s)$ and $a_2(s)$ will also be polynomials, and moreover $a_1(s)a_2(s) = 1$. This can only happen if $a_1(s)$ and $a_2(s)$ are both scalars. Therefore $W(s)$ must be unimodular. ∎

Hermite Forms Via Row or Column Operations.   By elementary operations we can convert polynomial matrices to several "standard" forms. Here we shall describe the *Hermite forms*, which are obtained by using only row (or only column) operations. Others will be described later—the *Popov echelon form* in Sec. 6.7.2 and, when both row and column operations are allowed, the *Smith form* below (Theorem 6.3-16).

Theorem 6.3-2.   Column Hermite Form [6]

Any $p \times m$ polynomial matrix of rank *r* can be reduced by elementary row operations (i.e., by premultiplication by a unimodular matrix) to a (lower or upper) *quasi-triangular* form in which

1. If $p > r$, the last $p - r$ rows are identically zero;
2. In column *j*, $1 \leq j \leq r$, the diagonal element is monic and of higher degree than any (nonzero) element above it;
3. In column *j*, $1 \leq j \leq r$, if the diagonal element is unity, then all elements above it are zero; and
4. If $m > r$, no particular statements can be made about the elements in the last $m - r$ columns and first *r* rows.

Remark 1.   By interchanging the roles of rows and columns, one can obtain a similar *row-Hermite form* the details of which we shall not spell out (just interchange *rows* and *columns* and replace *pre* by *post* in the statement of the theorem).

Remark 2.   When the matrix is square, we can define a unique Hermite form in the sense that $P(s)$ and $U(s)P(s)$, where $U(s)$ is any unimodular matrix, will have the same Hermite form. This will be proved in Sec. 6.7.1.

Proof.   We use the following obvious construction. By a row interchange, bring to the (1, 1) position the element of lowest degree in the first column. Call it $\tilde{p}_{11}(s)$.

Now by the Euclidean (division) algorithm† every other element in this column can be written as a multiple of $\tilde{p}_{11}(s)$ plus a remainder of lower degree than $\tilde{p}_{11}(s)$. Now by elementary row operations, we can subtract from every entry the appropriate multiple of $\tilde{p}_{11}(s)$ so as to leave only remainders of lower degree than $\tilde{p}_{11}(s)$. Now repeat the operation with a new (1, 1) element of lower degree than $\tilde{p}_{11}(s)$ and continue until all the elements in the first column except the (1, 1) element are zero. (We assume, of course, that in the original matrix the first column is not identically zero.)

Now consider the second column of the resulting matrix and, temporarily ignoring the first row, repeat the above procedure until all the entries below the (2, 2) element are zero. If the (1, 2) entry does not have lower degree than the (2, 2) entry, the division algorithm and an elementary row operation can be used to replace the (1, 2) entry by a polynomial of lower degree than the diagonal or (2, 2) entry. Continuing this procedure with the third column, fourth column, and so on finally gives the desired Hermite form. ∎

Example 6.3-1. Reduction to Hermite Form

The following steps are self-explanatory:

$$\begin{bmatrix} s^2 & 0 \\ 0 & s^2 \\ 1 & s+1 \end{bmatrix} \longrightarrow \begin{bmatrix} 1 & s+1 \\ 0 & s^2 \\ s^2 & 0 \end{bmatrix} \longrightarrow \begin{bmatrix} 1 & s+1 \\ 0 & s^2 \\ 0 & -s^2(s+1) \end{bmatrix} \longrightarrow \begin{bmatrix} 1 & s+1 \\ 0 & s^2 \\ 0 & 0 \end{bmatrix}$$

The corresponding unimodular matrix is (notice the order)

$$\begin{bmatrix} 1 & 0 & 0 \\ 0 & 1 & 0 \\ 0 & s+1 & 1 \end{bmatrix} \begin{bmatrix} 1 & 0 & 0 \\ 0 & 1 & 0 \\ -s^2 & 0 & 1 \end{bmatrix} \begin{bmatrix} 0 & 0 & 1 \\ 0 & 1 & 0 \\ 1 & 0 & 0 \end{bmatrix} = \begin{bmatrix} 0 & 0 & 1 \\ 0 & 1 & 0 \\ 1 & s+1 & s^2 \end{bmatrix}$$

Greatest Common Divisors. A greatest common right divisor (gcrd) of two matrices $\{N(s), D(s)\}$ with the same number of columns is any matrix $R(s)$ with the following properties:

1. $R(s)$ is a right divisor of $N(s)$ and $D(s)$; i.e., there exist polynomial matrices $\bar{N}(s)$, $\bar{D}(s)$ such that

$$N(s) = \bar{N}(s)R(s), \qquad D(s) = \bar{D}(s)R(s)$$

2. If $R_1(s)$ is any other right divisor of $N(s)$ and $D(s)$, then $R_1(s)$ is a right divisor of $R(s)$; i.e., there exists a polynomial matrix $W(s)$ such that $R(s) = W(s)R_1(s)$.

---

†As the reader may recall from elementary algebra (see also Sec. 2.4.4), this says that given any two polynomials $\{a(s), b(s)\}$ there exist *unique* polynomials $\{q(s), r(s)\}$ such that $a(s) = q(s)b(s) + r(s)$, deg $r(s) <$ deg $b(s)$. (It is possible to reduce a matrix to Hermite form without using the Euclidean division algorithm; see, e.g., [9, p. 15].)

Greatest common left divisors (gclds) can be defined with the obvious changes. For convenience, we shall henceforth talk only of gcrds, leaving the reader to make the appropriate changes for gclds, when necessary.

Lemma 6.3-3.   A Construction for a gcrd [6]

Given $m \times m$ and $p \times m$ polynomial matrices $D(s)$ and $N(s)$, form the matrix $[D'(s) \ N'(s)]'$, and find elementary row operations [or equivalently find a unimodular matrix $U(s)$] such that at least p of the bottom rows on the right-hand side are identically zero:

$$\underbrace{\begin{bmatrix} \overset{m}{U_{11}(s)} & \overset{p}{U_{12}(s)} \\ U_{21}(s) & U_{22}(s) \end{bmatrix}}_{U(s)} \begin{bmatrix} \overset{m}{D(s)} \\ N(s) \end{bmatrix} = \begin{bmatrix} \overset{m}{R(s)} \\ 0 \end{bmatrix}\begin{smallmatrix} m \\ p \end{smallmatrix} \tag{1}$$

(In particular we could use the construction for getting the Hermite form, but this is more than is necessary.) Then the square matrix denoted $R(s)$ in (1) will be a gcrd of $D(s)$ and $N(s)$.

Proof.   Let

$$\begin{bmatrix} U_{11}(s) & U_{12}(s) \\ U_{21}(s) & U_{22}(s) \end{bmatrix}^{-1} = \begin{bmatrix} V_{11}(s) & V_{12}(s) \\ V_{21}(s) & V_{22}(s) \end{bmatrix} \tag{2}$$

Then

$$\begin{bmatrix} D(s) \\ N(s) \end{bmatrix} = \begin{bmatrix} V_{11}(s) & V_{12}(s) \\ V_{21}(s) & V_{22}(s) \end{bmatrix}\begin{bmatrix} R(s) \\ 0 \end{bmatrix} \tag{3}$$

so that

$$D(s) = V_{11}(s)R(s), \qquad N(s) = V_{21}(s)R(s) \tag{4}$$

and $R(s)$ is a right divisor of $D(s)$ and $N(s)$. Next notice that we also have

$$R(s) = U_{11}(s)D(s) + U_{12}(s)N(s) \tag{5}$$

so that if $R_1(s)$ is another right divisor, say,

$$D(s) = D_1(s)R_1(s), \qquad N(s) = N_1(s)R_1(s)$$

then

$$R(s) = [U_{11}(s)D_1(s) + U_{12}(s)N_1(s)]R_1(s)$$

so that $R_1(s)$ is a right divisor of $R(s)$, and $R(s)$ is a gcrd.   ∎

Gcrds are Not Unique.   Notice that by carrying out the elementary operations in (1) in different orders we may get different matrices $U(s)$ and hence different gcrds. However, any two gcrds, $R_1(s)$ and $R_2(s)$, say, must be related (by definition) as

$$R_1(s) = W_2(s)R_2(s), \qquad R_2(s) = W_1(s)R_1(s), \qquad W_i(s) \text{ polynomial} \tag{6}$$

Since we can then write

$$R_1(s) = W_2(s)W_1(s)R_1(s) \tag{7}$$

it follows that

1. If $R_1(s)$ is nonsingular, then the $\{W_i(s), i = 1, 2\}$ must be unimodular, and hence the gcrd $R_2(s)$ is also nonsingular. *That is, if one gcrd is nonsingular, then all gcrds must be so, and they can only differ by a unimodular (left) factor.*
2. *If a gcrd is unimodular, then all gcrds must be unimodular.*

Lemma 6.3-4.   Nonsingular gcrds
   If

$$[D'(s) \quad N'(s)]'\text{ has full column rank} \tag{8}$$

then *all* gcrds of $\{N(s), D(s)\}$ must be nonsingular and can differ only by unimodular (left) factors.

Remark: We recall that a polynomial matrix will have full column rank if no nontrivial linear combination of its columns, with either rational or polynomial coefficients, is identically (i.e., for all s) equal to zero. Therefore the rank will be full for almost all s. If the rank is full for *all* s, the (not necessarily square) matrix will be called *irreducible*.

Proof.   From (1) we see that since elementary operations cannot change the rank of a matrix, $[R'(s) \ 0]'$ must also have full rank; i.e., $R(s)$ must be nonsingular. The rest follows from (6) and (7). ∎

The rank assumption (8) is very natural. If the rank of $[D'(s) \ N'(s)]$ is not full, then we can have gcrds of $\{N(s), D(s)\}$ with some elements of arbitrarily high degree. For example, if

$$[D'(s) \quad N'(s)] = \begin{bmatrix} s & s & 1 & 1 \\ 0 & 0 & 0 & 0 \end{bmatrix}$$

then for any α, the following are both gcrds,

$$R_1(s) = \begin{bmatrix} 1 & 0 \\ 1 & 0 \end{bmatrix} \quad \text{and} \quad R_2(s) = \begin{bmatrix} s^\alpha & 0 \\ 1 & 0 \end{bmatrix}$$

Therefore in the sequel, unless stated otherwise, we shall assume that (8) *holds*, a condition that is often met by having $D(s)$ nonsingular. In any case, as long as (8) is true, by elementary row operations we can always convert $[D'(s) \ N'(s)]$ to another matrix $[\tilde{D}'(s) \ \tilde{N}'(s)]$ in which $\tilde{D}(s)$ is nonsingular.

Relatively Prime (or Coprime) Matrices.   We shall say that two polynomial matrices with the same number of columns are *relatively right prime* or are *right coprime* if all *their gcrds are unimodular.*

Lemma 6.3-5.   Simple Bezout Identity [6]

$N(s)$ and $D(s)$ will be right coprime if and only if there exist polynomial matrices $X(s)$ and $Y(s)$ such that

$$X(s)N(s) + Y(s)D(s) = I \qquad (9)$$

Proof.   Given $N(s)$, $D(s)$ we can write [cf. (5) and (6)] any gcrd as

$$R(s) = \hat{X}(s)N(s) + \hat{Y}(s)D(s)$$

But if $N(s)$, $D(s)$ are coprime, then $R(s)$ must be unimodular, so that $R^{-1}(s)$ will be polynomial. Therefore, we can write

$$I = X(s)N(s) + Y(s)D(s)$$

where $X(s) = R^{-1}(s)\hat{X}(s)$ is polynomial, and similarly for $Y(s)$.

Conversely, suppose there exist $X(s)$, $Y(s)$ such that (9) holds, and let $R(s)$ be any gcrd,

$$N(s) = \bar{N}(s)R(s), \qquad D(s) = \bar{D}(s)R(s)$$

Then we can write

$$I = [X(s)\bar{N}(s) + Y(s)\bar{D}(s)]R(s)$$

so that

$$R^{-1}(s) = X(s)\bar{N}(s) + Y(s)\bar{D}(s) = \text{a polynomial matrix}$$

Therefore $R(s)$ must be unimodular, or, equivalently, $\{N(s), D(s)\}$ must be right coprime.   ∎

Lemma 6.3-6.   Rank Criterion for Relative Primeness [6]

$N(s)$ and $D(s)$ will be right coprime if and only if $[D'(s)\ N'(s)]'$ has full rank for *every* s (i.e., if and only if $[D'(s)\ N'(s)]'$ is irreducible).

Remark: Recall that by our standing assumption (8) we know that $[D'(s)\ N'(s)]'$ has full rank for almost all s.

Proof.   From (1) we see that $[D'(s)\ N'(s)]'$ has full rank for all s if and only if $R(s)$ is unimodular, i.e., if and only if all gcrds of $N(s)$ and $D(s)$ are unimodular, which by definition means that $N(s)$ and $D(s)$ are relatively prime.   ∎

For a matrix polynomial $P(s)$, we shall say that $s = \lambda$ is an eigenvalue of $P(s)$ if the equation

$$P(\lambda)p(\lambda) = 0 \qquad (10)$$

has a nonzero solution $p(\lambda)$. Such a solution $p(\lambda)$ is often called a *latent vector* of $P(s)$, and λ is called a *latent root* of $P(s)$. [To avoid degeneracies, these definitions are generally only used when $P(s)$ has full column rank.] In this language we can state the following result, which is a generalization of the obvious scalar version.

Lemma 6.3-7.   Latent-Roots Characterization of Relative Primeness

$N(s)$ and $D(s)$ will be (right) coprime if and only if they have no common latent vectors and associated common latent roots.

Proof.   We can have a nonzero solution $p(s)$ to

$$\begin{bmatrix} D(s) \\ N(s) \end{bmatrix} p(s) = 0 \tag{11}$$

if and only if $[D'(s)\ N'(s)]$ does not have full column rank for some s. Hence . . . . ∎

The above are among the oldest and best known tests for relative primeness. More recently the impetus of ideas from system theory (just as in Sec. 2.4.4 in the scalar case) has led to some new criteria. What appears to be one of the simplest and most useful of these will be briefly noted in Exercise 6.3-19.

Gcrds of Several Matrices and Irreducible Matrices.   So far we have spoken of gcrds for two matrices (with the same number of columns). However, the definition is easily extended.

A gcrd of a set of matrices $\{A_i(s), i = 1, \ldots, L\}$, each having the same number of columns, is any matrix $R(s)$ such that

1. $R(s)$ is a right divisor of the $\{A_i(s)\}$, i.e., $A_i(s) = \bar{A}_i(s)R(s)$, $i = 1, \ldots,$ L, and
2. If $R_1(s)$ is any other right divisor, then $R_1(s)$ divides $R(s)$ in the sense that $R(s) = W(s)R_1(s)$ for some polynomial matrix $W(s)$.

The previously given criteria for pairs of matrices can readily be extended to this more general case (see Exercise 6.3-3). We shall say that

A full column rank polynomial matrix $P(s)$
is *irreducible* if its rows are right coprime

The reader should check that an irreducible matrix has full (column) rank for all s (cf. the remark below Lemma 6.3-4). Therefore an irreducible square matrix is unimodular; also, any subset of columns of an unimodular matrix will be irreducible (with of course a similar statement, with obvious changes, for rows). As a final example, let us note that if $\{N(s), D(s)\}$ are right coprime then the matrix $[D'(s)\ N'(s)]'$ will be irreducible. Conversely, if a full column rank matrix $P(s)$ is irreducible, note that we can partition it, perhaps after some row operations, so that $P'(s) = [D'(s)\ N'(s)]$, $D(s)$ is nonsingular, and $N(s)D^{-1}(s)$ is an irreducible right MFD.

Irreducible (Left) MFDs from (Right) MFDs.   The construction described earlier for finding gcrds (cf. (1)–(3)) has some interesting special features when $D(s)$ is nonsingular. We recall that the basic operation is to find a unimodular matrix $U(s)$ such that

$$\begin{array}{c}m\\p\end{array}\begin{bmatrix} U_{11}(s) & U_{12}(s) \\ U_{21}(s) & U_{22}(s) \end{bmatrix}\begin{bmatrix} D(s) \\ N(s) \end{bmatrix}\begin{array}{c}m\\p\end{array} = \begin{bmatrix} \overset{m}{R(s)} \\ 0 \end{bmatrix}\begin{array}{c}m\\p\end{array} \tag{12}$$

Lemma 6.3-8.   Left MFDs from Right MFDs

Referring to (12), when $D(s)$ is nonsingular,

a. $U_{22}(s)$ will be nonsingular $\qquad\qquad\qquad\qquad\qquad\qquad\qquad (13)$

b. $N(s)D^{-1}(s) = -U_{22}^{-1}(s)U_{21}(s)$ $\qquad\qquad\qquad\qquad\qquad\qquad (14)$

c. $\{U_{21}(s), U_{22}(s)\}$ will be left coprime, i.e., $U_{22}^{-1}(s)U_{21}(s)$ will be an *irreducible left MFD.* $\qquad\qquad\qquad\qquad\qquad\qquad\qquad\qquad\qquad\qquad\qquad (15)$

d. If $\{N(s), D(s)\}$ are coprime, then

$$\deg \det D(s) = \deg \det U_{22}(s) \tag{16}$$

In other words, from a right MFD $N(s)D^{-1}(s)$ we can obtain an irreducible left MFD in the process of finding a gcrd of the elements of the right MFD. Similar results of course hold if we start with an arbitrary left MFD. (More can be said about the actual computational aspects of this calculation, see Sec. 6.7.2.)

Proof.   The relation (15) follows from the fact that the big matrix $U(s)$ in (12) is unimodular, so that any subset of its rows is irreducible.

To show that $U_{22}(s)$ is nonsingular, we first define a matrix $V(s)$ by

$$V(s)U(s) \triangleq \begin{bmatrix} V_{11}(s) & V_{12}(s) \\ V_{21}(s) & V_{22}(s) \end{bmatrix}\begin{bmatrix} U_{11}(s) & U_{12}(s) \\ U_{21}(s) & U_{22}(s) \end{bmatrix} = \begin{bmatrix} I & 0 \\ 0 & I \end{bmatrix}$$

Then applying $V(s)$ to (12), we can write

$$D(s) = V_{11}(s)R(s)$$

so that the nonsingularity of $D(s)$ will imply that $R(s)$ and $V_{11}(s)$ are nonsingular. Now by well-known formulas for block matrices, the determinant of the nonsingular (unimodular) matrix $U(s)$ is (cf. Exercises A.22 and A.11)

$$\frac{\det U_{22}(s)}{\det V_{11}(s)} = \det U(s) \neq 0 \tag{17}$$

which proves that $\det U_{22}(s) \neq 0$, i.e., that $U_{22}(s)$ is nonsingular. Now (14) follows from the equation [cf. (12)]

$$U_{21}(s)D(s) + U_{22}(s)N(s) = 0$$

Finally, *if* $\{N(s), D(s)\}$ *are coprime*, then $R(s)$ is unimodular, and

$$\deg \det D(s) = \deg \det V_{11}(s)$$

But since $U(s)$ is unimodular, (17) then shows that

$$\deg \det D(s) = \deg \det U_{22}(s) \quad \blacksquare$$

Another very useful consequence of having $D(s)$ nonsingular is the following generalization of Lemma 6.3-5, first shown to the author by S. Kung in 1976.

Lemma 6.3-9.   Generalized Bezout Identity

Let $\{N_R(s), D_R(s)\}$ be right coprime, with $D_R(s)$ nonsingular. Then there exist polynomial matrices $\{X(s), Y(s), X^*(s), Y^*(s)\}$ such that

$$\begin{bmatrix} -X(s) & Y(s) \\ D_L(s) & N_L(s) \end{bmatrix}\begin{bmatrix} -N_R(s) & X^*(s) \\ D_R(s) & Y^*(s) \end{bmatrix} = \begin{bmatrix} I & 0 \\ 0 & I \end{bmatrix} \tag{18a}$$

and

$$\begin{bmatrix} -N_R(s) & X^*(s) \\ D_R(s) & Y^*(s) \end{bmatrix}\begin{bmatrix} -X(s) & Y(s) \\ D_L(s) & N_L(s) \end{bmatrix} = \begin{bmatrix} I & 0 \\ 0 & I \end{bmatrix} \tag{18b}$$

Moreover, the block matrices in (18) will all be unimodular. We shall call (18a) the *forward Bezout identity* and (18b) the *reversed Bezout identity*.

Proof.   By Lemma 6.3-5, we know that $\{N_R(s), D_R(s)\}$ being right coprime implies that there exist polynomials $\{X(s), Y(s)\}$ such that $X(s)N_R(s) + Y(s)D_R(s) = I$. By Lemma 6.3-8, we know also that there will exist left coprime polynomials $\{N_L(s), D_L(s)\}$ such that $D_L^{-1}(s)N_L(s) = N_R(s)D_R^{-1}(s)$. Furthermore, the left coprimeness of $\{N_L(s), D_L(s)\}$ implies the existence of polynomials $\{\bar{X}(s), \bar{Y}(s)\}$ such that $D_L(s)\bar{X}(s) + N_L(s)\bar{Y}(s) = I$. Writing these relations in matrix form gives

$$\begin{bmatrix} -X(s) & Y(s) \\ D_L(s) & N_L(s) \end{bmatrix}\begin{bmatrix} -N_R(s) & \bar{X}(s) \\ D_R(s) & \bar{Y}(s) \end{bmatrix} = \begin{bmatrix} I & Q(s) \\ 0 & I \end{bmatrix} \qquad \bullet \tag{19}$$

where $Q(s) \triangleq Y(s)\bar{Y}(s) - X(s)\bar{X}(s)$. Therefore, postmultiplying both sides by the inverse of the block triangular matrix on the right-hand side gives (dropping the argument s for simplicity)

$$\begin{bmatrix} -X & Y \\ D_L & N_L \end{bmatrix}\begin{bmatrix} -N_R & N_R Q + \bar{X} \\ D_R & -D_R Q + \bar{Y} \end{bmatrix} = \begin{bmatrix} I & 0 \\ 0 & I \end{bmatrix} \tag{20}$$

which yields (18a) when we define $X^* = N_R Q + \bar{X}$, $Y^* = -D_R Q + \bar{Y}$. Next, (18b) follows by using the matrix fact (prove this) that $CD = I$ implies that $DC = I$ when $\{C, D\}$ are square (constant or polynomial) matrices. Finally, the unimodularity follows from Lemma 6.3-1.   ∎

6.3.2   Column- and Row-Reduced Matrices and Some Applications

A rational transfer function matrix $H(s)$ is said to be *proper* if

$$\lim_{s \to \infty} H(s) < \infty$$

and *strictly proper* if

$$\lim_{s \to \infty} H(s) = 0.$$

Unless otherwise specified, we shall assume that $H(s)$ is a $p \times m$ matrix.

In the scalar case, a transfer function is proper if the degree of the numerator polynomial is less than or equal to the degree of the denominator polynomial. The situation is not so simple in the matrix case, as we shall now explore in a series of lemmas.

We shall first define

$$\left.\begin{array}{l}\text{the degree of a}\\ \text{polynomial vector}\end{array}\right\} = \begin{array}{l}\text{the highest degree of}\\ \text{all the entries of the vector.}\end{array}$$

Then the following result is not unexpected.

Lemma 6.3-10.   Column Degrees of Proper Transfer Functions

If $H(s)$ is a *strictly proper* (proper) transfer function, and

$$H(s) = N(s)D^{-1}(s)$$

then every column of $N(s)$ has degree strictly less than (less than or equal to) that of the corresponding column of $D(s)$.

Proof.   We have

$$N(s) = H(s)D(s)$$

and therefore for the jth column we can write (in an obvious notation)

$$n_{ij}(s) = \sum_{k=1}^{m} H_{ik}(s)d_{kj}(s), \qquad i = 1, \ldots, p$$

But every element $H_{ik}(s)$ is strictly proper (proper) so that all elements $n_{ij}(s)$ must have degree less than (less than or equal to) that of the highest-degree polynomial in the jth column of $D(s)$. Hence . . . .   ∎

However, somewhat surprisingly, the converse of the result is not always true. For example, if

$$N(s) = [2s^2 + 1 \quad 2], \qquad D(s) = \begin{bmatrix} s^3 + s & s \\ s^2 + s + 1 & 1 \end{bmatrix}$$

the degrees of the columns of $N(s)$ are less than those of the corresponding columns of $D(s)$, but

$$N(s)D^{-1}(s) = \begin{bmatrix} \dfrac{-s^2 + s}{s^2 + s - 1} & \dfrac{s^3 + s - 1}{s^2 + s - 1} \end{bmatrix}$$

is not proper.

To obtain necessary and sufficient conditions for the properness of $N(s)D^{-1}(s)$ is not entirely trivial and requires the introduction of further concepts, especially those of *column reduced matrices* and of *poles and zeros at infinity*. We shall introduce the first concept here, while the latter will be studied in Section 6.5.3.

First, as a standard notation, we shall let

$$k_i \triangleq \text{the degree of the } i\text{th column of } D(s) \tag{21}$$

Then it is easy to see that

$$\deg \det D(s) \le \sum_{1}^{m} k_i$$

Inequality may hold because of possible cancellations. However, if $D(s)$ is such that equality holds in the above, we shall say that $D(s)$ is *column reduced*, a name suggested by Heymann [19]. The concept was first introduced into system theory by Wolovich [20], who used the perhaps somewhat less suggestive name *column proper*. An early mathematical treatment can be found in Wedderburn [8, Chap. 4]. [The physical significance of the concept is that column-reduced matrices have no *zeros* at infinity, as we shall explain in Section 6.5.3.]

To examine this concept further, consider the matrix

$$D(s) = \begin{bmatrix} s^3 + s & s + 2 \\ s^2 + s + 1 & 1 \end{bmatrix} = \begin{bmatrix} 1 & 1 \\ 0 & 0 \end{bmatrix} \begin{bmatrix} s^3 & 0 \\ 0 & s \end{bmatrix} + \begin{bmatrix} s & 2 \\ s^2 + s + 1 & 1 \end{bmatrix}$$

where the highest-degree terms in each column have been displayed separately. The sum of the column degrees is 4, but the determinantal degree is less than this because the coefficient matrix of the highest-degree terms in each column is singular.

In general, we can always write

$$D(s) = D_{hc}S(s) + L(s) \tag{22}$$

where

$$S(s) \triangleq \text{diag }\{s^{k_i}, i = 1, \ldots, m\} \tag{22a}$$

$D_{hc} = $ the *highest-column-degree coefficient matrix*, or more compactly, the *leading (column) coefficient matrix*, of $D(s)$ (22b)

a matrix whose ith column comprises the coefficients of s^{k_i} in the ith column of $D(s)$. Finally, $L(s)$ denotes the *remaining* terms and is a polynomial matrix with column degrees strictly less than those of $D(s)$. Then

$$\det D(s) = (\det D_{hc})s^{\sum_i k_i} + \text{terms of lower degree in } s \tag{23}$$

and therefore it follows that

> A nonsingular polynomial matrix is column reduced
> if and only if its leading (column) coefficient
> matrix is nonsingular   (24)

Here we can use this result to establish a correct converse of Lemma 6.3-10.

Lemma 6.3-11.   Properness of $N(s)D^{-1}(s)$ When $D(s)$ Is Column Reduced

If $D(s)$ is column-reduced, then $H(s) = N(s)D^{-1}(s)$ is strictly proper (proper) if and only if each column of $N(s)$ has degree less than (less than or equal to) the degree of the corresponding column of $D(s)$.

Proof.   For simplicity we shall only consider the strictly proper case, the modifications for the proper case being obvious.

For the "if" part, we have to show that with the above conditions on $\{N(s), D(s)\}$, every entry $h_{ij}(s)$ of $H(s) = N(s)D^{-1}(s)$ is strictly proper. We proceed directly, by using (the transposed form of) Cramer's rule (see Exercise A.8) to solve for $h_{ij}(s)$:

$$h_{ij}(s) = \frac{\det D^{ij}(s)}{\det D(s)}$$

where $D^{ij}(s)$ is the matrix obtained by replacing the jth row of $D(s)$ by the ith row of $N(s)$.

Now $D(s)$ is given by (22), while $D^{ij}(s)$ may be written similarly as

$$D^{ij}(s) = (D^{ij})_{hc}S(s) + L^{ij}(s)$$

$(D^{ij})_{hc}$ is the same as D_{hc} except for the jth row, which is now zero since each entry of the jth row of $N(s)$ is of lower degree than the corresponding entry of the jth row of $D(s)$. Hence, $(D^{ij})_{hc}$ is singular, while D_{hc} is nonsingular. It follows that

$$\deg \det D(s) = \deg \det S(s) = \sum_1^m k_i$$

whereas

$$\deg \det D^{ij}(s) < \sum_1^m k_i.$$

Therefore $h_{ij}(s)$ is strictly proper and hence so is the matrix $H(s)$.

For the "only if" part, we need only appeal to Lemma 6.3-10, which states that if $N(s)D^{-1}(s)$ is proper, the column degrees of $N(s)$ are less than those of the corresponding columns of $D(s)$ (whether or not $D(s)$ is column reduced).   ∎

For Lemma 6.3-11 to be a useful test, we need to know if a given matrix $D(s)$ can be made column-reduced by elementary column operations, say $D(s)W(s) = \bar{D}(s)$, $\bar{D}(s)$ column reduced. For then we could write $H(s) = \bar{N}(s)\bar{D}^{-1}(s)$, with $\bar{N}(s) = N(s)W(s)$ being a polynomial matrix, and apply Lemma 6.3-11 to $\{\bar{N}(s), \bar{D}(s)\}$.

Reduction to Column-Reduced Form.   In fact, the basic definitions suggest that any polynomial matrix can be made column reduced by using elementary column operations to successively reduce the individual column degrees until column-reducedness is achieved. The procedure is best seen via a simple example.

Example 6.3-2.   Obtaining a Column-Reduced Matrix

Let us consider the non-column-reduced matrix $D_3(s)$ of Example 6.2-1,

$$D_3(s) = \begin{bmatrix} (s+1)^2(s+2)^2 & -(s+1)^2(s+2) \\ 0 & s+2 \end{bmatrix} \tag{25}$$

Examining just the highest-order terms in each column, we see that we can take

$$\begin{bmatrix} s^4 & -s^3 \\ 0 & 0 \end{bmatrix} \longrightarrow \begin{bmatrix} 0 & -s^3 \\ 0 & 0 \end{bmatrix}$$

by the column operation of postmultiplying by the matrix with first column $[1 \ s]'$ and second column $[0 \ 1]'$. Applying this to $D_3(s)$ gives

$$D_3(s)\begin{bmatrix} 1 & 0 \\ s & 1 \end{bmatrix} = \begin{bmatrix} 2s^3 + 8s^2 + 10s + 4 & -(s^3 + 4s^2 + 5s + 2) \\ s^2 + 2s & s+2 \end{bmatrix}$$

which is still not column reduced. But postmultiplying again as shown gives

$$D_3(s)\begin{bmatrix} 1 & 0 \\ s & 1 \end{bmatrix}\begin{bmatrix} 1 & 0 \\ 2 & 1 \end{bmatrix} = \begin{bmatrix} 0 & s^3 + 4s^2 + 5s + 2 \\ s^2 + 4s + 4 & s+2 \end{bmatrix}$$

which is reduced. In this problem, the required steps could also have been seen by direct examination of $D_3(s)$ to be the addition to the first column of $s+2$ times the second column.   ∎

We should note that elementary row operations can also be used to make a matrix column-reduced—just use them to put the matrix in column-Hermite form (cf. Theorem 6.3-2), which is obviously column reduced.

An Important Equation.   It turns out that several systems problems lead us to ask if the equation

$$H(s)D(s) = N(s)$$

where $N(s)$ and $D(s)$ are *given* $p \times r$ and $m \times r$ polynomial matrices, with $D(s)$ of full column rank $r \leq m$, has a proper rational solution $H(s)$. A criterion can be obtained by using an extended definition [8]:

A full column rank matrix will be said to be column reduced
if its leading (column) coefficient matrix has full column rank.   (26)

We ask the reader to show that an equivalent characterization is that the sum of the column degrees equals the highest degree of all the $m \times m$ minors (m being the rank of the matrix) and also that any full column rank matrix can be transformed to column-reduced form by means of elementary column (or row) operations (cf. Example 6.3-2). We now have the following result [18].

Theorem 6.3-12.   Proper Solutions of a Rational Matrix Equation

Consider the equation $H(s)D(s) = N(s)$, where $N(s)$ and $D(s)$ are given $p \times r$ and $m \times r$ polynomial matrices, and $D(s)$ has full column rank $r \le m$. Form the matrix $F(s) \triangleq [D'(s) \ N'(s)]'$ and use elementary column operations to transform it to $\bar{F}(s) = [\bar{D}'(s) \ \bar{N}'(s)]$, where $\bar{D}(s)$ is column reduced. Then a proper $H(s)$ exists if and only if each column of $\bar{N}(s)$ has degree less than or equal to that of each corresponding column of $\bar{D}(s)$.

Proof.   The necessity (or "only if" part) follows from writing $H(s)\bar{d}_i(s) = \bar{n}_i(s)$, where $\{\bar{d}_i(s), \bar{n}_i(s)\}$ denote the ith columns of $\{\bar{D}(s), \bar{N}(s)\}$ respectively. It is clear that if $\deg n_i(s) > \deg d_i(s)$, then $H(s)$ must contain an improper entry. For sufficiency, select the rows of $\bar{D}(s)$ that contain a nonsingular minor of $\bar{D}_{hc}$, say $\tilde{D}(s) = [I \ \ 0]\bar{D}(s)$. Then $\tilde{D}(s)$ has the same column degrees as $\bar{D}(s)$ and is column reduced. Now we can verify that one proper solution is $H(s) = N(s)\tilde{D}^{-1}(s)[I \ \ 0]$. ∎

Further study of column-reduced matrices can clearly be helpful. Here is another important (in fact, characterizing) property, discovered by Forney [15] (see also [61, p. 37]).

Theorem 6.3-13.   The Predictable-Degree Property of Column-Reduced Matrices

Let $D(s)$ be a polynomial matrix of full column rank, and for any polynomial vector $p(s)$, let

$$q(s) = D(s)p(s) \tag{27a}$$

Then $D(s)$ is column reduced if and only if

$$\deg q(s) = \max_{i:\, p_i(s) \neq 0} [\deg p_i(s) + k_i] \tag{27b}$$

where $p_i(s)$ is the ith entry of $p(s)$ and k_i is the degree of the ith column of $D(s)$.

Proof.   Let

$$d = \max_{i:\, p_i(s) \neq 0} [\deg p_i(s) + k_i]$$

It is obvious that $\deg q(s) \le d$, and *the point is to show that equality holds*, i.e., that $q_0 \neq 0$ in

$$q(s) = q_0 s^d + q_1 s^{d-1} + \cdots + q_d$$

Now, by the definition of d, the $\{p_i(s)\}$ must have the form

$$p_i(s) = \alpha_i s^{d-k_i} + \text{(terms of lower degree)}$$

and, furthermore, not all the α_i can be zero. Then note that

$$q_0 = D_{hc}\alpha, \qquad \alpha' \triangleq [\alpha_1 \quad \cdots \quad \alpha_m]$$

so that if $D(s)$ is column reduced, D_{hc} will be of full rank (nonsingular), and therefore $q_0 \neq 0$.

Conversely, if Eqs. (27) hold, then $q_0 \neq 0$ for all $\alpha \neq 0$; i.e., D_{hc} must have full rank. ∎

Here is a first application of this result.

Lemma 6.3-14. **Invariance of Column Degrees of Column-Reduced Matrices**
[8, p. 49]

Let $D(s)$ and $\bar{D}(s)$ be column-reduced polynomial matrices, with column degrees arranged in order, say ascending. Then, if

$$D(s) = \bar{D}(s)U(s), \qquad U(s) \text{ unimodular}$$

$D(s)$ and $\bar{D}(s)$ will have the same column degrees.

Proof. By contradiction. Suppose the column degrees of $D(s)$ and $\bar{D}(s)$ are

$$\{k_1 \leq k_2 \leq \cdots \leq k_m\} \quad \text{and} \quad \{\bar{k}_1 \leq \bar{k}_2 \leq \cdots \leq \bar{k}_m\}$$

respectively, and that for some $l \leq m$,

$$k_i = \bar{k}_i, \qquad i \leq l - 1, \qquad \text{but} \quad k_l < \bar{k}_l$$

Let us write out $D(s) = \bar{D}(s)U(s)$ as

$$D(s) = \bar{D}(s) \begin{bmatrix} U_{11}(s) & U_{12}(s) \\ U_{21}(s) & U_{22}(s) \end{bmatrix} {}^{l-1}_{m-l+1}$$
$$\phantom{D(s) = \bar{D}(s)} \underset{l \quad\quad m-l}{}$$

The predictable-degree property of Theorem 6.3-13 shows that the $(p - l + 1) \times l$ submatrix $U_{21}(s)$ must be identically zero. Therefore the rank of $[U'_{11}(s) \quad U'_{21}(s)]'$ cannot exceed $l - 1$, and hence the rank of the whole $m \times m$ matrix $U(s)$ cannot exceed $(l - 1) + (m - l) = m - 1$; i.e., $U(s)$ must be singular and cannot be unimodular as assumed above. ∎

Division Theorems for Polynomial Matrices. An important application of column-reduced matrices is in the problem of extracting the strictly proper part of a given MFD. In the scalar case, we can do this by dividing one polynomial into another, and this motivates us to consider the corresponding results for matrices.

Given matrices $\{N(s), D(s)\}$ our aim is to find polynomial matrices $Q(s)$ and $R(s)$ such that

$$N(s) = Q(s)D(s) + R(s)$$

$$R(s)D^{-1}(s) \text{ is strictly proper} \tag{25}$$

One way of doing this is to form the rational matrix $H(s) = N(s)D^{-1}(s)$ and decompose it as

$$H(s) = N(s)D^{-1}(s) = H_{sp}(s) + P(s)$$

where $P(s)$ is a polynomial matrix and $H_{sp}(s)$ is strictly proper. Then we see that

$$H(s)D(s) = N(s) = P(s)D(s) + R(s)$$

where

$$R(s) = H_{sp}(s)D(s)$$

is clearly polynomial [it equals $N(s) - P(s)D(s)$] and is such that $R(s)D^{-1}(s)$ is strictly proper.

Now we know by Lemma 6.3-10 that

$$\deg \{i\text{th column of } R(s)\} < \deg \{i\text{th column of } D(s)\} \tag{26}$$

which might suggest that we can replace the condition (25) by (26), but as we noted in the discussion after Lemma 6.3-10, this may not be true unless $D(s)$ obeys further conditions, e.g., that it is column reduced.

It will be useful to formalize this discussion into a theorem, because we shall often refer to it in several other problems as well (see, e.g., Secs. 6.6, 7.5, 8.1, and 8.2).

Theorem 6.3-15.   Division Theorem for Polynomial Matrices

Let $D(s)$ be an $m \times m$ nonsingular polynomial matrix. Then, for any $p \times m$ polynomial matrix $N(s)$, there exist unique polynomial matrices $\{Q(s), R(s)\}$ such that $N(s) = Q(s)D(s) + R(s)$ and $R(s)D^{-1}(s)$ is strictly proper. [If $D(s)$ is also column reduced, then uniqueness will be ensured if the columns of $R(s)$ have degree strictly less than the corresponding column degrees of $D(s)$.]

Remark: It is worth noting that most mathematics books (e.g., [7]) only give this theorem for *regular* matrices $D(s)$, i.e., $D(s)$ such that its leading coefficient matrix, D_0 in $D(s) = D_0 s^r + \cdots + D_r$, $r =$ the highest degree of s appearing in $D(s)$, is nonsingular. As in this theorem, many results using regular matrices have a good chance of also holding with column- or row-reduced matrices.

Proof.   Everything has been proved but the uniqueness. For this, suppose there are two pairs such that

$$Q(s)D(s) + R(s) = \tilde{Q}(s)D(s) + \tilde{R}(s)$$

Then

$$Q(s) - \tilde{Q}(s) = [\tilde{R}(s) - R(s)]D^{-1}(s)$$

Now the left-hand side is polynomial (or constant), while, by hypothesis, the right-hand side is strictly proper (or zero). In any case, both sides must be zero for equality to hold, which proves the uniqueness. ∎

A very useful special case of Theorem 6.3-15 is when $D(s) = sI - A$, which is both row and column reduced and will yield a constant remainder matrix. The following result is classical [7].

Corollary.   Division by $sI - A$

Let
$$F(s) = F_0 s^n + F_1 s^{n-1} + \cdots + F_n$$

be a matrix polynomial, and let

$$F_r(A) \triangleq F_0 A^n + F_1 A^{n-1} + \cdots + F_n I$$
$$F_l(A) \triangleq A^n F_0 + A^{n-1} F_1 + \cdots + I F_n$$

be the right and left "values" of $F(s)$ at $s = A$.

Then we have

$$F(s) = Q_r(s)(sI - A) + R_r, \qquad R_r \triangleq F_r(A)$$
$$F(s) = (sI - A)Q_l(s) + R_l, \qquad R_l \triangleq F_l(A)$$

Proof.   We leave this to the reader, with the hint that

$$Q_r(s) = F_0 s^{n-1} + (F_0 A + F_1)s^{n-2} + \cdots + (F_0 A^{n-1} + \cdots + F_{n-1}) \quad \blacksquare$$

Row-Reduced Matrices.   Our discussions for column reducedness, column degrees, etc., have obvious counterparts which may be called row reducedness, row degrees, etc. There is no need to explicitly repeat all the previous discussions.

6.3.3   The Smith Form and Related Results

Earlier we presented the Hermite form, into which any polynomial matrix could be transformed by elementary row (or column) operations. We shall now consider what can be achieved when *both* row and column operations are permitted [7].

Theorem 6.3-16.   Smith Form

For any $p \times m$ polynomial matrix $P(s)$ we can find elementary row and column operations, or corresponding unimodular matrices $\{U(s), V(s)\}$, such that

$$U(s)P(s)V(s) = \Lambda(s) \tag{28a}$$

where

$$\Lambda(s) = \begin{bmatrix} \lambda_1(s) & & & \vdots & \\ & \ddots & & \vdots & 0 \\ & & \lambda_r(s) & \vdots & \\ \cline{1-5} & 0 & & \vdots & 0 \end{bmatrix} \begin{matrix} r \\ \\ \\ p\text{-}r \end{matrix} \tag{28b}$$
$$\qquad\qquad\quad\ \ \underbrace{}_{r} \quad \underbrace{}_{m\text{-}r}$$

$$r = \text{the (normal) rank of } P(s)$$

and the $\{\lambda_i(s)\}$ are unique monic polynomials obeying a *division property*

$$\lambda_i(s) \,|\, \lambda_{i+1}(s), \qquad i = 1, \dots, r-1 \tag{28c}$$

Moreover, if we define

$$\Delta_i(s) = \text{the gcd of all } i \times i \text{ minors of } P(s) \tag{29a}$$

then we can identify

$$\lambda_i(s) = \frac{\Delta_i(s)}{\Delta_{i-1}(s)}, \qquad \Delta_0(s) = 1 \tag{29b}$$

The matrix $\Lambda(s)$ is called the *Smith form* of $P(s)$.

Remark: The $\{\Delta_i(s)\}$ will be called the *determinantal divisors* of $P(s)$ and the $\{\lambda_i(s)\}$ the *invariant polynomials* of $P(s)$.

Proof.   We shall indicate how to form the matrix $\Lambda(s)$. First bring to the $(1, 1)$ position the element of least degree in the matrix $P(s)$. Now by elementary row operations, make all entries below the first one in the first column zero, as in the construction of the Hermite form. Next, by column operations, make all entries in the first row zero except for the first one. In the course of doing this, nonzero entries may reappear in the first column, and we should then repeat the above steps until we finally get all zeros in the first row and column except for the $(1, 1)$ element. [We ask the reader to show that the degree of the $(1, 1)$ element is reduced at each iteration, so that the algorithm is finite.]

Now if this $(1, 1)$ element does not divide every other entry in the whole matrix, by using the division algorithm and row and column interchanges we can bring a lower-degree element to the $(1, 1)$ position and then repeat the above steps to zero all other elements in the first column and first row. Proceeding in this way will give a matrix of the form

$$\begin{bmatrix} \lambda_1(s) & 0 & 0 & \cdots & 0 \\ 0 & & & & \\ \vdots & & P_1(s) & & \\ \vdots & & & & \\ 0 & & & & \end{bmatrix}, \qquad \lambda_1(s) \text{ divides every element of } P_1(s)$$

Now repeat the above operations on $P_1(s)$. Proceeding in this way gives the Smith form $\Lambda(s)$.

By using the Binet-Cauchy theorem (see Exercise 6.3-11) it can be seen that $\Delta_i(s) \triangleq$ the gcd of all $i \times i$ minors of $P(s)$ depends only on $P(s)$ and is independent of elementary row and column operations on $P(s)$. Therefore we can also identify

$$\Delta_i(s) = \text{the gcd of all } i \times i \text{ minors of } \Lambda(s)$$

The form of $\Lambda(s)$ then shows that

$$\Delta_1 = \lambda_1, \Delta_2 = \lambda_1\lambda_2, \ldots, \Delta_i = \lambda_1\lambda_2\ldots\lambda_i$$

or (with $\Delta_0 = 1$)

$$\lambda_1 = \Delta_1, \lambda_2 = \frac{\Delta_2}{\Delta_1}, \ldots, \lambda_i = \frac{\Delta_i}{\Delta_{i-1}}$$

Moreover, the uniqueness of the $\{\Delta_i\}$ implies that of the $\{\lambda_i\}$. This completes the proof. ∎

Remark: Although $\Lambda(s)$ is unique, the unimodular matrices $U(s)$, $V(s)$ are highly nonunique.

We shall write

$$P_1(s) \underset{S}{=} P_2(s) \tag{30}$$

to denote that $P_1(s)$ and $P_2(s)$ have the same Smith form, i.e., that they are *equivalent under elementary row and column operations*. The significance of being an equivalence relation is that we have the (easily verified) properties

$$P_1(s) \underset{S}{=} P_2(s) \Longrightarrow P_2(s) \underset{S}{=} P_1(s) \tag{31a}$$

$$P_1(s) \underset{S}{=} P_1(s) \tag{31b}$$

$$P_1(s) \underset{S}{=} P_2(s), \qquad P_2(s) \underset{S}{=} P_3(s) \Longrightarrow P_1(s) \underset{S}{=} P_3(s) \tag{31c}$$

The Smith form can be used to give a useful theoretical criterion for coprimeness.

Lemma 6.3-17.   Smith Form Test for Relative Primeness

$N(s)$ and $D(s)$ are right coprime if and only if the Smith form of $[N'(s) \; D'(s)]'$ is $[I \; 0]'$.

Proof.   By the construction for a gcrd we know that there exists a unimodular $U(s)$ such that

$$U(s)\begin{bmatrix} D(s) \\ N(s) \end{bmatrix} = \begin{bmatrix} R(s) \\ 0 \end{bmatrix} = \begin{bmatrix} I \\ 0 \end{bmatrix} R(s)$$

Now if $N(s)$ and $D(s)$ are coprime, any gcrd $R(s)$ must be unimodular, so that $R^{-1}(s)$ will be polynomial and we shall have

$$U(s)\begin{bmatrix} D(s) \\ N(s) \end{bmatrix} R^{-1}(s) = \begin{bmatrix} I \\ 0 \end{bmatrix} = \text{the Smith form}$$

We leave the proof of the converse to the reader. ∎

Example 6.3-3.

1. If P is unimodular, its Smith form is I, since we can write

$$P = P \cdot I \cdot I$$

2. If A_c is a companion matrix in controller form [i.e., with the coefficients of the characteristic polynomial $a(s)$ along the top row], then the Smith form of $sI - A$ is diag$\{1, 1, \ldots, 1, a(s)\}$. The reason is that there is a submatrix I_{n-1}, so $\Delta_{n-1} = 1$, and since $\Delta_i | \Delta_{i+1}$, we must have $\Delta_1 = 1$, $\Delta_2 = 1, \ldots, \Delta_{n-2} = 1$. Also, clearly, $\Delta_n = a(s)$. Therefore $\lambda_1 = \lambda_2 = \cdots = \lambda_{n-1} = 1$, and $\lambda_n = a(s)$. See also Exercise 6.3-15.

3. Suppose that

$$A = \text{block diag}\{C_i, i = 1, \ldots, r\}$$

where

$$C_i = \text{a companion matrix of the polynomial } \lambda_i(s)$$

and $n = \sum_1^r n_i$, $\lambda_i(s) | \lambda_{i+1}(s)$, $i = 1, \ldots, r - 1$. The Smith form of $sI - A$ is diag$\{I_{n-r}, \lambda_1(s), \ldots, \lambda_r(s)\}$. ∎

The converse of the last result is a basic theorem of matrix analysis, which we shall note now.

Theorem 6.3-18.   Rational Canonical Form

Suppose A is an $n \times n$ real-valued matrix and that $sI - A$ has $r \leq n$ *nonunity* invariant polynomials

$$\lambda_1(s) | \lambda_2(s) | \cdots | \lambda_r(s)$$

[The sum of the degrees of the $\{\lambda_i(s)\}$ must equal n. Why?] Then there exists a nonsingular matrix T such that

$$\bar{A} \triangleq T^{-1}AT = \text{block diag}\{C_i, i = 1, \ldots, r\}$$

Proof.   We can verify (see part 3 of Example (6.3-3)) that the invariant polynomials of $sI - \bar{A}$ are also $\{\lambda_1(s), \ldots, \lambda_r(s)\}$. The proof is completed by appealing to Lemma 6.3-19. ∎

Lemma 6.3-19.   Equivalence and Similarity

$sI - A$ and $sI - B$ will have the same Smith form if and only if A and B are similar.

Proof.   The main steps are outlined in Exercise 6.3-9, along with mention of an alternative criterion. ∎

6.3.4   *Linearizations, Matrix Pencils, and Kronecker Forms*

Let

$$P(s) = P_0 s^d + P_1 s^{d-1} + \cdots + P_d \tag{32}$$

be a $p \times m$ polynomial matrix. It is interesting and useful that we can use elementary row *and* column operations to transform an extended form of $P(s)$ to the *linear* form $sE - A$.

Lemma 6.3-20.   Linearization

We can write

$$\mathcal{P}(s) = U(s)[sE - A]V(s) \tag{33}$$

where

$$\mathcal{P}(s) = \text{block diag}\{\underbrace{I_m, \ldots, I_m, P(s)}_{d \text{ terms}}\}$$

$$P(s) = \text{any polynomial matrix as in (32)}$$

$\{E, A\}$ are *constant* matrices given by

$$E = \begin{bmatrix} P_0 & & & \\ & I_m & & \\ & & \ddots & \\ & & & I_m \end{bmatrix}, \quad A = \begin{bmatrix} -P_1 & -P_2 & \cdots & -P_d \\ I & & & \\ & \ddots & & \\ & & I & 0 \end{bmatrix} \tag{34}$$

and $\{U(s), V(s)\}$ are *unimodular* matrices whose exact form will be defined in the proof [cf. (35)–(37)].

Proof.   Let

$$V(s) = \begin{bmatrix} I & sI & \cdots & s^{d-1}I \\ & \ddots & \ddots & \vdots \\ & & \ddots & sI \\ & & & I \end{bmatrix}, \quad \text{a unimodular block Toeplitz matrix} \tag{35}$$

Then we can check that

$$(sE - A)V(s) = \begin{bmatrix} B_1(s) & B_2(s) & \cdots & B_{d-1}(s) & P(s) \\ -I & & & & 0 \\ & \ddots & & & \vdots \\ & & \ddots & & \vdots \\ & & & -I & 0 \end{bmatrix}$$

where

$$B_{i+1}(s) = sB_i(s) + P_{i+1}$$

and

$$B_1(s) = sP_0 + P_1 \tag{36}$$

Finally, we can verify that premultiplication of the above by the unimodular matrix

$$
U(s) = \begin{bmatrix}
0 & -I & & & \\
\cdot & & \cdot & & \\
\cdot & & & \cdot & \\
\cdot & & & & \\
0 & & & & -I \\
I & B_1(s) & \cdots & & B_{d-1}(s)
\end{bmatrix}
$$

yields (33). ∎

Corollary 6.3-21.

When $P(s)$ is a *regular* polynomial matrix, i.e., when $p = m$ and P_0 is nonsingular, a constant transformation will yield the familiar block companion form

$$
sE - A \longrightarrow \begin{bmatrix}
sI + \hat{P}_1 & \hat{P}_2 & \cdots & \hat{P}_d \\
-I & sI & & \\
& \cdot & \cdot & \\
& & \cdot & \cdot \\
& & & -I & sI
\end{bmatrix} \tag{37}
$$

where $\hat{P}_i = P_0^{-1} P_i$. ∎

Such linearization results have a long history (see, e.g., K.-H. Förster, *Math. Z.*, **95**, pp. 251–258, 1967) but were brought to prominence again by the work of Gohberg and his colleagues (see [21] and [22] and the references therein), which is again turning mathematical attention to polynomial matrices.

It follows from Lemma 6.3-20 that $P(s)$ and $sE - A$ have the same non-unity invariant polynomials. Van Dooren [23] has developed a reliable numerical algorithm for finding the invariant polynomials of matrices of the (linear) form $sE - A$.

Actually such matrices (with $E \neq I$) arise in various important problems in system theory—we have already encountered them in the PBH rank tests for controllability and observability (Sec. 6.2), and we shall encounter them again in studying the *zeros* of multivariable systems in state-space form (Sec. 6.5.3) and in other problems (e.g., in Chapter 8). It is therefore worthwhile to examine them more closely, following the work of Kronecker (as described in [7]).

Matrix Pencils and Kronecker Forms.   Linear matrix polynomials of the form $sE - A$ are known as (linear) matrix pencils. When $sE - A$ is square *and* $\det(sE - A) \not\equiv 0$, the pencil is said to be *regular*; otherwise it is called *singular*.

In trying to find a canonical form for linear pencils, it is natural to restrict oneself to *constant* elementary row and column operations,

$$U(sE - A)V = s\bar{E} - \bar{A} \tag{38}$$

where det $U \neq 0 \neq$ det V. Pencils related in this way are said to be *strictly equivalent*. Kronecker showed that U and V could be chosen so that

$$U(sE - A)V = \text{block diag } \{L_{\mu_1}, \ldots, L_{\mu_\alpha}, \tilde{L}_{\nu_1}, \ldots, \tilde{L}_{\nu_\beta}, sJ - I, sI - F\} \tag{39}$$

where $\{F, J, \{L_{\mu_i}\}, \{\tilde{L}_{\nu_i}\}\}$ are unique matrices such that

1. F is in Jordan (or in rational) form;
2. J is a nilpotent Jordan matrix, i.e., a matrix in Jordan form with all zero eigenvalues;
3. L_μ is a $\mu \times (\mu + 1)$ matrix of the form

$$\begin{bmatrix} s & -1 & & & \\ & \ddots & \ddots & & \bigcirc \\ & & \ddots & \ddots & \\ \bigcirc & & & s & -1 \end{bmatrix} \tag{40}$$

4. $\tilde{L}_\nu$ is a $(\nu + 1) \times \nu$ matrix with s along the diagonal and -1 along the first subdiagonal.

This is known as the *Kronecker canonical form* of the pencil.

An example will be given below, but first let us note that the matrices $\{L_\mu, \tilde{L}_\nu\}$ will not appear if the pencil is regular. These matrices capture the *singularity* of the pencil in the following way: For $\tilde{L}_\nu$ there exists a polynomial row vector such that

$$[1 \quad s \quad \cdots \quad s^\nu]\tilde{L}_\nu = [0 \quad \cdots \quad 0] \tag{41a}$$

while for L_μ there exists a column vector such that

$$L_\mu[1 \quad s \quad \cdots \quad s^\mu]' = [0 \quad \cdots \quad 0]' \tag{41b}$$

The sizes of the $\{L_\mu, \tilde{L}_\nu\}$ characterize them completely, and therefore they are given special names. The $\{\mu_i\}$ will be called the *right Kronecker indices* and the $\{\nu_i\}$ the *left Kronecker indices*.[†] We should note that the indices may be zero, corresponding to constant (degree zero) nulling vectors.

[†]These are often also known as *minimal indices*, for reasons best presented later (Sec. 6.5.4).

Example 6.3-4.

The following pencil is in Kronecker canonical form:

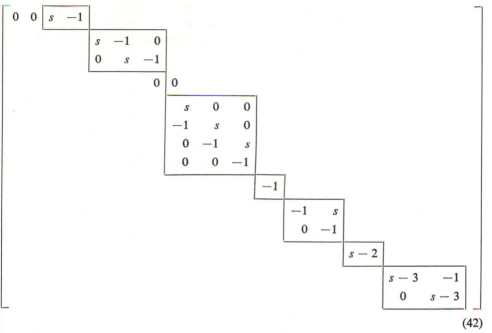

$$(42)$$

It has $\mu_1 = 0$, $\mu_2 = 0$, $\mu_3 = 1$, $\mu_4 = 2$, $\nu_1 = 0$, $\nu_2 = 3$, while

$$
J = \begin{bmatrix} 0 & \\ \hline & 0 & 1 \\ & 0 & 0 \end{bmatrix}, \qquad
F = \begin{bmatrix} 2 & \\ \hline & 3 & 1 \\ & 0 & 3 \end{bmatrix}
\qquad\blacksquare
$$

The matrices $\{L_{\mu_i}, \tilde{L}_{\nu_i}\}$ will not appear in regular pencils. The term $\{sJ - I\}$ will not appear if *E is nonsingular*, because then we can write

$$U(sE - A)V = U_1(sI - \bar{A})V, \qquad \bar{A} = E^{-1}A$$

In this case the notion of strict equivalence is the same as that of similarity (show this). The properties of regular pencils with E nonsingular are completely determined by the eigenvalues of the matrix $E^{-1}A$. When *E is singular*, this information has to be supplemented by the information in the matrix $sJ - I$. Rewriting this matrix as

$$
\begin{aligned}
sJ - I &= -s(s^{-1}I - J) \\
&= -s(\lambda I - J), \qquad \lambda = s^{-1}
\end{aligned}
\tag{43}
$$

and recalling that J is *nilpotent*, we may say that $sJ - I$ describes the singularity at $s = \infty$ (i.e., the eigenstructure at ∞) of the pencil $sE - F$. [This very heuristic argument will be made more precise in Sec. 6.5.3.]

It has been found that the eigenstructure at ∞ is important in several problems, e.g., in studying the properties of system zeros, multivariable root loci, improper systems, and interconnected systems, and the fact that the essential aspects of this structure at infinity can be reliably determined for pencils (see [23]) is an important reason for studying how to reduce arbitrary polynomial (and rational) matrices to pencil form. The linearization procedure of Lemma 6.3-20 is, however, not adequate for this purpose: While $P(s)$ and $sE - F$ as defined there have the same properties for all finite s, the unimodular matrices $U(s)$ and $V(s)$ can modify the behavior at $s = \infty$ [note that $\det U(s) = $ constant means that the unimodular matrix has *singular* behavior only at $s = \infty$]. The results for pencils suggest that we should study the structure at $s = \infty$ of $P(s)$ by examining the structure at $\lambda = 0$ of $P(\lambda^{-1})$. But $P(\lambda^{-1})$ is a rational rather than a polynomial matrix, and the techniques of this section are not directly applicable. We shall therefore postpone consideration of this point (and of the Kronecker indices of polynomial and rational matrices) to Sec. 6.5, where rational matrices will be studied in some detail.

Exercises

6.3-1.   *Further Properties of Unimodular Matrices*

a. Prove that every unimodular polynomial matrix with real or complex coefficients can be written as a product of elementary matrices. *Hint:* Reduce the unimodular matrix to Hermite form.

b. If $\{P(s), U(s)\}$ are polynomial matrices, with $U(s)$ unimodular, show that the pair $\{P(s), U(s)\}$ is always coprime (left or right, as the case may be).

6.3-2.   *Testing for coprimeness*

Check in several different ways whether the following pairs of matrices are right coprime.

a. $\begin{bmatrix} s & 0 \\ -s & s^2 \end{bmatrix}$ and $\begin{bmatrix} 0 & -(s+1)^2(s+2) \\ (s+2)^2 & (s+2) \end{bmatrix}$

b. $\begin{bmatrix} s & s \\ -s(s+1)^2 & -s \end{bmatrix}$ and $\begin{bmatrix} (s+1)^2(s+2)^2 & 0 \\ 0 & (s+2)^2 \end{bmatrix}$

c. $\begin{bmatrix} 2s+1 & s^2+1 \\ (s+1)^2 & s^2+2s \end{bmatrix}$ and $\begin{bmatrix} 2s^2+3s+5 & s^2+4s+1 \\ s^2+5s-1 & s^2+s-1 \end{bmatrix}$

Try also to find a gcrd if they are not coprime.

6.3-3.   *Coprimeness of Several Polynomials*

A set $\{A_i(s), i = 1, \ldots, r\}$ of polynomial matrices, all with the same number (say m) of columns, will be said to be right coprime if their right divisors are all unimodular. Show that the $A_i(s)$ are coprime if and only if

a. Rank $[A_1'(s) \cdots A_r'(s)]' = m$ for all s, and

b. There exist r polynomial matrices $X_i(s)$ such that $\sum_1^r X_i(s)A_i(s) = I$. Find the analogs of the other characterizations of coprimeness that were given in Sec. 6.3.1.

6.3-4. *Alternative Proof of (6.3-13)*

Prove that $U_{22}(s)$ in (6.3-12) is nonsingular by using contradiction. That is, assume there exists $\alpha(s)$ such that $\alpha(s)U_{22}(s) \equiv 0$. Then prove using $D(s)$ nonsingular that $\alpha(s)U_{21}(s) \equiv 0$ and hence that $U(s)$ is not unimodular. (Reference: W. A. Coppel, *Bull. Australian Math. Soc.*, **11**, pp. 89–113, 1974.)

6.3-5. *An Extension of Lemma 6.3-8*

Let $H(s) = N(s)D^{-1}(s)M(s)$, where N, D, M are polynomial matrices. By (6.3-17), we know that there exist polynomial matrices $\{U_{22}, U_{21}\}$ such that $ND^{-1} = -U_{22}^{-1}U_{21}$, $\{U_{22}, U_{21}\}$ left coprime. Let $A = U_{22}$, $B = -U_{21}M$, so that $H(s) = A^{-1}(s)B(s)$. Show that if $\{D, M\}$ are left coprime, the same will be true of $\{A, B\}$. *Hint:* By hypothesis there exist polynomial matrices $\tilde{X}, \tilde{Y}$ such that $M\tilde{X} + D\tilde{Y} = I$. Let $P = V_{22} - N\tilde{Y}V_{12}$, $Q = -\tilde{X}V_{12}$, where the V_{ij} are as defined in the proof of Lemma 6.3-8. Now show that $AP + BQ = U_{22}V_{22} + U_{21}V_{21} \triangleq I$. (Reference: W. A. Coppel, *Bull. Australian Math. Soc.*, **11**, pp. 89–113, 1974.)

6.3-6. *Another Proof of the Cayley-Hamilton Theorem* [7]

If $a(s) = \det(sI - A)$, we have $(sI - A)\operatorname{Adj}(sI - A) = a(s)I$. Show by using this relation and the corollary to Theorem 6.3-15 that $a(A) = 0$.

6.3-7. *Resolvent Formulas* [7]

Let $B(s) \triangleq \operatorname{Adj}(sI - A)$, and let $\det(sI - A) = a(s) = s^n + a_1s^{n-1} + \cdots + a_n$. Let

$$\delta(\lambda, \mu) \triangleq \frac{a(\lambda) - a(\mu)}{\lambda - \mu} = \lambda^{n-1} + (\mu + a_1)\lambda^{n-2}$$
$$+ (\mu^2 + a_1\mu + a_2)\lambda^{n-3} + \cdots$$

Show that this yields the resolvent formulas of Exercise A.23 by making the substitutions $\{\lambda = sI, \mu = A\}$ and $\{\lambda = A, \mu = sI\}$.

6.3-8.

Let $\{P(s), E, A\}$ be as in Lemma 6.3-20, and suppose that $P(s)$ is square and nonsingular. Show that $P^{-1}(s) = [I \ 0 \ \cdots \ 0][sE - A]^{-1}[I \ 0 \ \cdots \ 0]'$.

6.3-9. *Proof of Lemma 6.3-19*

a. Show that if $U(s)(sI - A) = (sI - B)V(s)$, where $U(s)$ and $V(s)$ are *unimodular*, then $U(sI - A) = (sI - B)V$, where U and V are constant matrices. [*Hint:* Use the division theorem and the fact that $(sI - B)^{-1}U$ is strictly proper.] Comparing terms in s now shows that $U = V$, so $UA = BU$.

b. Show that U is nonsingular. [*Hint:* Let W be the constant remainder when $U^{-1}(s)$ is divided on the left by $sI - A$; show that $WU = I$.] It follows that A and B are similar. The converse is easy.

Remark: C. I. Byrnes and M. A. Gauger (*J. Linear and Multilinear Algebra*, pp. 153–158, 1977) have derived an alternative (more easily verifiable) criterion for the similarity of two matrices A_1 and A_2: $\det(sI - A_1) = \det(sI - A_2)$ and $\operatorname{rank}(A_1 \otimes I - I \otimes A_1) = \operatorname{rank}(A_2 \otimes I - I \otimes A_2) = \operatorname{rank}(A_1 \otimes I - I \otimes A_2)$.

6.3-10.   Column-Reduced Matrices

Suppose $N(s)D^{-1}(s)$ is proper. Show that $[D'(s)\ N'(s)]'$ will be column reduced if and only if $D(s)$ is column reduced.

6.3-11.   GCDs of Minors of Equivalent Matrices

Suppose $P_1(s)$ and $P_2(s)$ are polynomial matrices related as $P_1(s) = U(s)P_2(s)V(s)$, where $\{U(s), V(s)\}$ are also polynomial.

a. Show that the gcd of all $i \times i$ minors of $P_2(s)$ divides the gcd of all $i \times i$ minors of $P_1(s)$ (see Exercise A.9 on the Binet-Cauchy theorem).

b. What can you say if $\{U(s), V(s)\}$ are unimodular?

6.3-12.   Minimal and Characteristic Polynomials

Show that the highest-degree invariant polynomial of $sI - A$ is also the minimal polynomial of A, i.e., it is the lowest degree polynomial, say $m(s)$, such that $m(A) = 0$.

Show that the characteristic polynomial of A is the product of all the invariant polynomials of $sI - A$.

6.3-13.   Smith Form of Real or Complex Matrices

If the elements of a matrix P are just real or complex numbers, show that the Smith form has $\lambda_1 = \lambda_2 = \cdots = \lambda_r = 1$, $\lambda_{r+i} = 0$, where r is the rank of P.

6.3-14.   Factorization of Matrices

Let R be a matrix with real elements. Use the result of Exercise 6.3-13 to prove the result (used in the Gilbert realization procedure of Sec. 6.1) that any matrix R of real or complex numbers can be written $R = BC$, where the number of columns of B is equal to the rank of R. (Another proof is given in Exercise A.17.)

6.3-15.   More on the Smith Form of a Companion Matrix

Let $a(s) = s^3 + a_1 s^2 + a_2 s + a_3$ and A_c be its associated *top-companion* matrix. Show that the unimodular matrices

$$U(s) = \begin{bmatrix} 0 & 0 & 1 \\ 0 & 1 & 0 \\ 1 & s + a_1 & s^2 + a_1 s + a_2 \end{bmatrix}, \quad V(s) = \begin{bmatrix} 0 & -1 & s^2 \\ -1 & 0 & s \\ 0 & 0 & 1 \end{bmatrix}$$

will convert $sI - A_c$ to its Smith form. Generalize to the $n \times n$ matrix case.

6.3-16.   Invariant Polynomials of Products

Let $P_i(s)$, $i = 1, 2$, have invariant polynomials $\{\psi_{i1}(s), \ldots, \psi_{ir_i}(s)\}$, $i = 1, 2$, where r_i is the rank of $P_i(s)$. Let $P(s) \triangleq P_1(s)P_2(s)$ have rank r ($r \leq \min\{r_1, r_2\}$), with invariant polynomials $\{\psi_1(s), \ldots, \psi_r(s)\}$. Show that $\psi_{ik}(s) | \psi_k(s)$, $i = 1, 2$, and $1 \leq k \leq r$.

6.3-17.   Simple Matrices

A polynomial matrix will be called *simple* if it has only one nonunity invariant polynomial. Show that a square $m \times m$ polynomial matrix $D(s)$ is simple if and only if it has rank greater than or equal to $m - 1$ for *all s*.

6.3-18.   Polynomial Solutions of Polynomial Equations

Let $F(s)$ be a full column rank polynomial matrix, and let $p(s)$ be a given polynomial vector. Show that the equation

$$F(s)q(s) = p(s)$$

will have a polynomial solution if and only if $F(s)$ is irreducible (i.e., it has full column rank for all s).

[*Remark:* Do this problem in at least two different ways].

6.3-19.   A Generalized Sylvester Resultant Matrix

a. Show that $\{N(s), D(s)\}$ will be right coprime if and only if we can find $\{A(s), B(s)\}$ left coprime with deg det $D(s) =$ deg det $A(s)$ and such that

$$A(s)N(s) + B(s)D(s) = 0. \tag{*}$$

b. Let $N(s) = N_0 s^L + N_1 s^{L-1} + \cdots + N_L$, with similar expressions for $\{D(s), B(s), A(s)\}$. Show that from (*) we can obtain the equations

$$[A_0 \quad B_0 \quad A_1 \quad B_1 \quad \cdots \quad A_{k-1} \quad B_{k-1}]S_k = 0, \qquad k = 1, 2, \ldots$$

where

$$S_k = \begin{bmatrix} N_0 & N_1 & \cdots & N_L & & \\ D_0 & D_1 & & D_L & & \\ 0 & N_0 & \cdots & N_{L-1} & N_L & \cdot \\ 0 & D_0 & & D_{L-1} & D_L & \cdot \\ & & & \cdot & & \\ & & & \cdot & & \\ & & & \cdot & & \\ & & & N_0 & & N_L \\ & & & D_0 & & D_L \end{bmatrix} \Bigg\} \; 2k \text{ block rows}$$

c. Let

$$r_0 = 0, \; r_i = \text{the rank of } S_i, \qquad i = 1, 2, \ldots$$
$$\alpha_0 = m + p, \; \alpha_i = r_i - r_{i-1}, \qquad i = 1, 2, \ldots$$

a. Show that $_i$ is a nonincreasing function of i

d. Let $n =$ the degree of det $D(s)$. Show that $\{N(s), D(s)\}$ will be right coprime if and only if $n = r_v - mv$, where v is the first integer such that $\alpha_{v+1} = m$. *Reference:* S. Kung, T. Kailath and M. Morf, "A Generalized Resultant Matrix," *Proc. 1976 IEEE Conference on Decision & Control*, pp. 892–895, Florida, December 1976. See also *IEEE Trans. Automat.*

Contr., **AC-23**, pp. 1043–1046, 1978. This reference also shows how to compute a gcrd of $\{N(s), D(s)\}$ and describes a way of exploiting the block Toeplitz structure of the $\bar{s}_k$ to speed up computations.

6.3-20.   *Intertwining Equivalence*

Show that the $p \times m$ polynomial matrices $P_1(s)$ and $P_2(s)$ are Smith equivalent if and only if there exist $p \times p$ and $m \times m$ polynomial matrices $\{M_1(s), M_2(s)\}$ such that $M_1(s)P_1(s) = P_2(s)M_2(s)$ and $\{P_1(s), M_2(s)\}$ are right coprime and $\{M_1(s), P_2(s)\}$ are left coprime. *Hint:* Use the generalized Bezout identities. *Note:* Such relations were first introduced by Fuhrmann and are especially important in studying problems where $P_1(s), P_2(s)$ have different dimensions provided $m_1 - m_2 = p_1 - p_2$ (see Sec. 8.2 for further discussion and references). In this case, the above relations will imply that $P_1(s)$ and $P_2(s)$ have the same nonunity invariant polynomials.

6.3-21.   *Skew-Prime Matrices*

a. Let $\{M_1(s), P_2(s)\}$ be left coprime, $\{P_1(s), M_2(s)\}$ be right coprime and such that $M_1(s)P_1(s) = P_2(s)M_2(s)$.

Show that there exist polynomial matrices $\{X(s), \bar{X}(s), Y(s), \bar{Y}(s)\}$ such that

$$\bar{X}M_1 + P_1 X = I = YP_2 + M_2\bar{Y}.$$

W. Wolovich [*IEEE Trans. Automat. Contr.*, **AC-23**, pp. 880–887, Oct. 1978] calls the matrices $\{M_1, P_1\}$ and $\{P_2, M_2\}$ *internally skew prime* (and $\{P_1, M_1\}$ and $\{M_2, P_2\}$ *externally skew prime*). *Hint:* Use the (reverse) Bezout identities.

b. If $\{M_1, P_1\}$ are internally skew prime, show that there exists $\{P_2, M_2\}$ such that $M_1P_1 = P_2M_2$ with $\{M_1, P_2\}$ left coprime and $\{P_1, M_2\}$ right coprime.

6.3-22.   *Generalized Polynomials*

Let $R[z]$ denote the (ring of) polynomials in z with real coefficients and let $R_\Lambda[z]$ denote the set of rational functions with no poles in some subset Λ of the complex plane.

$R_\Lambda[z]$ has properties analogous to $R[z]$ and therefore it has been called the ring of generalized polynomials by Pernebo [16].

a. Invertible elements in a ring are known as units. Show that the units in $R_\Lambda[z]$ are all nonzero rational functions with no poles or zeros in Λ.

b. Show that every nonzero $a \in R_\Lambda[z]$ can be *uniquely* factored as $a = pr$, where p is a monic polynomial with all its zeros in Λ and r is a unit in $R_\Lambda[z]$.

c. The Λ-degree, say $\deg_\Lambda a$, is defined as the degree of the polynomial p in the above factorization. Show that for all $\{a, b\}$ in $R_\Lambda[z]$, with $b \not\equiv 0$, there exist $\{q, r\}$ in $R_\Lambda[z]$ such that

$$a = bq + r, \quad \text{with} \quad \deg_\Lambda r < \deg_\Lambda b$$

d. Show that $R_\Lambda[z]$ is a principal ideal domain (cf. Sec. 5.1.3).

As shown, for example, in [6] and [9], almost all the results on polynomial matrices in Sec. 6.3 (and rational matrices in Sec. 6.5) apply to matrices with elements that are generalized polynomials. By choosing Λ to be the right half plane, and to include the point at ∞, we can ensure that our calculations will yield only proper stable transfer functions, see e.g. [13], [16], and [17].

6.4 SOME BASIC STATE-SPACE REALIZATIONS

In this section, we shall extend the realization procedures of Sec. 2.1 to obtain certain controller-, observer-, controllability-, and observability-type forms for a given MFD. We shall also describe how to transform given state-space realizations to one or the other of these special forms.

The ideas and results developed in this section will be fundamental for most of the rest of this book; we shall see that they provide a very concrete tool for understanding many aspects of multivariable system theory.

We shall restrict ourselves to strictly proper transfer functions, with the changes necessary to handle the more general case being straightforward— also see Chapter 8.

The methods we shall follow will be natural generalizations of those used in Sec. 2.1.1 for the scalar case, and therefore a quick review of that material will be useful. In particular, as in the scalar case, the key idea will be to obtain *controller-* and *observer*-form realizations of the basic MFDs $I_p(s)D^{-1}(s)$ and $D^{-1}(s)I_m(s)$.

6.4.1 Controller-Form Realizations from Right MFDs

Given a strictly proper right MFD

$$H(s) = N(s)D^{-1}(s),$$

we shall show how to construct a controllable realization of order equal to deg det $D(s)$. (The reader will recognize that this construction will be the proof of our previously stated Claim A in Sec. 6.2.3.)

We shall follow the lines of the procedure introduced in Sec. 2.1 to obtain controller-form realizations of a scalar transfer function. The final construction was first obtained (by other arguments) by Wang [24] and Wolovich [20] (see also Wolovich and Falb [25]).

Thus, given

$$H(s) = N(s)D^{-1}(s)$$

let us rewrite it as

$$D(s)\xi(s) = u(s) \tag{1a}†$$

$$y(s) = N(s)\xi(s) \tag{1b}$$

---

†$\xi(s)$ has been called the (transform of the) *partial state*. For notational convenience, we shall not use capital letters to denote transforms.

Now (1) corresponds to a *system of coupled* differential equations. In Sec. 2.1.1, we tackled the scalar version of this by rearranging the equation

$$a_0\xi^{(n)} + a_1\xi^{(n-1)} + \cdots + a_n\xi = u, \qquad a_0 \neq 0$$

as

$$\xi^{(n)} = -\frac{1}{a_0}(a_1\xi^{(n-1)}\cdots + a_n\xi) + \frac{u}{a_0} \tag{2}$$

Then, first assuming $\xi^{(n)}$ was available, we integrated it n times to obtain the lower-order derivatives, which were used, along with u, to synthesize $\xi^{(n)}$ according to (2). This was then used to *close the loop*, resulting in a simulation of the given equation.

We shall attempt to follow a similar path here, beginning with a rearrangement of $D(s)\xi = u$ into a more suitable and suggestive form for simulation. We have

$$\begin{bmatrix} d_{11}(s) & \cdots & d_{1m}(s) \\ \cdot & & \cdot \\ \cdot & & \cdot \\ \cdot & & \cdot \\ d_{m1}(s) & \cdots & d_{mm}(s) \end{bmatrix} \begin{bmatrix} \xi_1 \\ \cdot \\ \cdot \\ \cdot \\ \xi_m \end{bmatrix} = \begin{bmatrix} u_1 \\ \cdot \\ \cdot \\ \cdot \\ u_m \end{bmatrix}$$

A reasonable first step is to determine the highest-order derivative of each ξ_i, which is clearly equal to the *degree*, say k_i, of the ith column of $D(s)$. To display these degrees more explicitly, we can write $D(s)$ as

$$D(s) = D_{hc}S(s) + D_{lc}\Psi(s) \tag{3}$$

where [cf. (6.3-22)]

$$S(s) = \text{diag}\{s^{k_1}, \ldots, s^{k_m}\}$$

and the k_i are the column degrees of $D(s)$, and D_{hc} is the highest-column-degree coefficient matrix of $D(s)$.

The term $D_{lc}\Psi(s)$ accounts for the remaining *lower-column-degree* terms of $D(s)$, with D_{lc} a matrix of coefficients, and

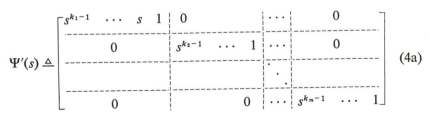

$$\tag{4a}$$

or, more compactly,

$$\Psi'(s) = \text{block diag } \{[s^{k_1-1}, \ldots, s, 1], \ldots, [s^{k_m-1}, \ldots, 1]\} \qquad (4b)$$

Analog-Computer Simulation.   Returning to Eq. (1) and following the pattern of the scalar case, we write it as

$$D_{hc}S(s)\xi(s) = -D_{lc}\Psi(s)\xi(s) + u(s) \qquad (5)$$

Then, *assuming that D_{hc} is invertible,* we have

$$S(s)\xi = -D_{hc}^{-1}D_{lc}\Psi(s)\xi + D_{hc}^{-1}u \qquad (6)$$

In the scalar case, the assumption $D_{hc} \neq 0$ means that we are really dealing with an nth-order equation. In the matrix case, the analogous assumption, with $n = \deg \det D(s)$, is that

$$D(s) \text{ is } column\ reduced.$$

As noted in Example 6.3-2, we can always reduce $D(s)$ to this form, without affecting the determinantal degree. Now we note that (6) displays on the left-hand side the highest-order derivatives of each ξ_i, in *decoupled* form. Therefore, we can apply Kelvin's technique: Assume temporarily that each $\xi_i^{(k_i)}$ is available and integrate each k_i times to obtain all the required lower-order derivatives. We shall then have *m chains,* with k_i integrators in the ith chain. We may assume without loss of generality† that the k_i are ordered such that

$$k_1 \leq k_2 \leq \cdots \leq k_m$$

Note that the outputs of the integrators are given by the entries of $\Psi(s)\xi(s)$. This gives us the so-called *core* realization with transfer function $\Psi(s)S^{-1}(s)$.

We now merely assemble these integrator outputs (states or lower-order derivatives) along with the input according to the prescription on the RHS (right-hand side) of (6) and thus generate the LHS (left-hand side) of (6). Closing the loop completes the realization of Eq. (1a)—see the symbolic diagram, Fig. 6.4-1.

The output equation (1b) is now easy to realize: We can write

$$y(s) = N(s)\xi(s) = N_{lc}\Psi(s)\xi(s) \qquad (7)$$

where N_{lc} is an appropriate matrix of coefficients, which shows clearly that y is obtained as weighted sums of the states (cf. Fig. 6.4-1).

†This corresponds to merely relabeling, i.e., permuting, the original equations.

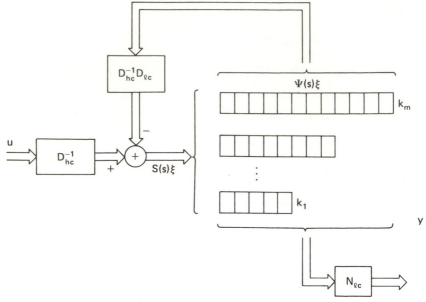

Figure 6.4-1. Schematic of controller-form realization $N(s)D^{-1}(s)$.

State-Space Equations. With an analog-computer realization in hand, we can readily describe it in state-space language. Recall that our first step was to set up m chains of k_i integrators each, with access to the input of the first integrator of each chain and to the outputs of every integrator of every chain. The corresponding system matrices for the realization of this core system are clearly

$$A_c^0 = \text{block diag}\left\{ \begin{bmatrix} 0 & & & \bigcirc \\ 1 & \cdot & & \\ & \cdot & \cdot & \\ \bigcirc & & 1 & 0 \end{bmatrix}, k_i \times k_i, i = 1, \ldots, m \right\} \tag{8a}\dagger$$

$$[B_c^0]' = \text{block diag} \{[1 \quad 0 \quad \cdots \quad 0], 1 \times k_i, i = 1, m\} \tag{8b}$$

and

$$C_c^0 = I_n, \quad n = \deg \det D(s) = \sum_1^m k_i \tag{8c}$$

It can also be checked by direct calculation that

$$(sI - A_c^0)^{-1}B_c^0 = C_c^0(sI - A_c^0)^{-1}B_c^0 = \Psi(s)S^{-1}(s) \tag{9}$$

†For reasons described in Sec. 6.4.6, we shall call the $\{k_i\}$ the *controllability indices* of (A_c, B_c, C_c); the reader may wish to check that the $\{k_i\}$ are the *right Kronecker* indices of the pencil $[sI - A_c \ B_c]$ (cf. Sec. 6.3.4).

The core realization is controllable and observable, as may be seen by direct calculation or more simply by looking at the simulation model: We have access to the first integrator of each chain (hence, controllability), and we observe the output of each integrator (hence, observability).

The next step is to assemble the states and the input according to the RHS of (6) and close the loop: This corresponds to state feedback† through a gain $D_{hc}^{-1} D_{lc}$ and to application of an input $D_{hc}^{-1} u$ (cf. Fig. 6.4-1).

The system matrices for the core realization modified in this way are

$$A_c = A_c^0 - B_c^0 D_{hc}^{-1} D_{lc}, \qquad B_c = B_c^0 D_{hc}^{-1} \qquad (10a)$$

while the output matrix, in accordance with (7), becomes

$$C_c = N_{lc} \qquad (10b)$$

The modifications required in (10) are easy to carry out: We merely replace certain rows of A_c^0 by the rows of $D_{hc}^{-1} D_{lc}$ and certain rows of B_c^0 by the rows of D_{hc}^{-1}. This should be clarified by the following simple example.

Example 6.4-1.   Controller-Form Realization of a Right MFD

  Let

$$N(s) = \begin{bmatrix} s & 0 \\ -s & s^2 \end{bmatrix}, \qquad D(s) = \begin{bmatrix} 0 & -(s^3 + 4s^2 + 5s + 2) \\ (s+2)^2 & s+2 \end{bmatrix}$$

which corresponds to the MFD (6.2-48) for the transfer function (6.2-44) in Example 6.2-1. Then we can see that

$$k_1 = 2, \qquad k_2 = 3, \qquad k_1 + k_2 = 5 = \deg \det D(s)$$

The highest-column-degree coefficient matrix is

$$D_{hc} = \begin{bmatrix} 0 & -1 \\ 1 & 0 \end{bmatrix} \quad \text{with} \quad D_{hc}^{-1} = \begin{bmatrix} 0 & 1 \\ -1 & 0 \end{bmatrix}$$

while

$$S(s) = \begin{bmatrix} s^3 & 0 \\ 0 & s^2 \end{bmatrix}, \qquad \Psi'(s) = \begin{bmatrix} s & 1 & 0 & 0 & 0 \\ 0 & 0 & s^2 & s & 1 \end{bmatrix}$$

$$D_{lc} = \begin{bmatrix} 0 & 0 & -4 & -5 & -2 \\ 4 & 4 & 0 & 1 & 2 \end{bmatrix}, \qquad N_{lc} = \begin{bmatrix} 1 & 0 & 0 & 0 & 0 \\ -1 & 0 & 1 & 0 & 0 \end{bmatrix}$$

Therefore,

$$D_{hc}^{-1} D_{lc} = \begin{bmatrix} 4 & 4 & 0 & 1 & 2 \\ 0 & 0 & 4 & 5 & 2 \end{bmatrix}$$

---

†The reader should check that the effect of state feedback $u = G[v - Kx]$ is to give a new realization $\{A - BGK, BG, C\}$.

and by (8)–(10),

$$A_c = \begin{bmatrix} -4 & -4 & 0 & -1 & -2 \\ 1 & 0 & 0 & 0 & 0 \\ \hline 0 & 0 & -4 & -5 & -2 \\ 0 & 0 & 1 & 0 & 0 \\ 0 & 0 & 0 & 1 & 0 \end{bmatrix}, \quad B_c = \begin{bmatrix} 0 & 1 \\ 0 & 0 \\ \hline -1 & 0 \\ 0 & 0 \\ 0 & 0 \end{bmatrix}, \quad C_c = N_{lc}$$

∎

In general, the A_c matrix will have $k_i \times k_i$ block diagonal elements in *top-companion* form, with the off-diagonal block elements being zero everywhere except possibly in the top row of each block. Similarly, the B_c matrix will have $k_i \times m$ blocks, with only the top row of each block being possibly nonzero. The C_c matrix will be full, in general.

When $m = 1$, we clearly have the *controller* (canonical) form. When $m > 1$, the reasons for the name controller form will be seen later (Sec. 7.1) in connection with the problem of pole shifting by state feedback. For reasons that will be explained in Sec. 6.4.5, we shall not call the form of this section a *canonical* controller form.

6.4.2   Some Properties of the Controller-Form Realization

Of course, the realization must be controllable, since it was obtained by state feedback around a controllable realization $\{A_c^0, B_c^0\}$ (just apply the PBH tests of Sec. 6.2.2).

Therefore,

$$\{A_c, B_c\} \text{ is controllable} \tag{11}$$

However, observability is not guaranteed, in general; it will depend on $N(s) = N_{lc}\Psi(s)$ and, in fact, on the relative primeness of $N(s)$ and $D(s)$ (cf. Sec. 6.5.1).

We must, of course, have the relation

$$N(s)D^{-1}(s) = H(s) = C_c(sI - A_c)^{-1}B_c \tag{12a}$$

Then from the expressions

$$N(s) = N_{lc}\Psi(s), \qquad N_{lc} = C_c \tag{12b}$$

and the fact that (12) is valid for every matrix N_{lc}, we would expect that

$$\Psi(s)D^{-1}(s) = [sI - A_c]^{-1}B_c \tag{13}$$

This useful result, which should be compared with (9), can also be proved by direct calculation. For example, by (8)–(10), we can write

$$(sI - A_c)\Psi(s) = (sI - A_c^0)\Psi(s) + B_c^0 D_{hc}^{-1} D_{lc}\Psi(s)$$

$$= B_c^0[S(s) + D_{hc}^{-1} D_{lc}\Psi(s)]$$

$$= B_c^0 D_{hc}^{-1}[D_{hc}S(s) + D_{lc}\Psi(s)] = B_c D(s)$$

The following lemma gives a useful and compact description of the construction of Sec. 6.4.1.

Lemma 6.4-1.

The controller-form realization $\{A_c, B_c, C_c\}$ of the column-reduced right MFD $N(s)D^{-1}(s)$ can be related as

$$\begin{bmatrix} sI - A_c & B_c \\ -C_c & 0 \end{bmatrix}\begin{bmatrix} \Psi(s) & 0 \\ 0 & I \end{bmatrix} = \begin{bmatrix} B_c & 0 \\ 0 & I \end{bmatrix}\begin{bmatrix} D(s) & I \\ -N(s) & 0 \end{bmatrix} \tag{14a}$$

where

$$\{\Psi(s), D(s)\} \text{ are right coprime} \tag{14b}$$

and

$$\{sI - A_c, B_c\} \text{ are left coprime} \tag{14c}$$

Proof. The relations (14a) follow immediately from (12)–(13). The fact (14c) follows from the controllability of $\{A_c, B_c\}$ [use the **PBH** test (Theorem 6.2-6) and Lemma 6.3-6]. Next, we note that the form of $\Psi(s)$ [cf. (4)] shows that $[\Psi'(s) \ D'(s)]'$ contains an $m \times m$ identity matrix, so that the rank will be m for all s. Then Lemma 6.3-6 shows that (14b) must hold. (The true significance of (14) will only appear gradually, but the active reader could profitably keep Exercise 6.3-20 in mind at this point.) ∎

Continuing with the properties of $\{A_c, B_c, C_c\}$, we recall that, in the scalar case, there was a direct relationship between the denominator polynomial $a(s)$ and the characteristic polynomial of the controller form, viz., that $a(s) = \det(sI - A_c)$. A similar relation holds in the multivariable case as well except for the fact that $\det(sI - A_c)$ is monic (i.e., has leading coefficient unity), while $\det D(s)$ is not (its leading coefficient is $\det D_{hc}$ [cf. (11)]). Therefore, we would expect that

$$\det(sI - A_c) = (\det D_{hc})^{-1} \det D(s) \tag{15}$$

and, in fact, we can prove this by direct calculation (Exercise 6.4-2).

However, here we shall use a more elegant method (see also Exercise 6.3-20), which will, in fact, prove substantially more. Thus, we shall show that (review Sec. 6.3.3)

$$sI - A_c \text{ and } D(s) \text{ have the same nonunity invariant polynomials} \tag{16}$$

and that the following matrices are *equivalent in the sense that they have the same Smith form* (cf. Theorem 6.3-16):

$$
\begin{bmatrix} sI - A_c & B_c \\ -C_c & 0 \end{bmatrix} \overset{S}{\simeq} \begin{bmatrix} I_n & 0 \\ 0 & N(s) \end{bmatrix}
\tag{17}
$$

These properties will be useful in Sec. 6.5.3 in defining the *poles* and *zeros* of a multivariable transfer function.

A closely related result is that the eigenvectors of A_c have the form

$$
p = \Psi(\lambda)q, \qquad \lambda = \text{an eigenvalue of } A_c
\tag{18a}
$$

where q is any vector such that

$$
D(\lambda)q = 0
\tag{18b}
$$

Note that (18) is a natural generalization of the result for the scalar case where A_c is a single companion matrix—see Exercise A.35.

Nice proofs of the results (15)–(18) (and several other facts) can be obtained by using the following result, which follows easily from Lemma 6.3-9.

Lemma 6.4-2.   Generalized Bezout Identities

Let $\{A_c, B_c\}$ be a controller-form realization of the right MFD $\Psi(s)D^{-1}(s)$, where $\Psi(s)$ is as in (4). Then there exist polynomial matrices $\{\tilde{X}(s), \tilde{Y}(s), X(s), Y(s)\}$ such that

$$
\begin{bmatrix} sI - A_c & B_c \\ -\tilde{X}(s) & \tilde{Y}(s) \end{bmatrix} \begin{bmatrix} X(s) & -\Psi(s) \\ Y(s) & D(s) \end{bmatrix} = \begin{bmatrix} I & 0 \\ 0 & I \end{bmatrix}
\tag{19}
$$

and

$$
\begin{bmatrix} X(s) & -\Psi(s) \\ Y(s) & D(s) \end{bmatrix} \begin{bmatrix} sI - A_c & B_c \\ -\tilde{X}(s) & \tilde{Y}(s) \end{bmatrix} = \begin{bmatrix} I & 0 \\ 0 & I \end{bmatrix}
\tag{20}
$$

Moreover, the three block matrices appearing in (19) and (20) are *all unimodular*. We shall refer to (19) as a *forward Bezout identity* and to (20) as a *reverse Bezout identity*.

Proof.   This is essentially a rearrangement of Lemma 6.3-9, but it will not hurt to repeat the arguments for this important result.

By (14b) and (14c), we know that $\{sI - A_c, B_c\}$ are left coprime and $\{\Psi(s), D(s)\}$ are right coprime. Therefore, by Lemma 6.3-5, we know that there exist polynomial matrices $\{X(s), Y(s)\}$ and $\{\bar{X}(s), \bar{Y}(s)\}$ such that

$$
(sI - A_c)X(s) + B_c Y(s) = I_n
\tag{21a}
$$

and

$$
\bar{X}(s)\Psi(s) + \bar{Y}(s)D(s) = I_m
\tag{21b}
$$

We can combine these (Bezout) identities and the earlier formula (13) in matrix form as

$$\begin{bmatrix} sI - A_c & B_c \\ -\bar{X}(s) & \bar{Y}(s) \end{bmatrix} \begin{bmatrix} X(s) & -\Psi(s) \\ Y(s) & D(s) \end{bmatrix} = \begin{bmatrix} I & 0 \\ Q(s) & I \end{bmatrix} \tag{22}$$

where

$$Q(s) \triangleq -\bar{X}(s)X(s) + \bar{Y}(s)Y(s), \qquad \text{a polynomial matrix}$$

Then premultiplying both sides by the inverse of the matrix on the right-hand side will yield (19) when we define

$$[-\tilde{X}(s) \quad \tilde{Y}(s)] \triangleq [-Q(s) \quad I] \begin{bmatrix} sI - A_c & B_c \\ -\bar{X}(s) & \bar{Y}(s) \end{bmatrix}$$

$$= [-\bar{X}(s) - Q(s)(sI - A_c) \quad \bar{Y}(s) - Q(s)B_c]$$

Next, we note that the block matrices in (19) are all square. Therefore, the rule for the determinant of a product of square matrices shows that all the block matrices in (19) must have determinants that are independent of s.

Finally, since a left inverse of a nonsingular matrix must also be a right inverse (prove this), (19) will also hold with the order of the matrices on the left-hand side interchanged. This completes the proof of the lemma. ∎

Now the results (15)–(18) can be deduced quite easily. We first note that from (19) we can write

$$\begin{bmatrix} sI - A_c & B_c \\ 0 & I \end{bmatrix} \begin{bmatrix} X(s) & -\Psi(s) \\ Y(s) & D(s) \end{bmatrix} = \begin{bmatrix} I & 0 \\ Y(s) & D(s) \end{bmatrix} \tag{23}$$

Now the second matrix in (23) being unimodular means that [cf. (6.3-30)]

$$\begin{bmatrix} sI - A_c & B_c \\ 0 & I \end{bmatrix} \overset{S}{=} \begin{bmatrix} I & 0 \\ Y(s) & D(s) \end{bmatrix}$$

i.e., that these matrices will have the same Smith form and hence the same nonunity invariant polynomials. Now elementary operations will not affect the Smith form, and therefore we shall have

$$\begin{bmatrix} sI - A_c & B_c \\ 0 & I \end{bmatrix} \overset{S}{=} \begin{bmatrix} sI - A_c & 0 \\ 0 & I \end{bmatrix}$$

and

$$\begin{bmatrix} I & 0 \\ Y(s) & D(s) \end{bmatrix} \overset{S}{=} \begin{bmatrix} I & 0 \\ 0 & D(s) \end{bmatrix}$$

The property (6.3-31c) of the equivalence relation $\underset{\sim}{S}$ then implies that

$$\begin{bmatrix} sI - A_c & 0 \\ 0 & I \end{bmatrix} \underset{\sim}{S} \begin{bmatrix} I & 0 \\ 0 & D(s) \end{bmatrix} \tag{24}$$

which immediately yields the claim (16) that $sI - A_c$ and $D(s)$ have the same nonunity invariant polynomials.

The characteristic polynomial of $sI - A_c$ is the product of its invariant polynomials, and therefore

$$\det (sI - A_c) \propto \det D(s)$$

The proportionality constant must clearly be $\det D_{hc}^{-1}$ in order to make the RHS monic. Equation (15) is thus verified.

To prove (17), we note that from (19) we can also write

$$\begin{bmatrix} sI - A_c & B_c \\ -C_c & 0 \end{bmatrix} \begin{bmatrix} X(s) & -\Psi(s) \\ Y(s) & D(s) \end{bmatrix} = \begin{bmatrix} I & 0 \\ -C_c X(s) & N(s) \end{bmatrix} \tag{25}$$

and now proceed as we did above to obtain (24) from (23).

To derive the eigenvector formula (18), we note from the relation

$$[sI - A_c \quad -B_c] \begin{bmatrix} \Psi(s) \\ D(s) \end{bmatrix} = 0 \tag{26}$$

that the rank m matrix $[\Psi'(s) \ D'(s)]$ spans† the null-space of $[sI - A_c \ -B_c]$. Now any eigenvector of A_c obeys $A_c p = \lambda p$, so that

$$[\lambda I - A_c \quad -B_c] \begin{bmatrix} p \\ 0 \end{bmatrix} = 0 \tag{27}$$

Therefore, $[p' \ 0]'$ lies in the null-space of $[\lambda I - A_c \ - B_c]$, and so we must have, for some vector q,

$$\begin{bmatrix} p \\ 0 \end{bmatrix} = \begin{bmatrix} \Psi(\lambda) \\ D(\lambda) \end{bmatrix} q \tag{28}$$

But this is just the previously stated characterization (18).

---

†We consider polynomial vectors as forming a linear vector space over the (field of) rational functions and use the fact (cf. Sec. 5.2) that $\dim \mathfrak{R}(A) + \dim \eta(A) = q$ for any $m \times q$ matrix A. In our case, $A = [sI - A_c \ - B_c]$, $q = m + n$, and $\dim \mathfrak{R}(A) = n$ because $[sI - A_c \ - B_c]$ has full rank.

6.4.3 Observer-Form Realizations from Left MFDs

We shall now consider how to obtain the *dual* observer-form realizations, starting from left MFDs

$$H(s) = D_L^{-1}(s)N_L(s)$$

The $D_L(s)$ and $N_L(s)$ here are $p \times p$ and $p \times m$ matrices, which are different from the $m \times m$ matrix $D_R(s)$ and the $p \times m$ matrix $N_R(s)$ of the right MFD. *However, we shall generally not use any special subscripts (L or R) to indicate this difference, relying on context instead.*

Perhaps the simplest way to obtain a realization is to first consider the transposed transfer function

$$H'(s) = N'(s)[D^{-1}(s)]' = N'(s)D^{-T}(s) \tag{29a}$$

and find a controller-form realization for it by the method of Sec. 6.4.1, say,

$$H'(s) = \bar{C}_c(sI - \bar{A}_c)^{-1}\bar{B}_c \tag{29b}$$

Then a realization for $H(s)$ clearly is, say,

$$H(s) = C_o(sI - A_o)^{-1}B_o \tag{30a}$$

where

$$C_o = \bar{B}'_c, \qquad A_o = \bar{A}'_c, \qquad B_o = \bar{C}'_c \tag{30b}$$

with the subscript o denoting the observer-type form. The only thing is to carefully identify $\{\bar{A}_c, \bar{B}_c, \bar{C}_c\}$ from $N'(s)D^{-T}(s)$ because, as the reader may expect, all the column-related properties of the controller realization will now depend on the rows of $N(s)$ and $D(s)$.

Thus, the significant degrees will be

$$l_i = \text{the degree of the } i\text{th } row \text{ of } D(s)$$

The matrix $D(s)$ must be *row reduced* (rather than column reduced), so that

$$\sum_1^p l_i = \deg \det D(s) \tag{31}†$$

and D_{hr}, the coefficient matrix of the highest-order terms in each row of $D(s)$, is nonsingular. Note that

$$D_{hr} = (D')_{hc}, \qquad N_{lr} = (N')_{lc}$$

†We shall call the $\{l_i\}$ the *observability indices* of $\{A_0, B_0, C_0\}$—cf. the footnote for (8a). The reader can check that they are also the *left Kronecker indices* of the pencil $[C' \ sI - A']'$ (cf. Sec. 6.3.4).

413

Pursuing this line of thought, we expect the *core-observer* realization to be given by [cf. (8)]

$$A_o^0 = \text{block diag}\left\{\begin{bmatrix} 0 & 1 & & & \\ & & \cdot & & \bigcirc \\ & & & \cdot & \\ & & & \cdot & 1 \\ \bigcirc & & & \cdot & 0 \end{bmatrix}, l_i \times l_i, i = 1, \ldots, p\right\} \qquad (32a)$$

$$C_o^0 = \text{block diag}\{[1 \quad 0 \quad \cdots \quad 0], 1 \times l_i, i = 1, \ldots, p\} \qquad (32b)$$

$$B_o^0 = I_n, \qquad n = \sum_1^p l_i = \deg \det D(s) \qquad (32c)$$

and the observer-form realization by [cf. (10)]

$$A_o = A_o^0 - D_{lr}D_{hr}^{-1}C_o^0 \qquad (33a)$$

$$C_o = D_{hr}^{-1}C_o^0, \qquad B_b = N_{lr} \qquad (33b)$$

Finally, the duality should yield the schematic diagram of Fig. 6.4-2. We have left out many details in this rapid sketch, which the reader should find useful to fill in. Here, however, we shall confirm the above results by a direct analysis (generalizing the scalar procedure of Sec. 2.1.1, Fig. 2.1-9), because familiarity with both types of arguments will be useful in further work.

Thus, given the left MFD

$$y(s) = D^{-1}(s)N(s)u(s)$$

define the *partial state* $\xi(s)$

$$D(s)y(s) = \xi(s) = N(s)u(s) \qquad (34)$$

Now we can write

$$D(s) = S_L(s)D_{hr} + \Psi_L(s)D_{lr} \qquad (35a)\dagger$$

where

$$S_L(s) = \text{diag}\{s^{l_i}, i = 1, \ldots, p\} \qquad (35b)$$

$$\Psi_L(s) = \text{block diag}\{[s^{l_i-1}, \ldots, s, 1], \qquad i = 1, \ldots, p\} \qquad (35c)$$

and therefore, using the assumption that $D(s)$ is row reduced, we obtain

$$y(s) = D_{hr}^{-1}S_L^{-1}(s)[\xi(s) - \Psi_L(s)D_{lr}y(s)] \qquad (36)$$

†The subscript L is introduced to distinguish these matrices from those introduced in Sec. 6.4.2 for *right* MFDs.

This equation can be implemented as shown in Fig. 6.4-2, where the integrator chains realize what may be called the *core-observer* form. The realization is completed by feeding the input into the integrator chains according to

$$\xi(s) = N(s)u(s) = \Psi_L(s)N_{lr}u(s) \qquad (37)$$

To write down the state-space equations, we can first check that the *core* transfer function $S_L^{-1}(s)\psi_L(s)$ is implemented with the triple $\{A_o^0, B_o^0, C_o^0\}$ specified by (32). Then we obtain the overall realization by using output feedback,

$$u^0(t) \longrightarrow u^0(t) - D_{lr}D_{hr}^{-1}y^0(t)$$

and also setting

$$u^0(t) = N_{lr}u(t), \qquad y^0(t) = D_{hr}y(t)$$

Putting these things together, we get the desired realization $\{A_o, B_o, C_o\}$ as described by (33).

The *nonfixed* columns of A_o will correspond directly to the columns of $D_{lr}D_{hr}^{-1}$, and those of C_o will correspond directly to the columns of D_{hr}^{-1}.

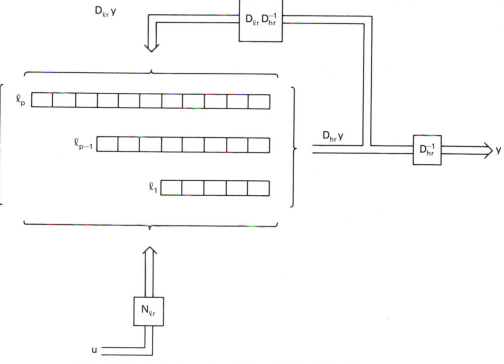

Figure 6.4-2. Symbolic representation of multivariable observer-form realizations of $D^{-1}(s)N(s)$.

As a last general remark, we should note the general form of the observer-form realization. From (33), we see that the matrix A_0 will have $l_t \times l_t$ block diagonal elements in left-companion form, with possibly nonzero elements in the first column, while the off-diagonal block elements will be zero everywhere except possibly in the first column. The C_o matrix will have $p \times l_t$ blocks, zero everywhere except possibly in the first column, while the B_o matrix will generally be full. We remark again that A_o' has the form of A_c (the A matrix of the controller form) except for the dimensions.

The reader will find it helpful to go through the steps described above in detail for the simple example given below; observe, for example, that the above procedure involves "cancelling differentiators with integrators," i.e., $\Psi_r(s)D_{lr}y(s)$ is obtained by differentiating $y(\cdot)$ and is fed in with $\xi(s)$—but the row properness of $D(s)$ enables the feed-in points to be shifted down the integrator chains, so that all differentiators may be eliminated—cf. the discussion of the scalar case in Sec. 2.1.2.

Example 6.4-2.   Observer-Form Realization of a Left MFD

We can form left MFDs of the transfer function of Example 6.4-1 as

$$H(s) = \begin{bmatrix} (s+1)^2(s+2)^2 & 0 \\ 0 & (s+2)^2 \end{bmatrix}^{-1} \begin{bmatrix} s & s(s+1)^2 \\ -s & -s \end{bmatrix}$$

$$= \begin{bmatrix} (s+1)^2(s+2) & s+2 \\ 0 & (s+2)^2 \end{bmatrix}^{-1} \begin{bmatrix} 0 & -s^2 \\ -s & -s \end{bmatrix}$$

The second MFD has determinantal degree 5 as compared to 6 for the first one, and we shall work with it. Note that

$$D(s) = \begin{bmatrix} s^3 & 0 \\ 0 & s^2 \end{bmatrix}\begin{bmatrix} 1 & 0 \\ 0 & 1 \end{bmatrix}$$

while

$$+ \begin{bmatrix} s^2 & s & 1 & 0 & 0 \\ 0 & 0 & 0 & s & 1 \end{bmatrix}\begin{bmatrix} 4 & 5 & 2 & 1 & 2 \\ 0 & 0 & 0 & 4 & 4 \end{bmatrix}'$$

$$N(s) = \begin{bmatrix} s^2 & s & 1 & 0 & 0 \\ 0 & 0 & 0 & s & 1 \end{bmatrix}\begin{bmatrix} 0 & 0 & 0 & -1 & 0 \\ 1 & 0 & 0 & -1 & 0 \end{bmatrix}'$$

The reader should now go through the details of the previously described realization procedure, suitably number the states from right to left along the integrator chains, and verify that

$$A_o = \begin{bmatrix} -4 & 1 & 0 & \vdots & 0 & 0 \\ -5 & 0 & 1 & \vdots & 0 & 0 \\ -2 & 0 & 0 & \vdots & 0 & 0 \\ \hline -1 & 0 & 0 & \vdots & -4 & 1 \\ -2 & 0 & 0 & \vdots & -4 & 0 \end{bmatrix}, \quad B_o = \begin{bmatrix} 0 & 1 \\ 0 & 0 \\ 0 & 0 \\ \hline -1 & -1 \\ 0 & 0 \end{bmatrix}$$

$$C_o = \begin{bmatrix} 1 & 0 & 0 & \vdots & 0 & 0 \\ 0 & 0 & 0 & \vdots & 1 & 0 \end{bmatrix}$$

Note when $D_{hr} = I$, as in this example, that the *nontrivial* columns of A_o (i.e., the ones whose structure is not fixed by A_o^0) have a very direct relationship to the columns of the matrix D_{lr}. ∎

The chief applications of the observer form appear in the design of state observers (Sec. 7.3). However, perhaps even more important than any direct application is that we now have two different ways of realizing the matrix transfer function $D^{-1}(s)$ (depending on whether it is column or row reduced). This can be exploited to get other useful realizations, not just for MFDs but also for transfer functions that arise (in many applications) in the form $H(s) = R(s)P^{-1}(s)Q(s) + W(s)$. We shall explore the possibilities for MFDs in Sec. 6.4.4 and the more general case in Sec. 8.2.

Properties of Observer-Form Realizations.   The properties of the controller-form realization, as described in Sec. 6.4.2, carry over with obvious changes to the observer-form realization. Therefore, here we note only the analog of the basic formulas (14) and (19), namely that the left MFD and the observer-form realization can be related as

$$\begin{bmatrix} \Psi_L(s) & 0 \\ 0 & I \end{bmatrix}\begin{bmatrix} sI - A_o & B_o \\ -C_o & 0 \end{bmatrix} = \begin{bmatrix} D_L(s) & N_L(s) \\ -I & 0 \end{bmatrix}\begin{bmatrix} C_o & 0 \\ 0 & I \end{bmatrix} \qquad (38)$$

where

$$\{\Psi_L(s), D_L(s)\} \text{ are left coprime and } \{C_o, sI - A_o\} \text{ are right coprime} \qquad (39)$$

The analog of (27) is

$$\begin{bmatrix} X(s) & Y(s) \\ \Psi_L(s) & D_L(s) \end{bmatrix}\begin{bmatrix} sI - A_o & -\tilde{X}(s) \\ -C_o & \tilde{Y}(s) \end{bmatrix} = \begin{bmatrix} I & 0 \\ 0 & I \end{bmatrix}$$

$$= \begin{bmatrix} sI - A_o & -\tilde{X}(s) \\ -C_o & -\tilde{Y}(s) \end{bmatrix}\begin{bmatrix} X(s) & Y(s) \\ \Psi_L(s) & D_L(s) \end{bmatrix} \qquad (40)$$

where all the block matrices are unimodular.

6.4.4   Controllability- and Observability-Form Realizations

Reviewing our controller-form realization procedure, we can say that what was done was first to realize the simple right MFD

$$\xi(s) = I_m D^{-1}(s)u(s) \qquad (41)$$

in *controller form* and then obtain the complete realization by *reading out* the partial states $\xi(s)$ via the equation

$$y(s) = N(s)D^{-1}(s)u(s) = N(s)\xi(s) \qquad (42)$$

However, we can clearly also write

$$y(s) = N(s)[D^{-1}(s)I_m]u(s) \qquad (43)$$

and first realize the simple left MFD, say,

$$D^{-1}(s)I_m u(s) = \bar{y}(s) \qquad (44a)$$

in *observer form* and then complete the realization by implementing

$$y(s) = N(s)\bar{y}(s) \qquad (44b)$$

This will give what we shall call a *controllability-form* realization of $N(s)D^{-1}(s)$. To actually carry out the above construction, we need to start with $D(s)$ in row-reduced form, which can be arranged by appropriate column operations on $N(s)$ and $D(s)$. Then we can proceed to implement (44a) in observer form by writing [cf. (34)]

$$D(s)\bar{y}(s) = \bar{\xi}(s) = u(s)$$

as

$$D_{hr}\bar{y}(s) = S^{-1}(s)[u(s) - \psi_L(s)D_{lr}\bar{y}(s)] \qquad (45)$$

A realization can then be set up as shown in schematic form in Fig. 6.4-3, where the output portion [yielding $y(s)$] should temporarily be ignored. Comparison with the block diagram of Fig. 6.4-2 [temporarily ignore the input $u(s)$] will show that they are consistent, except for the arrangement of the integrator chains—these are shown *reversed* in order to indicate that (as in the scalar case) we shall number the states (integrator outputs) from left to right rather than from right to left as in the observer form (compare Fig. 2.1-10 and 2.1-7 in Chapter 2). This means that the A matrix will now be given by [cf. (33)], say,

$$\tilde{I}A_o\tilde{I} = A_{co} \qquad (46)$$

where

$$\tilde{I} = \text{block diag } \{\tilde{I}_{l_i}, i = 1, \ldots, m\}$$

The effect of premultiplication by $\tilde{I}$ is to interchange the *block rows*, and postmultiplication will interchange the *block columns*. Therefore, we can see that A_{co} will have $l_i \times l_i$ block diagonal elements that are right-companion matrices, with possibly nonzero elements in the last column, while the off-diagonal block elements will be zero everywhere except possibly in the last column. Moreover, since the numerator matrix in (44) is I_m, we can see that [cf. (33b)]

$$B'_{co} = B'_o\tilde{I} = \text{block diag } \{[1 \quad 0 \quad \cdots \quad 0], 1 \times l_i, i = 1, \ldots, p\} \qquad (47)$$

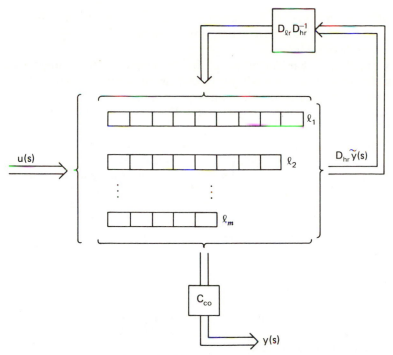

Figure 6.4-3. Schematic of the controllability-form realization of a right MFD.

Note that B_{co} is simpler in form than in the controller form, where [cf. (10)] $B_c = B_{co} D_{hc}^{-1}$.

The reason for the name *controllability form* can now already be made clear. Notice from (46) and (47) that (the subscript co has been suppressed for convenience)

$$\{b_1 \quad Ab_1 \quad \cdots \quad A^{l_1-1}b_1, b_2, \ldots, b_m, \ldots, A^{l_m-1}b_m\} = I_n \qquad (48)$$

the $n \times n$ identity matrix. Therefore, the controllability matrix is clearly of full rank and has a simple form—it is easy to solve equations involving $\mathcal{C}_{co}$ as the coefficient matrix by focusing on the special columns in (48).

Thus, in particular, let us show how to complete the realization by computing the output matrix C_{co}. We know that

$$C_{co}[B_{co} \quad A_{co}B_{co} \quad \cdots \quad A_{co}^{n-1}B_{co}] = [h_1 \quad h_2 \quad \cdots \quad h_n] \qquad (49)$$

where the $\{h_i\}$ are the ($p \times m$ block) Markov parameters; i.e.,

$$H(s) = \sum_{i=1}^{\infty} h_i s^{-i}$$

But using the property of the special columns (48), we see that *the columns of C_{co} can be obtained directly as the corresponding columns of* $[h_1 \ h_2 \ \cdots \ h_n]$ (see Example 6.4-3 below).

A question that might be raised by the above construction is whether using the transfer function will mean that the unobservable parts of the system $\{N(s), D(s)\}$ will be lost. The answer is no, basically because of the property that $C_1(sI - A)^{-1}B = C_2(sI - A)^{-1}B$ with $\{A, B\}$ controllable implies that $C_1 = C_2$.

We should also ask how the output matrix C_{co} can be directly obtained from $\{N(s), D(s)\}$ without going to the transfer function and the Markov parameters. For this we note that the observer-form realization of $D^{-1}(s)$ gives us

$$D^{-1}(s) = C_o(sI - A_o)^{-1}B_o = C_o\check{I}(sI - A_{co})^{-1}B_{co} \qquad (50)$$

Therefore we shall have

$$N(s)D^{-1}(s) = N(s)C_o\check{I}(sI - A_{co})^{-1}B_{co} \qquad (51)$$

The right-hand side is apparently nonproper even if $N(s)D^{-1}(s)$ is strictly proper, as we always assume. To reconcile these facts, we use the division Theorem 6.3-15 to write, say,

$$\bar{N}(s) \triangleq N(s)C_o\check{I} = Q(s)(sI - A_{co}) + C$$

where C is a unique constant matrix, which can moreover be found as (cf. the Corollary to Theorem 6.3-15),

$$C = \bar{N}(A_{co}) \qquad (52)$$

Therefore

$$N(s)D^{-1}(s) = C(sI - A_{co})^{-1}B_{co} + Q(s)B_{co}$$

Because $N(s)D^{-1}(s)$ is strictly proper, we must have $Q(s)B_{co} \equiv 0$, and C must equal the output matrix of the controllability form, viz., $C = C_{co}$. Both these methods of calculating C_{co} are illustrated in Example 6.4-3.

Example 6.4-3.   Controllability-Form Realization of a Right MFD

The right MFD of Example 6.4-1 can be realized in controllability form by first finding an observer-type realization of

$$D^{-1}(s) = \begin{bmatrix} 0 & -(s^3 + 4s^2 + 5s + 2) \\ (s + 2)^2 & s + 2 \end{bmatrix}^{-1}$$

and then suitably transposing rows and columns. For the observer realization, we see that $D(s)$ is row reduced (and column reduced) and that

$$D_{lr}D_{hr}^{-1} = \begin{bmatrix} 0 & 0 & 0 & 4 & 4 \\ -4 & -5 & -2 & 1 & 2 \end{bmatrix} \begin{bmatrix} 0 & -1 \\ 1 & 0 \end{bmatrix}^{-1} = \begin{bmatrix} 4 & 5 & 2 & -1 & -2 \\ 0 & 0 & 0 & 4 & 4 \end{bmatrix}'$$

Therefore

$$A_{co} = \begin{bmatrix} 0 & 0 & -2 & 0 & 0 \\ 1 & 0 & -5 & 0 & 0 \\ 0 & 1 & -4 & 0 & 0 \\ \hline 0 & 0 & 2 & 0 & -4 \\ 0 & 0 & 1 & 1 & -4 \end{bmatrix} \tag{53a}$$

while

$$B_{co}' = \begin{bmatrix} 1 & 0 & 0 & 0 & 0 \\ \hline 0 & 0 & 0 & 1 & 0 \end{bmatrix} \tag{53b}$$

Note that

$$[b_1 \quad Ab_1 \quad A^2b_1 \quad b_2 \quad Ab_2] = I_5 \tag{53c}$$

Now, if we know the Markov parameters $\{h_i, i = 1, 2, 3\}$, we can calculate C_{co}. We have [recall that $H(s)$ is given by (6.2-44)]

$$h_1 = \lim_{s \to \infty} sH(s) = \begin{bmatrix} 0 & 1 \\ -1 & -1 \end{bmatrix}$$

$$h_2 = \lim_{s \to \infty} s[sH(s) - h_1] = \begin{bmatrix} 0 & -4 \\ 4 & 4 \end{bmatrix}$$

$$h_3 = \lim_{s \to \infty} s[s^2 H(s) - sh_1 - h_2] = \begin{bmatrix} 1 & 12 \\ -12 & -12 \end{bmatrix}$$

and hence

$$C_{co}[b_1 \quad b_2 \quad Ab_1 \quad Ab_2 \quad A^2b_1 \quad A^2b_2] = \begin{bmatrix} 0 & 1 & 0 & -4 & 1 & 12 \\ -1 & -1 & 4 & 4 & -12 & -12 \end{bmatrix}$$

so that, using (53c), we can read off

$$C_{co} = \begin{bmatrix} 0 & 0 & 1 & 1 & -4 \\ -1 & 4 & -12 & -1 & 4 \end{bmatrix} \tag{53d}$$

Notice that we do not need to know $n = 5$ Markov parameters but only 3, where 3 is the length of the longest integrator chain in the realization; i.e., 3 is the highest row degree of $D(s)$.

Another often quicker method of finding C_{co} is via the polynomial matrix formula (52). For this we note that

$$C_o \tilde{I} = D_{hr}^{-1} \begin{bmatrix} 0 & 0 & 1 & 0 & 0 \\ 0 & 0 & 0 & 0 & 1 \end{bmatrix} = \begin{bmatrix} 0 & 0 & 0 & 0 & 1 \\ 0 & 0 & -1 & 0 & 0 \end{bmatrix}$$

and

$$\bar{N}(s) = \begin{bmatrix} 0 & 0 & 0 & 0 & s \\ 0 & 0 & -s^2 & 0 & -s \end{bmatrix}$$

$$= \begin{bmatrix} 0 & 0 & 0 & 0 & 0 \\ 0 & 0 & -1 & 0 & 0 \end{bmatrix} s^2 + \begin{bmatrix} 0 & 0 & 0 & 0 & 1 \\ 0 & 0 & 0 & 0 & -1 \end{bmatrix} s$$

Therefore

$$C_{co} = \bar{N}(A_{co})$$

$$= \begin{bmatrix} 0 & 0 & 0 & 0 & 0 \\ 0 & 0 & -1 & 0 & 0 \end{bmatrix} A_{co}^2 + \begin{bmatrix} 0 & 0 & 0 & 0 & 1 \\ 0 & 0 & 0 & 0 & -1 \end{bmatrix} A_{co}$$

$$= \begin{bmatrix} 0 & 0 & 0 & 0 & 0 \\ 0 & -1 & 4 & 0 & 0 \end{bmatrix} A_{co} + \begin{bmatrix} 0 & 0 & 1 & 1 & -4 \\ 0 & 0 & -1 & -1 & 4 \end{bmatrix}$$

$$= \begin{bmatrix} 0 & 0 & 1 & 1 & -4 \\ -1 & 4 & -12 & -1 & 4 \end{bmatrix}$$

as before.   ∎

The Observability Form.   By now, it will relieve both author and reader if we just briefly note that the *observability* form can be obtained by a *duality* argument or directly by writing a left MFD description as

$$y(s) = D^{-1}(s)N(s)u(s)$$
$$= [I_p D^{-1}(s)]N(s)u(s)$$

and first implementing $I_p D^{-1}(s)$ in controller form.

Rather than go into all the details, it may be more useful to remind the reader that the scalar realization procedures introduced in Sec. 2.1.1 can perhaps be better understood now—the key is that $D^{-1}(s)$ can be realized in two basic forms and that each of these forms can then be associated with $N(s)$ either as $N(s)D^{-1}(s)$ or $D^{-1}(s)N(s)$ (with proper attention to dimensions). The fact that the scalar versions of $N(s)$ and $D(s)$ always commute was perhaps a block to understanding rather than an aid.

6.4.5 Canonical State-Space Realizations and Canonical MFDs

In Sec. 6.4.1–6.4.4, we have shown how to generalize the scalar procedures of Sec. 2.1.1 to obtain several different realizations of a matrix transfer function. While the basic concepts of Sec. 2.1.1 have thus been (often nicely) extended, there is one feature that is still missing. With any nth-order controllable realization of a nominal scalar transfer function

$$H(s) = \frac{b(s)}{a(s)}, \qquad \deg a(s) = n$$

we could associate a *unique* state-space realization of order n, called the controller canonical form, a unique controllability canonical form, and so on. Here, however, this has not been achieved. It is true that given a right MFD

$$H(s) = N(s)D^{-1}(s)$$

we described a procedure in Sec. 6.4.1 for obtaining a controller-form realization of order equal to $\deg \det D(s)$, but this realization can be different for MFDs that vary only *trivially* from $N(s)D^{-1}(s)$. Thus let

$$H(s) = \tilde{N}(s)\tilde{D}^{-1}(s)$$

where

$$\tilde{N}(s) = N(s)U(s), \qquad \tilde{D}(s) = D(s)U(s), \qquad U(s) \text{ unimodular}$$

Then the procedure of Sec. 6.4.1 will generally yield a different controller-form realization of $H(s)$. *This lack of uniqueness is why we do not call the realizations of Sec. 6.4.1–6.4.4 canonical.*

Example 6.4-4.

Consider the MFD studied in Example 6.4-1, which had

$$D(s) = \begin{bmatrix} 0 & -(s^3 + 4s^2 + 5s + 2) \\ (s + 2)^2 & s + 2 \end{bmatrix} \tag{54a}$$

and gave the controller-form matrices

$$A_c = \begin{bmatrix} -4 & -4 & 0 & -1 & -2 \\ 1 & 0 & 0 & 0 & 0 \\ \hline 0 & 0 & -4 & -5 & -2 \\ 0 & 0 & 1 & 0 & 0 \\ 0 & 0 & 0 & 1 & 0 \end{bmatrix}, \quad B_c = \begin{bmatrix} 0 & 1 \\ 0 & 0 \\ \hline -1 & 0 \\ 0 & 0 \\ 0 & 0 \end{bmatrix} \tag{54b}$$

Now we can replace $D(s)$ by

$$\tilde{D}(s) = D(s) \begin{bmatrix} 1 & 2s \\ 0 & 2 \end{bmatrix}$$

$$= \begin{bmatrix} 0 & -2(s^3 + 4s^2 + 5s + 2) \\ (s + 2)^2 & 2s^3 + 8s^2 + 10s + 4 \end{bmatrix} \tag{55a}$$

which has not changed the column degrees or the determinantal degree. However, the controller-form realization of

$$H(s) = \tilde{N}(s)\tilde{D}^{-1}(s), \qquad \tilde{N}(s) \triangleq N(s)\begin{bmatrix} 1 & 2s \\ 0 & 2 \end{bmatrix}$$

will be different, and we can check that it will be

$$\tilde{A}_c = \begin{bmatrix} -4 & -4 & 0 & 0 & 0 \\ 1 & 0 & 0 & 0 & 0 \\ \hline 0 & 0 & -4 & -5 & -2 \\ 0 & 0 & 1 & 0 & 0 \\ 0 & 0 & 0 & 1 & 0 \end{bmatrix}, \qquad B_c = \begin{bmatrix} 1 & 1 \\ 0 & 0 \\ \hline -\frac{1}{2} & 0 \\ 0 & 0 \\ 0 & 0 \end{bmatrix} \qquad (55b)$$

which is a different realization from $\{A_c, B_c\}$ given earlier. ∎

The reason for the lack of uniqueness is clearly that in the matrix case we can have nontrivially different (right) MFDs for a given transfer function. However, if we can specify *unique* (right) MFDs for a given transfer function, then these MFDs and the associated state-space realizations can be called *canonical*. (As a matter of interest, it will be shown in Sec. 6.7.3 that this eminently reasonable decision is in fact consistent with the proper mathematical definition of what *canonical* should mean.)

The first unique MFDs were described by Rosenbrock [12], whose results were based on the Hermite forms of Theorem 6.3-2 and will be described in Sec. 6.7.1. In Sec. 6.4.6 we shall solve similar problems for state-space realizations. That is, given a family, $\{T^{-1}AT, T^{-1}B, CT, T \text{ any nonsingular matrix}\}$, of controllable (observable) realizations, we shall show how to associate with it some unique controllability- and controller- (observability- and observer-) type forms. It must be said, however, that our concern with the uniqueness question is only secondary (the uniqueness is not often relevant), and our main interest in Sec. 6.4.6 is how to convert an arbitrary state-space realization to one or the other of the four forms of Sec. 6.4.1–6.4.4.

6.4.6 Transformations of State-Space Realizations

In the previous sections we showed how to obtain state-space realizations in certain special forms starting with given MFDs of a transfer function. However, in many problems, we start not with a transfer function but with some given set of state-space equations

$$\dot{x}(t) = Ax(t) + Bu(t), \qquad y(t) = Cx(t)$$

Therefore we shall describe some methods for transforming a given matrix realization $\{A, B, C\}$, with $\{A, B\}$ controllable, to the controllability- and controller-type forms. Similar (dual) arguments will apply to the corresponding results for observable realizations and observability- and observer-type forms, and so these will not be examined here.

We start with the controllability matrix of the given realization

$$\mathcal{C}(A, B) = [B \quad AB \quad \cdots \quad A^{n-1}B]$$

which by assumption is of full rank. Therefore it contains n linearly independent columns, which we can use as a new basis for the state space (cf. Sec. 5.2). Equivalently, we use these n columns to form a nonsingular transformation matrix T and then compute a new realization as

$$\bar{A} = T^{-1}AT, \qquad \bar{B} = T^{-1}B, \qquad \bar{C} = CT$$

Unlike the scalar case, the selection of linearly independent vectors from $\mathcal{C}(A, B)$ can now be made in many ways, but we shall need to impose certain restrictions to get the controllability- and controller-type forms described earlier.

To get the controllability forms, we shall require that the selected vectors be such that they can be arranged in *chains* as

$$T = [b_1, Ab_1, \ldots, A^{r_1-1}b_1, b_2, \ldots, A^{r_2-1}b_2, \ldots, A^{r_m-1}b_m] \qquad (56)$$

where the $\{r_i\}$ are nonnegative integers and the chain beginning with b_i is omitted if $r_i = 0$. Then we can check that the matrices

$$\bar{A} = T^{-1}AT, \qquad \bar{B} = T^{-1}B$$

will in fact be in the *controllability form* of Sec. 6.4.4, namely

$$\bar{A}_{co} = \begin{bmatrix} 0 & 0 & 0 & x & & x & & x \\ 1 & 0 & 0 & x & & x & & x \\ 0 & 1 & 0 & x & \bigcirc & x & \bigcirc & x \\ 0 & 0 & 1 & x & & x & & x \\ & & & x & 0 & 0 & x & & x \\ \bigcirc & & & x & 1 & 0 & x & \bigcirc & x \\ & & & x & 0 & 1 & x & & x \\ & & & x & & & x & 0 & 0 & x \\ \bigcirc & & & x & \bigcirc & & x & 1 & 0 & x \\ & & & x & & & x & 0 & 1 & x \end{bmatrix}$$

$$\bar{B}'_{co} = \begin{bmatrix} 1 & 0 & 0 & 0 & 0 & 0 & 0 & 0 & 0 & 0 \\ 0 & 0 & 0 & 0 & 1 & 0 & 0 & 0 & 0 & 0 \\ 0 & 0 & 0 & 0 & 0 & 0 & 0 & 1 & 0 & 0 \end{bmatrix}$$

This follows by using the rule (cf. Sec. 5.2.2) that

the ith column of $\bar{A}$ equals the representation of
$A \cdot (i$th column of $T)$ in terms of the new basis
represented by the columns of T

However, even with this restriction, there are many ways of selecting an appropriate transformation matrix T, and correspondingly there is no *unique* controllability form (similar statements can be made for the controller form) in the multivariable case. Still, as Luenberger [26] pointed out, there are certain natural choices of matrix T that should have special interest, and we shall study them here.

It will be convenient, following Kalman [27], to use a so-called *crate* or *Young* diagram as an aid to describing all selection procedures.

The crate is (cf. Fig. 6.4-4) a table with m columns, representing the columns of the input matrix $B = [b_1 \ b_2 \ \cdots \ b_m]$, and n rows corresponding to the powers $\{I, A, \ldots, A^{n-1}\}$; the (i, j)th cell of the crate then represents the column vector $A^{i-1}b_j$, and choosing n linearly independent columns of $\mathcal{C}(A, B)$ corresponds to selecting n cells of the crate. This can clearly be done in many ways, but at least two procedures (meeting the requirement mentioned earlier) suggest themselves immediately.

Scheme I. Search the Crate by Columns. We first select b_1 and indicate this by putting an x in the $A^0 b_1$ cell. Now if Ab_1 is linearly independent of b_1, we put an x in this cell as well and continue down the first column of the crate until we either x in all the cells or we find a vector, say $A^{l_1}b_1$, that is linearly *dependent* on the earlier vectors in the column. We denote this fact by putting a 0 in the corresponding $(l_1, 1)$ cell; then we note (by a now-

b_1	b_2	b_3	b_4	
x	x	x	0	$A^0 = I$
x	x	0		A
x	0			A^2
x				A^3
				A^4
				A^5
				A^6

Figure 6.4-4. Typical crate diagram filled in by *searching by columns*. We have $l_1 = 4, l_2 = 2, l_3 = 1, l_4 = 0$.

familiar argument) that when this happens all the remaining vectors in that column will be linearly dependent on the previously selected vectors. We indicate this by leaving the corresponding cells blank. If we have not found n linearly independent elements in the first column, we go to the second column. If b_2 is linearly independent of *all* previously selected vectors $\{b_1, Ab_1, \ldots, A^{l_1-1}b_1\}$, we put an x in the corresponding cell. Now we repeat this procedure with Ab_2 and continue in this way with successive columns if necessary until n linearly independent vectors have been found. With this scheme, the crate diagram will have the general form shown in Fig. 6.4-4. The cells with the 0s correspond to the vectors $\{A^{l_i}b_i, i = 1, \ldots, m\}$. The pattern depends heavily on the order in which the inputs are arranged, since the tendency is to have a few long chains of xs and not all inputs may be called upon. A more uniform treatment of the system inputs is provided by another natural search procedure.

Scheme II. Search the Crate by Rows.   Now we search the rows until we find a vector, say $A^{k_i}b_i$, that is linearly dependent on *all* the previously selected vectors. We put a 0 in the corresponding cell and note again that all vectors below it in the same column will also be linearly dependent on the already-selected vectors (prove this). Therefore we leave all the corresponding cells blank and go on if necessary to the next linearly independent vector encountered in the row search. (We may remark that searching the crate by rows corresponds to searching the columns of the controllability matrix from left to right.)

A typical crate diagram produced by this scheme will appear as in Fig. 6.4-5—the tendency now is to have several chains of nearly equal lengths

	b_1	b_2	b_3	b_4	
	x	x	x	0	I
	x	0	x		A
	x		x		A^2
	0		0		A^3
					A^4
					A^5
					A^6

Figure 6.4-5.   Typical crate diagram filled in by *searching by rows*. We have $k_1 = 3, k_2 = 1, k_3 = 3$. The set of lengths $\{3, 1, 3\}$ will be the same even if the order of the $\{b_i\}$ is permuted.

$\{k_1, \ldots, k_m\}$. It can be shown (see Exercise 6.4-11) that the lengths we get here will remain the same even if the columns are permuted.

Some Canonical Controllability Forms. The controllability forms corresponding to these two schemes will have the following forms, which we shall show for the patterns given by Figs. 6.4-4 and 6.4-5.

Scheme I. Corresponding to Fig. 6.4-4, the transformation matrix T will be

$$T_1 = [b_1 \quad Ab_1 \quad A^2b_1 \quad A^3b_1 \quad b_2 \quad Ab_2 \quad b_3]$$

and correspondingly (by the rules of Sec. 5.2)

$$
A_{1\dot{c}o} =
\left[
\begin{array}{cccc|cc|c}
0 & 0 & 0 & x & 0 & x & x \\
1 & 0 & 0 & x & 0 & x & x \\
0 & 1 & 0 & x & 0 & x & x \\
0 & 0 & 1 & x & 0 & x & x \\
\hline
0 & 0 & 0 & 0 & 0 & x & x \\
0 & 0 & 0 & 0 & 1 & x & x \\
\hline
0 & 0 & 0 & 0 & 0 & 0 & x
\end{array}
\right]
\begin{array}{l} \\ \\ {\scriptstyle l_1} \\ \\ {\scriptstyle l_2} \\ \\ {\scriptstyle l_3} \end{array}
,
\quad
B_{1co} =
\left[
\begin{array}{cccc}
1 & 0 & 0 & x \\
0 & 0 & 0 & x \\
0 & 0 & 0 & x \\
0 & 0 & 0 & x \\
\hline
0 & 1 & 0 & x \\
0 & 0 & 0 & x \\
\hline
0 & 0 & 1 & x
\end{array}
\right]
$$

$$\qquad\qquad {\scriptstyle l_1 \qquad\qquad l_2 \quad l_3}$$

where the xs denote possibly nonzero values, representing the dependence of the vectors $\{A^{l_i}b_i\}$ on the elements in T_1.

In more detail, the xs are equal to the coefficients $\{\beta_{ijk}\}$ in the following relations, written with reference to Fig. 6.4-4:

$$A^4b_1 = \beta_{110}b_1 + \beta_{111}Ab_1 + \beta_{112}A^2b_1 + \beta_{113}A^3b_1$$

$$A^2b_2 = \beta_{210}b_1 + \beta_{211}Ab_1 + \beta_{212}A^2b_1 + \beta_{213}A^3b_1 + \beta_{220}b_2$$
$$\qquad + \beta_{221}Ab_2$$

and similarly for Ab_3, b_4. With this pattern in mind the reader can see that for any Scheme I selection the expressions will be of the type

$$A^{l_i}b_i = \sum_{j=1}^{i}\sum_{k=0}^{l_j-1} \beta_{ijk}A^kb_j, \qquad l_i > 0 \tag{57a}$$

$$b_i = \sum_{j=1}^{i-1}\sum_{k=0}^{l_j-1} \beta_{ijk}A^kb_j, \qquad l_i = 0 \tag{57b}$$

[This is not so formidable as it may look: The 0 in the ith column of the crate represents $A^{l_i}b_i$, and we write this as a combination of the terms $\{A^kb_j, k = 0, \ldots, l_j - 1\}$ in the jth column for $j = 0, 1, \ldots$ up to i (or $i - 1$ if $l_i = 0$).]

Note that because of the particular selection scheme of searching the crate by columns, the transformed system consists of subsystems of order l_i represented by the diagonal blocks, with only a one-sided or *triangular* coupling allowed between the subsystems. However, the block sizes are highly dependent on the order in which the inputs are arranged.

[It is worth noting that once an order of the $\{b_i\}$ has been chosen, the parameters $\{l_i, \beta_{ijk}\}$ are the *same* for all pairs $\{\bar{A}, \bar{B}\}$ that are similar to $\{A, B\}$. The reason is that if $\bar{A} = T^{-1}AT$, $\bar{B} = T^{-1}B$, then $\bar{A}^i\bar{b}_j = T^{-1}(A^ib_j)$, so that all linear dependencies between the $\{A^ib_j\}$ are just transferred to the $\{\bar{A}^i\bar{b}_j\}$. The fact that all pairs $\{T^{-1}AT, T^{-1}B, \det T \neq 0\}$ have the same controllability form $\{A_{1co}, B_{1co}\}$ is the reason for calling this form *canonical* (see also Sec. 6.7.3).]

We have not mentioned the output matrix so far, but it is easy to bring it in. We shall have $C_{1co} = CT_1$. The entries of C_{1co} will not have any particular form, but they will also be invariant under similarity transformation. For suppose that instead of $\{A, B, C\}$ we had $\{T^{-1}AT, T^{-1}B, CT\}$. Then the transformation matrix would be $T^{-1}T_1$ rather than T_1, and therefore $C_{1co} = CT \cdot T^{-1}T_1 = CT_1$, as before.

Controllability Form with Scheme II.   Let us now study the forms obtained from Scheme II. The crate is now searched by rows and gives rise to dependency relations of the form (written for Fig. 6.4-5)

$$A^3b_1 = \alpha_{110}b_1 + \alpha_{111}Ab_1 + \alpha_{112}A^2b_1$$
$$+ \alpha_{120}b_2$$
$$+ \alpha_{130}b_3 + \alpha_{131}Ab_3 + \alpha_{132}A^2b_3 \tag{58a}$$

$$Ab_2 = \alpha_{210}b_1 + \alpha_{211}Ab_1$$
$$+ \alpha_{220}b_2$$
$$+ \alpha_{230}b_3 \tag{58b}$$

$$A^3b_3 = \alpha_{330}b_1 + \alpha_{331}Ab_1 + \alpha_{312}A^2b_1$$
$$+ \alpha_{320}b_2$$
$$+ \alpha_{330}b_3 + \alpha_{331}Ab_3 + \alpha_{332}A^2b_3 \tag{58c}$$

and finally

$$b_4 = \alpha_{410}b_1 + \alpha_{420}b_2 + \alpha_{430}b_3$$

With these relations and with

$$T_2 = [b_1 \quad Ab_1 \quad A^2b_1 \quad b_2 \quad b_3 \quad Ab_3 \quad A^2b_3]$$

$\{T_2^{-1}AT_2, T_2^{-1}B\}$ will have the form

$$A_{2co} = \begin{bmatrix} 0 & 0 & \alpha_{110} & \alpha_{210} & 0 & 0 & \alpha_{310} \\ 1 & 0 & \alpha_{111} & \alpha_{211} & 0 & 0 & \alpha_{311} \\ 0 & 1 & \alpha_{112} & 0 & 0 & 0 & \alpha_{312} \\ 0 & 0 & \alpha_{120} & \alpha_{220} & 0 & 0 & \alpha_{320} \\ 0 & 0 & \alpha_{130} & \alpha_{230} & 0 & 0 & \alpha_{230} \\ 0 & 0 & \alpha_{131} & 0 & 1 & 0 & \alpha_{331} \\ 0 & 0 & \alpha_{132} & 0 & 0 & 1 & \alpha_{332} \end{bmatrix}$$

$$B_{2co} = \begin{bmatrix} 1 & 0 & 0 & \alpha_{410} \\ 0 & 0 & 0 & 0 \\ 0 & 0 & 0 & 0 \\ 0 & 1 & 0 & \alpha_{420} \\ 0 & 0 & 1 & \alpha_{430} \\ 0 & 0 & 0 & 0 \\ 0 & 0 & 0 & 0 \end{bmatrix} \tag{59}$$

(Again we may note that the $\{k_i, \alpha_{ijk}\}$ are the same for all pairs similar to $\{A, B\}$, and therefore $\{A_{2co}, B_{2co}\}$ is again a *canonical representation*.) The output is uniquely determined as

$$C_{2co} = CT_2$$

Note that this (canonical) form is different from that obtained via Scheme I. We again have subsystems defined by the diagonal blocks, but the coupling is no longer *one-sided*—subsystem 2 is coupled not just to subsystem 3 but also to subsystem 1 and so on. However, notice also that the coupling is not completely arbitrary but is determined by the crate diagram—thus subsystem 2 does not feed into all states of subsystems 1 and 3, because of the *fixed* zeros corresponding to the fact (see Fig. 6.4-5) that Ab_2 depends only on $\{b_1, b_2, b_3, Ab_1\}$ and not on $\{A^2b_1, Ab_3, A^2b_3\}$.

The presence of these fixed zeros was first noted by Denery [28] and Popov [29]; the significance of this fact is in the identification problem, where one goal is to estimate the parameters $\{\alpha_{ijk}, k_i\}$ of a unique representative of a set of similar realizations (see, e.g., [30] and [31]). The integers $\{k_i, \sum k_i = m\}$ and the entries in $\{A_{2co}, B_{2co}\}$ that are not fixed to be zero or unity (i.e., the $\{\alpha_{ijk}\}$) are *independent* in the sense that we can give them arbitrary values and still be sure that there is a controllable pair $\{A, B\}$ whose controllability canonical form will be $\{A_{2co}, B_{2co}\}$ with the given parameters. This is obvious—just enter the parameters into matrices A and B as shown in (59) and verify that Scheme II gives us back the same $\{k_i, \alpha_{ijk}\}$. The same independence property also holds for the parameters $\{l_i, \beta_{ijk}\}$ defined earlier for Scheme I canonical forms. However, careful attention to the zero locations is necessary for these results to be true (see Example 6.4-5 below).

The Controllability Indices.   We have mentioned before that there can be several different ways of searching the crate diagram and each of them will yield an associated set of controllability canonical forms. In particular, each scheme will have a special associated set of indices, describing the lengths of the *chains* in each column of the crate diagram.

However, the indices $\{k_i\}$ that arise in connection with Scheme II have a special significance. They arise from searching the crate by rows or, equivalently, by searching the controllability matrix

$$\mathcal{C}(A, B) = [B \quad AB \quad \cdots \quad A^{n-1}B]$$

from left to right. We have previously defined the smallest integer μ such that

$$\operatorname{rank} [B \quad AB \quad \cdots \quad A^{\mu-1}B] = n$$

to be the *controllability index* of $[A, B]$. We can now see that

$$\mu = k_{\max}, \qquad \begin{array}{l} \text{the length of the longest chain} \\ \text{in the crate diagram for Scheme II} \end{array}$$

In view of this, it is reasonable that

$$\{k_i\} = \text{the controllability indices of } \{A, B\}$$

We may note that the set of controllability indices is invariant under similarity transformation (see Exercise 6.4-11).

Example 6.4-5.
   Let $n = 4$, $m = 2$, $k_1 = 3$, $k_2 = 1$. Then is the pair

$$A_1 = \begin{bmatrix} 0 & 0 & 1 & 0 \\ 1 & 0 & 2 & 0 \\ 0 & 1 & 3 & \textcircled{1} \\ \hline 0 & 0 & -21 & 5 \end{bmatrix}, \qquad B_1 = \begin{bmatrix} 1 & 0 \\ 0 & 0 \\ 0 & 0 \\ \hline 0 & 1 \end{bmatrix}$$

in controllability canonical form for Scheme II? Repeat for a pair $\{A_2, B_2\}$ where $B_2 = B_1$ and $A_2 = A_1$ except in the circled $(3, 4)$ entry, which should now be set equal to 0.

Solution. The answer is no. One way (the longer one) is to calculate the controllability canonical form of $\{A_1, B_1\}$. We shall do this as a review. The first step is to form

$$
\mathcal{C}(A_1, B_1) =
\begin{array}{c}
\begin{array}{cccccccc}
b_1 & b_2 & Ab_1 & Ab_2 & A^2b_1 & A^2b_2 & A^3b_1 & A^3b_2
\end{array} \\
\begin{bmatrix}
1 & 0 & 0 & 0 & 0 & 1 & 1 & 8 \\
0 & 0 & 1 & 0 & 0 & 2 & 2 & 17 \\
0 & 0 & 0 & 1 & 1 & 8 & 3 & 30 \\
0 & 1 & 0 & 5 & 0 & 4 & -21 & -14
\end{bmatrix}
\end{array}
$$

Searching the columns from left to right,† we find the basis

$$T_2 = [b_1 \quad Ab_1 \quad b_2 \quad Ab_2]$$

and note that

$$A^2b_1 = Ab_2 - 5b_2$$
$$A^2b_2 = b_1 + 2Ab_1 + 8Ab_2 - 36b_2$$

Therefore the controllability canonical form for $\{A_1, B_1\}$ is

$$
A_{co} =
\left[\begin{array}{cc|cc}
0 & 0 & 0 & 1 \\
1 & 0 & 0 & 2 \\
\hline
0 & 1 & 0 & -36 \\
0 & -5 & 1 & 8
\end{array}\right] \neq A_1, \qquad
B_{co} =
\left[\begin{array}{c|c}
1 & 0 \\
0 & 0 \\
\hline
0 & 1 \\
0 & 0
\end{array}\right] \neq B_1
$$

On the other hand, for the pair $\{A_2, B_2\}$ as specified in the problem, the same procedure will show that this pair is already in canonical controllability form.

Actually, both these answers could have been obtained more directly by using the crate diagram. Since $k_1 = 3$, $k_2 = 1$, this must have the form

b_1	b_2	
x	x	I
x	0	A
x		A^2
0		A^3

which shows that while A^3b_1 can depend on $\{b_1, b_2, Ab_1, A^2b_1\}$, Ab_2 can only depend on $\{b_1, b_2, Ab_1\}$. Therefore in the controllability canonical form the third column of A_{co} can be quite arbitrary, but the fourth column must have a zero in the third row.

We leave it to the reader to check whether $\{A_1, B_1\}$ and $\{A_2, B_2\}$ can be controllability canonical forms for Scheme I. ∎

†We note that this corresponds to searching the crate by rows.

Example 6.4-6.   Constructing Canonical Controllability Forms

We shall construct the controllability canonical forms for the pairs $\{A_c, B_c\}$ and $\{\tilde{A}_c, \tilde{B}_c\}$ that we obtained in Example 6.4-4 as two different (noncanonical) realizations of the same transfer function matrix. We start with the pair $\{A_c, B_c\}$ as in (54b) and form

$$
\mathcal{C}(A_c, B_c) = \begin{matrix}
b_1 & b_2 & Ab_1 & Ab_2 & A^2b_1 & A^2b_2 & A^3b_1 \\
\begin{bmatrix}
0 & 1 & 0 & -4 & 1 & 12 & -6 & \cdot \\
0 & 0 & 0 & 1 & 0 & -4 & 1 & \cdot \\
-1 & 0 & 4 & 0 & -11 & 0 & 26 & \cdot \\
0 & 0 & -1 & 0 & 4 & 0 & -11 & \cdot \\
0 & 0 & 0 & 0 & -1 & 0 & 4 & \cdot
\end{bmatrix}
\end{matrix} \tag{60}
$$

where we stop because $\{b_1, b_2, Ab_1, Ab_2, A^2b_1\}$ are already linearly independent, and

$$
A^2b_2 = -4b_2 - 4Ab_2
$$
$$
A^3b_1 = -2b_1 - 5Ab_1 - 4A^2b_1 + 2b_2 + Ab_2
$$

Therefore the *Scheme II controllability canonical* form can now be written down by inspection as

$$
A_{2co} = \begin{bmatrix}
0 & 0 & -2 & 0 & 0 \\
1 & 0 & -5 & 0 & 0 \\
0 & 1 & -4 & 0 & 0 \\
\hline
0 & 0 & 2 & 0 & -4 \\
0 & 0 & 1 & 1 & -4
\end{bmatrix}, \quad
B_{2co} = \begin{bmatrix}
1 & 0 \\
0 & 0 \\
0 & 0 \\
\hline
0 & 1 \\
0 & 0
\end{bmatrix} \tag{61}
$$

As a partial check that this is indeed a *canonical* form, we shall show that we get exactly the same pair $\{A_{2co}, B_{2co}\}$ if we start with the other realization (55b) of Example 6.4-4, for which the controllability matrix turns out to be

$$
\mathcal{C}(\tilde{A}_c, \tilde{B}_c) = \begin{matrix}
\tilde{b}_1 & \tilde{b}_2 & \tilde{A}\tilde{b}_1 & \tilde{A}\tilde{b}_2 & \tilde{A}^2b_1 & \tilde{A}^2b_2 & \tilde{A}^3b_1 \\
\begin{bmatrix}
1 & 1 & -4 & -4 & 12 & 12 & 32 & \cdot \\
0 & 0 & 1 & 1 & -4 & -4 & 12 & \cdot \\
-\frac{1}{2} & 0 & 2 & 0 & -\frac{11}{2} & 0 & 13 & \cdot \\
0 & 0 & -\frac{1}{2} & 0 & 2 & 0 & -\frac{11}{2} & \cdot \\
0 & 0 & 0 & 0 & -\frac{1}{2} & 0 & 2 & \cdot
\end{bmatrix}
\end{matrix}
$$

where again $\{\tilde{b}_1, \tilde{b}_2, \tilde{A}\tilde{b}_1, \tilde{A}\tilde{b}_2, \tilde{A}^2\tilde{b}_1\}$ are linearly independent. Furthermore, we find that

$$
\tilde{A}^2\tilde{b}_2 = -4\tilde{b}_2 - 4\tilde{A}\tilde{b}_2
$$

and

$$
\tilde{A}^3\tilde{b}_1 = -2\tilde{b}_1 - 5\tilde{A}\tilde{b} - 4\tilde{A}^2\tilde{b}_1 + 2\tilde{b}_2 + \tilde{A}\tilde{b}_2
$$

which are exactly the same relationships as for the $\{A_c, B_c\}$ pair. Therefore the two pairs will have the same *canonical* controllability form.

To determine the Scheme I controllability *canonical* form associated with searching the crate diagram by columns (Scheme I), we return to $\mathcal{C}(A, b)$ as computed in (60) and note that $\{b_1, Ab_1, A^2b_1, A^3b_1, b_2\}$ are linearly independent. We shall also find that

$$A^4b_1 = -4b_1 - 12Ab_1 - 13A^2b_1 - 6A^3b_1$$

$$Ab_2 = 2b_1 + 5Ab_1 + 4A^2b_1 + A^3b_1 - 2b_2$$

which will then yield

$$A_{1co} = \begin{bmatrix} 0 & 0 & 0 & -4 & 2 \\ 1 & 0 & 0 & -12 & 5 \\ 0 & 1 & 0 & -13 & 4 \\ 0 & 0 & 1 & -6 & 1 \\ \hline 0 & 0 & 0 & 0 & -2 \end{bmatrix}, \quad B_{1co} = \begin{bmatrix} 1 & 0 \\ 0 & 0 \\ 0 & 0 \\ 0 & 0 \\ \hline 0 & 1 \end{bmatrix} \qquad (62)$$

We shall leave it to the reader to check that the same *canonical* matrices (62) will be obtained if we start with the pair $\{\tilde{A}_c, \tilde{B}_c\}$. We should point out that if we had interchanged the inputs b_2 and b_1 and then used Scheme I the linearly independent vectors we would obtain would be $\{b_2, Ab_2, b_1, Ab_1, A^2b_1\}$ with corresponding matrices

$$\tilde{A}_{1co} = \begin{bmatrix} 0 & -4 & 0 & 0 & 2 \\ 1 & -4 & 0 & 0 & 1 \\ \hline 0 & 0 & 0 & 0 & -2 \\ 0 & 0 & 1 & 0 & -5 \\ 0 & 0 & 0 & 1 & -4 \end{bmatrix}, \quad \tilde{B}_{1co} = \begin{bmatrix} 1 & 0 \\ 0 & 0 \\ \hline 0 & 1 \\ 0 & 0 \\ 0 & 0 \end{bmatrix} \qquad (63)$$

We see that with Scheme I the subblock dimensions can depend on the order in which the inputs are arranged, whereas Scheme II is indifferent to this order. ∎

Some Canonical Controller Forms. As is already clear from the scalar case, the controller-type forms are somewhat less easy to obtain by similarity transformations from an arbitrary controllable pair $\{A, B\}$ than were the controllability forms. Brunovsky [32], Luenberger [26], and Tuel [33] were the first to describe general methods for doing this, though their procedures did not clearly specify *canonical* controller forms, for reasons we shall see. Then we shall describe a method due to Popov [29] for obtaining a canonical controller form when Scheme II is used.

First, however, consider the case where n linearly independent vectors have been obtained (by some scheme of searching the crate diagram) and they can be arranged in chains as

$$Q = [b_1 \quad Ab_1 \quad \cdots \quad A^{p_1-1}b_1 \quad \cdots \quad b_m \quad \cdots \quad A^{p_m-1}b_m] \qquad (64a)$$

Let us assume that $\{p_i > 0, i = 1, \ldots, m\}$ and define

$$\sigma_1 = p_1, \qquad \sigma_2 = p_1 + p_2, \ldots, \sigma_m = \sum_1^m p_i = n \qquad (64\text{b})$$

Now let

$$q_i = \text{the } \sigma_i\text{th row of } Q^{-1} \qquad (64\text{c})$$

and form

$$T^{-1} = [(A')^{p_1-1}q_1' \quad \cdots \quad q_1' \quad (A')^{p_2-1}q_2' \quad \cdots \quad q_m']' \qquad (64\text{d})$$

(It can be shown that T^{-1} is nonsingular—do this.) Then we can verify that

$$\bar{A} = T^{-1}AT, \qquad \bar{B} = T^{-1}B \qquad (64\text{e})$$

are such that $\{\bar{A}, \bar{B}\}$ is in controller-type form; e.g., for $n = 7$, $m = 3$,

$$
\bar{A} = \begin{bmatrix}
x & x & x & x & x & x & x \\
1 & 0 & 0 & 0 & 0 & 0 & 0 \\
0 & 1 & 0 & 0 & 0 & 0 & 0 \\
\hline
x & x & x & x & x & x & x \\
\hline
x & x & x & x & x & x & x \\
0 & 0 & 0 & 0 & 1 & 0 & 0 \\
0 & 0 & 0 & 0 & 0 & 1 & 0
\end{bmatrix}, \qquad
\bar{B} = \begin{bmatrix}
x & x & x \\
0 & x & x \\
0 & x & x \\
\hline
0 & x & x \\
\hline
x & x & x \\
x & x & 0 \\
x & x & 0
\end{bmatrix} \qquad (65)
$$

where the xs denote possibly nonzero elements. In general, however, it is hard to say whether any of them should definitely be zero. Therefore $\{\bar{A}, \bar{B}\}$ cannot be said to be in *canonical* form, in the sense that we cannot be sure that a pair $\{\tilde{A}, \tilde{B}\}$ similar to $\{A, B\}$ will give rise to the *same* controller form by using the above construction.

However, we can be more specific if Scheme II is used, and this was shown by Denery [28] and Popov [29]. Actually for Scheme II, Popov showed that a more illuminating construction than the one just described can be obtained; we shall describe it here via an example, for simplicity of notation.

Example 6.4-7.   Canonical Controller Forms

We shall construct the canonical version of the controller-form pair $\{A_c, B_c\}$ of Example 6.4-1.

The controllability matrix was already worked out in Example 6.4-6 [cf. (60)] as also were the dependency relations for a Scheme II search,

$$A^3b_1 = 2b_2 + Ab_2 - 2b_1 - 5Ab_1 - 4A^2b_1$$
$$A^2b_2 = -4b_2 - 4Ab_2$$

The Popov construction is to take the first relation, put all terms with nonzero powers of A on the left, and then factor out A as

$$A[\underbrace{A^2 b_1 + 5b_1 + 4Ab_1 - b_2}_{e_{13}}] = -2b_1 + 2b_2$$

Now repeat this operation with (the factor) e_{13},

$$A[\underbrace{Ab_1 + 4b_1}_{e_{12}}] = e_{13} - 5b_1 + b_2$$

and then with e_{12},

$$Ab_1 = e_{12} - 4b_1, \qquad e_{11} \triangleq b_1$$

Do the same for the next chain

$$A[\underbrace{Ab_2 + 4b_2}_{e_{22}}] = -4b_2$$

$$Ab_2 = e_{22} - 4b_2, \qquad e_{21} \triangleq b_2$$

Now it can be seen that the vectors $\{e_{11} \; e_{12} \; e_{13} \; e_{21} \; e_{22}\}$ form a basis, with respect to which (the canonical controller form is)

$$A_{2c} = \begin{bmatrix} -4 & -5 & -2 & 0 & 0 \\ 1 & 0 & 0 & 0 & 0 \\ 0 & 1 & 0 & 0 & 0 \\ \hdashline 0 & 1 & 2 & -4 & -4 \\ 0 & 0 & 0 & 1 & 0 \end{bmatrix}, \qquad B_{2c} = \begin{bmatrix} 1 & 0 \\ 0 & 0 \\ 0 & 0 \\ \hdashline 0 & 1 \\ 0 & 0 \end{bmatrix} \qquad (66)$$

which is different from the two *noncanonical* controller-type forms $\{A_c, B_c\}$ and $\{\tilde{A}_c, \tilde{B}_c\}$ that we had obtained in Example 6.4-4.

Another important fact, which may perhaps be missed in this simple example, is that unlike the $\bar{B}$ matrix in (65), the B_{2c} matrix in (66) will have (fixed) zeros in all rows except the first, $(\sigma_1 + 1)$th, $(\sigma_2 + 1)$th, etc.—the reader should try to show this.

Let us note that we can also find the controller canonical form corresponding to Scheme I, *but with considerably more labor*. Starting from (60), we follow the procedure of (64), which yields $\sigma_1 = 4$, $\sigma_2 = 5$, and (we show some intermediate steps for the benefit of the active student)

$$Q = \begin{bmatrix} 0 & 0 & 1 & -6 & 1 \\ 0 & 0 & 0 & 1 & 0 \\ -1 & 4 & -11 & 26 & 0 \\ 0 & -1 & 4 & -11 & 0 \\ 0 & 0 & -1 & 4 & 0 \end{bmatrix}, \qquad Q^{-1} = \begin{bmatrix} 0 & 2 & -1 & -4 & -5 \\ 0 & 5 & 0 & -1 & -4 \\ 0 & 4 & 0 & 0 & -1 \\ 0 & 1 & 0 & 0 & 0 \\ 1 & 2 & 0 & 0 & 1 \end{bmatrix}$$

$$
T^{-1} = \begin{bmatrix} 12 & 16 & -1 & 2 & 8 \\ -4 & -4 & 0 & -1 & -2 \\ 1 & 0 & 0 & 0 & 0 \\ 0 & 1 & 0 & 0 & 0 \\ 1 & 2 & 0 & 0 & 1 \end{bmatrix}, \qquad T = \begin{bmatrix} 0 & 0 & 1 & 0 & 0 \\ 0 & 0 & 0 & 1 & 0 \\ -1 & -2 & 0 & 0 & 4 \\ 0 & -1 & -2 & 0 & -2 \\ 0 & 0 & -1 & -2 & 1 \end{bmatrix}
$$

and, finally, the Scheme I canonical form,

$$
A_{1c} = \left[\begin{array}{cccc:c} -6 & -13 & -12 & -4 & 0 \\ 1 & 0 & 0 & 0 & 0 \\ 0 & 1 & 0 & 0 & 0 \\ 0 & 0 & 1 & 0 & 0 \\ \hdashline 0 & 1 & 2 & 0 & 0 \end{array}\right], \qquad B_{1c} = \left[\begin{array}{cc} 1 & 12 \\ 0 & -4 \\ 0 & 1 \\ 0 & 0 \\ \hdashline 0 & 1 \end{array}\right] \qquad (67) \qquad ∎
$$

A significant fact about Scheme I forms, besides the greater labor involved in obtaining them, is that the B matrix is not so simple as in the Scheme II forms—in B_{1c}, the xs can occur in all rows (cf. the discussion of B_{2c} above). This feature makes Scheme II forms more useful in certain applications, e.g., in the mode-shifting problem to be studied in Sec. 7.1. We shall also see in Sec. 6.7.2 that the MFDs naturally associated with Scheme II forms have a nice (*polynomial-echelon*) structure.

As a final remark here, however, let us note that from a computational point of view, the actual numerical calculations of this section could be organized more efficiently—see [34]–[36] for some studies in this direction.

Exercises

6.4-1.

Find controller-, observer-, and controllability-type realizations of

$$
H(s) = \begin{bmatrix} \dfrac{1}{(s-1)^2} & \dfrac{1}{(s-1)(s+3)} \\[2ex] \dfrac{-6}{(s-1)(s+3)^2} & \dfrac{s-2}{(s+3)^2} \end{bmatrix}
$$

6.4-2.

Find observer- and controllability-type realizations of

$$
H(s) = \begin{bmatrix} \dfrac{1}{s+1} & \dfrac{2}{s+1} \\[2ex] \dfrac{-1}{(s+1)(s+2)} & \dfrac{1}{s+2} \end{bmatrix}
$$

6.4-3.

Suppose

$$
A = \begin{bmatrix} 1 & 0 & 0 \\ 0 & -1 & 0 \\ 0 & 0 & -3 \end{bmatrix}, \qquad B = \begin{bmatrix} 0 \\ 1 \\ 1 \end{bmatrix}, \qquad C' = \begin{bmatrix} 1 & 0 \\ -1 & 2 \\ 0 & 1 \end{bmatrix}
$$

a. Is the realization controllable? Observable?

b. Try to write $C(sI - A)^{-1}B = D_L^{-1}(s)N_L(s) = N_R(s)D_R^{-1}(s)$, where $D_L(s)$, $D_R(s)$ have determinants of degree 3.

c. Repeat part b if the determinants are to be of degree 2.

6.4-4.

Suppose $D(s)$ is column reduced and has the form

$$D(s) = \begin{bmatrix} D_c(s) & D_{c\bar{c}}(s) \\ 0 & D_{\bar{c}}(s) \end{bmatrix}$$

Does the controller-type realization of $N(s)D^{-1}(s)$ have any special features? Explain why or why not.

6.4-5.

a. Prove by direct calculation that $(sI - A_c^o)^{-1}B_c^o = \Psi(s)S^{-1}(s) =$ transfer function from $u(s)$ to $\xi(s)$.

b. If we apply feedback and make an input transformation as in $u(s) = D_{hc}^{-1}[v(s) - D_{lc}\xi(s)]$, find the new transfer function from $u(s)$ to $\xi(s)$.

6.4-6. *Obtaining the Controllability Form*

In the text, we obtained the controllability form by relabeling the states of the observer-form realization of $D^{-1}(s)I_m$. Show that a more direct method is to start by writing $D(s) = S(s)D_{hr} + \tilde{\Psi}(s)D_{lr}$, where $\tilde{\Psi}(s) =$ block diag $\{[1\ s\ \cdots\ s^{l_i-1}]\}$. [Note that the order of the powers of s is reversed as compared to $\Psi(s)$ or $\Psi_L(s)$.] Now let $D_{lr}D_{hr}^{-1} = \bar{D}$, so that $D(s) = [S(s) + \tilde{\Psi}(s)\bar{D}]D_{hr}$. Also define $C_o =$ block diag $\{[0\ \underbrace{\cdots\cdots\ 1}_{l_1}]\}$.

6.4-7.

Show that the problem of setting up a desired state vector can be solved *by inspection* if a controllability form realization is used. For a discrete-time system, what is the shortest number of steps necessary to set up an arbitrary state?

6.4-8.

Suppose the $n \times n$ matrix $sI - A$ has only q nonunity invariant polynomials, $q < n$. Show that $\{C, A\}$ will be nonobservable for all $p \times n$ matrices C with $p < q$. Deduce a similar statement for the controllability of $\{A, B\}$. *Hint:* Use the rational canonical form for A (Theorem 6.3-18) and the PBH rank test (Theorem 6.2-6).

6.4-9.

Suppose that $\{A, b\}$ is controllable and that q_n is the last row of $\mathcal{C}(A, b)$. Let $T = [(A')^{n-1}q_n'\ \cdots\ q_n']'$.

a. Show that T is nonsingular. *Hint:* For part a, show that $T\mathcal{C}$ is a triangular matrix with ones on the main diagonal.

·b. Show that $\{TAT^{-1}, Tb\}$ are in controller canonical form.

c. Extend the proof in part a to show that the matrix T in Eq. (6.4-64d) is nonsingular.

6.4-10.

Show that the general form of the Scheme II dependency relations is [cf. (58)]

$$A^{k_i}b_i = \sum_{j=1}^{i-1} \sum_{k=0}^{K_1} \alpha_{ijk} A^k b_j + \sum_{j=i}^{m} \sum_{k=0}^{K_2} \alpha_{ijk} A^k b_j$$

where $K_1 = \min(k_i, k_j - 1)$, $K_2 = \min(k_i, k_j) - 1$.

6.4-11.

Show that the set of chain lengths obtained in a Scheme II search is invariant under a similarity transformation. *Hint:* Check the dimensions of the range spaces of $\{B\}$, $\{B, AB\}$. . . .

6.5   SOME PROPERTIES OF RATIONAL MATRICES

In Sec. 2.4.1 we developed, for scalar systems, the relationships among the three concepts of minimal realizations, jointly controllable and observable realizations, and irreducible MFDs. In Sec. 6.2 we obtained the multivariable analogs of the relations between the first two concepts, and we can now extend that discussion by showing in Sec. 6.5.1 the significance of irreducible MFDs. In Sec. 6.5.2 we shall present the important Smith-McMillan canonical representation of transfer functions and use it in Sec. 6.5.3 to study the finite and infinite poles and zeros of multivariable systems. In Sec. 6.5.4, we shall show how the null-space structure of rational matrices can be displayed by using the so-called minimal bases of rational vector spaces; this allows us to extend to arbitrary rational matrices the concept of Kronecker indices introduced in Sec. 6.3.4 for linear pencils.

6.5.1   Irreducible MFDs and Minimal Realizations

The constructions of Sec. 6.4 show how to obtain from an MFD $N(s)D^{-1}(s)$ a controllable realization of order equal to the determinantal degree of $D(s)$. To obtain the minimal degree, it therefore seems reasonable to reduce the MFD to "lowest" terms by extracting a gcrd of $N(s)$ and $D(s)$. We shall justify this hope here by proving the following theorem.

Theorem 6.5-1.

Any realization of an MFD with order equal to the degree of the determinant of the denominator matrix will be a minimal (equivalently, a controllable and observable) realization if and only if the MFD is irreducible.

This theorem is an obvious generalization of Theorem 2.4-4 for scalar transfer functions and can in fact be proved in a very similar way. Thus we first establish the following.

Lemma 6.5-2.

If there exists one controllable and observable realization of $N(s)D^{-1}(s)$, with order $n = \deg \det D(s)$, then all realizations of order n will also be controllable and observable.

Proof.   We use the facts that minimality is equivalent to joint controllability and observability (Theorem 6.2-3) and that for any two realizations $\{A_i, B_i, C_i, i = 1, 2\}$ of the same order

$$\mathcal{O}_1 \mathcal{C}_1 = \mathcal{O}_2 \mathcal{C}_2$$

Now if realization $\{A_1, B_1, C_1\}$ is minimal,

$$\text{rank } \mathcal{O}_1 \mathcal{C}_1 \triangleq \rho(\mathcal{O}_1 \mathcal{C}_1) = n$$

and hence $\rho(\mathcal{O}_2 \mathcal{C}_2) = n$. But then because

$$\rho(\mathcal{O}_2, \mathcal{C}_2) \leq \min \{\rho(\mathcal{O}_2), \rho(\mathcal{C}_2)\}$$

we have $\rho(\mathcal{O}_2) = n = \rho(\mathcal{C}_2)$, so that $\{A_2, B_2, C_2\}$ is controllable and observable. ∎

Now all we have to do is find one minimal realization, and this will be provided by the controller-form realization of Sec. 6.4.1.

Lemma 6.5-3.

A controller-form realization of $N(s)D^{-1}(s)$, of order equal to deg det $D(s)$, will also be observable (and hence minimal) if and only if the MFD is irreducible.

Proof.†   Assume that the controller-form realization of Sec. 6.4-1, $\{A_c, B_c, C_c\}$, is not observable. Then by the PBH test (Theorem 6.2-5) there is an eigenvector of A_c that is orthogonal to the rows of C_c. Now in (6.4-18) we showed that any eigenvector of A_c had the form $e = \Psi(\lambda)p$, where $\Psi(\lambda)$ is as in (6.4-4) and p is a solution of $D(\lambda)p = 0$. But if this eigenvector is orthogonal to $C_c \triangleq N_{lc}$, we shall also have $N_{lc}e = 0$, i.e., $N_{lc}\Psi(\lambda)p = 0$ or $N(\lambda)p = 0$. Therefore $N(s)$ and $D(s)$ cannot be relatively prime (Lemma 6.3-7), which contradicts our hypothesis. This proves sufficiency. Necessity of the condition follows from the fact that a realization of a *reducible* MFD (of order equal to the determinantal degree of the denominator) cannot be minimal, for by extracting a gcd we obtain an MFD of lower determinantal degree and can make a realization of this lower order. ∎

Theorem 6.5-1 then follows from these two lemmas. With this theorem and Theorem 6.2-3, we shall have obtained for the matrix case all the inter-relations noted in Sec. 2.4.1 among the notions of minimality, irreducibility, and joint controllability and observability.

We also proved in Theorem 6.2-4 that any two minimal realizations can be related by a similarity transformation. There is an important analogous result for MFDs.

---

†Compare with the proof in Sec. 2.4.1.

Theorem 6.5-4.

Suppose $\{N_i(s)D_i^{-1}(s), i = 1, 2\}$ are two irreducible MFDs. Then there exists a *unimodular* matrix $U(s)$ such that

$$D_1(s) = D_2(s)U(s), \qquad N_1(s) = N_2(s)U(s)$$

Proof.   Since $\{N_i(s)D_i^{-1}(s), i = 1, 2\}$ are both realizations, we have

$$\dot{N}_1(s) = N_2(s)D_2^{-1}(s)\acute{D}_1(s)$$

and a candidate for the unimodular matrix is

$$U(s) \triangleq D_2^{-1}(s)D_1(s)$$

We shall show that $U(s)$ is unimodular by proving that $D_2^{-1}(s)D_1(s)$ and its inverse $D_1^{-1}(s)D_2(s)$ are both polynomial matrices (Lemma 6.3-1). To do this, first note that because $N_1(s)$ and $D_1(s)$ are coprime there exist polynomial matrices $X(s)$ and $Y(s)$ such that (Lemma 6.3-5) $X(s)N_1(s) + Y(s)D_1(s) = I$. Therefore

$$X(s)N_2(s)D_2^{-1}(s)D_1(s) + Y(s)D_2(s)D_2^{-1}(s)D_1(s) = I$$

or

$$[X(s)N_2(s) + Y(s)D_2(s)]U(s) = I$$

which shows that $U^{-1}(s)$ is polynomial. But interchanging the subscripts 1 and 2 in the above argument will show that $U(s)$ is polynomial, thus completing the proof. ∎

The following result is a natural extension of Theorem 6.5-4.

Lemma 6.5-5.

If $\{N(s), D(s)\}$ is any MFD of $H(s)$ and $\{\bar{N}(s), \bar{D}(s)\}$ is an *irreducible* MFD of $H(s)$, then there exists a polynomial matrix $R(s)$, not necessarily unimodular, such that $N(s) = \bar{N}(s)R(s)$, $D(s) = \bar{D}(s)R(s)$.

Proof.   If $\{N(s), D(s)\}$ are not coprime, let $R_1(s)$ be a gcrd, so that

$$N(s) = N_1(s)R_1(s), \qquad D(s) = D_1(s)R_1(s)$$

and $H(s) = N_1(s)D_1^{-1}(s)$ is irreducible. But then by Theorem 6.5-4, since $\bar{N}(s)\bar{D}^{-1}(s)$ is also irreducible, there exists a unimodular matrix $U(s)$ such that

$$N_1(s) = \bar{N}(s)U(s), \qquad D_1(s) = \bar{D}(s)U(s)$$

and defining $R(s) = U(s)R_1(s)$ completes the proof. ∎

So far we have talked only of right MFDs, but clearly all our statements have obvious analogs for left MFDs. However, let us explicitly record the following result, which removes a possible ambiguity in Theorem 6.5-1.

Lemma 6.5-6.

The determinantal degree of any irreducible right MFD of $H(s)$ is equal to the determinantal degree of any irreducible left MFD of $H(s)$.

Proof.   Otherwise by using Lemma 6.5-3 (and its dual) we would have jointly controllable and observable and hence minimal (Theorem 6.2-3) realizations of different orders. This is impossible.   ∎

Theorem 6.5-1 gives one answer to the question posed in Sec. 6.1 as to the minimal order of any realization of $H(s)$: It is the determinantal degree of the denominator of any irreducible MFD of $H(s)$.

Here is another useful expression for the minimal order, independently discovered by Ho ([37] and [38]), Silverman [39], and Youla and Tissi [40].

Lemma 6.5-7.

Let $\{h_i, i = 1, \ldots\}$ be the Markov parameters of a strictly proper matrix transfer function $H(s)$; i.e., let

$$H(s) = \sum_1^\infty h_i s^{-i}$$

Then the minimal order of any state-space realization of $H(s)$ is given by

$$n_{\min} = \text{rank of } \mathbf{M}$$

where $\mathbf{M}$ is the infinite (block) Hankel matrix

$$\mathbf{M} = \begin{bmatrix} h_1 & h_2 & h_3 & \cdots \\ h_2 & h_3 & h_4 & \cdots \\ h_3 & h_4 & h_5 & \cdots \\ \cdot & \cdot & \cdot & \\ \cdot & \cdot & \cdot & \\ \cdot & \cdot & \cdot & \end{bmatrix}$$

Proof.   If $\{A, B, C\}$ is a *minimal* realization of order n, then

$$h_i = CA^{i-1}B, \qquad i = 1, 2, \ldots$$

and

$$M[n, n] = \mathcal{O}(C, A)\mathcal{C}(A, B)$$

where $M[i, j]$ is the leading submatrix of $\mathbf{M}$ formed by the first i block rows and the first j block columns. The minimality of $\{A, B, C\}$ means that $\mathcal{O}$ and $\mathcal{C}$ have rank n and therefore rank $M[n, n] = n$. Then by virtue of the Cayley-Hamilton theorem, we shall have

$$\text{rank } M[n + i, n + 1] = n, \qquad i \geq 0$$

This proves the lemma.   ∎

Important Remark: To really use Lemma 6.5-7, we need an upper bound, say n_0, on the minimal degree. For a transfer function matrix $H(s)$, one upper bound is $\min(m, p) \cdot \deg d(s)$, the least common multiple of the denominators of the entries in $H(s)$ (show this). To obtain better bounds, we should study in more detail the properties of $H(s)$ as a matrix of rational functions, and we shall begin to do this now.

One important consequence will be an alternative theoretically important formula for $n_{\min}$ [see (7) below].

6.5.2   The Smith-McMillan Form of H(s)

The reader should review the Smith form discussed in Sec. 6.3.3. Given a rational matrix $H(s)$, whose entries are assumed to be in reduced form, let us write it as

$$H(s) = \frac{N(s)}{d(s)}$$

where

$$d(s) = \text{the } monic \text{ least common multiple of the}$$
$$\text{denominators of the entries of } H(s)$$

Then $d(s)H(s) = N(s)$ is a polynomial matrix, so that we can write, say,

$$d(s)H(s) = N(s) = U_1(s)\Lambda(s)U_2(s) \tag{1}$$

where the $\{U_i(s)\}$ are unimodular matrices and $\Lambda(s)$ is in Smith form. Now write

$$U_1^{-1}(s)H(s)U_2^{-1}(s) = \frac{\Lambda(s)}{d(s)} = \text{diag}\left\{\frac{\lambda_i(s)}{d(s)}\right\}$$

and reduce the elements of the rational matrix $\Lambda(s)/d(s)$ to *lowest* terms. That is, let, say,

$$\frac{\lambda_i(s)}{d(s)} = \frac{\epsilon_i(s)}{\psi_i(s)} \tag{2a}$$

where

$$\{\epsilon_i(s), \psi_i(s)\} \text{ are coprime, } i = 1, \ldots, r \tag{2b}$$

and

$$r = \text{the (normal) rank of } H(s) \tag{2c}$$

Then we can write

$$H(s) = U_1(s)M(s)U_2(s) \tag{3a}$$

where

$$M(s) = \begin{bmatrix} \text{diag } \{\epsilon_i(s)/\psi_i(s)\} & 0 \\ 0 & 0 \end{bmatrix} \tag{3b}$$

We also note the properties

$$\psi_{i+1}(s)\,|\,\psi_i(s) \quad \text{(i.e., } \psi_{i+1} \text{ divides } \psi_i), \quad i = 1, \ldots, r-1 \tag{4a}$$

$$\epsilon_i(s)\,|\,\epsilon_{i+1}(s), \quad i = 1, \ldots, r-1 \tag{4b}$$

$$d(s) = \psi_1(s) \tag{5}$$

Properties (4a) and (4b) are reasonably obvious. For (5), we note that if $\psi_1(s) \neq d(s)$, then from $\epsilon_1(s)/\psi_1(s) = \lambda_1(s)/d(s)$, it must be that $d(s)$ and $\lambda_1(s)$ have a common factor. But then so will $d(s)$ and all the elements of $N(s)$, which would contradict the definition of $d(s)$.

Example 6.5-1.

Let $H(s)$ be as in Example 6.2-1,

$$H(s) = \frac{1}{d(s)}\begin{bmatrix} s & s(s+1)^2 \\ -s(s+1)^2 & -s(s+1)^2 \end{bmatrix} \tag{6a}$$

$$d(s) = (s+1)^2(s+2)^2$$

Then we can check that

$$H(s) = \begin{bmatrix} 1 & 0 \\ -(s+1)^2 & 1 \end{bmatrix}\begin{bmatrix} s/(s+1)^2(s+2)^2 & 0 \\ 0 & s^2/(s+2) \end{bmatrix}\begin{bmatrix} 1 & (s+1)^2 \\ 0 & 1 \end{bmatrix} \tag{6b}$$

We should note that even though $H(s)$ is strictly proper, its Smith-McMillan form may not be so [cf. the $(2, 2)$ element $s^2/(s+2)$]. The reason for this discrepancy is that the unimodular matrices $\{U(s), V(s)\}$ can introduce additional poles and zeros at $s = \infty$, as will be explained in more detail in Sec. 6.5-3. ∎

As might be expected from the construction, the matrix $M(s)$ is unique, while the $\{U_i(s)\}$ can be chosen in many ways. This result was first developed by McMillan [41] in 1952, following the above construction via the Smith form of a related polynomial matrix. Therefore $M(s)$ is called the *Smith-McMillan* form of $H(s)$. $M(s)$ is a general canonical form for rational matrices and can in fact be defined directly without references to polynomial matrices, as we shall see in Sec. 6.5.3. We shall find the Smith-McMillan form to be a valuable conceptual and theoretical tool, though like the Smith form it will not generally be convenient for actual numerical computations.

Historically, the Smith-McMillan form was used by Kalman ([42] and [43]) to give a generalization of Gilbert's technique of Sec. 6.1, which we recall worked only when the *denominator polynomial* $d(s)$ had distinct roots. Kalman used the Smith-McMillan form to construct a controllable and observable realization of $M(s)$ using only

$$n_{\min} = \sum \deg \psi_i(s) \tag{7}$$

integrators. This alternative expression for the minimal order turned out to coincide with what McMillan in 1952 [42] had called the *degree* of $H(s)$ and had shown to be the minimum number of inductors and capacitors required to realize a *passive* impedance matrix $H(s)$. Kalman's result (7) shows that if we allow integrators (op amps) as building blocks, then the McMillan degree is the minimal order of any proper $H(s)$, whether it is a passive impedance or not.

Here we shall not go into Kalman's construction (see, however, Exercise 6.5-1), because by using the results of Sec. 6.5.1 we shall directly establish the truth of formula (7).

Thus note that from (3) that we can write $M(s)$ as a right or left MFD,

$$M(s) = \mathcal{E}(s)\psi_R^{-1}(s) = \psi_L^{-1}(s)\mathcal{E}(s) \tag{8}$$

where

$$\mathcal{E}(s) = \begin{bmatrix} \text{diag } \{\epsilon_i(s)\} & 0 \\ \hline 0 & 0 \end{bmatrix}, \qquad p \times m \tag{9}$$

$$\psi_R(s) = \begin{bmatrix} \text{diag } \{\psi_i(s)\} & 0 \\ \hline 0 & I_{m-r} \end{bmatrix}, \qquad m \times m \tag{10}$$

and

$$\psi_L(s) = \begin{bmatrix} \text{diag } \{\psi_i(s)\} & 0 \\ \hline 0 & I_{p-r} \end{bmatrix} \tag{11}$$

The fact that $\{\epsilon_i(s), \psi_i(s)\}$, $i = 1, \ldots, r$, are coprime clearly implies that the matrices $\{\mathcal{E}(s), \psi_R(s)\}$ are coprime, and hence so are

$$N_0(s) \triangleq U_1(s)\mathcal{E}(s), \qquad D_0(s) \triangleq U_2^{-1}(s)\psi_R(s)$$

where $\{U_1(s), U_2(s)\}$ are unimodular matrices, so that we can now write $H(s)$ as an *irreducible* MFD [cf. (3)], say,

$$
\begin{aligned}
H(s) &= [U_1(s)\mathcal{E}(s)][\psi_R^{-1}(s)U_2(s)] \\
&= N_0(s)D_0^{-1}(s)
\end{aligned}
\tag{12}
$$

But then the claim (7) follows immediately by Theorem 6.5-1,

$$
\begin{aligned}
n_{\min} &= \text{deg det } D_0(s) = \text{deg det } \psi_R(s) \\
&= \sum \text{deg } \psi_i(s), \qquad \text{the } \textit{McMillan degree of } H(s)
\end{aligned}
\tag{13}
$$

Some Related Results.   In Theorem 6.5-4, we showed that

$$N_1(s)D_1^{-1}(s) = N_2(s)D_2^{-1}(s) \qquad \text{(both irreducible)}$$

implies that

$$D_1(s) = D_2(s)U_R(s), \qquad N_1(s) = N_2(s)U_R(s), \qquad U_R(s) \text{ unimodular} \quad (14)$$

Taking $N_2(s) = \mathcal{E}(s)$, we see that the Smith form of any right numerator is $\mathcal{E}(s)$ as defined in (9). By a similar argument $\mathcal{E}(s)$ is also the Smith form of the left numerator of any irreducible MFD of $H(s)$. This means we have proved the following result [12], which will be used presently to study the zeros of $H(s)$.

Lemma 6.5-8.

The (right or left) numerators of irreducible MFDs of $H(s)$ all have the same Smith form.

We cannot make a similar claim for the (right or left) denominators because they may not have the same size. But by using (10) and (11), we have the following result [12].

Lemma 6.5-9.

The denominators of irreducible MFDs of $H(s)$ all have the same nonunity invariant polynomials.

In particular, of course, this means they will have the same determinant, a fact proved differently in Lemma 6.5-6. [In fact, the results of Lemma 6.5-6 and Lemma 6.5-9 also follow easily from (6.4-16), as we ask the reader to show.]

6.5.3   *Poles and Zeros of Multivariable Transfer Functions*

Poles and zeros of multivariable systems can be defined in several (not all equivalent) ways, but the definitions that yield the most significant consequences are the following [12]:

the *zeros* are the roots of the (nonzero) numerator polynomials $\{\epsilon_i(s)\}$ in the Smith-McMillan form of $H(s)$ $\qquad$ (15)

the *poles* are the roots of the denominator polynomials
in the Smith-McMillan form of $H(s)$ $\qquad$ (16)

The transfer function $H(s)$ of Example 6.5-1 has Smith-McMillan form

$$M(s) = \text{diag} \left\{ \frac{s}{(s+1)^2(s+2)^2}, \frac{s^2}{s+2} \right\}$$

so that $H(s)$ has three zeros at $s = 0$, two poles at $s = -1$, and three poles at $s = -2$.

We should note that multivariable transfer functions can have poles and zeros at the same location: Although each pair $\{\epsilon_i(s), \psi_i(s)\}$ is coprime, an $\epsilon_i(s)$ and a $\psi_j(s), j \neq i$, can still have common roots. For example, we could have

$$M(s) = \text{diag } \{(s - s_0)^{-2}, 1, (s - s_0)^3, (s - s_0)^3\}$$

To reflect this fact, and for other reasons, it will be useful to rewrite the (nontrivial part of the) Smith-McMillan form as

$$\text{diag } \left\{ \frac{\epsilon_i(s)}{\psi_i(s)} \right\} = \prod_\alpha M_\alpha(s) \tag{17}$$

where α ranges over the set of poles and zeros of $H(s)$, and each $M_\alpha(s)$ has the form

$$M_\alpha(s) = \text{diag } \{(s - \alpha)^{\sigma_1}, \ldots, (s - \alpha)^{\sigma_r}\} \tag{18a}$$

The divisibility properties of the $\{\epsilon_i(s), \psi_i(s)\}$ imply that the indices $\{\sigma_i\}$ form a nondecreasing sequence

$$\sigma_1 \leq \sigma_2 \leq \cdots \leq \sigma_r \tag{18b}$$

The $\{\sigma_i\}$ will of course depend on α, but for simplicity we shall often not indicate this explicitly. There will be a pole at α only if some σ_i is negative, and we shall say that the *order* of the pole is $-\sigma_1$ (if $\sigma_1 < 0$), while the sum of all the negative indices will be called the *degree* of the pole at α. Similarly, the order of the zero at α is σ_r (if $\sigma_r > 0$), while its degree is the sum of all the positive indices. We shall call the set $\{\sigma_1, \ldots, \sigma_r\}$ the *structural* indices of $H(s)$ at α.

To build up our intuition for these definitions of the poles and zeros of a matrix transfer function $H(s)$, let us examine them more closely in the cases where we may have irreducible MFDs and minimal state-space realizations for $H(s)$.

Irreducible MFDs.   Suppose first that we have an irreducible right MFD

$$H(s) = N(s)D^{-1}(s)$$

Then it follows immediately from Lemma 6.5-9 that

> the *poles* of $H(s)$ are the roots of det $D(s) = 0$, where $D(s)$ is the denominator of any irreducible MFD of $H(s)$

For the zeros, we need a little more thought because $H(s)$ may not be square. Note first that by Lemma 6.5-8 we immediately have the following result:

> if $H(s)$ is square and nonsingular, then its zeros are the roots of det $N(s) = 0$, where $N(s)$ is the numerator in any irreducible MFD of $H(s)$

More generally, Lemma 6.5-8 shows that

> the zeros are the roots of the invariant polynomials
> of the numerator of any irreducible MFD of $H(s)$ (19)

Another way of saying this is that

> the zeros are the frequencies at which the
> rank of $N(s)$ drops below its normal rank (20)

For example, if

$$N(s) = \begin{bmatrix} s+1 & 0 \\ 0 & s-2 \end{bmatrix}$$

there are zeros at $s = -1$ and $s = 2$, since the rank of $N(s)$ drops at these frequencies. On the other hand, if

$$N(s) = \begin{bmatrix} s+1 & 0 & s-2 \\ 0 & s-2 & s+1 \end{bmatrix}$$

there are no zeros. This is in fact a *typical* or *generic* situation—rectangular (or nonsquare) systems generally have no zeros because it is unlikely that all minors of size less than or equal to the normal rank will be simultaneously zero.

Poles and Zeros of Minimal State-Space Realizations. Suppose that the multivariable system is given in state-space form, with $\{A, B, C\}$ minimal. Then we recall the useful fact, proved in Sec. 6.4 [cf. (6.4-17)], that

$$\begin{bmatrix} sI - A_c & B_c \\ -C_c & 0 \end{bmatrix} \stackrel{S}{\sim} \begin{bmatrix} I_n & 0 \\ 0 & N(s) \end{bmatrix} \tag{21}$$

and note that since (similarity and) equivalence transformations do not affect the Smith form, we can also write (again using Lemma 6.5-8)

$$\begin{bmatrix} sI - A & B \\ -C & 0 \end{bmatrix} \stackrel{S}{\sim} \begin{bmatrix} I_n & 0 \\ 0 & N(s) \end{bmatrix} \stackrel{S}{\sim} \begin{bmatrix} I_n & 0 \\ 0 & \mathcal{E}(s) \end{bmatrix} \tag{22}$$

Therefore the zeros of a minimal realization can be computed as the zeros of the invariant polynomials of the block (so-called *system*) matrix on the left-hand side of (22).

Dynamical Interpretation of Zeros ([44] to [46]). This characterization allows us to give a physical interpretation of the zeros. If s_0 is a zero frequency, then all matrices in (22) will lose rank at $s = s_0$, and there will exist a vector $[x'_0 \ u'_0]'$ such that

$$
\begin{bmatrix} s_0 I - A & B \\ -C & 0 \end{bmatrix} \begin{bmatrix} x_0 \\ u_0 \end{bmatrix} = 0 \tag{23a}
$$

But this means that if we have an input

$$
u(t) = u_0 e^{s_0 t}, \qquad t \geq 0 \tag{23b}
$$

then there exists an initial state x_0 such that the response is

$$
y(t) \equiv 0, \qquad t > 0 \tag{23c}\dagger
$$

This *transmission-blocking* property is well known for scalar systems (cf. Exercise 2.2-18) and further confirms the reasonableness of our definition (15) of multivariable zeros.

Apart from definitions (of which there are still others; see, e.g., [45] and [47]) we should ask if the multivariable zeros have other properties and implications similar to those known for scalar systems. In fact several results are available in this direction, see especially [45] and [46].

Thus, it can be shown that the zeros are the poles of the inverse of $H(s)$ [minimal‡ left inverse if $H(s)$ is not square but has full column rank—see Exercise 6.5-14]. Also, all minimal (left) inverses will be stable only when the zeros of $H(s)$ are in the left half plane. For scalar systems, we know that the zero frequencies are just the modes (of a minimal realization) that can be made unobservable by using state feedback to "move a pole under a zero" (see Example 3.3-6). A similar statement is true for left-invertible multivariable systems, but for non-left-invertible systems there is a degeneracy in that *any* frequency can be made into an unobservable mode by state feedback. These phenomena will be explored in Sec. 7.6 after we have introduced state feedback for multivariable systems.

Poles and Zeros at Infinity.   It is important to note that the definitions we have used apply only to poles and zeros at *finite* points in the complex s plane, because the (highly nonunique) unimodular matrices used to get the unique Smith-McMillan form destroy information about the behavior at $s = \infty$—the point is that, as we shall see, unimodular matrices can have both poles and zeros at ∞.

There are several problems in which it is important to keep track of the behavior at $s = \infty$. Poles at $s = \infty$ correspond to nonproper systems (or

---

†This discussion is really only relevant when the system matrix in (23a) has full column rank; otherwise, we can force $y(t) \equiv 0, t > 0$, for any frequency s_0. (Why?) (See also Sec. 7.6.)

‡That is, one with lowest finite polar degree.

systems with differentiators), as may arise in constructing inverse systems, while the zeros at ∞ are important, for example, in studying the asymptotic behavior of multivariable root loci. Recall that for scalar systems with a numerator of degree m, $m < n$, m of the closed-loop poles will converge toward the m finite zeros as the feedback gain goes to infinity, while the remaining $n - m$ poles will converge to the $n - m$ *zeros at infinity* (cf. Sec. 3.1.1). The properties and in fact the proper definition of multivariable root loci are only beginning to be studied, but it is clear that the concept of poles and zeros at infinity will be important in this study (see, e.g., [48] and the references therein).

To obtain the pole-zero structure at infinity, we can proceed as follows (cf. the procedure used in the study of linear pencils in Sec. 6.3.4): Make a change of variables $s \rightarrow \lambda^{-1}$, and form the Smith-McMillan form of $H(\lambda^{-1})$. Then the pole-zero structure of $H(\lambda^{-1})$ at $\lambda = 0$ will give the pole-zero structure of $H(s)$ at $s = \infty$.

Example 6.5-2.
Consider the simple unimodular matrix

$$U(s) = \begin{bmatrix} 1 & s \\ 0 & 1 \end{bmatrix} \tag{24}$$

Here

$$U(\lambda^{-1}) = \begin{bmatrix} 1 & \lambda^{-1} \\ 0 & 1 \end{bmatrix} \tag{25}$$

whose Smith-McMillan form can be seen to be

$$\begin{bmatrix} \lambda^{-1} & 0 \\ 0 & \lambda \end{bmatrix} \tag{26}$$

which shows that $U(s)$ has a (first-order) pole and zero at ∞. ∎

The justification of this procedure is the following [41]: Because unimodular matrices can introduce (or delete) poles and zeros at infinity (but not at any finite point), the Smith-McMillan form will only accurately reflect the behavior of $H(s)$ at all finite s but not for $s = \infty$. To correctly determine the structure at $s = \infty$, we can make any transformation of the form

$$s = \frac{a\lambda + b}{c\lambda + d} \tag{27}$$

where $c \neq 0$ and $ad - bc \neq 0$, which will merely transform the complex s plane into itself and in particular will move the point at $s = \infty$ to the point $\lambda = -d/c$. The Smith-McMillan form for $H(\lambda)$ will accurately reflect the behavior of $H(\lambda)$ [and therefore of $H(s)$] for all points except those at $\lambda = \infty$.

In particular, the Smith-McMillan structure at $\lambda = -d/c$ will accurately reflect that of $H(s)$ at $s = \infty$. Of the many possible choices of $\{a, b, c, d\}$ it seems simplest to choose $a = 0 = d$, $b = c = 1$.

By this means we can define structural indices $\{\sigma_i\}$ not only at the finite poles and zeros of $H(s)$ but also at the infinite poles and zeros.

Example 6.5-3.

The matrix

$$H(s) = \text{diag}\left\{\frac{s}{s-1}, \frac{1}{s-1}, (s-1)^2\right\}$$

has the Smith-McMillan form

$$M(s) = \text{diag}\left\{(s-1)^{-1}, (s-1)^{-1}, s(s-1)^2\right\}$$

and the finite structural indices are seen to be the following:

at $s = 0$: $\sigma_1 = 0$, $\sigma_2 = 0$, $\sigma_3 = 1$

at $s = 1$: $\sigma_1 = -1$, $\sigma_2 = -1$, $\sigma_3 = 2$

For the point at infinity, we transform to

$$H(\lambda^{-1}) = \text{diag}\left\{\frac{-1}{\lambda-1}, \frac{-\lambda}{\lambda-1}, \frac{(\lambda-1)^2}{\lambda^2}\right\}$$

which has the Smith-McMillan form

$$M(\lambda) = \text{diag}\left\{\frac{1}{\lambda^2(\lambda-1)}, \frac{1}{\lambda-1}, \lambda(\lambda-1)^2\right\}$$

so that for the structural indices of $H(s)$ we have

at $s = \infty$: $\sigma_1 = -2$, $\sigma_2 = 0$, $\sigma_3 = 1$ ∎

Valuations and a Direct Characterization of the Smith-McMillan Form.
It is now worthwhile to point out that it is not really necessary to go through a separate calculation of the Smith-McMillan form for $H(s^{-1})$. We shall establish this by showing how to directly compute the Smith-McMillan form of $H(s)$ at each pole and zero, finite or infinite [50]. Recall first that for the finite poles and zeros we wrote [cf. (17)–(18)] the nonzero part of the Smith-McMillan form as the product of terms of

$$M_\alpha(s) = \text{diag}\left\{(s-\alpha)^{\sigma_1(\alpha)}, \ldots, (s-\alpha)^{\sigma_r(\alpha)}\right\} \qquad (28a)$$

where

$$\sigma_1(\alpha) \leq \sigma_2(\alpha) \leq \cdots \leq \sigma_r(\alpha) \qquad (28b)$$

For any scalar rational function $g(s)$, suppose we can write, for some integer v_α,

$$g(s) = (s - \alpha)^{v_\alpha} p(s)/q(s), \qquad -\infty < \alpha < \infty \tag{29a}$$

where $\{p(s), q(s)\}$ are coprime and not divisible by $s - \alpha$. Then we call†

$$v_\alpha \triangleq \text{the } s - \alpha \text{ valuation of } g(s) \tag{29b}$$

For consistency in later formulas, we shall define

$$v_\alpha = \infty \qquad \text{if } g(s) \equiv 0 \tag{29c}$$

We extend this definition to rational matrices $H(s)$ by writing, as in [15],

$$v_\alpha^{(i)}(H) \triangleq \text{the } i\text{th valuation of } H(s) \text{ at } s - \alpha$$
$$= \min \{v_\alpha(|H|^{(i)})\} \tag{30a}$$

where the minimum is taken over all

$$|H|^{(i)} \triangleq \text{an } i \times i \text{ minor of } H(s) \tag{30b}$$

That is, $v_\alpha^{(i)}(H)$ is the (algebraically) *smallest* of the $s - \alpha$ valuations of all the $i \times i$ minors of $H(s)$‡. Then by using the Binet-Cauchy theorem (cf. Exercise A.9) we can show that (cf. Exercise 6.5-6)

$$v_\alpha^{(i)}(H) = v_\alpha^{(i)}(M) = v_\alpha^{(i)}(M_\alpha) \tag{31}$$

where M and M_α are as in (3) and (17). .
The special form of $M(s)$ then shows that

$$\sigma_1(\alpha) = v_\alpha^{(1)}(H)$$
$$\sigma_2(\alpha) = v_\alpha^{(2)}(H) - v_\alpha^{(1)}(H)$$
$$\cdot$$
$$\cdot$$
$$\cdot$$
$$\sigma_r(\alpha) = v_\alpha^{(r)}(H) - v_\alpha^{(r-1)}(H) \tag{32‡}$$

where we recall again that r is the normal rank of H. The formula (32) holds

---

†Some authors write $v_{s-\alpha}[g(s)]$ instead of $v_\alpha[g(s)]$—this is helpful when working over finite fields where $s - \alpha$ is replaced by a so-called primitive polynomial (see, e.g., [13] and [15]). As mentioned before, we shall confine ourselves to the complex field.

‡It is worth noting that

$$v_\alpha^{(r)}(H) = \left\{ \begin{array}{l} \text{the number of} \\ \text{zeros of } H \text{ at } \alpha \end{array} \right\} - \left\{ \begin{array}{l} \text{the number of} \\ \text{poles of } H \text{ at } \alpha \end{array} \right\}$$

for any α, but if α is not a pole or zero of $H(s)$, then all the α-valuations will be zero.

We should note that the present direct construction of the Smith-McMillan form, building it up via valuations of H, shows clearly the uniqueness of the form (see also Exercise 6.5-7).

Example 6.5-1.   (Continued)

Let us consider again the rational matrix $H(s)$ in (6a) of Example 6.5-1. Then we can calculate the valuations as shown:

	s	s + 1	s + 2
$\nu^{(1)}$	1	−2	−2
$\nu^{(2)}$	3	−2	−3

It follows that

$$\sigma_1(0) = 1, \qquad \sigma_2(0) = 3 - 1 = 2$$
$$\sigma_1(-1) = -2, \qquad \sigma_2(-1) = -2 + 2 = 0$$
$$\sigma_1(-2) = -2, \qquad \sigma_2(-2) = -3 + 2 = -1$$

and therefore that

$$M(s) = \begin{bmatrix} s & 0 \\ 0 & s^2 \end{bmatrix} \begin{bmatrix} (s+1)^{-2} & 0 \\ 0 & 1 \end{bmatrix} \begin{bmatrix} (s+2)^{-2} & 0 \\ 0 & (s+2)^{-1} \end{bmatrix}$$

$$= \begin{bmatrix} s/(s+1)^2(s+2)^2 & 0 \\ 0 & s^2/(s+2) \end{bmatrix}$$

as we found before.   ∎

Example 6.5-3.   (Continued)

For the matrix

$$H(s) = \text{diag}\left\{ \frac{s}{s-1}, \frac{1}{s-1}, (s-1)^2 \right\}$$

we see that

$$v_0^{(1)} = 0, \qquad v_0^{(2)} = 0, \qquad v_0^{(3)} = 1$$
$$v_1^{(1)} = -1, \qquad v_1^{(2)} = -2, \qquad v_1^{(3)} = 0$$

so that

$$\sigma_1(0) = 0, \qquad \sigma_2(0) = 0, \qquad \sigma_3(0) = 1$$
$$\sigma_1(1) = -1, \qquad \sigma_2(1) = -1, \qquad \sigma_3(1) = 2$$

and the Smith-McMillan form is

$$M(s) = \begin{bmatrix} 1 & & \\ & 1 & \\ & & s \end{bmatrix} \begin{bmatrix} (s-1)^{-1} & & \\ & (s-1)^{-1} & \\ & & (s-1)^2 \end{bmatrix}$$

as we found earlier.   ∎

Behavior at Infinity.   To handle the point at infinity, we shall define† for any scalar rational function $g(s)$

$$v_\infty(g) \triangleq \text{the } \infty \text{ valuation of } g(s)$$
$$= \text{denominator degree} - \text{numerator degree} \tag{33}$$

For example,

$$v_\infty\left[\frac{s}{(s-1)^2}\right] = 1$$

The definition is extended to matrices as before; namely,

$$v_\infty^{(i)}(H) = \text{the algebraically smallest } \infty \text{ valuation of all} \\ \text{the } i \times i \text{ minors of } H(s) \tag{34}$$

Then for the Smith-McMillan form at $s = \infty$, we can write

$$M_\infty(s) = \text{diag}\{s^{\sigma_1(\infty)}, \ldots, s^{\sigma_r(\infty)}\} \tag{35a}$$

where

$$\sigma_1(\infty) = v_\infty^{(1)}, \sigma_2(\infty) = v_\infty^{(2)} - v_\infty^{(1)}, \ldots \tag{35b}$$

Example 6.5-3.   (Continued)
  For

$$H(s) = \text{diag}\left\{\frac{s}{s-1}, \frac{1}{s-1}, (s-1)^2\right\}$$

we have

$$v_\infty^{(1)} = -2, \qquad v_\infty^{(2)} = -2, \qquad v_\infty^{(3)} = -1$$

so that

$$\sigma_1(\infty) = -2, \qquad \sigma_2(\infty) = 0, \qquad \sigma_3(\infty) = 1$$

which is what we found before by determining the Smith-McMillan form of $H(\lambda^{-1})$. ∎

  We should note that the point at infinity has to be kept separate from the rest of the Smith-McMillan form so that we do not *mix up* the behavior at $s = 0$ and $s = \infty$. Thus, note that in the above continuation of Example 6.5-3 we cannot combine the forms obtained for finite s and infinite s as

$$\begin{bmatrix} 1 & & \\ & 1 & \\ & & s \end{bmatrix}\begin{bmatrix} (s-1)^{-1} & & \\ & (s-1)^{-1} & \\ & & (s-1)^2 \end{bmatrix}\begin{bmatrix} s^{-2} & & \\ & 1 & \\ & & s \end{bmatrix} \neq M(s)$$

†We could also write $v_\infty(g)$ as $v_{s^{-1}}(g)$, which would make more sense when working over finite fields where ∞ has no meaning.

What we can do is define (in an obvious notation)

$$M\left(s, \frac{1}{\lambda}\right) = M_{\text{fin}}(s)M_\infty\left(\frac{1}{\lambda}\right)$$

so that in Example 6.5-3 we have

$$M\left(s, \frac{1}{\lambda}\right) = \begin{bmatrix} 1 & & \\ & 1 & \\ & & s \end{bmatrix} \begin{bmatrix} (s-1)^{-1} & & \\ & (s-1)^{-1} & \\ & & (s-1)^2 \end{bmatrix} \begin{bmatrix} \lambda^{+2} & & \\ & 1 & \\ & & \lambda^{-1} \end{bmatrix}$$

This suggests that one way to handle the problem is to work with rational functions in two variables of the type $H(s/\lambda)$. This can in fact be done and with some profit, but it will require an extension of the notions of equivalence, coprimeness, etc., and therefore here we shall only refer to [49].

Another way is to make the transformation (cf. (27)) $\lambda = 1/(s + \beta)$, where β is a point that is not a pole or a zero of $H(s)$, see, e.g., [16]. Finally we remark that just as finite poles of $H(s)$ correspond to exponential terms in the free response, so also the infinite poles correspond to impulsive terms in the free response to *arbitrary* initial conditions at $t = 0-$. This physical interpretation is useful in understanding various mathematical properties of the poles and zeros at ∞ (see [46]).

6.5.4 Nullspace Structure; Minimal Polynomial Bases and Kronecker Indices

A rational $p \times m$ matrix $H(s)$ is singular if it is not square and invertible. We shall see that the nature of the singularity can be characterized by certain so-called *left* and *right minimal* (or *Kronecker*) *indices*.

We note first that the set of all rational $m \times 1$ vectors $\{f(s)\}$ such that

$$H(s)f(s) = 0$$

is a vector space (over the field of scalar rational functions) called the *right null-space* of $H(s)$. Every element of this null-space is orthogonal to the rows of $H(s)$, and the dimension of this null-space is $\alpha = m - r$ (cf. Sec. 5.2). If we had a vector space over the real or complex numbers, its dimension would essentially characterize it. But for our more general situation of a rational vector space, it turns out that there is a richer structure, first noted by Kronecker (in connection with linear pencils, cf. Sec. 6.3.4) and exposed in detail by Wedderburn [8] and later, independently, by Forney [15]. Here we follow the slightly different development in [46] and [60] (see also [61, Sec. 5]).

This structure is captured by the notion of a *minimal polynomial basis* for the (right null-) space, which can be obtained as follows.

First note that we can, without loss of generality, restrict ourselves to polynomial bases—just multiply through by the least common multiple of the denominators. Now among all polynomial vectors in the space, which in our case are polynomial solutions of

$$H(s)f(s) = 0$$

choose a nonzero one, $f_1(s)$, of least degree μ_1. Then among all solutions linearly independent of $f_1(s)$, choose one, say $f_2(s)$, of least degree μ_2, $\mu_2 \geq \mu_1$. Continuing in this way, we shall get $\alpha = m - r$ polynomial vectors

$$F(s) = [f_1(s) \quad f_2(s) \quad \cdots \quad f_\alpha(s)] \tag{36a}$$

with degrees

$$\mu_1 \leq \mu_2 \leq \cdots \leq \mu_\alpha \tag{36b}$$

Of course there can be many (nontrivially different) sets of polynomial vectors of this sort, but it turns out that they must all have the same set of degrees $\{\mu_1, \ldots, \mu_\alpha\}$.

For suppose we obtained another set of vectors $\bar{F}(s) = [\bar{f}_1(s) \quad \cdots \quad \bar{f}_\alpha(s)]$ with column degrees $\{\bar{\mu}_1, \ldots, \bar{\mu}_\alpha\}$. Suppose now that

$$\mu_1 = \cdots = \mu_{n_1} < \mu_{n_1+1} = \cdots = \mu_{n_2} < \cdots$$

and similarly

$$\bar{\mu}_1 = \cdots = \bar{\mu}_{\bar{n}_1} < \bar{\mu}_{\bar{n}_1+1} = \cdots = \bar{\mu}_{\bar{n}_2} < \cdots$$

Obviously $\mu_1 = \bar{\mu}_1$, and we shall now prove that $n_1 = \bar{n}_1$ and $\mu_{n_1+1} = \bar{\mu}_{\bar{n}_1+1}$. Note first that every column $\bar{f}_i(s)$, $1 \leq i \leq \bar{n}_1$, must be a linear combination of the $\{f_i(s), 1 \leq i \leq n_1\}$; otherwise we could replace $f_{n_1+1}(s)$ by an $\bar{f}_i(s)$ of lower degree $\bar{\mu}_1 = \mu_1$. Therefore $\bar{n}_1 \leq n_1$. On the other hand, reversing roles, every $f_i(s)$, $1 \leq i \leq n_1$, must be a linear combination of the $\{\bar{f}_i(s),$ $1 \leq i \leq \bar{n}_1\}$; otherwise we could replace $\bar{f}_{\bar{n}_1+1}(s)$ by an $f_i(s)$ of lower degree $\mu_1 = \bar{\mu}_1$. Therefore $n_1 \leq \bar{n}_1$, and hence $n_1 = \bar{n}_1$. Moreover, since the $\{f_i(s)\}$ and $\{\bar{f}_i(s)\}$ span the same space, we must have $\mu_{n_1+1} = \bar{\mu}_{\bar{n}_1+1}$, and so on.

Therefore the above method of selecting a basis specifies a *unique* set of indices for the right null-space of $H(s)$, which we shall call the *right minimal indices* of $H(s)$. More generally, any set of polynomial vectors, say,

$$F(s) = [f_1(s) \quad \cdots \quad f_\alpha(s)]$$

selected in the sequential fashion described above and forming a basis for a rational vector space will be called a *minimal (polynomial) basis* for the space.

To get some further insight into these concepts, let us consider the special case where $H(s)$ is a linear pencil

$$H(s) = sE - A$$

Then it turns out, perhaps not surprisingly, that in this case the right minimal indices concide with the *right Kronecker indices* as defined in Sec. 6.3.4. That discussion is worth rereading at this point, because we can give a nice explicit formula for the right minimal basis of a pencil in Kronecker canonical form.

For this, we recall from Eq. (6.3-41b) that

$$L_\mu[1 \quad s \quad \cdots \quad s^\mu]' = [0 \quad \cdots \quad 0]'$$

so that for the pencil (6.3-42) in Example 6.3-4 we can see that a basis for the right null-space is

$$F(s) \triangleq \begin{bmatrix} 1 & 0 & 0 & 0 & 0 & 0 & 0 & 0 & 0 & \cdots & 0 \\ 0 & 1 & 0 & 0 & 0 & 0 & 0 & 0 & 0 & \cdots & 0 \\ 0 & 0 & 1 & s & 0 & 0 & 0 & 0 & 0 & \cdots & 0 \\ 0 & 0 & 0 & 0 & 1 & s & s^2 & 0 & 0 & \cdots & 0 \end{bmatrix}'$$

corresponding to the indices $\{0, 0, 1, 2\}$.

It is easy in this case to see that $F(s)$ is a minimal basis, but furthermore we can get some clues from it as to how to recognize a minimal basis in more complicated situations. Note first that for any strictly equivalent pencil

$$U(sE - A)V$$

where $\{U, V\}$ are constant nonsingular matrices, the right minimal basis changes to

$$\bar{F}(s) = V^{-1}F(s)$$

The basis is now more complicated, but we note that the column degrees of $\bar{F}(s)$ are still the same, viz., $\{0, 0, 1, 2\}$.

Moreover, we may notice that both $F(s)$ and $\bar{F}(s)$ are

1. *Column reduced*, i.e., the leading column coefficient matrix has full rank, and
2. *Irreducible*, i.e., the rank is full for all (finite) values of s.

It turns out that these two properties in fact completely characterize a minimal polynomial basis for any rational vector space,† as we shall shortly prove. Yet another characterization is obtained if we define, as in [15], the *order* of a polynomial basis as the sum of the column degrees of the vectors in the basis; it then turns out that a polynomial basis is minimal if and only if it has minimal order.

We shall collect these facts into the following theorem [46] and [60].

---

†It is also possible to define rational minimal bases (cf. [15] and [46]).

Theorem 6.5-10.   Minimal Bases

Consider a full column rank polynomial matrix

$$F(s) = [f_1(s) \quad \cdots \quad f_\alpha(s)],$$

with column degrees $\{\mu_1 \leq \mu_2 \cdots \leq \mu_\alpha\}$. Then the following statements are equivalent:

1. The columns of $F(s)$ form a minimal basis for the rational vector space they generate.
2. $F(s)$ is column-reduced and irreducible.
3. $F(s)$ has minimal order.

Proof. $1 \Rightarrow 2, 3 \Rightarrow 2$: We demonstrate that if $F(s)$ is not column reduced or not irreducible, then it is possible to find another polynomial basis $\{\bar{f}_i(s)\}$ for the same space with the same degrees in all but a single column, say the jth, and with $\bar{\mu}_j < \mu_j$. It is easy to see that this contradicts both the hypothesis that $\{f_i(s)\}$ is a minimal basis and the hypothesis that the basis has minimal order.

If $F(s)$ is not column reduced, by an elementary column operation, we can obtain a new basis with the degree of just one column reduced from the degree of that column in $F(s)$ and with the other column degrees unchanged.

If $F(s)$ is not irreducible, then its columns are dependent for some value of s, say $s = a$, so there exist constants $\{c_i\}$, not all zero, such that

$$\sum_{i=1}^{\alpha} c_i f_i(a) = 0 \tag{37}$$

It follows that

$$\bar{f}(s) \triangleq \sum_{i=1}^{\alpha} \frac{c_i f_i(s)}{s - a} \tag{38}$$

is a polynomial vector. Now suppose $f_j(s)$ has highest degree among all $f_i(s)$ for which $c_i \neq 0$; then

$$\deg \bar{f}(s) = \deg f_j(s) - 1 < \deg f_j(s)$$

On replacing $f_j(s)$ by $\bar{f}(s)$ in the original basis, we still have a polynomial basis for the same space, but the degree of one column is now reduced from the corresponding degree in $F(s)$.

$2 \Rightarrow 1, 2 \Rightarrow 3$: For the converse proofs, we use the predictable-degree property of a column-reduced matrix (Theorem 6.3-13) and the fact (see Exercise 6.3-18) that if $F(s)$ is irreducible and $p(s)$ is a polynomial vector, then the solution $q(s)$ of

$$p(s) = F(s)q(s)$$

must be polynomial. [The latter result is also immediately obtained on using the Bezout characterization of the irreducibility of $F(s)$ (cf. Lemma 6.3-5), namely that there exists a polynomial matrix $X(s)$ such that $X(s)F(s) = I$.]

Suppose now that $F(s)$ is column reduced and irreducible but is not a minimal basis or not a minimal-order basis. Let $P(s)$ denote either a minimal basis or a minimal-order basis,

$$P(s) = [p_1(s) \quad \cdots \quad p_\alpha(s)]$$

with column degrees $m_1 \le m_2 \le \cdots \le m_\alpha$. In either case, then, it is easy to see that there must be some j such that

$$\mu_1 \le m_1, \ldots, \mu_{j-1} \le m_{j-1} \quad \text{but} \quad \mu_j > m_j \tag{39}$$

Now since the $\{f_i(s)\}$ form a basis for the space, we must have

$$p_k(s) = \sum_i f_i(s) q_i(s), \qquad k = 1, \ldots, j$$

By the irreducibility of $F(s)$, the $q_i(s)$ must be polynomial. Then by the predictable-degree property and the inequalities in (39), the summation index i can only take values $i = 1, \ldots, j - 1$. This contradicts the independence of the $\{p_k(s), k = 1, \ldots, j\}$. Hence $F(s)$ must be both a minimal basis and a minimal-order basis. The theorem is thus completely proved. ∎

We see that a minimal basis of a rational vector space has a unique associated set of minimal indices, which for reasons explained earlier might also be called *Kronecker indices*. Therefore the null-spaces of a rational transfer function matrix can be characterized by a set of left and right Kronecker (or minimal) indices.

Example 6.5-4.

We note that the matrix

$$H(s) = \begin{bmatrix} s^{-1} & 0 & s^{-1} & s \\ 0 & (s+1)^2 & (s+1)^2 & 0 \\ -1 & (s+1)^2 & s^2 + 2s & -s^2 \end{bmatrix}$$

can be written

$$H(s) = \begin{bmatrix} 1 & 0 & 0 \\ 0 & 1 & 0 \\ -s & 1 & 1 \end{bmatrix} \begin{bmatrix} s^{-1} & 0 & 0 & 0 \\ 0 & (s+1)^2 & 0 & 0 \\ 0 & 0 & 0 & 0 \end{bmatrix} \begin{bmatrix} 1 & 0 & 1 & s^2 \\ 0 & 1 & 1 & 0 \\ 0 & 0 & 1 & 0 \\ 0 & 0 & 0 & 1 \end{bmatrix}$$

$$= U(s) M(s) V(s), \text{ say}$$

We see that $H(s)$ has normal rank 2 and that its Smith-McMillan form is $M(s)$. We see also that

$$U^{-1}(s) H(s) = \begin{bmatrix} s^{-1} & 0 & s^{-1} & s \\ 0 & (s+1)^2 & (s+1)^2 & 0 \\ 0 & 0 & 0 & 0 \end{bmatrix}$$

which shows that the last row of $U^{-1}(s)$ is a polynomial basis for the left null-space of $H(s)$, and so also the last two columns of $V^{-1}(s)$ can be seen to be a polynomial basis for the right null-space of $H(s)$. These bases are

$$[s \quad -1 \quad 1] \qquad \text{for the left null-space}$$

and

$$\begin{bmatrix} -1 & -1 & 1 & 0 \\ -s^2 & 0 & 0 & 1 \end{bmatrix}' \qquad \text{for the right null-space}$$

We see also that these happen to be minimal, by statement 2 of Theorem 6.5-10. Therefore the left and right Kronecker indices are $\{1\}$ and $\{0, 2\}$, respectively. ∎

I. C. Gohberg pointed out to the author that the significance of minimal bases was perhaps first realized by J. Plemelj in 1908 and then substantially developed in 1943 by N. I. Mushkelishvili and N. P. Vekua (see the discussion in [61, Sec. 5]). These authors were studying the so called Riemann-Hilbert problem, which was later shown to be closely related to the theory of Wiener-Hopf integral equations, as described for example in the definitive paper of Gohberg and Krein [62]. Certain so called "factorization indices" play an important role in this theory, and it is therefore not surprising that these are closely related to the Kronecker indices, as has recently been shown in [63] and [64]. We shall make the relationship more explicit in Sec. 7.1.3.

The Defect of a Matrix.   We recall that an interesting role of the infinite frequency poles and zeros in the *scalar* case is to help ensure the following balance:

the total (finite and infinite) number of poles
= the total number of (finite or infinite) zeros.

However this nice relation does not carry over to the matrix case. For example, consider the matrix

$$H(s) = \begin{bmatrix} I \\ G(s) \end{bmatrix}, \qquad G(s) \text{ rational}$$

The poles of $H(s)$ are exactly those of $G(s)$ but $H(s)$ has no zeros, because it has full rank for all s. (Another proof follows by the extended Bezout criterion of Exercise 6.5-15—take $X(s) = [I \ 0]$). However it turns out that in this case, the total number of poles (finite and infinite) equals the sum of the Kronecker indices of $H(s)$. We can in fact prove a more general result [50], [51].

If $H(s)$ is a rational matrix of rank r, we shall define (extending a concept of Forney's [15])

$$\text{def } H(s) \triangleq \text{the } \textit{defect} \text{ of } H(s).$$
$$= -\sum_{\alpha \in \mathbb{C}} v_{\alpha}^{(r)}(H), \tag{40}$$

where $\mathbb{C}$ is the entire complex plane (including ∞) and where we recall that

$$v_\alpha^{(r)}(H) = \text{the } r\text{th valuation of } H(s) \text{ at } s - \alpha$$
$$= \text{the minimum of the valuations at } \alpha \text{ of all the}$$
$$r \times r \text{ minors of } H(s).$$

From the properties of the Smith-McMillan form of $H(s)$, we can see that (cf. (32))

$$v_\alpha^{(r)}(H) = \text{the difference between the number of} \atop \text{zeros and poles at } s = \alpha \tag{41}$$

and we note that this property is true for finite *and* infinite values of α. Therefore we have that

$$\text{def } H(s) = \text{(the total (finite and infinite) number} \atop \text{of poles of } H(s)) - \text{(the total number of} \atop \text{zeros of } H(s)). \tag{42}$$

The following result then holds [50], [51].

Theorem 6.5-11.   Defects and Kronecker Indices

For any rational matrix $H(s)$,

$$\text{def } H(s) = \text{the sum of the right and left Kronecker indices of } H(s). \tag{43}$$

Proof.   Note that we can always, and in many ways, write

$$H(s) = H_1(s)H_2(s)$$

where $H_1(s)$ has full column rank r and $H_2(s)$ has full row rank r. Then

$$v_\alpha^{(r)}(H) = \min v_\alpha \{\text{all } r \times r \text{ minors of } H_1H_2\}$$
$$= \min v_\alpha \{\text{all } (r \times r \text{ minor of } H_1)(r \times r \text{ minor of } H_2)\}$$

Now there is one product in which the minimum of each factor occurs, so that

$$v_\alpha^{(r)}(H) = \min v_\alpha \{\text{all } r \times r \text{ minors of } H_1\}$$
$$+ \min v_\alpha \{\text{all } r \times r \text{ minors of } H_2\}$$
$$= v_\alpha^{(r)}(H_1) + v_\alpha^{(r)}(H_2)$$

Therefore, summing over all α, we have

$$\text{def } H = \text{def } H_1 + \text{def } H_2$$

Now it is easy to see that with $H_1(s)$ and $H_2(s)$ as above, the left [right] null space of $H(s)$ is the same as that of $H_1(s)$ [$H_2(s)$]. Let $G_1(s)$ be an irreducible, row-reduced polynomial basis for the left null space of $H_1(s)$, hence of $H(s)$. Its row degrees $\{l_i\}$ are then the left Kronecker indices of $H(s)$, and the sum of these is simply given by (see Exercise 6.5-19)

$$\sum l_i = \text{def } G_1$$

Also, by Exercise 6.5-25,

$$\text{def } G_1 = \text{def } H_1$$

so that

$$\sum l_i = \text{def } H_1$$

Similarly, if $\{k_i\}$ are the right Kronecker indices of $H_2(s)$, and hence of $H(s)$, we can show that

$$\sum k_i = \text{def } H_2$$

The desired result then follows easily from the earlier decomposition of def H. ∎

Special Solutions of Rational Equations. There are several problems that reduce ultimately to determining solutions $H(s)$ to the equation

$$H(s)A(s) = B(s) \tag{44}$$

where $A(s)$ and $B(s)$ are specified $m \times r$ and $p \times r$ matrices, with $A(s)$ of full column rank $r \leq m$. In particular it is of interest to have tests, involving $A(s)$ and $B(s)$, that determine whether certain special types of solution exist, e.g., proper solutions, stable solutions, or feedforward solutions (i.e., those with all poles at the origin, of interest in discrete-time systems because of their 'deadbeat' behaviour). This general problem, which arises in several contexts (e.g., system inversion, model matching, etc.), was formulated by Wang & Davison [52], and subsequently studied by Forney [15] and Kung & Kailath [50] among others. Here we shall present a simple condition for determining when there exists a solution with no poles at some specified frequency α.

First write

$$F(s) \triangleq \begin{bmatrix} B(s) \\ A(s) \end{bmatrix} \triangleq \begin{bmatrix} \bar{B}(s) \\ \bar{A}(s) \end{bmatrix} Q(s) \triangleq \bar{F}(s)Q(s), \tag{45}$$

where $\bar{F}(s)$ has no poles or zeros at α and $Q(s)$ is some non-singular matrix (which evidently has the same pole-zero structure as $F(s)$). Note that $\bar{A}(s)$ and $\bar{B}(s)$ will have no poles at α. Now (44) may be written as

$$H(s)\bar{A}(s) = \bar{B}(s)$$

If $\bar{A}(s)$ has *no zeros* at α, then we know that it possesses a left inverse, $\bar{A}^{-L}(s)$, that has no poles at α (cf. Exercise 6.5-14). It is in this case easily verified that

$$H(s) = \bar{B}(s)\bar{A}^{-L}(s)$$

constitutes a solution of (44) that has no poles at α. We shall now demonstrate the *converse*: If $\bar{A}(s)$ *has* zeros at α, then every solution $H(s)$ must contain poles at α. For this, write

$$\begin{bmatrix} H(s) \\ I \end{bmatrix} \bar{A}(s) = \begin{bmatrix} \bar{B}(s) \\ \bar{A}(s) \end{bmatrix} = \bar{F}(s).$$

If $\bar{A}(s)$ has zeros at α then $\bar{A}(\alpha)$ is a matrix with finite entries (since $\bar{A}(s)$ has no poles at α) that has rank less than r, which is possible only if $H(\alpha)$ has infinite entries (since $\bar{F}(\alpha)$ is a finite matrix of rank r). Thus a solution $H(s)$ of (44) with no poles at α exists if and only if $\bar{A}(s)$ has no zeros at α.

To interpret the above condition in terms of the given matrices $A(s)$ and $B(s)$, note that since $\bar{A}(s)$ has no poles at α, by (41) we have that

$$v_\alpha^{(r)}(\bar{A}(s)) = \text{the total number of zeros of } \bar{A}(s) \text{ at } \alpha \tag{46}$$

Since $\bar{A}(s) = A(s)Q^{-1}(s)$, we can see that

$$v_\alpha^{(r)}(\bar{A}) = v_\alpha^{(r)}(A) - v_\alpha^{(r)}(Q) \tag{47a}$$

$$= v_\alpha^{(r)}(A) - v_\alpha^{(r)}(F) \tag{47b}$$

Therefore we have the result, first noted in [50], that Eq. (44) has a solution with no poles at α if and only if

$$v_\alpha^{(r)}(A) = v_\alpha^{(r)}(F) \tag{48}$$

A noteworthy feature of the above result is that it is independent of whether $A(s)$ and $B(s)$ are polynomial or rational, and that it holds for any α, finite or infinite. In particular, applying it to $\alpha = \infty$, we have the result that Eq. (44) has a proper solution if and only if

$$v_\infty^{(r)}(A) = v_\infty^{(r)}(F) \tag{49}$$

Suppose now that we applied this criterion to a *polynomial* matrix $F(s)$ that had first been made column reduced, which can be achieved by column operations on (44) that leave $H(s)$ unaffected. Then (cf. Exercise 6.5.19) $-v_\infty^{(r)}(F)$ will be the sum of the column degrees of $F(s)$. Now $v_\infty^{(r)}(A)$ will attain this value if and only if the first m rows of the leading column coefficient matrix of $F(s)$ have full rank. This yields another simple test for properness [58], alternative to the one given in Theorem 6.13-12.

We see that stable solutions will be obtained if and only if (47) holds for all α with Re $\alpha \geq 0$; in discrete-time, feedforward solutions will be obtained if and only if (47) holds for all α except $\alpha = 0$. Other results can also be obtained (see [50] and also Sec. 6.7.2).

In the above discussions, we assumed that $A(s)$ had full rank, which allowed consideration of an arbitrary $B(s)$. If $A(s)$ is not of full rank, then (44) will not have a solution unless $B(s)$ is in the column range space of $A(s)$, i.e., if and only if

$$\text{rank } A(s) = \text{rank } F(s) \tag{50}$$

If this simple necessary condition is satisfied then it may be proved by a straightforward modification of the previous proof that (48) still holds; the only change in the proof is that in the factorization of (45), $\bar{F}(s)$ and $Q(s)$ must be required to have full rank.

Exercises

6.5-1. *Kalman's Generalization of the Gilbert Realization Procedure* [42], [43]

Let $H(s) = U_1(s)M(s)U_2(s)$, where $M(s)$ is the unique Smith-McMillan form of $H(s)$.

a. Show that we can write $H(s) = \sum_1^r [c_i(s)b_i(s)]/\psi_i(s) + W(s)$, where r and $\{\psi_i(s)\}$ are as defined in Sec. 6.5.3, $\{c_i(s)\}$ and $\{b_i(s)\}$ are polynomial row and column vectors, respectively, and $W(s)$ is a polynomial matrix.

b. If $H(s)$ is *strictly proper*, show that we can write $H(s) = \sum_1^r H_i(s)$, $H_i(s) = [c_i(s)b_i(s)] \bmod \psi_i(s)/\psi_i(s)$.

c. Then a state-space realization of $H(s)$ can be obtained as the parallel combination of realizations of the $\{H_i(s)\}$. Use the matrix identity given in Exercise A.33(d) to find a state-space realization of $H_i(s)$.

d. Prove that the realization of $H(s)$ obtained in part c [with order equal to the McMillan degree, $\sum \deg \psi_i(s)$] is minimal.

6.5-2.

Let $N_R D_R^{-1} = D_L^{-1} N_L$ define two irreducible MFDs. If $D_R^{-1}C$ is also irreducible, show that so is $D_L^{-1}N_L C$. *Hint:* By contradiction; assume $D_L^{-1}N_L C = \bar{N}\bar{D}^{-1}$, $\deg \det \bar{D} < \deg \det D_L$.

6.5-3. *(Rosenbrock)*

Let $H(s) = C(sI - A)^{-1}B = N(s)D^{-1}(s)$ with $\{A, B, C\}$ minimal and $\{N(s), D(s)\}$ right coprime. Then if $H_1(s) = C_1(sI - A)^{-1}B$, show that there exists a polynomial matrix $N_1(s)$ such that $H_1(s) = N_1(s)D^{-1}(s)$. Also show that, conversely, if $N_1(s)$ is any polynomial matrix such that $N_1(s)D^{-1}(s)$ is strictly proper, then there exists a C_1 such that $H_1(s) = C_1(sI - A)^{-1}B$.

6.5-4. *Zeros of State-Space Realizations*

a. Show that the zeros of a minimal realization $\{A, B, C\}$ are unaffected by any state or output feedback that does not affect the minimality.

b. Let $\{A, B, C\}$ be a minimal realization with an equal number of inputs and outputs. Show that $\det \{C \text{ Adj } (sI - A)B\} = [\det(sI - A)]^{m-1}\Pi_1^r \epsilon_i(s)$, where the $\epsilon_i(s)$ are the nonzero numerator polynomials in the Smith-McMillan form of $C(sI - A)^{-1}B$.

6.5-5. *Poles, Zeros and Minors*

a. Show that a rational matrix will have a pole at α if and only if some minor has a pole at α.

b. Show that it will have a zero at α if *every* minor of a certain order has a zero at α, but that this condition is *not necessary*.

6.5-6.   Valuations Under Elementary Transformations

Given the $p \times m$ matrix $H(s)$, there are $\binom{p}{i}$ and $\binom{m}{i}$ ways, respectively, of selecting i rows and i columns of $H(s)$. Let $|H|_{kl}^{(i)}$ denote the $i \times i$ minor of $H(s)$ formed by making the k-th and l-th selections, respectively, of i rows and columns from the above possibilities. If now

$$H(s) = U(s)M(s)V(s)$$

then the Binet-Cauchy theorem (Exercise A.9) enables us to assert that

$$|H|_{kl}^{(i)} = \sum_{p,q} |U|_{kp}^{(i)} |M|_{pq}^{(i)} |V|_{lq}^{(i)}$$

a. If $U(s)$, $V(s)$ are unimodular, so that neither they nor their inverses have finite poles, show (by using the above relations and their reverses) that

$$\min_{k,l} \{v_\alpha(|H|_{kl}^{(i)})\} = \min_{k,l} \{v_\alpha(|M|_{kl}^{(i)})\}$$

where $v_\alpha(\ )$ denotes the valuation at the finite point α. In other words, using the notation of (30a), show that $v_\alpha^{(i)}(H) = v_\alpha^{(i)}(M)$.

b. Show that for this result to hold at some α (including $\alpha = \infty$) it suffices that $U(s)$ and $V(s)$ have no poles at α, and that they possess left and right inverses, respectively, that also have no poles at α.

6.5-7.   The Invariant Rational Functions [49]

Given a rational matrix $H(s)$ of rank r, let $d_k(s)$ be the least common multiple of the denominators of all the $k \times k$ minors, say $\{|H|_{j,l}^{(k)}\}$, of $H(s)$. Let $n_k(s)$ be the greatest common divisor of all the (numerator) polynomials $\{d_k(s)|H|_{j,l}^{(k)}\}$. Define $g_k(s) = n_k(s)/d_k(s)$, reduced to lowest terms, and define $m_k(s) = g_k(s)/g_{k-1}(s)$, $k = 1, 2, \ldots, r$, with $g_0(s) = 1$. Show that the $\{m_k(s)\}$ are the nonzero elements of the Smith-McMillan form of $H(s)$. *Remarks:* Comparing the above with the definition in Sec. 6.3.3 of the invariant polynomial matrix, we see that the $\{m_k(s)\}$ may be called the *invariant rational functions* of $H(s)$. Incidentally, this also proves the uniqueness of the Smith-McMillan form.

6.5-8.   The Minimal, Characteristic, and Zero Polynomials of $H(s)$

a. Refer to Eq. (6.5-5) and show that $\psi_1(s)$ is the minimal polynomial of the matrix A in any minimal realization $\{A, B, C\}$ of $H(s)$.

b. Show that the characteristic polynomial $a(s)$ of a minimal realization of $H(s)$ is the least common multiple of the denominators of the minors (of all orders) of $H(s)$.

c. Let $|H|^{(r)} =$ an $r \times r$ minor of $H(s)$, $r = \operatorname{rank} H(s)$, and let

$$b(s) = \text{the gcd of all the polynomials } \{a(s)|H|_i^{(r)}\},$$
$$\text{where } i \text{ ranges over the set of all the}$$
$$r \times r \text{ minors of } H(s).$$

Show that the zeros of $b(s)$ are the same in location and multiplicity as the finite zeros of $H(s)$. [$b(s)$ may be called the zero polynomial of $H(s)$.]

6.5-9.   Some Properties of the McMillan Degree

We defined the McMillan degree δ_M of a proper rational matrix as (cf. (6.5-7)) the sum of the degrees of the denominator polynomials in its Smith-McMillan form. For a general rational matrix $H(s)$, decompose it as

$$H(s) = H_{sp}(s) + D(s)$$

where $H_{sp}(s)$ is strictly proper and $D(s)$ is polynomial. Define its McMillan degree as

$$\delta_M(H(s)) = \delta_M(H_{sp}(s)) + \delta_M(D(s^{-1})).$$

Show that

1. $\delta_M(H)$ = sum of polar degrees at *all* poles, those at infinity included
2. If $H(s)$ is nonsingular,

$$\delta_M(H(s)) = \delta_M(H^{-1}(s))$$

3. If G is a constant matrix, $\delta(G) = 0$, and if G is also nonsingular

$$\delta_M(GH(s)) = \delta_M(H(s))$$

4. $\delta_M(H_1(s) + H_2(s)) \leq \delta_M(H_1(s)) + \delta_M(H_2(s))$ with equality if $H_1(s)$ and $H_2(s)$ have no common poles (finite or infinite).

Remark: Lévy [49] has shown that if we form $H(s/\lambda)$ and rewrite it as $H(s/\lambda) = N(s, \lambda)D^{-1}(s, \lambda)$, then $\delta_M(H(s)) = \deg | D(s, \lambda)|$, if $\{N(s, \lambda), D(s, \lambda)\}$ are coprime in the sense that

a. $N(s, 1)$ and $D(s, 1)$ are right coprime, and b. $N(1, \lambda)$ and $D(1, \lambda)$ are right coprime or b . [$N'(1, 0)$, $D'(1, 0)$] is full rank

and where the degree of a polynomial in s and λ is the highest degree of any of the monomials $s^\alpha \lambda^\beta$ in it. [Example: $\deg (s^3\lambda^2 + s^2\lambda + s\lambda^6) = 7$].

Use this definition to derive the properties 1 to 4 above.

6.5-10.

Suppose $\{H(s), M(s)\}$ are rational matrices related as $H(s) = U(s)M(s)V(s)$, where $\{U(s), V(s)\}$ are unimodular.

1. For every polynomial $(s - \alpha)$, show that the (algebraically) lowest power of $(s - \alpha)$ that can be factored out of any of the $i \times i$ minors of $H(s)$ and $M(s)$ is the same. *Note.* If $(s - \alpha)$ is not a pole or zero of $H(s)$, the lowest power will be 0.
2. Extend this result to $\alpha = \infty$, by showing that the highest relative order (denominator degree $-$ numerator degree) among the $i \times i$ minors of $H(s)$ and $M(s)$ is the same.

6.5-11.   Orders and Degrees of Poles and Zeros

Let $H(s)$ have a pole (zero) at $s = \alpha$. Show that

1. The (algebraically) lowest negative power of $s - \alpha$ that occurs in any *entry* of $H(s)$ is the *order* of the pole at $s - \alpha$.
2. The lowest negative power of $s - \alpha$ that occurs in *any* minor of $H(s)$ is the *degree* of the pole at $s = \alpha$.
3. Denote the highest *positive* power of $s - \alpha$ that occurs in *all* the $i \times i$ minors of $H(s)$ by w_i (with $w_i = 0$ if no positive power occurs in the $i \times i$ minors). Then show that the *order* of the zero at α is given by $w_r - w_{r-1}$, where r is the rank of $H(s)$ and that the *degree* of this zero is given by $w_r +$ the polar degree at α.

6.5-12.

a. Show that the pole- (zero-) structure of a nonsingular matrix is identical to the zero- (pole-) structure of its inverse.

b. Show that a non-singular matrix has as many poles as zeros (when those at infinity are included).

6.5-13. *Existence of Inverse Systems*

1. Let $H(s)$ be a $p \times m$ rational transfer function, $p \geqq m$. Show that $H(s)$ has a "left" inverse (not necessarily proper) if and only if $H(s)$ has full column rank, i.e., rank $[H(s)] = m$. [A left inverse $H^{-L}(s)$ is such that $H^{-L}(s)H(s) = I$, so that if $y(s) = H(s)u(s)$, then $u(s) = H^{-L}(s)y(s)$.]
2. If $N(s)D^{-1}(s)$ is an irreducible MFD of $H(s)$, show that $H(s)$ will have a left inverse (not necessarily proper) if and only if $N(s)$ has rank m. *Hint:* Recall the properties of linear equations and rank.
3. Use the identity

$$\begin{bmatrix} sI - A & B \\ -C & 0 \end{bmatrix}\begin{bmatrix} I & -(sI - A)^{-1}B \\ 0 & I \end{bmatrix} = \begin{bmatrix} sI - A & 0 \\ -C & H(s) \end{bmatrix}$$

to find a condition that $\{A, B, C\}$ defines a left-invertible realization.
4. State and prove the corresponding results for right inverses.

6.5-14. *Properties of Left Inverses*

1. Use the Smith-McMillan form to characterize all left inverses of a rational matrix $H(s)$ with full column rank, and show that the zeros of $H(s)$ will be among the poles of all such inverses.
2. Show how to find a "minimal" left inverse and show that its poles are exactly the zeros of $H(s)$.

6.5-15. *Extended Bezout Test*

Show that a rational matrix $H(s)$ of full column rank has no zeros at α, finite or infinite, if and only if there exists a rational matrix $X(s)$ with no poles at α such that

$$X(s)H(s) = I$$

[This is a generalization [46] of a well-known test of Bezout, see Lemma 6.3-5.]

6.5-16. *Diagonal Decompositions* [46]

Let a rational matrix $H(s)$ be related to a diagonal (but *not* necessarily Smith-McMillan) form via unimodular matrices:

$$H(s) = U_1(s)\mathfrak{D}(s)U_2(s),$$

where $\mathfrak{D}(s) = \text{diag}\{d_1(s), \dots, d_r(s)\}$.

1. Show that the $\{d_j(s)\}$ have respective pole or zero multiplicities (counted negative for poles and positive for zeros) that are simply a reordering of the structural indices, $\{\sigma_i(\alpha)\}$, of $H(s)$ (cf. Eq. (6.5.18)).
2. Show that this result holds even when $U_1(s)$ and $U_2(s)$ are not necessarily unimodular but merely possess *no* poles or zeros at α, with $U_1(s)$ being perhaps only left invertible and $U_2(s)$ right invertible.

6.5-17. *Structural Indices from Laurent Expansions*

Suppose we have available a Laurent-series expansion of a matrix transfer function $H(s)$ about the point $s = \alpha$, say $H(s) = R_{-i}(s - \alpha)^{-i} + \cdots + R_0 + R_1(s - \alpha) + \cdots$. Define the Toeplitz matrices $\{T_i\}$ to be upper triangular Toeplitz matrices with first row $[R_{-i} \dots R_i]$. Take $T_{-i-1} = 0$.

1. If n_i denotes the *number* of structural indices of $H(s)$ at α whose value is less or equal to $i - l$ show that the $\{n_i\}$ are given by

$$n_i = \text{rank } T_{i-l} - \text{rank } T_{i-l-1}, \qquad i = 0, 1, \dots$$

Hint: First justify the procedure for $H(s)$ in the Smith-McMillan form.

2. Show that the structural indices may be uniquely reconstructed from the $\{n_i\}$. *Remark:* These results, as well as an efficient numerical algorithm, can be found in P. Van Dooren, P. Dewilde, and J. Vandewalle, *IEEE Trans. Circuits and Systems*, **CAS-26**, pp. 180–189, March 1979. See also [46, Ch. II].

6.5-18. *A Formula for the Minimal Order*

Show that the minimal order of any realization of

$$H(s) = R_{-i}(s - \alpha)^{-i} + \cdots + \frac{R_{-1}}{s - \alpha} + R_0$$

is equal to the rank of an upper-triangular Hankel matrix with first row $[R_{-1} \ R_{-2} \ \cdots \ R_{-i}]$. Note that this generalizes Gilbert's formula (cf. Eqs. (6.1.4)) for the case simple roots ($l = 1$).

6.5-19. *Column Reduced Polynomial Matrices*

Let $D(s)$ be a full column rank matrix that is also column reduced (i.e., its leading column coefficient matrix has full rank).

1. Show that the infinite structural indices of $D(s)$ are $\{-k_1, \dots, -k_m\}$, where the $\{k_i\}$ are the column degrees of $D(s)$ arranged in descending order.
2. This shows that column (or row) reduced matrices have no zeros at infinity. Show by example that the converse statement is not true.

3. Show that a pencil $sE - F$ with full column (row) rank has no zeros at infinity if and only if there exists a constant column (row) transformation that will make the pencil column (row) reduced.
   Hint: Consider the pencil in its Kronecker canonical form.

6.5-20.   *Column Reduced Rational Matrices* [46], [60]
   Assume $H(s)$ has full column rank m, and make a Laurent expansion at α:

$$H(s) = R_{-l}(s - \alpha)^{-l} + \cdots + R_0 + R_1(s - \alpha) + \cdots$$
$$= [W_0 + W_1(s - \alpha) + \cdots]D_\alpha(s) = W_\alpha(s)D_\alpha(s),$$

where

$$D_\alpha(s) = \text{diag}\{(s - \alpha)^{k_1}, \ldots, (s - \alpha)^{k_m}\}$$

and

$$k_i = \text{the (algebraically) lowest degree of } s - \alpha$$
$$\text{occurring in the } i\text{th column of } H(s).$$

We shall say that $H(s)$ is column reduced at α if W_0 has full column rank.

1. Generalize to '$\alpha = \infty$' to show that this matches with our usage for polynomial matrices.
2. Show that if W_0 has full column rank the $\{k_i\}$ are simply the structural indices $\{\sigma_i\}$ of $H(s)$.

Remark: This result shows that the pole-zero structure of a column reduced $H(s)$ is given by that of its column vectors taken separately. (A column reduced matrix is, partly for the above reason, sometimes said to have *orthogonal* columns, see [15] and [46]).

6.5-21.   *A Strict Linearization of a Rational Matrix* ([46], [51])
   Let $H(s)$ be a possibly singular rational matrix, decomposed into its strictly proper and polynomial parts as $H(s) = H_{sp}(s) + D(s)$. Let $\{A_f, B_f, C_f\}$ be a minimal realization of $H_{sp}(s)$, so that $H_{sp}(s) = C_f(sI_{\delta_f} - A_f)^{-1}B_f$. Let $\{A_\infty, B_\infty, C_\infty, D\}$ be a minimal realization of $s^{-1}D(s^{-1})$, so that $s^{-1}D(s^{-1}) = C_\infty(sI_{\delta_\infty} - A_\infty)^{-1}B_\infty + D$ or $D(s) = C_\infty(I_{\delta_\infty} - sA_\infty)^{-1}B_\infty + D$. Then show that the following linear pencil has the same zero structure [finite, infinite, and singular (i.e., Kronecker index)] as $H(s)$:

$$\begin{bmatrix} sI_{\delta_f} - A_f & 0 & B_f \\ 0 & I_{\delta_\infty} - sA_\infty & B_\infty \\ \hline -C_f & -C_\infty & D \end{bmatrix}$$

6.5-22.   *(Kung)*
   Let $G(s)$ and $F(s)$ be $\bar{p} \times (p + m)$ and $(p + m) \times r$ polynomial matrices with $\bar{p} = p + m - r$, $r \leq \min(p, m)$. Assume that $G(s)$ is *left irreducible* [i.e., its columns are left coprime, or, equivalently, $G(s)$ has full row rank for all s], that $F(s)$ is *right irreducible*, and that $G(s)F(s) = 0$. Then if these matrices

are partitioned as follows,

$$G = [\underset{m}{G_1} \quad \underset{p}{G_2}], \qquad F' = [\underset{m}{F'_1} \quad \underset{p}{F'_2}]$$

prove that
    a. $G_1(s)$ and $F_2(s)$ have the same invariant polynomials, and
    b. $G_2(s)$ and $F_1(s)$ have the same invariant polynomials.

6.5-23. (*Kung*)

Assume that $\{G(s), F(s)\}$ are as in Exercise 6.5-22 except that now $G(s)$ is *row reduced* rather than left irreducible and $F(s)$ is *column reduced* rather than right irreducible. Let $G_{1hr} \triangleq$ the leading row coefficient matrix of $G_1(s)$, and similarly define G_{2hr}, F_{1hc}, and F_{2hc}, respectively. Show that
    a. The rank deficiency of $G_{1hr} = $ the rank deficiency of F_{2hc}, and
    b. The rank deficiency of $G_{2hr} = $ the rank deficiency of F_{1hc}.

6.5-24.

If $H(s)$ is a rational matrix of full column rank, and $T(s)$ is a non-singular rational matrix of appropriate dimension, show that def $HT = $ def H.

6.5-25. (*Forney* [15])

Given a matrix $H(s)$ of full column rank, and letting $G(s)$ be any basis for its (left) null space, i.e., $G(s)H(s) = 0$, show that def $H = $ def G.
Hint: Show that $H(s)$ can be brought to the form

$$\bar{H}(s) = \begin{bmatrix} I \\ W(s) \end{bmatrix}$$

without altering its defect, and that $G(s)$ can correspondingly be made $\bar{G}(s) = [W(s) \quad -I]$ without altering its defect.

*6.6 NERODE EQUIVALENCE FOR MULTIVARIABLE SYSTEMS

We have spent considerable time in this chapter on the relations between state-space descriptions and MFDs. A key role in this was played by the development in Sec. 6.4 of the multivariable analogs of the four special scalar realizations of Sec. 2.1.1. There we obtained these realizations by starting with Kelvin's technique for analog-computer simulations, and these procedures were generalized in Sec. 6.4. However, in the scalar case in Sec. 5.1 we also showed how the concept of *Nerode equivalence* could be used to obtain the special realizations in a general and abstract way, without any reference to analog-computer techniques.

In this section, we shall show how MFDs can be used to obtain the multivariable analog of the results of Sec. 5.1, a review of which will be helpful at this time.

The key idea of Nerode equivalence is that the state space can be identified with a space of equivalence classes of input sequences. Thus for a linear time-invariant *discrete-time* (for simplicity) system, we can identify the *zero* state at some time,

$t = 1$ say, with the class, say $\mathfrak{M}$, of all *polynomial* inputs (i.e., inputs terminating at $t = 0$) that produce a zero output for $t \geq 1$.

In the scalar case, we showed that all elements in this special class $\mathfrak{M}$ had the form $p(z)a(z)$, with $a(z)$ being the denominator polynomial of the scalar transfer function $H(z) = b(z)/a(z)$. Then we showed that the state at $t = 1$ set up by any polynomial input $u(z)$ could be identified with the remainder $r(z)$ obtained via division by $a(z)$,

$$u(z) = q(z)a(z) + r(z), \qquad \deg r(z) < \deg a(z)$$

This remainder $r(z)$ provides a unique representative of every input that will leave the system at $t = 1$ in the same state as would be set up by the given $u(z)$ [i.e., all these inputs will yield the same output $y(\cdot)$ for $t \geq 1$]. Since $\deg a(z) = n$, $r(z)$ is a polynomial of degree less than or equal to $n - 1$ and its coefficients (or linear combinations thereof) can be represented by a point in an n-dimensional state space. This characterization then led us easily to the state equations. We shall see that the use of the MFD allows us to generalize the above construction in a nice way.

The first step is to find a characterization of the space $\mathfrak{M}$ of all polynomial input vectors that yield a zero response for $t \geq 1$ or equivalently that yield a polynomial output $y(z)$. Suppose therefore that we have a transfer function with an *irreducible* right MFD

$$H(z) = N(z)D^{-1}(z)$$

where $D(z)$ and $N(z)$ are known $m \times m$ and $p \times m$ polynomial matrices. Then we can readily establish the following result.

Lemma 6.6-1.

Let $u(z) \in \mathfrak{M}$, the space of all $m \times 1$ polynomial (input) vectors such that

$$y(z) \triangleq H(z)u(z)$$
$$= \text{a } p \times 1 \text{ polynomial (output) vector}$$

Then all such $u(z)$ will have the form

$$u(z) = D(z)f(z)$$

for some $m \times 1$ polynomial vector $f(z)$.

Proof.   Since $N(z)D^{-1}(z)$ is irreducible, $N(z)$ and $D(z)$ are right coprime, and therefore there exist polynomial matrices $X(z)$ and $Y(z)$ such that (Lemma 6.3-5)

$$X(z)N(z) + Y(z)D(z) = I$$

Now if $u(z)$ is such that

$$H(z)u(z) = y(z), \qquad \text{a polynomial}$$

then

$$X(z)N(z)D^{-1}(z)u(z) = X(z)y(z)$$

so that

$$[I - Y(z)D(z)]D^{-1}(z)u(z) = X(z)y(z)$$

or

$$D^{-1}(z)u(z) = Y(z)u(z) + X(z)y(z)$$
$$= \text{a polynomial vector, say } f(z)$$

That is, $u(z)$ has the form $D(z)f(z)$, as was to be proved. ∎

Having identified the space $\mathfrak{M}$ as all inputs of the form $D(z)f(z)$ for polynomial $f(z)$, we may proceed to construct the abstract state space as in the scalar case. We first pick some input $u_1(z)$ and define the (Nerode equivalence) class of inputs $\{u_1(z) + \mathfrak{M}\}$ to be a particular state. We then pick an input $u_2(z)$ that is not in this class and form the class $\{u_2(z) + \mathfrak{M}\}$, and so on.

We now pick one representative from each class and, as in the scalar case, a natural representative is the unique element $r_i(z)$ of the class $\{u_i(z) + \mathfrak{M}\}$ for which $D^{-1}(z)r_i(z)$ is strictly proper. The proof that such an $r_i(z)$ exists, and that it is unique, is simple: By the division theorem (Theorem 6.3-15), we can write

$$u_i(z) = D(z)q(z) + r_i(z)$$

where $D^{-1}(z)r_i(z)$ is strictly proper. Then $r_i(z) \in \{u_i(z) + \mathfrak{M}\}$, since $D(z)q(z) \in \mathfrak{M}$, so existence is proved. If there exist $r(z)$, $s(z) \in \{u_i(z) + \mathfrak{M}\}$ we must have

$$r(z) - s(z) = D(z)q(z)$$

for some polynomial $q(z)$, but then it is impossible for $D^{-1}(z)r(z)$ and $D^{-1}(z)s(z)$ to both be strictly proper unless $r(z) = s(z)$, so uniqueness is proved.

It is trivial to see, furthermore, that *every* $r(z)$ such that $D^{-1}(z)r(z)$ is strictly proper is the representative of some Nerode equivalence class. It then follows from the above that the state space dimension is precisely the dimension of the space of vectors $r(z)$ for which $D^{-1}(z)r(z)$ is strictly proper. We now prove that this dimension equals the determinantal degree of $D(z)$:

Let $V(z)$ be a unimodular matrix such that $\bar{D}(z) \triangleq V(z)D(z)$ is row-reduced with row degrees $\{\rho_i\}$. Then

$$D^{-1}(z)r(z) = \bar{D}^{-1}(z)p(z) \tag{1a}$$

where

$$p(z) = V(z)r(z) \tag{1b}$$

Now the right hand side of (1a) is strictly proper if and only if the row degrees of $p(z)$ are less than the corresponding row degrees of $\bar{D}(z)$, by Lemma 6.3-11, so the space of $\{p(z)\}$ such that $\{\bar{D}^{-1}(z)p(z)\}$ are strictly proper has dimension

$$n = \sum \rho_i = \deg \det \bar{D}(z) = \deg \det D(z) \tag{2}$$

and, since $V(z)$ in (1b) is nonsingular, this is also the dimension of the space of $r(z)$ for which $D^{-1}(z)r(z)$ is strictly proper.

The state-space matrices $\{A, B, C\}$ can now be computed by generalizing the procedure described in Example 5.2-5 for the scalar case. Let us consider a simple example.

Example 6.6-1.

Let

$$H(z) = N(z)D^{-1}(z)$$

where

$$N(z) = [1 \quad -1] \quad \text{and} \quad D(z) = \begin{bmatrix} 2z + 1 & 1 \\ z - 2 & z^2 \end{bmatrix}$$

It is easy to check that $N(z)$ and $D(z)$ are coprime and that $D(z)$ is row reduced with

$$\det D(z) = 2z^3 + z^2 - z + 2$$

and

$$\rho_1 = 1, \qquad \rho_2 = 2$$

The remainders will have the form $r(z) = [r_1(z) \quad r_2(z)]'$, with $r_i(z)$ of degree ρ_i. Now the 2×1 polynomial remainder vectors $r(z)$ obviously form a linear vector space (over the field of real or complex numbers), and a *natural basis* for this space is given by the vectors

$$\begin{bmatrix} 1 \\ 0 \end{bmatrix}, \quad \begin{bmatrix} 0 \\ 1 \end{bmatrix}, \quad \begin{bmatrix} 0 \\ z \end{bmatrix}$$

Clearly, any 2×1 vector $r(z)$, with $r_1(z)$ constant and $r_2(z)$ of degree 1, can be represented by a linear combination (with real or complex coefficients) of these vectors so that the space of $r(z)$-vectors has dimension 3. The evolution of the state when a new input is applied will be described by the expression

$$[zr(z) + u(1)] \bmod D(z) = zr(z) \bmod D(z) + u(1) \bmod D(z)$$

Now to evaluate $zr(z) \bmod D(z)$ it suffices to evaluate

$$z \begin{bmatrix} 1 \\ 0 \end{bmatrix} \bmod D(z) = \begin{bmatrix} z \\ 0 \end{bmatrix} \bmod D(z)$$

$$z \begin{bmatrix} 0 \\ 1 \end{bmatrix} \bmod D(z) = \begin{bmatrix} 0 \\ z \end{bmatrix} \bmod D(z)$$

$$z \begin{bmatrix} 0 \\ z \end{bmatrix} \bmod D(z) = \begin{bmatrix} 0 \\ z^2 \end{bmatrix} \bmod D(z)$$

To do this, let us write

$$D(z) = \begin{bmatrix} 2z + 1 & 1 \\ z - 2 & z^2 \end{bmatrix} = \begin{bmatrix} z & 0 \\ 0 & z^2 \end{bmatrix} \begin{bmatrix} 2 & 0 \\ 0 & 1 \end{bmatrix} + \begin{bmatrix} 1 & 1 \\ z - 2 & 0 \end{bmatrix}$$

$$= S(z)D_{hr} + L(z), \text{ say}$$

Therefore

$$S(z) = D(z)D_{hr}^{-1} - L(z)D_{hr}^{-1}$$

or

$$\begin{bmatrix} z & 0 \\ 0 & z^2 \end{bmatrix} = D(z)\begin{bmatrix} \frac{1}{2} & 0 \\ 0 & 1 \end{bmatrix} - \begin{bmatrix} \frac{1}{2} & 1 \\ (z-2)/2 & 0 \end{bmatrix}$$

It follows that

$$\begin{bmatrix} z \\ 0 \end{bmatrix} \bmod D(z) = -\begin{bmatrix} \frac{1}{2} \\ (z-2)/2 \end{bmatrix} = -\frac{1}{2}\begin{bmatrix} 1 \\ 0 \end{bmatrix} + 1\begin{bmatrix} 0 \\ 1 \end{bmatrix} - \frac{1}{2}\begin{bmatrix} 0 \\ z \end{bmatrix} \quad (2a)$$

$$\begin{bmatrix} 0 \\ z \end{bmatrix} \bmod D(z) = \begin{bmatrix} 0 \\ z \end{bmatrix} \quad (2b)$$

$$\begin{bmatrix} 0 \\ z^2 \end{bmatrix} \bmod D(z) = -\begin{bmatrix} 1 \\ 0 \end{bmatrix} \quad (2c)$$

Now we can translate these results into state-space language as follows. Suppose that

$$r(z) = \begin{bmatrix} r_1(z) \\ r_2(z) \end{bmatrix} = \begin{bmatrix} r_{11} \\ r_{21}z + r_{22} \end{bmatrix}$$

and define the *state vector* at $t = 1$ associated with $r(z)$ as

$$x(1) = [r_{11} \quad r_{22} \quad r_{21}]'$$

Then the translation of (2) into state-space language is that the state matrix A is such that

$$A\begin{bmatrix} 1 \\ 0 \\ 0 \end{bmatrix} = \begin{bmatrix} -\frac{1}{2} \\ 1 \\ -\frac{1}{2} \end{bmatrix}, \quad A\begin{bmatrix} 0 \\ 0 \\ 1 \end{bmatrix} = \begin{bmatrix} 0 \\ 0 \\ 1 \end{bmatrix}, \quad A\begin{bmatrix} 0 \\ 1 \\ 0 \end{bmatrix} = \begin{bmatrix} -1 \\ 0 \\ 0 \end{bmatrix}$$

or equivalently that

$$A = \begin{bmatrix} -\frac{1}{2} & 0 & -1 \\ 1 & 0 & 0 \\ -\frac{1}{2} & 1 & 0 \end{bmatrix}$$

Similarly,

$$u(1) = u_1(1)\begin{bmatrix} 1 \\ 0 \end{bmatrix} + u_2(1)\begin{bmatrix} 0 \\ 1 \end{bmatrix}$$

translates to

$$B = \begin{bmatrix} 1 & 0 \\ 0 & 1 \\ 0 & 0 \end{bmatrix}$$

This realization is in the controllability form of Sec. 6.4.4. Therefore the output matrix can be computed from the Markov parameters of $H(z)$. We see that, in an obvious notation,

$$I = [b_1 \quad b_2 \quad Ab_2]$$

so that

$$C = [Cb_1 \quad Cb_2 \quad CAb_2]$$

Now

$$H(z) = [\tfrac{1}{2} \quad 0]z^{-1} + [\tfrac{1}{4} \quad -1]z^{-2} + \cdots$$

which then shows that

$$C = [\tfrac{1}{2} \quad 0 \quad -1] \quad \blacksquare$$

It should be clear how to carry out the above procedure in the general case, and therefore we see that the use of the MFD $N(z)D^{-1}(z)$, with $D(z)$ row reduced and $\{N(z), D(z)\}$ right coprime, has enabled a natural generalization of the scalar construction to the multivariable case.

However, there are some features particular to the matrix case. To get the observability-form realization (cf. Sec. 5.1.1) we must begin with an irreducible *left* MFD, $H(z) = D^{-1}(z)N(z)$ with $D(z)$ *column reduced*. It will be a useful exercise for the reader to go through the construction for this case, where we now identify the states with output sequences rather than equivalence classes of input sequences. Moreover, with somewhat more effort we can also use this abstract framework to obtain controller-form realizations of right MFDs [with $D(z)$ column reduced] and observer-form realizations of left MFDs [with $D(z)$ row reduced].

Realization from Markov Parameters. In Sec. 5.1.2, we also showed how the basic Nerode idea could be applied to the problem of realization from Markov parameters. Very similar arguments can be used in the matrix case, so we shall not pursue them here. We note that certain fast (generalized Lanczos and Euclidean division) realization algorithms can be developed, generalizing the results of Berlekamp, Massey and Rissanen referenced in Chapter 5 (see [53, Chap. 4]). A natural extension of scalar concepts to the matrix case is provided by the use of certain polynomial echelon forms to be described in Sec. 6.7.2 (see Example 6.7-7).

6.7 CANONICAL MATRIX-FRACTION AND STATE-SPACE DESCRIPTIONS

In Sec. 6.4.5, which should be reread at this point, we drew attention to the possible nonuniqueness in associating MFDs with transfer functions and to the consequent nonuniqueness of the state-space realizations obtained via these MFDs.

Given a transfer function $H(s)$, we can obtain a unique MFD $\{N(s), D(s)\}$ where

$$H(s) = N(s)D^{-1}(s)$$

by specifying either $D(s)$ or $N(s)$ uniquely. It is more common to require that the (always square) matrix $D(s)$ be uniquely specified, and in Sec. 6.7.1 and 6.7.2 we shall describe two important cases and some applications. Finally in Sec. 6.7.3 we shall give the formal algebraic definition of a *canonical form*.

6.7.1   Hermite-Form MFDs and Scheme I Realizations

The first special form for polynomial matrices, under elementary row or column operations, that we described in this book was the Hermite form of Theorem 6.3-2. That description will be recalled now, though it simplifies considerably if we restrict it to square nonsingular polynomial matrices.

Given an $m \times m$ nonsingular polynomial matrix, by elementary column operations we can convert it to a *unique* so-called *Hermite-form* $D_H(s)$, where

> $D_H(s)$ is upper triangular, with each diagonal element monic and of higher degree than any other element in the same row (if a diagonal element is 1, all other entries in that row will be zero)

The construction of $D_H(s)$ is fairly simple and was described in the proof of Theorem 6.3-2. It remains only to check the uniqueness, which means checking that any two matrices

$$D(s) \text{ and } \bar{D}(s) = D(s)U(s), \qquad U(s) \text{ unimodular}$$

will have the same Hermite form.

Proof of Uniqueness [6].   Since Hermite forms are obtained by elementary column operations, the forms for $D(s)$ and $\bar{D}(s) = D(s)U(s)$, $U(s)$ unimodular, must be related as

$$\bar{D}_H(s) = D_H(s)W(s), \qquad W(s) \text{ unimodular} \tag{1}$$

But we shall prove that the only such $W(s)$ is

$$W(s) = I \tag{2}$$

For this we note first that since D_H and $\bar{D}_H$ are nonsingular and upper triangular,

$$W(s) = D_H^{-1}(s)\bar{D}_H(s) \text{ is upper triangular}$$

Therefore since $W(s)$ is unimodular, its diagonal elements must be constant and in fact unity [to keep the $\{\bar{d}_{ii}(s), d_{ii}(s)\}$ monic].

But now we can write

$$\bar{d}_{11}(s) = d_{11}(s)$$
$$\bar{d}_{12}(s) = d_{11}(s)w_{12}(s) + d_{12}(s) \cdot 1$$

But by construction

$$\deg \bar{d}_{12}(s) < \deg \bar{d}_{11}(s) = \deg d_{11}(s)$$

and so we must have

$$w_{12}(s) = 0$$

Similarly, from

$$\bar{d}_{13}(s) = d_{11}(s)w_{13}(s) + d_{12}(s)w_{23}(s) + d_{13}(s)$$

we see that $w_{13}(s) = 0$ and so on for $\{w_{14}(s), \ldots, w_{1m}(s)\}$.
Now start with the second row,

$$\bar{d}_{22}(s) = d_{22}(s), \qquad \bar{d}_{23}(s) = d_{22}(s)w_{23}(s) + d_{23}(s)$$

which shows that $w_{23}(s) = 0$ and so on. Therefore all the off-diagonal elements of $W(s)$ must be zero, which proves our claim that $W(s) = I$ and that the Hermite form of $D(s)$ is unique.   ∎

With a unique denominator $D_H(s)$, we could of course associate a unique numerator $N_H(s) \triangleq H(s)D_H(s)$, and the Hermite MFD $N_H(s)D_H^{-1}(s)$ is a unique or canonical MFD of $H(s)$.

We note also that by using elementary row operations in place of elementary column operations we can obtain a unique row-Hermite form for $D(s)$ and thereby a unique Hermite left MFD.

Associated State-Space Realizations.   The matrix $D_H(s)$ is row reduced, and we can readily associate a controllability-form state-space realization with $D_H^{-1}(s)$ and $N_H(s)D_H^{-1}(s)$ by following the procedure of Sec. 6.4.4. It is an interesting fact ([54] and [55]) that the realization so obtained will be in the Scheme I controllability canonical form as described in Sec. 6.4.6.

It will be useful to first consider an example.

Example 6.7-1.

Consider the two matrices in (6.4-54a) and (6.4-55a) of Example 6.4-4,

$$D(s) = \begin{bmatrix} 0 & -(s^3 + 4s^2 + 5s + 2) \\ (s+2)^2 & s+2 \end{bmatrix}$$

and

$$\tilde{D}(s) = \begin{bmatrix} 0 & -2(s^3 + 4s^2 + 5s + 2) \\ (s+2)^2 & 2s^3 + 8s^2 + 10s + 4 \end{bmatrix}$$

The upper triangular Hermite form for $D(s)$ is readily seen to be, say,

$$D(s)\begin{bmatrix} 1 & 0 \\ -(s+2) & 1 \end{bmatrix} = \begin{bmatrix} s^4 + 6s^3 + 13s^2 + 12s + 4 & -(s^3 + 4s^2 + 5s + 2) \\ 0 & s + 2 \end{bmatrix}$$

$$= D_H(s) \tag{3}$$

and we can verify that the Hermite form for $\tilde{D}(s)$ is the same.

With the unique row-reduced Hermite form we can associate a unique *controllability-form realization* for $D_H^{-1}(s)$ by following the procedure of Sec. 6.4.4. We have

$$D_{hr} = I, \qquad D'_{lr} = \begin{bmatrix} 6 & 13 & 12 & 4 & 0 \\ -1 & -4 & -5 & -2 & 2 \end{bmatrix}$$

so that

$$A_o = \begin{bmatrix} -6 & 1 & 0 & 0 & \vdots & 1 \\ -13 & 0 & 1 & 0 & \vdots & 4 \\ -12 & 0 & 0 & 1 & \vdots & 5 \\ -4 & 0 & 0 & 0 & \vdots & 2 \\ \hdashline 0 & 0 & 0 & 0 & \vdots & -2 \end{bmatrix}, \qquad C'_o = \begin{bmatrix} 1 & 0 \\ 0 & 0 \\ 0 & 0 \\ 0 & 0 \\ \hdashline 0 & 1 \end{bmatrix}$$

and by properly reordering variables the final result is

$$A_{co} = \begin{bmatrix} 0 & 0 & 0 & -4 & \vdots & 2 \\ 1 & 0 & 0 & -12 & \vdots & 5 \\ 0 & 1 & 0 & -13 & \vdots & 4 \\ 0 & 0 & 1 & -6 & \vdots & 1 \\ \hdashline 0 & 0 & 0 & 0 & \vdots & -2 \end{bmatrix}, \qquad B_{co} = \begin{bmatrix} 1 & 0 \\ 0 & 0 \\ 0 & 0 \\ 0 & 0 \\ \hdashline 0 & 1 \end{bmatrix} \tag{4}$$

Not unexpectedly, this realization is *different* from the one calculated in Sec. 6.4.4 (cf. (6.4-53)) for the matrix $D^{-1}(s)$ rather than $D_H^{-1}(s)$; however, the reader should check that it does turn out to be the same as the *canonical* controllability realization obtained by applying the Scheme I construction of Sec. 6.4.6 to the noncanonical realization (6.4-53)—this was in fact already shown in Example 6.4-6; cf. Eq. (6-4-62). ∎

Example 6.7-1 suggests that there must be a close relationship between the $\{\beta_{ljk}\}$ parameters giving the dependency relations for Scheme I controllability forms [cf. (6.4-57)] and the entries of the matrix $D_H(s)$ of the Hermite-form MFD.

In fact, let

$$D_H(s) = \left[d_{ij}(s) = \sum_{l=0}^{L} d_{ijl} s^l \right] \tag{5a}$$

where if $\{l_i, i = 1, \ldots, m\}$ are the degrees of the diagonal elements of $D_H(s)$, then L is the largest such degree. We shall show that

$$d_{ijl} = \begin{cases} -\beta_{jil}, & l \neq l_i \\ 1, & l = l_i \end{cases} \tag{5b}$$

The reader should verify that this is so for the realizations studied in Example 6.7-1 [for which the dependency relations were worked out above (6.4-62)].

For a general proof, we first note that for any controllable pair $\{A, B\}$ there will exist some MFD of $(sI - A)^{-1}B$, say

$$(sI - A)^{-1}B = N(s)D^{-1}(s) \tag{6}$$

From this expression we shall derive a relationship between the columns of the controllability matrix $\mathcal{C}(A, B)$ and the entries of $D(s)$. Therefore special structures of $D(s)$ will be reflected in special relations between the columns of $\mathcal{C}(A, B)$, and this will lead us to the desired result. Let us write

$$D(s) = D_L + D_{L-1}s + \cdots + D_0 s^L \tag{7}$$

where L is the highest power of s that occurs in $D(s)$. Then by comparing the coefficient of s^{-1} on both sides of the equation

$$(sI - A)^{-1}BD(s) = N(s)$$

or

$$s^{-1}(I + As^{-1} + A^2 s^{-2} + \cdots)B(D_0 s^L + \cdots + D_L) = \text{a polynomial} \tag{8}$$

we can write

$$BD_L + ABD_{L-1} + \cdots + A^L BD_0 = 0$$

or

$$[B \quad AB \quad \cdots \quad A^L B]\begin{bmatrix} D_L \\ \cdot \\ \cdot \\ \cdot \\ D_0 \end{bmatrix} = 0 \tag{9}$$

Therefore the coefficients of $D(s)$ express certain linear dependence relations among the columns of $\mathcal{C}(A, B)$.

This clearly describes certain ways of going through the (crate diagram of the) columns of the controllability matrix $\mathcal{C}(A, B)$. *Different structures of $D(s)$ will correspond to different ways of exploring the crate diagram and vice versa.*

Thus for the matrix $D_H(s)$ of Example 6.7-1, we see that

$$[D_4' \quad D_3' \quad \cdots \quad D_o'] = \begin{bmatrix} 4 & 0 & 12 & 0 & 13 & 0 & 6 & 0 & 1 & 0 \\ -2 & 2 & -5 & 1 & -4 & 0 & -1 & 0 & 0 & 0 \end{bmatrix} \tag{10}$$

In this case, the dependence relations (9) can be read off as

$$4b_1 + 12Ab_1 + 13A^2b_1 + 6A^3b_1 + 1A^4b_1 = 0 \tag{11a}$$

and

$$-2b_1 + 2b_2 - 5Ab_1 + Ab_2 - 4A^2b_1 - A^3b_1 = 0 \tag{11b}$$

In a crate diagram this would yield a picture of the form

	b_1	b_2
I	x	x
A	x	0
A²	x	0
A³	x	0
A⁴	0	0

which clearly corresponds to searching by columns. (Note also that because there is a zero in the left-hand corner of every block D_i [(10) shows the D_i'] the search goes first through $\{b_1, Ab_1, \ldots, A^ib_1, \ldots\}$. For the detailed calculations in the crate, see Example 6.4-6, especially Eq. (6.4-62).)

It should now be clear how to complete the proof of (6). If

$$D(s) = \left\{ \sum_o^L d_{ijl}s^l \right\}$$

is in Hermite form, then we must have, with

$$l_i \triangleq \text{ the degree of the } i\text{th row of } D_H(s)$$

the properties

1. $d_{ijl} = 0, i > j$ [$D_H(s)$ is upper triangular],
2. $d_{jjl} = 1$ (the diagonal entries are monic), and
3. $d_{ijl} = 0, l > l_i$ (the diagonal entry has the highest degree in its row).

Then the dependency equation (9), written for the jth column, becomes, if $l_j > 0$,

$$A^{l_j}b_j = -\sum_{i=1}^{j} \sum_{l=0}^{l_i-1} A^l b_i d_{ijl} \tag{12a}$$

while if $l_j = 0$, it is

$$b_j = -\sum_{i=1}^{j-1} \sum_{l=0}^{l_i-1} A^l b_i d_{ijl} \tag{12b}$$

Comparing these with the Scheme I dependency relations of Eq. (6.4-57) shows immediately that $d_{ijl} = -\beta_{jil}$, as claimed earlier in (5).

6.7.2 Popov or Polynomial-Echelon MFDs and Scheme II Realizations

The analysis in Sec. 6.7.1 brought out the one-to-one relationship between Hermite-form MFDs and Scheme I for searching by columns the crate diagram of the controllability matrix of their associated state-space realizations. This suggests that with Scheme II for searching the crate diagram by rows (cf. Sec. 6.4.6) we should be able to associate a unique MFD. In fact the corresponding MFDs are quite interesting and will be studied in this section. They were first introduced by Popov [11] and then studied in more detail by Morf [56], Forney [15], Eckberg [57], and Kung and Kailath [58].

An active reader can try to deduce the nature of the special structure of $D(s)$ associated with row searching of the crate by some experimentation with the dependency relation (9),

$$[B \quad AB \quad \cdots \quad A^L B]\mathfrak{D} = 0$$
$$\mathfrak{D}' = [D'_L \quad \cdots \quad D'_0]$$

which will show that $\mathfrak{D}'$ will tend to have chains of nonzero entries in a certain *echelon* structure. In fact, with some effort the reader could try to deduce the relationship [analogous to (6)] between the coefficients of $D(s)$ and the Popov parameters $\{\alpha_{ijk}\}$ of any associated realization as defined by Eq. (6.4-58).

Here we shall take the less motivated but algebraically simpler route of proceeding by definition.

We shall say that a polynomial matrix $D(s)$ is in *Popov form* or in *polynomial-echelon form* if it has the following characteristics:

1. It is column reduced, with its column degrees arranged in ascending order, say

$$k_1 \leq k_2 \leq \cdots \leq k_m \tag{13}$$

2. For column j, $1 \leq j \leq m$, there is a so-called *pivot index* p_j such that

    a. $d_{p_j j}(s)$ has degree k_j (14a)

    b. $d_{p_j j}(s)$ is monic   (14b)

    c. $d_{p_j j}(s)$ is the last (or lowest) entry of degree k_j in the jth column; i.e., deg $d_{ij}(s) < k_j$ if $i > p_j$ (14c)

d. if $k_i = k_j$ and $i < j$, then $p_i < p_j$; i.e., the
    pivot indices are arranged to be increasing   (14d)

e. $d_{p_j i}(s)$ has degree less than k_j if $i \neq j$ (14e)

Example 6.7-2

For ease of comparison with earlier results, let us take a matrix much studied in Sec. 6.4,

$$D(s) = \begin{bmatrix} 0 & -(s^3 + 4s^2 + 5s + 2) \\ (s+2)^2 & s+2 \end{bmatrix} \tag{15a}$$

This is already column reduced, with pivot indices $p_1 = 2$, $p_2 = 1$ and all other characteristics of the echelon form except that $d_{p_2 2}(s)$ is not monic. It is easy to arrange this, so that the echelon form is

$$D_E(s) = \begin{bmatrix} 0 & s^3 + 4s^2 + 5s + 2 \\ (s+2)^2 & -s - 2 \end{bmatrix} \tag{15b}$$

We shall give less trivial examples later, but here let us note that if we write down, using the method of Sec. 6.4.4, the controllability-form realization $\{A_{co}, B_{co}\}$ associated with $D_E^{-1}(s)$ we shall find that

$$A_{co} = \begin{bmatrix} 0 & 0 & -2 & 0 & 0 \\ 1 & 0 & -5 & 0 & 0 \\ 0 & 1 & -4 & 0 & 0 \\ \hline 0 & 0 & 2 & 0 & -4 \\ 0 & 0 & 1 & 1 & -4 \end{bmatrix}, \quad B_{co} = \begin{bmatrix} 1 & 0 \\ 0 & 0 \\ 0 & 0 \\ \hline 0 & 1 \\ 0 & 0 \end{bmatrix} \tag{15c}$$

Moreover, the calculations in Example 6.4-6 show that this is also the *canonical* Scheme II controllability form associated with any fifth-order controllable realization of $D_E^{-1}(s)$. ∎

The general form of the result noted in Example 6.7-2 is that if

$$D_E(s) = \left[d_{ij}(s) = \sum_{l=0}^{k_j} d_{ijl} s^l \right] \tag{16}$$

where $\{k_1, \ldots, k_m\}$ are the column degrees, then

$$d_{ijl} = \begin{cases} 1, & l = k_j \\ -\alpha_{p_j i l}, & l \leq k_{j-1} \end{cases} \tag{17}$$

where the $\{\alpha_{ijk}\}$ are the so-called Popov parameters describing the Scheme II (crate search by rows) dependency relations for the columns of the controllability matrix of any controllable realization of $D_E^{-1}(s)$.

The reader can verify that (17) holds for our simple example (15). For a general proof, we can follow the same pattern as for the Hermite form in Sec. 6.7-1—however, the algebra (especially the subscripts) is more laborious here, particularly because the pivot elements do not lie along the diagonal so that $D_{hc} \neq I$ (as compared to $D_{hr} = I$ for the Hermite form).

Therefore we shall not give a detailed proof of (17) here—it will be more useful to examine more closely the matrix $[D'_L, \ldots, D'_0]$ that would arise in the proof [through the relation (9)]. Such an examination will in fact explain the reason for the special name *polynomial-echelon* form; it will also indicate that a closely related quasi-canonical form can often be as useful as the canonical echelon form (see e.g., [53] and [58]).

Example 6.7-3.   A Polynomial-Echelon Matrix
    The following matrix

$$D_E(s) = \begin{bmatrix} 5s + 1 & s^2 + 3s + 2 & 4s + 5 \\ 3s + 4 & 2s + 1 & s^3 + s^2 + 2 \\ s + 7 & 3 & 2 \end{bmatrix} \tag{18}$$

can readily be seen to be in what we have called polynomial-echelon form (with column degrees $k_1 = 1, k_2 = 2, k_3 = 3$ and pivot indices $p_1 = 3, p_2 = 1, p_3 = 2$). However, it is not very clear why this matrix deserves the name polynomial-echelon form or even just echelon form. To get some insight into this we shall, following Morf [56], write

$$D_E(s) = D_3 + D_2 s + D_1 s^2 + D_0 s^3$$

and form the *coefficient matrix*

$$\mathfrak{D}'_E \triangleq [D'_3 \quad D'_2 \quad D'_1 \quad D'_0]$$

$$= \begin{bmatrix} 1 & 4 & 7 & 5 & 3 & ① & 0 & 0 & 0 & 0 & 0 & 0 \\ 2 & 1 & 3 & 3 & 2 & 0 & ① & 0 & 0 & 0 & 0 & 0 \\ 5 & 2 & 2 & 4 & 0 & 0 & 0 & 1 & 0 & 0 & ① & 0 \end{bmatrix} \tag{19}$$

We can see that this is in fact the usual echelon form for real matrices. However, there is an additional constraint on the *pivot* indices [shown circled in (19)]: We see that if α_i is the column index of the pivot entry of the *i*th row of $\mathfrak{D}'_E$, then the integers $\{\alpha_i, i = 1, \ldots, m\}$ are *distinct* in the strong sense that

$$\alpha_i \neq \alpha_j \bmod m \qquad \text{if } i \neq j \tag{20}$$

where *m* is the number of rows in $\mathfrak{D}'_E$. In our example (19), $\alpha_1 = 6, \alpha_2 = 7, \alpha_3 = 11$, and $m = 3$, and we see that (20) holds. ∎

It is this distinctive feature that leads us to use the name polynomial-echelon form; the reader should check that this special property is a consequence of our requirement (14e) on the polynomial-echelon form. Other consequences of this requirement are that in the coefficient matrix $\mathfrak{D}'$

> 1. the pivot entries are the only nonzero entries
>   in their respective columns   (21a)
>
> 2. The columns $\alpha_i + qm$, q any positive integer,
>   are identically zero   (21b)

We shall describe presently how to convert an arbitrary polynomial matrix to this special form by elementary column operations. However, let us note first that the polynomial-echelon form is *unique* or *canonical* in the sense that

$$D(s) \text{ and } D(s)U(s), \ U(s) \text{ unimodular, have the same}$$
$$\text{polynomial-echelon form} \tag{22}$$

A direct algebraic proof of (22) is not difficult, but it is laborious and will not be given here. We may only mention that an indirect proof follows from the fact noted above that a matrix in polynomial-echelon form can be uniquely associated with a pair $\{A, B\}$ in Scheme II controllability form, and we showed in Sec. 6.4.6 that such forms were unique (canonical).

Actually, as mentioned in Sec. 6.4.6, in many applications (e.g., pole shifting) the canonical feature of the Scheme II forms is not significant, and one could work just as well with noncanonical controller-type forms. A similar statement holds for the polynomial-echelon forms: One can often work just as well with certain noncanonical so-called quasi-echelon forms. In particular, we shall find that the requirement (14e), that all entries in a row containing a pivot element have degrees lower than that of the pivot element, can often be replaced by the weaker constraint that

$$\text{the pivot indices } \{p_i\} \text{ are distinct} \tag{14e'}$$

Example 6.7-4.   A Quasi-Canonical Polynomial-Echelon Form
Consider the matrix

$$\tilde{D}_E(s) = \begin{bmatrix} 5s + 1 & s^2 + 3s + 2 & 2s^3 + 3s^2 + 4s + 5 \\ 3s + 4 & 2s + 1 & s^3 + s^2 + 2 \\ s + 7 & 3 + 4s & 2 + 5s + 3s^2 \end{bmatrix} \tag{23}$$

As compared to the canonical polynomial-echelon matrix (18) of Example 6.7-3, the matrix $\tilde{D}_E(s)$ can be seen to satisfy all the assumptions (13)–(14) except for (14e). However, it does satisfy the new assumption (14e'). To see the effect of this change, let us again form the coefficient matrix

$$\tilde{D}'_E = \begin{bmatrix} 1 & 4 & 7 & 5 & 3 & \boxed{1} & 0 & 0 & 0 & 0 & 0 & 0 \\ 2 & 1 & 3 & 3 & 2 & 4 & \boxed{1} & 0 & 0 & 0 & 0 & 0 \\ 5 & 2 & 2 & 4 & 0 & 5 & 3 & 1 & 3 & 2 & \boxed{1} & 0 \end{bmatrix} \tag{24}$$

This is again in the usual echelon form for real-valued matrices. Moreover, by virtue of (14e'), we see that we still have the special property (20) that the pivot indices are strongly distinct. However, the features (21a) and (21b) are now lost, and because of this, the forms obeying (13), (14a)–(14d), and (14e') are not unique (canonical). It is not hard to see how to reintroduce the uniqueness [viz., replace (14e') by (14e)], but as noted before, in many applications it may not be necessary to go all the way to the canonical form. ∎

This is a good place to explain how to transform a given matrix $D(s)$ to echelon (canonical and quasi-canonical) form; we follow a method suggested by B. Lévy (personal communication, 1977).

Transformation to Polynomial-Echelon Form. The first step is to make $D(s)$ column reduced, and we shall assume that this has been done. The next thing is to find a unimodular matrix $U(s)$ such that

$$D(s)U(s) = E(s) \tag{25}$$

where $E(s)$ is in canonical or quasi-canonical-echelon form. Let us rewrite the above relation as

$$[D(s) \quad -I_m] \begin{bmatrix} U(s) \\ E(s) \end{bmatrix} = 0 \tag{26}$$

and let us write

$$\begin{bmatrix} U(s) \\ E(s) \end{bmatrix} = \begin{bmatrix} U_0 \\ E_0 \end{bmatrix} s^L + \begin{bmatrix} U_1 \\ E_1 \end{bmatrix} s^{L-1} + \cdots + \begin{bmatrix} U_L \\ E_L \end{bmatrix} \tag{27}$$

where

$$L = \text{the highest power of } s \text{ in } D(s)$$

(We leave it to the reader to show that L is also the highest power of s in $[U'(s) \ E'(s)]'$. *Hint*: Consider $D^{-1}(s)$.) Then we can rewrite (26) as

$$[D(s) \quad sD(s) \quad \cdots \quad s^L D(s) \quad -I_m \quad \cdots \quad -s^L I_m]\alpha = 0 \tag{28}$$

where α is the real-valued matrix

$$\alpha' = [U'_L \quad \cdots \quad U'_0 \quad E'_L \quad \cdots \quad E'_0]$$

The point now is to find $\mathcal{C}$ in echelon form (canonical or quasi-canonical) with all its pivot elements occurring in the lower half of $\mathcal{C}$. Then, recalling our earlier discussions on the relation between the coefficient matrix and the polynomial matrix, we shall see that $E(s)$ will be in echelon form.

To find $\mathcal{C}$ in echelon form, we should search for a set of m so-called *primary dependent columns* of the polynomial matrix in (28), viz.,

$$[D(s) \quad \cdots \quad s^L D(s) \quad -I \quad \cdots \quad -s^L I] = \mathcal{B}(s)$$

A dependent column of $\mathcal{B}(s)$ will be defined as a column that is a linear combination of the preceding columns of $\mathcal{B}(s)$. A set of dependent columns of $\mathcal{B}(s)$ is called *primary* if their column locations, say $\{\alpha_1, \ldots, \alpha_m\}$, are strongly distinct in the sense that

$$\alpha_i \neq \alpha_j \bmod m \qquad \text{if } i \neq j$$

As the reader will recall from our earlier discussions, this strong distinctness property characterizes the echelon form. We should note also that since $D(s)$ is nonsingular its right null-space is zero, and consequently all the (primary) dependent columns will be in the right half of $\mathcal{B}(s)$. A numerical example will help to clarify the above.

Example 6.7-5.   Construction of Echelon Forms

To keep the algebra down, we shall take a very simple example. Transform

$$D(s) = \begin{bmatrix} -3s & s + 2 \\ 1 - s & 1 \end{bmatrix} \tag{29}$$

to echelon form. We first write out $\mathcal{B}(s)$ as

$$\begin{bmatrix} -3s & s + 2 & -3s^2 & s^2 + 2s & -1 & 0 & -s & 0 \\ 1 - s & 1 & s - s^2 & s & 0 & -1 & 0 & -s \end{bmatrix}$$

and now search for dependent columns. The first dependent column turns out to be B_7, the seventh column,

$$B_7 + B_6 + 2B_5 + B_2 = 0 \tag{30}$$

The second dependent column is B_8,

$$B_8 + 3B_7 - B_6 - B_1 = 0 \tag{31}$$

Since $8 \neq 7 \bmod 2$, B_8 is primary. Therefore $\mathcal{C}$ is now determined as

$$\mathcal{C}' = \begin{bmatrix} 0 & 1 & 0 & 0 & 2 & 1 & \boxed{1} & 0 \\ -1 & 0 & 0 & 0 & 0 & -1 & 3 & \boxed{1} \end{bmatrix} \tag{32}$$

and

$$E'(s) = \begin{bmatrix} 2 & 1 \\ 0 & -1 \end{bmatrix} + \begin{bmatrix} 1 & 0 \\ 3 & 1 \end{bmatrix} s$$

so that

$$E(s) = \begin{bmatrix} s+2 & 3s \\ 1 & s-1 \end{bmatrix} \tag{33}$$

This is in the quasi-canonical echelon form. To obtain the true canonical form we must eliminate the entry 3 that lies in the same row as the first pivot index. This is easily done by adding -3 times the first column of $\mathcal{Q}$ to the second column, yielding, say

$$\tilde{\mathcal{Q}}' = \begin{bmatrix} 0 & 1 & 0 & 0 & 2 & 1 & 1 & 0 \\ -1 & -3 & 0 & 0 & -6 & -4 & 0 & 1 \end{bmatrix} \tag{34}$$

and now we have a canonical echelon form

$$\tilde{E}'(s) = \begin{bmatrix} s+2 & 1 \\ -6 & s-4 \end{bmatrix} \tag{35}$$

The reader should check that going from $\mathcal{Q}$ to $\tilde{\mathcal{Q}}$ is really achieved by substituting for B_7 from (30) into (31), so that (31) becomes

$$B_8 - 4B_6 - 6B_5 - B_2 - B_1 = 0 \tag{36}$$

which corresponds to the second column of $\tilde{\mathcal{Q}}$. It is also easy to see the nature of the nonuniqueness in the quasi-canonical form (e.g., we could replace $3B_7$ by $2B_7 - B_6 - 2B_5 - B_2$, and so on). We note also that for many purposes it suffices to determine the quasi-canonical form.

Though the above example is too simple to really demonstrate this, we should point out that the special 'shift' structure of $\mathcal{B}(s)$ makes it possible to obtain a fast algorithm for finding the dependent columns (see [58]). ∎

More on the Equation $H(s)D(s) = N(s)$.   At the end of Sec. 6.5.4, we discussed conditions for the existence of proper solutions $H(s)$ to equations of the form

$$H(s)D(s) = N(s) \tag{37}$$

where $\{N(s), D(s)\}$ are given $p \times r$ and $m \times r$ polynomial matrices and with $D(s)$ of full column rank r. It turns out that by finding a *left minimal basis* (LMB) of

$$F(s) \triangleq [D'(s) \quad N'(s)]' \tag{38}$$

in polynomial echelon form, we can efficiently determine a *minimal* proper solution. Here minimal means of minimal McMillan degree.

Theorem 6.7-1.   Minimal Proper Solutions

Let $\bar{E}(s)$ be a (quasicanonical) echelon form LMB of $F(s)$ (with $p + m - r$ rows). Then there will exist a proper solution to (37) if and only if there are p rows of $\bar{E}(s)$ that have pivot indices greater than m.

Furthermore suppose p such rows exist and denote the matrix they form as

$$E(s) = [-N_l(s) \quad D_l(s)]_p \qquad (39)$$
$${}_m \qquad\qquad {}_p$$

Then $H(s) = D_l^{-1}(s)N_l(s)$ is a minimal proper solution of (37).

Proof.   For properness, combine the test noted below (6.5-49) with the result of Exercise 6.5-23. For minimality, use the predictable degree property to show that no other choice can decrease the order. See [15] for more details.   ∎

The key step is to determine an LMB in echelon form, which problem we shall now consider, following [58].

Finding an LMB in Echelon Form.   We follow the lines of the method described above to find the echelon form of a matrix. We first write

$$\bar{E}(s)F(s) = 0$$

in expanded form as

$$\mathcal{E}\mathcal{F}(s) = 0 \qquad (40a)$$

where

$$\mathcal{E} = [E_\nu \ \cdots \ E_o], \qquad \bar{E}(s) = E_o z^\nu + \cdots + E_\nu \qquad (40b)$$

and

$$\mathcal{F}(s) = [F'(s) \quad sF'(s) \quad \ldots \quad s^\nu F'(s)]' \qquad (40c)$$

The integer ν is usually not known *a priori* but will be determined in the course of the construction. If $\bar{E}(s)$ is in polynomial echelon form, then we see that the rows of $\mathcal{E}$ will define certain dependencies between the rows of $\mathcal{F}(s)$. And in particular, the pivot locations in $\mathcal{E}$ will define certain *primary* dependent rows in $\mathcal{F}(s)$, while there will be several other *secondary* dependent rows as well (cf. our earlier discussion of the structure of the polynomial echelon form).

Now these primary dependent rows can be determined by using the following selection rule: Find the first row, say f_1, in $\mathcal{F}(s)$ that is dependent on the preceding rows. Then remove from consideration all later rows k such that $k = f_1 \bmod q$, where q is the number of rows of $F(s)$. Now search for the next dependent row, say f_2, and remove from consideration all rows with indices equal to $f_2 \bmod q$. Continue in this way till $\bar{p}$ rows have been found, where $\bar{p}$ is the number of rows in the LMB ($\bar{p} = q - r$, where r is the rank of the full column rank matrix $F(s)$). Reference [58] describes some numerically efficient ways of organizing this sequential search procedure.

Example 6.7-6.

Find a left minimal basis of $F'(s) = [D'(s)\ N'(s)]$, where

$$D(s) = \begin{bmatrix} s^4 + 2s^3 + 0 & - & 5s - 4 & 2s^3 + 2s^2 + 2s - 8 \\ -s^4 + 7s^3 + 7s^2 + 14s + 6 & & -2s^4 - 5s^3 + s^2 + 3s \\ -2s^3 - s^2 - 17s - 9 & & 2s^4 + 3s^3 - s^2 - s - 2 \end{bmatrix}$$

$$N(s) = (2s^4 + 3s^3 - s^2 - 9s - 4)\begin{bmatrix} 1 & 0 \\ 0 & 1 \end{bmatrix}$$

Solution.   Let $\{f_1, f_2, \dots\}$ denote the rows of $\mathcal{F}(s) \triangleq [F'(s)\ sF'(s) \dots]'$. We find that the first dependent row is f_8,

$$f_8 + f_7 + f_6 - f_5 - 3f_4 + 2f_3 + f_2 = 0$$

The next is f_9, which is also primary,

$$f_9 - 2f_6 + 2f_5 + f_4 - f_1 = 0$$

The next is f_{14}, which is secondary, and must be ignored. The next is f_{15},

$$f_{15} - f_{13} - 2f_6 + f_5 - 2f_3 - 3f_2 = 0$$

These three relations yield the matrix

$$\mathcal{E} = \begin{bmatrix} 0 & 1 & 2 & -3 & -1 & 1 & 1 & \circled{1} & 0 & 0 & 0 & 0 & 0 & 0 & 0 \\ -1 & 0 & 0 & 1 & 2 & -2 & 0 & 0 & \circled{1} & 0 & 0 & 0 & 0 & 0 & 0 \\ 0 & -3 & -2 & 0 & 1 & -2 & 0 & 0 & 0 & 0 & 0 & 0 & -1 & 0 & \circled{1} \end{bmatrix}$$

which corresponds to an LMB in quasicanonical echelon form

$$\bar{E}(s) = \begin{bmatrix} s & s+1 & s+2 & -3 & -1 \\ -2s-1 & 0 & 0 & s+1 & 2 \\ -2s & -3 & -s^2-2 & 0 & s^2+1 \end{bmatrix}$$

In the notation of Theorem 6.7-1, we have $p = 2$, $m = 3$ and $r = 2$. We see also that rows 2 and 3 have row pivot indices greater than 3 (as shown immediately by the circled entries in $\mathcal{E}$). Therefore a minimal proper solution of (37) can be found as

$$H(s) = \begin{bmatrix} s+1 & 2 \\ 0 & s^2+1 \end{bmatrix}^{-1} \begin{bmatrix} 2s+1 & 0 & 0 \\ 2s & 3 & s^2+2 \end{bmatrix} \quad \blacksquare$$

Some Problems Reducible to Finding an LMB.   We conclude this section by noting some problems that can be reduced to finding an LMB.

One is the problem of finding a minimal left MFD, say $A^{-1}B$, from a given right MFD, ND^{-1} (or vice versa). For this we must solve the equation

$$A^{-1}B = ND^{-1}$$

or

$$[-B \quad A]F(s) = 0$$

where

$$F'(s) = [D'(s) \quad N'(s)]$$

Then the minimal left MFD is completely determined by finding an LMB of $F(s)$. [This redeems a promise we made in Lemma 6.3-8].

Another is the problem of finding inverse systems. Thus suppose we have

$$y(s) = N(s)D^{-1}(s)u(s)$$

where $N(s)$ is a $p \times m$ matrix with full column rank $m \leq p$. Then a left inverse exists and can be found by seeking a pair $\{A(s), B(s)\}$ such that

$$u(s) = A^{-1}(s)B(s)y(s)$$

i.e., such that

$$A^{-1}(s)B(s)N(s)D^{-1}(s) = I$$

or equivalently such that

$$[B(s) \quad -A(s)]\begin{bmatrix} N(s) \\ D(s) \end{bmatrix} = 0$$

Again Theorem 6.7-1 can be applied to find a proper minimal inverse if one exists.

Without going into too much detail, (for which see [53]) we might also note that the problem of finding a minimal realization to match a given set of Markov parameters, say $\{h_1, \ldots, h_N\}$, is essentially one (cf. Sec. 5.1.2) of finding a polynomial matrix

$$D(z) = D_L z^L + \cdots + D_o$$

such that

$$(41)$$

where the entries marked X (corresponding to the unknown Markov parameters $h_i, i > N$) can be suitably chosen to achieve the desired equalities. Now we can find $[D'_0 \ldots]'$ in echelon form by searching sequentially for primary dependent columns in the Hankel matrix denoted $\mathbf{M}$ in (41).

Example 6.7-7. A Minimal Partial Realization Problem

Find a minimal partial realization of the matrix sequence

$$\begin{bmatrix} -1 & 1 \\ 0 & 0 \end{bmatrix}, \begin{bmatrix} 0 & 1 \\ 0 & 0 \end{bmatrix}, \begin{bmatrix} 1 & 1 \\ -1 & 1 \end{bmatrix}, \begin{bmatrix} 1 & 2 \\ 0 & 1 \end{bmatrix}, \begin{bmatrix} 1 & 4 \\ 1 & 1 \end{bmatrix}, \begin{bmatrix} 2 & 7 \\ 1 & 2 \end{bmatrix}$$

Solution. We first form an (incomplete) Hankel matrix

$$\mathbf{M} = \begin{bmatrix} -1 & 1 & 0 & 1 & 1 & 1 & 1 & 2 & 1 & 4 & 2 & 7 \\ 0 & 0 & 0 & 0 & -1 & 1 & 0 & 1 & 1 & 1 & 1 & 2 \\ 0 & 1 & 1 & 1 & 1 & 2 & 1 & 4 & 2 & 7 & x & x \\ 0 & 0 & -1 & 1 & 0 & 1 & 1 & 1 & 1 & 2 & x & x \\ 1 & 1 & 1 & 2 & 1 & 4 & 2 & 7 & x & x & x & x \\ -1 & 1 & 0 & 1 & 1 & 1 & 1 & 2 & x & x & x & x \\ 1 & 2 & 1 & 4 & 2 & 7 & x & x & x & x & x & x \\ 0 & 1 & 1 & 1 & 1 & 2 & x & x & x & x & x & x \\ 1 & 4 & 2 & 7 & x & x & x & x & x & x & x & x \\ 1 & 1 & 1 & 2 & x & x & x & x & x & x & x & x \\ 2 & 7 & x & x & x & x & x & x & x & x & x & x \\ 1 & 2 & x & x & x & x & x & x & x & x & x & x \end{bmatrix}$$

To find the solution $\mathbf{D}(z)$, let us note that the first dependent column is M_4, in the sense that by suitably choosing the x's, we can have

$$-M_1 - 2M_2 + M_3 + M_4 = 0$$

where M_i denotes the ith column of $\mathbf{M}$.

The second dependent column (not necessarily primary) is M_6, since (in the same sense as above)

$$-M_3 - 2M_4 + M_5 + M_6 = 0$$

However, because $6 = 4 \bmod 2$, M_6 is *not* a primary dependent column, and hence should be omitted from our selection. It can be shown that M_7 is the second primary dependent column, since (in the same sense as above)

$$M_1 + 2M_2 - M_5 - M_6 + M_7 = 0$$

Now we have obtained all the primary dependent columns and we can identify

$$D_0 = \begin{bmatrix} -1 & 1 \\ -2 & 2 \end{bmatrix}, \qquad D_1 = \begin{bmatrix} 1 & 0 \\ 1 & 0 \end{bmatrix}, \qquad D_2 = \begin{bmatrix} 0 & -1 \\ 0 & -1 \end{bmatrix}, \qquad D_3 = \begin{bmatrix} 0 & 1 \\ 0 & 0 \end{bmatrix}$$

with $D_4 = D_5 = 0$. Then

$$D(z) = \begin{bmatrix} z - 1 & z^3 - z^2 + 1 \\ z - 2 & - z^2 + 2 \end{bmatrix}$$

and the numerator matrix $N(z)$ can be found via **M** as

$$N(z) = \begin{bmatrix} 0 & -z^2 \\ 0 & -1 \end{bmatrix}$$

A state-space realization can be readily obtained by the methods of Sec. 6.4.1 as

$$A = \begin{bmatrix} -1 & -1 & 0 & 1 \\ -2 & -1 & 0 & 2 \\ 0 & 1 & 0 & 0 \\ 0 & 0 & 1 & 0 \end{bmatrix} \qquad B = \begin{bmatrix} 0 & 1 \\ 1 & -1 \\ 0 & 0 \\ 0 & 0 \end{bmatrix} \qquad C = \begin{bmatrix} 0 & -1 & 0 & 0 \\ 0 & 0 & 0 & -1 \end{bmatrix}$$

For more details, including the justification of this procedure, and ways of exploiting the special Hankel structure to speed up the computations, we refer to [53, Ch. 4]. ∎

Finally we note that certain control problems of disturbance rejection can also be reduced to finding special solutions of rational equations (i.e., essentially to finding an LMB), as has recently been shown by Emre and Hautus [18].

*6.7.3   The Formal Definition of Canonical Form†

In many situations, we have a set, say X, of object $x \in X$, along with an equivalence relation, denoted $\sim$, that allows us to regard two equivalent objects as being the same for some purpose. For example, X may be a set of matrices and we may wish to consider as being equivalent all matrices *similar* to a given matrix A, i.e., all matrices of the form $\{T^{-1}AT, T$ arbitrary but nonsingular$\}$. Or we may regard as being equivalent all matrices obtained from a given matrix by applying elementary row operations to it. And so on. The basic (defining) properties of an equivalence relation are

1. Transitivity: $x \sim y$ and $y \sim z$ implies $x \sim z$;
2. Symmetry: $x \sim y$ implies $y \sim x$; and
3. Reflexivity: $x \sim x$.

†This section is included to show that the liberally used term *canonical* does in fact have a formal mathematical definition.

Now, given an equivalence relation on a set X, we can decompose X into subsets each composed of elements equivalent to each other. And doing this gives us a new so-called quotient set $X/\sim$ of subsets of X. A set in the quotient set can be characterized by any of its elements, since any element in the set will be equivalent to any other. A *canonical* description of X under $\sim$ can then be regarded as being a nice choice of characteristic element for each set in the quotient set $X/\sim$. But *nice* is a subjective word, and all we can ask is that we have a *unique* description of each set in $X/\sim$. More formally, we can give the following definition (see, e.g., the standard algebra textbook [59, pp. 287–288]): Given a set X and an equivalence relation $\sim$, a subset C of X will be said to be a set of *canonical forms for X under* $\sim$ if for every $x \in X$ there exists one and only one $c \in C$ such that $x \sim c$.

Let X be the set of $n \times n$ matrices A and $\sim$ be similarity. Then the set of Jordan forms is a set of canonical forms for X. We can take this statement as known, perhaps by virtue of the fact that we have heard it asserted so often or by noting that the Jordan form of a matrix is indeed the same for all matrices similar to the given matrix.

It will be useful, in less familiar examples, to have a formal way of checking when a given set C is a set of canonical forms for X under $\sim$. This is provided by the following: Let

$$f: X \longrightarrow C$$

denote the map that associates to each $x \in X$ a unique image $c = f(x)$ such that $c \sim x$. Then we shall show that the map f is what is called a *complete invariant*.

A mapping f from a set X to any other set T is said to be *invariant* under $\sim$ on X if

$$x \sim y \Longrightarrow f(x) = f(y)$$

Example 6.7-6.

The map that associates with any $n \times n$ matrix A its characteristic polynomial $\det(sI - A)$ is an invariant under similarity. However, the converse need not be true—nonsimilar matrices can have the same characteristic polynomial, e.g.,

$$A_1 = \begin{bmatrix} 1 & 0 \\ 0 & 1 \end{bmatrix} \quad \text{and} \quad A_2 = \begin{bmatrix} 1 & 1 \\ 0 & 1 \end{bmatrix}$$

Our next definition is meant to handle this situation. ∎

A mapping f from a set X to any other set T is said to be a *complete invariant* under $\sim$ on X if

$$f(x) = f(y) \Longleftrightarrow x \sim y$$

Example 6.7-7.

It is proved in books on matrix theory (e.g., [6] or [7]) that the map that associates with any $n \times n$ matrix A its invariant polynomials (cf. Sec. 6.3) (or equivalently its Jordan form) is a complete invariant under similarity. ∎

Theorem. If C is a set of canonical forms for X under $\sim$, then the map $f: X \longrightarrow C$ that associates to each $x \in X$ a unique $c \in C$ such that $c \sim x$ is a complete invariant. Note also that

$$f(c) = c$$

Proof. To prove that f is invariant, note that if $f(x) = c_1, c_1 \sim x$, and $f(y) = c_2, c_2 \sim y$, and $x \sim y$, then $c_1 = c_2$. For if $x \sim y$, then $c_2 \sim y \sim x$ and $c_1 \sim x$, which is a contradiction if $c_1 \neq c_2$ because by definition there is only one $c \in C$ such that $c \sim x$. Second, to prove that f is complete, suppose $f(x) = c$ and $f(y) = c$. Then $c \sim x$ and $c \sim y$ so that $x \sim y$.

Finally, let $c \in C$; since $c \sim c$ and $f(c) \sim c$, by uniqueness we must have $c = f(c)$. ∎

The Nonfixed Parameters of the State-Space Canonical Forms. We would expect that the so-called nonfixed (not necessarily equal to 0 or 1) parameters of the controllability and controller *canonical* forms of Sec. 6.4.6 should also have certain nice properties.

We have already noted that the $\{l_i, \beta_{ijk}\}$ and the $\{k_i, \alpha_{ijk}\}$ are

1. *Invariant*, i.e., all similar pairs have the same values of these parameters, and
2. *Independent*, i.e., given arbitrary $\{l_i, \beta_{ijk}\}, \{k_i, \alpha_{ijk}\}$ (subject of course to $\sum l_i = n = \sum k_i$), there exist pairs $\{A, B\}$ that have these parameter sets.

These parameters are also *complete* in the sense that if they are the same for two pairs $\{A, B\}$ and $\{\bar{A}, \bar{B}\}$ then these pairs must be similar. The reason is that both pairs will have the same controllability canonical form.

Exercises

6.7-1.

Suppose $D_H(s)$ has row degrees $l_1 = 3, l_2 = 0, l_3 = 1$. Indicate the possible form of the $[D_3' \; D_2' \; D_1' \; D_0']'$ that can be associated with $D_H(s)$ [cf. Eq. (6.7-9)] and also of the crate diagram that can be associated with the pair $\{A, B\}$ of any controllable realization of $N(s)D_H^{-1}(s)$. Repeat for $D_E(s)$ with $p_1 = 2, p_2 = 3, p_3 = 1$ and $k_1 = 2 = k_2, k_3 = 3$.

6.7-2.

Let

$$D(s) = \begin{bmatrix} 6s + 8 & s^2 + 3s + 2 \\ s^3 + 4 & 7s + 5 \end{bmatrix}$$

Find the Popov parameters $\{\alpha_{ijk}\}$ uniquely associated with any Scheme II controllability canonical state-space realization $N(s)D^{-1}(s)$.

REFERENCES

1. E. G. GILBERT, "Controllability and Observability in Multivariable Control Systems," *SIAM J. Control*, 1, pp. 128–151, 1963.

2. R. E. KALMAN, "Mathematical Description of Linear Dynamical Systems," *SIAM J. Control*, **1**, pp. 152–192, 1963.

3. B. NOBLE, *Applied Linear Algebra*, Prentice-Hall, Englewood Cliffs, N.J., 1969; 2nd rev. ed., 1977.

4. G. STRANG, *Linear Algebra and Its Applications*, Academic Press, New York, 1976.

5. C. LANCZOS, *Linear Differential Operators*, Van Nostrand Reinhold, New York, 1959.

6. C. C. MACDUFFEE, *The Theory of Matrices*, Springer, Berlin, 1933; reprinted by Chelsea Publishing Co. New York, 1950.

7. F. R. GANTMAKHER, *Theory of Matrices*, Vols. 1 and 2, Chelsea Publishing Co. New York, 1959.

8. J. H. M. WEDDERBURN, *Lectures on Matrices, SIAM J. Control and Optim.*, 1980.

9. M. NEWMAN, *Integral Matrices*, Academic Press, New York, 1972.

10. V. BELEVITCH, "On Network Analysis by Polynomial Matrices," in *Recent Developments in Network Theory* (S. R. Deards, ed.), Pergamon, Elmsford, N.Y., 1963, pp. 19–30. See also V. BELEVITCH, *Classical Network Theory*, Holden-Day, San Francisco, 1968.

11. V. M. POPOV, "Some Properties of Control Systems with Irreducible Matrix Transfer Functions," in *Lecture Notes in Mathematics*, Vol. 144, Springer, Berlin, 1969, pp. 169–180.

12. H. H. ROSENBROCK, *State-Space and Multivariable Theory*, Wiley, New York, 1970.

13. G. D. FORNEY, JR., "Convolutional Codes I: Algebraic Structure," *IEEE Trans. Inf. Theory*, **IT-16**, pp. 720–738, 1970. See also **IT-17**, p. 360, 1971.

14. W. A. WOLOVICH, *Linear Multivariable Systems*, Springer-Verlag, New York, Inc., 1974.

15. G. D. FORNEY, JR., "Minimal Bases of Rational Vector Spaces, with Applications to Multivariable Linear Systems," *SIAM J. Control*, **13**, pp. 493–520, 1975.

16. L. PERNEBO, "Algebraic Control Theory for Linear Multivariable Systems," Ph.D. Dissertation, Dept. of Automatic Control, Lund Institute of Technology, Lund, Sweden, 1978.

17. N. T. HUNG and B. D. O. ANDERSON, "Triangularization Technique for the Design of Multivariable Control Systems," *IEEE Trans. Automat. Contr.*, **AC-24**, pp. 455–460, Jan. 1979.

18. E. EMRE and M. L. J. HAUTUS, "A Polynomial Characterization of (A,B)-invariant and Reachability Subspaces," Memorandum COSOR 78-19, Dept. of Mathematics, Eindhoven Institute of Technology, Eindhoven, The Netherlands, Oct. 1978.

19. M. HEYMANN, "Structure and Realization Problems in the Theory of Dynamical Systems," CISM Courses and Lectures—No. 204, Springer-Verlag New York, Inc., New York, 1975.

20. W. A. WOLOVICH, "The Determination of State-Space Representations for Linear Multivariable Systems," in *Proceedings of the Second IFAC Symposium on Multivariable Technical Control Systems, Dusseldorf, Oct. 1971.*

21. I. C. GOHBERG, P. LANCASTER and L. RODMAN, "Spectral Analysis of Matrix Polynomials, I and II," *J. Linear Alg. Appl.*, **20**, pp. 1–44, 65–88, 1978.

22. I. C. GOHBERG and L. RODMAN, "On Spectral Analysis of Non-Monic Matrix and Operator Polynomials, I and II," *Israel J. Math.*, **30**, pp. 133–151, 1978.

23. P. VAN DOOREN, "The Computation of Kronecker's Canonical Form of a Singular Pencil," *J. Linear Alg. Appl.*, **27**, pp. 103–140, Oct. 1979.

24. S-H. WANG, "Design of Linear Multivariable Systems," *Memo. No. ERL-M309*, Electronics Research Laboratory, University of California, Berkeley, Oct. 1971.

25. W. A. WOLOVICH and P. L. FALB, "On the Structure of Multivariable Systems," *SIAM J. Control*, **7**, no. 3, pp. 437–451, Aug. 1969.

26. D. G. LUENBERGER, "Canonical Forms for Linear Multivariable Systems," *IEEE Trans. Autom. Control*, **AC-12**, pp. 290–293, 1967.

27. R. E. KALMAN, "Kronecker Invariants and Feedback," in *Ordinary Differential Equations*, (L. Weiss, ed.), Academic Press, New York, 1972, pp. 459–471.

28. D. DENERY, "An Identification Algorithm that is Insensitive to Initial Parameter Estimates," *AIAA J.*, **9**, pp. 371–377, 1971.

29. V. M. POPOV, "Invariant Description of Linear, Time-Invariant Controllable Systems, *SIAM J. Contr.*, **10**, pp. 252–264, May 1972.

30. K. GLOVER and J. C. WILLEMS, "Parametrizations of Linear Dynamical Systems: Canonical Forms and Identifiability," *IEEE Trans. Autom. Control*, **AC-19**, no. 6, pp. 640–645, Dec. 1974. See also M. J. DENHAM, *ibid.*, pp. 646–655.

31. J. M. C. CLARK, "The Consistent Selection of Local Coordinates in Linear System Identification," *Proceedings of the Joint Automatic Control Conference*, Purdue University, Lafayelte, Indiana, June 1976.

32. P. BRUNOVSKY, "A Classification of Linear Controllable Systems," *Kybernetika (Praha)*, **6**, pp. 173–187, 1970.

33. W. TUEL, "Canonical Forms for Linear Systems—I," *Res. Rept. No. RJ375*, IBM Research Laboratory, San Jose, Calif., 1966.

34. D. Q. MAYNE, "An Elementary Derivation of Rosenbrock's Minimal Realization Algorithm," *IEEE Trans. Autom. Control*, **AC-18**, no. 3, pp. 306–307, June 1973.

35. J. D. APLEVICH, "Direct Computation of Canonical Forms for Linear Systems by Elementary Matrix Operations," *IEEE Trans. Autom. Control*, **AC-19**, no. 2, pp. 124–126, April 1974.

36. K. C. DALY, "The Computation of Luenberger Canonical Forms Using Elementary Similarity Transformations," *Int. J. Syst. Sci.*, **7**, pp. 1–15, 1976.

37. B. L. HO, "An Effective Construction of Realizations from Input/Output Descriptions," Ph.D. dissertation, Stanford University, Stanford, Calif., 1966.

38. B. L. Ho and R. E. Kalman, "Effective Construction of Linear, State-Variable Models from Input/Output Functions," *Regelungstechnik*, **14**, pp. 545–548, 1966.

39. L. M. Silverman, "Structural Properties of Time-Variable Linear Systems," Ph.D. dissertation, Dept. of Electrical Engineering, Columbia University, New York, 1966.

40. D. C. Youla and P. Tissi, "*N*-Port Synthesis Via Reactance Extraction, Part I," *IEEE Int. Conv. Rec.*, **14**, Pt. 7, pp. 183–205, 1966.

41. B. McMillan, "Introduction to Formal Realization Theory," *Bell Syst. Tech. J.*, **31**, pp. 217–279, 541–600, 1952.

42. R. E. Kalman, "Irreducible Realizations and the Degree of a Rational Matrix," *SIAM J. Appl. Math.*, **13**, pp. 520–544, June 1965.

43. R. E. Kalman, "On Structural Properties of Linear Constant Multivariable Systems," Paper 6A, in *Proceedings of the Third IFAC Congress, London, 1966.*

44. C. A. Desoer and J. D. Schulman, "Zeros and Poles of Matrix Transfer Functions and Their Dynamical Interpretation," *IEEE Trans. Circuits Syst.*, **CAS-21**, pp. 3–8, 1974.

45. A. G. J. MacFarlane and N. Karcanias, "Poles and Zeros of Linear Multivariable Systems: A Survey of the Algebraic, Geometric and Complex Variable Theory," *Int. J. Control.*, **24**, pp. 33–74, 1976.

46. G. Verghese, "Infinite-Frequency Behavior in Generalized Dynamical Systems," Ph.D. dissertation, Dept. of Electrical Engineering, Stanford University, Stanford, Calif., Dec. 1978.

47. A. C. Pugh, "Transmission and System Zeros," *Int. J. Control*, **26**, pp. 315–324, Aug. 1977.

48. D. H. Owens, *Feedback and Multivariable Systems*, IEE Control Engineering Series # 7, Peter Peregrinus Press, London, England, 1978.

49. B. Lévy, "Algebraic Structure of 2-D Systems," Ph.D. dissertation, Dept. of Electrical Engineering, Stanford University, Stanford, Calif., Aug. 1980.

50. S. Kung and T. Kailath, "Some Notes on Valuation Theory in Linear Systems," *Proc. 1978 IEEE Conference on Decision and Control*, San Diego, Calif. pp. 515–517.

51. G. Verghese, P. Van Dooren and T. Kailath, "Properties of the System Matrix of a Generalized State-Space System," *Int. J. Control,* **30**, pp. 235–243, Aug. 1979.

52. S. H. Wang and E. J. Davison, "A Minimization Algorithm for the Design of Linear Multivariable Systems," *IEEE Trans. Autom. Control*, **AC-18**, pp. 220–225, 1973.

53. S. Kung, "Multivariable and Multidimensional Systems: Analysis and Design," Ph.D. Dissertation, Department of Electrical Engineering, Stanford University, Stanford, Calif. June 1977.

54. B. W. Dickinson, T. Kailath, and M. Morf, "Canonical Matrix Fraction and State-Space Descriptions for Deterministic and Stochastic Linear Systems," *IEEE Trans. Autom. Control*, **AC-19**, pp. 656–667, Dec. 1974.

55. B. W. DICKINSON, "Properties and Applications of Matrix-Fraction Descriptions of Linear Systems," Ph.D. dissertation, Dept. of Electrical Engineering, Stanford University, Stanford, Calif., Aug. 1974.

56. M. MORF, "Fast Algorithms for Multivariable Systems," Ph.D. dissertation, Dept. of Electrical Engineering, Stanford University, Stanford, Calif., Aug. 1974.

57. A. E. ECKBERG, "Algebraic System Theory with Application to Decentralized Control," Ph.D. dissertation, Dept. of Electrical Engineering, M.I.T., Cambridge, Mass., June 1973.

58. S. KUNG, T. KAILATH and M. MORF, "Fast and Stable Algorithms for Minimal Design Problems," pp. 97–104, *Proc. Fourth IFAC International Symposium on Multivariable Technological Systems*, ed. D. P. Altherton, Pergamon Press, London, 1977.

59. S. MACLANE and G. BIRKHOFF, *Algebra*, Macmillan, New York, 1967.

60. G. VERGHESE and T. KAILATH, "Rational Matrix Structure," *Proceedings of the 1979 IEEE Conference on Decision and Control*, Florida, Dec. 1979.

61. N. P. VEKUA, *Systems of Singular Integral Equations*, Noordhoff, The Netherlands, 1967. (Russian original, 1950)

62. I. C. GOHBERG and M. G. KREIN, "Systems of Integral Equations on a Half Line with Kernel Depending on the Difference of Arguments," *Amer. Math. Soc. Transl.*, **14**, pp. 217–287, 1960. (Russian original, 1958)

63. I. C. GOHBERG and L. LERER, "Factorization Indices and Kronecker Indices of Matrix Polynomials," *Integ. Equations and Operator Theory*, **2**, No. 2, 1979.

64. P. A. FUHRMANN and J. C. WILLEMS, "Factorization Indices at Infinity for Rational Matrix Functions," *Integ. Equations and Operator Theory*, **2**, No. 3, 1979.

STATE FEEDBACK AND COMPENSATOR DESIGN

7

7.0 INTRODUCTION

In Chapters 3 and 4, we discussed at some length the problem of shifting the poles of a scalar (single-input, single-output) system by means of separately designed controllers and observers or by direct transfer function calculations. In this chapter we shall discuss more briefly the multivariable analogs of these problems.

The multivariable problem of course has a much richer structure, and there are still several open questions. However, we shall see that using a combination of state-space and transfer function methods can be highly effective here.

7.1 STATE-SPACE ANALYSIS OF LINEAR STATE-FEEDBACK

The problem is the following: Given

$$\dot{x}(t) = Ax(t) + Bu(t) \tag{1}$$

find a *nonsingular* $m \times m$ matrix G and an $m \times n$ *gain* matrix K such that under the state-feedback operation

$$u(t) \longrightarrow Gv(t) - Kx(t) \tag{2}$$

499

the new state equation

$$\dot{x}(t) = (A - BK)x(t) + BGv(t) \tag{3}$$

has the characteristic polynomial

$$\det{(sI - A + BK)} = \alpha(s), \quad \text{an arbitrary monic polynomial of degree } n \tag{4}$$

To solve the problem, we shall have to assume that

$$\{A, B\} \text{ is controllable} \tag{5}$$

Otherwise certain modes cannot be shifted. We ask the reader to prove this by using either the standard form of Sec. 6.2.2 for a noncontrollable pair or by using the PBH eigenvector test (Theorem 6.2-5). Here we shall concentrate on proving the *sufficiency* of condition (5).

This problem can be solved in several different ways, and unlike its scalar analog in Sec. 3.2, in the multivariable case there can be many gain matrices K yielding the same characteristic polynomial $\alpha(s)$.

In the scalar case (Sec. 3.2.1), we first described a direct solution method (Bass-Gura) and then a method using the controller canonical form (Rissanen). In the matrix case, it is simpler to first study the problem with $\{A, B\}$ in the controller form (Sec. 7.1.1) and then to try a direct method (Sec. 7.1.2). Still other methods will be described in Sec. 7.2.1.

7.1.1   Controller-Form Method

The first step will be to transform the given controllable pair $\{A, B\}$ to controller form by Scheme II of Sec. 6.4.6. That is, we search the columns of the controllability matrix $\mathcal{C}(A, B)$ from left to right until we find n linearly independent vectors, which we then rearrange in the form

$$\{b_1 \quad Ab_1 \quad \cdots \quad A^{k_1-1}b_1 \quad b_2 \quad \cdots \quad A^{k_m-1}b_m\} \tag{6}$$

Then by suitable recombinations of these vectors we can find a new basis

$$T = \{e_{11} \quad \cdots \quad e_{1k_1} \quad e_{21} \quad \cdots \quad e_{m1} \quad \cdots \quad e_{mk_m}\} \tag{7}$$

with respect to which the pair $\{A, B\}$ is in controller form, i.e.,

$$T^{-1}AT = A_c, \quad T^{-1}B = B_c \tag{8}$$

where A_c and B_c have the forms (shown for $n = 10$, $m = 3$, $k_1 = k_2 = 3$, $k_3 = 4$; the unmarked entries are zeros):

$$
A_c = \begin{bmatrix}
x & x & x & x & x & x & x & x & x & x \\
1 & & & & & & & & & \\
 & 1 & & & & & & & & \\
\hline
x & x & x & x & x & x & x & x & x & x \\
 & & & 1 & & & & & & \\
 & & & & 1 & & & & & \\
\hline
x & x & x & x & x & x & x & x & x & x \\
 & & & & & & 1 & & & \\
 & & & & & & & 1 & & \\
 & & & & & & & & 1 &
\end{bmatrix}, \qquad
B_c = \begin{bmatrix}
1 & x & x \\
 & & \\
\hline
 & 1 & x \\
 & & \\
\hline
 & & 1
\end{bmatrix}
\qquad (9)\dagger
$$

Under the transformation

$$ x(t) = Tx_c(t) $$

the state equation (3) goes over into

$$ \dot{x}_c(t) = (A_c - B_c K_c)x_c(t) + B_c Gv(t) \tag{10} $$

where $\{A_c, B_c\}$ are as in (8) and

$$ K_c \triangleq KT \tag{11} $$

The first step in the pole-shifting algorithm will be to perform elementary column operations on B_c to *zero out* the entries marked x in the $\{$1st, $(k_1 + 1)$th, $(k_1 + k_2 + 1)$th, . . .$\}$ rows of B_c. This can always be done by elementary transformations because of the appropriately located 1s in these rows. Let us choose the (nonsingular) matrix G to represent these elementary transformations; i.e., we choose G such that

$$ B_c G = \text{block diag } \{[1 \quad 0 \quad \cdots \quad 0]', k_i \times 1, i = 1, \ldots, m\} $$
$$ = B_o^{\circ}, \text{ say} \tag{12} $$

Let us also define

$$ \tilde{K}_c = G^{-1}K_c, \qquad K_c = G\tilde{K}_c $$

so that we shall have

$$ B_c K_c = B_c G\tilde{K}_c = B_o^{\circ}\tilde{K}_c = \begin{bmatrix} \tilde{k}_{11} & \cdots & \tilde{k}_{1n} \\ \tilde{k}_{21} & \cdots & \tilde{k}_{2n} \\ \vdots & & \\ \vdots & & \\ \tilde{k}_{m1} & \cdots & \tilde{k}_{mn} \end{bmatrix} \tag{13} $$

†We have assumed that B has full rank. If not, e.g., if b_3 is linearly dependent on $\{b_1, b_2\}$, we ask the reader to check that the input u_3 will play no role in the succeeding arguments.

It then follows that we can make

$$A_c - B_0^\circ \tilde{K}_c = \text{a matrix with arbitrary elements in rows}$$
$$\{1, k_1 + 1, \ldots, \textstyle\sum_1^{m-1} k_i + 1\} \text{ and the other rows}$$
$$\text{just as in } A_c \tag{14}$$

It is now clear that we can arrange $A_c - B_0^\circ \tilde{K}_c$ in various convenient forms. For example, we could choose it so that

$$A_c - B_0^\circ \tilde{K}_c = A_o^\circ = \text{block diag} \left\{ \begin{bmatrix} 0 & & & \\ 1 & \cdot & & \\ & \cdot & \cdot & \\ & & \cdot & \cdot \\ & & & 1 & 0 \end{bmatrix}, k_i \times k_i \right\} \tag{15}$$

which will have the characteristic polynomial

$$\det(sI - A_o^\circ) = s^n \tag{16}$$

(We may note that $\{A_o^\circ, B_o^\circ\}$ is just the *core* realization we set up in Sec. 6.4.1—more on this in Sec. 7.1.3.)

A method of getting a desired characteristic polynomial

$$\alpha(s) = s^n + \alpha_1 s^{n-1} + \cdots + \alpha_n$$

is the following:

1. Choose the $\{2\text{nd}, 3\text{rd}, \ldots\}$ rows of $\tilde{K}_c$ to make all elements in the $\{(k_1 + 1), (k_1 + k_2 + 1), \ldots\}$ rows zero except for the $\{\{k_1 + 1, k_1\}, \{k_1 + k_2 + 1, k_1 + k_2\}, \ldots\}$ elements, which should be made equal to 1. The resulting matrix will be in *top-companion* form.
2. Now choose the first row of $\tilde{K}_c$ to make the first row of $A_c - B_0^\circ \tilde{K}_c$ equal to $[\alpha_1 \cdots \alpha_n]$.

Then clearly we shall have

$$\alpha(s) = \det(sI - A_c + B_0^\circ \tilde{K}_c) = \det(sI - A_c + B_c G \tilde{K}_c)$$
$$= \det(sI - A + BG\tilde{K}) = \det(sI - A + BK) \tag{17}$$

That is, we have proved that by a suitable choice of input transformation G and feedback gain matrix K we can arrange for a controllable pair $\{A, B\}$ to have an arbitrary (nth-degree) characteristic polynomial. We have thus generalized the scalar result of Sec. 3.2.1. However, it is worth noting that our procedure (cf. step 1 above) was to use a preliminary feedback operation to first reduce the pair $\{A_c, B_c\}$ to a *single-companion-block* (so-called *cyclic*) form, so that the new A matrix and the first column of the new B matrix formed a scalar controllable system. Note also that though only the input $u_1(\cdot)$ is used at the final stage, all the inputs $\{u_i\}$ enter in setting up the cyclic form.

The significance of these remarks will become apparent once again as we examine the possibility of a direct solution to the problem.

7.1.2   A Direct Method

Although the solution just given is quite elegant, it relies of course on a preliminary transformation to controller form, which may be quite laborious in the matrix case. It is therefore of interest to try to obtain a more direct solution, as in the Bass-Gura method of Sec. 3.2.1.

In this method, we first write

$$\det (sI - A + BK) = \det (sI - A) \det [I + (sI - A)^{-1}BK] \qquad (18)$$

but unlike the scalar case, we cannot simplify this any further because

$$(sI - A)^{-1}BK$$

may not be of rank 1. However, let us try to find K in the form

$$K = qk \qquad (19)$$

where k and q are $m \times 1$ and $1 \times n$ matrices to be suitably chosen. Now if we define $Bq = \tilde{b}$, we can write

$$\det [I + (sI - A)^{-1}Bqk] = 1 + k(sI - A)^{-1}\tilde{b} \qquad (20)$$

which shows that provided

$$\{A, Bq\} = \{A, \tilde{b}\} \text{ is controllable} \qquad (21)$$

we shall have reduced the matrix problem to a scalar problem, and k can be found by any of the formulas of Sec. 3.2.1.

The only question is whether if $\{A, B\}$ is controllable there always exists a vector q that will make $\{A, Bq\}$ controllable. Unfortunately in general the answer is no. A simple example is provided by (applying the PBH rank test to)

$$A = \begin{bmatrix} 0 & 1 & 0 \\ 0 & 0 & 0 \\ 0 & 0 & 0 \end{bmatrix}, \qquad B = \begin{bmatrix} 0 & 0 \\ 1 & 0 \\ 0 & 1 \end{bmatrix}.$$

The point actually is that $\{A, B\}$ can never be controllable if $sI - A$ has more invariant polynomials than the number of inputs (cf. Exercise 6.4-8), and therefore if $\{A, Bq\}$ is to be controllable, $sI - A$ must have only one *nonunity* invariant polynomial. Polynomial matrices with this property will be called *simple*. A real matrix A such that $sI - A$ is *simple* will be called *cyclic*. From the discussion of the Smith form and the rational canonical form (Theorems 6.3-16–6.3-18) we may characterize a cyclic matrix as one that can be transformed to a (*one-block*) *companion* matrix by a similarity transformation.

While the fact that $sI - A$ is simple is a necessary condition for the rank 1 feedback method to work, it still remains to prove that for such matrices A we can find a suitable vector q. We shall prove the following result.

Lemma 7.1-1.   ([1] and [2])

If $\{A, B\}$ is controllable and A is cyclic, there exists at least one vector q such that $\{A, Bq\}$ is controllable. In fact, it holds that "almost any" $m \times 1$ vector q will suffice.

In other words, if we pick m real numbers at random, this will almost always give a satisfactory q. More insight into this fact will follow from the proof of the lemma (given in Sec. 7.2.3).

However, we have the restriction that A be cyclic. Wonham [1] and Gopinath [2] also proved that a preliminary state feedback could be used to convert a noncyclic A matrix to a cyclic one. (Recall that this was what was done in step 1 of our first solution method.) Actually, Brasch and Pearson [3] gave a much more convenient form of this result.

Lemma 7.1-2.   [3]

If $\{A, B, C\}$ is a minimal realization, then almost any (constant) output feedback F will make $A - BFC$ cyclic.

These two lemmas then provide the following design procedure:

1. By a preliminary and "almost arbitrary" state or output feedback, we make A cyclic.
2. By an almost arbitrary choice of q, we make $\{A, Bq\}$ controllable.
3. Now apply the scalar design formulas to find a gain matrix k to make

$$\det(sI - A + Bqk) = \alpha(s)$$

The proofs of Lemmas 7.1-1 and 7.1-2 given in the original references [1]–[3] are based on state-space methods and are fairly lengthy. It turns out that simpler proofs can be obtained by working with the transfer functions $(sI - A)^{-1}B$ and $C(sI - A)^{-1}B$. But once we go to this, we could just as easily handle transfer functions given in MFD form, $N(s)D^{-1}(s)$, and therefore in Sec. 7.2.1 we shall examine the pole-shifting problem in this context. We shall find that this approach not only gives us an alternative description of the methods of this section but also leads to some new insights and results. Section 7.2.3 will contain the proofs of Lemmas 1 and 2 above.

We close this section with some further remarks on state-space *canonical* forms.

7.1.3   The Brunovsky Canonical Form, Kronecker and Factorization Indices

In Sec. 7.1.1, we used the controller form obtained via Scheme II of Sec. 6.4.6, i.e., searching the crate by rows. The resulting form was actually a canonical controller form, and we shall explore some further *canonical* properties here.

First, however, we may ask why we specified that the controller form was to be obtained by Scheme II rather than Scheme I or the other schemes noted in Sec. 6.4.6. The reason is that with these other schemes, the B matrix will not have the property that nonzero elements occur only in rows $\{1, k_1 + 1, k_1 + k_2 + 1, \ldots\}$; we should recall from Sec. 7.1.1 that it is this special property that makes it easy to *reset* the xs in the A_c matrix and thus solve the mode-shifting problem.

Now when $\{A_c, B_c\}$ is obtained by Scheme II, their form is completely determined by the canonical parameters

$$\{k_i, \alpha_{ijk}\}$$

and it is of interest to examine the behavior of these parameters under the transformations we have used.

First, by definition the above parameters are invariant under similarity transformation,

$$\{A, B\} \longrightarrow \{T^{-1}AT, T^{-1}B\}$$

However, if we also allow input transformations and state feedback, so that

$$\{A, B\} \longrightarrow \{T^{-1}(A - BGK)T, T^{-1}BG\}, \qquad \det G \neq 0 \tag{22}$$

then the parameters $\{\alpha_{ijk}\}$ can be zeroed out, and the only remaining invariants are the *controllability indices* $\{k_i\}$. [It can be seen that according to the definitions of Sec. 6.7.3 the indices $\{k_i\}$ are a set of complete invariants for $\{A, B\}$ under operations (22), i.e., for all $T, G, K, \{T^{-1}(A - BGK)T, T^{-1}BG\}$ have the same indices $\{k_i\}$, which is the invariance property, and if two pairs $\{A_i, B_i, i = 1, 2\}$ have the same indices, then there must exist T, G, K such that

$$A_1 = T^{-1}(A_2 - B_2GK)T, \qquad B_1 = T^{-1}B_2G$$

which is the completeness property.]

The controller-form pair corresponding to all the $\{\alpha_{ijk}\} = 0$ is the *core* realization [cf. Sec. 6.4.1 and also Eqs. (12) and (15) above] $\{A_o^\circ, B_o^\circ\}$. This is a *unique* form in the sense that two realizations that are related by similarity, input transformations, and state feedback, i.e., as

$$A_1 = T^{-1}(A_2 - B_2GK)T, \qquad B_1 = T^{-1}B_2G$$

will have the same core realization $\{A_o^\circ, B_o^\circ\}$. Therefore this form is a canonical form under the specified operations; it is often called the *Brunovsky canonical form* after its discoverer [4].

The significance of the controllability indices $\{k_i\}$ and of Scheme II by which they were defined stands out clearly in this form.

Rosenbrock [5] and Kalman [6] recognized that these indices occur in the work of Kronecker (1890) as briefly described in Sec. 6.5.4. Exercise 7.1-1 provides some elaboration of this connection; see also References [7], [8], and [9] and Sec. 7.2.2. As mentioned in Sec. 6.5.4, Ref. [9] brought out a connection between these Kronecker indices and the so called Wiener-Hopf factorization indices. More explicitly, suppose that $U^{-1}(s)$ is a unimodular matrix such that $D(s)U^{-1}(s)$ is column-reduced. We can write this, in the notation of Eq. 6.4-3, as

$$D(s)U(s) = D_{hc}S(s) + L(s), \quad S(s) = \text{diag}\{s^{k_i}\}$$

or

$$D(s) = (D_{hc} + L(s)S^{-1}(s))S(s)U(s)$$

where the nonsingularity of D_{hc} and the strict properness of $L(s)S^{-1}(s)$ ensures that the first factor on the left is bounded at ∞, together with its inverse. According to Gohberg and Krein, this identifies the Kronecker indices $\{k_i\}$ as the left factorization indices at infinity of $D(s)$. Reference [9] develops some implications of this connection for the control structure theorem of Sec. 7.2.2 below.

Exercise

7.1-1. *Kronecker Equivalence and State Feedback*

Following Kronecker, in Sec. 6.3.4, we defined two pencils $[zI - A \ B]$ and $[zI - \hat{A} \ \hat{B}]$ to be equivalent if and only if $[zI - A \ B] = L[zI - \hat{A} \ \hat{B}]R$, for constant nonsingular matrices L and R.

a. Show that a suitable L and R exist if $\hat{A} = T^{-1}(A - BGK)T$, $\hat{B} = T^{-1}BG$ for G nonsingular and K arbitrary.

b. Conversely if a suitable L and R exist, show that they must have the form

$$L = T, \qquad R = \begin{bmatrix} T^{-1} & 0 \\ -K & G \end{bmatrix}$$

for some nonsingular G and an arbitrary $m \times n$ matrix K.

c. What can you conclude about the relation between Kronecker equivalence and the transformations used to obtain the Brunovsky canonical form?

7.2 TRANSFER FUNCTION ANALYSIS
OF LINEAR STATE FEEDBACK

In a brief discussion at the end of Sec. 3.2.1 and in more detail in Sec. 4.5 we saw that a suitable transfer function point of view could be helpful in studying the state-feedback problem. This turns out to be even more true in the multi-variable case. Thus from Sec. 7.1.1 it is clear that in the multivariable case there is a great deal of freedom in the choice of the feedback-gain matrix to achieve a given set of closed-loop modes. In Sec. 7.2.1 we shall show how a transfer function analysis leads us naturally to a unique characterization of the feedback gain matrix in terms of the eigenvalues *and* eigenvectors of the closed-loop system. This characterization can be exploited to special effect in the analysis of the multivariable quadratic regulator problem (cf. Sec. 7.4).

Another possibility for flexibility in the multivariable case is that the closed-loop state matrix has nontrivial invariant polynomials (cf. Sec. 6.5.2) other than just its characteristic polynomial. In Sec. 7.2.2 we shall prove an important *control structure theorem* of Rosenbrock [5] which shows how the controllability (Kronecker) indices set lower bounds on the size of the companion block structure of the closed-loop state matrix.

Finally in Sec. 7.2.3 we shall use transfer function methods to give simple proofs of the two lemmas that we quoted in Sec. 7.1.2 in studying a direct approach to state feedback.

7.2.1   *Alternative Formulas for the Feedback Gain Matrix*

We saw in Sec. 6.4.2 that there was a direct relationship between the controller canonical form and a right MFD

$$H(s) = N(s)D^{-1}(s), \qquad D(s) \text{ column reduced} \tag{1}$$

Referring to the schematic block diagram for the controller form (Fig. 6.4-1), we note that the states of the realization are completely determined by the partial state $\xi(s)$ and its derivatives $\Psi(s)\xi(s)$. Therefore state feedback through a gain matrix K_c can be schematically represented as in Fig. 7.2-1 (cf. also the analogous discussion of the scalar case in Sec. 4.5.3).

From Fig. 7.2-1 we can see that the closed-loop transfer function is, say,

$$H_K(s) = N(s)D_K^{-1}(s) \tag{2}$$

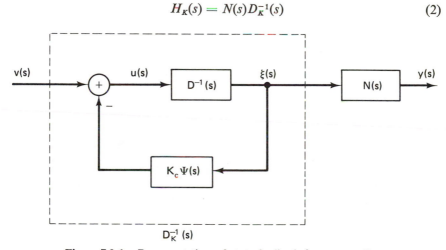

Figure 7.2-1.   Representation of state feedback for a controller-form realization of $H(s) = N(s)D^{-1}(s)$.

where

$$D_K(s) \triangleq D(s) + K_c \Psi(s) \tag{3}$$

If we now recall the representation [cf. Eq. (6.4-3)]

$$D(s) = D_{hc} S(s) + D_{lc} \Psi(s)$$

where

$$S(s) \triangleq \text{diag} \{s^{k_i}, i = 1, \ldots, m\}$$

and

$$\Psi(s) = \text{block diag} \{[s^{k_i-1} \quad \cdots \quad 1], i = 1, \ldots, m\}$$
$$k_i = \text{the degree of the } i\text{th column of } D(s)$$

we have

$$D_K(s) = D_{hc} S(s) + (D_{lc} + K_c) \Psi(s) \tag{4}$$

The significant facts to be noted from (2)–(4) are that state feedback

1. does not alter the numerator polynomial $N(s)$
2. does not alter D_{hc}, the highest-degree-coefficient matrix of $D(s)$ (5)
3. can completely change D_{lc}, the lower-degree-coefficient matrix of $D(s)$ (6)
4. cannot change the column degrees of $D(s)$ (7)

Constraint (5) can be relaxed by allowing a nonsingular input transformation G, which will give

$$H_{K,G}(s) \triangleq N(s)[D_{hc} S(s) + (D_{lc} + K_c) \Psi(s)]^{-1} G$$
$$= N(s)[G^{-1} D_{hc} S(s) + G^{-1}(D_{lc} + K_c) \Psi(s)]^{-1} \tag{8}$$

A convenient choice, and one we shall make here, is that

$$G = D_{hc} \tag{9}$$

which the reader may wish to check is equivalent to the choice we made in the design method of Sec. 7.1 [cf. (7.1.2)].

The fact that we can alter D_{lc} at will by proper choice of K_c indicates that we should be able to make the characteristic polynomial of the closed-loop system equal to an arbitrary monic polynomial, say,

$$\alpha(s) = s^n + \alpha_1 s^{n-1} + \cdots + \alpha_n \tag{10}$$

Of course we already proved this in Sec. 7.1.1, but a direct transfer function proof can also be given [10].

Thus, assume that the column degrees of $D(s)$ are arranged as

$$k_1 \leq k_2 \cdots \leq k_m, \qquad \sum_{1}^{m} k_i = n \qquad (11)$$

and define polynomials $\{\alpha_i(s), i = 1, \ldots, m\}$ by

$$\alpha(s) = s^n + \alpha_1(s)s^{n-k_1} + \alpha_2(s)s^{n-k_1-k_2} + \cdots + \alpha_m(s) \qquad (12)$$

where $\alpha_i(s)$ is a polynomial of degree less than k_i. For example, if $k_1 = 1$, $k_2 = 3$, $k_3 = 4$, and

$$\alpha(s) = s^8 + 2s^7 + 3s^6 + s^5 + s^3 + s + 1$$

then

$$\alpha_1(s) = 2, \qquad \alpha_2(s) = 3s^2 + s, \qquad \alpha_3(s) = s^3 + s + 1$$

Now we can check that†

$$\det \begin{bmatrix} s^{k_1} + \alpha_1(s) & \alpha_2(s) & \cdots & \alpha_m(s) \\ -1 & s^{k_2} & & \\ & \cdot & \cdot & \\ & & \cdot & \cdot \\ & & -1 & s^{k_m} \end{bmatrix} = \alpha(s) \qquad (13)$$

Now from (8) and (9) we have

$$D_{K,G}(s) \triangleq S(s) + D_{hc}^{-1}(D_{lc} + K_c)\Psi(s)$$

so that if we choose K_c so that

$$K_c \Psi(s) = D_{hc} \begin{bmatrix} \alpha_1(s) & \cdots & \alpha_m(s) \\ -1 & & \\ & \cdot & \\ & \cdot & \\ & -1 & \\ & & -1 & 0 \end{bmatrix} - D_{lc}\Psi(s) \qquad (14)$$

then we shall have [cf. (6.4-15)]

$$\det (sI - A_c + B_c GK_c) = \det D_{K,G}(s) = \alpha(s) \qquad (15)$$

†This is a generalized form of the (companion) matrix $sI - A_c$.

It will be an interesting exercise for the student to reconcile the expression for K_c obtained from (14) with the one obtained by the method of Sec. 7.1.1.

We note also that, as in Sec. 7.1, the choice of K_c is not unique because we can define several different matrices $D_{K,G}(s)$ with the same characteristic polynomial $\alpha(s)$.

This freedom of choice in the selection of the gain matrix should ideally be used to achieve some other desirable properties of the closed-loop system besides control over its modes. Unfortunately there have been few studies of systematic ways of doing this, except in the multivariable quadratic regulator theory (cf. Sec. 3.4 and 7.4).

In this connection, it will be useful to note some alternative characterizations of the feedback gain matrix in terms not only of the desired closed-loop eigenvalues but also certain specified sets of closed-loop eigenvectors.

Eigenvector Formulas for the Feedback Gain Matrix ([11] and [12]). The closed-loop system is described by

$$H_{K,G}(s) = N(s)D_{K,G}^{-1}(s)$$

where [cf. (8) and (9)]

$$D_{K,G}(s) = S(s) + D_{hc}^{-1}[D_{lc} + K_c\Psi(s)]$$
$$= D_{hc}^{-1}[D(s) + K_c\Psi(s)] \tag{16}$$

The closed-loop eigenvalues will be the roots, say $\{\mu_i, i = 1, \ldots, n\}$, of the equation [cf. (6.4-16)]

$$\det D_{K,G}(\mu) = 0 \tag{17}$$

The eigenvectors of the state matrix of the associated controller-form realization of $H_{K,G}(s)$ will be given by [cf. formula (6.4-18)]

$$f_i = \Psi(\mu_i)p_i \tag{18}$$

where p_i is such that

$$D_{K,G}(\mu_i)p_i = 0 \tag{19}$$

But by (16), this implies that

$$D(\mu_i)p_i + K_c\Psi(\mu_i)p_i = 0 \tag{20}$$

which with (18), yields

$$K_cf_i = -D(\mu_i)p_i = -g_i, \text{ say} \tag{21}$$

Now if the $\{\mu_i\}$ are distinct, the eigenvectors $\{f_i\}$ will be linearly independent.

Then we can write

$$K_c[f_1 \quad \cdots \quad f_n] = -[g_1 \quad \cdots \quad g_n]$$

and solve uniquely for K_c as

$$K_c = -[g_1 \quad \cdots \quad g_n][f_1 \quad \cdots \quad f_n]^{-1} \tag{22}$$

A unique solution can also be obtained when the $\{\mu_i\}$ are not distinct, but we shall have to use *generalized eigenvectors* (cf. Exercise A.51). We shall not pursue this generalization here, since our interest is really to reverse the above arguments. That is, we can proceed as follows in the state-feedback problem:

1. Choose a desired set of distinct closed-loop eigenvalues $\{\mu_i\}$.
2. Choose a set of eigenvectors $\{f_i\}$ such that
   a. $f_i \in \Re[\Psi(\mu_i)]$, the range space of $\Psi(\mu_i)$,
   b. The $\{f_i\}$ are linearly independent, and
   c. If f_i is associated with μ_i, then the complex conjugate of f_i must be associated with the complex conjugate of μ_i.
3. Since $\Psi(s)$ is always full rank, there will exist a unique m-vector p_i such that $f_i = \Psi(\mu_i)p_i$. Then define $g_i = D(\mu_i)p_i$.

Now if we choose K_c as in (22), then it follows from the above analysis that the controller form associated with the closed-loop transfer function

$$N(s)D_K^{-1}(s), \qquad D_K(s) = D(s) + K_c\Psi(s)$$

will have eigenvalues $\{\mu_i\}$ and eigenvectors $\{f_i\}$, as desired.

Therefore specifying eigenvalues and *a* set of eigenvectors completely specifies the feedback gain matrix and the only question is how to choose the $\{\mu_i\}$ and $\{f_i\}$. Note that the $\{f_i\}$ are not completely arbitrary but must satisfy the constraints noted above under step 2. In particular, since $\Psi(\cdot)$ is an $n \times m$ matrix, each eigenvector f_i must lie in the m-dimensional subspace spanned by the columns of $\Psi(\mu_i)$, and we can only choose any vector in this subspace, subject furthermore to the requirements 2b and 2c.

Moore [11] gives some examples of how this freedom of choice may be used to improve the transient response characteristic of multivariable non-zero-set-point control systems (whose scalar analog is discussed in Sec. 3.2.2). In Sec. 7.4 we shall see how the *optimal multivariable quadratic regulator* also specifies a particular set of eigenvalues and eigenvectors. In this connection, however, we might note that our formulas above have all been for systems described by MFDs or their controller-form state-space realizations. While an arbitrary controllable pair $\{A, B\}$ can always be converted to controller form, a more direct treatment is often desirable [12].

For this, consider the closed-loop system shown in Fig. 7.2-2. Assume

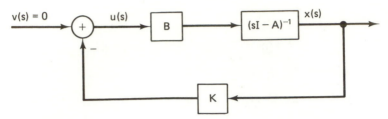

Figure 7.2-2. Schematic representation of the closed-loop system.

that $V(s) \equiv 0$ (no external input) and that the closed-loop system supports a motion at the single frequency μ_i, so that

$$x(t) = X_i e^{\mu_i t}, \qquad u(t) = U_i e^{\mu_i t} \tag{23a}$$

Here the vectors $\{X_i, U_i\}$ are such that

$$(\mu_i I - A)X_i = BU_i \tag{23b}$$

and

$$U_i = -KX_i \tag{23c}$$

We see that

$$(\mu_i I - A + BK)X_i = 0 \tag{23d}$$

so that X_i must be an eigenvector of the closed-loop system, associated with the eigenvalue μ_i. If we assume that the closed-loop eigenvalues $\{\mu_i\}$ are distinct, then the corresponding eigenvectors $\{X_i\}$ will be linearly independent, and in this case the feedback gain matrix can be written as

$$K = -[U_1 \quad \cdots \quad U_n][X_1 \quad \cdots \quad X_n]^{-1} \tag{24}$$

which is a general version of the controller-form formula (22).

To use formula (24) as a basis for choosing K, we must specify the $\{U_i, X_i\}$ subject to constraints (23). One way is to specify the $\{U_i\}$ in some way and then to calculate $\{X_i\}$ from (23b). (For a unique solution, we must assume that μ_i is not an eigenvalue of A, i.e., that the closed-loop eigenvalues are different from the open-loop eigenvalues.) An example of this route will be given in Sec. 7.4 when we study the multivariable quadratic regulator.

On the other hand, if we choose X_i first, then U_i can be found from (23b) as (assuming B has full rank)

$$U_i = (B'B)^{-1}B'(\mu_i I - A)X_i \tag{25}$$

However this expression will satisfy (23b) only if this X_i was consistent in the first place, i.e., only if

$$X_i \in \Re[(\mu_i I - A)^{-1} B] \tag{26}$$

To clarify this, suppose that $\{A, B\}$ is in controller form, so that

$$(\mu_i I - A)^{-1} B = \Psi(\mu_i) D^{-1}(\mu_i)$$

Then condition (26) is equivalent to the previously noted requirement 2a on the eigenvectors. Therefore, provided (26) is met as well as the conditions that the $\{X_i\}$ are linearly independent and that if $\mu_i = \mu_j^*$ we choose $X_i = X_j^*$, then K can be calculated via (24) and (25).

*7.2.2   Rosenbrock's Control Structure Theorem

In Sec. 7.2.1 we explored in some detail the consequences of properties (5) and (6) of state feedback. Here we shall examine the significance of property (7) that

the column degrees of $D_K(s)$ = the column degrees of $D(s)$

We recall that the column degrees [of a column-reduced $D(s)$, as assumed in (1)] are also the *controllability indices* of any minimal realization of $\Psi(s) D^{-1}(s)$.

We denote the column degrees by

$$k_1 \leq k_2 \leq \cdots \leq k_m$$

The sum of the column degrees is the degree of the characteristic polynomial of any minimal realization $\{A, B, C\}$ of $\Psi(s) D^{-1}(s)$. Now by referring to the Smith form (cf. Sec. 6.3.3) of $sI - A$, say,

$$sI - A \sim \text{diag}\{\psi_i(s)\}$$

where the invariant polynomials $\{\psi_i(s)\}$ are such that

$$\psi_{i+1}(s) \,|\, \psi_i(s), \qquad i = 1, \ldots, n \tag{27}\dagger$$

we see that the characteristic polynomial

$$\det(sI - A) = \prod_1^n \psi_i(s)$$

We showed in Sec. 7.2.1 that properties (5) and (6) of state feedback enabled us to obtain any monic polynomial of degree

$$n = \sum_1^m k_i = \deg \det(sI - A) = \deg \det D(s)$$

$\dagger$In Sec. 6.3.3 we denoted the invariant polynomials by $\lambda_i(s)$ and had them arranged so that $\lambda_i(s) \,|\, \lambda_{i+1}(s)$. The reason for the change of notation here is that $sI - A$ and $D(s)$ are now *denominator* matrices of a transfer function, whose Smith-McMillan form (cf. Sec. 6.5.2) has *denominator* polynomials $\psi_i(s)$ arranged so that $\psi_{i+1}(s) \,|\, \psi_i(s)$.

as the characteristic polynomial after feedback. This leads us to expect that the individual k_i should be related to the degrees of the separate invariant polynomials of $sI - A$ and the closely related matrix $D(s)$.

To pursue this connection, we note first that there are only m column degrees (controllability indices), while there are n invariant polynomials of $sI - A$ and m invariant polynomials of $D(s)$. These facts are partly reconciled by the following results.

Lemma 7.2-1

There are at most m *nonunity* invariant polynomials of $sI - A$, and they coincide with the nonunity invariant polynomials $\{\psi_i(s)\}$ of $D(s)$.

Proof.   This was already shown in Sec. 6.4.2 [cf. (6.4-16)].   ∎

Lemma 7.2-2.

The invariant polynomials $\{\psi_i(s)\}$ of the (column-reduced) matrix $D(s)$ satisfy the degree constraints

$$1. \quad \sum_{i=1}^{r} \deg \psi_{m+1-i}(s) \leq \sum_{i=1}^{r} k_i, \qquad r = 1, \ldots, m \tag{28}$$

or, equivalently,

$$2. \quad \sum_{i=1}^{r} \deg \psi_i(s) \geq \sum_{i=1}^{r} k_{m+1-i}, \qquad r = 1, \ldots, m \tag{29}$$

There is equality when $r = m$.

Proof.   From the discussion of the Smith form in Sec. 6.3.3 we know that

$$\psi_m(s) = \text{gcd of all } 1 \times 1 \text{ minors of } D(s)$$

and that

$$\prod_{i=0}^{r-1} \psi_{m-i}(s) = \text{gcd of all } r \times r \text{ minors of } D(s), r = 1, 2, \ldots, m$$

Now since $D(s)$ is column reduced, it follows that

the least degree of any 1×1 minor is k_1

the least degree of any 2×2 minor is $k_1 + k_2$

and so on.

Therefore the gcd of all 1×1 minors cannot have degree greater than k_1, i.e.,

$$\deg \psi_m(s) \leq k_1$$

The gcd of all 2×2 minors cannot have degree greater than $k_1 + k_2$, i.e.,

$$\deg \psi_m(s)\psi_{m-1}(s) \leq k_1 + k_2$$

Proceeding in this way gives all the relations (28). Since $D(s)$ is column reduced, equality holds when $r = m$, and therefore the inequalities in constraint 1 can be reversed to give those in constraint 2.

(It will be a useful exercise to obtain the result of Lemma 7.2-2 by deriving the Smith form of $sI - A_c$.) ∎

Reference to Theorem 6.3-18 will show that a matrix with invariant factors $\{\psi_m(s), \ldots, \psi_1(s)\}$ can be taken by a similarity transformation to the so-called *rational* canonical form,

$$R = \text{block diag}\,\{A_i, A_i = \text{a companion matrix with det }(sI - A_i) = \psi_i(s)\}$$

Therefore Lemma 7.2-2 sets lower bounds on the size of the (so-called *cyclic*) blocks in the rational canonical form of a matrix A obtained from a realization of $\Psi(s)D^{-1}(s)$. Now the state-feedback constructions of the previous sections (especially Sec. 7.1.1) can give us different rational canonical forms for the A matrix. For example, we can readily get one with block sizes equal to the $\{k_i\}$—just use the feedback to null out all *off-block-diagonal* elements in the *coefficient* rows and make the elements in each diagonal block equal to the coefficients of $\{\psi_m(s), \psi_{m-1}(s), \ldots, \psi_1(s)\}$. However, we can get other forms as well; e.g., in Sec. 7.1.1 we used feedback to give us just a single-block companion matrix.

How much freedom do we have? The simple Lemma 7.2-2 shows that the controllability indices put bounds on the sizes of the blocks in the rational canonical form of $A_c - B_c K_c$. Moreover, Rosenbrock discovered that the *converse* of Lemma 7.2-2 is also true, though it is not quite so easy to prove. We first note an important preliminary result.

Lemma 7.2-3.   [5, p. 184]

Given a set of polynomials $\{\psi_i(s), i = 1, \ldots, m\}$ satisfying

$$\psi_{i+1}(s) \,|\, \psi_i(s), \qquad i = 1, \ldots, m$$

and the degree constraints (28), there is a column-reduced matrix $\tilde{D}(s)$ with column degrees $k_1 \leq k_2 \leq \cdots \leq k_m$ and invariant polynomials $\{\psi_i(s)\}$. Moreover, the highest-column-degree coefficient matrix of $\tilde{D}(s)$ can be made equal to an arbitrary nonsingular matrix.

Proof.   We form a matrix

$$\bar{D}(s) = \text{diag}\,\{\psi_{m+1-i}(s), i = 1, \ldots, m\}$$

$\bar{D}(s)$ is clearly column reduced with degrees δ_i, say, where

$$\delta_i = \deg \psi_{m+1-i}(s), \qquad i = 1, \ldots, m$$

Then we shall show that by elementary row and column operations we can transform $\bar{D}(s)$ to another column-reduced matrix with degrees equal to the $\{k_i\}$. The idea is the following.

First make the normalization

$$\hat{D}(s) = \bar{D}_{hc}^{-1}\bar{D}(s)$$

so that $\hat{D}(s)$ has highest-column-degree coefficient matrix equal to I_m. Now suppose that $\delta_1 < k_1$ so that $\delta_j > k_j$ for some $j = 2, \ldots, m$. Then add s times row 1 to row j. Now, in the resulting matrix, let β be the leading coefficient of the (j, j) element and set $\alpha = \delta_j - \delta_1 - 1$. Then subtract βs^α times (the new) column 1 from (the new) column j. The resulting matrix will still be column proper, and let it be normalized (as above) so that its highest-column-degree coefficient matrix is I_m. Then it turns out that in the resulting matrix,

$$\deg (\text{column } 1) = \delta_1 + 1$$
$$\deg (\text{column } j) = \delta_j - 1$$
$$\deg (\text{column } i) = \delta_i, \qquad i \neq 1 \quad \text{or} \quad j$$

(The reader may find it useful at this stage to consider a simple example, as given below in Example 7.2-1.) To complete the proof we just repeat the steps until $\delta_1 = k_1$. Then we proceed similarly with the next degree that is not equal to the corresponding controllability index. And so on. We can multiply the final matrix on the left by any nonsingular real matrix in order to obtain a desired $\tilde{D}_{hc}$. Call the resulting matrix $\tilde{D}(s)$. Since all the elementary row and column operations used above do not affect the invariant factors, $\tilde{D}(s)$ meets all the conditions of the lemma. ∎

The role of state feedback now becomes clear. Once we have obtained a matrix $\tilde{D}(s)$ as in Lemma 7.2-3, then we can write

$$\tilde{D}(s) = D_{hc}S(s) + \tilde{D}_{lc}\Psi(s)$$

and given $D(s)$, with the same D_{hc} and the same column degrees as $\tilde{D}(s)$, we can choose a state-feedback matrix K_c to make

$$\begin{aligned} D_K(s) &= [D(s) + K_c\Psi(s)] \\ &= [D_{hc}S(s) + (D_{lc} + K_c)\Psi(s)] \\ &= [D_{hc}S(s) + \tilde{D}_{lc}\Psi(s)] = \tilde{D}(s) \end{aligned}$$

We thus have proved the following important theorem of Rosenbrock, cast in state-space language.

Theorem 7.2-4.   The Control Structure Theorem [5]

Let $\{A, B\}$ be a controllable pair with controllability indices $k_1 \leq k_2 \leq \cdots \leq k_m$. Let $\psi_i(s)$, $i = 1, \ldots, q, q \leq m$, be any set of monic polynomials satisfying the divisibility conditions $\psi_{i+1}(s) | \psi_i(s)$, $i = 1, \ldots, q - 1$ and $\sum_1^q \deg \psi_i = n$. Then there is a feedback gain matrix K such that the given polynomials are the nonunity invariant polynomials of $sI - A + BK$ if and only if

$$\sum_{i=1}^{r} \deg \psi_{n+1-i}(s) \leq \sum_1^r k_i, \qquad i \leq r \leq q$$

Proof.   Already given above. We note that the proof used transfer function methods. A direct state space proof can be given but is less simple. ∎

Corollary 7.2-5.

Let $\{A, B\}$ be controllable. Then A can always be made cyclic, with arbitrary characteristic polynomial, by linear state feedback. Moreover, this result is true for systems defined over finite fields [e.g., $GF(2)$ as in Example 2.3-8] as well as the usual real field.

Proof. For real fields, this is the result proved several times before. However, the proofs in Secs. 7.1.1 and 7.1.2 will not work in the finite field case, because (as shown by Heymann [13]) rank 1 feedback is not sufficient in this case. The proof via Rosenbrock's theorem is quite general,† however, and goes as follows: Choose $\psi_1(s) = \alpha(s)$, the desired characteristic polynomial, and choose $\psi_2(s) = \cdots = \psi_m(s) = 1$. This choice meets all the conditions of the theorem. ∎

Example 7.2-1.   [5]

A nontrivial example over the real field of the full generality of Rosenbrock's theorem is when $k_1 = k_2 = 3$ and

$$\psi_1(s) = (s^2 + 1)^2, \qquad \psi_2(s) = s^2 + 1$$

The characteristic polynomial is $\psi_1(s)\psi_2(s)$ and we could obtain a realization with two invariant factors if we could split $\psi_1(s)\psi_2(s)$ into two third-degree factors, so that we could then use the methods of Sec. 7.1.1 to obtain a block diagonal matrix with each block having one of the factors as the characteristic polynomial. However, the only third-degree factors of $\psi_1(s)\psi_2(s)$ will have imaginary coefficients, and so we cannot proceed in this way. By Rosenbrock's theorem we form diag $\{(s^2 + 1)^2,$ $s^2 + 1\}$ and use the methods of Lemma 7.2-3 to transform it:

$$\begin{bmatrix} s & 1 \\ 1 & 0 \end{bmatrix}\begin{bmatrix} (s^2 + 1) & 0 \\ 0 & (s^2 + 1)^2 \end{bmatrix}\begin{bmatrix} 1 & s \\ 0 & -1 \end{bmatrix} = \begin{bmatrix} s^3 + s & -s^2 - 1 \\ s^2 + 1 & s^3 + s \end{bmatrix}$$

Suppose the given $\{A, B\}$ pair is

$$A = \begin{bmatrix} -2 & 0 & -6 & -3 & 0 & -1 \\ 1 & 0 & 0 & 0 & 0 & 0 \\ 0 & 1 & 0 & 0 & 0 & 0 \\ -4 & -1 & -7 & -2 & -1 & 0 \\ 0 & 0 & 0 & 1 & 0 & 0 \\ 0 & 0 & 0 & 0 & 1 & 0 \end{bmatrix}, \quad B = \begin{bmatrix} 1 & 0 \\ 0 & 0 \\ 0 & 0 \\ 0 & 1 \\ 0 & 0 \\ 0 & 0 \end{bmatrix}$$

which has MFD

$$(sI - A)^{-1}B = \Psi(s)D^{-1}(s)$$

with

$$\Psi(s) = \begin{bmatrix} s^2 & s & 1 & 0 & 0 & 0 \\ 0 & 0 & 0 & s^2 & s & 1 \end{bmatrix}'$$

$$D(s) = \begin{bmatrix} s^3 + 2s^2 + 6 & 3s^2 + 1 \\ 4s^2 + s + 7 & s^3 + 2s + 1 \end{bmatrix}$$

†The proof given in Sec. 7.2.1 also applies to finite fields.

By comparing

$$\tilde{D}(s) = D(s) + K\Psi(s)$$

we find that

$$K = \begin{bmatrix} -2 & 1 & -6 & -4 & 0 & 2 \\ -3 & -1 & -6 & 0 & -1 & -1 \end{bmatrix}$$

is a feedback gain matrix that will make $sI - A + BK$ have invariant polynomials $\{\psi_1(s), \psi_2(s)\}$ as specified. ∎

*7.2.3 Two Useful Theorems on State and Output Feedback

In Sec. 7.1.2 we used two results, Lemma 7.1-1 (Wonham) and Lemma 7.1-2 (Brasch and Pearson), that were helpful in effectively reducing the solution of the multivariable mode-shifting problem to that of the scalar case studied in Sec. 3.2.1. However, we noted in Sec. 7.1.2 that though these results were originally derived in a state-space context easier proofs could be given in a transfer function framework. This will be done in this section, following [14], where we shall restate and then actually prove the transfer function versions of the previously mentioned lemmas. A quick review of the material in Sec. 6.3.3 on the Smith form (see also Exercise 6.3-17) will be helpful at this point.

We begin with the new version of (the dual of) Wonham's Lemma 7.2-1.

Theorem 7.2-6.

Suppose $\{N(s), D(s)\}$ are coprime $p \times m$ and $m \times m$ matrices, with $D(s)$ nonsingular and *simple*.† Then $\{cN(s), D(s)\}$ will be coprime for almost all constant row vectors c.

Proof.   Let $U(s)$ and $V(s)$ be unimodular matrices such that

$$U(s)D(s)V(s) = \tilde{D}(s), \qquad \text{the Smith form of } D(s)$$
$$= \text{diag}\{a(s), 1, \ldots, 1\}$$

since $D(s)$ is simple. Also let the $p \times 1$ vector

$$f(s) = \text{the first column of } \tilde{N}(s) \triangleq N(s)V(s)$$

Since

$$\begin{bmatrix} \tilde{D}(s) \\ \tilde{N}(s) \end{bmatrix} = \begin{bmatrix} U(s) & 0 \\ 0 & I_p \end{bmatrix} \begin{bmatrix} D(s) \\ N(s) \end{bmatrix} V(s) \tag{30}$$

$\{\tilde{D}(s), \tilde{N}(s)\}$ will also be coprime, and so $a(s)$ and $f(s)$ can have no common factors. If $a(s)$ has $\bar{n}$ distinct roots $\{\lambda_i\}$, this means that

$$f(\lambda_i) \neq 0, \qquad i = 1, \ldots, \bar{n}$$

Now a randomly chosen $1 \times p$ vector c almost surely does not lie in any of the

---

†We may recall that in Sec. 7.1.2 we defined a polynomial matrix to be simple if it has only one nonunity invariant polynomial.

$(p - 1)$-dimensional hyperplanes orthogonal to the $f(\lambda_i)$, so that almost surely

$$cf(\lambda_i) \neq 0, \qquad i = 1, \ldots, \bar{n}$$

and therefore $\{a(s), cf(s)\}$ will be coprime polynomials. But then we see that

$$\text{rank}\begin{bmatrix} a(s) & & & \\ & 1 & & \bigcirc \\ \bigcirc & & \cdot & \\ & & & 1 \\ cf(s) & x & x & x \end{bmatrix} = m \qquad \text{for all } s \qquad (31)\dagger$$

so that $\{\tilde{D}(s), c\tilde{N}(s)\}$ are coprime and [by use of (30)] so also $\{D(s), cN(s)\}$.

We turn now to the result of Brasch and Pearson (Lemma 7.1-2) that if $\{A, B, C\}$ is minimal, then (almost any) output feedback matrix K will make $\{A + BKC\}$ *cyclic* (or nonderogatory). Such matrices can be characterized in various ways. Here we note that a matrix A is cyclic if the polynomial matrix $sI - A$ is *simple*, i.e., if it has only one nonunity invariant polynomial. This leads to the transfer function version of Lemma 7.1-2. The proof will use the Hermite form described in Sec. 6.3.1.

Theorem 7.2-7.

If $\{D(s), N(s)\}$ are coprime $m \times m$ and $p \times m$ polynomial matrices, with $D(s)$ nonsingular, then

$$D_K(s) = D(s) + KN(s)$$

will be simple for almost all constant (output-feedback) matrices K.

Proof.   We first assume that by elementary operations the matrix $D(s)$ has been transformed to lower triangular Hermite form (cf. Theorem 6.3-2), say,

$$D(s) = \begin{bmatrix} d_{11}(s) & 0 & 0 \\ d_{21}(s) & d_{22}(s) & 0 \\ d_{31}(s) & d_{32}(s) & d_{33}(s) \end{bmatrix}$$

where we have assumed $m = 3$ for simplicity. Now $d_{33}(s)$ is obviously simple; we shall show that by suitable output feedback to the *second input* we can obtain a new matrix

$$\tilde{D}_1(s) = D(s) + K_1\alpha N(s)$$

$$= \begin{bmatrix} d_{11}(s) & 0 & 0 \\ \tilde{d}_{21}(s) & \tilde{d}_{22}(s) & \delta_1(s) \\ d_{31}(s) & d_{32}(s) & d_{33}(s) \end{bmatrix} \qquad (32)\ddagger$$

†The xs denote the entries of the other columns of $c\tilde{N}(s)$, whose exact value is of no relevance here.

‡Here $K_1 = [0\ 1\ 0]'$, and α is almost any constant row vector.

where the indicated 2×2 submatrix of $\tilde{D}_1(s)$ is simple. Then by a further output feedback to the *first input* we shall obtain

$$\tilde{D}_2(s) = \tilde{D}_1(s) + K_2\beta N(s) \tag{33}†$$

as a simple matrix. (And so on, if $m > 3$.) To show how to find α, let us first write the $p \times m$ ($m = 3$) matrix $N(s)$ in terms of its columns as

$$N(s) = [N_1(s) \quad N_2(s) \quad N_3(s)]$$

Then since $\{D(s), N(s)\}$ are assumed coprime, we can see that $\{d_{33}(s), N_3(s)\}$ must also be coprime.

Since $\{d_{33}(s), N_3(s)\}$ are coprime, by Theorem 7.2-6 for almost all vectors α it must be true that $\{d_{33}(s), \alpha N_3(s)\}$ are coprime. Now with one such vector α, use output feedback into the second input, so that

$$\tilde{D}_1(s) = D(s) + K_1\alpha N(s), \qquad K_1 = [0 \quad 1 \quad 0]'$$

$$= \begin{bmatrix} d_{11}(s) & 0 & 0 \\ d_{21}(s) & \tilde{d}_{22}(s) & \alpha N_3(s) \\ d_{31}(s) & d_{32}(s) & d_{33}(s) \end{bmatrix}, \qquad \text{say.}$$

Now we claim that the 2×2 submatrix of $\tilde{D}_1(s)$ indicated by the dashed lines is simple. The reason is that the pair $\{d_{33}(s), \alpha N_3(s)\}$ is coprime and therefore this 2×2 submatrix must have rank greater than or equal to 1 for all s. By the division property of the invariant polynomials, at least one invariant polynomial must be unity, and therefore the 2×2 submatrix must be simple.

Moreover, this 2×2 submatrix and the columns $\{N_2(s), N_3(s)\}$ must be coprime because otherwise $\{\tilde{D}_1(s), N(s)\}$ would not be coprime, and since elementary operations do not affect rank, $\{D(s), N(s)\}$ would not be coprime, contrary to assumption. Now we proceed as before, feeding back $\beta N(s)$ into the first input, where β is one of the "almost all" row vectors that make

$$\left\{ \begin{bmatrix} d_{22} & \alpha N_3 \\ d_{32} & d_{33} \end{bmatrix}, \ \beta[N_2 \quad N_3] \right\} \text{ right coprime}$$

A proof by induction can be developed for a general $m \times m$ matrix $\tilde{D}(s)$, though we shall be content with the above sketch here. ∎

Example 7.2-2.

Consider a system with

$$D(s) = \begin{bmatrix} (s+1)(s+2) & 0 \\ s+1 & s+1 \end{bmatrix}, \qquad N(s) = \begin{bmatrix} s & 1 \\ 2 & 2 \end{bmatrix}$$

†Here $K_2 = [1 \ 0 \ 0]'$, and β is almost any constant row vector.

By computing the gcds of the 1×1 and 2×2 minors, the invariant factors of $D(s)$ are readily seen to be $s + 1$ and $(s + 1)(s + 2)$, so that $D(s)$ is not *simple*.

This can also be seen by noting that the rank of $D(s)$ at $s = -1$ is zero, whereas if $D(s)$ were simple the rank would be at least unity.

However, if we use constant output feedback, then we can change $D(s)$ to

$$D_K(s) = D(s) + KN(s)$$

and by Theorem 7.2-7, $D_K(s)$ should be simple for almost all K. In fact, the reader should check that $D_K(s)$ will be simple for all K such that

$$K \neq \begin{bmatrix} \dfrac{\alpha(2\alpha - 1)}{\alpha + 1} & -\dfrac{\alpha(2\alpha - 1)}{2\alpha + 1} \\ 0 & \alpha \end{bmatrix}, \quad \alpha \neq 1$$

For example, if we choose

$$K = \begin{bmatrix} 1 & 1 \\ -1 & 0.5 \end{bmatrix}$$

then

$$D_K(s) = \begin{bmatrix} s^2 + 4s + 4 & 3 \\ 0 & s + 4 \end{bmatrix}$$

which is simple [since the rank of $D_K(s)$ is at least unity for all s].

For this $D_K(s)$, we also remark that by Theorem 7.2-6 $\{[\alpha \ \beta]N(s), D_K(s)\}$ will be right coprime for almost all $\{\alpha, \beta\}$, including $\{\alpha = 1, \beta = 0\}$ and $\{\alpha = 0, \beta = 1\}$. On the other hand, the reader should check that the pair $\{[\alpha \ \beta]N(s), D(s)\}$ was not coprime for any choice of $\{\alpha, \beta\}$. ∎

Exercises

7.2-1.

Suppose t and r are integers such that $n = tm + r, 0 \leq r < m$, and that the controllability indices of $\{A, B\}$ are $k_1 = k_2 = \cdots = k_r = t + 1, k_{r+1} = \cdots = k_m = t$. Then show that *any* set of invariant polynomials $\{1, \ldots, 1, \psi_m(s), \ldots, \psi_1(s)\}$ can be achieved by state feedback around $\{A, B\}$. *Remark:* This distribution of values of the controllability indices is *generic* in that almost all pairs $\{A, B\}$ will have this distribution (for some value of r, including $r = 0$). See [9] for several related results.

7.2-2.

In using formula (22), it may happen that the $\{f_i, g_i\}$ are complex valued, which might give rise to a complex-valued K_c matrix. Show that this can be avoided by replacing the columns $\{f_i, f_i^*\}$ by $\{\text{Re } f_i, \text{Im } f_i\}$ and correspondingly for the columns $\{g_i, g_i^*\}$.

7.2-3. *Generalization of Theorem 7.2-6* [14]

Let us define the *simplicity* of any full rank (not necessarily square) polynomial matrix as the number of its nonunity invariant polynomials. Then a matrix is *simple* if and only if it has simplicity 1. Now show that if $\{N(s), D(s)\}$ are right coprime and $D(s)$ has simplicity $q \geq 1$, then $\{CN(s), D(s)\}$ will be right coprime for almost all $q \times p$ constant matrices C. *Hint:* Note that the rank defect of $D(\lambda_i)$ is less than or equal to q, and follow the proof used in Sec. 7.2.3.

7.2-4. *Generalization of Theorem 7.2-7* [14]
    Consider an MFD$\{N(s), D(s)\}$ that is not right coprime but is such that $[D'(s) \; N'(s)]'$ has simplicity $\sigma \geq 1$ (i.e., it has σ nonunity invariant polynomials; the case $\sigma = 0$ corresponds to coprimeness).
    a. Show that $\sigma \leq q$, where q is the simplicity of $D(s)$.
    b. Show that for almost all $p \times m$ output-feedback matrices K the matrix $D(s) + KN(s)$ will have simplicity $\sigma + 1$.

7.3 DESIGN OF STATE OBSERVERS

We suppose now that the states of the system

$$\dot{x}(t) = Ax(t) + Bu(t), \qquad x(0) = x_0$$
$$y(t) = Cx(t)$$

are not directly measurable. In view of the detailed discussions in the scalar case (Sec. 4.1) we proceed directly to form the equations of the asymptotic observer,

$$\dot{\hat{x}}(t) = A\hat{x}(t) + Bu(t) + L[y(t) - C\hat{x}(t)], \qquad \hat{x}(0) = \hat{x}_0 \qquad (1)$$

where $\hat{x}_0$ can be chosen arbitrarily, e.g., $\hat{x}_0 = 0$. The $n \times p$ matrix L is to be chosen so that the error,

$$\tilde{x}(t) = x(t) - \hat{x}(t)$$

which obeys the equation

$$\dot{\tilde{x}}(t) = (A - LC)\tilde{x}(t), \qquad \tilde{x}(0) = x_0 - \hat{x}_0 \qquad (2)$$

decays to zero at any desired rate. Therefore the question is whether we can find L such that

$$\det (sI - A + LC) = \text{an arbitrary monic } n\text{th-degree polynomial}$$

Since

$$\det (sI - A + LC) = \det (sI - A' + C'L')$$

we see that this is similar to the problem as solved in Sec. 7.1, with $\{A, B\}$ instead of $\{A', C'\}$. Therefore, provided $\{A', C'\}$ is controllable, or equivalently,

$$\{C, A\} \text{ is observable}$$

we can find L by one of the methods discussed in Sec. 7.1. We shall not pursue the details here.

Combined Observer-Controller Systems ([15]–[17])

The observer can be combined with the controller as (cf. Fig. 4.2-1) leading to a system with state equations

$$\begin{bmatrix} \dot{x}(t) \\ \dot{\hat{x}}(t) \end{bmatrix} = \begin{bmatrix} A & -BK \\ LC & A - BK - LC \end{bmatrix} \begin{bmatrix} x(t) \\ \hat{x}(t) \end{bmatrix} + \begin{bmatrix} BG \\ BG \end{bmatrix} u(t) \tag{3a}$$

$$y(t) = [C \quad 0][x'(t) \quad \hat{x}'(t)]' \tag{3b}$$

The characteristic polynomial of this system is easily seen to be

$$a_{o-c}(s) = \det(sI - A + BK)\det(sI - A + LC) \tag{4}$$

Therefore as in the scalar case the modes of the combined system are completely under our control, and stability may be guaranteed. Moreover, the controller modes can be set independently of the observer modes, and K and L can be separately computed.

As in the scalar case, we can show that the transfer function of the overall system is

$$H_{o-c}(s) = C(sI - A + BK)^{-1}BG \tag{5a}$$

where G is the input matrix used in the feedback law

$$u(t) \longrightarrow Gv(t) - K\hat{x}(t) \tag{5b}$$

Therefore, as shown in the scalar case by Example 4.2-1, except for greater initial transients, the overall system will behave like one with pure state feedback.

The observer, of course, introduces the need for additional states, but since the observer may often be implemented electronically no matter what form the given physical plant has (electronic or mechanical or hydraulic or hybrid), this fact may not be very burdensome. Nevertheless, there is at least some theoretical insight to be gained in trying to understand the minimal number of additional states that is required, and we shall study two aspects of this problem here.

Reduced-Order Observers ([16] and [2])

In the scalar case (Sec. 4.3) we noted that the output equation

$$y(t) = cx(t)$$

gave us a known linear relation between the n state variables, implying that only $n - 1$ of them needed to be estimated. In the matrix case, the equation

$$y(t) = Cx(t), \qquad C = \text{a } p \times n \text{ matrix}$$

gives us p linear relationships between the n state variables, so that we should need to estimate only $n - p$ of them. This can be proved by a straightfoward extension of the argument used in the scalar case, which we shall leave to the reader as a useful review.

Observer for Linear Functionals of the States ([16]–[18])

A different kind of reduction in the number of observer states can be obtained when the states are to be used in conjunction with a feedback controller. Suppose we use a scheme, e.g., one of those of Sec. 7.1, in which K has rank 1, e.g., $K = qk$, where k is a row vector. In this case, it may be unnecessary to estimate all the n states separately since we only need an estimate of $kx(t)$, a particular linear combination of the states.

In fact Luenberger [16] (with a small correction by Wonham and Morse [18]) showed that an observer can always be constructed so that its poles are freely assignable and its dimension is $v - 1$, where v is the observability index of (C, A); i.e., v is the smallest integer such that the rows of $[C', A'C', \ldots,$ $(A')^{v-1}C']'$ have rank n. Another characterization is that v is the length of the longest *chain* of the form $\{c, cA, \ldots, cA^{v-1}\}$, where c is a row of C. For an arbitrary $\{C, A\}$ pair, we would expect the chains to be of roughly equal length [there is no reason one row of C, i.e., one of the p outputs $\{y_1(\cdot), \ldots, y_p(\cdot)\}$, should be favored over the others], so that we might expect

$$v \doteq [n/p]$$

where $[n/p]$ denotes the smallest integer larger than n/p. When $p = 1$, $v = n$, but v can be substantially less than n or $n - p$ when $p > 1$. Therefore taking into account the fact that only a single linear functional has to be estimated may be quite worthwhile, depending again of course on the mode of implementation and the cost.

When we have to observe more than one linear functional, the minimum number of states required for the observer is not known, though some special results are available in the literature. For example, suppose the characteristic polynomial of the observer is not left completely free but is chosen as a product of factors of degree $l_{p+1-i} - 1$, $1 \le i \le m$ (m = the number of inputs), where $\{l_1 \le l_2 \ldots \le l_p\}$ are the observability indices of $\{C, A\}$. Then it has been shown [19] that we can observe m linear functionals with no more than $\sum_1^m (l_{p+1-i} - 1)$ states in the observer. (This number may or may not be less than $n - p$.) Another (weaker) result, which we shall justify in Sec. 7.5, is that we can always manage with $m(v - 1)$ states.

The problem of determining the minimum number of observer states (when $m > 1$) seems to be a difficult analytic problem (except in special cases); however, as pointed out before, since the observer can be implemented with electronic logic (which gets cheaper almost by the day), the question of a minimal-order observer may not be especially significant.

Exercises

7.3-1.   *A Standard Form*

Consider a realization $\dot{x}(t) = Ax(t) + Bu(t)$, $y = Cx(t)$, with $\{C, A\}$ in observer canonical form (cf. Sec. 6.4.3) and with observability indices $\{v_i, i = 1, \ldots, p\}$.

a. Show that by a nonsingular transformation of $y(\cdot)$ we can transform C to

$$
C_0^\circ = \begin{bmatrix} 1 & 0 & \cdots & 0 & & & & & \\ & & & & 1 & 0 & \cdots & 0 & \\ & & & & & & & & \ddots \\ & & & & & & & & 1 & 0 & 0 \end{bmatrix}
$$

b. Show that the transformed system can be regarded as comprising p single-output, v_i-state systems of the form $\dot{x}^{(i)} = F^{(i)}x^{(i)} + G_1^{(i)}y + G_2^{(i)}u$, $y_i = [1 \ 0 \ \cdots \ 0]x^{(i)}$, $i = 1, \ldots, p$.

c. Use this fact to give an alternative proof that we need only an $(n - p)$-dimensional state observer.

7.3-2.   *Estimation of a single linear functional*

Suppose that we make the observers for the p subsystems in the representation of Exercise 7.3-1 all have the same $v - 1$ poles, where $v = \max_i v_i =$ the observability index of C, A. (This can be done by adding observable but uncontrollable states to subsystems for which $v_i < v$.) Show that this scheme allows us to estimate a linear functional $kx(t)$ with a $(v - 1)$-dimensional estimator.

7.3-3.   [15]

Consider two systems $\dot{x}(t) = Ax(t) + Bu(t)$, $y(t) = Cx(t)$ and $\dot{z}(t) = Fz(t) + Hy(t) + Gu(t)$ and suppose there exists a matrix T such that $TA - FT = HC$.

a. Show how to choose G and H so that we can write $z(t) = Tx(t) + e^{Ft}[z(0) - Tx(0)]$. *Hint:* Consider $\dot{x}(t) - T\dot{z}(t)$ and also see Exercise A.40.

b. Give conditions on $\{A, F\}$ such that the above matrix T is unique.

c. Is it necessary that $x(\cdot)$ and $z(\cdot)$ have the same number of elements?

d. Explain how this provides another approach to observer design.

7.4   A BRIEF LOOK AT THE MULTIVARIABLE QUADRATIC REGULATOR

In this section, we shall discuss the multivariable version of the quadratic regulator problem of Sec. 3.4. Our treatment will be brief because this problem has been extensively discussed in the literature (see, e.g., [20]–[22]). Our approach, however, will be somewhat novel, being based on the result established in Sec. 7.2.1 that the feedback gain matrix can be completely specified by giving not only the closed-loop eigenvalues but also an allowable set of closed-loop eigenvectors.

The quadratic regulator problem is to find a control input $u(\cdot)$ to minimize the cost functional

$$J = \int_0^\infty [x'(t)Qx(t) + u'(t)Ru(t)] \, dt \tag{1}$$

for the system

$$\dot{x}(t) = Ax(t) + Bu(t), \qquad x(0) = x_0 \tag{2}$$

where R and Q are given matrices, $R > 0$ (positive definite) and $Q \geq 0$ (positive semidefinite). We can, without loss of generality, then write

$$Q = C'C$$

and define

$$y(t) = Cx(t) \tag{3}$$

As mentioned in Sec. 3.4.2, this problem can be tackled in many ways, but here we only note that by, for example, the Euler-Lagrange technique of Sec. 3.4.1 the optimal trajectory $x(\cdot)$ and the optimal control $u(\cdot)$ can be determined via the two-point boundary-value equations [cf. (3.4-24)]

$$\begin{bmatrix} \dot{x}(t) \\ \dot{\lambda}(t) \end{bmatrix} = \mathbf{M} \begin{bmatrix} x(t) \\ \lambda(t) \end{bmatrix}, \qquad \begin{bmatrix} x(0) \\ \lambda(\infty) \end{bmatrix} = \begin{bmatrix} x_0 \\ 0 \end{bmatrix} \tag{4a}$$

where $\mathbf{M}$ is the Hamiltonian matrix

$$\mathbf{M} = \begin{bmatrix} A & -BR^{-1}B' \\ -C'C & -A' \end{bmatrix} \tag{4b}$$

and

$$u(t) = -R^{-1}B'\lambda(t) \tag{5}$$

It is interesting to represent these equations in the block diagram form of Fig. 7.4-1, which shows that the optimal input can be represented as the response of a system [the so-called *adjoint* system (cf. Sec. 9.3)] with transfer function $-H'(-s)$ placed in the output-feedback loop. However, this form is not directly implementable since we have a *noncausal* constraint $\lambda(\infty) = 0$ on the *final* (rather than initial) state of $\lambda(\cdot)$.

Nevertheless, the solution (4)–(5) is actually a state-feedback solution, which will follow by showing (as in Sec. 3.4 for the scalar case) that $\lambda(\cdot)$ and $x(\cdot)$ can be related as

$$\lambda(t) = \bar{P}x(t) \tag{6}$$

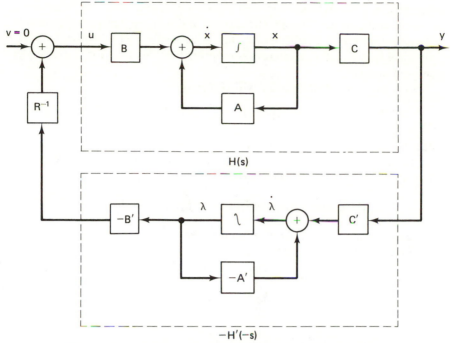

Figure 7.4-1. The optimal configuration [i.e., the optimal $u(\cdot)$] is obtained by putting $-H'(-s)$ in a feedback loop around $H(s)$. Notice the interesting dualities in the above.

where $\bar{P}$ can be found, under the assumptions

$$\{A, B\} \text{ stabilizable and } \{C, A\} \text{ detectable} \tag{7}$$

as the unique non-negative-definite solution of the algebraic Riccati equation (ARE)

$$0 = A'\bar{P} + \bar{P}A - \bar{P}BR^{-1}B'\bar{P} + C'C \tag{8}$$

From (5) and (6), we see that the optimal control is indeed of state-feedback type,

$$u(t) = -\bar{K}x(t), \qquad \bar{K} = R^{-1}B'\bar{P} \tag{9}$$

The analysis of the ARE (8) can be carried out as in Sec. 3.4.3 (see also [20]–[21]), but we shall not do so here. Instead we shall obtain an alternative formula for $\bar{K}$ without reference to the ARE [12]. [In fact we can derive the ARE formulas (8) and (9) from the new formula for K.]

For this, we recall that in Sec. 7.2.1 we showed that $\bar{K}$ is determined by (and, conversely, determines) the eigenvalues and eigenvectors of the closed-loop system, while here we find that the feedback gain matrix is fixed by $\bar{P}$ or, more basically, fixed by the (choice of Q and R in the) optimality criterion. It follows that the optimality criterion must specify a set of optimal closed-loop eigenvalues and eigenvectors (or optimal eigensystem for short) under assumptions (7).

Now, the *optimal eigenvalues* will be the stable roots of

$$\det (sI - M) = 0 \tag{10a}$$

a fact that can be proved exactly as in the scalar case (cf. Sec. 3.4.2). We note that there will be $2n$ roots in all—if μ_i is a root, then so is $-\mu_i$. We shall assume, for simplicity of analysis, that

$$\text{the stable roots, say } \{\mu_1, \ldots, \mu_n\}, \text{ are all distinct} \tag{10b}$$

To determine the *optimal eigenvectors*, let us first redraw Fig. 7.4-1 as in Fig. 7.4-2. Then we note that with no external input $[V(s) \equiv 0]$ and with X_i being the eigenvector associated with the closed-loop eigenvalue μ_i, we have

$$x(t) = X_i e^{\mu_i t}, \qquad u(t) = U_i e^{\mu_i t} \tag{11}$$

where

$$(\mu_i I - A)X_i = BU_i = -B\bar{K}X_i \tag{12}$$

or

$$(\mu_i I - A + B\bar{K})X_i = 0 \tag{13}$$

For optimality of the system, we must have [cf. (5)]

$$u(t) = -R^{-1}B'\lambda(t)$$

so that when (11) holds, we must also have

$$\lambda(t) = \Lambda_i e^{\mu_i t}, \qquad \text{with } U_i = -R^{-1}B'\Lambda_i \tag{14}$$

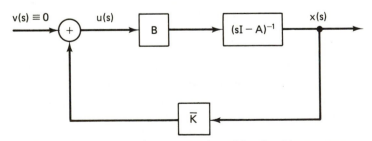

Figure 7.4-2.   Schematic representation of the closed-loop system.

By (4a) and (4b) we must also have

$$\mu_i \begin{bmatrix} X_i \\ \Lambda_i \end{bmatrix} = \begin{bmatrix} A & -BR^{-1}B' \\ -C'C & -A' \end{bmatrix} \begin{bmatrix} X_i \\ \Lambda_i \end{bmatrix}, \qquad i = 1, \ldots, n \qquad (15)$$

Therefore the optimal (closed-loop) eigenvectors $\{X_i\}$ (and also the eigenvectors $\{\Lambda_i\}$ of the closed-loop adjoint system) can be obtained from the eigenvectors of the Hamiltonian matrix. We saw earlier that the optimal closed-loop eigenvalues were the stable eigenvalues of the same matrix.

Now assumption (10b) that the closed-loop eigenvalues are all distinct will mean that the corresponding closed-loop eigenvectors $\{X_i, i = 1, \ldots, n\}$ will be linearly independent. And it follows from this that the optimal feedback gain $\bar{K}$ can be calculated from the relations

$$-R^{-1}B'\Lambda_i = U_i = -\bar{K}X_i, \qquad i = 1, \ldots, n \qquad (16)$$

as

$$\bar{K} = -R^{-1}B'[\Lambda_1, \ldots, \Lambda_n][X_1, \ldots, X_n]^{-1} \qquad (17)$$

(The gain matrix $\bar{K}$ can also be found when the $\{\mu_i\}$ are not distinct, but we shall have to introduce *generalized* eigenvectors, etc., as in the Jordan form; this case is examined in [12].)

The result (17) is usually derived via a study of the Algebraic Riccati Equation (9) (see e.g., Sec. 3.4.3 and Exercises 3.4-9, 10). The present analysis is not only more elementary, but, as we shall see, it can be easily adapted to problems where the given data are not in state-space form. Thus suppose that we are given

$$H(s) \triangleq C(sI - A)^{-1}B \qquad (18)$$

as a matrix of rational fractions. For simplicity (see [12]) let us also assume that the

$$\{\pm \mu_i\} \text{ are not eigenvalues of A} \qquad (19)$$

The $\{\pm \mu_i\}$ are the closed-loop eigenvalues of the system of Fig. 7.4-1, and can therefore be computed as the roots of

$$\det [R + H'(-s)H(s)] = 0 \qquad (20)$$

From (14)–(15) we find that the U_i may be specified as the solutions to the equation

$$[R + H'(-\mu_i)H(\mu_i)]U_i = 0, \qquad i = 1, \ldots, n \qquad (21)$$

for left-half-plane μ_i. The corresponding eigenvectors are given by

$$X_i = (\mu_i I - A)^{-1}BU_i, \qquad i = 1, \ldots, n \qquad (22)$$

whence $\bar{K}$ can be found as

$$\bar{K} = -[U_1, \ldots, U_n][X_1, \ldots, X_n]^{-1} \tag{23}$$

A simple numerical example will be given below.
    Alternatively, suppose that we have an MFD description

$$H(s) = N(s)D^{-1}(s) \tag{24}$$

Then we shall show that the equation

$$[D'(-s)RD(s) + N'(-s)N(s)]\xi(s) = 0 \tag{25}$$

where $\xi(s)$ is the partial state associated with the MFD, will have nontrivial solutions, ξ_i say, only when $s = \mu_i$, $i = 1, \ldots, n$, and also that the feedback gain matrix for a controller-form realization of $N(s)D^{-1}(s)$ can be formed as

$$\bar{K}_c = -[D(\mu_1)\xi_1 \cdots D(\mu_n)\xi_n][\Psi(\mu_1)\xi_1 \cdots \Psi(\mu_n)\xi_n]^{-1} \tag{26}$$

where $\Psi(s)$ is the matrix of chains of powers of s, defined, e.g., by (6.4-4).
    To justify this recipe, we just substitute $H(s) = N(s)D^{-1}(s)$ into (19) and use the fact that

$$y(s) = N(s)\xi(s), \qquad D(s)\xi(s) = u(s)$$

to obtain formula (25). Then (26) follows by using the characterization (6.4-18) of the eigenvectors of a controller-form realization of $N(s)D^{-1}(s)$.

Example 7.4-1.
    Consider the controllable system

$$\dot{x} = \begin{bmatrix} 0 & -1 \\ 0 & 0 \end{bmatrix} x + \begin{bmatrix} 1 & 0 \\ 0 & 1 \end{bmatrix} u$$

and let

$$R = I, \qquad Q = \begin{bmatrix} 4 & 2 \\ 2 & 1 \end{bmatrix} = \begin{bmatrix} 2 \\ 1 \end{bmatrix} [2 \quad 1]$$

We shall use formulas (21)–(24), for which we first choose

$$C = [2 \quad 1]$$

and then calculate the transfer function as

$$H(s) = C(sI - A)^{-1}B = [2 \quad 1] \begin{bmatrix} 1/s & -1/s^2 \\ 0 & 1/s \end{bmatrix} = [2/s \quad (s-2)/s^2]$$

We now find the solutions of

$$[R + H'(-s)H(s)]U(s) = 0$$

i.e.,

$$\frac{1}{s^4}\left[\begin{bmatrix} s^4 & 0 \\ 0 & s^4 \end{bmatrix} + \begin{bmatrix} -2s \\ -s-2 \end{bmatrix}\begin{bmatrix} 2s & s-2 \end{bmatrix}\right]U(s) = 0$$

or

$$\frac{1}{s^4}\begin{bmatrix} s^4 - 4s^2 & -2s^2 + 4s \\ -2s^2 - 4s & s^4 - s^2 + 4 \end{bmatrix}U(s) = 0$$

Taking the determinant of the matrix, we find the optimal modes to be the left-half-plane roots of

$$s^4 - 5s^2 + 4 = (s^2 - 1)(s^2 - 4) = 0$$

which are just $\mu_1 = -1$ and $\mu_2 = -2$. The corresponding U_i are then found easily, say,

$$s = -1: \quad \begin{bmatrix} -3 & -6 \\ 2 & 4 \end{bmatrix}U_{-1} = 0 \Longrightarrow U_{-1} = \begin{bmatrix} 2 \\ -1 \end{bmatrix}$$

$$s = -2: \quad \begin{bmatrix} 0 & -16 \\ 0 & 16 \end{bmatrix}U_{-2} = 0 \Longrightarrow U_{-2} = \begin{bmatrix} 2 \\ 0 \end{bmatrix}$$

The eigenvectors corresponding to these U_i are determined by

$$X_i = (\mu_i I - A)^{-1} B U_i$$

so that

$$X_{-1} = \begin{bmatrix} -1 & -1 \\ 0 & -1 \end{bmatrix}\begin{bmatrix} 2 \\ -1 \end{bmatrix} = \begin{bmatrix} -1 \\ 1 \end{bmatrix}$$

and

$$X_{-2} = \begin{bmatrix} -\frac{1}{2} & -\frac{1}{4} \\ 0 & -\frac{1}{2} \end{bmatrix}\begin{bmatrix} 2 \\ 0 \end{bmatrix} = \begin{bmatrix} -1 \\ 0 \end{bmatrix}$$

Then we can calculate

$$\bar{K} = -[U_{-1} \quad U_{-2}][X_{-1} \quad X_{-2}]^{-1} = \begin{bmatrix} 2 & 0 \\ 0 & 1 \end{bmatrix}$$

As a check, note that

$$A - B\bar{K} = \begin{bmatrix} -2 & -1 \\ 0 & -1 \end{bmatrix}$$

which has eigenvalues -1 and -2 and corresponding eigenvectors $[-1 \ 1]'$ and $[-1 \ 0]'$.

If we were only interested in obtaining eigenvalues -1 and -2, then many choices of K are possible; e.g.,

$$K = \begin{bmatrix} 2 & -1 \\ 0 & 1 \end{bmatrix}, \quad \text{yielding } A - BK = \begin{bmatrix} -2 & 0 \\ 0 & -1 \end{bmatrix}$$

The eigenvectors associated with this solution are different. ∎

Asymptotic Root Loci.   In the scalar case, we studied the asymptotic behavior of the roots of the characteristic equation as the cost of control went to zero or infinity (cf. Sec. 3.4.1). A similar analysis can be carried out in the multivariable case. For example, if $H(s)$ is nonsingular, $R = rI$, and $r \to 0$, then the poles move to the left-half-plane zeros, finite *and* infinite, of $H'(-s)H(s)$. Those that move to infinity (providing quicker regulation to zero) group into Butterworth patterns of different orders and radii. The detailed derivations are somewhat complicated, and here we shall only refer to [23] and [24]. (We may mention that for singular $H(s)$ there are still details to be clarified.)

7.5 TRANSFER FUNCTION DESIGN OF COMPENSATORS

In Sec. 4.5, we showed for scalar systems how transfer function analysis could be used to design compensators without introducing any state-space notions. These methods allowed us not only to obtain results exactly equivalent to those obtained by use of an observer-controller pair but also suggested some more general compensation schemes. Here we shall develop the multivariable analogs of those results.

The Observer-Controller-Type Compensator.   In this section we shall, following Wolovich [10], extend the transfer function compensator design procedure of Sec. 4.5.3 to the multivariable case. Thus, given a controllable and observable system with the strictly proper transfer function

$$H(s) = N(s)D^{-1}(s) \tag{1}$$

with $\{N(s), D(s)\}$ left coprime, we wish to design a compensator to make the overall system have a strictly proper transfer function, say,

$$H_c(s) = N(s)P^{-1}(s) \tag{2}$$

with all modes of the system freely assignable.

From our discussions in Sec. 4.5.3 we know that what is required is to first reconstruct the partial state

$$\xi(s) = D^{-1}(s)u(s) \tag{3}$$

and then feed an appropriate polynomial (matrix) multiple of it, say $M(s)\xi(s)$, back to the input [see Fig. 7.5-1(a)].

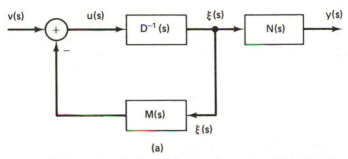

(a)

Figure 7.5-1(a). Feedback of the partial state. The choice $M(s) = P(s) - D(s)$ will give a new transfer function $N(s)^{-1}P(s)$.

We then have

$$v(s) = [D(s) + M(s)]\xi(s)$$

and hence

$$y(s) = N(s)[D(s) + M(s)]^{-1}v(s) \tag{4}$$

Thus to obtain a desired denominator $P(s)$ we choose

$$M(s) = P(s) - D(s) \tag{5}$$

To reconstruct the partial state, we again, as in Sec. 4.5.3, make use of the Bezout identity relating the right coprime matrices $N(s)$ and $D(s)$,

$$X(s)N(s) + Y(s)D(s) = I \tag{6}$$

which immediately yields the configuration of Fig. 7.5-1(b). As in the scalar case, to make this scheme realizable, we introduce a denominator polynominal (matrix), $\Delta(s)$.
Then

$$[\Delta(s)M(s)X(s)]N(s) + [\Delta(s)M(s)Y(s)]D(s) = \Delta(s)M(s) \tag{7}$$

which yields the setup of Fig. 7.5-1(c). We still need to *reduce* the numerators $\Delta(s)M(s)X(s)$ and $\Delta(s)M(s)Y(s)$ to $N_u(s)$ and $N_y(s)$, respectively, so that $\Delta^{-1}(s)N_u(s)$ and $\Delta^{-1}(s)N_y(s)$ are realizable (i.e., proper). The reduction process closely parallels the scalar case, except for obvious changes. In the matrix case, we first determine a minimal left factorization of $H(s)$, say,

$$H(s) = A^{-1}(s)B(s), \; A(s) \text{ row reduced} \tag{8a}$$

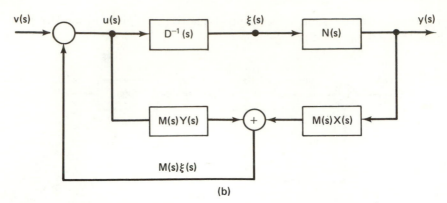

(b)

Figure 7.5-1(b).   Reconstruction of the partial state based on the Bezout identity $X(s)N(s) + Y(s)D(s) = I.$

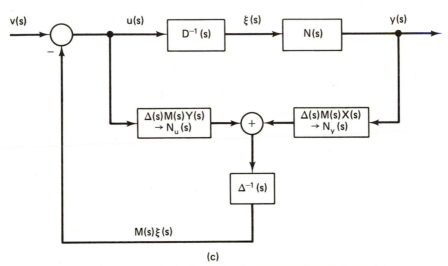

(c)

Figure 7.5-1(c).   Physically realizable reconstruction of the partial state.

so that

$$B(s)D(s) = A(s)N(s) \tag{8b}$$

Then we arrange the relations (7)–(8) in matrix form as

$$\begin{bmatrix} F(s) & G(s) \\ B(s) & -A(s) \end{bmatrix} \begin{bmatrix} D(s) \\ N(s) \end{bmatrix} = \begin{bmatrix} \Delta(s)M(s) \\ 0 \end{bmatrix} \tag{9}$$

where $F(s) = \Delta(s)M(s)Y(s)$ and $G(s) = \Delta(s)M(s)X(s)$. Now elementary row operations can be used to reduce (9) to the form

$$\begin{bmatrix} N_u(s) & N_y(s) \\ B(s) & -A(s) \end{bmatrix}\begin{bmatrix} D(s) \\ N(s) \end{bmatrix} = \begin{bmatrix} \Delta(s)M(s) \\ 0 \end{bmatrix} \tag{10}$$

where $N_y(s)$ is such that

$$\text{column degrees of } N_y(s) < \text{column degrees of } A(s) \tag{11}$$

[The elementary row operations are of course just subtracting from the first row of (9), $L(s)$ times the second row, where $L(s)$ is obtained by division as $G(s) = L(s)A(s) + N_y(s)$.] The first relation in (10) can also be rewritten, using (5), as [cf. (4.5–35)]

$$[N_u(s) + \Delta(s)]D(s) + N_y(s)N(s) = \Delta(s)P(s) \tag{12}$$

It can be shown that the modes of the configuration of Fig. 7.5-1(c) are exactly the zeros of $\det \Delta(s) \det P(s)$, and we leave this as an exercise.

We still have to show how to choose $\Delta(s)$ such that $\Delta^{-1}(s)N_u(s)$ and $\Delta^{-1}(s)N_y(s)$ are realizable and $\det \Delta(s)$ is arbitrary. For this, we need bounds on the row degrees of $N_u(s)$ and $N_y(s)$. Now since $A^{-1}(s)B(s)$ is a minimal left MFD and $A(s)$ is row reduced, the highest *row* degree of $A(s)$ must be ν, the observability index of the given (minimal) system (see Sec. 6.2.1). This must then also be its highest column degree, and therefore [cf. (11)] the highest column degree of $N_y(s)$ can at most be $\nu - 1$, which will be an upper bound on the *row* degrees of $N_y(s)$ as well. If we therefore choose

$$\Delta(s) = \begin{bmatrix} s^{\nu-1} + \beta_1(s) & -1 & & & \bigcirc \\ \beta_2(s) & s^{\nu-1} & \cdot & & \\ \cdot & & \cdot & \cdot & \\ \cdot & & & \cdot & -1 \\ \beta_m(s) & \bigcirc & & \cdot & s^{\nu-1} \end{bmatrix} \tag{13}$$

where the $\beta_i(s)$ all have degree less than $\nu - 1$, then by Lemma 6.3-11 $\Delta^{-1}(s)N_y(s)$ will be proper. In addition [cf. Eq. (7.2-13)] we shall have

$$\det \Delta(s) = s^{m(\nu-1)} + \beta_1(s)s^{(m-1)(\nu-1)} + \cdots + \beta_m(s) \tag{14}$$

which may be made into an arbitrary polynomial, $\beta(s)$, of degree $m(\nu - 1)$. To see under what conditions, with this choice of $\Delta(s)$, $\Delta^{-1}(s)N_u(s)$ is not improper, we note from (12) that

$$[\Delta^{-1}(s)N_y(s)][N(s)P^{-1}(s)] + [I + \Delta^{-1}(s)N_u(s)]D(s)P^{-1}(s) = I \tag{15a}$$

Now since $\Delta^{-1}(s)N_y(s)$ is proper and $N(s)P^{-1}(s)$ is strictly proper, we may ignore the first of the two terms on the left-hand side at $s = \infty$. We then see that

$$\Delta^{-1}(s)N_u(s) \text{ is not improper if and only if}$$
$$D(s)P^{-1}(s) \text{ is } exactly \ proper; \text{ i.e.,}$$
$$D(s)P^{-1}(s)|_{s=\infty} \text{ is finite and nonsingular} \tag{15b}$$

It can be shown now that the poles of the configuration of Fig. 7.5-1(c) are exactly the roots of det $\Delta(s)P(s)$, and we leave this as an exercise to the reader.

To interpret the condition (15b), let us, for now, restrict ourselves to devising a compensation scheme corresponding to the observer-controller scheme of Sec. 7.3 in which we used only (estimated) state feedback. From Sec. 7.2.1, we know that, assuming now that $D(s)$ is column reduced,

the $P(s)$ obtainable by state feedback (with possible input transformations) are restricted to have the same column degrees as $D(s)$ and to be column reduced $\tag{16}$

From Lemma 6.3-11, it follows that in this case we have $D(s)P^{-1}(s)$ and $P(s)D^{-1}(s)$ both proper, i.e., $D(s)P^{-1}(s)$ exactly proper. Hence $\Delta^{-1}(s)N_u(s)$ is proper (strictly proper if $D_{hc} = P_{hc}$), and the scheme of Fig. 7.5-1(c) is thus realizable.

We now have, therefore, a complete transfer function equivalent of the state-space observer-controller design of earlier sections. In fact, we can now also shed more light on a question left unresolved in our state-space analysis. Namely, in the last part of Sec. 7.3 we raised the question of the minimum order of an observer required to estimate not all the states but only m linear functionals of the states.

The present analysis shows that an observer can be designed using the transfer functions $\Delta^{-1}(s)N_u(s)$ and $\Delta^{-1}(s)N_y(s)$, where, by (13), $\Delta(s)$ can be chosen so that

$$\deg \det \Delta(s) = m(v - 1)$$

This will not in general be the minimal order that might be possible, but as noted in Sec. 7.3, (state-space) analysis of this problem is not easy. A special case in which the above result does happen to give the minimal order is that of single-input systems where $m = 1$; now the compensation of (13) will have order $v - 1$, which is the result obtained (by state-space methods) by Luenberger [16] and Wonham and Morse [18] (cf. the discussion at the end of Sec. 7.3 and also Exercise 7.3-2).

More General Nominal Transfer Functions. As in the scalar case (cf. Sec. 4.5.2), the frequency-domain solution suggests certain immediate extensions. For example, we may wish to make the $m(v - 1)$ modes of the compensator [the roots of det $\Delta(s)$] appear in the transfer function. This can be achieved, as in the scalar case, by modifying (5) to the following:

$$M(s) = \Delta^{-1}(s)P(s) - D(s) \tag{17}$$

which will yield an overall transfer function†

$$H_c(s) = N(s)P^{-1}(s)\Delta(s) \tag{18}$$

Here $P(s)$ is no longer restricted by (15)–(16), but the condition (15b) for realizability is now replaced quite naturally by the condition that

$$D(s)P^{-1}(s)\Delta(s) \text{ is exactly proper} \tag{19}$$

Thus if we wish to obtain a transfer function of the type (18) and if (19) is satisfied, then our compensation scheme works as before, except that the calculations now use $M(s)$ as defined in (17). (Although this $M(s)$ is rational, the quantity $\Delta(s)M(s)$ that is used in our calculations [see (7)–(12)] is still polynomial.) The modes of the resulting system are the roots of det $P(s)$.

The condition (19) is the analog of the constraint in the scalar case (cf. Sec. 4.5.2) that the above compensation schemes cannot change the *relative order* of the original transfer function. To relax this constraint, we can use a somewhat more general compensation scheme, with both state feedback and some additional series compensation of the input. Wolovich ([25] and [26]) has made some studies along these lines, though certain points remain unsettled in the analysis.

MFD techniques have also been successfully applied to the multivariable versions of the basic control system problems of output (and state) regulation with disturbance rejection. The general problem here can be described via the block diagram of Fig. 7.5-2. The inputs to the system G are control inputs, $u(\cdot)$, and disturbance (or noise) inputs, $e(\cdot)$. The outputs are the variables to be controlled, $y(\cdot)$, and the variables that can be measured, $z(\cdot)$. The measured variables may be the states of a realization of G, and of course, the

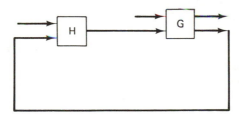

Figure 7.5-2. Block diagram of a general control configuration.

†The description of a transfer function by a *triple* as in (18) will be discussed at greater length in Chapter 8.

variables $y(\cdot)$ and $z(\cdot)$ may have some components in common. A general control problem is to find a proper controller H such that the closed-loop system is internally stable and

1. The input-output relation between a command input $u_c(\cdot)$ and the output $y(\cdot)$ should satisfy certain (so-called *servo*) requirements, e.g., that $y(\cdot)$ should be equal to $u_c(\cdot)$.
2. The input-output relation between the disturbance $e(\cdot)$ and the output $y(\cdot)$ should satisfy certain (so-called *regulator*) requirements, e.g., that step changes in $e(\cdot)$ should give no steady-state error in $y(\cdot)$ or that unstable components of $e(\cdot)$ should not appear in $y(\cdot)$.
3. The above relationships should be *robust* or insensitive to small changes in the parameters of G.

There have been many papers on these problems, using a variety of methods. Here we shall only mention the papers of Bengtsson [27], and Cheng and Pearson [28], which give an MFD analysis motivated by a certain geometric state-space approach, as developed chiefly by Wonham and his colleagues (see, e.g., [29] to [31]). Some concepts from this geometric theory will be encountered in the next section. Other state-space based approaches have been pursued by several others (see, e.g., [32] and the surveys [33] and [34]). A more complete and elegant MFD solution of the servo and regulator problems has been given by Pernebo [35], using techniques and results from polynomial and rational matrix theory as developed in Chapter 6. However we shall refrain from pursuing this topic further here, since apart from space limitations our concern is more with introducing some of the basic concepts of system structure than with studying specific control problems. We mention in particular our almost total neglect of multivariable stability analyses, referring only to the reprint volume [36] for reviews of the methods of Mac-Farlane (characteristic loci), Rosenbrock (inverse Nyquist), Mayne (sequentical return difference), and others, which have to be considered alongside the quadratic regulator method briefly noted in Sec. 7.4.

Exercises

7.5-1.

Show that the modes of the configuration in Fig. 7.5-1(c) are the zeros of $|P(s)| \cdot |\Delta(S)|$.

7.5-2.

Show that if, in the scheme of Fig. 7.5-1(c), the highest-column-degree coefficient matrices of $D(s)$ and $P(s)$ are equal, $D_{hc} = P_{hc}$, then $\Delta^{-1}(s)N_u(s)$ is *strictly* proper.

7.5-3.

Construct a compensator dual to that of Fig. 7.5-1(c), beginning with a plant whose transfer function is described by $H(s) = A^{-1}(s)B(s)$. [This structure involves injecting a signal at the plant *output*, which is generally an undesirable feature, since the output is usually a high level (energy, power, voltage, etc.) signal. However, if the *plant* is actually a low-level-output *sensor* whose response characteristics we wish to modify, such a compensator structure may be useful.]

7.5-4.

Given a nonsingular column reduced matrix $D(s)$, so that $D^{-1}(s)$ is proper, show that $[D(s) + M(s)]^{-1}$ is proper if the column degrees of $M(s)$ are less than the corresponding column degrees of $D(s)$.

Show that this is not necessarily true if $D^{-1}(s)$ is proper but not column reduced. *Hint: Consider*

$$D(s) = \begin{bmatrix} s^2 & s^3 \\ 0 & s^2 \end{bmatrix}, \qquad M(s) = \begin{bmatrix} 1 & 0 \\ 0 & 1 - s^2 \end{bmatrix}$$

7.5-5.

This exercise outlines a compensation scheme that realizes the transfer function $N(s)P^{-1}(s)$ from $N(s)D^{-1}(s)$ under the constraint that $D(s)P^{-1}(s)$ is only proper, not exactly proper as in (15b) or (19). Assume $P(s)$, but not necessarily $D(s)$, is column-reduced.

1. Use the division theorem to write $P(s) = Q(s)D(s) + M(s)$, with $M(s)D^{-1}(s)$ strictly proper. Prove that $M(s)P^{-1}(s)$ is strictly proper, and that $Q^{-1}(s)$ is proper.
2. From the above derive a suitable compensation scheme that is identical in form to Fig. 7.5-1(a) except that the block $D^{-1}(s)$ is preceded by a block $Q^{-1}(s)$. Show that this may be realized even if the partial state is not directly available, by a scheme such as in Fig. 7.5-1(c).

7.5-6.   *Parametrizing Stabilizing Feedback Compensators* [*Antsaklis*]

We are given a plant

$$H(s) = B(s)A^{-1}(s)$$

and a feedback input

$$u(s) = -F(s)y(s)$$

where the feedback compensator is

$$F(s) = P^{-1}(s)Q(s)$$

1. Show that the system transfer function is given by $B(PA + QB)^{-1}P$ (with the argument s dropped for notational convenience) so that the modes of the closed-loop system are determined by the zeros of $D(s) \triangleq PA + QB$.
2. The Bezout relationships for BA^{-1} give (as in Lemma 6.3-9)

$$\begin{bmatrix} -B_1 & A_1 \\ X_1 & Y_1 \end{bmatrix}\begin{bmatrix} -Y & A \\ X & B \end{bmatrix} = \begin{bmatrix} I & 0 \\ 0 & I \end{bmatrix} = \begin{bmatrix} -Y & A \\ X & B \end{bmatrix}\begin{bmatrix} -B_1 & A_1 \\ X_1 & Y_1 \end{bmatrix}$$

Show that the whole class of stabilizing feedback compensators is obtained from

$$[P \quad Q] = [N \quad D] \begin{bmatrix} -B_1 & A_1 \\ X_1 & Y_1 \end{bmatrix}$$

where D is any nonsingular matrix that is stable (i.e., det D has its roots in the left-half-plane) and N is any polynomial matrix (appropriately dimensioned, of course).

3. Let $K(s) \triangleq D^{-1}N$, and use the above relationships to show that

$$F(s) = (X_1 - KB_1)^{-1}(Y_1 + KA_1)$$
$$= (Y + AK)(X - BK)^{-1}$$

and that

$$K(s) = (FB + A)^{-1}(FX - Y)$$
$$= (X_1F - Y_1)(B_1F + A_1)^{-1}$$

Remark: This parametrization of stabilizing compensators was first given by D. Youla, H. Jabr and J. Bongiorno (*IEEE Trans. Automat. Contr.*, **AC-21**, pp. 319–338, 1976) and then further developed by L. Cheng and J. B. Pearson [28]; the above simple MFD approach is developed from a report by P. J. Antsaklis [*Tech. Rept. 7806,* Dept. of Elec. Engg, Rice University, Houston, Texas, 1978; see also *IEEE Trans. Automat. Contr., AC-24*, Aug. 1979].

7.6   OBSERVABILITY UNDER FEEDBACK, AND INVARIANT ZEROS; {A,B}-INVARIANT AND MAXIMALLY UNOBSERVABLE SUBSPACES

For scalar systems we proved in Section 3.2 that state-feedback did not affect the controllability of a realization. On the other hand, since state-feedback can be used to cancel out zeros, the observability of a scalar realization can be affected by state-feedback (see Example 3.3.6). The modes affected by such pole-zero cancellations become unobservable and *maximal unobservability* is achieved by using state-feedback to cancel out all the zeros. The observable part of the resulting realization will clearly have no zeros and therefore its observability will be unaffected by state-feedback. We should note also that this *strong* observability of (scalar) systems without zeros is equivalent to the property that the states of any minimal realization of such a system can be determined purely from the output $y(\cdot)$ without any knowledge of $u(\cdot)$—this property is sometimes also called *perfect observability*.

In this section, we shall see that similar properties also hold for multivariable systems, but under certain extra conditions that help to further elaborate the properties of *multivariable zeros* (cf. Section 6.5.3). It will be conceptually helpful to consider discrete-time systems here, though the mathematics applies equally to continuous-time systems.

The fact that controllability is not affected by state-feedback can be proved as in the scalar case and we leave this to the reader. For observability, a longer analysis will be needed. First it will be convenient to note that one charact-erization of an unobservable realization $\{C, A\}$ is that there exists a nontrivial 'unobservable' subspace of states, say $\mathcal{V}$, such that $\mathcal{V}$ is A-invariant ($A\mathcal{V} \subset \mathcal{V}$) and $\mathcal{V} \subset \mathcal{N}(C)$, the null space of C. [Under the action of the system any $x_0 \in \mathcal{V}$ will remain in $\mathcal{V}$ and therefore in $\mathcal{N}(C)$, i.e., it will give a zero output $y(\cdot)$.] With this characterization in mind, we can describe a subspace $\mathcal{V}$ as being *unobservable with unknown inputs* if $\mathcal{V} \subset \mathcal{N}(C)$ and if given any x_0 in $\mathcal{V}$, there exists an input u_0 such that $Ax_0 + Bu_0$ remains in $\mathcal{V}$ and hence in $\mathcal{N}(C)$. This can be written more formally as

$$\mathcal{V} \text{ unobservable with unknown inputs} \iff A\mathcal{V} \subset \mathcal{B} + \mathcal{V}, \mathcal{V} \subset \mathcal{N}(C) \quad (1)$$

where $\mathcal{B}$ denotes the range space of the input matrix B. Such subspaces were called "$\{A, B\}$-invariant subspaces in $\mathcal{N}(C)$" by Wonham and Morse (see, [29] and [30]) who have made an extensive study of their properties. A basic result relates such subspaces to state-feedback.

Lemma 7.6-1.

$\mathcal{V}$ is an $\{A, B\}$-invariant subspace if and only if there exists a (nonunique) feed-back-gain matrix K such that $\mathcal{V}$ is $(A - BK)$-invariant. In other words

$$A\mathcal{V} \subset \mathcal{B} + \mathcal{V} \iff \text{there exists } K \text{ such that } (A - BK)\mathcal{V} \subset \mathcal{V} \quad (2)$$

Proof. The $\Longleftarrow$ part is obvious. For $\Longrightarrow$, let V and B be bases for the subspaces, $\mathcal{V}$ and $\mathcal{B}$. Then

$$A\mathcal{V} \subset \mathcal{B} + \mathcal{V} \Longrightarrow AV = B\tilde{K} + VL \quad (3)$$

for some matrices $\{\tilde{K}, L\}$. Now V is a matrix whose columns are independent n-vectors, and therefore it will have a left inverse matrix V^f such that

$$V^f V = I$$

Therefore we can write (3) as

$$VL = AV - B\tilde{K} = (A - BK)V, \qquad \tilde{K} = KV$$

which is the same as saying that

$$(A - BK)\mathcal{V} \subset \mathcal{V}$$

Finally we note that K is not unique, because bases can be chosen in many ways.   ∎

In words, we know that $A\mathcal{V} \subset B + \mathcal{V}$ means that for any initial state $x_0 \in \mathcal{V}$, we can find some input sequence $\{u_i, i \geq 0\}$ such that $x_{i+1} = Ax_i +$

Bu_i, $i \geq 0$, will remain in $\mathcal{V}$. Lemma 7.6.1 shows that such a sequence $\{u_i\}$ can also be obtained by state-feedback!

Next we note that if $\mathcal{V}_1$ and $\mathcal{V}_2$ are $\{A, B\}$-invariant then so is their sum $\mathcal{V}_1 + \mathcal{V}_2$. Therefore there must be a unique "maximal" $\{A, B\}$-invariant subspace† in $\mathfrak{N}(C)$, which we shall denote by $\mathcal{V}^*$. By Lemma 7.6.1, this is also the maximal unobservable subspace under state feedback.

A Standard Form to Display $\mathcal{V}^*$.   As with unobservable systems in Sections 2.4.2 and 6.2, the unobservable space $\mathcal{V}^*$ can be clearly displayed in a certain standard form.

Theorem 7.6-2.

Suppose $\{A, B, C\}$ is a minimal realization and that (nonunique) state-feedback, $u = -Kx$, has been used to make $\{A - BK, B, C\}$ maximally unobservable (i.e., with an $\{A, B\}$-invariant subspace in $\mathfrak{N}(C)$ of largest possible dimension). Then by suitable choice of input and state bases, the resulting realization can be described by equations of the form

$$\begin{bmatrix} x_{i+1}^1 \\ x_{i+1}^2 \\ \hline y_i \end{bmatrix} = \left[\begin{array}{cc|cc} A_{11} + B_1K_{11} & 0 & B_1 & 0 \\ A_{21} + B_2K_{21} & A_{22} + B_2K_{22} & 0 & B_2 \\ \hline C_1 & 0 & 0 & 0 \end{array}\right] \begin{bmatrix} x_i^1 \\ x_i^2 \\ \hline u_i \end{bmatrix} \tag{4}$$

where K_{11}, K_{21}, K_{22} are arbitrary, and the realization $\{A_{11} + B_1K_{11}, B_1, C_1\}$ is perfectly observable, i.e., x_0^1 can be determined from knowledge of

$$\{y_i, i \geq 0\}$$

irrespective of what the inputs $\{u_i\}$ are.

Relation to Zeros.   We shall defer the fairly straightforward proofs of these results in order to pursue the more interesting question of whether (as in the scalar case) the unobservable modes (i.e., the eigenvalues of $A_{22} + B_2K_{22}$) are the "zeros" of the original realization $\{A, B, C\}$.

We would expect this to be true, except that the presence of the feedback-term K_{22} should give us pause. For if $\{A_{22}, B_2\}$ is controllable, the eigenvalues of $\{A_{22} + B_2K_{22}\}$ can be arbitrarily fixed by choice of K_{22}. (This cannot happen in the scalar case because there $B_2 \equiv 0$). We would not want to term these "zeros" of $\{A, B, C\}$. The only hope is that $\{A_{22}, B_2\}$ is not completely controllable. In that case, by a suitable choice of basis, we can arrange that

$$A_{22} = \begin{bmatrix} \bar{A}_{22} & 0 \\ \bar{A}_{32} & \bar{A}_{33} \end{bmatrix}, \qquad B_2 = \begin{bmatrix} 0 \\ \bar{B}_3 \end{bmatrix} \tag{5}$$

where $\{\bar{A}_{33}, \bar{B}_3\}$ is completely controllable. If we partition K_{22} correspond-

---

†A subspace is maximal if there is no other subspace that strictly includes it.

ingly as

$$K_{22} = [\bar{K}_{22} \quad \bar{K}_{23}] \tag{6}$$

the unobservable modes can be divided into those of $\{\bar{A}_{33} + \bar{B}_3\bar{K}_{23}\}$, which may be *arbitrary*, and those of $\bar{A}_{22}$, which are *fixed*.

We shall see that $\{\bar{A}_{33}, \bar{B}_3\} \not\equiv 0$ if and only if the 'system matrix'

$$\mathbf{P}(z) = \begin{bmatrix} zI - A & B \\ C & 0 \end{bmatrix} \tag{7}$$

is of less than full column rank for *all* values of z. [This will explain the reason for the restriction imposed in the footnote for Eq. (6.5.23c).] Moreover we shall find that the modes of $\{\bar{A}_{22}\}$ will in fact be the zeros of $\{A, B, C\}$, i.e., by Section 6.5.3, they will be values of z for which $\mathbf{P}(z)$ has less than normal rank.

This discussion explains our remark in the introduction about the expected relation between zeros and observability under feedback only holding under certain conditions. We know from Sec. 6.5.3 that information about zeros can also be couched in MFD terms, and therefore we can develop transfer function characterizations of $\{A, B\}$-invariant subspaces and of the decomposition (4). We shall not pursue this here,† but shall conclude with proofs of the various results mentioned above. Our arguments generally follow those of Molinari [38] (see also [39] and [40] and the references therein).

Proof of Theorem 7.6.2. Choose bases such that $\mathcal{V}^*$, the largest $\{A, B\}$-invariant subspace in $\mathfrak{N}(C)$ has the form

$$\mathcal{V}^* = \mathfrak{R}\left\{\begin{bmatrix} 0 \\ I \end{bmatrix}\right\},$$

and partition $\{A, B, C\}$ correspondingly as

$$\begin{bmatrix} A_{11} & A_{12} \\ A_{21} & A_{22} \end{bmatrix}, \quad \begin{bmatrix} B_1 \\ B_2 \end{bmatrix}, \quad [C_1 \quad 0]$$

Thus, after application of appropriate feedback $u = -Kx$ that will introduce the unobservability corresponding to $\mathcal{V}^*$, the realization will have the form

$$A - BK = \begin{bmatrix} \tilde{A}_{11} & 0 \\ \tilde{A}_{21} & \tilde{A}_{22} \end{bmatrix}, \quad \begin{bmatrix} B_1 \\ B_2 \end{bmatrix}, \quad [C_1 \quad 0]$$

[Note that $\mathcal{V}^*$ is $(A - BK)$-invariant and in $\mathfrak{N}(C)$.]

Let $\tilde{B}_1$ be a basis for $\mathcal{B}_1 = \mathfrak{R}(B_1)$. The controllability of $\{A, B\}$ guarantees that $\tilde{B}_1$ is nontrivial. Then an input transformation will yield

$$B = \begin{bmatrix} B_1 \\ B_2 \end{bmatrix} \longrightarrow \begin{bmatrix} \tilde{B}_1 & 0 \\ \tilde{B}_{21} & \tilde{B}_2 \end{bmatrix},$$

†Refs. 37a, b do this, but in a more abstract way.

and using a suitable state-space transformation we can arrange to have

$$B \longrightarrow \begin{bmatrix} \tilde{B}_1 & 0 \\ 0 & \tilde{B}_2 \end{bmatrix}, \tag{8}$$

while leaving the structures of $(A - BK)$, and C unchanged. $\tilde{B}_1$ has full column rank by construction and $\tilde{B}_2$ will have full column rank if we assume that the original matrix B has full column rank.

Next we note that to preserve maximal unobservability, any further feedback we may use must have gain matrices of the form

$$\tilde{K} = \begin{bmatrix} \tilde{K}_{11} & 0 \\ \tilde{K}_{21} & \tilde{K}_{22} \end{bmatrix}. \tag{9}$$

Otherwise a nonzero entry would be introduced in the $(1, 2)$-block of $A - BK$. Combining these results and dropping the tilde yields the formula (4). Now the perfect observability of $\{A_{11} - B_1 K_{11}, B_1, C_1\}$ is equivalent to that of $\{A_{11}, B_1, C_1\}$. By definition, $\{A_{11}, B_1, C_1\}$ will be perfectly observable if (and only if) there is no $\{A_{11}, B_1\}$-invariant subspace in $\mathfrak{N}(C_1)$. Now if there were such a subspace, we could perform a decomposition on it of the type described in the first part of the proof above, which would mean that $\{A, B, C\}$ would exhibit a larger $\{A, B\}$-invariant subspace in $\mathfrak{N}(C)$ than $\mathcal{V}^*$. This is a contradiction.

An immediate consequence of the perfect observability, and one that will be needed shortly, is the following:

$$\begin{bmatrix} zI - A_{11} & B_1 \\ C_1 & 0 \end{bmatrix} \text{ has full column rank for all } z \tag{10}$$

To prove this, assume the contrary. Then

$$\begin{bmatrix} z_0 I - A_{11} & B_1 \\ C_1 & 0 \end{bmatrix} \begin{bmatrix} x_0 \\ u_0 \end{bmatrix} = 0, \qquad \begin{bmatrix} x_0 \\ u_0 \end{bmatrix} \neq 0. \tag{11}$$

Since B_1 has full column rank by construction, it is easy to see that we cannot have $x_0 = 0$. Now the input sequence $\{u_0, z_0 u_0, z_0^2 u_0, \ldots\}$ applied to the system with initial state x_0 can be seen to result in the nontrivial state sequence $\{x_0, z_0 x_0, z_0^2 x_0, \ldots\}$ for which the output remains zero. This contradicts the perfect observability of $\{A_{11}, B_1, C_1\}$.

The Smith Form of P(z) and Proof of Claims on Zeros. The standard form of (4) with the additional decomposition of (5) is obtained from the given system by appropriate feedback $u = -Kx$, and by changes of input- and state-space bases, all of which can be represented as real invertible operations on the system matrix $\mathbf{P}(z)$:

$$\begin{bmatrix} zI - A + BK & B \\ C & 0 \end{bmatrix} = \begin{bmatrix} zI - A & B \\ C & 0 \end{bmatrix} \begin{bmatrix} I & 0 \\ K & I \end{bmatrix}, \tag{11a}$$

$$\begin{bmatrix} zI - A & BG \\ C & 0 \end{bmatrix} = \begin{bmatrix} zI - A & B \\ C & 0 \end{bmatrix} \begin{bmatrix} I & 0 \\ 0 & G \end{bmatrix}, \tag{11b}$$

$$\begin{bmatrix} zI - T^{-1}AT & T^{-1}B \\ CT & 0 \end{bmatrix} = \begin{bmatrix} T^{-1} & 0 \\ 0 & I \end{bmatrix} \begin{bmatrix} zI - A & B \\ C & 0 \end{bmatrix} \begin{bmatrix} T & 0 \\ 0 & I \end{bmatrix} \tag{11c}$$

Since these constant operations do not modify the Smith form of $\mathbf{P}(z)$, we find

$$\mathbf{P}(z) \sim \begin{bmatrix} zI - A_{11} & & & B_1 & 0 \\ -\bar{A}_{21} & zI - \bar{A}_{22} & & 0 & 0 \\ -\bar{A}_{31} & -\bar{A}_{32} & zI - \bar{A}_{33} & 0 & \bar{B}_3 \\ C_1 & 0 & 0 & 0 & 0 \end{bmatrix} \tag{12}$$

By simple permutation, we obtain

$$\mathbf{P}(z) \sim \left[\begin{array}{cc:c:cc} zI - A_{11} & B_1 & & & \\ C_1 & 0 & \bigcirc & & \bigcirc \\ \hdashline -\bar{A}_{21} & 0 & zI - \bar{A}_{22} & & \bigcirc \\ \hdashline -\bar{A}_{31} & 0 & -\bar{A}_{32} & zI - \bar{A}_{33} & \bar{B}_3 \end{array} \right] \tag{13}$$

Now since $\{\bar{A}_{33}, \bar{B}_3\}$ is controllable, $[zI - \bar{A}_{33} \ \bar{B}_3]$ always has full row rank and

$$[zI - \bar{A}_{33} \ \ \bar{B}_3] \sim [I \ \ 0]. \tag{14}$$

Also, from (10) we see that

$$\begin{bmatrix} zI - A_{11} & B_1 \\ C_1 & 0 \end{bmatrix} \sim \begin{bmatrix} I \\ 0 \end{bmatrix} \tag{15}$$

Consequently

$$\mathbf{P}(z) \sim \left[\begin{array}{c:c:c} I & & \\ 0 & \bigcirc & \bigcirc \\ \hdashline \bigcirc & zI - \bar{A}_{22} & \bigcirc \\ \hdashline \bigcirc & \bigcirc & I \quad 0 \end{array} \right] \tag{16}$$

The earlier claims, that $\mathbf{P}(z)$ has less than full column rank for all z if and only if $\{\bar{A}_{33}, \bar{B}_3\} \neq 0$, and that the eigenvalues of $\bar{A}_{22}$ are the values of z for which $\mathbf{P}(z)$ has less than normal rank, can now be immediately verified from (16). In fact, we clearly also have the stronger result that the (finite) zero *structure* of $\mathbf{P}(z)$ is identical to that of $zI - \bar{A}_{22}$.

It may be shown that the zero structure of $\mathbf{P}(z)$ at $z = \infty$ is isomorphic to that of the left-hand matrix in (15) at $z = \infty$. This result is of interest because the infinite zero structure of $\mathbf{P}(z)$ is known to be isomorphic to that of $C(sI - A)^{-1}B$, see [41] and [42].

REFERENCES

1. W. M. WONHAM, "On Pole Assignment in Multi-Input Controllable Linear Systems," *IEEE Trans. Autom. Control*, **AC-12**, pp. 660–665, Dec. 1967.

2. B. GOPINATH, "On the Control of Linear Multiple Input-Output Systems," *Bell Syst. Tech. J.*, **50**, pp. 1063–1081, May 1971.

3. F. M. BRASCH and J. B. PEARSON, "Pole Placement Using Dynamic Compensators," *IEEE Trans. Autom. Control*, **AC-15**, pp. 34–43, Feb. 1970.

4. P. BRUNOVSKY, "A Classification of Linear Controllable Systems," *Kybernetika (Praha)*, **3**, pp. 173–187, 1970.

5. H. H. ROSENBROCK, *State-Space and Multivariable Theory*, Wiley, New York, 1970.

6. R. E. KALMAN, "Kronecker Invariants and Feedback," in *Ordinary Differential Equations* (L. Weiss, ed.), Academic Press, New York, 1972, pp. 459–471.

7. F. R. GANTMAKHER, *Theory of Matrices*, Chelsea Publishing Co., New York, 1959.

8. J. S. THORP, "The Singular Pencil of a Linear Dynamical System," *Int. J. Control*, **18**, pp. 577–596, 1973.

9. I. C. GOHBERG and L. LERER, "Factorization Indices and Kronecker Indices of Matrix Polynomials," *Integ. Equations and Operator Theory*, **2**, no. 2, 1979. See also Ref. 64 of Ch. 6.

10. W. A. WOLOVICH, "The Differential Operator Approach to Linear System Analysis and Design," *J. Franklin Inst.*, **301**, nos. 1 & 2, pp. 27–47, Jan.–Feb. 1976.

11. B. C. MOORE, "On the Flexibility Offered by State Feedback in Multivariable Systems Beyond Closed Loop Eigenvalue Assignment," *IEEE Trans. Autom. Control*, **AC-21**, pp. 689–692, 1976. See also **AC-22**, pp. 140–141, 1977.

12. G. VERGHESE and T. KAILATH, "Fixing the State-Feedback Gain by Choice of Closed-Loop Eigensystem," in *Proceedings of the 1977 IEEE Conference on Decision and Control*, New Orleans, Dec. 1977, pp. 1245–1248.

13. M. HEYMANN, "On the Input and Output Reducibility of Multivariable Linear Systems," *IEEE Trans. Autom. Control*, **AC-15**, pp. 563–569, Oct. 1970.

14. S. KUNG, personal communication, May 1976.

15. D. G. Luenberger, "Observing the State of a Linear System," *IEEE Trans. Mil. Electron.*, **MIL-8**, pp. 74–80, 1964.

16. D. G. Luenberger, "Observers for Multivariable Systems," *IEEE Trans. Autom. Control*, **AC-11**, pp. 190–199, 1966.

17. D. G. Luenberger, "An Introduction to Observers," *IEEE Trans. Autom. Control*, **AC-16**, pp. 596–602, 1971.

18. W. M. Wonham and A. S. Morse, "Feedback Invariants of Linear Multivariable Systems," *Automatica*, **8**, pp. 93–100, Jan. 1972.

19. T. E. Fortmann and D. Williamson, "Design of Low-Order Observers for Linear Feedback Control Laws," *IEEE Trans. Autom. Control*, **AC-17**, pp. 301–308, 1972.

20. B. D. O. Anderson and J. B. Moore, *Linear Optimal Control*, Prentice-Hall, Englewood Cliffs, N.J., 1971.

21. H. Kwakernaak and R. Sivan, *Linear Optimal Control Systems*, Wiley, New York, 1972.

22. E. A. Rynaski and R. F. Whitbeck, "Theory and Application of Linear Optimal Control," *Cornell Aero. Lab. Rept. 1H-1943-F1, 193*, Cornell Aero. Lab., Buffalo, N.Y. (Also Rept. AFFDL-TR-65-28, Flight Dynamics Lab. Wright-Patterson Air Force Base, Dayton, Ohio.) Oct. 1965.

23. H. Kwakernaak, "Asymptotic Root Loci of Multivariable Linear Optimal Regulators," *IEEE Trans. Autom. Control*, **AC-21**, pp. 378–382, 1976.

24. B. Kouvaritakis, "The Optimal Root Loci of Linear Multivariable Systems," *Int. J. Control*, **28**, no. 1, pp. 33–62, Jul. 1978.

25. W. A. Wolovich, "On the Synthesis of Multivariable Systems," *IEEE Trans. Autom. Control*, **AC-18**, pp. 46–50, 1973.

26. W. A. Wolovich, *Linear Multivariable Systems*, Springer-Verlag New York, Inc., 1974.

27. G. Bengtsson, "Output Regulation and Internal Models—A Frequency Domain Approach," *Automatica*, **13**, pp. 33–345, July 1977.

28. L. Cheng and J. B. Pearson, "Frequency Domain Synthesis of Multivariable Linear Regulators," *IEEE Trans. Autom. Control*, **AC-23**, pp. 3–15, February 1978.

29. G. Basile and G. Marro, "Controlled and Conditioned Invariant Subspaces in Linear System Theory," *J. Opt. Theory Appl.*, **3**, pp. 306–315, 1969. See also *Int. J. Control*, **17**, pp. 931–943, 1973.

30. W. M. Wonham, "Linear Multivariable Control: A Geometric Approach, second edition," Springer-Verlag, Berlin, 1979.

31. W. M. Wonham, "Geometric State-Space Theory in Linear Multivariable Control: A Status Report," *Automatica*, **15**, pp. 5–13, 1979.

32. E. J. Davison, "The Robust Decentralized Control of a General Servomechanism Problem," *IEEE Trans. Autom. Control*, **AC-21**, pp. 14–24, 1976.

33. C. A. Desoer and Y. T. Wang, "Linear Time-Invariant Robust Servomechanism Problem: A Self-Contained Exposition," in *Advances in Control and*

Dynamic Systems, vol. XVI, ed. by C. T. Leondes, Academic Press, New York, 1979.

34. H. G. KWATNY and K. C. KALNITSKY, "On Alternative Methodologies for the Design of Robust Linear Multivariable Regulators," *IEEE Trans. Autom. Control*, AC-23, pp. 930–933, October 1978.

35. L. PERNEBO, "Algebraic Control Theory for Linear Multivariable Systems," Ph.D. Dissertation, Dept. of Automat. Control, Lund Institute of Technology, Lund, Sweden, 1978.

36. A. G. J. MACFARLANE, ed., *Frequency-Response Methods in Automatic Control*, IEEE Reprint Book, IEEE Press, New York, 1979. See also *IEEE Trans. Autom. Control*, AC-24, pp. 250–265, April 1979.

37a. E. EMRE and M. L. J. HAUTUS, "A Polynomial Characterisation of (A,B)-invariant and Reachability Subspaces," *SIAM J. Control and Optim.,* 1980.

37b. P. A. FUHRMANN and J. C. WILLEMS, "A Study of {A, B}-Invariant Subspaces via Polynomial Models," Math. Report 218, University of the Negev, Beer Sheva, Israel, May 1979.

38. B. P. MOLINARI, "Zeros of the System Matrix," *IEEE Trans. Autom. Control*, AC-21, pp. 795–797, Oct. 1976. See also ibid, pp. 761–764.

39. B. P. MOLINARI, "Structural Invariants of Linear Multivariable Systems," *Int. J. Control*, 28, pp. 493–510, Oct. 1978.

40. N. SUDA and E. MUTSUYOSHI, Invariant zeros and input-output structure of linear time-invariant systems, *Intern. J. Control*, 28, pp. 525–535, Oct. 1979.

41. G. VERGHESE, P. VAN DOOREN and T. KAILATH, "Properties of the System Matrix of a Generalized State-Space System," *Int. J. Control*, 1979.

42. G. VERGHESE, "Infinite-Frequency Behavior in Generalized Dynamical Systems," Ph.D. Dissertation, Department of Electrical Engineering, Stanford University, Ca., December 1978.

GENERAL DIFFERENTIAL SYSTEMS AND POLYNOMIAL MATRIX DESCRIPTIONS

8

8.0 INTRODUCTION

So far in this book we have considered systems described by state-space equations

$$\dot{x}(t) = Ax(t) + Bu(t), x(t_0) = x_0 \tag{1a}$$

$$y(t) = Cx(t) \tag{1b}$$

or by matrix-fraction descriptions of the form

$$D_L(d/dt)y(t) = N_L(d/dt)u(t) \tag{2}$$

or

$$D_R(d/dt)\xi(t) = u(t) \tag{3a}$$

$$y(t) = N_R(d/dt)\xi(t) \tag{3b}$$

where $D_L(\cdot)$ and $D_R(\cdot)$ are nonsingular (polynomial) matrices.

Now in many physical problems, or in manipulation of MFDs (cf. Section 7.5), we can encounter more general equations of the form

$$P(d/dt)\xi(t) = Q(d/dt)u(t) \tag{4a}$$

$$y(t) = R(d/dt)\xi(t) + W(d/dt)u(t) \tag{4b}$$

where to ensure a unique solution we assume that $P(\cdot)$ is a nonsingular

549

(polynomial) matrix. Such descriptions will be called *polynomial matrix descriptions* (or PMDs).

In view of our extensive results in earlier chapters, it would seem that to analyze and understand the properties of PMDs, we should have convenient ways of obtaining "equivalent" state-space or matrix-fraction descriptions. However several questions arise in doing this.

First of all, we should ensure that all significant information is preserved in the transformation to state-space or matrix-fraction form. But what does this requirement mean? Preservation of the transfer function? But we know that the transfer function only gives a complete description of "minimal" (state-space) or "irreducible" (matrix-fraction) systems, and we have to introduce certain controllability and observability and coprimeness notions to study more general state-space realizations and MFDs. What are the corresponding concepts for general differential systems? Would they allow us to work directly with the coefficients $\{P, Q, R, W\}$ without prior transformation to state-space or matrix-fraction form? If so, what are good (canonical) $\{P, Q, R, W\}$ descriptions? Also, what new problems would such results allow us to solve? Or what old problems can be better solved with such tools? And so on.

Such questions were first raised by Rosenbrock in the mid-sixties (cf. the books [1]-[2] and the references therein) and through his extensive contributions, and the related work of several others, especially Wolovich [3], many powerful and elegant results have now been obtained.

However some important questions are still open and a complete treatment is not yet available. Nevertheless, in this chapter we shall, by judicious use of earlier results (especially those of Section 6.4 on realizations of MFDs), be able to reproduce the major results of Rosenbrock and others in a quite simple and direct way.

In Sec. 8.1, we introduce some basic notations and operations. Then in Sec. 8.2 the ideas of Sec. 6.4 are adapted to derive a simple state-space realization for a given PMD. This realization can be described in a compact matrix form, which in turn suggests a useful matrix transformation between PMDs. Pursuit of this notion leads to a deeper study of equivalence relations between PMDs. The notion of irreducible PMDs is introduced in Sec. 8.3 and related to the minimality of the associated state-space realization. Redundancy in PMDs is explored and related to the concepts of controllability and observability for certain associated realizations. Finally we show some applications of the concepts of system equivalence in defining the zeros and poles of general PMDs and in characterizing the properties of certain interconnected networks.

8.1 POLYNOMIAL MATRIX DESCRIPTIONS AND SYSTEM MATRICES

We shall define the quadruple introduced in (8.0.4)

$$\{P(s),\ Q(s),\ R(s),\ W(s)\} \tag{1}$$

as a *polynomial matrix description (PMD)* of a given system. Then the right and left *matrix-fraction descriptions (MFDs)* of (8.0.2-3) can be written as

$$\{D_R(s),\ I,\ N_R(s),\ 0\} \tag{2}$$

and

$$\{D_L(s),\ N_L(s),\ I,\ 0\} \tag{3}$$

respectively. Similarly, the *state-space description* will be

$$\{sI - A,\ B,\ C,\ J(s)\} \tag{4}$$

These various possible descriptions of a given system must clearly be related. In particular, of course, the (reduced) transfer functions as calculated from the different descriptions must all be the same. For the general differential system, Laplace transformation (with zero initial conditions) will give

$$\begin{aligned} P(s)\xi(s) &= Q(s)u(s) \\ y(s) &= R(s)\xi(s) + W(s)u(s) \end{aligned} \tag{5}$$

so that the transfer function will be

$$H(s) = R(s)P^{-1}(s)Q(s) + W(s) \tag{6}$$

Therefore we must have

$$\begin{aligned} R(s)P^{-1}(s)Q(s) + W(s) &= N_R(s)D_R^{-1}(s) \\ &= D_L^{-1}(s)N_L(s) = C(sI - A)^{-1}B + J(s) \end{aligned}$$

However we know by now that the transfer function does not reflect all the aspects of a given nonminimal state-space realization or a given non-irreducible MFD. Clearly we should explore this point for general PMDs. For example, what does noncontrollability of a state-space realization correspond to for PMDs?

Even more fundamentally, we have to explore the following question. Different investigators may write different sets of equations for the same system, e.g.,

$$\ddot{y}(t) + y(t) = u(t)$$

or

$$\begin{bmatrix} \dot{x}_1(t) \\ \dot{x}_2(t) \end{bmatrix} = \begin{bmatrix} 0 & 1 \\ -1 & 0 \end{bmatrix} \begin{bmatrix} x_1(t) \\ x_2(t) \end{bmatrix} + \begin{bmatrix} 0 \\ 1 \end{bmatrix} u(t)$$

In what sense can these descriptions be said to be equivalent? Only if we can transform one description to the other in some natural "information-preserving" way. Such ways are easy to determine in the above simple example, but can become much less so for more complicated systems of equations, e.g., those with mixed algebraic and differential equations, or for systems of the form

$$A\ddot{x}(t) + B\dot{x}(t) + Cx(t) = u(t)$$

In such cases, and in the more general ones encompassed by the equations (8.0.4), there may be no entirely obvious way of going to an MFD or a state-space realization, and in particular there may be differences as to the nature of the information required to be preserved.

To begin the study of these questions, it will be useful to first consider some of the basic operations that might be applied to simplify a given PMD or to transform a given PMD to some other form.

Operations on PMDs.   Given a PMD $\{P, Q, R, W\}$† corresponding to the transfer function

$$H = RP^{-1}Q + W$$

we know from the analysis in Chapter 6 that the following operations and combinations thereof will be useful.

1. *Operations on P:* Operations of the form

$$P \longrightarrow M_1 \bar{P} M_2 \tag{7}$$

where the $\{M_i\}$ are polynomial matrices, are used to make $P(s)$ row or column reduced or to convert it to some special canonical form, etc. The $\{M_i\}$ are often constant or unimodular matrices, but they need not be so and may in fact be rectangular if P and $\bar{P}$ have different dimensions.

2. *Reductions Modulo P:* To be able to apply the state-space realization methods of Sec. 6.4, it will be necessary to arrange that $P^{-1}Q$ and/or RP^{-1} be strictly proper. As explained in Sec. 6.3.2, this can be done by using the division theorem to write

$$Q = PY + \bar{Q} \tag{8a}$$

where

$$\text{the row degrees of } \bar{Q} < \text{the row degrees of } P \tag{8b}$$

†Unless necessary for clarity, we shall often not explicitly show the arguments.

Then if P is row reduced, we know that $P^{-1}\bar{Q}$ will always be strictly proper. Similarly, we can write

$$R = XP + \bar{R} \tag{9a}$$

with

the column degrees of $\bar{R}$ < the column degrees of P (9b)

Then if P is column reduced, we know that $\bar{R}P^{-1}$ will always be strictly proper.

The System Matrix.   Rosenbrock has noted that the above operations on the PMD can be conveniently represented by operations on the

$$\text{system matrix, } \mathbf{P} \triangleq \begin{bmatrix} P(s) & Q(s) \\ -R(s) & W(s) \end{bmatrix} \tag{10}$$

Thus operation (7) can be represented as

$$\begin{bmatrix} M_1 & 0 \\ 0 & I_p \end{bmatrix} \begin{bmatrix} P & Q \\ -R & W \end{bmatrix} \begin{bmatrix} M_2 & 0 \\ 0 & I_m \end{bmatrix} = \begin{bmatrix} M_1 P M_2 & M_1 Q \\ -R M_2 & W \end{bmatrix}$$

The reduction operations (8) can be represented as

$$\begin{bmatrix} P & Q \\ -R & W \end{bmatrix} \begin{bmatrix} I & -Y \\ 0 & I \end{bmatrix} = \begin{bmatrix} P & \bar{Q} \\ -R & W + RY \end{bmatrix}$$

and the reduction operations (9) as

$$\begin{bmatrix} I & 0 \\ X & I \end{bmatrix} \begin{bmatrix} P & Q \\ -R & W \end{bmatrix} = \begin{bmatrix} P & Q \\ -\bar{R} & W + XQ \end{bmatrix}$$

We can combine these operations into the form

$$\begin{bmatrix} \bar{P} & \bar{Q} \\ -\bar{R} & \bar{W} \end{bmatrix} = \begin{bmatrix} M_1 & 0 \\ X & I_p \end{bmatrix} \begin{bmatrix} P & Q \\ -R & W \end{bmatrix} \begin{bmatrix} M_2 & -Y \\ 0 & I_m \end{bmatrix} \tag{11}$$

where $\{M_1, M_2, X, Y\}$ are some polynomial matrices.

The Special Case of Unimodular M_i.   The case of unimodular matrices $\{M_i\}$ is particularly important, because then the transformation (11) can be readily seen to preserve several important features of the PMDs. For example, in this case, we can see easily that the transformation will preserve the transfer function,

$$\bar{R}\bar{P}^{-1}\bar{Q} + \bar{W} = (R - XP)M_2 M_2^{-1} P^{-1} M_1^{-1} M_1 (Q - PY)$$
$$+ W + XQ + RY - XPY$$
$$= RP^{-1}Q + W \tag{12}$$

[The reason for the minus sign on the R term in the system matrix can also be seen from the above calculations.] Because the $\{M_i\}$ are unimodular the transformation (11) implies that (note $\bar{P} = M_1 P M_2$)

$$P \text{ and } \bar{P} \text{ will have the same determinantal degree} \qquad (13)$$

and, more generally, that

$$P \text{ and } \bar{P} \text{ will have the same Smith form} \qquad (14)$$

Similarly, by noting from (11) that

$$[\bar{P} \quad \bar{Q}] = M_1 [P \quad Q] \begin{bmatrix} M_2 & -Y \\ 0 & I \end{bmatrix}$$

and using the unimodularity of M_1 and M_2, we can say that

$$[\bar{P} \quad \bar{Q}] \text{ and } [P \quad Q] \text{ have the same Smith form} \qquad (15a)$$

So also we can prove that

$$\begin{bmatrix} \bar{P} \\ \bar{R} \end{bmatrix} \text{ and } \begin{bmatrix} P \\ R \end{bmatrix} \text{ have the same Smith form} \qquad (15b)$$

and in fact that the system matrices

$$\bar{\mathbf{P}} \text{ and } \mathbf{P} \text{ have the same Smith form} \qquad (15c)$$

Moreover, under this assumption of unimodular $\{M_i\}$, we can prove readily that the transformation (11) is what is called an *equivalence transformation* or *equivalence relation*. That is, let us write

$$\mathbf{P}_1 \sim \mathbf{P}_2 \qquad (16a)$$

to denote the relation

$$\begin{bmatrix} M_1 & 0 \\ X & I \end{bmatrix} \begin{bmatrix} P_1 & Q_1 \\ -R_1 & W_1 \end{bmatrix} \begin{bmatrix} M_2 & -Y \\ 0 & I \end{bmatrix} = \begin{bmatrix} P_2 & Q_2 \\ -R_2 & W_2 \end{bmatrix} \qquad (16b)$$

where $\{X, Y\}$ are arbitrary polynomial matrices, while

$$M_1, M_2 \text{ are unimodular polynomial matrices} \qquad (16c)$$

It then is easy to check that $\sim$ obeys the following (characterizing) properties of an equivalence relation:

1. *symmetry:* $\mathbf{P}_1 \sim \mathbf{P}_2 \Longrightarrow \mathbf{P}_2 \sim \mathbf{P}_1$ (17a)

2. *reflexivity:* $\mathbf{P} \sim \mathbf{P}$ (17b)

3. *transitivity:* $\mathbf{P}_1 \sim \mathbf{P}_2$ and $\mathbf{P}_2 \sim \mathbf{P}_3 \Longrightarrow \mathbf{P}_1 \sim \mathbf{P}_3$ (17c)

The transformation (11) or (16) thus has many nice properties—*the only difficulty is that often we may not be able to choose the $\{M_i\}$ to be unimodular.* A simple case is when the $\{P_i\}$ do not have the same dimensions. However, even having equal dimensions may not be enough.

Example 8.1-1.   Do Transformations Always Exist?

Let us try to find a relation of the form (16) between the system matrices

$$\begin{bmatrix} (s+2)^2 & 1 \\ -(s+1) & 0 \end{bmatrix} \quad \text{and} \quad \begin{bmatrix} (s+2)^2 & s+1 \\ -1 & 0 \end{bmatrix}$$

both of which correspond to the transfer function

$$H(s) = (s+1)(s+2)^{-2} = (s+2)^{-2}(s+1)$$

Now if (16) holds, we must have M_1 and M_2 being nonzero constants (with product unity) and

$$M_2(s+1) - M_2(s+2)^2 X(s) = 1$$

But since the right-hand side is constant, all the coefficients of $X(s)$ (which multiply s^2) must be zero, so $X(s) \equiv 0$ and then $M_2 = 0$. This is a contradiction, so that there is no transformation of the form (16) that will relate the above system matrices.

This is disappointing, because everyone would expect some kind of equivalence relation between the two system matrices, which arise from obviously equivalent ways of writing the transfer function. ∎

We must therefore look for a more powerful equivalence relation than (16), and for this we must apparently relax the condition that the $\{M_i\}$ must be unimodular; in fact, we should also perhaps explicitly recognize that they need not be square unless the $\{P_i\}$ are artificially made to have the same dimensions. Some reflection will show that one way of doing this is to rewrite the transformation (7)

$$P \longrightarrow M_1 \bar{P} M_2$$

in the more general form

$$M_1 P \longrightarrow \bar{P} M_2 \tag{18}$$

Correspondingly, the transformation (11) or (16) should be modified to

$$\begin{bmatrix} M_1 & 0 \\ X & I_p \end{bmatrix} \begin{bmatrix} P & Q \\ -R & W \end{bmatrix} = \begin{bmatrix} \bar{P} & \bar{Q} \\ -\bar{R} & \bar{W} \end{bmatrix} \begin{bmatrix} M_2 & -Y \\ 0 & I_m \end{bmatrix} \tag{19}$$

[though, of course, M_2 and Y will not in general be the same matrices as in (16)].

This more general transformation still preserves the transfer function [cf. (12)],

$$
\begin{aligned}
\bar{R}\bar{P}^{-1}\bar{Q} + \bar{W} &= \bar{R}\bar{P}^{-1}(M_1 Q + \bar{P}Y) + W + XQ - \bar{R}Y \\
&= \bar{R}\bar{P}^{-1}M_1 Q + W + XQ \\
&= \bar{R}M_2 P^{-1}Q + W + XQ \\
&= (R - XP)P^{-1}Q + W + XQ \\
&= RP^{-1}Q + W
\end{aligned} \tag{20}
$$

But unless we put more restrictions on the $\{M_i\}$ it is not clear how we might get the other nice properties (13)–(17) of the earlier transformation. For example, it is not clear when (19) will define an equivalence relationship—in particular try to prove (17a) for it.

To get more insight, it will be useful to study the transformation of a given PMD to state-space form (19); since we know the natural equivalence relation between state-space realizations, namely similarity, we should then be in a better position to understand the restrictions on (19). We shall pursue these thoughts in Sec. 8.2 and in particular resolve the difficulty raised in Example 8.1-1.

A final remark is in order here. We have seen that the system matrix seems to be helpful in our analysis, and this will be reinforced as we proceed. It is therefore useful to note that the system matrix can be associated with the equations [cf. (5)]

$$
\begin{bmatrix} P(s) & Q(s) \\ -R(s) & W(s) \end{bmatrix} \begin{bmatrix} \xi(s) \\ -u(s) \end{bmatrix} = \begin{bmatrix} 0 \\ -y(s) \end{bmatrix} \tag{21}
$$

An application of this fact will be provided by Exercise 8.1-2. In Exercise 8.1-3 we shall present some of the relations between the system matrix and the transfer function.

Exercises

8.1-1.   *Rank of* $\mathbf{P}(s)$.

If $\{P, Q, R, W\}$ is a PMD realizing $H = RP^{-1}Q + W$, show that we can write

$$
\mathbf{P}(s) = \begin{bmatrix} P(s) & 0 \\ -R(s) & I_p \end{bmatrix} \begin{bmatrix} P^{-1}(s) & 0 \\ 0 & H(s) \end{bmatrix} \begin{bmatrix} P(s) & Q(s) \\ 0 & I_m \end{bmatrix}
$$

Therefore show that the rank of $\mathbf{P}(s)$ is the sum of the rank of $P(s)$ and that of $H(s)$.

8.1-2. *System Matrices and Transfer Functions* [1]

If $H(s) = R(s)P^{-1}(s)Q(s) + W(s)$,

a. Show that

$$H_{ij}(s) \triangleq \text{the } (i,j)\text{th element of } H(s) = \mathbf{P}\begin{Bmatrix} 1, 2, \ldots, r, r+i \\ 1, 2, \ldots, r, r+j \end{Bmatrix} \div \det P$$

where the special symbol denotes the minor of **P** formed by rows $\{1, \ldots, r, r+i\}$ and columns $\{1, \ldots, r, r+j\}$; and

b. With an obvious extension of the above notation, show that

$$H\begin{Bmatrix} i_1, \ldots, i_q \\ j_1, \ldots, j_q \end{Bmatrix} = \mathbf{P}\begin{Bmatrix} 1, \ldots, r, r+i_1, \ldots, r+i_q \\ 1, \ldots, r, r+j_1, \ldots, r+j_q \end{Bmatrix} \div \det P$$

c. If $H(s)$ is nonsingular, find a relation between $\det H(s)$ and $\det P(s)$ (also note Exercise 8.1-1).

8.1-3. *Inverse Systems* [1]

If a PMD $\{P, Q, R, W\}$ realizes an invertible $m \times m$ transfer function $H(s)$, show that a system matrix for $H^{-1}(s)$ is

$$\mathbf{P}_I = \begin{bmatrix} P & Q & 0 \\ -R & W & -I_m \\ 0 & I_m & 0 \end{bmatrix}$$

Hint: Note that $\mathbf{P}_I[\zeta'(s)\ -u'(s)\ -y'(s)]' = [0\ 0\ -u'(s)]'$ and compare with (8.1-21).

Show that the following matrix can also characterize $H^{-1}(s)$,

$$\mathbf{P}_M = \begin{bmatrix} I & R & -W \\ 0 & P & Q \\ 0 & 0 & I \end{bmatrix}$$

Try to find interpretations for $\mathbf{P}_I^{-1}$ and $\mathbf{P}_M^{-1}$.

8.2 STATE-SPACE REALIZATIONS OF PMDS AND SOME CONCEPTS OF SYSTEM EQUIVALENCE

We shall extend the methods used for MFDs in Sec. 6.4 to obtain state-space realizations for PMDs. This will lead us to certain natural equivalence relations between PMDs.

We begin with a PMD

$$\{P(s), Q(s), R(s), W(s)\} \tag{1}$$

with the associated transfer function

$$H(s) = R(s)P^{-1}(s)Q(s) + W(s) \tag{2}$$

where $\{P, Q, R, W\}$ are known $l \times l$, $l \times m$, $p \times l$, and $p \times m$ polynomial matrices.

From the discussion in Sec. 8.1, we know that by unimodular operations we can arrange that

$$P(s) \text{ is row reduced} \tag{3a}$$

with row degrees, say,

$$\rho_i, \qquad i = 1, \ldots, l, \qquad \sum_1^l \rho_i = n \tag{3b}$$

Then if $P^{-1}Q$ is not strictly proper, let us divide (by Theorem 6.3-15) to obtain

$$Q(s) = P(s)Y(s) + \bar{Q}(s) \tag{4a}$$

where

$$P^{-1}(s)\bar{Q}(s) \text{ will be strictly proper} \tag{4b}$$

Now we can use the method of Sec. 6.4.3 to obtain an *observer-form* realization of $P^{-1}(s)\bar{Q}(s)$ as

$$C_o(sI - A_o)^{-1} = P^{-1}(s)\Psi_L(s) \tag{5a}$$

and

$$C_o(sI - A_o)^{-1}B_o = P^{-1}(s)\bar{Q}(s) \tag{5b}$$

where

$$\Psi_L(s) = \text{block diag } \{[s^{\rho_i - 1} \quad \cdots \quad s \quad 1]', i = 1, \ldots, l\} \tag{6a}$$

$$P(s) = S(s)P_{hr} + \Psi_L(s)P_{lr}, \qquad S(s) = \text{diag } \{s^{\rho_i}\} \tag{6b}$$

$$A_o = A_o^\circ - P_{lr}P_{hr}^{-1}C_o^\circ \tag{6c}$$

$$C_o^\circ = \text{block diag } \{[1 \quad 0 \quad \cdots \quad 0], 1 \times \rho_i\} \tag{6d}$$

$$C_o = P_{hr}^{-1}C_o^\circ \tag{6e}$$

and B_o is such that $\Psi_L(s)B_o = \bar{Q}(s)$. Also, of course,

$$\text{the dimension of } A_o = \text{deg det } P(s) = n \tag{7}$$

This means that we now have achieved the result

$$\begin{aligned} H(s) &= R(s)P^{-1}(s)Q(s) + W(s) \\ &= R(s)P^{-1}(s)\bar{Q}(s) + [W(s) + R(s)Y(s)] \\ &= R(s)C_o(sI - A_o)^{-1}B_o + [W(s) + R(s)Y(s)] \end{aligned} \tag{8}$$

This is not yet in state-space form because the term $R(s)C_o(sI - A_o)^{-1}B_o$ may not be a proper rational matrix. Clearly, the thing to do is to divide $R(s)C_o$ by $sI - A_o$ to obtain, say,

$$R(s)C_o = \Lambda(s)(sI - A_o) + C \tag{9}$$

where C is a unique constant matrix and $C(sI - A_o)^{-1}$ is strictly proper.†
Then (8) can be rewritten as

$$H(s) = C(sI - A_o)^{-1}B_o + J(s) \tag{10a}$$

where

$$J(s) \triangleq W(s) + R(s)Y(s) + \Lambda(s)B_o \tag{10b}$$

In other words $\{A_o, B_o, C, J(s)\}$ is a *state-space realization* of the given PMD $\{P(s), Q(s), R(s), W(s)\}$.

Of course, the state-space realization will only be proper if $J(s)$ is a constant matrix, which will be true if and only if $H(s)$ is proper (i.e., $\lim_{s \to \infty} H(s)$ is a constant matrix). We see from (10) that it is not easy to say from the PMD whether or not $H(s)$ is proper—$W(s) = 0$ is not a necessary condition. Now, nonproper systems may arise in the course of analysis, especially in the study of inverse systems and of interconnected systems. For example, if RL circuits are regarded as interconnections of R and sL subsystems, then we have some nonproper subsystems. Therefore since the analysis is not particularly simplified by the assumption that $H(s)$ is proper, we shall study the general case here.

A Compact Description of the Realization Algorithm.   The above realization algorithm can be described in a compact way as

$$\begin{bmatrix} \Psi_L(s) & 0 \\ -\Lambda(s) & I_p \end{bmatrix}\begin{bmatrix} sI - A_o & B_o \\ -C & J(s) \end{bmatrix} = \begin{bmatrix} P(s) & Q(s) \\ -R(s) & W(s) \end{bmatrix}\begin{bmatrix} C_o & -Y(s) \\ 0 & I_m \end{bmatrix} \tag{11}$$

From (5)–(6) and Sec. 6.4.3, we also note the important properties that

$$\Psi_L(s) \text{ and } P(s) \text{ are left coprime} \tag{12a}$$

and

$$sI - A_o \text{ and } C_o \text{ are right coprime} \tag{12b}$$

(Let us briefly recall one proof of these facts. For (12a), we just have to note from the form of $\Psi_L(s)$ that $[\Psi_L(s) \ P(s)]$ contains an $l \times l$ identity matrix

---

†Moreover, by the corollary to Theorem 6.3-15, $C = R(A_o)C_o$. We note that such a procedure was used in Sec. 6.4.4 to obtain the output matrix of controllability-type forms. As shown there, C can also be expressed in terms of the Markov parameters of $R(s)P^{-1}(s)$.

and therefore has rank l for all s. A similar argument can be applied to $\{C_o, sI - A_o\}$. Or, alternatively, we proved in Sec. 6.5.1 that nth-order state-space realizations of $P^{-1}(s)\Psi_L(s)$, where $n = \deg \det P(s)$, are controllable and observable, which means [cf. (5a)] that $\{C_o, sI - A_o\}$ must be coprime, as stated in (12b).)

It will be useful to generalize (11) somewhat by noting that we can make the substitutions

$$\{A_o, B_o, C_o\} \longrightarrow \{T^{-1}A_oT, T^{-1}B_o, C_oT\} \tag{13}$$

and

$$\{P(s), Q(s), R(s)\} \longrightarrow \{V_1(s)P(s)V_2(s), V_1(s)Q(s), R(s)V_2(s)\} \tag{14}$$

to take into account similarity transformations of the state-space realization and to account for any unimodular transformations that may have been used to make $P(s)$ row reduced. These transformations will modify (11) to the more general form

$$\begin{bmatrix} M_1(s) & 0 \\ X(s) & I_p \end{bmatrix} \begin{bmatrix} sI - A & B \\ -C & J(s) \end{bmatrix} = \begin{bmatrix} P(s) & Q(s) \\ -R(s) & W(s) \end{bmatrix} \begin{bmatrix} M_2(s) & -Y(s) \\ 0 & I_m \end{bmatrix} \tag{15a}$$

where now

$$\{M_1(s), P(s)\} \text{ are left coprime} \tag{15b}$$

$$\{sI - A, M_2(s)\} \text{ are right coprime} \tag{15c}$$

We note of course that the $\{M_i\}$ in (15a) will not in general be unimodular (or even square), so that we have an example of the more general transformation (8.1-19) noted at the end of Sec. 8.1. Moreover, unlike the situation there, we now have some additional constraints (15b) and (15c), which will allow us to show that (15a) is indeed an equivalence relation.

First, however, let us note that the transformation (15) does allow us to resolve the difficulty encountered in Example 8.1-1.

Example 8.2-1.

The system matrices of Example 8.1-1 will be related by (15) if we can find polynomials $\{M_1(s), M_2(s)\}$ (not constants, as in Example 8.1-1) and $\{X(s), Y(s)\}$ obeying (15b) and (15c) and such that

$$\begin{bmatrix} M_1(s) & 0 \\ X(s) & 1 \end{bmatrix} \begin{bmatrix} (s+2)^2 & 1 \\ -(s+1) & 0 \end{bmatrix} = \begin{bmatrix} (s+2)^2 & s+1 \\ -1 & 0 \end{bmatrix} \begin{bmatrix} M_2(s) & -Y(s) \\ 0 & 1 \end{bmatrix}$$

Some simple algebra will show that the choice

$$M_1(s) = (s+1) = M_2(s), \qquad X(s) = 0 = Y(s)$$

will be satisfactory. ∎

Properties of the Transformation (15).   Returning to (15), however, we would now like to see if it is indeed an equivalence relation. The first thing to check is that the transformation is symmetric, i.e., if (15a) implies that there exist polynomial matrices $\{M_3, M_4, E, F\}$ such that

$$\begin{bmatrix} sI - A & B \\ -C & J \end{bmatrix}\begin{bmatrix} M_4 & -E \\ 0 & I \end{bmatrix} = \begin{bmatrix} M_3 & 0 \\ F & I \end{bmatrix}\begin{bmatrix} P & Q \\ -R & W \end{bmatrix} \tag{16a}$$

and

$$\{sI - A, M_3\} \text{ are right coprime} \tag{16b}$$

$$\{M_4, P\} \text{ are left coprime} \tag{16c}$$

To show this, we must clearly use the given properties (15b) and (15c) in some symmetric way. This is provided by the generalized Bezout identities obtained as in Lemma 6.4-2 (see also Lemma 6.3-9),

$$\begin{bmatrix} -M_2 & X^* \\ sI - A & Y^* \end{bmatrix}\begin{bmatrix} -\tilde{X} & \tilde{Y} \\ P & M_1 \end{bmatrix} = \begin{bmatrix} I & 0 \\ 0 & I \end{bmatrix} = \begin{bmatrix} -\tilde{X} & \tilde{Y} \\ P & M_1 \end{bmatrix}\begin{bmatrix} -M_2 & X^* \\ sI - A & Y^* \end{bmatrix} \tag{17}$$

Motivated by this, and some algebraic exploration, we can check that (16) will be satisfied by choosing

$$M_4 = \tilde{X}, \qquad M_3 = Y^*, \qquad E = \tilde{Y}B - \tilde{X}Y, \qquad F = RX^* - XY^* \tag{18}$$

It is now easy to check the other (reflexivity, transitivity) properties [cf. (8.1-17)] to show that the transformation (15) is indeed an equivalence relation.

We note also that the natural (similarity) equivalence relation between state-space realizations is also of the form (15), viz.,

$$\begin{bmatrix} T & 0 \\ 0 & I \end{bmatrix}\begin{bmatrix} sI - A_1 & B_1 \\ -C_1 & J_1 \end{bmatrix} = \begin{bmatrix} sI - A_2 & B_2 \\ -C_2 & J_2 \end{bmatrix}\begin{bmatrix} T & 0 \\ 0 & I \end{bmatrix} \tag{19}$$

for any nonsingular constant matrix T. (The coprimeness conditions are trivially met).

This suggests that two PMDs may be said to be equivalent if they can each be transformed by relations of the form (15) to the same state-space realization. This definition was first proposed by Fuhrmann ([4] and [5]), who came to it in a much more abstract way; Wolovich [3] had a similar result, though expressed in a much less compact form (cf. Exercise 8.2-8).

Fuhrmann's System Equivalence (FSE).   We shall say that

$$\begin{matrix} {}^{r_1} \\ {}_p \end{matrix}\begin{bmatrix} \overset{r_1}{P_1} & \overset{m}{Q_1} \\ -R_1 & W_1 \end{bmatrix} \overset{F}{\sim} \begin{matrix} {}^{r_2} \\ {}_p \end{matrix}\begin{bmatrix} \overset{r_2}{P_2} & \overset{m}{Q_2} \\ -R_2 & W_2 \end{bmatrix} \quad \text{or} \quad \mathbf{P_1} \overset{F}{\sim} \mathbf{P_2} \tag{20a}$$

if there exist polynomial matrices $\{M_1, X, M_2, Y\}$, *of appropriate dimensions,* such that

$$\begin{bmatrix} M_1 & 0 \\ X & I_p \end{bmatrix}\begin{bmatrix} P_1 & Q_1 \\ -R_1 & W_1 \end{bmatrix} = \begin{bmatrix} P_2 & Q_2 \\ -R_2 & W_2 \end{bmatrix}\begin{bmatrix} M_2 & -Y \\ 0 & I_m \end{bmatrix} \qquad (20b)$$

and

$$\{M_1, P_2\} \text{ are left coprime} \qquad (20c)$$

$$\{P_1, M_2\} \text{ are right coprime} \qquad (20d)$$

The symbol $\overset{F}{\sim}$ will be read as "equivalent in the Fuhrmann sense" or "Fuhrmann system equivalent."

We have already shown [cf. (8.1-20)] that any transformation of the form (20b) preserves the transfer function,

$$\mathbf{P}_1 \overset{F}{\sim} \mathbf{P}_2 \Longrightarrow R_1 P_1^{-1} Q_1 + W_1 = R_2 P_2^{-1} Q_2 + W_2 \qquad (21)$$

We should also note that as a consequence of (20c) we have the generalized Bezout identities

$$\begin{bmatrix} -\tilde{X} & \tilde{Y} \\ P_2 & M_1 \end{bmatrix}\begin{bmatrix} -M_2 & X^* \\ P_1 & Y^* \end{bmatrix} = \begin{bmatrix} I & 0 \\ 0 & I \end{bmatrix} = \begin{bmatrix} -M_2 & X^* \\ P_1 & Y^* \end{bmatrix}\begin{bmatrix} -\tilde{X} & \tilde{Y} \\ P_2 & M_1 \end{bmatrix} \qquad (22)$$

from which we can conclude that (20) implies

$$\begin{bmatrix} P_1 & Q_1 \\ -R_1 & W_1 \end{bmatrix}\begin{bmatrix} \tilde{X} & -E \\ 0 & I \end{bmatrix} = \begin{bmatrix} X^* & 0 \\ F & I \end{bmatrix}\begin{bmatrix} P_2 & Q_2 \\ -R_2 & W_2 \end{bmatrix} \qquad (23a)$$

where

$$\{P_1, X^*\} \text{ are left coprime} \qquad (23b)$$

and

$$\{\tilde{X}, P_2\} \text{ are right coprime} \qquad (23c)$$

and

$$E \triangleq \tilde{Y}Q_1 - \tilde{X}Y, \qquad F \triangleq R_2 X^* - XY^* \qquad (23d)$$

Using (20) and (22)–(23), we can prove that FSE is indeed an equivalence relation. The next result shows that FSE specializes naturally for state-space realizations.

Theorem 8.2-1. FSE for State-Space Realizations

The state-space realizations $\{A_i, B_i, C_i, J(s)\}$ are FSE if and only if they are similar.

Proof. The "only if" part is immediate from (19). The "if" part will take more work. We now assume that the realizations satisfy a relation of the form

$$\begin{bmatrix} M_1(s) & 0 \\ X(s) & I \end{bmatrix} \begin{bmatrix} sI - A_1 & B_1 \\ -C_1 & J_1(s) \end{bmatrix} = \begin{bmatrix} sI - A_2 & B_2 \\ -C_2 & J_2(s) \end{bmatrix} \begin{bmatrix} M_2(s) & -Y(s) \\ 0 & I \end{bmatrix} \quad (24)$$

Since $sI - A$ is row reduced, by the division theorem (Theorem 6.3-15), we can write

$$M_1(s) = (sI - A_2)U(s) + T, \; T = \text{a constant matrix} \quad (25)$$

But by (24) $M_1(s)(sI - A_1) = (sI - A_2)M_2(s)$, so that we can write

$$T(sI - A_1) = (sI - A_2)T_0$$

where

$$T_0 \triangleq M_2(s) - U(s)(sI - A_1).$$

Note that T_0 must be a constant matrix because $T_0(sI - A_1)^{-1} = (sI - A_2)^{-1}T$, a strictly proper matrix. But it now follows that

$$T = T_0 \quad \text{and} \quad TA_1 = A_2T$$

Next from (24), we also have

$$M_1(s)B_1 = -(sI - A_2)Y(s) + B_2$$

which by using (25) gives the relation

$$(sI - A_2)(U(s)B_1 + Y(s)) + (TB_1 - B_2) = 0$$

or

$$U(s)B_1 + Y(s) = (sI - A_2)^{-1}(TB_1 - B_2)$$

The matrix on the left hand side is polynomial, while that on the right is strictly proper, so that we may conclude that $B_2 = TB_1$. We can prove similarly that $C_1 = C_2T$. To complete the proof that we have a similarity relation, it remains only to prove that T is nonsingular. We have not yet used the fact that by the hypothesis of FSE, $M_1(s)$ and $sI - A_2$ are left coprime. Hence by (25) we can say that (use Lemma 6.3-5)

$$T \text{ and } sI - A_2 \text{ are left coprime}$$

Combining this fact with the identity

$$T(sI - A_1) = (sI - A_2)T$$

we can write, for some polynomial matrices $\{\Delta_1(s), \Delta_2(s)\}$ that

$$[sI - A_2 \quad T] \begin{bmatrix} \Delta_1(s) & T \\ \Delta_2(s) & -sI + A_1 \end{bmatrix} = [I \quad 0]$$

By elementary column operations, we can ensure that $\Delta_2(s)$ is a constant matrix, so that

$$(sI - A_2)\Delta_1(s) + (T\Delta_2 - I) = 0$$

But then,

$$\Delta_1(s) = (sI - A_2)^{-1}(T\Delta_2 - I)$$

where the right hand side is strictly proper; this identity can only hold if $T\Delta_2 = I$. Hence T is nonsingular and this completes the proof. ∎

Theorem 8.2-1 shows that FSE is *a* generalization of the natural notion of equivalence for *state-space realizations*. The natural equivalence notion for *MFD*s was described in Sec. 6.5 (Lemma 6.5-5), and we have a result similar to Theorem 8.2-1 for it.

Theorem 8.2-2.

Two right MFDs $\{N_i(s), D_i(s)\}$, $i = 1, 2$, are FSE if and only if there exists a unimodular matrix $U(s)$ such that

$$N_1(s) = N_2(s)U(s), \qquad D_1(s) = D_2(s)U(s)$$

A similar result holds for left MFDs.

Proof.   Let

$$\begin{bmatrix} M_1 & 0 \\ X & I \end{bmatrix} \begin{bmatrix} D_1 & I \\ -N_1 & 0 \end{bmatrix} = \begin{bmatrix} D_2 & I \\ -N_2 & 0 \end{bmatrix} \begin{bmatrix} M_2 & -Y \\ 0 & I \end{bmatrix}$$

Then

$$M_1 = I - D_2 Y \quad \text{and} \quad M_1 D_1 = D_2 M_2 \tag{26}$$

combining which we get

$$D_1 = D_2(M_2 + YD_1) = D_2U, \text{ say}$$

So also from

$$XD_1 - N_1 = -N_2M_2, \qquad X = N_2 Y$$

we get

$$N_1 = N_2(M_2 + YD_1) = N_2U$$

The unimodularity of $U(s)$ may now be simply proved. We use (26) and the coprimeness conditions of FSE (that $\{M_1, D_2\}$ are left coprime, and $\{D_1, M_2\}$ are right coprime) to write the Bezout identities (cf. Lemma 6.3-9)

$$\begin{bmatrix} \tilde{X} & \tilde{Y} \\ D_2 & M_1 \end{bmatrix} \begin{bmatrix} -M_2 & Y \\ D_1 & I \end{bmatrix} = \begin{bmatrix} I & 0 \\ 0 & I \end{bmatrix}$$

where all the block matrices are unimodular. Hence

$$\begin{bmatrix} -M_2 & Y \\ D_1 & I \end{bmatrix}\begin{bmatrix} I & 0 \\ -D_1 & I \end{bmatrix} = \begin{bmatrix} -(M_2 + YD_1) & Y \\ 0 & I \end{bmatrix}$$

is unimodular, and so also $U = M_2 + YD_1$. The converse part is easy. ∎

Our realization procedure and Theorems 8.2-1 and 8.2-2 show that FSE has some nice properties. However, there are ways in which FSE is not so convenient an equivalence relation as we would like. For example, as may be expected from the realization algorithm [cf. (7)], it is true that $\mathbf{P}_1 \overset{F}{\sim} \mathbf{P}_2$ implies that P_1 and P_2 have the same determinantal degree. However, a direct proof of this fact from the definition (20) is not so easy. The same goes for the consequence that, e.g., P_1 and P_2 have the same Smith form or rather, since P_1 and P_2 may have different dimensions, the same, nonunity invariant polynomials. When we recall how easily such results could be obtained for the special equivalence relation (8.1-16) described in Sec. 8.1 we see that the difficulty arises from the fact that in FSE the matrices $\{M_i\}$ are not unimodular.

A Modification of FSE [6].   Let us consider how we may ameliorate this situation. We know that

$$\{M_1, P_2\} \text{ are left coprime}$$
$$\{P_1, M_2\} \text{ are right coprime}$$

Therefore, using also the relation

$$M_1 P_1 = P_2 M_2$$

we know that there exist polynomial matrices $\{\tilde{X}, \tilde{Y}, X^*, Y^*\}$ such that [cf. (22)]

$$\underbrace{\begin{bmatrix} -\tilde{X} & \tilde{Y} \\ P_2 & M_1 \end{bmatrix}}_{\text{unimodular}}\begin{bmatrix} -M_2 & X^* \\ P_1 & Y^* \end{bmatrix} = \begin{bmatrix} I & 0 \\ 0 & I \end{bmatrix} \tag{27}$$

This suggests that we *extend* M_1 to the unimodular matrix in (27) and correspondingly modify the FSE relation (20a) to the form (P_1 is $l_1 \times l_1$ and P_2 is $l_2 \times l_2$)

$$\begin{bmatrix} -\tilde{X} & \tilde{Y} & 0 \\ P_2 & M_1 & 0 \\ 0 & X & I_1 \end{bmatrix}\begin{bmatrix} I_{l_2} & 0 & 0 \\ 0 & P_1 & Q_1 \\ 0 & -R_1 & W_1 \end{bmatrix} = \begin{bmatrix} I_{l_1} & 0 & 0 \\ 0 & P_2 & Q_2 \\ 0 & -R_2 & W_2 \end{bmatrix}\begin{bmatrix} -\tilde{X} & \tilde{Y}P_1 & \tilde{Y}Q_1 \\ I & M_2 & -Y \\ 0 & 0 & I \end{bmatrix} \tag{28}$$

Now note that

$$\begin{bmatrix} I & \tilde{X} \\ 0 & I \end{bmatrix}\begin{bmatrix} -\tilde{X} & \tilde{Y}P_1 \\ I & M_2 \end{bmatrix} = \begin{bmatrix} 0 & I \\ I & M_2 \end{bmatrix} \tag{29}$$

which shows that all the matrices in (29) are unimodular.

Therefore if $\mathbf{P}_1 \overset{F}{\sim} \mathbf{P}_2$, we have shown that we can find *expanded* matrices such that

$$\begin{bmatrix} M_{1e} & 0 \\ X_e & I_p \end{bmatrix}\begin{bmatrix} P_{1e} & Q_{1e} \\ -R_{1e} & W_1 \end{bmatrix} = \begin{bmatrix} P_{2e} & Q_{2e} \\ -R_{2e} & W_2 \end{bmatrix}\begin{bmatrix} M_{2e} & -Y_e \\ 0 & I_m \end{bmatrix} \tag{30}$$

where the subscript e is used (temporarily) to denote *expanded* matrices as in (28). The point of doing this is that since

$$M_{1e}, M_{2e} \text{ are unimodular}$$

it is easy to see that, for example,

$$\deg \det P_{1e} = \deg \det P_{2e}$$

from which it is evident [cf. (28)] that

$$\deg \det P_1 = \deg \det P_2$$

This was a property that was not obvious from the original assumption that $\mathbf{P}_1 \overset{F}{\sim} \mathbf{P}_2$. Similarly since P_{1e} and P_{2e} will clearly have the same Smith form, we can see from (28) that P_1 and P_2 will have the same nonunity invariant polynomials. So also for $[P_1 \ Q_1]$ and $[P_2 \ Q_2]$ and so on. In other words, by this means we have proved the following result.

Theorem 8.2-3.   Properties of FSE

FSE is an equivalence relation on the matrices $\{P, Q, R, W\}$ that preserves the
1. Determinantal degree of P and
2. The nonunity invariant polynomials of P, $[P \ Q]$, $[P' \ -R']'$, and **P**.

Rosenbrock's System Equivalence (RSE).   Extended matrices of the type used above in (28) were first introduced by Rosenbrock in his approach to state-space realization of PMDs. Therefore we shall define two PMDs to be *equivalent in the Rosenbrock sense* (or RSE)

$$\{P_1, Q_1, R_1, W_1\} \overset{R}{\sim} \{P_2, Q_2, R_2, W_2\} \tag{31a}$$

if for some integer

$$q \leq \max(n_1, n_2), \qquad n_i = \deg \det P_i \tag{31b}$$

there exist polynomial matrices X, Y, M_{1e}, M_{2e} (of appropriate dimensions) such that

$$
\begin{bmatrix} M_{1e} & 0 \\ \hline X & I_p \end{bmatrix} \underbrace{\begin{bmatrix} I_{q-l_1} & 0 & 0 \\ 0 & P_1 & Q_1 \\ \hline 0 & -R_1 & W_1 \end{bmatrix}}_{\mathbf{P}_1} = \begin{bmatrix} I_{q-l_2} & 0 & 0 \\ 0 & P_2 & Q_2 \\ \hline 0 & -R_2 & W_2 \end{bmatrix} \begin{bmatrix} M_{2e} & -Y \\ \hline 0 & I_m \end{bmatrix} \tag{31c}
$$

and

$$ M_{1e}, M_{2e} \text{ are unimodular} \tag{31d} $$

For simplicity, *we often shall not explicitly show the expansion in (31c)* and shall use the simpler notation

$$
\begin{bmatrix} M_1 & 0 \\ X & I \end{bmatrix} \begin{bmatrix} P_1 & Q_1 \\ -R_1 & W_1 \end{bmatrix} = \begin{bmatrix} P_2 & Q_2 \\ -R_2 & W_2 \end{bmatrix} \begin{bmatrix} M_2 & -Y \\ 0 & I \end{bmatrix} \tag{32}
$$

where it is understood that the $\{P_i, Q_i, R_i\}$ have been expanded as appropriate. We shall write

$$ \mathbf{P}_1 \overset{R}{\sim} \mathbf{P}_2 $$

where the $\{\mathbf{P}_i, i = 1, 2\}$ are the system matrices (in expanded form or not—it does not matter) to denote Rosenbrock's system equivalence [1][2].

Theorem 8.2-4.   RSE is Equivalent to FSE

$$ \mathbf{P}_1 \overset{R}{\sim} \mathbf{P}_2 \Longleftrightarrow \mathbf{P}_1 \overset{F}{\sim} \mathbf{P}_2 $$

Proof.   We essentially introduced RSE by proving that

$$ \mathbf{P}_1 \overset{F}{\sim} \mathbf{P}_2 \Longrightarrow \mathbf{P}_1 \overset{R}{\sim} \mathbf{P}_2 $$

For the converse, we can proceed as follows [6].

If $\mathbf{P}_1 \overset{R}{\sim} \mathbf{P}_2$, (31c) must hold, and by appropriate partitioning we can rewrite it as (watch the subscripts)

$$
\begin{array}{c} \\ {\scriptstyle q-l_2} \\ {\scriptstyle l_2} \\ {\scriptstyle p} \end{array}
\begin{bmatrix} \overset{q-l_1}{M_{11}} & \overset{l_1}{M_{12}} & \overset{p}{0} \\ M_{13} & M_{14} & 0 \\ \hline X_1 & X_2 & I_p \end{bmatrix}
\begin{bmatrix} I_{q-l_1} & 0 & 0 \\ 0 & P_1 & Q_1 \\ \hline 0 & -R_1 & W_1 \end{bmatrix}
$$

$$
= \begin{bmatrix} I_{q-l_2} & 0 & 0 \\ 0 & P_2 & Q_2 \\ \hline 0 & -R_2 & W_2 \end{bmatrix}
\begin{bmatrix} \overset{q-l_2}{M_{21}} & \overset{l_2}{M_{22}} & \overset{m}{-Y_1} \\ M_{23} & M_{24} & -Y_2 \\ \hline 0 & 0 & I_m \end{bmatrix} \begin{array}{l} {\scriptstyle q-l_1} \\ {\scriptstyle l_1} \\ {\scriptstyle m} \end{array} \tag{33}
$$

From this, we see that we can write

$$
\begin{bmatrix} M_{14} & 0 \\ X_2 & I_p \end{bmatrix}\begin{bmatrix} P_1 & Q_1 \\ -R_1 & W_1 \end{bmatrix} = \begin{bmatrix} P_2 & Q_2 \\ -R_2 & W_2 \end{bmatrix}\begin{bmatrix} M_{24} & -Y_2 \\ 0 & I_m \end{bmatrix}
$$

which is of the form of Fuhrmann's system equivalence, provided we can prove that $\{M_{14}, P_2\}$ are left coprime, and $\{P_1, M_{24}\}$ are right coprime. We can do this by noting that from (33) we also have the relationship

$$
M_{13} = P_2 M_{23}, \qquad M_{12} P_1 = M_{22}
$$

Therefore we can rewrite M_1 and M_2 as

$$
M_1 = \begin{bmatrix} M_{11} & M_{12} \\ P_2 M_{23} & M_{14} \end{bmatrix}, \qquad M_2 = \begin{bmatrix} M_{21} & M_{12} P_1 \\ M_{23} & M_{24} \end{bmatrix}
$$

This shows that P_2 and M_{14} must be left coprime, because otherwise M_1 would have a nonunimodular factor. Similarly, the unimodularity of M_2 implies that P_1 and M_{24} must be right coprime. This completes the proof that RSE $\Rightarrow$ FSE and hence of the theorem. ∎

While mathematically RSE and FSE are thus completely equivalent, RSE is generally more useful for simplifying a given PMD. This follows from the fact that because of the unimodularity of M_2, we can write [cf. (32)]

$$
\mathbf{P}_1 \overset{R}{\sim} \mathbf{P}_2 \Longleftrightarrow \begin{bmatrix} M_1 & 0 \\ X & I \end{bmatrix}\begin{bmatrix} P_1 & Q_1 \\ -R_1 & W_1 \end{bmatrix}\begin{bmatrix} M_2 & -Y \\ 0 & I \end{bmatrix} = \begin{bmatrix} P_2 & Q_2 \\ -R_2 & W_2 \end{bmatrix} \tag{34}
$$

for some possibly expanded $\{P_i, Q_i, R_i, W_i\}$ and unimodular $\{M_i\}$. Therefore, for example, we can apply elementary row and column operations to simplify $\mathbf{P}_1$—some applications will be given in Sec. 8.3.3. FSE is not so useful in this regard because the allowed transformations depend on both $\mathbf{P}_1$ and $\mathbf{P}_2$—but it is often simpler for proving things because no expansion is required.

Relations Between the Partial States.   So far we have discussed equivalence entirely in terms of the coefficient matrices without reference to the physical variables $\{\xi, y, u\}$. To bring this in, note that if $\mathbf{P}_1 \overset{F}{\sim} \mathbf{P}_2$, by postmultiplying both sides of (20b) by $[\xi'\ -u']'$,

$$
\begin{bmatrix} M_1 & 0 \\ X & I \end{bmatrix}\begin{bmatrix} P_1 & Q_1 \\ -R_1 & W_1 \end{bmatrix}\begin{bmatrix} \xi_1 \\ -u \end{bmatrix} = \begin{bmatrix} P_2 & Q_2 \\ -R_2 & W_2 \end{bmatrix}\begin{bmatrix} M_2 & -Y \\ 0 & I \end{bmatrix}\begin{bmatrix} \xi_1 \\ -u \end{bmatrix} \tag{35a}
$$

and using the defining equations [cf. (8.1-21)], we obtain

$$
\begin{bmatrix} P_2 & Q_2 \\ -R_2 & W_2 \end{bmatrix}\begin{bmatrix} M_2 \xi_1 + Yu \\ -u \end{bmatrix} = \begin{bmatrix} M_1 & 0 \\ X & I \end{bmatrix}\begin{bmatrix} 0 \\ -y \end{bmatrix} = \begin{bmatrix} 0 \\ -y \end{bmatrix}
$$

$$
= \begin{bmatrix} P_2 & Q_2 \\ -R_2 & W_2 \end{bmatrix}\begin{bmatrix} \xi_2 \\ -u \end{bmatrix} \tag{35b}
$$

Therefore a possible relation between the partial states is

$$\xi_2(t) = M_2(D)\xi_1(t) + Y(D)u(t), \qquad D = \frac{d}{dt} \tag{36a}$$

By a similar argument applied to (23a), we can write

$$\xi_1(t) = \tilde{X}(D)\xi_2(t) + E(D)u(t) \tag{36b}$$

Therefore we may expect that the partial states of two equivalent (FSE or RSE) PMDs can be related by an invertible transformation.

Actually both this and its converse are true. If (36a) and (36b) hold for some polynomial matrices $\{M_2, Y, \tilde{X}, E\}$, then we can prove that $\mathbf{P}_1 \overset{F}{\sim} \mathbf{P}_2$. (see Exercises 8.2-10 and 8.2-11). These results are essentially due to Pernebo [7] to whose paper we refer for more details (see also [8]).

Extended Transfer Functions and Morf's System Equivalence [9]. Let us return to the general differential equations

$$P(D)\xi(t) = Q(D)u(t), \, D = d/dt,$$
$$y(t) = R(D)\xi(t) + W(D)u(t)$$

and take Laplace transforms to get, say,

$$P(s)\xi(s) = Q(s)u(s) + \alpha_0(s)$$
$$y(s) = R(s)\xi(s) + W(s)u(s) + \beta_0(s)$$

where $\alpha_0(s), \beta_0(s)$ are polynomial vectors in s whose coefficients are determined by the initial values of $\{\xi(\cdot), u(\cdot)\}$ and their derivatives (cf. Exercise 8.2-2).

The transfer function is obtained when initial conditions are zero. Thus

$$y(s) = H(s)u(s)\Big|_{\alpha_0(s)=0=\beta_0(s)}$$

with

$$H(s) = R(s)P^{-1}(s)Q(s) + W(s)$$

However, we could also introduce an *extended or generalized transfer function* as

$$\begin{bmatrix} y(s) \\ \xi(s) \\ u(s) \end{bmatrix} = \mathbf{H}(s)\begin{bmatrix} \beta_0(s) \\ \alpha_0(s) \\ u(s) \end{bmatrix}, \qquad \mathbf{H}(s) = \begin{bmatrix} I_p & R(s)P^{-1}(s) & H(s) \\ 0 & P^{-1}(s) & P^{-1}(s)Q(s) \\ 0 & 0 & I_m \end{bmatrix} \tag{37}$$

Note that the entries of $\mathbf{H}(s)$ have interpretations similar to those for the usual transfer function $H(s)$; e.g., we have

$$y(s) = R(s)P^{-1}(s)\alpha_0(s)\big|_{u(s)=0=\beta_0(s)}$$

An important feature of the expanded representation (37) is that, unlike $H(s)$, *the generalized transfer function is invertible.*† In fact, the inverse is

$$\mathbf{H}^{-1}(s) = \begin{bmatrix} I_p & -R(s) & -W(s) \\ 0 & P(s) & -Q(s) \\ 0 & 0 & I_m \end{bmatrix} \tag{38}$$

as can be checked by multiplication of $\mathbf{H}(s)$ and $\mathbf{H}^{-1}(s)$ or more directly by noting that we can write (cf. Exercise 8.1-3)

$$\begin{bmatrix} \beta_0(s) \\ \alpha_0(s) \\ u(s) \end{bmatrix} = \underbrace{\begin{bmatrix} I_p & -R(s) & -W(s) \\ 0 & P(s) & -Q(s) \\ 0 & 0 & I_m \end{bmatrix}}_{\mathbf{P}_M(s), \text{ say}} \begin{bmatrix} y(s) \\ \xi(s) \\ u(s) \end{bmatrix} \tag{39}$$

A striking thing is that the matrix $\mathbf{H}^{-1}(s) = \mathbf{P}_M(s)$, say, is a *polynomial matrix* [$H(s)$ is rational, of course]. Moreover, it compares favorably with the system matrix $\mathbf{P}(s)$ introduced in Sec. 8.1 via the equations [cf. (8.1-21) adapted to nonzero initial conditions]

$$\begin{bmatrix} \alpha_0(s) \\ -y(s) \end{bmatrix} = \underbrace{\begin{bmatrix} P(s) & Q(s) \\ -R(s) & W(s) \end{bmatrix}}_{\mathbf{P}(s)} \begin{bmatrix} \xi(s) \\ -u(s) \end{bmatrix} - \begin{bmatrix} 0 \\ \beta_0(s) \end{bmatrix} \tag{40}$$

Thus $\mathbf{P}_M(s)$ is invertible, unlike $\mathbf{P}(s)$; also, it is more natural to consider transformations between the *given* variables $\{\beta_0(s), \alpha_0(s), u(s)\}$ and the *response* variables $\{y(s), \xi(s), u(s)\}$ [cf. (37) and (39)] than between the mixed sets $\{\alpha_0(s), y(s)\}$ and $\{\xi(s), u(s)\}$ [cf. (40)].

To define an equivalence (MSE) between these extended system matrices $\{\mathbf{P}_{Mi}, i = 1, 2\}$, we shall write

$$\{\mathbf{P}_{M1} \overset{M}{\sim} \mathbf{P}_{M2}\} \tag{41a}$$

if there exist polynomial matrices $\{X, Y, K, L\}$ such that

$$\begin{bmatrix} K & 0 \\ X & I \end{bmatrix} \mathbf{P}_{M1} = \mathbf{P}_{M2} \begin{bmatrix} I & 0 \\ Y & L \end{bmatrix} \tag{41b}$$

where

$$\{K, \mathbf{P}_2\} \text{ are left coprime} \tag{42a}$$

$$\{\mathbf{P}_1, L\} \text{ are right coprime} \tag{42b}$$

---

†It may be noted that the function $\mathbf{H}(s)$ also arises in a scattering-theory approach to linear systems (see [10] and [11]).

With this definition, it can be shown (see [6] and Exercise 8.2-13) that

$$\mathbf{P}_{M1} \overset{M}{\sim} \mathbf{P}_{M2} \Longleftrightarrow \mathbf{P}_1 \overset{F}{\sim} \mathbf{P}_2 \tag{43}$$

[Morf's original definition [9] required that (as in RSE) $\{K, L\}$ be unimodular. With this condition, we can prove that MSE $\Longrightarrow$ FSE, but the converse is not clear.] There are several features of MSE that seem worthy of further investigation, especially the connections with scattering descriptions [10][11].

The significance of any definition, of course, resides in its consequences and applications, and so we turn to such questions in Sec. 8.3.

Exercises

8.2-1.

Find the matrices that will relate the system matrices

$$\begin{bmatrix} (s+2)^2 & -1 \\ s+1 & 0 \end{bmatrix} \quad \text{to} \quad \begin{bmatrix} (s+2)^2 & s+1 \\ -1 & 0 \end{bmatrix}$$

by (a) RSE and (b) MSE (cf. Example 8.1-1).

8.2-2.   *Initial Conditions in PMDs*

If $D = d/dt$ and $F(\cdot)$ is a polynomial, say, $F(\lambda) = F_0 \lambda^n + F_1 \lambda^{n-1} + \cdots + F_n$, the $\{F_i, i = 1, \ldots, n\}$ being constant matrices, show that the Laplace transform of $F(D)\xi(t)$ can be written as $\mathcal{L}\{F(D)\xi(t)\} = F(s)\xi(s) - \xi_0(s)$, where $\xi_0(s) = [Is^{n-1} \ \cdots \ I]\mathbf{T}(F_i)[\xi'(0) \ \cdots \ \xi'^{(n-1)}(0)]'$ and $\mathbf{T}(F_i)$ denotes a lower triangular Toeplitz matrix with $[F_0' \ \cdots \ F_{n-1}']'$ as its first column.

8.2-3.   *More on FSE*

a. Let $\mathbf{P}_k = $ block diag $\{I_k, \mathbf{P}\}$, $k = 0, 1, 2, \ldots$. Show that the $\mathbf{P}_k$ are all FSE to each other.

b. Refer to (20), and show that it implies the left coprimeness of the matrices

$$\begin{bmatrix} M_1 & 0 \\ X & I \end{bmatrix} \quad \text{and} \quad \begin{bmatrix} P_2 & Q_2 \\ -R_2 & W_2 \end{bmatrix}$$

8.2-4.

Given a system matrix

$$\left[\begin{array}{cc|c} s+1 & s^3 & 0 \\ 0 & s+1 & 1 \\ \hline -1 & 0 & 0 \end{array}\right]$$

show how to reduce it to state-space form

$$\left[\begin{array}{cc|c} s+1 & -1 & -3 \\ 0 & s+1 & 1 \\ \hline -1 & 0 & 2-s \end{array}\right]$$

by elementary row and column operations.

8.2-5.

Find a transformation of FSE to relate

$$\begin{bmatrix} (s+2)^2 & s+1 \\ -1 & 0 \end{bmatrix} \quad \text{and} \quad \begin{bmatrix} (s+2)^2 & 1 \\ -(s+1) & 0 \end{bmatrix}$$

Compare with the result in Example 8.2-1. Why is there a difference?

8.2-6.

a. If $\{P_i, Q_i, R_i, W_i, i = 1, 2\}$ are FSE, in the text we proved that deg det P_1 = deg det P_2 by relating FSE to RSE. Obtain a direct proof by using the Bezout identity (8.2-22).

b. If $\{P_1, Q_1, R_1, W_1\} \overset{F}{\sim} \{P_2, Q_2, R_2, W_2\}$ and $\{P_2, Q_2, R_2, W_2\} \overset{F}{\sim} \{P_3, Q_3, R_3, W_3\}$, show (without, as in the text, using the concept of RSE) that $\{P_1, Q_1, R_1, W_1\} \overset{F}{\sim} \{P_3, Q_3, R_3, W_3\}$.

8.2-7.   *Relating Left and Right MFDs*

Let $\{N_R, D_R\}$ and $\{D_L, N_L\}$ be irreducible right and left MFDs. Show that they can be related as

$$\begin{bmatrix} N_L & 0 \\ 0 & I \end{bmatrix}\begin{bmatrix} D_R & I \\ -N_R & 0 \end{bmatrix} = \begin{bmatrix} D_L & N_L \\ -I & 0 \end{bmatrix}\begin{bmatrix} N_R & 0 \\ 0 & I \end{bmatrix}$$

and

$$\begin{bmatrix} D_R & I \\ -N_R & 0 \end{bmatrix}\begin{bmatrix} \tilde{X} & \tilde{Y} \\ 0 & I \end{bmatrix} = \begin{bmatrix} X^* & 0 \\ -Y^* & I \end{bmatrix}\begin{bmatrix} D_L & N_L \\ -I & 0 \end{bmatrix}$$

where $\{\tilde{X}, \tilde{Y}, X^*, Y^*\}$ are polynomial matrices determined by the Bezout identity

$$\begin{bmatrix} N_L & -D_L \\ \tilde{Y} & \tilde{X} \end{bmatrix}\begin{bmatrix} X^* & D_R \\ -Y^* & N_R \end{bmatrix} = \begin{bmatrix} I & 0 \\ 0 & I \end{bmatrix}$$

8.2-8.   *Wolovich's definition of equivalence*

Wolovich [3] defines a PMD $\{P, Q, R, W\}$ to be equivalent to a state-space realization $\{A, B, C, J\}$, with the size of A being equal to deg det $P(s)$, if the following three conditions are met:

a. The partial state $\xi(\cdot)$ corresponding to a given input $u(\cdot)$ and given initial conditions on $\xi(\cdot)$ and its derivatives, $P(D)\xi(t) = Q(D)u(t)$, can be expressed as

$$\xi(t) = C_o x(t) + Y(D)u(t) \qquad (*)$$

for some constant matrix C_o and some polynomial matrix $Y(D)$, with $x(\cdot)$ being the solution to $\dot{x}(t) = Ax(t) + Bu(t)$, for the same $u(\cdot)$ and a *unique* initial state $x(0)$.

b. Conversely, if $\xi(t) = C_o x(t) + Y(D)u(t)$, $\dot{x}(t) = Ax(t) + Bu(t)$, $x(0) = x_o$, for some known $u(\cdot)$ and $x(0)$, then $\xi(\cdot)$ also satisfies $P(D)\xi(t) = Q(t)u(t)$

for the same $u(\cdot)$ and some appropriate *unique* initial conditions on $\zeta(\cdot)$ and its derivatives.

c. Inserting expression $(*)$ for $\zeta(\cdot)$ into the output equation for the PMD gives, say,

$$y(t) = R(D)[C_o x(t) + Y(D)u(t)] + W(D)u(t)$$
$$= [\Lambda(D)(DI - A) + \bar{C}]x(t) + [R(D)Y(D) + W(D)]u(t)$$
$$= \bar{C}x(t) + [\Lambda(D)B + R(D)Y(D) + W(D)]u(t)$$
$$= \bar{C}x(t) + \bar{J}(D)u(t)$$

The third condition is that we must have $\bar{C} = C, \bar{J}(D) = J(D)$. Show that this definition of equivalence is the same as FSE (or RSE).

8.2-9.   *Relations Between The Partial States* [7]

Let $P_i(D)\zeta_i(t) = Q_i(D)u(t), y(t) = R_i(D)\zeta_i(t), i = 1, 2$, be two PMDs such that for a given $u(\cdot)$ there is a one-to-one relationship between the two partial states of the form $\zeta_1(t) = F(D)\zeta_2(t) + G(D)u(t)$ for some polynomial matrices $F(\cdot)$ and $G(\cdot)$. If $\{P_2(D), F(D)\}$ are right coprime, show that there exist polynomial matrices $\{X(\cdot), Y(\cdot)\}$ such that $\zeta_2(t) = X(D)\zeta_1(t) + Y(D)u(t)$.

8.2-10.   *A Useful Lemma* [6]

Consider two proper differential systems $P(D)\zeta_1(t) = Q(D)u(t), A(D)\zeta_2(t) = B(D)u(t)$, where $P(\cdot)$ is square and nonsingular, while $A(\cdot)$ may be rectangular with more rows than columns. Let $\mathcal{A}_i, i = 1, 2$, denote the set of solutions $\{\zeta_i(\cdot)\}$ to the above systems.

a. Show that $\mathcal{A}_1 \subseteq \mathcal{A}_2$ if and only if there exists a polynomial matrix $C(\cdot)$ such that $[A \ B] = C[P \ Q]$.

b. Show that $\mathcal{A}_1 = \mathcal{A}_2$ if and only if $C(\cdot)$ is unimodular.

Remark: A weaker form of this result was first given by Blomberg [8].

8.2-11.   *FSE and The Partial States* [6][7]

Suppose the partial states of two PMDs are related by an invertible mapping $\zeta_2(t) = M_2(D)\zeta_1(t) + Y(D)u(t)$ [with inverse $\zeta_1(t) = \tilde{X}(D)\zeta_2(t) + E(D)u(t)$; cf. (8.2-36a) and (8.2-36b)]. Show that the PMDs must be FSE. *Hint:* Note that $[P_1 \ Q_1][\zeta_1' \ -u']' = 0$ and

$$\left\{ \begin{bmatrix} P_2 & Q_2 \\ -R_2 & W_2 \end{bmatrix} \begin{bmatrix} M_2 & -Y \\ 0 & I \end{bmatrix} - \begin{bmatrix} 0 & 0 \\ -R_1 & W_1 \end{bmatrix} \right\} \begin{bmatrix} \zeta_1 \\ -u \end{bmatrix} = 0$$

Then use the result of Exercise 8.2-10 to conclude that the 2×2 block matrix in the braces must have the form

$$\begin{bmatrix} M_1 \\ X \end{bmatrix} [P_1 \ Q_1]$$

8.2-12.   *Equivalence of Inverse Systems* [6]

Refer to Exercise 8.1-3 for the system matrix $\{P_I\}$ of the inverse of a PMD. Show that $\mathbf{P}_{I1} \overset{F}{\sim} \mathbf{P}_{I2} \Leftrightarrow \mathbf{P}_1 \overset{F}{\sim} \mathbf{P}_2$; i.e., the system matrices of the inverse systems will be FSE if and only if the same is true of the system matrices of the original system. Reference: H. H. Rosenbrock and A. J. Van der Weiden, *Inter. J. Control*, **25**, pp. 389–392, 1977.

8.2-13.   *MSE and FSE* [6]

Prove that MSE as defined by (41)–(42) is equivalent to FSE. *Hint:* Note that $\mathbf{P}_M$ and $\mathbf{P}_I$ as in Exercise 8.2-12 are related by certain row and column operations. (See also Exercise 8.1-3).

8.2-14.   *Realization via Nerode Equivalence*

Given an MFD, $N(s)D^{-1}(s)$, not necessarily irreducible, show how to extend the Nerode equivalence approach of Section 6.6 to construct a state-space realization of order equal to the determinantal degree of $D(s)$. *Hint:* Apply the arguments of Section 6.6 to $D^{-1}(s)$ and then incorporate the effects of $N(s)$. *Remark:* For a discussion couched in more abstract terms, see Reference [5].

8.3   SOME PROPERTIES AND APPLICATIONS OF SYSTEM EQUIVALENCE

We shall illustrate the significance of the equivalence concepts introduced in Sec. 8.2 in a number of different problems.

We shall begin in Sec. 8.3.1 with the extension to PMDs of the important results obtained in Secs. 2.4.1, 6.2, and 6.5 on the relationships between minimal state-space realizations and irreducible MFDs. The results extend in a natural way, but an interesting fact is that the proofs are actually somewhat simpler than in the earlier sections, because we now have available the powerful equivalence results of Sec. 8.2. In Sec. 8.3.2, we shall demonstrate the value of these equivalence results in studying the zeros and poles of general PMDs.

In Sec. 8.3.3, we shall show another application to developing controllability and observability criteria for certain interconnections of minimal state-space systems. The results here provide a nice generalization of the scalar results of Example 2.4-4.

8.3.1   Some Properties of Irreducible PMDs

A PMD $\{P, Q, R, W\}$ will be said to be *irreducible* if

$$\{P \text{ and } Q\} \text{ are left coprime} \tag{1a}$$

$$\{R \text{ and } P\} \text{ are right coprime} \tag{1b}$$

By the construction of Sec. 8.2, we can associate with any PMD $\{P, Q, R, W\}$ a state-space realization [cf. (8.2-11)]

$$\{A_o, B_o, C, J(s)\} \tag{2a}$$

of order

$$n = \deg \det P(s) \tag{2b}$$

Now if the PMD $\{P, Q, R, W\}$ is irreducible, this special realization (2) will be minimal, as follows from identity (8.2-11),

$$\begin{bmatrix} \Psi_L(s) & 0 \\ -\Lambda(s) & I_p \end{bmatrix} \begin{bmatrix} sI - A_o & B_o \\ -C & J(s) \end{bmatrix} = \begin{bmatrix} P(s) & Q(s) \\ -R(s) & W(s) \end{bmatrix} \begin{bmatrix} C_o & -Y(s) \\ 0 & I_m \end{bmatrix}$$

which shows that (cf. Theorem 8.2-3)

$$\{P, Q\} \text{ coprime} \Longleftrightarrow \{sI - A_o, B_o\} \text{ coprime} \tag{3a}$$

$$\{R, P\} \text{ coprime} \Longleftrightarrow \{C, sI - A_o\} \text{ coprime} \tag{3b}$$

But if the special realization (2) is minimal, then by Lemma 6.5-2, all nth-order realizations [recall $n = \deg \det P(s)$] of $\{P, Q, R, W\}$ must also be minimal. Therefore, we have essentially proved the following result, which generalizes Theorem 6.5-1.

Theorem 8.3-1.

All state-space realizations of a PMD $\{P, Q, R, W\}$, of order $n = \deg \det P(s)$, will be minimal if and only if the PMD is irreducible.

Proof.   The *if* part was proved above. For the *only if* part, we just reverse the arguments. That is, if $\{A, B, C, D\}$ is a minimal realization of $\{P, Q, R, W\}$ with $n = \deg \det P(s)$, then the special realization (2) must also be minimal because it has the same order. Therefore, $\{sI - A_o, B_o\}$ will be left coprime and $\{C, sI - A_o\}$ will be right coprime. Then identity (3) shows that the same will be true of $\{P, Q\}$ and $\{R, P\}$, respectively.   ∎

We showed in Sec. 6.2.1 that all minimal state-space realizations of a given transfer function were related by similarity transformations. Similarly, in Sec. 6.5.1, we proved that all irreducible MFDs were related by unimodular transformations (cf. Theorem 6.5-4). There is, of course, a corresponding result for PMDs.

Theorem 8.3-2.

All irreducible PMDs of a given transfer function are equivalent (RSE or FSE).

Proof.   By the construction of Sec. 8.2, we know that

$$\{P_i, Q_i, R_i, W_i\} \overset{F}{\sim} \{C_i, A_{oi}, B_{oi}, J_i(s)\}, \qquad i = 1, 2$$

Since the PMDs are irreducible, the state-space realizations will be minimal (by Theorem 8.3-1). Now all minimal state-space realizations are similar to each other (cf. Theorem 6.2-4) and therefore also FSE or RSE. Then, by the transitivity property of equivalence relations, the PMDs must be equivalent (FSE or RSE).   ∎

The construction of Sec. 8.2, which is summarized by (3) above, shows how to associate a state-space realization with a PMD. It should be clear that we can also associate an MFD with a given PMD. This can be done by going through the state-space realization, provided this is either controllable *or* observable (or both). For example, if $\{A_o, B_o\}$ is controllable, we can find T such that $\{T^{-1}A_oT, T^{-1}B_o\}$ is in controller form and

$$\{P(s), Q(s), R(s), W(s)\} \overset{F}{\sim} \{T^{-1}A_oT, T^{-1}B_o, CT, J\}$$

But by the construction of Sec. 6.4.3 [cf. (6.4-37)] we know that

$$\{T^{-1}A_oT, T^{-1}B_o, CT, J\} \overset{F}{\sim} \{D(s), I, \bar{N}(s), J(s)\}$$

Since it can easily be verified that

$$\{D(s), I, \bar{N}(s), J(s)\} \overset{F}{\sim} \{D(s), I, N(s), 0\}$$

where

$$N(s) \triangleq \bar{N}(s) + J(s)D(s)$$

the above procedure shows how to rewrite the PMD in MFD form, while keeping the same transfer function

$$RP^{-1}Q + W = (\bar{N} + JD)D^{-1}$$

as well as the other properties (Theorem 8.3-2) that are invariant under FSE or RSE. It is worthwhile to record this result as a theorem.

Theorem 8.3-3.   *From PMDs to MFDs*

Given $\{P, Q, R, W\}$ with $\{P, Q\}$ left coprime, there exist polynomial matrices $\{N_R(s), D_R(s)\}$ such that

$$\begin{bmatrix} P & Q \\ -R & W \end{bmatrix} \overset{F}{\sim} \begin{bmatrix} D_R & I \\ -N_R & 0 \end{bmatrix}$$

On the other hand, if we have $\{R, P\}$ right coprime, we can find matrices $\{N_L(s), D_L(s)\}$ such that

$$\begin{bmatrix} P & Q \\ -R & W \end{bmatrix} \overset{F}{\sim} \begin{bmatrix} D_L & N_L \\ I & 0 \end{bmatrix}$$

If $\{P, Q, R, W\}$ is irreducible, then so are the MFDs $\{N_R, D_R\}$ and $\{N_L, D_L\}$.

Proof. The above results have all essentially been proved in the preceding discussion. An alternative proof, not using state-space realizations, has been given by Coppel [12] and is outlined in Exercise 6.3-5. ∎

8.3.2 Poles and Zeros of PMDs: Transmission and Decoupling Zeros

In Sec. 6.5.3 we showed how to define poles and zeros for irreducible MFDs and minimal state-space realizations. Here we shall extend those results to reducible as well as irreducible PMDs.

The Transmission Poles and Zero of a PMD. The (finite) poles and zeros of a transfer function $H(s)$ are most naturally defined, as we did in Sec. 6.5.3, through the Smith-McMillan form of $H(s)$,

$$H(s) = U_L(s)\mathcal{E}(s)\Psi_R^{-1}(s)U_R(s) \tag{4}$$

where

$U_L(s)$, $U_R(s)$ are unimodular

$\mathcal{E}(s) = \text{diag}\{\epsilon_i(s)\}, \qquad \Psi_R(s) = \text{diag}\{\psi_i(s)\}$

$\{\mathcal{E}(s), \Psi_R(s)\}$ are right coprime

Then (cf. Sec. 6.5.3) the

$$\text{poles of } H(s) \triangleq \text{ the roots of the } \{\psi_i(s)\} \tag{5a}$$

$$\text{zeros of } H(s) \triangleq \text{ the roots of the } \{\epsilon_i(s)\} \tag{5b}$$

Now the transfer function associated with a PMD $\{P, Q, R, W\}$ is

$$H(s) = R(s)P^{-1}(s)Q(s) + W(s) \tag{6}$$

and we might ask if the poles and zeros cannot be defined directly in terms of the PMD. In Sec. 6.5.3, we showed that this could be done for irreducible PMDs given in state-space form or in MFD form. For example, we showed that, given

$$H(s) = C(sI - A)^{-1}B, \qquad \{A, B, C\} \text{ minimal} \tag{7}$$

we could write [cf. (6.5-22)]

$$\begin{bmatrix} sI - A & B \\ -C & 0 \end{bmatrix} \overset{s}{\sim} \begin{bmatrix} I & 0 \\ 0 & \mathcal{E}(s) \end{bmatrix} \tag{8}$$

Note that, though this result was proved for strictly proper systems in Sec. 6.5.3, the arguments used there carry over without change to the more general case

$$H(s) = C(sI - A)^{-1}B + J(s), \qquad \{A, B, C\} \text{ minimal} \tag{9}$$

to yield

$$\begin{bmatrix} sI - A & B \\ -C & J(s) \end{bmatrix} \overset{s}{\sim} \begin{bmatrix} I & 0 \\ 0 & \mathcal{E}(s) \end{bmatrix} \tag{10}$$

But now, if

$$\{P, Q, R, W\} \text{ is irreducible} \tag{11}$$

and has the same transfer function as $\{A, B, C, J(s)\}$, then by Theorem 8.3-2 we know that

$$\begin{bmatrix} P(s) & Q(s) \\ -R(s) & W(s) \end{bmatrix} \overset{F}{\sim} \begin{bmatrix} sI - A & B \\ -C & J(s) \end{bmatrix} \tag{12}$$

By the properties of FSE (Theorem 8.2-3), this means, of course, that

$$\begin{bmatrix} P(s) & Q(s) \\ -R(s) & W(s) \end{bmatrix} \overset{s}{\sim} \begin{bmatrix} I & 0 \\ 0 & \mathcal{E}(s) \end{bmatrix} \tag{13}$$

In other words, *whenever* $\{P, Q, R, W\}$ *is irreducible, the invariant polynomials of the system matrix define the zeros of* $H(s)$.

It is of value to give a direct proof of (13), without appealing to (6.5-22). For this, we note that the Smith-McMillan form yields

$$\{\Psi_R(s), U_R(s), U_L(s)\mathcal{E}(s), 0\}$$

as an irreducible PMD of $H(s)$ [cf. (4)]. Then, if $\{P, Q, R, W\}$ is another irreducible PMD, Theorem 8.3-2 implies that

$$\begin{bmatrix} P & Q \\ -R & W \end{bmatrix} \overset{F}{\sim} \begin{bmatrix} \Psi_R(s) & U_R(s) \\ -U_L(s)\mathcal{E}(s) & 0 \end{bmatrix} \tag{14}$$

Now elementary transformations can be used to reduce the right hand matrix to the Smith equivalent form

$$\begin{bmatrix} \Psi_R & I \\ \mathcal{E} & 0 \end{bmatrix} \overset{s}{\sim} \begin{bmatrix} 0 & I \\ \mathcal{E} & 0 \end{bmatrix} \overset{s}{\sim} \begin{bmatrix} I & 0 \\ 0 & \mathcal{E} \end{bmatrix} \tag{15}$$

The properties of FSE show that both block matrices in (14) have the same Smith form, so that (14) and (15) together again prove (13).

Let us note also that relations (12) and (14) show also that for irreducible PMDs,

$$\text{the finite poles of } H(s) = \text{the roots of } \det P(s) \tag{16}$$

Decoupling Zeros.   If a state-space realization $\{A, B, C\}$ is not minimal, we know that it has noncontrollable and/or nonobservable modes, which can be displayed by suitable similarity transformations (cf. Sec. 6.2). The noncontrollable and/or nonobservable parts of the system in this special transformed form are redundant and make no contribution to the *transfer function* between the given inputs and outputs (i.e., to the response of the system when the initial state is zero). The uncontrollable modes cannot be excited from the input (though they may appear due to noise effects in the system, which can set up initial conditions that excite all the modes). Similarly, unobservable modes cannot be seen in the specified outputs of the realization.

Let us now explore the analogous phenomena for PMDs. A natural way to begin is to associate a state-space realization $\{A, B, C, J\}$ with a given PMD $\{P, Q, R, W\}$. This can be done in different ways, but if we use the one described in Sec. 8.2 we have the advantage that we can write

$$
\begin{bmatrix} sI - A & B \\ -C & J(s) \end{bmatrix} \overset{F}{\sim} \begin{bmatrix} P(s) & Q(s) \\ -R(s) & W(s) \end{bmatrix}
\tag{17}
$$

We have already shown (Theorem 8.2.3) that (17) implies the following Smith equivalences (*which we shall relax to meaning preservation of the nonunity invariant polynomials*)

$$
sI - A \overset{S}{\sim} P(s)
\tag{18a}
$$

$$
[sI - A \quad B] \overset{S}{\sim} [P(s) \quad Q(s)]
\tag{18b}
$$

$$
[sI - A' \quad C']' \overset{S}{\sim} [P'(s) \quad R'(s)]'
\tag{18c}
$$

as well as the Smith equivalence of the system matrices of (17) themselves. These relationships will be used to study the controllability-observability properties of a PMD via those of an equivalent state-space system.

We first note that the presence of the 'direct transmission' term $J(s)$ in a state-space realization does not alter the discussions of, or results on, the controllability and observability of such systems that we presented in Sec. 2.4.1; we leave this for the reader to verify.

Now bring the state-space system of (17) to the standard form system of Sec. 2.4.2, using similarity transformations (thereby maintaining the system equivalence). Then

$$
\begin{bmatrix} P(s) & Q(s) \\ -R(s) & W(s) \end{bmatrix} \overset{F}{\sim}
\begin{bmatrix}
sI - A_{co} & 0 & -A_{13} & 0 & B_1 \\
-A_{21} & sI - A_{c\bar{o}} & -A_{23} & -A_{24} & B_2 \\
0 & 0 & sI - A_{\bar{o}\bar{c}} & 0 & 0 \\
0 & 0 & -A_{43} & sI - A_{\bar{o}\bar{o}} & 0 \\
\hline
-C_1 & 0 & -C_2 & 0 & J(s)
\end{bmatrix}
$$

$$
\overset{\triangle}{=} \begin{bmatrix} sI - \bar{A} & \bar{B} \\ -\bar{C} & J(s) \end{bmatrix}
\tag{19}
$$

where, for instance, the subsystem

$$\left\{ \begin{bmatrix} A_{co} & 0 \\ A_{21} & A_{c\bar{o}} \end{bmatrix}, \begin{bmatrix} B_1 \\ B_2 \end{bmatrix}, [C_1 \quad 0] \right\} \tag{20}$$

is controllable. Hence (by the PBH test, for example)

$$\begin{bmatrix} sI - A_{co} & 0 & \vdots & B_1 \\ -A_{21} & sI - A_{c\bar{o}} & \vdots & B_2 \end{bmatrix} \overset{S}{\sim} [I \quad 0] \tag{21}$$

It follows quite straightforwardly from (18b), (19) and (21) that

$$[P(s) \quad Q(s)] \overset{S}{\sim} [sI - \bar{A} \quad \bar{B}] \overset{S}{\sim} \begin{bmatrix} I & 0 & \vdots & 0 \\ \vdots & \vdots & ------ \\ 0 & \vdots & sI - A_{o\bar{c}} & 0 \\ & \vdots & -A_{43} & sI - A_{\bar{o}\bar{c}} \end{bmatrix} \tag{22}$$

from which the (Smith) *zeros* of $[sI - \bar{A} \quad \bar{B}]$ and $[P(s) \quad Q(s)]$ are seen to be precisely those of $sI - A_{o\bar{c}}$ and $sI - A_{\bar{o}\bar{c}}$. [In general, however, the invariant *polynomials* of $[sI - \bar{A} \quad \bar{B}]$ and $[P(s) \quad Q(s)]$ are not those of $sI - A_{o\bar{c}}$ and $sI - A_{\bar{o}\bar{c}}$ taken separately because the term A_{43} essentially causes a regrouping of the zeros. For example,

$$\begin{bmatrix} s - 1 & 0 \\ 0 & s - 1 \end{bmatrix}$$

is already in Smith form, with invariant polynomials $\{s - 1, s - 1\}$, but

$$\begin{bmatrix} s - 1 & 0 \\ 1 & s - 1 \end{bmatrix}$$

has Smith form

$$\begin{bmatrix} 1 & 0 \\ 0 & (s - 1)^2 \end{bmatrix}$$

and so has the single invariant polynomial $(s - 1)^2$.]

The (Smith) zeros of $[P(s) \quad Q(s)]$ thus correspond to uncontrollable modes of the equivalent state-space realization. Following Rosenbrock [1], we shall term these the *input-decoupling* (or *i.d.*) *zeros* of the PMD. A dual development identifies the Smith zeros of $[P'(s) \quad R'(s)]'$ with the unobservable modes of the equivalent state-space system, and these are termed the *output-decoupling* (or *o.d.*) *zeros* of the PMD.

Suppose now that $\{P(s), Q(s)\}$ have a nontrivial (i.e., nonunimodular) greatest common left divisor (gcld), denoted by $Z_{id}(s)$ say, corresponding to the i.d. zeros of the PMD system. Then we may write

$$P(s) = Z_{id}(s)P_{id}(s), \qquad Q(s) = Z_{id}(s)Q_{id}(s) \tag{23}$$

where now $\{P_{id}(s), Q_{id}(s)\}$ are coprime. Extracting i.d. zeros from a PMD, i.e., effecting the change

$$\{P, Q, R, W\} \longrightarrow \{P_{id}, Q_{id}, R, W\} \tag{24}$$

(which of course leaves the transfer function unchanged), results in a system with no i.d. zeros, but with possibly some remaining o.d. zeros, corresponding to the zeros of $[P_{id}(s)\ R(s)]$. The relationship in (19) indicates that in fact the zeros of $sI - A_{\delta\delta}$ disappear when the i.d. zeros are extracted, and the remaining o.d. zeros are those of $sI - A_{c\delta}$. The o.d. zeros of the *original* system that disappear when the i.d. zeros are extracted, we shall term the *input-output decoupling* (or *i.o.d.*) *zeros* of the PMD. By identifying these with the zeros of $sI - A_{\varepsilon\delta}$ we see in fact that the i.o.d. zeros have an identity that is independent of the particular extraction procedure used. For example, the i.o.d. zeros are also obtained as the remaining i.d. zeros after extraction of all the o.d. zeros first, by cancelling a GCRD, $Z_{od}(s)$, from $P(s)$ and $R(s)$.

The i.d. and o.d. zeros may be given direct dynamical interpretations. For example, an o.d. zero at $s = \beta$ corresponds to a (partial) state motion of the form $\xi(t) = ke^{\beta t}$ (with $u(t) \equiv 0$) such that

$$y(t) \equiv 0, \ t \geq 0.$$

Finally, let us show that the set of decoupling zeros that must be extracted from a PMD $\{P, Q, R, W\}$ to obtain an irreducible PMD with the same transfer function can be written as

$$\left\{ \begin{matrix} \text{decoupling} \\ \text{zeros} \end{matrix} \right\} = \{\text{i.d. zeros}\} + \{\text{o.d. zeros}\} - \{\text{i.o.d. zeros}\} \tag{25}$$

To show this, suppose we have first extracted a gcld, $Z_{id}(s)$, from $P(s)$ and $Q(s)$ to get

$$P(s) = Z_{id}(s)P_{id}(s), \qquad Q(s) = Z_{id}(s)Q_{id}(s) \tag{26}$$

where $\{P_{id}(s), Q_{id}(s)\}$ are now left coprime. Suppose further that $\{R(s), P_{id}(s)\}$ have a gcrd $Z_{oi}(s)$, so that, say,

$$R(s) = R_{oi}(s)Z_{oi}(s), \qquad P_{id}(s) = P_{oi}(s)Z_{oi}(s) \tag{27}$$

where now $\{R_{oi}(s), P_{oi}(s)\}$ are right coprime.

The process of removing the i.d. zeros and then the o.d. zeros of the resulting PMD can then be described by the equations

$$
\begin{bmatrix} P & Q \\ -R & W \end{bmatrix} = \begin{bmatrix} Z_{td} & 0 \\ 0 & I \end{bmatrix} \begin{bmatrix} P_{td} & Q_{td} \\ -R & W \end{bmatrix}
$$

$$
= \begin{bmatrix} Z_{td} & 0 \\ 0 & I \end{bmatrix} \begin{bmatrix} P_{oi} & Q_{td} \\ -R_{oi} & W \end{bmatrix} \begin{bmatrix} Z_{oi} & 0 \\ 0 & I \end{bmatrix} \tag{28a}
$$

where the 'reduced' PMD

$$
\{R_{oi}, P_{oi}, Q_{td}, W\} \text{ is irreducible} \tag{28b}
$$

From (28) we see that the set of o.d. zeros that disappears after the i.d. zeros are eliminated from the PMD is given by

$$
\{\text{i.o.d. zeros}\} \triangleq \{\text{o.d. zeros}\} - \{\text{zeros of det } Z_{oi}(s)\} \tag{29}
$$

Now the poles of the *transfer function* will be just the roots of det $P_{oi}(s)$, while the PMD $\{P, Q, R, W\}$ has certain *additional* modes given by the roots of (cf. (28))

$$
\frac{\det P(s)}{\det P_{oi}(s)} = \det Z_{td}(s) \det Z_{oi}(s) \tag{30}
$$

These are precisely the decoupling zeros, and we see that

$$
\{\text{decoupling zeros}\} = \{\text{i.d. zeros}\} + \{\text{zeros of det } Z_{oi}(s)\}
$$

which combined with (29) establishes the relation (25).

Motivated by the above results, Rosenbrock [2] defines the poles and zeros of a general PMD as

$$
\{\text{zeros of a PMD}\} = \{(\text{transfer function) zeros}\}
$$
$$
+ \{\text{decoupling zeros}\} \tag{31}
$$
$$
\{\text{poles of a PMD}\} = \{(\text{transfer function) poles}\}
$$
$$
+ \{\text{decoupling zeros}\} \tag{32}
$$

To distinguish them from the decoupling zeros, the transfer function zeros are often called the *transmission zeros*, as we noted in Sec. 6.5.3.

Example 8.3-1.   Poles and Zeros of a PMD
The PMD associated with the system matrix

$$
\mathbf{P} = \begin{bmatrix}
1 & 0 & 0 & 0 & \vdots & 0 \\
0 & 1 & 0 & 0 & \vdots & 0 \\
0 & 0 & s^2(s+1) & s(s+2) & \vdots & -s \\
0 & 0 & 0 & s+2 & \vdots & 1 \\
\hdashline
0 & 0 & 0 & -1 & \vdots & 0
\end{bmatrix}
\tag{33}
$$

has

$$\{\text{i.d. zeros}\} = \{0\}, \qquad \{\text{o.d. zeros}\} = \{0, 0, -1\}$$
$$\{\text{i.o.d. zeros}\} = \{0\}, \qquad \{\text{decoupling zeros}\} = \{0, 0, -1\}$$

To see this, note first that $s = 0$ will make the third row of $\mathbf{P}(s)$ zero, so that $[P(0) \ Q(0)]$ will have rank $3 < 4$. Therefore, $s = 0$ is an i.d. zero. The only other possible i.d. zeros are at $s = -1$ or $s = -2$, but the rank of $[P(s) \ Q(s)]$ is full (i.e., 4) at these values, and therefore $s = 0$ is the only i.d. zero. On the other hand, we see that the values $\{s = 0, s = 0, s = -1\}$ will make $[P'(s) \ R'(s)]'$ lose rank, and therefore these will be the o.d. zeros. Next, if we remove the i.d. zero (at $s = 0$), the resulting system matrix will be

$$
\mathbf{P_1} = \begin{bmatrix}
1 & 0 & 0 & 0 & \vdots & 0 \\
0 & 1 & 0 & 0 & \vdots & 0 \\
0 & 0 & s(s+1) & s+2 & \vdots & -1 \\
0 & 0 & 0 & -1 & \vdots & 1 \\
\hdashline
0 & 0 & 0 & -1 & \vdots & 0
\end{bmatrix}
$$

Now there are, of course, no i.d. zeros, but we can see also that one of the o.d. zeros at $s = 0$ has disappeared. Therefore, $s = 0$ is an i.o.d. zero. On the other hand, if we first remove the o.d. zeros, we shall have

$$
\mathbf{P_2} = \begin{bmatrix}
1 & 0 & 0 & 0 & \vdots & 0 \\
0 & 1 & 0 & 0 & \vdots & 0 \\
0 & 0 & 1 & s(s+2) & \vdots & -s \\
0 & 0 & 0 & s+2 & \vdots & 1 \\
\hdashline
0 & 0 & 0 & -1 & \vdots & 0
\end{bmatrix}
$$

This has no i.d. zeros anymore, showing again that the disappearing i.d. zero $s = 0$ is the only i.o.d. zero. The decoupling zeros are

$$\{0\} + \{0, 0, -1\} - \{0\} = \{0, 0, -1\}$$

The poles of the PMD are the roots of $\det P(s) = 0$, namely $\{0, 0, -1, -2\}$. Removing the decoupling zeros $\{0, 0, -1\}$ leaves the one pole at $s = -2$. This can be verified by computing

$$H(s) = R(s)P^{-1}(s)Q(s) = \frac{1}{s+2}$$

or by examining the system matrix $\mathbf{P}_2(s)$, which has no i.d. zeros or o.d. zeros and therefore corresponds to an irreducible PMD. Note also that there is no value of s for which $\mathbf{P}_2(s)$ loses rank, and therefore there are no transmission zeros (cf. Sec. 6.5.3), consistent again with the fact that $H(s) = 1/(s+2)$.

Transmission Zeros of Reducible PMDs. In Sec. 8.3.2, we showed that, for an *irreducible* PMD,

$$\mathbf{P}(s) = \begin{array}{c} {}_l \\ {}_p \end{array} \begin{bmatrix} \overset{l}{P(s)} & \overset{m}{Q(s)} \\ -R(s) & W(s) \end{bmatrix} \overset{s}{\underset{\sim}{}} \begin{bmatrix} I & 0 \\ 0 & \mathcal{E}(s) \end{bmatrix} \tag{34}$$

Therefore, as in Sec. 6.5.3 for MFDs and state-space triples, the transmission zeros of a general irreducible PMD can be found as the roots of the invariant polynomials of the system matrix $\mathbf{P}(s)$, or equivalently, as the values of s for which the rank of the system matrix is less than its *normal* rank. It is of interest to ask whether the zeros of the invariant polynomials of a general (not irreducible) system matrix would give all the transmission *and* decoupling zeros of the PMD. The answer is that this is always true when $H(s)$ is non-singular but may not be true otherwise.

In the nonsingular case, we need only refer to (28), which will show that (Exercise 8.1-1 shows that $\mathbf{P}$ is nonsingular)

$$\det \mathbf{P} = \det Z_{id} \det \mathbf{P}_{red} \det Z_{oi} \tag{35}$$

so that the roots of $\det \mathbf{P}$ will be the transmission zeros (i.e., the roots of $\det \mathbf{P}_{red}$) and the decoupling zeros (i.e., the roots of $\det Z_{id}(s) \det Z_{oi}(s) = 0$). However, $\det \mathbf{P}(s)$ is also the product of the invariant polynomials of $\mathbf{P}(s)$. Hence ....

This equality breaks down when $H(s)$ is singular, as is readily shown by the following simple example: Let $l = 3, p = 2, m = 1$, and

$$\mathbf{P}(s) = \left[\begin{array}{ccc|c} s-1 & 0 & 0 & 0 \\ 0 & s+1 & 0 & 1 \\ 0 & 0 & s+3 & 1 \\ \hline 1 & -1 & 0 & 0 \\ 0 & 2 & 1 & 0 \end{array} \right]$$

We have an i.d. zero at $s = 1$ because at this value of s the first three rows of $\mathbf{P}(s)$ (i.e., $[P(s) \; Q(s)]$) lose rank. (There are no other i.d. zeros and no o.d. zeros or i.o.d. zeros.) On the other hand, because there are more outputs than inputs ($p > m$), even though we lose rank at $s = 1$ in the first three rows, we do not lose rank in the whole matrix $\mathbf{P}(s)$ at this value of s. Therefore the i.d. zero at $s = 1$ is not a zero of any invariant polynomial of $\mathbf{P}(s)$.

Similar phenomena can occur with o.d. zeros—loss of rank in the first l columns of $\mathbf{P}(s)$ need not imply loss of rank in the whole matrix when $m > p$.

Nevertheless, in view of their transmission-blocking properties (which follow as in Sec. 6.5.3), the Smith form zeros of $\mathbf{P}(s)$, also often called the *invariant zeros* of the system, are a significant feature of the system and play a role in several systems problems (see, e.g., [13] to [15] and also Sec. 7.6).

8.3.3   Controllability and Observability of Interconnected Systems

It is often helpful to know when the series or parallel or feedback connection of controllable and/or observable realizations is still controllable and/or observable. Some results were given for scalar systems in Sec. 2.4 (cf. Example 2.4-5 and Exercises 2.4-7 and 2.4-8). The multivariable case is more difficult, though it was already studied (and resolved for diagonalizable systems) in the early paper of Gilbert [16].

In this section, we shall illustrate how the notion of system equivalence and the use of MFDs can provide some nice answers to these problems.

The case of series interconnection will illustrate the major ideas. (For a more comprehensive treatment, see References [17] to [19]). Consider the series connection of

$$H_1(s) \text{ followed by } H_2(s)$$

where

$$H_i(s) = C_i(sI - A_i)^{-1}B_i, \qquad i = 1, 2$$
$$\{A_i, B_i, C_i\} \text{ minimal}, \qquad \dim A_i = n_i$$

A natural set of state equations for the combined realization will be

$$\begin{bmatrix} \dot{x}_1(t) \\ \dot{x}_2(t) \end{bmatrix} = \begin{bmatrix} A_1 & 0 \\ B_2C_1 & A_2 \end{bmatrix} \begin{bmatrix} x_1(t) \\ x_2(t) \end{bmatrix} + \begin{bmatrix} B_1 \\ 0 \end{bmatrix} u(t) \tag{36a}$$

$$y(t) = [0 \quad C_2][x_1'(t) \quad x_2'(t)]' \tag{36b}$$

We can write down the controllability and observability matrices for this system, but it is not easy to deduce from them any simple conditions on the $\{A_i, B_i, C_i\}$ that will ensure the controllability and/or observability of the overall system.

Rosenbrock [1] suggested that the use of MFDs and system matrices would be useful. Thus, suppose

$$H_1(s) = D_{L1}^{-1}(s)N_{L1}(s), \qquad \deg \det D_L = n_1$$
$$H_2(s) = N_{R2}(s)D_{R2}^{-1}(s), \qquad \deg \det D_R = n_2$$

where both the representations are assumed to be irreducible, i.e., $\{D_{L1}, N_{L1}\}$ are left coprime and $\{N_{R2}, D_{R2}\}$ are right coprime. Then the series connection obeys

$$y = N_{R2}D_{R2}^{-1}D_{L1}^{-1}N_{L1}u = N_{R2}(D_{L1}D_{R2})^{-1}N_{L1}u \qquad (37)$$

This suggests that the condition

$$N_{R2} \text{ and } D_{L1}D_{R2} \text{ are right coprime} \qquad (38)$$

might be necessary and sufficient for the observability of any state-space realization (of order $n = \deg \det D_{L1} \det D_{R2}$) of the series combination of the minimal realizations $H_1(s)$ and $H_2(s)$. However, this is not necessarily true. All we can say is that by the procedure of Sec. 8.2 we can construct one realization $\{A_o, B_o, C, J(s)\}$ of $\{D_{L1}D_{R2}, N_{L1}, N_{R2}, 0\}$ and that, by virtue of (8.2-11), this realization will be observable if and only if N_{R2} and $D_{L1}D_{R2}$ are coprime. It need not follow from this that the state-space realization (36) that we started with will be observable. However, if we can show that the special realization is equivalent (say RSE) to the original state-space realization, then both realizations will be observable or unobservable together and the condition (38) will apply to the realization (36).

To carry out this scheme, let us start with two systems with PMDs described by the system matrices

$$\mathbf{P}_i = \begin{bmatrix} P_i & Q_i \\ -R_i & W_i \end{bmatrix}, \qquad i = 1, 2$$

Then a natural PMD for the series combination of system 1 followed by system 2 is obtained by imposing the interconnection constraints

$$y_1 = u_2, \qquad y = y_2, \qquad u = u_1$$

on the subsystems, which obey the relations

$$y_i = R_i \xi_i + W_i u_i, \qquad P_i \xi_i = Q_i u_i, \qquad i = 1, 2.$$

This leads to the PMD

$$\mathbf{P}_{ser} = \begin{bmatrix} \overbrace{\begin{matrix} 0 & P_1 & 0 \\ P_2 & 0 & Q_2 \\ 0 & -R_1 & -I \end{matrix}}^{P_{ser}} & \begin{matrix} Q_1 \\ 0 \\ W_1 \end{matrix} \\ \hline \begin{matrix} -R_2 & 0 & W_2 \end{matrix} & 0 \end{bmatrix} \qquad (39)$$

By a simple calculation, we can see that (with P_{ser} as in (39))

$$\det P_{ser} = \det P_1 \cdot \det P_2 \qquad (40)$$

It is worth noting that if the $\mathbf{P}_i$ were originally in state-space form, then $\mathbf{P}_{ser}$ as just defined in (39) will no longer be in state-space form! However it can be reduced to such form by transformations of RSE. To see this, note first that if we make RSE transformations of the original PMDs,

$$\begin{bmatrix} M_{1i} & 0 \\ X_i & I \end{bmatrix} \begin{bmatrix} P_i & Q_i \\ -R_i & W_i \end{bmatrix} \begin{bmatrix} M_{2i} & Y_i \\ 0 & I \end{bmatrix} = \begin{bmatrix} \tilde{P}_i & \tilde{Q}_i \\ -\tilde{R}_i & \tilde{W}_i \end{bmatrix}, \qquad i = 1, 2$$

then we can verify that the $\mathbf{P}_{ser}$ transforms accordingly,

$$\tilde{\mathbf{P}}_{ser} = \begin{bmatrix} M_{11} & 0 & 0 & 0 \\ 0 & M_{12} & 0 & 0 \\ X_1 & 0 & I & 0 \\ \hline 0 & X_2 & 0 & I \end{bmatrix} \mathbf{P}_{ser} \begin{bmatrix} M_{22} & 0 & Y_2 & 0 \\ 0 & M_{21} & 0 & Y_1 \\ 0 & 0 & I & 0 \\ \hline 0 & 0 & 0 & I \end{bmatrix} \qquad (41)$$

Now for our original state-space problem, we can use the above results as follows. Let us define

$$\{sI - A_i, B_i, C_i, J_i\} \triangleq \{P_i, Q_i, R_i, W_i\}$$

[Notice that we are actually studying a generalization of (36), where the $\{J_i\}$ were assumed for simplicity to be zero.] Then notice that by elementary operations we can transform

$$\mathbf{P}_{ser} = \begin{bmatrix} 0 & P_1 & 0 & Q_1 \\ P_2 & 0 & Q_2 & 0 \\ 0 & -R_1 & -I & W_1 \\ \hline -R_2 & 0 & W_2 & 0 \end{bmatrix} \overset{R}{\sim} \begin{bmatrix} 0 & P_1 & 0 & Q_1 \\ P_2 & -Q_2R_1 & 0 & Q_2W_1 \\ 0 & -R_1 & -I & W_1 \\ \hline -R_2 & 0 & W_2 & 0 \end{bmatrix}$$

$$\overset{R}{\sim} \begin{bmatrix} 0 & P_1 & 0 & Q_1 \\ P_2 & -Q_2R_1 & 0 & Q_2W_1 \\ 0 & 0 & -I & W_1 \\ \hline -R_2 & -W_2R_1 & W_2 & 0 \end{bmatrix} \overset{R}{\sim} \begin{bmatrix} P_1 & 0 & 0 & Q_1 \\ -Q_2R_1 & P_2 & 0 & Q_2W_1 \\ 0 & 0 & I & 0 \\ \hline -W_2R_1 & -R_2 & 0 & W_2W_1 \end{bmatrix}$$

We may identify the last matrix as the system matrix associated with the state-space description (cf. (36), which must be extended to handle nonzero J_i). Now we assume that the state-space descriptions $\{A_i, B_i, C_i, J_i(s)\}$

are minimal and that we have the alternative irreducible matrix-fraction descriptions

$$H_1(s) = C_1(sI - A_1)^{-1}B_1 + J_1(s) = D_{L1}^{-1}(s)N_{L1}(s)$$
$$H_2(s) = C_2(sI - A_2)^{-1}B_2 + J_2(s) = N_{R1}(s)D_{R2}^{-1}(s)$$

By Theorem 8.3-2, we can write

$$\begin{bmatrix} sI - A_1 & B_1 \\ -C_1 & J_1(s) \end{bmatrix} \overset{R}{\sim} \begin{bmatrix} D_{L1} & N_{L1} \\ -I & 0 \end{bmatrix}$$

$$\begin{bmatrix} sI - A_2 & B_2 \\ -C_2 & J_2(s) \end{bmatrix} \overset{R}{\sim} \begin{bmatrix} D_{R2} & I \\ -N_{R2} & 0 \end{bmatrix}$$

Therefore,

$$\mathbf{P}_{\text{ser}} \overset{R}{\sim} \tilde{\mathbf{P}}_{\text{ser}} = \begin{bmatrix} 0 & D_{L1} & 0 & \vdots & N_{L1} \\ D_{R2} & 0 & I & \vdots & 0 \\ 0 & -I & -I & \vdots & 0 \\ \hdashline -N_{R2} & 0 & 0 & \vdots & 0 \end{bmatrix} \tag{42}$$

Now by elementary operations consistent with RSE, we can transform

$$\tilde{\mathbf{P}}_{\text{ser}} \overset{R}{\sim} \begin{bmatrix} 0 & D_{L1} & 0 & \vdots & N_{L1} \\ D_{R2} & -I & 0 & \vdots & 0 \\ 0 & 0 & -I & \vdots & 0 \\ \hdashline -N_{R2} & 0 & 0 & \vdots & 0 \end{bmatrix}$$

$$\overset{R}{\sim} \begin{bmatrix} D_{L1}D_{R2} & D_{L1} & 0 & \vdots & N_{L1} \\ 0 & -I & 0 & \vdots & 0 \\ 0 & 0 & -I & \vdots & 0 \\ \hdashline -N_{R2} & 0 & 0 & \vdots & 0 \end{bmatrix} \overset{R}{\sim} \begin{bmatrix} D_{L1}D_{R2} & 0 & 0 & \vdots & N_{L1} \\ 0 & I & 0 & \vdots & 0 \\ 0 & 0 & I & \vdots & 0 \\ \hdashline -N_{R2} & 0 & 0 & \vdots & 0 \end{bmatrix}$$

Therefore, we have proved

$$\begin{bmatrix} sI - A_1 & 0 & \vdots & 0 \\ -B_2C_1 & sI - A_2 & \vdots & 0 \\ \hdashline W_2R_1 & -R_2 & \vdots & W_2W_1 \end{bmatrix} \overset{R}{\sim} \begin{bmatrix} D_{L1}D_{R2} & \vdots & N_{L1} \\ \hdashline -N_{R2} & \vdots & 0 \end{bmatrix} \tag{43}$$

from which it is obvious that the series combination will be observable if and only if $\{N_{R2}, D_{L1}D_{R2}\}$ are right coprime and controllable if and only if $\{D_{L1}D_{R2}, N_{L1}\}$ are left coprime.

These results, due to Rosenbrock [1, p. 127], are simpler than any expressed in terms of state-space descriptions, but they are still not as nice as we might expect from knowledge of the scalar condition derived in Example 2.4-4, which was, for example, that observability held if and only if

$$b_2(s) \text{ and } a_1(s) \text{ were coprime}$$

where $H_i(s) = b_i(s)/a_i(s)$. There is, in fact, a multivariable analog of this condition, as first shown by Callier and Nahum [20]. Assume that

$$H_i(s) = D_{Li}^{-1}(s)N_{Li}(s), \qquad i = 1, 2$$

Then the series combination of $H_1(s)$ followed by $H_2(s)$ will be observable if and only if

$$N_{L2} \text{ and } D_{L1} \text{ are right coprime} \qquad (44)$$

[No simple condition for controllability is available when left MFDs are used for both $H_1(s)$ and $H_2(s)$, but, of course, there will be a *dual* result when right MFDs are used for both $H_1(s)$ and $H_2(s)$—see Exercise 8.3-11.] To establish (44), we note only that, by suitable elementary operations

$$\tilde{\mathbf{P}}_{\text{ser}} = \begin{bmatrix} 0 & D_{L1} & 0 & \vdots & N_{L1} \\ D_{L2} & 0 & N_{L2} & \vdots & 0 \\ 0 & -I & -I & \vdots & 0 \\ \hline -I & 0 & 0 & \vdots & 0 \end{bmatrix} \overset{R}{\sim} \begin{bmatrix} 0 & D_{L1} & 0 & \vdots & N_{L1} \\ 0 & N_{L2} & 0 & \vdots & 0 \\ 0 & 0 & I & \vdots & 0 \\ \hline I & 0 & 0 & \vdots & 0 \end{bmatrix}$$

and for observability to hold the rank of the first three block columns must be full for all s, which will be true if and only if condition (44) holds. Several related results and extensions are noted in the exercises and in [17]–[21].

Exercises

8.3-1.

Suppose $\{P, Q, R, W\}$ is a PMD. Show that there exists a pair $\{N, D\}$ such that $\{P, Q, R, W\} \overset{R}{\sim} \{D, N, I, W\}$ (and hence, in particular, that D and P have the same determinantal degree) if and only if R and P are right coprime.

8.3-2. [1]

If $\{P, Q, R, W\}$ is irreducible, show that the matrix

$$\begin{bmatrix} P^2 & PQ \\ -RP & -RQ \end{bmatrix} \quad \text{has the Smith form} \quad \begin{bmatrix} I_n & 0 \\ 0 & 0 \end{bmatrix}$$

Hint: The given matrix has an obvious factorization as a block column matrix times a block row matrix. Now use Lemma 6.3-9.

8.3-3.   *Inverse Systems*

Let $H(s)$ be an $m \times m$ transfer function given by $H(s) = R(s)P^{-1}(s)Q(s) + W(s)$, det $W(s) \not\equiv 0$.

a. Show that det $H(s) \not\equiv 0$ if and only if det $[P(s) + Q(s)W^{-1}(s)R(s)] \not\equiv 0$.

b. When $H(s)$ is nonsingular, show that we can write $H^{-1}(s) = \tilde{R}(s)\tilde{P}^{-1}(s)\tilde{Q}(s) + \tilde{W}(s)$, where $\tilde{W} = W^{-1}$, $\tilde{Q} = QW^{-1}$, $\tilde{R} = -W^{-1}R$, $\tilde{P} = P + QW^{-1}R$.

c. If $W(s)$ is not invertible, show that we can write $H^{-1}(s) = \hat{R}(s)\hat{P}^{-1}(s)\hat{Q}(s) + \hat{W}(s)$ for suitable $\{\hat{P}, \hat{Q}, \hat{R}, \hat{W}\}$. (See also Exercise 8.1-3.)

d. Show that $\{\tilde{P}, \tilde{Q}, \tilde{R}, \tilde{W}\}$ and $\{\hat{P}, \hat{Q}, \hat{R}, \hat{W}\}$ will be irreducible if and only if the same is true of $\{P, Q, R, W\}$.

8.3-4.

Refer to Exercise 8.3-3 and show that $\{\tilde{P}, \tilde{Q}, \tilde{R}, \tilde{W}\}$, $\{\hat{P}, \hat{Q}, \hat{R}, \hat{W}\}$, and $\{P, Q, R, W\}$ all have the same decoupling zeros.

8.3-5.   *Feedback Connections*

Suppose we use feedback $u(t) = v(t) - F(D)y(t)$ around a PMD $\{P, Q, R, W\}$.

a. Show that a system matrix for the overall closed-loop system is

$$
\left[
\begin{array}{ccc:c}
P & Q & 0 & 0 \\
-R & W & I & 0 \\
0 & -I & F(s) & I \\
\hdashline
0 & 0 & -I & 0
\end{array}
\right]
$$

b. Note that for part a we have to prove also that the 3×3 block matrix is nonsingular, which will follow by showing that its determinant is proportional to det $P(s)$ det $[I + F(s)H(s)]$, where $H(s) = R(s)P^{-1}(s)Q(s) + W(s)$.

c. Show that $\{P, Q, R, W\}$ irreducible implies that the closed-loop system matrix is also irreducible.

d. Hence prove that the ratio of the closed-loop to the open-loop characteristic polynomials is det $[I + F(s)H(s)]$. This formula was first derived by Popov in 1964—see [2, Sec. 3.1] for generalizations.

8.3-6.   *Smith Form of $P(s)$* [1]

Suppose that $\{P, Q, R, W\}$ is an irreducible PMD with a transfer function $H(s) = RP^{-1}Q + W$ that has the Smith-McMillan form diag $\{\epsilon_i/\psi_i\}$.

a. Show that the Smith form of $P(s)$ is diag $\{1, 1, \ldots, 1, \psi_m, \ldots, \psi_1\}$.

b. Let

$$
\mathbf{P}\begin{Bmatrix} 1, 2, \ldots, l, i_1, \ldots, i_k \\ 1, 2, \ldots, l, j_1, \ldots, j_k \end{Bmatrix} = \begin{array}{l} \text{the minor formed by the rows} \\ = \{1, 2, \ldots, l, i_1, \ldots, i_k\} \text{ and the} \\ \text{columns } \{1, 2, \ldots, l, j_1, \ldots, j_k\} \\ \text{of } \mathbf{P} \end{array}
$$

Show that the greater common divisor of all such minors is proportional to $\epsilon_1\epsilon_2 \ldots \epsilon_k \psi_{k+1} \psi_{k+2} \ldots \psi_q$, where $k = 1, 2, \ldots, q$ and $q = \min(m, p)$.

8.3-7.

Show that all the zeros (transmission plus decoupling) of a PMD can be obtained as the roots of the gcd of the largest-size nonzero minors of the form

$$\mathbf{P} \begin{Bmatrix} 1, \ldots, l, i_1, \ldots, i_k \\ 1, \ldots, l, j_1, \ldots, j_k \end{Bmatrix}$$

Hint: Use (8.3-25) and the result of Exercise 8.3-6. Reference: H. H. Rosenbrock, Int. J. Control, **20**, pp. 525–527, 1974.

8.3-8.

a. Show that every i.o.d. zero is both an i.d. zero and an o.d. zero.

b. Show that the i.d. zeros, o.d. zeros, and i.o.d. zeros of a PMD are invariant under FSE or RSE but not necessarily under state or output feedback.

c. Show that the *invariant* zeros of $\{A, B, C, J(s)\}$ are unchanged by output feedback. Show that the number of i.d. zeros of the PMD $\{sI - A, B, C\}$ is $n - r$. State and prove the corresponding result when $\{C, A\}$ is nonobservable.

8.3-9.   *Series Connections* [20]

a. Establish the necessary and sufficient conditions shown in the table in the accompanying figure for the controllability of $H_1(s)H_2(s)$.

b. Make up a similar table for observability.

H_2 \ H_1	$= N_{R2}D_{R2}^{-1}$	$= D_{L1}^{-1}N_{L1}$
$= N_2 D_{R2}^{-1}$	$\{N_{R1}, D_{R2}\}$ are left coprime	$\{D_{L1}D_{L2}, N_{L1}\}$ are left coprime
$= D_{L2}^{-1}N_{L2}$	$\{N_{L2}N_{R1}, D_{R2}\}$ are left coprime	No nice condition known

8.3-10.   *Parallel Connections*

Suppose $\{A_i, B_i, C_i, J_i(s)\}$ are two minimal realizations—connected in parallel, i.e., $u_1 = u_2$ and $y = y_1 + y_2$. Again, the use of MFDs gives some convenient conditions.

a. Show that the parallel connection will be controllable if and only if $\{D_{R1}, D_{R2}\}$ are left coprime.

b. Find and prove a similar result for observability.

c. Make up a table similar to the ones described for the series connection in Exercise 8.3-11.

Reference: P. A. Fuhrmann, IEEE Trans. Circuits Syst., **CAS-22**, p. 57, 1975.

REFERENCES

1. H. H. ROSENBROCK, *State-Space and Multivariable Theory*, Wiley, New York, 1970.

2. H. H. ROSENBROCK, *Computer-Aided Control System Design*, Academic Press, New York, 1974.

3. W. A. WOLOVICH, *Linear Multivariable Systems*, Springer-Verlag New York, Inc., New York, 1974.

4. P. A. FUHRMANN, "Algebraic System Theory: An Analyst's Point of View," *J. Franklin Inst.*, **301**, pp. 521–540, 1976.

5. P. A. FUHRMANN, "On Strict System Equivalence and Similarity," *Int. J. Control*, **25**, pp. 5–10, 1977.

6. B. LÉVY, S. KUNG, T. KAILATH, and M. MORF, "A Unification of System Equivalence Definitions," in *Proceedings of the 1977 IEEE Conference on Decision and Control*, New Orleans, Dec. 1977, pp. 795–800.

7. L. A. PERNEBO, "Notes on Strict System Equivalence," *Int. J. Control*, **25**, pp. 21–38, 1977.

8. H. BROMBERG and R. YLINEN, "An Operator Algebra for Analysis and Synthesis of Feedback and Other Systems," Rept. B31, Systems Theory Lab., Helsinki Institute of Technology, Helsinki, Norway, 1976.

9. M. MORF, "Extended System and Transfer Function Matrices and System Equivalence," in *Proceedings of the 1975 IEEE Conference on Decision and Control*, Houston, Dec. 1975, pp. 199–206.

10. J. W. HELTON, "Discrete-Time Systems, Operator Models, and Scattering Theory," *J. Functional Anal.*, **16**, pp. 15–38, May 1974.

11. J. W. HELTON, "Systems with Infinite-Dimensional State Space: The Hilbert Space Approach," *Proc. IEEE*, **64**, pp. 145–160, 1976.

12. W. A. COPPEL, "Matrices of Rational Functions," *Bull. Aust. Math. Soc.*, **11**, pp. 89–113, 1974.

13. A. G. MACFARLANE and N. KARCANIAS, "Poles and Zeros of Linear Multivariable Systems: A Survey," *Int. J. Control*, **24**, pp. 33–74, 1976; see also **26**, pp. 157–163, 1977.

14. N. SUDA and E. MUTSUYOSHI, "Invariant Zeros and Input-Output Structure of Linear Time-Invariant Systems," *Int. J. Control*, **28**, pp. 525–535, 1978.

15. N. KARCANIAS and B. KOUVARITAKIS, "The Use of Frequency Transmission Concepts in Linear Multivariable System Analysis," *Int. J. Control*, **28**, pp. 197–240, 1978.

16. E. G. GILBERT, "Controllability and Observability in Multivariable Linear Systems," *SIAM J. Control*, **1**, pp. 128–151, 1963.

17. H. H. ROSENBROCK and A. C. PUGH, "Contributions to a Hierarchical Theory of Systems," *Int. J. Contr.*, **19**, pp. 845–867, 1974.

18. G. VERGHESE, B. LÉVY and T. KAILATH, "Generalized State-Space for Degenerate Systems," *IEEE Trans. on Automat. Control*, **AC-25**, 1980.

19. G. VERGHESE, *Infinite-Frequency Behavior in Generalized Dynamical Systems*, Ph.D. Dissertation, Dept. of Electrical Engineering, Stanford University, Ca. December 1978.

20. F. M. CALLIER and C. D. NAHUM, "Necessary and Sufficient Conditions for the Complete Controllability and Observability of Systems in Series," *IEEE Trans. Circuits Syst.*, **CAS-22**, pp. 90–95, 1975.

21. H. BART, I. GOHBERG, M. A. KAASHOEK and P. VAN DOOREN, "Factorizations of Transfer Functions," Rept. No. 107, Dept. of Mathematics, Vrije Universiteit, Amsterdam, The Netherlands, Jun. 1979.

SOME RESULTS
FOR TIME-VARIANT SYSTEMS

9

9.0 INTRODUCTION

In the earlier chapters we have shown that a lot can be learned about linear systems without having any explicit formula for the time-domain solution of the state-space equations. The point is that one rarely needs an explicit solution, and if numerical values are desired, these are best found by using the currently best available computer routine. Rather than explicit formulas for the solution, what one often really uses is some convenient representation of the solution. The differential equation itself is often just such a representation. For constant-parameter state equations, the Laplace transform and the matrix exponential often provide somewhat more explicit representations. For time-variant equations, one can use the so-called fundamental matrix or an important special case, the *state-transition matrix* (already studied in the time-invariant case in Sec. 2.5.1). In Sec. 9.1 we shall develop the properties of the state-transition matrix in more detail and describe its use in representing the solution of the state equations. Then applications will be made in Sec. 9.2 to studying state controllability and observability for time-variant equations. The discussion of observability is used to bring up certain dualities, extending those discussed for constant systems in earlier chapters. We shall show that the notion of *adjoint system* is more significant for time-variant systems and shall give some examples of the usefulness of this concept in Sec. 9.3. We shall also explore the close relationship between adjoints and the notion of reciprocity, familiar from electric circuits and electromagnetic theory.

594

9.1 TIME-VARIANT STATE EQUATIONS; STATE-TRANSITION MATRICES

In studying the solution of the time-invariant equation

$$\dot{x}(t) = Ax(t) + bu(t)$$

we found no difficulty in guessing the results by extending results and concepts from the scalar case to the vector or matrix case. However, the chief way in which matrices differ from scalars is that matrix multiplication is not commutative; i.e., AB may not be equal to BA. In fact, we find that

$$e^{At} \cdot e^{Bt} \neq e^{(A+B)t} \qquad \text{in general}$$

Moreover, as we may expect, equality will hold if $AB = BA$ and in fact only if $AB = BA$ (try to show this).

The problem with noncommutativity comes immediately to the fore when we treat equations with time-variant coefficients, where we shall see that the well-known solution for scalar time-variant linear equations has no direct matrix analog in general. Thus, the *scalar* equation

$$\dot{x}(t) = a(t)x(t), \qquad t \geq 0, \qquad x(0) = x_0, \qquad x(\cdot) \text{ scalar}$$

has the solution (show this)

$$x(t) = \left[\exp \int_0^t a(s)\, ds \right] x_0, \qquad t \geq 0$$

By analogy with the constant (time-invariant) case, we might expect that the vector differential equation

$$\dot{x}(t) = A(t)x(t), \qquad x(0) = x_0, \qquad t \geq 0$$

will have the solution

$$\left[\exp \int_0^t A(s)\, ds \right] x_0 \tag{1}$$

where

$$\exp \int_0^t A(s)\, ds \triangleq I + \int_0^t A(s)\, ds + \tfrac{1}{2} \int_0^t A(\sigma)\, d\sigma \int_0^t A(s)\, ds + \cdots$$

However, this is not true since

$$\frac{d}{dt} \exp \int_0^t A(s)\, ds = A(t) + \frac{A(t)}{2} \int_0^t A(s)\, ds + \int_0^t A(\sigma)\, d\sigma \frac{A(t)}{2} + \cdots$$

$$\neq A(t) \exp \int_0^t A(s)\, ds$$

595

However, if $A(t)$ and $\int_0^t A(s)\,ds$ commute, then we can see that equality will hold and (1) will be a solution of the differential equation. Unfortunately $A(\cdot)$ and its integral do not always commute, which is why our attempted solution breaks down. Therefore we shall have to abandon any approach via analogy and proceed by a more fundamental route, with some attention to the questions of existence and uniqueness of solutions.

Since we shall generally have to resort to a computer to solve the differential equations we encounter, one may wonder if attention to the *theory* of differential equations is relevant. Actually it is in such cases that the theory is really necessary: We can always get some numbers from the computer by writing

$$x(t + \Delta) = [I + A(t)\Delta]x(t), \qquad x(0) = x_0$$

and plugging in $t = 0, \Delta, 2\Delta, \ldots$ with Δ sufficiently small. But these numbers may be of no significance if the equation we are trying to solve has no solution or does not have a unique solution!

We shall therefore first briefly note in a general context some results on existence and uniqueness of solutions of ordinary differential equations, whether linear or nonlinear, time-invariant or time-variant.

Fundamental Results on Differential Equations [1]–[4]

Consider the vector differential equation

$$\dot{x}(t) = f[x(t), t], \qquad a \le t \le b$$

with the *initial* condition

$$x(t_0) = x_0, \qquad t_0 = \text{any point in } [a, b]$$

A solution will be written $\phi(\cdot)$ or $\phi(\cdot\,; x_0, t_0)$.

Such equations may exhibit a variety of behaviors. We shall consider some aspects.

1. *It may be that there is no solution for arbitrary a and b:*

Example A: Consider the equation

$$\dot{x}(t) = (t - 1)^{-1}, \qquad x(0) = 0$$

A solution is $\ln(1 - t)$, $0 \le t < 1$, but there is no solution for $t \ge 1$.

Example B: *A Riccati Equation*: The nonlinear equation

$$\dot{x}(t) = x^2(t) + 1, \qquad x(0) = 0$$

has no solution on the interval $0 \leq t \leq 2$, but it does have one on $0 \leq t < \pi/2$.

2. *Even if there is a solution, there may not be a unique solution:*

Example C: The equation

$$\dot{x}(t) = 2\sqrt{|x(t)|}, \qquad x(0) = 0$$

has an infinity of solutions

$$\phi(t;0,0) = \begin{cases} 0, & 0 \leq t \leq c \leq \infty \\ 2(t-c)^2, & c \leq t \end{cases}$$

one for each value of the arbitrary number c, $0 \leq c \leq \infty$.

3. *Uniqueness is guaranteed if $f(x, t)$ satisfies a so-called Lipschitz condition over $[a, b]$*, i.e., if

$$\|f(x, t) - f(\tilde{x}, t)\| \leq k(t) \|x - \tilde{x}\|, \qquad a \leq t \leq b \qquad (2)$$

where $k(\cdot)$ is a piecewise continuous function on $[a, b]$ and $\|\cdot\|$ denotes the norm. This important result has a fairly simple proof, which the reader can look up in many books on differential equations. Note that the function $\sqrt{x}$ in Example C is not Lipschitz. On the other hand, $f(\cdot, \cdot)$ will be Lipschitz if it is sufficiently smooth, e.g., if the functions $\{\partial f_i/\partial x_j\}$ are continuous over (a, b).

When $f(x, t)$ obeys a Lipschitz condition, existence is also guaranteed in addition to uniqueness.

We should note that the Lipschitz condition is only a *sufficient* condition, and unique solutions may exist even when it is not met. Weaker existence and uniqueness conditions are discussed in [1, Chap. 2].

4. *An Important Device.* In many problems the difficulty is not with existence but with uniqueness; i.e., we often have two or more solutions or we have an "obvious" solution but are not sure if it is the only one. A major application of the uniqueness result is to prove various identities in cases where certain associated differential equations may not have explicit closed-form solutions. The method is the following: To check if

$$F(t) \stackrel{?}{=} G(t), \qquad a \leq t \leq b$$

try to show that

1. $F(t_0) = G(t_0)$ for some t_0 in $[a, b]$ (3a)

2. $F(\cdot)$ and $G(\cdot)$ obey the same differential equations   (3b)

3. These equations are known to have unique solutions   (3c)

If this can be done, then the conjectured equality is true.
    We shall see several examples of this method in this chapter.

Linear Time-Variant Equations

With this background, we shall return to the linear equation

$$\dot{x}(t) = A(t)x(t), \qquad x(t_0) = x_0, \qquad t_0 \in [a, b] \tag{4}\dagger$$

where we shall assume that $A(\cdot)$ is piecewise continuous. (Weaker conditions, e.g., integrability, will also do; see [1, p. 67].) Now over every *finite* interval we note that

$$\| A(t)x(t) - A(t)\tilde{x}(t) \| \le \max \| A(\cdot) \| \| x(t) - \tilde{x}(t) \|$$

so that the Lipschitz condition is satisfied. Therefore there exists a unique solution $\phi(t; x_0, t_0)$ going through x_0 at $t = t_0$.

The State-Transition Matrix.   There is a *representation* for the solution that is useful in establishing various results.
    Let

$$\phi_i(\cdot; \tau) = \text{the solution of (4) with value at } \tau \text{ equal to}$$
$$e_i, \text{ the } i\text{th column of the identity matrix } I_n \tag{5a}$$

and form the matrix

$$\Phi(\cdot, \tau) = [\phi_1(\cdot, \tau) \quad \cdots \quad \phi_n(\cdot, \tau)] \tag{5b}$$

Then by linearity we see that

$$\sum_1^n \phi_i(\cdot, t_0)x_{i0} \triangleq \Phi(\cdot, t_0)x_0$$

is the (unique) solution of the differential equation (4). With the usual abuse of notation we shall write

$$x(t) = \Phi(t, t_0)x_0, \qquad t, t_0 \in [a, b] \tag{6}$$

---

†It may be worth recalling that an important way in which time-variant realizations arise is as a result of linearizing a nonlinear time-invariant system (cf. the discussion at the end of Sec. 2.2.2).

and stress that t_0 can be any point in the interval of definition of the differential equation. This fact leads to the name *state-transition matrix* for $\Phi(\cdot, \cdot)$. It follows from this interpretation [or directly from (5)] that

$$\Phi(t, t) = I, \qquad t \in [a, b] \tag{7}$$

We can also establish the important property that

$$\Phi(t_1, t_2) = \Phi(t_1, t_3)\Phi(t_3, t_2) \qquad \text{for any points } \{t_1, t_2, t_3\} \text{ in } [a, b] \tag{8}$$

a useful special case of which is

$$\Phi^{-1}(t_1, t_2) = \Phi(t_2, t_1) \tag{9}$$

The so-called *semigroup property* (8) can be proved by comparing the expressions

$$x(t_1) = \Phi(t_1, t_2)x(t_2) \tag{10}$$

and

$$x(t_1) = \Phi(t_1, t_3)x(t_3), \qquad x(t_3) = \Phi(t_3, t_2)x(t_2) \tag{11}$$

Here is another proof that illustrates a technique mentioned above: let

$$R(t) = \Phi(t, t_3)\Phi(t_3, t_2) \text{ and let } L(t) = \Phi(t, t_2) \tag{12}$$

Then

$$\frac{dR(t)}{dt} = A(t)\Phi(t, t_3)\Phi(t_3, t_2) = A(t)R(t) \tag{13a}$$

and

$$\frac{dL(t)}{dt} = A(t)\Phi(t, t_2) = A(t)L(t) \tag{13b}$$

Therefore $L(\cdot)$ and $R(\cdot)$ obey the same differential equation. Moreover, at $t = t_3$,

$$R(t_3) = \Phi(t_3, t_3)\Phi(t_3, t_2) = \Phi(t_3, t_2) = L(t_3) \tag{13c}$$

Therefore by the uniqueness theorem for solutions of linear differential equations $L(\cdot) \equiv R(\cdot)$, which is relation (8).

It is useful to note that $\Phi(\cdot, \tau)$ can be characterized as the unique solution of the matrix differential equation

$$\frac{dM(t)}{dt} = A(t)M(t), \qquad M(\tau) = I \tag{14}$$

a fact that is just a restatement of (5). We note also that

$$\Phi(t, \tau) = \exp \int_{\tau}^{t} A(\sigma)\, d\sigma \quad \text{if } A(t) \text{ commutes with } \int_{0}^{t} A(\sigma)\, d\sigma \quad (15)$$

This can happen when $A(\cdot)$ is diagonal or when $A(\cdot)$ is such that [5]

$$A(t_1)A(t_2) = A(t_2)A(t_1), \quad \text{all } t_1, t_2 \text{ in } [a, b] \quad (16)$$

In particular this includes the case of a constant A matrix, in which case we showed in Sec. 2.5.1 that $\Phi(t, \tau) = \exp A(t - \tau)$. The arguments (4)–(8) of Sec. 2.5 have certain generalizations to the time-variant case involving the so-called *product* or *multiplicative* integral of A (see [5, pp. 131–135]). However, we shall not pursue this topic here. The matrix exponential can be computed by Laplace transform and other techniques; for time-variant systems, simulation or direct numerical evaluation are often the only means of determining $\Phi(\cdot, \cdot)$. Nevertheless, we can obtain several useful results without knowing $\Phi(\cdot, \cdot)$ explicitly. Here is a first example.

The Forced or Inhomogeneous Equation: Superposition Integral. The equation

$$\dot{x}(t) = A(t)x(t) + B(t)u(t), \quad x(t_0) = x_0 \quad (17)$$

satisfies a Lipschitz condition if the unforced or homogeneous equation does, since

$$\| [A(t)x(t) + B(t)u(t)] - [A(t)\tilde{x}(t) + B(t)u(t)] \|$$
$$= \| A(t)x(t) - A(t)\tilde{x}(t) \| \quad (18)$$

We shall solve the equation by first determining the impulse response of the system represented by the equation and then using the superposition property to handle a general input. Later we shall describe certain other methods as well.

To calculate the impulse response, we first consider the single-input equation

$$\dot{x}(t) = A(t)x(t) + b(t)u(t), \quad x(t_{0-}) = 0, \quad u(t) = \delta(t - t_0) \quad (19)$$

The solution of this equation for $t \geq t_{0+}$ is the same as the solution of the homogeneous equation

$$\dot{x}(t) = A(t)x(t), \quad x(t_{0+}) = b(t_0)$$

Therefore the impulse response is

$$h(t, t_0) = \Phi(t, t_0)b(t_0)$$

and the response to an arbitrary input $u(\cdot)$ is given (for zero initial state) by the superposition integral,

$$x(t) = \int_{t_0}^{t} \Phi(t, \tau)b(\tau)u(\tau) \, d\tau \tag{20a}$$

Now if we have m inputs rather than just one, the use of superposition will show that

$$x(t) = \sum_{1}^{m} \int_{t_0}^{t} \Phi(t, \tau)b_i(\tau)u_i(\tau) \, d\tau$$

$$= \int_{t_0}^{t} \Phi(t, \tau)B(\tau)u(\tau) \, d\tau \tag{20b}$$

Finally, the effect of nonzero initial conditions at $t = t_{0-}$ is easily accommodated by combining (6) and (20) to obtain the complete solution

$$x(t) = \Phi(t, t_0)x_0 + \int_{t_0}^{t} \Phi(t, \tau)B(\tau)u(\tau) \, d\tau \tag{21}$$

A rapid check of the correctness of the final result can be obtained again using our fundamental device (3) for checking expressions involving solutions of time-variant equations. Thus we just verify that the solution takes the value x_0 at $t = t_0$ and that its derivative is

$$A(t)\Phi(t, t_0)x_0 + \int_{t_0}^{t} A(t)\Phi(t, \tau)b(\tau)u(\tau) \, d\tau + \Phi(t, t)b(t)u(t)$$

$$= A(t)x(t) + b(t)u(t)$$

Input-Output Relations. For the realization

$$\begin{aligned} \dot{x}(t) &= A(t)x(t) + B(t)u(t), \qquad x(t_0) = x_0 \\ y(t) &= C(t)x(t) + D(t)u(t) \end{aligned} \tag{22}$$

the impulse response from $u(\cdot)$ to $y(\cdot)$ is, using (21),

$$h(t, \tau) = C(t)\Phi(t, \tau)B(\tau) + D(t)\delta(t - \tau) \tag{23}$$

An important (but not unexpected) property of the impulse-response function is that it is invariant under certain state transformations.

Let $P(\cdot)$ be an $n \times n$ matrix such that

$$P^{-1}(t) \text{ and } \dot{P}(t) \text{ exist for all } t \tag{24}$$

and let

$$x(t) = P(t)\xi(t), \qquad \xi(t) = P^{-1}(t)x(t) \tag{25}$$

Then it can be checked that the realizations (22) and

$$\dot{\xi}(t) = [P^{-1}(t)A(t)P(t) - P^{-1}(t)\dot{P}(t)]\xi(t) + P^{-1}(t)B(t)u(t)$$
$$y(t) = C(t)P(t)\xi(t) + D(t)u(t), \qquad \xi(t_0) = P^{-1}(0)x_0$$
(26)

have the same impulse-response function.

An important special case is obtained by choosing $P(\cdot)$ so that

$$P^{-1}(t)A(t)P(t) - P^{-1}(t)\dot{P}(t) \equiv 0 \tag{27a}$$

or, equivalently,

$$\dot{P}(t) = A(t)P(t) \tag{27b}$$

Any nonsingular $n \times n$ matrix $P(\cdot)$ that satisfies (27b) is said to be a *fundamental matrix* of $A(\cdot)$ (see also Example 9.1-3 and Exercise 9.1-4). In particular we may set $P(t_0) = I$, in which case

$$P(t) = \Phi(t, t_0), \qquad \text{the state-transition matrix of } A(\cdot) \tag{28}$$

Then the resulting realization is

$$\dot{\xi}(t) = \Phi(t_0, t)B(t)u(t), \qquad \xi(t_0) = x_0$$
$$y(t) = C(t)\Phi(t, t_0)\xi(t) + D(t)u(t)$$
(29)

which, it should be noted, does not have any *state-variable* feedback (cf. Fig. 9.1-1); its state-transition matrix is identity. Therefore the form (29) is often useful in simplifying calculations.

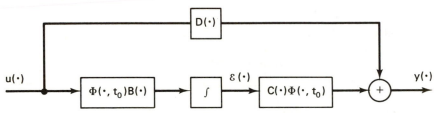

Figure 9.1-1   Realization of Eq. (29).

Thus we see immediately (just integrate) that

$$\xi(t) = x_0 + \int_{t_0}^{t} \Phi(t_0, \tau)B(\tau)u(\tau)\, d\tau \tag{30}$$

Therefore, using (25) and (28), we can write

$$x(t) = \Phi(t, t_0)\xi(t) = \Phi(t, t_0)x_0 + \int_{t_0}^{t} \Phi(t, t_0)\Phi(t_0, \tau)B(\tau)u(\tau)\, d\tau$$
$$= \Phi(t, t_0)x_0 + \int_{t_0}^{t} \Phi(t, \tau)B(\tau)u(\tau)\, d\tau$$

which is the result (21) we obtained somewhat more laboriously earlier. [In differential equation theory, the method just used is called Lagrange's method of *variation of constants*.] This simplifying transformation can be useful in other calculations as well.

However, it should be pointed out that the transformation (24)–(25) can adversely affect the internal properties of the realization. For example, consider the effect of the transformation (29) on a constant realization $\{A, B, C\}$—some of the entries of the transformed matrices $\Phi(t_0, \cdot)B$ and $C\Phi(\cdot, t_0)$ can grow exponentially. However, if we require that the transformation matrix $P(\cdot)$ be such that

1. $P(\cdot)$ has a continuous derivative on $[t_0, \infty)$ (31a)

2. $P(\cdot)$ and $\dot{P}(\cdot)$ are bounded in $[t_0, \infty)$ (31b)

3. There exists a constant m such that
$$0 < m < |\det P(t)|, \qquad t \geq t_0 \tag{31c}$$

then it can be shown that the transformation will preserve the internal stability of the realization [5]. Transformations obeying the above conditions are known as *Lyapunov transformations*.

We shall now present some worked examples to illustrate the results and methods of this section.

Example 9.1-1.

1. Show that

$$P = \int_0^\infty e^{At} Q e^{A't} \, dt, \qquad A \text{ stable}$$

satisfies the Lyapunov equation $AP + PA' + Q = 0$

2. Is it true that

$$P = e^{At} P e^{A't} + \int_0^t e^{As} Q' e^{A's} \, ds \ ?$$

Solution

1.

$$AP + PA' = \int_0^\infty [Ae^{At} Q e^{A't} + e^{At} Q e^{A't} A'] \, dt$$

$$= \int_0^\infty \frac{d}{dt}[e^{At} Q e^{A't}] \, dt = e^{At} Q e^{A't} \Big|_0^\infty$$

But at $t = \infty$, the integrand is zero because A is stable (i.e., its eigenvalues have negative real parts). Therefore

$$AP + PA' = 0 - e^{At} Q e^{A't} \big|_{t=0} = -Q$$

2. The two sides have the same value P when $t = 0$. The LHS obeys the differential equation

$$\frac{d[\text{LHS}]}{dt} = 0, \qquad t \geq 0$$

As for the RHS, we have

$$\frac{d}{dt}[\text{RHS}] = Ae^{At}Pe^{A't} + e^{At}Pe^{A't}A' + e^{At}Qe^{A't}$$

But

$$Ae^{At} = e^{At}A, \qquad e^{A't}A' = A'e^{A't}$$

so that

$$\frac{d}{dt}[\text{RHS}] = e^{At}[AP + PA' + Q]e^{A't} = e^{At} \cdot 0 \cdot e^{A't} = 0$$

by the result of part 1. Therefore the LHS and RHS obey the same equation,

$$\frac{dM(t)}{dt} = 0, \qquad M(0) = P$$

which clearly has a unique solution $[M(t) = P]$. Therefore the conjectured equality is true. ∎

Example 9.1-2.

Show that a $p \times m$ impulse-response matrix function $h(t, \tau)$ has a finite-dimensional state-variable realization if and only it can be written in the form

$$h(t, \tau) = D(t)\delta(t - \tau) + M(t)N(\tau), \qquad t \geq \tau$$

where $D(\cdot)$, $M(\cdot)$, $N(\cdot)$ are matrices of conformable dimensions.

Solution.   A finite-dimensional realization

$$\dot{x}(t) = A(t)x(t) + B(t)u(t)$$
$$y(t) = C(t)x(t) + D(t)u(t), \qquad a \leq t \leq b$$

has impulse response

$$h(t, \tau) = C(t)\Phi(t, \tau)B(\tau) + D(t)\delta(t - \tau)$$

where $\Phi(\cdot, \cdot)$ is the state-transition matrix of $A(\cdot)$. If we let, for any $t_0 \in [a, b]$,

$$M(t) = C(t)\Phi(t, t_0), \qquad N(t) = \Phi(t_0, \tau)B(\tau)$$

then $h(t, \tau)$ has the form specified in the problem. This proves the *only if* part of the problem.

For the *if* part, given $h(t, \tau)$ as in the problem, we note that

$$\dot{x}(t) = N(t)u(t)$$
$$y(t) = M(t)x(t) + D(t)u(t)$$

is a finite-dimensional realization of $h(t, \tau)$.   ∎

Example 9.1-3.

1. Show that the equation

$$\ddot{y}(t) + \frac{4}{t}\dot{y}(t) + \frac{2}{t^2}y(t) = u(t)$$

has a realization

$$A(t) = \begin{bmatrix} -4/t & -2/t^2 \\ 1 & 0 \end{bmatrix}, \quad B = \begin{bmatrix} 1 \\ 0 \end{bmatrix}$$
$$C = [0 \quad 1]$$

2. Show that two linearly independent solutions of $\dot{x}(t) = A(t)x(t)$ are

$$\begin{bmatrix} -1/t^2 \\ 1/t \end{bmatrix} \quad \text{and} \quad \begin{bmatrix} 2/t^3 \\ -1/t^2 \end{bmatrix}$$

3. Hence find the state-transition matrix.

Solution.

1. We can use Kelvin's method (cf. Sec. 2.1.1) as shown in the figure. Therefore

$$\begin{bmatrix} \dot{x}_1(t) \\ \dot{x}_2(t) \end{bmatrix} = \begin{bmatrix} -4/t & -2/t^2 \\ 1 & 0 \end{bmatrix}\begin{bmatrix} x_1(t) \\ x_2(t) \end{bmatrix} + \begin{bmatrix} 1 \\ 0 \end{bmatrix}u(t)$$
$$y(t) = [0 \quad 1][x_1(t) \quad x_2(t)]'$$

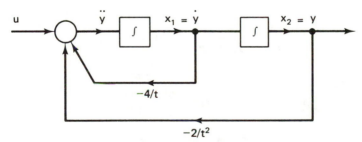

2. We note that $x_2(t) = y(t)$, $x_1(t) = \dot{y}(t)$ and that

$$y(t) = t^{-1} \quad \text{and} \quad y(t) = -t^{-2}$$

are two linearly independent solutions of the homogeneous equation $(u(\cdot) \equiv 0)$.

Therefore

$$\psi_1(t) = \begin{bmatrix} -1/t^2 \\ 1/t \end{bmatrix} \quad \text{and} \quad \psi_2(t) = \begin{bmatrix} 2/t^3 \\ -1/t^2 \end{bmatrix}$$

are linearly independent solutions of $\dot{x}(t) = A(t)x(t)$.

3. To find $\Phi(t, \tau)$ we need two linearly independent solutions $\{\phi_i(\cdot, \tau)\}$ such that

$$\phi_1(\tau, \tau) = \begin{bmatrix} 1 \\ 0 \end{bmatrix}, \quad \phi_2(\tau, \tau) = \begin{bmatrix} 0 \\ 1 \end{bmatrix}$$

We know that

$$\psi_1(\tau) = \begin{bmatrix} -\tau^{-2} \\ \tau^{-1} \end{bmatrix}, \quad \psi_2(\tau) = \begin{bmatrix} 2\tau^{-3} \\ -\tau^{-2} \end{bmatrix}$$

which suggests that suitable choices can be

$$\phi_1(t, \tau) = \tau^2\psi_1(t) + \tau^3\psi_2(t)$$
$$\phi_2(t, \tau) = 2\tau\psi_1(t) + \tau^2\psi_2(t)$$

Therefore

$$\Phi(t, \tau) = \begin{bmatrix} -\tau^2 t^{-2} + 2\tau^3 t^{-3} & -2\tau t^{-2} + 2\tau^2 t^{-3} \\ \tau^2 t^{-1} - \tau^3 t^{-2} & 2\tau t^{-1} - \tau^2 t^{-2} \end{bmatrix}$$

[We can check this by computing $d\Phi(t, \tau)/dt$ and verifying that it is equal to $A(t)\Phi(t, \tau)$.]

Another method of finding $\Phi(t, \tau)$ is to first form a so-called fundamental matrix $M(t)$ by putting together n ($n = 2$ in this case) linearly independent solutions of $\dot{x}(t) = A(t)x(t)$. Then we can verify that

$$\Phi(t, \tau) = M(t)M^{-1}(\tau)$$

In our problem we can take

$$M(t) = [\psi_1(t) \quad \psi_2(t)] = \begin{bmatrix} -t^{-2} & 2t^{-3} \\ t^{-1} & -t^{-2} \end{bmatrix}$$

Then

$$M^{-1}(\tau) = \begin{bmatrix} \tau^2 & 2\tau \\ \tau^3 & \tau^2 \end{bmatrix}$$

and we can verify that $M(t)M^{-1}(\tau) = \Phi(t, \tau)$. ∎

Exercises

9.1-1.

Show that $X(t) = e^{At}Ce^{Bt}$ is the unique solution of the equation $\dot{X}(t) = AX(t) + X(t)B$, $X(0) = C$. What would be the solution if A and B were time-variant matrices?

9.1-2.

a. Show that $(d/dt)P^{-1}(t) = -P^{-1}(t)\dot{P}(t)P^{-1}(t)$. *Hint:* Differentiate the identity $I = P(t)P^{-1}(t)$.

b. Use this result to show that if $\Phi(t, \tau)$ is the state-transition matrix of $A(t)$, then $\Phi'(\tau, t)$ is the state-transition matrix of $-A'(t)$.

9.1-3.

a. If

$$A(t) = \begin{bmatrix} A_{11}(t) & A_{12}(t) \\ 0 & A_{22}(t) \end{bmatrix}$$

show that we can write

$$\Phi(t, t_0) = \begin{bmatrix} \Phi_{11}(t, t_0) & \Phi_{12}(t, t_0) \\ 0 & \Phi_{22}(t, t_0) \end{bmatrix}, \quad \frac{d}{dt}\Phi_{ii}(t, t_0) = A_{ii}\Phi_{ii}(t, t_0), \quad i = 1, 2$$

b. Check that

$$A(t) = \begin{bmatrix} -1 & e^{2t} \\ 0 & -1 \end{bmatrix}$$

yields

$$\Phi(t, 0) = \begin{bmatrix} e^{-t} & (e^t - e^{-t})/2 \\ 0 & e^{-t} \end{bmatrix}$$

Note that though $A(t)$ has eigenvalues $(-1, -1)$ for all t, the system $\dot{x}(t) = A(t)x(t)$ is not stable in that there can be initial conditions for which $x(t) \longrightarrow \infty$ as $t \longrightarrow \infty$. Therefore there is no significance to the concept of "time-variant natural frequencies."

9.1-4.   *Fundamental Matrices and Wronskians*

A *fundamental matrix* of $A(\cdot)$ is any solution of the matrix differential equation $dM(t)/dt = A(t)M(t)$, det $M(t_0) \neq 0$, $a \leq t_0 \leq b$.

a. Prove that $M(t)$ is nonsingular for any value of t.

b. Prove that $\Phi(t, \tau) = M(t)M^{-1}(\tau)$.

c. Show that det $M(t) = $ det $M(t_0) \cdot \exp \int_{t_0}^t \text{tr } A(\tau)d\tau$. Det $M(t)$ is known as the *Wronskian* of the differential equation.

9.1-5.

Verify that the equation $\dot{x}(t) = A(t)x(t)$, with

$$A(t) = \begin{bmatrix} 2 & -e^t \\ e^{-t} & 1 \end{bmatrix}$$

has the state-transition matrix

$$\Phi(t, t_0) = \begin{bmatrix} e^{2(t-t_0)} \cos(t - t_0) & -e^{(2t-t_0)} \sin(t - t_0) \\ e^{(t-2t_0)} \sin(t - t_0) & e^{(t-t_0)} \cos(t - t_0) \end{bmatrix}$$

9.1-6.

Find the state-transition matrix for the equation $\dot{x}(t) = e^{-At}Ce^{At}x(t)$. A and C are constant square matrices. *Hint:* Find a differential equation for $z(t) = e^{At}x(t)$.

9.1-7.

If A and C are stability matrices, show that

$$\int_0^\infty e^{A't}Ce^{At}\,dt = \frac{1}{2\pi j}\int_{-j\infty}^{j\infty}(-sI - A')^{-1}C(sI - A)^{-1}\,ds$$

9.1-8.

a. Show that $\det[I + A(t)\Delta] = 1 + \Delta\ \text{tr}\ A(t) + $ higher-order terms in Δ.

b. Use this result to show that $\det\Phi(t, \tau) = \exp\int_\tau^t \text{tr}\ [A(s)]\,ds$.

9.1-9.   *Periodic Systems*

Let $\dot{x}(t) = A(t)x(t)$, $A(t + T) = A(t)$. Define a constant matrix R via the equation $e^{RT} = \Phi(T, 0)$, $\Phi(\cdot, \cdot) = $ state-transition matrix of $A(\cdot)$ and let $P(t) = e^{Rt}\Phi(0, t), \xi(t) = P(t)x(t)$.

a. Show that $\dot{\xi}(t) = R\xi(t)$ and $\Phi(t, t_0) = P^{-1}(t)e^{R(t-t_0)}P(t_0)$ with $P(t + T) = P(t)$.

b. Determine R and $P(\cdot)$ when

$$A(t) = \begin{bmatrix} -1 + \cos t & 0 \\ 0 & -2 + \cos t \end{bmatrix}$$

Notice that by this so-called *Floquet method* we have reduced the study of the original periodic system to that of a constant system.

9.1-10.   (*De Russo, Roy, Close*)

Show that the differential equation $\ddot{y}(t) - 2t\dot{y}(t) - (2 - t^2)y(t) = 0$ can be realized by state equations with

(a) $A_1(t) = \begin{bmatrix} 0 & 1 \\ 2 - t^2 & 2t \end{bmatrix}$

or

(b) $A_2(t) = \begin{bmatrix} t & 1 \\ 1 & t \end{bmatrix}$

Which choice is more convenient for computation of $\Phi(t, \tau)$?

9.1-11.   *Some Formulas for Two-Point Boundary-Value Equations*

Consider the two-point system of linear differential equations

$$\begin{bmatrix} \dot{x}_1(t) \\ \dot{x}_2(t) \end{bmatrix} = \underbrace{\begin{bmatrix} A_{11}(t) & A_{12}(t) \\ A_{21}(t) & A_{22}(t) \end{bmatrix}}_{A(t)}\begin{bmatrix} x_1(t) \\ x_2(t) \end{bmatrix}$$

with the boundary conditions $x_1(t_0) = x_{10}, x_2(t_f) = x_{2f}$. Let $\Phi(t, t_0)$ be the state-transition matrix of $A(\cdot)$, and partition it correspondingly as

$$\Phi(t, t_0) = \begin{bmatrix} \Phi_{11}(t, t_0) & \Phi_{12}(t, t_0) \\ \Phi_{21}(t, t_0) & \Phi_{22}(t, t_0) \end{bmatrix}$$

Assume that $\Phi_{22}(t, t_0)$ is nonsingular;

a. We can write

$$\begin{bmatrix} x_1(t) \\ x_2(t_0) \end{bmatrix} = \begin{bmatrix} P_{11}(t, t_0) & P_{12}(t, t_0) \\ P_{21}(t, t_0) & P_{22}(t, t_0) \end{bmatrix} \begin{bmatrix} x_1(t_0) \\ x_2(t) \end{bmatrix}$$

where (with arguments omitted for simplicity) $P_{11} \triangleq \Phi_{11} - \Phi_{12}\Phi_{22}^{-1}\Phi_{21}$, $P_{12} \triangleq \Phi_{12}\Phi_{22}^{-1}, P_{21} = -\Phi_{22}^{-1}\Phi_{21}, P_{22} = \Phi_{22}^{-1}$; and

b. The P_{ij} obey the differential equations

$$\dot{P}_{12} = A_{11}P_{12} - P_{12}A_{22} - P_{12}A_{21}P_{12} + A_{12}$$
$$\text{(a Riccati differential equation)}$$
$$\dot{P}_{11} = [A_{11} - P_{12}A_{21}]P_{11} \quad , \quad \dot{P}_{22} = -P_{22}[A_{22} + A_{21}P_{12}]$$
$$\dot{P}_{21} = -P_{22}A_{21}P_{11}$$

with initial conditions $P_{ij}(t_0, t_0) = I\delta_{ij}, i, j = 1, 2$. *Note:* We see that $P_{12}(\cdot, t_0)$ is a basic quantity since all the others are specified once $P_{12}(\cdot, t_0)$ is known. Note that $P_{12}(\cdot, t_0)$ obeys a Riccati-type differential equation, which by proper identification of the $A_{ij}(\cdot)$ matrices can be made identical to the Riccati equation encountered in the optimal quadratic regulator problem (cf. Sec. 3.4.3). The other three functions $\{P_{11}, P_{21}, P_{22}\}$ also play a role in understanding the properties of the Riccati variable $P_{12}(\cdot, t_0)$, as is brought out by considering a scattering-theory framework for least-squares estimation and control problems (see, e.g., G. Verghese, B. Friedlander, and T. Kailath, "Scattering Theory and Least-Squares Estimation," Pt. III., *IEEE Trans. Automat. Contr.*, **AC-25**, 1980, and the references therein).

9.2   CONTROLLABILITY AND OBSERVABILITY PROPERTIES

In Sec. 2.3 we showed that if we permitted impulsive inputs we could move a controllable realization $\{A, b\}$ from an initial state x_0 to a new arbitrary state. Moreover, because of the impulsive nature of the input this change could be made in *zero* time (viz., in $0-$ to $0+$). We can now investigate the problem under the constraint of nonimpulsive finite energy inputs, in which case of course we cannot expect the change of state to be achievable in zero time (Sec. 9.2.1).

In Sec. 9.2.1, we shall present various equivalent conditions for the solution of the state controllability for time-variant systems and shall show how they specialize to the time-invariant case. In Sec. 9.2.2, we shall consider the analogous results for the state-observability problem. We shall also examine the duality between these concepts and show that the so-called *adjoint* realization $\{-A'(\cdot), C'(\cdot), B'(\cdot)\}$ is more appropriate in the time-variant case than the dual realization $\{A', C', B'\}$ used in the constant-parameter case.

9.2.1   The Controllability Gramian

Suppose we have a realization

$$\dot{x}(t) = A(t)x(t) + B(t)u(t), \qquad x(t_0) = x_0 \tag{1}$$

and we wish to choose a finite energy input $u(\cdot)$ that will bring the state to zero by a desired time, say t_f. We shall show that this can be done if and only if [6]

the rows of $\Phi(t_0, \cdot)B(\cdot)$ are linearly independent on $[t_0, t_f]$ \qquad (2)

where $\Phi(\cdot, \cdot)$ is the state-transition matrix of $A(\cdot)$. That is, there must exist no constant nonzero vector q such that

$$q'\Phi(t_0, t)B(t) \equiv 0, \qquad t_0 \le t \le t_f \tag{3}$$

A well-known and more explicit test for linear independence of any functions $\{l_i(\cdot), i = 1, \ldots, n\}$ is that their *Gramian* matrix

$$G = [G_{ij}], \qquad G_{ij} = \int_{t_0}^{t_f} l_i(\tau)l_j(\tau)\, d\tau$$

be nonsingular (the simple proof is outlined in Exercise 9.2-1). In our case, this means that a necessary and sufficient condition for controllability of (1) to the origin is that the following *controllability Gramian* be nonsingular [7]:

$$\mathcal{C}(t_0, t_f) = \int_{t_0}^{t_f} \Phi(t_0, \tau)B(\tau)B'(\tau)\Phi'(t_0, \tau)\, d\tau \tag{4}$$

Some properties and an alternative expression for computing $\mathcal{C}(t_0, t_f)$ are given in Exercise 9.2-2. Here we shall show how conditions (2) and (4) arise.

Given (1), our task is to find $u(\cdot)$ so that

$$0 = x(t_f) = \Phi(t_f, t_0)x_0 + \int_{t_0}^{t_f} \Phi(t_f, \tau)B(\tau)u(\tau)\, d\tau \tag{5}$$

or

$$-\Phi(t_f, t_0)x_0 = \int_{t_0}^{t_f} \Phi(t_f, \tau)B(\tau)u(\tau)\, d\tau$$

By using Eqs. (9.1-8–9.1-9), this can be rewritten as

$$-x_0 = \int_{t_0}^{t_f} L(\tau)u(\tau)\,d\tau, \qquad L(\tau) = \Phi(t_0, \tau)B(\tau) \tag{6}$$

How can we solve (6), which is an *integral equation* because the unknown function $u(\cdot)$ appears under an integral sign?

We can try to get some insight into this question by replacing the *infinite* number of equations in (6) (one for each value of τ) by an *approximate* finite set

$$-x_0 = \sum_{1}^{N-1} L(\tau_i)u(\tau_i)\Delta \tag{7a}$$

where

$$\tau_i = t_0 + i\Delta, \qquad i = 0, \dots, N, \qquad N\Delta = t_f - t_0 \tag{7b}$$

$$L(\tau_i) = \Phi(t_0, \tau_i)B(\tau_i) \tag{7c}$$

Let

$$\mathcal{U} = [u(\tau_0) \quad \cdots \quad u(\tau_{N-1})]' \tag{8a}$$

$$\mathcal{L} = [L(\tau_0)\Delta \quad \cdots \quad L(\tau_{N-1})\Delta], \qquad \text{an } n \times N \text{ matrix} \tag{8b}$$

Then (7a) can be rewritten as

$$\mathcal{L}\mathcal{U} = -x_0 \tag{8c}$$

a set with more unknowns than equations. [Such equations are studied in Secs. A.4 and A.5 (see also Sec. 6.2.1), a review of which would be helpful at this time.] Clearly, there can be a solution $\mathcal{U}$ if and only if $-x_0 \in \mathcal{R}(\mathcal{L})$; i.e., $-x_0$ is a linear combination of the columns of $\mathcal{L}$. Since x_0 is an arbitrary n-vector, there can be a solution if and only if at least n columns of $\mathcal{L}$ are linearly independent. But the fact that row rank equal column rank means that there will be a solution if and only if

$$\text{the rows of } \mathcal{L} \text{ are linearly independent} \tag{9}$$

i.e., if and only if $\mathcal{L}$ has full rank. Clearly, condition (9) makes plausible the fact that (6) will have a solution if and only if

$$\text{the rows of } L(\tau) \text{ are linearly independent}$$

which was condition (2). A direct proof will be given presently, but let us continue with the heuristic discussion a little longer.

Because there are more unknowns than equations, there will be more than one solution of the approximate equations (8c) when (9) is met. We ask the reader to prove this and go on here to note that one particular solution is

$$\mathcal{U}_* = -\mathcal{L}'(\mathcal{L}\mathcal{L}')^{-1}x_0 \tag{10}$$

as can be verified by plugging into Eq. (8c). We may ask why $\mathcal{L}\mathcal{L}'$ is invertible. This can be proved by using Sylvester's equality (Exercise A.18) or by adapting the method indicated in Exercise 9.2-1. Now from (7)–(10) we see that

$$\mathcal{L}\mathcal{L}' = \sum_0^{N-1} L(\tau_i)L'(\tau_i)\Delta^2$$

$$= \Delta \sum_0^{N-1} \Phi(t_0, \tau_i)B(\tau_i)B'(\tau_i)\Phi'(t_0, \tau_i)\Delta$$

and that

$$u_*(\tau_i) = -B'(\tau_i)\Phi'(t_0, \tau_i)\Delta(\mathcal{L}\mathcal{L}')^{-1}x_0$$

$$= -B'(\tau_i)\Phi'(t_0, \tau_i)\left[\sum_0^{N-1} \Phi(t_0, \tau_i)B(\tau_i)B'(\tau_i)\Phi'(t_0, \tau_i)\Delta\right]^{-1} x_0$$

Now taking limits as

$$\Delta \longrightarrow 0, \qquad N \longrightarrow \infty, \qquad N\Delta = t_f - t_0, \qquad \tau_i = \tau$$

we get

$$u_*(\tau) = -B'(\tau)\Phi'(t_0, \tau)\mathcal{C}^{-1}(t_0, t_f)x_0 \tag{11}$$

where $\mathcal{C}(t_0, t_f)$ is the controllability Gramian introduced above in (4), as the solution of the integral equation (6). Actually this can be easily confirmed by noting that

$$\int_{t_0}^{t_f} L(\tau)u_*(\tau)\, d\tau = -\int_{t_0}^{t_f} \Phi(t_0, \tau)B(\tau)B'(\tau)\Phi'(t_0, \tau)\mathcal{C}^{-1}(t_0, t_f)x_0\, d\tau$$

$$= -\mathcal{C}(t_0, t_f)\mathcal{C}^{-1}(t_0, t_f)x_0 = -x_0$$

In other words, if we assume that the controllability Gramian is nonsingular, then there is a solution to the controllability-to-the-origin problem, and we could have proved this by just *guessing* that (11) was a solution and checking it. However, the heuristic discussion serves to explain the origins of $u_*(\cdot)$. The careful reader will recall from Sec. 6.2.1 that this particular solution $u_*(\cdot)$ should have a certain optimality property—this is pursued in Exercises 9.2-10 and 9.2-11.

Let us, however, first complete the present discussion by proving directly, as promised above, that the condition (2) that $L(\cdot)$ have full rank is a necessary condition for the solvability of (6). Indeed, if this were not so, there would be a linear combination of the rows of $L(\cdot)$ that would add to zero, say,

$$q'L(\tau) \equiv 0, \qquad t_0 \leq \tau \leq t_f$$

Now suppose

$$x_0 = q$$

Then if there existed a $u(\cdot)$ that would satisfy (6) we would have

$$-q = \int_{t_0}^{t_f} L(\tau)u(\tau)\, d\tau$$

so that

$$-q'q = \int_{t_0}^{t_f} q'L(\tau)u(\tau)\, d\tau = \int 0 \cdot u(\tau)\, d\tau = 0$$

But this is impossible unless (every component of) q is zero, contrary to our initial assumption.

We may remark that the above conditions (2) and (4) for controllability to the origin are also necessary and sufficient for state controllability to any other state. We leave the proof to the reader. However, we should note that if we are not required to go from *arbitrary* initial states to *arbitrary* final states, then the above conditions need not apply. The reader should be able to show that to go from x_0 at $t = t_0$ to x_f at $t = t_f$ it is necessary and sufficient that

$$\Phi(t_0, t_f)x_f - x_0 \text{ be a linear combination of the columns of } \mathcal{C}(t_0, t_f) \qquad (12)$$

Specialization to Time-Invariant Systems. When $\{A(\cdot), B(\cdot)\}$ are constant or time invariant, we can write

$$\Phi(t_0, \tau) = \exp A(t_0 - \tau)$$

and we can also, without loss of generality, take $t_0 = 0$. Then the basic controllability conditions (2)–(3) can be written as follows: The pair $\{A, B\}$ is controllable (to the origin) if and only if the rows of $e^{-A\tau}B$ are linearly independent over $(0, t_f)$ or, equivalently, if and only if

$$qe^{-A\tau}B \equiv 0, \qquad 0 \leq \tau \leq t_f, \qquad \Longrightarrow q = 0 \qquad (13)$$

Now it can be seen in several ways (show this) that

$$qe^{-A\tau}B \equiv 0, \qquad 0 \leq \tau \leq t_f$$

implies that

$$qA^iB = 0, \qquad i = 0, 1, \ldots, n-1 \qquad (14)$$

The converse is also true, by virtue of the Cayley-Hamilton theorem. Therefore we can say that $\{A, B\}$ will be controllable if and only if

$$q[B \quad AB \quad \cdots \quad A^{n-1}B] = 0 \Longrightarrow q = 0$$

But this in turn will be true if and only if the matrix

$$\mathcal{C} \triangleq [B \quad AB \quad \cdots \quad A^{n-1}B] \text{ has full rank } n$$

which was the condition developed in Secs. 2.3.2 and 6.2 for *instantaneous* controllability to the origin. We see in fact that the value of t_f plays no role in the above analysis, and so the interval length can be arbitrarily small for time-invariant systems.

One might ask if there is an analogous possibility for time-variant systems. In fact, it is not difficult to show (see [9]) that one can take any state $x(t-)$ to an arbitrary state at $t+$ by using impulsive inputs if and only if rank $\mathcal{C}_k(t) = n$, where

$$\mathcal{C}_k(t) \triangleq [B(t) \quad (A(t) - d/dt)B(t) \ldots (A(t) - d/dt)^{k-1}B(t)] \qquad (15)$$

for some k (possibly greater than n); it is assumed that $\{A(\cdot), b(\cdot)\}$ are sufficiently differentiable that $\mathcal{C}_k(t)$ makes sense. The example

$$\dot{x}(t) = \sin t \, u(t), \qquad t \geq 0$$

shows why the condition $k > n$ may be necessary. The matrix $\mathcal{C}_n(t)$ has some interesting theoretical relationships with the controllability Gramian (see Exercise 9.2-12).

Example 9.2-1.

Let

$$A(t) = \begin{bmatrix} 0 & 0 \\ 0 & -1 \end{bmatrix}, \qquad b_1(t) = \begin{bmatrix} 1 \\ e^{-2t} \end{bmatrix}, \qquad b_2(t) = \begin{bmatrix} 1 \\ e^{-t} \end{bmatrix}$$

1. Find $\exp At$.
2. Determine the controllability or otherwise over $[0, 1]$ of $\{A(\cdot), b_1(\cdot)\}$.
3. Repeat for $\{A(\cdot), b_2(\cdot)\}$.

Solution.

1.

$$e^{At} = \mathcal{L}^{-1} \begin{bmatrix} s & 0 \\ 0 & s+1 \end{bmatrix}^{-1} = \begin{bmatrix} 1 & 0 \\ 0 & e^{-t} \end{bmatrix} = \Phi(t, 0)$$

2. The rows of

$$\Phi(0, t)b_1(t) = e^{-At}b_1(t) = \begin{bmatrix} 1 & 0 \\ 0 & e^{t} \end{bmatrix} \begin{bmatrix} 1 \\ e^{-2t} \end{bmatrix} = \begin{bmatrix} 1 \\ e^{-t} \end{bmatrix}$$

are clearly linearly independent over [0, 1], and therefore $\{A(\cdot), b_1(\cdot)\}$ is controllable over [0, 1] or, in fact, over any interval.

3. The rows of

$$e^{-At}b_2(t) = \begin{bmatrix} 1 & 0 \\ 0 & e^{+t} \end{bmatrix} \begin{bmatrix} 1 \\ e^{-t} \end{bmatrix} = \begin{bmatrix} 1 \\ 1 \end{bmatrix}$$

are linearly dependent over [0, 1], and therefore $\{A(\cdot), b_2(\cdot)\}$ is not controllable over [0, 1] or over any interval.

Note that linear independence or dependence refers to combinations of vectors with *constant* (real or complex) coefficients; weights that are functions of t are not permitted. Note also that the ranks of the matrices

$$[b_1(t) \quad Ab_1(t)] \quad \text{or} \quad [b_1(t) \quad e^{At}b_1(t)]$$

are not relevant to the question of controllability (however, see Exercise 9.2-12).

The controllability Gramian tests are relevant but we shall leave them to the reader to compute. In simple problems, the Gramian tests are often more complicated than the others, but they might be more reliable for actual numerical computation and analysis; we note that the Gramians can be recursively computed via certain linear matrix differential equations (see Exercise 9.2-2 and also C. Van Loan, *IEEE Trans. Automat. Contr.*, **AC-23**, pp. 395–404, 1978).

9.2.2   The Observability Gramian and a Duality

Consider a realization

$$\dot{x}(t) = A(t)x(t) + B(t)u(t)$$
$$y(t) = C(t)x(t), \qquad t_0 \leq t \leq t_f$$

where $A(\cdot)$ and $C(\cdot)$ are continuous. As long as $B(\cdot)u(\cdot)$ is known its effect can be eliminated, so that without loss of generality we can assume $u(\cdot) \equiv 0$. Such a realization will be said to be *observable* on $[t_0, t_f]$ if $x(t_0)$ can be deduced from knowledge of $\{y(t), t_0 \leq t \leq t_f\}$ [and of $A(\cdot)$ and $C(\cdot)$]. We shall show [cf. (2)–(4)] that this will be true if and only if

the columns of $C(\cdot)\Phi(\cdot, t_0)$ are linearly independent over $[t_0, t_f]$ (16)

or, equivalently, if and only if

$$C(t)\Phi(t, t_0)p \equiv 0, \qquad t_0 \leq t \leq t_f, \qquad \Longrightarrow p \equiv 0 \tag{17}$$

or, equivalently, if and only if the Gramian matrix

$$\Theta(t_0, t_f) = \int_{t_0}^{t_f} \Phi'(\tau, t_0)C'(\tau)C(\tau)\Phi(\tau, t_0) \, d\tau \tag{18}$$

is nonsingular. This matrix will be called the *observability Gramian* of $\{C(\cdot), A(\cdot)\}$.

To establish these results, we have to consider how to obtain x_0 unambiguously from the equation

$$y(t) = C(t)x(t) = C(t)\Phi(t, t_0)x_0, \qquad t_0 \le t \le t_f \tag{19}$$

Now, as compared to the situation in the controllability problem, we have n unknowns and an infinite number of equations, one for each t in $[t_0, t_f]$. This *dual* problem can be attacked by first working with an *approximate* finite set of equations

$$y(\tau_i) = C(\tau_i)\Phi(\tau_i, t_0)x_0, \qquad i = 0, \ldots, N$$

where

$$\tau_i = t_0 + i\Delta, \qquad N\Delta = t_f - t_0$$

Such equations were also studied in Sec. 6.2.1, and by following that discussion we would be led to conditions (16)–(18). However, we shall leave that approach as a useful exercise for the reader and give a direct proof here (following the direct proofs given in Sec. 9.2.1).

Thus returning to (19), we first note that if the columns of $C(\cdot)\Phi(\cdot, t_0)$ are linearly independent over $[t_0, t_f]$, so that the observability Gramian is nonsingular, then we can obtain x_0 from (19) as

$$x_0 = \Theta^{-1}(t_0, t_f) \int_{t_0}^{t_f} \Phi'(t, t_0)C'(t)y(t) \, dt$$

On the other hand, if the columns are linearly dependent so that

$$C(t)\Phi(t, t_0)q \equiv 0, \qquad t_0 \le t \le t_f, \qquad q \ne 0$$

then, clearly, choosing $x_0 = q$ will give us zero output over $[t_0, t_f]$, so that $x_0 = q$ cannot be *observed*. The other equivalences in (16)–(18) are now easy to obtain. We see also by arguments similar to those used at the end of Sec. 9.2.1 that conditions (16)–(18) all reduce in the time-invariant case to the familiar condition that

$$\text{rank } \{\Theta'(C, A) \triangleq [C' \quad A'C' \quad \cdots \quad A'^{n-1}C']\} = n$$

We note also that, as for controllability, under suitable differentiability assumptions we can introduce the matrix

$$\Theta_k'(t) \triangleq [C'(t) \quad \cdots \quad (A'(t) + d/dt)^{k-1}C'(t)]$$

which when $k = n$ has certain interesting connections with the observability Gramian (cf. Exercise 9.2-12).

By comparing the criteria for controllability and observability the reader will notice obvious *dualities*:

$$\text{rows} \longleftrightarrow \text{columns}$$
$$B(\cdot) \longleftrightarrow C'(\cdot)$$
$$C(\cdot) \longleftrightarrow B'(\cdot) \tag{20a}$$
$$\Phi(t_0, \cdot) \longleftrightarrow \Phi'(\cdot, t_0)$$

It is natural to ask from these correspondences whether

$$A(\cdot) \overset{?}{\longleftrightarrow} A'(\cdot) \tag{20b}$$

The answer is *no*. The proper correspondence is

$$A(\cdot) \longleftrightarrow -A'(\cdot) \tag{20c}$$

We can see this as follows. Let $\Psi(\cdot, t_0)$ be the state-transition matrix of $-A'(\cdot)$; i.e.,

$$\frac{d\Psi(t, t_0)}{dt} = -A'(t)\Psi(t, t_0), \qquad \Psi(t_0, t_0) = I$$

Now

$$(d/dt)\Psi^{-1}(t, t_0) = -\Psi^{-1}(t, t_0)\dot{\Psi}(t, t_0)\Psi^{-1}(t, t_0)$$
$$= +\Psi^{-1}(t, t_0)A'(t)$$

Therefore, taking transposes and using [cf. (9.1-9)]

$$\Psi^{-1}(t, t_0) = \Psi(t_0, t)$$

we shall have

$$\frac{d\Psi'(t_0, t)}{dt} = A(t)\Psi'(t_0, t), \qquad \Psi(t_0, t_0) = I$$

But then by uniqueness [of the solution to (9.1-14)] we must have

$$\Psi(t_0, \cdot) = \Phi'(\cdot, t_0) \tag{21}$$

where $\Phi(\cdot, t_0)$ is the state-transition matrix of $A(\cdot)$. Hence the correspondence is

$$\Phi(t_0, \cdot) \longleftrightarrow \Phi'(\cdot, t_0) = \Psi(t_0, \cdot) \quad \text{or} \quad A(\cdot) \longleftrightarrow -A'(\cdot)$$

Therefore the complete correspondence between time-variant realizations is given by

$$
\left.\begin{aligned}
\dot{x}(t) &= A(t)x(t) + B(t)u(t) \\
y(t) &= C(t)x(t), \quad t_0 \leq t \leq t_f
\end{aligned}\right\} \longleftrightarrow
\begin{cases}
\dot{x}(t) = -A'(t)x(t) + C'(t)u(t) \\
y(t) = B'(t)x(t), \quad t_0 \leq t \leq t_f
\end{cases}
$$

$$(22)$$

It is also easy to verify that the controllability Gramian of $\{A(\cdot), B(\cdot)\}$ is the observability Gramian of $\{B'(\cdot), -A'(\cdot)\}$ and that the observability Gramian of $\{C(\cdot), A(\cdot)\}$ is the controllability Gramian of $\{-A'(\cdot), C'(\cdot)\}$. For example, we see that

$$
\mathcal{O}[B'(\cdot), -A'(\cdot)] = \int_{t_0}^{t_f} \Psi'(\tau, t_0) B(\tau) B'(\tau) \Psi(\tau, t_0) \, d\tau
$$

$$
= \int_{t_0}^{t_f} \Phi(t_0, \tau) B(\tau) B'(\tau) \Phi'(t_0, \tau) \, d\tau
$$

$$
= \mathcal{C}[A(\cdot), B(\cdot)] \tag{23}
$$

The realization $\{-A'(\cdot), C'(\cdot), B'(\cdot)\}$ is said to be the *adjoint* of the realization $\{A(\cdot), B(\cdot), C(\cdot)\}$. Now we should note that for time-invariant systems we defined a *dual* realization of $\{A, B, C\}$ as $\{A', C', B'\}$, whereas the adjoint is $\{-A', C', B'\}$. However, it is easy to see that $\mathcal{O}(C', -A')$ is non-singular if and only if $\mathcal{O}(C', A)$ is nonsingular and so also for $\mathcal{C}(A', C')$ and $\mathcal{C}(-A', C')$. In other words, insofar as our applications in Chapters 2–8 were concerned, it makes little difference whether we use the *adjoint* or the *dual* realization. However, in the time-variant case, it is the adjoint that is more significant.

To emphasize the difference between the adjoint and the dual, we may note that for single input, single output time-invariant systems, the dual $\{A', c', b'\}$ realizes the same transfer function as $\{A, b, c\}$, while the adjoint $\{-A', c', b'\}$ realizes a different transfer function.

We shall study the adjoint system in more detail in Sec. 9.3.

Joint Observability and Controllability. To conclude this section, we might ask to examine the significance of joint observability and controllability for time-variant realizations. Unfortunately the situation here is more complicated, chiefly because causality of the impulse response imposes certain somewhat artificial constraints on the analysis (see [8] and [6]).

Here we shall only note that the most useful theorems are obtained by restricting attention to the class of so-called *uniform realizations* [10]. A realization is uniform if its coefficients are continuous and bounded and the realization is *uniformly controllable* in the sense that for all t, there exists a δ and a constant α, depending on δ but not on t, such that the controllability Gramian over $[t - \delta, t]$ is strictly positive definite

$$
\mathcal{C}(t - \delta, t) \geq \alpha(\delta)I > 0 \tag{24}
$$

and *uniformly observable* in the sense that for all t there exists a δ such that

$$\mathcal{O}(t, t + \delta) \geq \beta(\delta)I > 0. \tag{25}$$

(Recall that $A > B\,(A \geq B)$ means that $A - B$ is positive (semi) definite). These definitions were originally introduced by Kalman [11] in a slightly more general form, without the restriction to bounded coefficients. Kalman showed that these conditions sufficed to ensure the asymptotic stability of the optimal time-variant quadratic regulator [11]. Other important facts are that external and internal stability are equivalent for such realizations [12] and that all uniform realizations of a given impulse response can be related by Lyapunov transformations, which we may recall from (9.1.31) are a special class of time-variant similarity transformations.

Another important class of so-called *balanced realizations* is obtained when the controllability and observability Gramians are equal and diagonal

$$\mathcal{C}(t_0, t_f) = \mathcal{O}(t_0, t_f) = \Lambda.$$

The point is that one can now make a meaningful assignment of "less controllable" and "less observable" states. Without the equality of the Gramians one could trade-off controllability for observability or vice versa, as can be seen by considering diagonal realizations. We now also have a reasonable way of constructing reduced-order models—form a balanced realization (see Exercise 9.2-15) and delete the part of the state-space that is, by some criterion, "least controllable," and therefore also "least observable." This proposal was made by Moore [13], who used singular value decompositions of the Gramians to help identify the "least" controllable (and observable) states. The rationale for this comes from the result of Exercise 9.2-16. Moore [13] also showed the important result that a reduced-order approximation of a stable balanced realization is also stable and balanced. For several interesting related results, see also [14] and [15].

Exercises

9.2-1.   *Gramian Test for Linear Independence*
    Show that n functions $\{l_i(\tau),\ a \leq \tau \leq b\}$ will be linearly independent if and only if the following matrix is positive definite: $\mathcal{G} = [\mathcal{G}_{ij}]$, $\mathcal{G}_{ij} = \int_a^b l_i(t)l_j(t)\,dt$. Hint: Note that $\alpha'\mathcal{G}\alpha = \int_a^b [\sum \alpha_i l_i(t)]^2\,dt$.

9.2-2.   *Properties of the Controllability Gramian*
    a. Show that $\mathcal{C}(t_0, t_f)$ is a symmetric and nonnegative-definite matrix.
    b. Show that $\mathcal{C}(t_0, t_f)$ satisfies the relations

$$\frac{d\mathcal{C}(t, t_f)}{dt} = A(t)\mathcal{C}(t, t_f) + \mathcal{C}(t, t_f)A'(t) - B(t)B'(t), \qquad \mathcal{C}(t_f, t_f) = 0$$

$$\mathcal{C}(t_0, t_f) = \mathcal{C}(t_0, t) + \Phi(t_0, t)\mathcal{C}(t, t_f)\Phi'(t_0, t)$$

c. Guess the corresponding properties for $\Theta(t_0, t_f)$.

9.2-3.

If $x(t) = T(t)z(t)$, $\dot{T}(\cdot)$ and $T^{-1}(\cdot)$ exist, show that $\mathcal{C}_z(t_0, t_f) = T^{-1}(t_0)\mathcal{C}_x(t_0, t_f)[T^{-1}(t_0)]'$.

9.2-4.

Show that $\dot{x} = Ax + bu$ is controllable if and only if the only solution of $\dot{\mathcal{E}}(t) = A'\mathcal{E}(t)$, $\mathcal{E}(0) = \mathcal{E}_0$, such that $\mathcal{E}'(t)b \equiv 0$ for all t is the one obtained when $\mathcal{E}_0 = 0$.

9.2-5.

Prove that $\{A, B\}$ is controllable if and only if $C(sI - A)^{-1}B \equiv 0 \Rightarrow C = 0$; i.e., the rows of $(sI - A)^{-1}B$ are linearly independent functions of s.

9.2-6.

If $\{A(\cdot), B(\cdot)\}$ is controllable over $[t_0, t_f]$, show that $\{A(\cdot) - B(\cdot)K(\cdot)$, $B(\cdot)\}$ will be also, where $K(\cdot)$ is an arbitrary matrix.

9.2-7.   (*Rosenbrock*)

Show that the assumption of continuous $C(\cdot)$ is necessary in studying the observability of time-variant systems by considering the example $\dot{x} = 0$, $y(t) = c(t)x(t)$, $t \geq 0$, $c(t) = 1 - 1(t)$, $x(0) = x_0$. Clearly, $x(t) = x_0$, $t \geq 0$, $y(0) = x_0$, so that the system is observable. However, show that $\Theta(0, t_f) = 0$ for all $t_f > 0$. Why is there not a corresponding example for controllability?

9.2-8.

From $y_0(t) = Ce^{At}x_0$, obtain n equations by *sampling* at the instants $0, \Delta, \ldots, (n - 1)\Delta$.

a. Show that these equations have a unique solution if and only if the pair $\{C, e^{A\Delta}\}$ is *observable*.

b. Let A be a 2×2 matrix. If $\{C, A\}$ is observable, show that we can always choose Δ so that $\{C, e^{A\Delta}\}$ is observable.

c. Generalize the result of part b to the case of matrix A of arbitrary size.

9.2-9.

We have a time-variant system $\dot{x}(t) = A(t)x(t) + b(t)u(t)$, $x(t_0) = x_0$ and wish to choose $u(\cdot)$ so as to minimize the *cost* $J = \int_{t_0}^{t_f} [x'(t)Q(t)x(t) + u'(t)R(t)u(t)]\, dt + x'(t_f)Q_f x(t_f)$, where $Q(\cdot)$ and $R(\cdot)$ are given positive-definite matrices. Show that if $[A(\cdot), b(\cdot)]$ is controllable, the optimum cost is finite for all t_f. Assume that the elements of $A(\cdot), b(\cdot), Q(\cdot), R(\cdot)$ are all bounded functions of t.

9.2-10.   *Optimality Property of $u_*(\cdot)$ [7]*

Show that the special solution $u_*(\cdot)$ defined by (9.2-11) has least energy among all controls that take an arbitrary initial state to the origin by time t_f. *Hint:* Read Sec. 6.2.1.

9.2-11.   *Closed-Loop Form for $u_*(\cdot)$*

a. Show that $u_*(\cdot)$ as given by (9.2-11) can be written in the *closed-loop* form $u_*(t) = -K(t)x_*(t)$, where $\dot{x}_*(t) = A(t)x_*(t) + B(t)u_*(t)$, $x_*(t_0) = x_0$, and $K(t) = -B'(t)\mathbb{C}^{-1}(t, t_f)$. Do this by using the principle of optimality (cf. Sec. 3.5.2) and also by direct algebraic manipulation.

b. Let $P(t) \triangleq \mathbb{C}^{-1}(t, t_f)$. Show that $P(\cdot)$ obeys a Riccati-type differential equation.

c. For parts a and b we apparently need to assume that $\mathbb{C}(t, t_f)$ is invertible for all $t \in [t_0, t_f]$, while for (9.2-11) it suffices for $\mathbb{C}(t_0, t_f)$ to be invertible. Does this mean that the closed-loop form is less general than the open-loop form (9.2-11)?

9.2-12. *Properties of* $\mathbb{C}_n(t)$ [9]

Consider the matrix $\mathbb{C}_n(t)$ as defined by (9.2-15).

1. Show that rank $\mathbb{C}_n(t) = n$ for *some* t in $[t_0, t_f]$ implies that rank $\mathbb{C}(t_0, t_f) = n$.
2. Show that rank $\mathbb{C}_n(t) = n$ for *almost all* t in $[t_0, t_f]$ if and only if rank $\mathbb{C}(t, \tau) = n$ for all subintervals (t, τ) in $[t_0, t_f]$. Such realizations may be said to be *totally controllable*.
3. Assume that we have a scalar (single input, single output) system. Show that rank $\mathbb{C}_n(t) = n$ for *all* t in $[t_0, t_f]$ if and only if we can transform the realization to the time-variant controllability form $\{A_{co}(\cdot), b_{co}(\cdot)\}$. [We note that $b'_{co}(\cdot) = [1 \ 0 \ \cdots \ 0]$ and $A_{co} = $ a right-companion matrix.]

9.2-13.

If $\dot{x}(t) = Ax(t) + Bu(t), t \geq 0$, show that the feedback law $u(t) = -B'W_c^{-1}(0, T)x(t) + v(t)$ gives a stable closed-loop system, where for any T,

$$W_c(0, T) \triangleq \int_0^T e^{-A\tau}BB'e^{-A'\tau}\,d\tau.$$ *Hint:* Use the Lyapunov criterion of Theorem 2.5-1. (*Reference:* D. Kleinman, *IEEE Trans. Autom. Control*, **AC-15**, p. 692, 1970.)

9.2.14. (*Lax Pairs*)

Let $L(t) = \Phi(t, 0)L(0)\Phi(0, t)$, where $\Phi(t, 0)$ is the state-transition matrix of $A(\cdot)$. Show that

1. The eigenvalues of $L(t)$ are the same for all t.
2. $dL(t)/dt = L(t)A(t) - A(t)L(t) \triangleq$ the commutator $[L, A]$ of $L(\cdot)$ and $A(\cdot)$.

9.2.15. *Balancing Transformations*

1. Show that the eigenvalues of the product $\mathbb{C}(t - \delta, t)\mathbb{O}(t, t + \delta)$ are invariant under similarity transformation.
2. Hence show that the eigenvalues of this product are nonnegative.
3. Find a transformation that will bring any realization to balanced form (Moore). *Hint:* First consider $\mathbb{C}(t - \delta, t)$ nonsingular, and use a symmetric square root of the positive definite matrix $\mathbb{C}(t - \delta, t)$ to transform to a realization with $\mathbb{C} = I$.
4. Show that for a time-invariant selfdual ($A = A', B' = C$) realization, $\mathbb{C}(t - \delta, t) = \mathbb{O}(t, t + \delta)$. Then show that the transformation to the balanced realization is orthogonal. Will this result extend to time-variant systems?

9.2-16. [13].

Show that the set of all states that can be reached in a time $t \leq t_1$, with inputs of energy less than or equal to unity over $[t_0, t_1]$ is the same as the set of vectors $\{R_2^{\frac{1}{2}}(t_0, t_1)p\}$, where $R_2^{\frac{1}{2}}$ is a square-root of the reachability gramian and p ranges over the set of all $n \times 1$ vectors of length n. The structure of this set can be well displayed by singular value analysis.

State and prove the corresponding result for observability.

9.2-17.

The open-loop form of the optimal control law for going to the origin is given by (9.2-11) as $u(t) = -B'(t)\Phi'(t_0, t)\mathfrak{C}^{-1}(t_0, t_f)x_0$. Show that the matrix $L(t) \triangleq B'(t)\Phi'(t_0, t)$ can be obtained by solving the *adjoint* system $\dot{\lambda}(t) = -A'(t)\lambda(t)$, $\lambda(t_0) = e_i$, $l_i(t) = B'(t)\lambda(t)$, where $e_i =$ the ith column of the $n \times n$ identity matrix and $l_i(\cdot) =$ the ith column of $L(\cdot)$. *Note:* Other applications of the adjoint system will be presented in Sec. 9.3.

9.3 ADJOINT SYSTEMS AND SOME APPLICATIONS

Consider a linear time-variant system with impulse response $h(t, \tau)$, the response measured at time t to a unit impulse input at time τ. Now there are often applications in which the final time, t_f say, is fixed, and we wish to study the behavior of

$$h(t_f, \tau) \text{ as a function of } \tau$$

This is not very easy to do unless we have an explicit formula for $h(t, \tau)$. If, as usually happens, the impulse response has to be determined by simulation, then we apparently have to proceed as follows: Put in impulses at various times $\{t_0 + k(t_f - t_0)/N, \; k = 0, 1, \ldots, N,\}$ and measure the responses $h[t_f, t_0 + k(t_f - t_0)/N]$. Then by cross-plotting we can approximately obtain $h(t_f, \tau)$ as a function of τ, with the approximation improving as $N \to \infty$. An example is illustrated in Fig. 9.3-1.

This method can clearly be quite expensive in many problems. In fact it was in connection with a rocket analysis program at Hughes Aircraft Company that R. Battin showed how one could use state-space descriptions and some simple differential equation theory to obtain $h(t_f, \tau)$ in a *single* simulation [16]. Actually, the basic idea comes from the so-called *principle of reciprocity*, which is familiar in many linear systems problems. For example, the current in loop i of an electric circuit arising from a voltage source in loop j is the same as the current in loop j that would be caused by putting the voltage source in loop i. Similar examples abound in electromagnetic theory and in mechanics. For example, the deflection of a stretched string at position y due to a point mass at position x is the same as the deflection at x that would be caused by the same point mass at y.

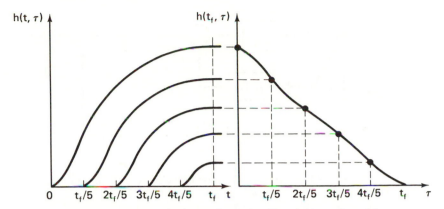

Figure 9.3-1   Cross-plotting to determine $h(t_f, \tau)$ (approximately).

Therefore it seems that we might obtain $h(t_f, \tau)$ by applying an impulse at time t_f to the output of the linear system and measuring the response at time τ. However, there seem to be two difficulties in actually doing this for any given system: How do we reverse the roles of input and output, and how do we reverse the direction of time (to measure the response at the time τ which precedes t_f)? As on so many previous occasions, it is helpful when faced with such questions to suppose that we have an internal description, or say an analog-computer simulation, of the given system. Then some exploration will show that the roles of input and output can be reversed just by reversing the arrows on the input and output leads of each of the elements (integrators, adders, and scalors) of the simulation and also replacing adders by nodes and nodes by adders—cf. Figs. 9.3-2(a) and 9.3-2(b). The resulting realization (system) is said to be *adjoint* to the original realization (system). Moreover, the problem that time τ is before time t_f can be overcome by making a change

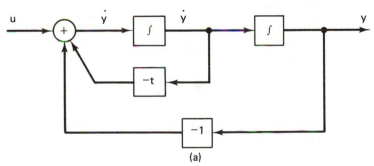

(a)

Figure 9.3-2(a)   Simulation of $\dfrac{d^2 y}{dt^2} + t\dfrac{dy}{dt} + y = u$.

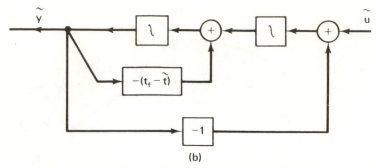

(b)

Figure 9.3-2(b)   The modified adjoint simulation of the system

$$\frac{d^2\tilde{y}}{d\tilde{t}^2} + \frac{d}{d\tilde{t}}[(t_f - \tilde{t})\tilde{y}] + \tilde{y} = \tilde{u}.$$

of variables

$$t \longrightarrow \tilde{t} \qquad \text{where } \tilde{t} = t_f - t$$

so that $\tilde{t} = 0$ when $t = t_f$. The resulting system is often called the *modified adjoint* system.

Example 9.3-1.

Suppose

$$\frac{d^2y}{dt^2} + t\frac{dy}{dt} + y = u$$

Then we shall find that

$$h(t, \tau) = \begin{cases} e^{-t^2/2} \int_\tau^t e^{r^2/2}\, dr, & t > \tau \\ 0, & t < \tau \end{cases}$$

A simulation is shown in Fig. 9.3-2(a), and the so-called *modified adjoint* simulation is shown in Fig. 9.3-2(b).

The impulse response of the modified adjoint realization is

$$\tilde{h}(\tilde{t}, \tau) = \begin{cases} e^{-(t_f - \tau)^2/2} \int_{t_f - \tilde{t}}^{t_f - \tau} e^{r^2/2}\, dr, & \tilde{t} > \tau \\ 0, & \tilde{t} < \tau \end{cases}$$

Note that

$h(t_f, \tau) =$ the response of the original system at $t = t_f$ to an impulse input at $t = \tau$

$ =$ the response of the modified adjoint at $\tilde{t} = t_f - \tau$ (i.e., $t = \tau$) to an impulse input at $\tilde{t} = 0$ (i.e., $t = t_f$)

$ = \tilde{h}(t_f - \tau, 0)$

In Fig. 9.3-3, we have shown again (cf. Fig. 9.3-1) the curves of $h(t, \tau)$ for six different values of τ. But instead of cross-plotting to obtain $h(t_f, \tau)$, a single simulation of the modified adjoint system will give the curve $\tilde{h}(\tilde{t}, 0)$, where $\tilde{t} = t_f - t$. Finally, a simple reversal of the time scale gives $h(t_f, \tau)$.

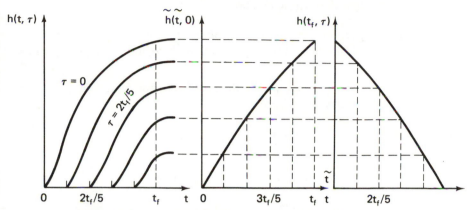

Figure 9.3-3   A single simulation of the modified adjoint system gives $\tilde{h}(\tilde{t}, 0)$, which is equivalent (after the time reversal $\tilde{t} = t_f - t$) to $h(t_f, \tau)$ as shown in Fig. 9.3-1.

Some General Results

In general, if we have a realization

$$\frac{dx(t)}{dt} = A(t)x(t) + B(t)u(t), \qquad x(t_0) = x_0$$
$$y(t) = C(t)x(t), \qquad\qquad t_0 \leq t \leq t_f \tag{1}$$

then the *modified adjoint* realization is defined by (with $\tilde{t} = t_f - t$)

$$\frac{d\tilde{\lambda}(t)}{dt} = +A'(t_f - t)\tilde{\lambda}(t) + C'(t_f - t)\tilde{\mu}(t), \qquad \tilde{\lambda}(0) = \lambda_0$$
$$\tilde{\eta}(t) = B'(t_f - t)\tilde{\lambda}(t), \qquad\qquad 0 \leq t \leq t_f - t_0 \tag{2}$$

The *adjoint realization* is one that is defined in *reverse* time, so that the independent variable is t and not $\tilde{t} = t_f - t$,

$$\frac{d\lambda(t)}{dt} = -A'(t)\lambda(t) + C'(t)\mu(t), \qquad \lambda(t_f) = \lambda_0$$
$$\eta(t) = B'(t)\lambda(t), \qquad\qquad t_0 \leq t \leq t_f \tag{3}$$

and its impulse response will be denoted $h_a(t, \tau)$.

Now we shall establish the following basic results. For $t \geq \tau$,

$$h(t, \tau) = C(t)\Phi(t, \tau)B(\tau) \tag{4}$$

$$\tilde{h}'(t, \tau) = C(t_f - \tau)\Phi(t_f - \tau, t_f - t)B(t_f - t) \tag{5}$$

$$h'_a(t, \tau) = C(\tau)\Phi(\tau, t)B(t) \tag{6}$$

For analytical purposes, the adjoint is often more convenient (see below), though the modified adjoint must be used if we have to actually carry out a simulation.

Proof of (4)–(6).   The key step in the proof is to note that if $\Phi(\cdot, *)$ is the state-transition matrix of $A(\cdot)$,

$$\frac{d\Phi(t, \tau)}{dt} = A(t)\Phi(t, \tau)$$

then $\Phi'(*, \cdot)$ is the state-transition matrix of $-A'(\cdot)$; i.e.,

$$\frac{d\Phi'(\tau, t)}{dt} = -A'(t)\Phi'(\tau, t) \tag{7}$$

To see this we note that

$$0 = \frac{d}{dt}I = \frac{d}{dt}[\Phi(t, \tau)\Phi(\tau, t)] = \left[\frac{d}{dt}\Phi(t, \tau)\right]\Phi(\tau, t) + \Phi(t, \tau)\frac{d}{dt}\Phi(\tau, t)$$

Therefore

$$\frac{d\Phi(\tau, t)}{dt} = -\Phi^{-1}(t, \tau)A(t)\Phi(t, \tau)\Phi(\tau, t) = -\Phi(\tau, t)A(t) \tag{8}$$

from which (7) follows easily.

Then the impulse response of the adjoint realization (3) is

$$h_a(t, \tau) = B'(t)\Phi'(\tau, t)C'(\tau) = [C(\tau)\Phi(\tau, t)B(t)]'$$

Finally, for the impulse response of the modified adjoint realization (2), we just note that an obvious change of variables in (7) yields

$$-\frac{d\Phi'(T - \tau, T - t)}{dt} = -A'(t - \tau)\Phi'(T - \tau, T - t) \tag{9}$$

Formulas (4)–(6), which we derived for systems with state-space realizations, also imply the more general input-output relationships

$$h(t, \tau) = h'_a(\tau, t) \tag{10}$$

$$= \tilde{h}'(t_f - \tau, t_f - t) \tag{11}$$

Relation (10) is a statement of the reciprocity theorem. For time-invariant systems, (10) yields the transfer function formula

$$H_a(s) = -H'(-s) \tag{12}$$

Adjoint systems have numerous applications in system and circuit theory, especially in studying the sensitivity of system behavior to parameter variations; see, e.g., [17]–[19]. We shall now give a brief indication of the nature of these applications.

The Adjoint Lemma

An important formula for adjoint systems is Eq. (13) below, which is a specialization to state-space equations of a general (defining) relation for adjoints [18].

Lemma 9.3-1.

   Let

$$\dot{x}(t) = A(t)x(t) + B(t)u(t)$$

and

$$\dot{\lambda}(t) = -A'(t)\lambda(t) + \Gamma(t)\mu(t)$$

Then

$$\lambda'(t_f)x(t_f) - \lambda'(t_0)x(t_0) = \int_{t_0}^{t_f} [\lambda'(t)B(t)u(t) + \mu'(t)\Gamma'(t)x(t)]\, dt \tag{13}$$

Proof.

$$\frac{d}{dt}\lambda'(t)x(t) = \lambda'(t)[A(t)x(t) + B(t)u(t)] + [-A'(t)\lambda(t) + \Gamma(t)\mu(t)]'x(t)$$
$$= \lambda'(t)B(t)u(t) + \mu'(t)\Gamma'(t)x(t) \tag{14}$$

Now integrating both sides from t_0 to t_f gives the result (13).   ∎

We shall now describe some problems that can be nicely resolved by using formula (13) with judicious choices of $\lambda(t_f)$ and $\Gamma(\cdot)\mu(\cdot)$.

Application 1: Midcourse Corrections.   Consider a system that is supposed to follow a nominal trajectory described by

$$\dot{x}_0(t) = f[x_0(t), u_0(t), t], \qquad t \geq t_0 \tag{15}$$

but because of errors actually follows one described by

$$\dot{x}(t) = f[x(t), u(t), t], \qquad t \geq t_0 \tag{16}$$

If the deviations, say,

$$x(t) - x_0(t) = \delta x(t), \qquad u(t) - u_0(t) = \delta u(t) \tag{17}$$

are small, then they will obey the equation

$$\frac{d}{dt}\,\delta x(t) = A(t)\delta x(t) + B(t)\delta u(t), \qquad t \geq t_0 \tag{18}$$

where (cf. Sec. 2.2.2)

$$A(t) = \left[\frac{\partial f_i}{\partial x_j}\right], \qquad B(t) = \left[\frac{\partial f_i}{\partial u_j}\right]$$

Suppose now we wish to choose $\delta u(\cdot)$ so that for some vector c

$$c\delta x(t_f) = 0 \tag{19}$$

For example, if the components of the state vector $x(\cdot)$ are position and velocity, we may wish one of the position coordinates of $\delta x(t_f)$ to be zero, while being indifferent to the values of the others; this can be achieved by taking $c = [1 \; 0 \; \cdots \; 0]$.

Now the equation for $\delta x(\cdot)$ tells us that

$$c\delta x(t_f) = c\Phi(t_f, t_0)\delta x(t_0) + c \int_{t_0}^{t_f} \Phi(t_f, \tau)B(\tau)\delta u(\tau)\,d\tau \tag{20}$$

where $\Phi(\cdot, \cdot)$ is the state-transition matrix of $A(\cdot)$.

For a direct attack on this problem, we would need to know how $\Phi(t_f, \tau)$ changes as τ changes, which in turn suggests consideration of the adjoint system.

In fact, by formula (13) of Lemma 9.3-1, we see that if we define

$$\dot\lambda(t) = -A'(t)\lambda(t), \qquad \lambda(t_f) = c' \tag{21}$$

then

$$c\delta x(t_f) - \lambda'(t_0)\delta x(t_0) = \int_{t_0}^{t_f} \lambda'(t)B(t)\delta u(t)\,dt \tag{22}$$

We could satisfy this equation by various choices of $\delta u(\cdot)$. For example, we could apply an impulsive correction at time τ, so that

$$\delta u(t) = k(\tau)\delta(t - \tau), \qquad t_0 \leq \tau \leq t_f \tag{23}$$

Then $k(\tau)$ will be determined by the equation

$$0 - \lambda'(t_0)\delta x(t_0) = \lambda'(\tau)B(\tau)k(\tau) \tag{24}$$

If $B(\cdot)$ is a column vector, then $u(\cdot)$ and $k(\cdot)$ will be scalar, and we can readily solve (24) to get

$$k(\tau) = -\frac{\lambda'(t_0)\delta x(t_0)}{\lambda'(\tau)B(\tau)} \tag{25}$$

$\lambda(\cdot)$ is obtained by a *single* simulation of the adjoint system (backward in time from $t = t_f$, or rather by simulating the modified adjoint system forward in time). We can then choose any value of τ that gives a reasonable (say smallest) value of $k(\tau)$.

Application 2: Iterative Optimization.   Another problem closely related to that in Application 1 is one where $\delta x(t_0)$ is zero, but we have to choose $\delta u(\cdot)$ so as to maximize

$$\frac{c\delta x(t_f)}{\int_{t_0}^{t_f} [\delta u(t)]^2 \, dt} \tag{26}$$

As will be explained below, this requirement can arise in certain iterative optimization schemes.

Now we note that by the Schwarz inequality

$$\frac{|c\delta x(t_f)|}{\left[\int_{t_0}^{t_f} [\delta u(t)]^2 \, dt\right]^{1/2}} = \frac{\left|\int_{t_0}^{t_f} \lambda'(t)B(t)\delta u(t)dt\right|}{\left[\int_{t_0}^{t_f} [\delta u(t)]^2 \, dt\right]^{1/2}}$$

$$\leq \left[\int_{t_0}^{t_f} |\lambda'(t)B(t)|^2 \, dt\right]^{1/2} \tag{27}$$

with equality being achieved by

$$\delta u(t) = k\lambda'(t)B(t), \qquad t_0 \leq t \leq t_f \tag{28}$$

where k is an arbitrary constant.

This maximization problem may arise as follows. We wish to find a control $u(\cdot)$ in a system obeying

$$\dot{x}(t) = f[x(t), u(t), t], \qquad t_0 \leq t \leq t_f$$

so as to maximize the value of a given payoff function $\phi[x(t)]$ at the final time t_f. Suppose further that we have, on some basis, found a trial solution $u_0(\cdot)$ which we wish to improve by using a "small" perturbation

$$u(\cdot) = u_0(\cdot) + \delta u(\cdot)$$

The resulting change in the value of the payoff function is, ignoring second-order terms,

$$\delta\phi \triangleq \phi[x_0(t_f) + \delta x(t_f)] - \phi[x_0(t_f)]$$

$$= c\delta x(t_f)$$

where

$$c = \nabla\phi[x_0(t_f)], \qquad \text{the gradient of } \phi \text{ evaluated at } x_0(t_f)$$

However, since the value of $|c\delta x(t_f)|$ can be arbitrarily increased by just multiplying $\delta u(\cdot)$ by a constant, we should impose a constraint on $\delta u(\cdot)$. This leads to the problem of maximizing the ratio in (26).

Finally, we may recall that adjoint equations and variables arise as Lagrange multipliers in the calculus of variations, as, for example, in the optimization problem discussed in Sec. 3.4.2.

Concluding Remarks

Most of the results in this chapter were already known by the late sixties. Much remains to be done to try to parallel the developments since then in the theory of constant multivariable systems.

Most progress has been made for systems with analytic coefficients, where there is now a fairly complete theory of realizability. The first results, obtained in 1966 (see the surveys [20]), were in terms of a Hankel matrix with entries

$$h_{ij}(t, \tau) = \left(\frac{\partial}{\partial t}\right)^{i-1} \left(-\frac{\partial}{\partial \tau}\right)^{j-1} h(t, \tau), \qquad i, j = 1, 2, \ldots$$

Later Kamen [21] (following some unpublished 1968 work of Kalman's) showed that it was sufficient to consider only the values of the $\{h_{ij}(t, \tau)\}$ along the diagonal, i.e., when $t = \tau$. This reduction to functions of one variable rather than two allows the use of certain algebraic results (especially a Bezout identity for coprimeness) to help construct minimal realizations, see [21]. (A different kind of algebraic approach to discrete-time systems is initiated in [22].)

Finally, we may note that a general method for constructing state-space models from input-output descriptions is given in Schumitzky [23], where it is then applied to the problem of representing open-loop control laws in state-feedback form.

REFERENCES

1. E. A. CODDINGTON and N. LEVINSON, *Theory of Ordinary Differential Equations*, McGraw-Hill, New York, 1955.

2. E. L. INCE, *Ordinary Differential Equations*, Dover, New York, 1956.

3. V. I. ARNOLD, *Ordinary Differential Equations*, MIT Press, Cambridge, Ma., 1973.

4. W. HIRSCH and S. SMALE, *Differential Equations and Linear Algebra*, Academic Press, New York, 1974.

5. F. R. GANTMAKHER, *Theory of Matrices*, Vol. I, Chelsea Publishing Co. New York, 1959.

6. L. WEISS, "On the Structure Theory of Linear Differential Systems," *SIAM J. Control*, **6**, pp. 659–680, 1968.

7. R. E. KALMAN, Y. C. HO, and K. S. NARENDRA, "Controllability of Linear Dynamical Systems," in *Contrib. Differ. Equations*, **1**, pp. 189–213, 1962.

8. D. C. YOULA, "The Synthesis of Linear Dynamical Systems from Prescribed Weighting Patterns," *SIAM J. Appl. Math.*, **14**, pp. 527–549, 1966.

9. L. M. SILVERMAN and H. E. MEADOWS, "Controllability and Observability in Time-Variable Linear Systems," *SIAM J. Control*, **5**, pp. 64–73, 1967.

10. L. M. SILVERMAN, "Synthesis of Impulse Response Matrices by Internally Stable and Passive Realizations", *IEEE Trans. Circuit Theory*, **CT-15**, pp. 238–245, September 1968.

11. R. E. KALMAN, "Contributions to the Theory of Optimal Control", *Bol. Soc. Matem. Mex.*, **5**, pp. 102–119, 1960.

12. L. M. SILVERMAN and B. D. O. ANDERSON, "Controllability, Observability and Stability of Linear Systems," *SIAM J. Control*, **6**, pp. 121–129, 1968.

13. B. C. MOORE, "Singular Value Analysis of Linear Systems," *IEEE Trans. Automat. Contr.*, 1980. Also available as Repts. 7801–7802, Systems Control Group, Dept. of Elec. Engg., University of Toronto, Canada, May 1978.

14. L. PERNEBO and L. M. SILVERMAN, "Model Reduction via Balanced State-Space Representations," *IEEE Trans. Automat. Contr.*, to appear.

15. E. VERRIEST, Ph.D. Dissertation, Dept. of Elec. Engg., Stanford Univ., Stanford, CA, 1979.

16. H. LANING and R. BATTIN, *Random Processes in Automatic Control*, McGraw-Hill, N.Y., 1956.

17. S. W. DIRECTOR and R. A. ROHRER, "The Generalized Adjoint Network and Network Sensitivities," *IEEE Trans. Circuit Theory*, **CT-16**, pp. 318–323, 1969.

18. D. G. LUENBERGER, *Optimization by Vector Space Methods*, J. Wiley, N.Y., 1969.

19. A. E. BRYSON and Y. C. HO, *Applied Optimal Control*, Hemisphere Publishing Co., Washington, D.C., 1975.

20a. L. M. SILVERMAN, "Realization of Linear Dynamical Systems," *IEEE Trans. Automat. Contr.*, **AC-16**, pp. 554–567, 1971.

20b. J. C. WILLEMS and S. K. MITTER, "Controllability, Observability, Pole Allocation and State Reconstruction," *IEEE Trans. Automat. Contr.*, **AC-16**, pp. 582–595, 1971.

21. E. W. KAMEN, "New Results in Realization Theory for Linear Time-Varying Analytic Systems," *IEEE Trans. Automat. Contr.*, **AC-24**, pp. 866–878, Dec. 1979.

22. E. W. KAMEN and K. M. HAFEZ, "Algebraic Theory of Linear Time-Varying Systems," *SIAM J. Contr. and Optim.*, **17**, pp. 500–510, July 1979.

23. A SCHUMITZKY, "State Feedback Control for General Linear Systems," *Proc. Fourth Symp. on the Math. Thy. of Networks and Systems*, pp. 194–200, Delft, Holland, July 1979.

SOME FURTHER READING   10

10.0 INTRODUCTION

It is perhaps natural to wonder, even at the end of such a long book, about further directions and further extensions of the concepts and results presented so far. In fact, there are several, but let us briefly consider two of them.

So far we have only studied systems that can be completely characterized by a finite set of coupled ordinary differential equations. But clearly there are many physical problems where such idealizations are inadequate and we have to model systems via sets of partial differential equations (e.g., transmission lines, networks, elastic deformations, moving fluids, etc.). There is now accumulating quite a large body of system-theoretical literature on such "distributed parameter" or "infinite-dimensional" systems. There is quite a diversity of approaches, each using different mathematical tools. The author has not had the occasion to really pursue these researches. But he is happy to report to interested readers that a preliminary exploration shows that the basic approach of this book can also be pursued in the case of infinite-dimensional systems. Of course there will be differences, both expected and unexpected and in both tools and results, from the finite-dimensional case, but it is satisfying that our basic framework—first trying to build up a relationship between state-space and transfer function (MFD and PMD) descriptions, and then interactively using these in pursuing the major systems concepts (of state- and modal-controllability and observability, irreducibility and minimality, the significance of poles and zeros, the problem of pole-shifting)—

can be carried over to this more general problem area. This is not to say that
it has all been done, or that there will not be major gaps, but a preliminary
investigation suggests that trying to ask the same questions in roughly the
same order that we did for finite-dimensional systems can be a useful guide
not only in sorting out and rearranging the results already available in the
literature but also in suggesting new questions for investigation.

These remarks apply also to the developing area of multidimensional
systems, which are useful in seeking a more fundamental approach to the processing of signals carried by two- and three-dimensional wave fields (as in
image deblurring systems, beamforming arrays, diagnostic imaging, etc.). The
transform analysis of such problems also turns out to be relevant to the
analysis of delay-differential equations, multivariable network synthesis with
lumped and distributed elements, the analysis of multi-step, multi-derivative
integration formulas, etc. (see the papers collected in [1]–[3]). Here the author
can say on the basis of more personal experience (see the theses [4], [5] and the
papers [6]–[8]) that the approach of this book can be a useful guide in exploring and developing a new branch of system theory.

We shall give a brief introduction to distributed parameter and multidimensional systems in Secs. 10.1 and 10.2, just enough to point out some
parallels with earlier results and to provide references for further study. The
transfer function analysis of 2-D systems leads naturally to the use of algebraic geometry techniques. In Sec. 10.3 we list some applications of algebraic
geometry to system theory, including some that lead to interesting implications for certain nonlinear systems. Finally, not to slight the many challenges
still left for finite-dimensional, 1-D, linear systems, Sec. 10.4 talks very briefly
about a striking mathematical result on model reduction that should provide
some opportunity for system-theoretic application.

10.1   DISTRIBUTED PARAMETER SYSTEMS

Distributed parameter systems are those that have to be modeled via partial
differential equations, or delay-differential equations, or integro-differential
equations.

For such systems, the state-space has to be an infinite-dimensional space
(e.g., the linear space of all continuous functions, or all square-integrable
functions, or even spaces of impulsive and generalized functions). For example, consider the realization

$$\dot{x}(t) = ax(t) + x(t - \tau) + bu(t)$$
$$y(t) = cx(t)$$

To compute $x(t)$, we need to know the values of $x(\cdot)$ in the τ seconds prior to
t, requiring that the "state-space" be the space of continuous functions on

$[-\tau, 0]$, an infinite-dimensional space. From another point of view, an analog computer realization will clearly require a delay line of length τ, which is an "infinite-dimensional" element. So also, the transfer function of the system is

$$H(s) = cb/(s - a - e^{-s\tau})$$

which is irrational and not just a ratio of finite sums of powers of s.

Another example arises from the one-dimensional controlled wave equation (see also Exercise 2.2-7)

$$\frac{\partial^2 \bar{w}}{\partial t^2} - \frac{\partial^2 \bar{w}}{\partial x^2} = b(x)u(t)$$

with boundary conditions being imposed at $x = 0$ and $x = 1$. This can be rewritten in state-space form as

$$\dot{w}(t) = A\,w(t) + B\,u(t) \qquad\qquad (1)$$

where $w' = [\partial\bar{w}/\partial t \quad \partial\bar{w}/\partial x]$ and

$$A = \begin{bmatrix} 0 & \partial/\partial x \\ \partial/\partial x & 0 \end{bmatrix}, \qquad B = \begin{bmatrix} b(x) \\ 0 \end{bmatrix}$$

Note that A is now an 'operator' on a space of differentiable functions; also the control $u(\cdot)$ is "distributed" over the interval $[x = 0, x = 1]$. In many problems, the control enters at points or curves on the boundary, e.g., if $b(x) = 0$ in the above, but $\alpha\partial\bar{w}(1, t)/\partial t + \partial\bar{w}(1, t)/\partial x = u(t)$. As part of a comprehensive survey paper, D.L. Russell [9, pp. 672–680] derives for (1) many results analogous to those of our Secs. 2.3.2 and 3.2.1; the analogy is fairly close and will be quite instructive for our readers. Extensions to higher-dimensional wave equations and to other equations are not so nice, nor all worked out, but we refer the reader to [9] for more discussion; for further results, especially on quadratic regulator problems, the reader can profit from the books of J.L. Lions [10] and of M. Delfour, A. Bensoussan and S.K. Mitter [11].

Direct generalization begins to break down when we come to the analogs of Sec. 2.4 on joint controllability and observability and on relations to the transfer function. Thus since $H(s) = c(sI - A)^{-1}b$ is irrational (recall that $\{A, b, c\}$ may be operators effecting differentiations or delays), it may have more complicated singularities than just poles (e.g., branch points or essential singularities). And these singularities may not have a unique relation to the 'spectral properties' of the operator A. As one consequence, it turns out that without further assumptions, one can have two different 'controllable and observable' realizations (of a given $H(s)$), one stable and the other not (see

Baras [12], Helton [13]–[14], Dewilde [15], Baras and Dewilde [16]). As in the finite-dimensional case, it turns out that a transfer function analysis can illuminate some of these issues and can be especially useful in design. We shall give a brief indication of how this can be done. We shall assume discrete-time systems for simplicity and can begin by thinking of $H(z)$ in the infinite-dimensional case as a ratio

$$H(z) = \sum_{0}^{\infty} b_i z^i \Big/ \sum_{0}^{\infty} a_i z^i = b(z)/a(z)$$

Of course, such "infinite-degree polynomials" $\{b(z), a(z)\}$ are meaningless unless we define (some mode of convergence for) them and for various reasons a natural first choice is as bounded analytic (so-called H^{∞}) functions in $|z| \leq 1$, see, e.g., Hoffman [17], Duren [18], and also [15]–[16]. One motivation is that a famous theorem of Bochner and Chandrasekharan asserts that a linear time-invariant system taking square-integrable discrete-time input functions to similar output functions, must have an H^{∞} transfer function. Another is that products of H^{∞} functions are in H^{∞}.

We could now ask, as in Sec. 2.1, for a controllable state-space realization of $b(z)/a(z)$. We cannot use Kelvin's method because $\{b(z), a(z)\}$ may not be given as power series in z. Here, however, the Nerode equivalence construction of Sec. 5.1 comes to the rescue, and allows us to construct a controllable realization by using some elementary properties of subspaces in Hilbert space.

We shall not describe this here, but shall go on to explore how the observability of this realization might be related to the coprimeness of $\{b(z), a(z)\}$. This issue brings up some new points. We shall say that $\{b(z), a(z)\}$ are strongly coprime if there exist H^{∞} functions $\{x(z), y(z)\}$ such that

$$x(z)a(z) + y(z)b(z) = 1$$

Clearly $\{b(z), a(z)\}$ with a common zero in $|z| < 1$ are not coprime, but the above condition is stronger because it does not even allow the limiting case of a sequence of $z_i \longrightarrow \lambda$ on the unit circle such that $a(z_i) \neq 0 \neq b(z_i)$ but $a(\lambda) = 0 = b(\lambda)$. [This is a consequence of a famous so-called Corona Theorem of L. Carleson according to which the above condition holds if and only if $\inf \{|b(z)| + |a(z)|\} \geq \delta > 0$.]

We could say that $\{a(z), b(z)\}$ are weakly coprime if they have no common zeros, but this needs some qualification. To extract the zeros of an H^{∞} function we could first write

$$a(z) = \prod_{1}^{\infty} (z - \alpha_i) \cdot \tilde{a}(z)$$

but this is not desirable because $\tilde{a}(z)$ will not in general be in H^{∞} and also the infinite product may not converge. This can be remedied by factoring $a(z)$ as

$$a(z) = B(z) \cdot a_0(z)$$

where, with $|\alpha_i| < 1$, and $\sum (1 - |\alpha_i|) < \infty$,

$$B(z) = \prod_1^\infty (z - \alpha_i)(1 - \alpha_i^* z)\alpha_i^*/|\alpha_i| \quad , \quad |\alpha_i| < 1, \sum (1 - |\alpha_i|) < \infty$$

is called a *Blaschke product*. $|B(z)| = 1$ when $|z| = 1$, so that it is in H^∞ and therefore so is $a_0(z)$. The above factorization is analogous to splitting out "all pass" factors from a rational transfer function to leave behind the "minimum-phase" part. For a general H^∞ function, there may be a further "singular" zero structure [17], so that the general factorization is

$$a(z) = a_{\text{in}}(z)a_{\text{out}}(z)$$

where all the zero structure is contained in the so-called *inner function* $a_{\text{in}}(z)$ and $a_{\text{out}}(z)$ is called the *outer function* part of $a(z)$.

This now allows us to define $\{a(z), b(z)\}$ as being *weakly coprime* if they have no common inner part.

Corresponding to these two kinds of coprimeness, we shall have two kinds of observability, strong and weak, and then dually two kinds of controllability. (The realization obtained by the Nerode construction turns out to be weakly controllable, except when $a(z)$ is an inner function.) Moreover it turns out that if a realization is weakly controllable and strongly observable, or vice versa, or strongly controllable and strongly observable, then the singularities of the transfer function coincide with the 'spectra' of the operator A in the state-space realization. For reasons of space, if nothing else, we cannot go further into these concepts here; an interested reader can refer to the book of Fuhrmann [19]. However this book, and in fact the present system theory literature on this subject (see [12]–[16]), is at a fairly high mathematical level and does not follow the route we have begun to describe here. (A partial report on progress with our approach can be found in [20].)

Except for the observation that the transfer-function methods of Sec. 4.5 can be used here as well, we shall stop our brief discussion of distributed-parameter systems here.

10.2   2-D SYSTEMS

We shall briefly consider linear shift-invariant discrete-time systems defined by an input-output-relation

$$y(i, j) = \sum_{k, l} h(i - k, j - l)u(k, l)$$

where the impulse response function $h(\cdot, \cdot)$ is often also called the *point-spread function.* In the transform domain, we can write this as

$$Y(z, \omega) = H(z, \omega)U(z, \omega)$$

where

$$H(z, \omega) \triangleq \sum_{i,j=-\infty}^{\infty} h(i, j)z^{-i}\omega^{-j}$$

is the transfer function.

An important special class of systems is obtained by restricting attention to rational transfer functions, say

$$H(z, \omega) = \frac{\sum_{I} b_{kl}z^{-k}\omega^{-l}}{1 + \sum_{I-\{0,0\}} a_{kl}z^{-k}\omega^{-l}} = \frac{b(z, \omega)}{a(z, \omega)}$$

where I is some set of points located about the point $(0, 0)$. These may be called "nearest-neighbor" models.

If I is all the points in one quadrant, the system is said to be *quarter plane causal*. For example, I may be the set of all $\{k, l: k \geq 0, l \geq 0\}$, in which case we shall have

$$H(z, \omega) = \sum_{i,j=0}^{\infty} h(i, j)z^{-i}\omega^{-j}$$

so that $y(i, j)$ will depend on $\{y(k, l), u(k, l), k \leq i, l \leq j\}$. This leads to a so-called SW (southwest) causal system. Other specifications can be given for I (to make for example the system be asymmetric half plane causal) but in many cases, certain coordinate transformations can be devised to reduce them to the quarter plane case (see [5, Ch. 1]). For convenience therefore we shall henceforth restrict ourselves to SW causal systems.

We may say that $H(z, \omega)$ is irreducible if $\{a(z, \omega), b(z, \omega)\}$ are coprime in the sense that they have no common factor $f(z, \omega)$ such that

$$a(z, \omega) = \bar{a}(z, \omega)f(z, \omega), \qquad b(z, \omega) = \bar{b}(z, \omega)f(z, \omega)$$

Note however that unlike the 1-D case, $a(z, \omega)$ and $b(z, \omega)$ may still have a common zero (take $a = z\omega - 1$, $b = z - \omega$). [In fact, two polynomials $a(z, \omega)$, $b(z, \omega)$ chosen 'at random' will almost always have common zeros. This is a result of Bezout, which generalizes the fact that two randomly chosen lines in a plane will almost always intersect each other. Therefore zero coprimeness is a strong form of coprimeness.] This implies that there can be two kinds of singularities for $H(z, \omega)$ depending on whether $\{a(z_0, \omega_0) = 0, b(z_0, \omega_0) \neq 0\}$ or whether $\{a(z_0, \omega_0) = 0 = b(z_0, \omega_0)\}$. These are called nonessential singularities of the first and second kind respectively. (It can be shown that no other singularities are possible for rational functions.) Singularities of the first kind generalize the 1-D notion of poles, but those of the second kind have no 1-D analog and cause various differences between 1-D and 2-D results.

For example, a striking fact, discovered by Goodman [21] is that if $H(z, \omega)$ has nonessential singularities of the second kind on the boundary $\{|z| = 1, |\omega| = 1\}$, then the numerator polynomial can play a role in the (input-output) stability of the system! Jury [22] gives a detailed review of stability problems and related questions (see also Willsky [23]).

This however is not the only problem where we encounter significant differences from the 1-D case. Returning to coprimeness, a famous theorem (Hilbert's Nullstellensatz) asserts that $\{a(z, \omega), b(z, \omega)\}$ are zero coprime if and only if there exist polynomials $\{x(z, \omega), y(z, \omega)\}$ such that

$$x(z, \omega)a(z, \omega) + y(z, \omega)b(z, \omega) = 1$$

This is exactly as in the Bezout identity for 1-D polynomials.

However, $\{a(z, \omega), b(z, \omega)\}$ are factor coprime if and only if there exist polynomials $\{x_i(z, \omega), y_i(z, \omega), i = 1, 2\}$ such that

$$x_1 a + y_1 b = r_1(\omega), \qquad x_2 a + y_2 b = r_2(z)$$

where r_1 depends only on ω, and r_2 only on z.

Corresponding to these two notions of coprimeness, different degrees of flexibility can be achieved in relocating the poles of $H(z, \omega)$—this can be analyzed by transfer-function methods exactly as in Secs. 4.5 and 7.5 (for the matrix case). Here however we shall only refer to [5] for the full story, including applications to delay-differential systems, which were first studied by state-space methods.

In fact, state-space realizations $\{A, B, C\}$ of $H(z, \omega)$ can be obtained [7] by seeking analog computer simulations as in Sec. 2.1, leading to state-space equations of a form originally proposed on different grounds by Roesser [24] and yielding a transfer function of the form

$$H(z, \omega) = C\left[\begin{pmatrix} zI & \\ & \omega I \end{pmatrix} - A\right]^{-1} B$$

Roesser and others (Attasi, Fornasini, and Marchesini) proposed definitions of state-controllability and observability, but these had controllable and observable realizations that were not minimal, and minimal realizations that were not controllable and observable.

However, if we use the factor comprimeness concepts to define (modal) controllability and observability, then the natural relationship to minimality is recovered [7]. We say $\{C, A\}$ is modally observable if

$$C \text{ and } \begin{pmatrix} zI \\ & \omega I \end{pmatrix} - A \text{ are left factor coprime}$$

with a similar definition for modal controllability of $\{A, B\}$. As in the PBH tests of Secs. 2.4.3 and 6.2, there are corresponding eigenvalue-eigenvector (or rather eigencurve-eigencone) characterizations, which explain the adjective modal.

We see that the properties of curves in 2-space are important here, so that the proper mathematical tool is algebraic geometry. Actually some very modern results of this subject are useful in system theory, e.g., the so-called Serre conjecture, the recent proof of which was one of the achievements for which D. Quillen received one of the 1978 Fields Medals in mathematics. On the other hand, the needs of system theory have led to the development of some new mathematical results, e.g., that so-called 'primitive' factorizations always exist for 2-D polynomial matrices (see [6]), and furthermore, that they may not in the N-D case, $N \geq 3$, as discovered by B. Lévy (1977) and then independently by Youla and Gnavi [25].

10.3 SOME OTHER APPLICATIONS OF ALGEBRAIC GEOMETRY; NONLINEAR SYSTEMS

We should mention that algebraic geometry also has a role to play in 1-D systems, especially in the study of invariants and canonical forms, for example [26], [27] and the surveys [28], [29]. We should also note the studies of Brockett and his colleagues on the geometry of the space of all rational functions ([30]–[32])—with results that have interesting implications for system identification and also for recent work on certain nonlinear systems (Toda Lattices and Korteweg de Vries equations) that have some striking properties, including certain solutions ("solitons") that have linear superposition properties (see, e.g., [33]–[35]). This appearance of linear ideas in some highly nonlinear problems leads us to note some references in which the groundwork for a nonlinear system theory is being laid. We list here the papers of Brockett [36], [37], Hermann and Krener [38], Sussman [39], [40], Gilbert [41], Mitzel, Clancy and Rugh [42], Tarn and Nonoyama [43], Sontag [44], and Fliess [45]. Again a variety of mathematical tools arises, especially differential geometry and Lie Algebras.

10.4 APPROXIMATION AND MODEL REDUCTION

We should not of course give the impression that there are no more interesting results for finite-dimensional 1-D linear systems. Here is one to whet your fancy: we know that the rank of the Hankel matrix formed from the impulse response coefficients

$$\mathbf{M} = \begin{bmatrix} h_1 & h_2 & h_3 & \cdots \\ h_2 & h_3 & & \cdots \\ h_3 & & & \\ \cdot & & & \\ \cdot & & & \\ \cdot & & & \end{bmatrix}$$

determines the order of a minimal realization. However, rank is very sensitive to perturbations in the element values, and with noisy measurements it is quite conceivable that $\mathbf{M}$ might have a very large (or infinite) rank. It is then important to try to find a realization of some fixed rank, say r, whose Hankel matrix, $\hat{\mathbf{M}}$ say, is such that

$$\|\mathbf{M} - \hat{\mathbf{M}}\| = \text{minimum}$$

(where $\|A\| \triangleq \sup \|Ax\|$ over all $\|x\| = 1$). It is fairly wellknown that a rank r matrix achieving this minimum can be found by keeping only the r largest singular values in the singular value decomposition of $\mathbf{M}$. The problem of course is that this approximating matrix will not in general be Hankel (and therefore will not correspond to a rational transfer function of degree r). However, Adamjan, Arov, and Krein [46], following some work of Nehari [47], were able to show that the best Hankel approximant achieved the same minimum value as the unconstrained approximant, and in fact they also give an explicit formula for the rational function corresponding to this Hankel approximant.

This striking result, pointed out to the author by P.Dewilde in July 1978, should have useful system-theoretic applications. Most present work in this area is based on less quantitative criteria for how much is lost when using a reduced model (see [48], [49]).

We could go on with suggestions for more reading on other topics, e.g., large-scale systems with decentralized control and observation, fast $O(n^2)$ and $O(n \log^2 n)$ algorithms for systems problems, more on numerical analysis and computer-aided design, generalized state-space systems and singular perturbations, stability analyses and system design, etc. (see, e.g. [50], [51]). However, we shall stop here, for as it was written long ago

> "*of the making of many books there is
> no end, and much study is a weariness
> of the flesh.*"

REFERENCES

1. N. K. Bose, ed., "Special Issue on Multidimensional Systems," *Proc. IEEE*, **65**, June 1977.

2. M. G. Ekstrom and S. K. Mitra, eds., *Two Dimensional Signal Processing*, Dowden, Hutchinson and Ross, Stroudsburg, PA, 1978.

3. N. K. Bose, ed., *Multidimensional Systems: Theory and Applications*, IEEE Press, N.Y., 1979.

4. S-Y. Kung, "Multivariable and Multidimensional Systems: Analysis and Design," Ph.D. Dissertation, Dept. of Electrical Engg., Stanford Univ., Stanford, CA 94305, June 1977.

5. B. Lévy, "Algebraic Structures for Two-Dimensional Systems," Ph.D. Dissertation, Dept. of Elec. Engg., Stanford Univ., Stanford, CA, August 1980.

6. M. Morf, B. Lévy and S-Y. Kung, "New Results on 2-D Systems Theory, Part I: 2-D Polynomial Matrices, Factorization and Coprimeness," *Proc. IEEE*, **65**, pp. 861–872, June 1977.

7. S-Y. Kung, B. Lévy, M. Morf and T. Kailath, "New Results in 2-D System Theory, Pt. II: State-Space Models, Realization and the Notions of Controllability, Observability and Minimality," *Proc. IEEE*, **65**, pp. 945–961, June 1977.

8. B. Lévy, M. Morf and S-Y. Kung, "New Results in 2-D System Theory, Pt. III: Recursive Realization and Estimation Algorithms," *Proc. 20th Midwest Symp. Circuits & Systems*, p. 178, Lubbock, TX, 1977.

9. D. L. Russell, "Controllability and Stabilizability Theory for Linear Partial Differential Equations: Recent Progress and Open Questions," *SIAM Review*, **20**, pp. 639–739, October 1978.

10. J. L. Lions, *Optimal Control of Systems Governed by Partial Differential Equations*, Springer-Verlag, N.Y., 1971.

11. M. Delfour, A. Bensoussan and S. K. Mitter, *Linear Infinite Dimensional Systems*, MIT Press, to appear.

12. J. S. Baras, "Intrinsic Models for Infinite Dimensional Linear Systems," Ph.D. Dissertation, Harvard Univ., 1973.

13. J. W. Helton, "Discrete Time Systems, Operator Models and Scattering Theory," *J. Func. Anal.*, **16**, pp. 15–38, 1974.

14. J. W. Helton, "Systems with Infinite Dimensional State Space: The Hilbert Space Approach," *Proc. IEEE*, **64**, pp. 145–160, Jan. 1976.

15. P. Dewilde, "Input-Output Description of Roomy Systems," *SIAM J. Contr. and Optim.*, **14**, pp. 712–736, 1976. (Based on a 1971 report, Dept. of Math., Univ. of California, Berkeley, CA.)

16. J. S. Baras and P. Dewilde, "Invariant Subspace Methods in Linear Multivariable-Distributed Systems and Lumped-Distributed Network Synthesis," *Proc. IEEE*, **64**, pp. 160–178, January 1976.

17. K. Hoffman, *Banach Spaces of Analytic Functions*, Prentice-Hall, Englewood Cliffs, N.J., 1962.

641

18. P. L. Duren, *Theory of H^p Spaces*, Academic Press, N.Y., 1970.

19. P. A. Fuhrmann, *Linear Operators and Systems in Hilbert Space*, McGraw-Hill, N.Y., 1980.

20. J. Baras and T. Kailath, "Transfer Function Analysis of Infinite-Dimensional Systems—A Simplified Approach," Tech. Rept., Univ. of Maryland, MD, 1979.

21. D. M. Goodman, "Some Stability Properties of Two-Dimensional Linear Shift-Invariant Digital Filters," *IEEE Trans. Circuits & Syst.*, **CAS-24**, pp. 201–208, April 1977.

22a. E. I. Jury, "Stability of Multidimensional Scalar and Matrix Polynomials," *Proc. IEEE*, **66**, pp. 1018–1047, September 1978.

22b. E. I. Jury, "Problems in Multidimensional System Theory," Rept. No. 78–06, Dept. of Automatic Control, ETH, Zurich, November 1978.

23. A. S. Willsky, "Relationships Between Digital Signal Processing and Control and Estimation Theory," *Proc. IEEE*, **66**, pp. 996–1017, September 1978.

24. R. P. Roesser, "A Discrete State-Space Model for Linear Image Processing," *IEEE Trans. Automat. Contr.*, **AC-20**, pp. 1–10, February 1975.

25. D. C. Youla and G. Gnavi, "Notes on N-Dimensional System Theory," *IEEE Trans. Circuits Syst.*, **CAS-26**, pp. 105–111, February 1979.

26. M. Hazewinkel and A. M. Perdon, "On the Theory of Families of Linear Systems," *Proc. Fourth Symp. on the Math. Thy. of Networks and Systems*, pp. 155–161, Delft, Holland, July 1979.

27. C. Martin and R. Hermann, "Application of Algebraic Geometry to Systems Theory, Pt. V: The McMillan Degree and Kronecker Indices as Topological and Holomorphic Invariants, *SIAM J. Control*, **16**, pp. 743–748, 1978.

28. C. Byrnes and P. Falb, "Applications of Algebraic Geometry in System Theory," *Amer. J. Math.*, **101**, pp. 337–363, 1979.

29. R. Hermann, "Algebro-Geometric and Lie-Theoretic Techniques," *Systems Theory*, Math. Sci. Press, Brookline, MA, 1977. (Also see review by W. M. Wonham, *IEEE Trans. Automat. Contr.*, **AC-24**, pp. 519–520, 1979.)

30. R. W. Brockett, "Some Geometric Questions in the Theory of Linear Systems," *IEEE Trans. Automat. Contr.*, **AC-21**, pp. 449–455, August 1976.

31. C. Byrnes, "On Certain Families of Rational Functions Arising in Dynamics," *Proc. 1978 IEEE Conf. on Decision & Control*, pp. 1002–1006, San Diego, CA, January 1979.

32. P. S. Krishnaprasad, "Symplectic Mechanics and Rational Functions," *Richerche di Automatica*, Jan. 1980.

33. M. J. Ablowitz, D. J. Kaup, A. C. Newell and H. Segur, "The Inverse Scattering Transform—Fourier Analysis for Nonlinear Problems," *Stud. Appl. Math.*, **53**, pp. 249–315, 1974.

34. J. Moser, ed., *Dynamical Systems: Theory and Applications*, Lecture Notes in Physics, vol. 38, Springer-Verlag, N.Y., 1975.

35. F. CALOGERO, ed., *Nonlinear Evolution Equations Solvable by the Spectral Transform*, Pitman, London, 1978.

36. R. W. BROCKETT, "Nonlinear Systems and Differential Geometry," *Proc. IEEE*, **64**, pp. 61–72, January 1976.

37. R. W. BROCKETT, "Volterra Series and Geometric Control Theory," *Automatica*, **12**, pp. 167–176, 1976.

38. R. HERMANN and A. J. KRENER, "Nonlinear Controllability and Observability," *IEEE Trans. Automat. Contr.*, **AC-22**, pp. 728–740, October 1977.

39. H. J. SUSSMAN, "Minimal Realizations of Nonlinear Systems," *Geometric Methods in Systems Theory*, D. Q. Mayne and R. W. Brockett, eds., D. Reidel, Holland, 1973.

40. H. J. SUSSMAN, "Existence and Uniqueness of Minimal Realizations of Nonlinear Systems," *Math. Syst. Theory*, **10**, pp. 263–284, 1976.

41. E. G. GILBERT, "Bilinear and 2-power Input-Output Maps: Finite Dimensional Realizations and the Role of Functional Series," *IEEE Trans. Automat. Contr.* **AC-23**, pp. 418–425, June 1978.

42. G. E. MITZEL, S. J. CLANCY and W. J. RUGH, "On Transfer Function Representations for Homogeneous Nonlinear Systems," *IEEE Trans. Automat. Contr.* **AC-24**, pp. 242–249, April 1979.

43. T. J. TARN and S. NONOYAMA, "An Algebraic Structure of Discrete-time Biaffine Systems," *IEEE Trans. Automat. Contr.* **AC-24**, pp. 211–221, April 1979.

44. E. D. SONTAG, "Realization Theory of Discrere-time Nonlinear Systems, Part I-The Bounded Case," *IEEE Trans. Circuits and Systems*, **CAS-26**, pp. 342–355, May 1979.

45. M. FLIESS, "A Functional Realization Theory for Linear and Nonlinear Recursive M-D Systems," *Proc. 1978 IEEE Conf. on Decision & Control*, pp. 661–666, San Diego, CA, January 1979.

46. V. M. ADAMJAN, D. Z. AROV, and M. G. KREIN, *Mat. USSR Sbornik*, **15**, pp. 31–73, 1971 (English Translation).

47. Z. NEHARI, "On Bounded Bilinear Forms," *Ann. Math.*, **65**, pp. 153–162, 1957.

48. H. P. ZEIGER and A. J. McEWEN, "Approximate Linear Realizations of Given Dimension via Ho's Algorithm," *IEEE Trans. Automat. Contr.*, **AC-19**, p. 153, April 1974.

49. S-Y. KUNG, "A New Low-Order Approximation Algorithm via Singular Value Decompositions," *Proc. IEEE Conf. Decision & Control*, Ft. Lauderdale, Florida, December 1979. Also *IEEE Trans. Automat. Contr.*, 1980.

50. M. ATHANS, ed, "Special Issue on Large-Scale Systems and Decentralized Control," *IEEE Trans. Automat. Contr.*, **AC-23**, April 1978.

51. M. SAIN, ed., "Special Issue on Linear Multivariable Control," *IEEE Trans. Automat. Contr.*, **AC-25**, Dec. 1980.

SOME FACTS FROM MATRIX THEORY

Appendix

We have collected together, in the form of groups of exercises, all the matrix facts that will be needed for the first part of this book. It is not necessary to actually work out any of the exercises, at least on a first reading. But you should try to do so after a result in the text provides some motivation for its significance and value. If the proof seems difficult, do not hesitate to look up the relevant topic in a book or a cited paper. Use your good sense on how far to pursue something—the goal is not the mathematics per se but to learn how to find and use mathematics for your needs. By using the Appendix this way, by the end of this course you will have learned a surprisingly large amount of matrix theory, and you will be better prepared and motivated to profit from regular courses in these subjects.

There are of course several good books on matrix theory and linear algebra, and we give an annotated list of some of our favorites ([1]–[19]) in the References. Note, however, that none of the books has everything we shall need, though [1] and [2] come very close.

1 BASIC OPERATIONS

A.1 Transformations and the Usual Rule for Matrix Multiplication

We have the relations

$$\left.\begin{array}{l} y_1 = a_{11}x_1 + a_{12}x_2 + a_{13}x_3 \\ y_2 = a_{21}x_1 + a_{22}x_2 + a_{23}x_3 \\ y_3 = a_{31}x_1 + a_{32}x_2 + a_{33}x_3 \end{array}\right\} \quad \text{or} \quad y_i = \sum_{j=1}^{3} a_{ij}x_j, \quad i = 1, 2, 3$$

and

$$\left.\begin{array}{l} z_1 = b_{11}y_1 + b_{12}y_2 + b_{13}y_3 \\ z_2 = b_{21}y_1 + b_{22}y_2 + b_{23}y_3 \end{array}\right\} \quad \text{or} \quad z_i = \sum_{j=1}^{3} b_{ij}y_j, \qquad i = 1, 2$$

Let x, y, z be column matrices with components $\{x_i\}$, $\{y_i\}$, $\{z_i\}$, respectively. Represent in matrix notation the relations between (1) y and x, (2) z and y, and (3) z and x. This example explains the origin of the usual rule for matrix multiplication. We say *usual* because for special applications other rules can be given, e.g., the so-called *Schur product*, defined by

$$[A \circ B]_{ij} = a_{ij}b_{ij}$$

or the *Lie product*, defined by

$$[A, B]_{ij} = \sum_{k=1}^{n} [a_{ik}b_{kj} - b_{ik}a_{kj}]$$

or the *Kronecker product*, defined by

$$A \otimes B = \begin{bmatrix} a_{11}B & a_{12}B & \cdots & a_{1n}B \\ \vdots & & & \\ \vdots & & & \\ a_{m1}B & & \cdots & a_{mn}B \end{bmatrix}$$

A.2   Vectors and Column Matrices

Let $\vec{x}$ be a vector in 2-space, with Ox and Oy the (standard) coordinate axes for the space. Let the components of $\vec{x}$ along these axes be x_1 and x_2. The column matrix x with elements x_1 and x_2 will be a *representation* of $\vec{x}$. Similarly, let y be a representation of $\vec{y}$.

1. Show that

$$\langle \vec{x}, \vec{y} \rangle \triangleq \text{scalar product of } \vec{x} \text{ and } \vec{y}$$
$$= x'y = y'x \triangleq \langle x, y \rangle$$
$$\|\vec{x}\| \triangleq \text{length of } \vec{x} = \sqrt{x'x} \triangleq \|x\|$$
$$\cos \theta \triangleq \text{cosine of angle between } \vec{x} \text{ and } \vec{y} = x'y/\|x\| \|y\|$$

If $x'y = 0$, we say the vectors $\vec{x}$ and $\vec{y}$, and the (column) matrices x and y, are *orthogonal*. (Note well: If $\{x, y\}$ have complex entries, we must use the Hermitian transpose above; i.e., we must also take complex conjugates when we transpose: otherwise we shall not have $\|x\|$ nonnegative. For notational convenience, we shall assume throughout this book that we are dealing with real-valued vectors, leaving the reader to make the appropriate adjustments as above when complex vectors are encountered, e.g., as eigenvectors.)

2. A vector $\vec{y}$ is obtained from $\vec{x}$ by rotating $\vec{x}$ through an angle α. Find the relation between α, y, and x.

A.3 The Cauchy-Schwarz Inequality

Let x and y be two N-component column matrices. Show that

$$|\langle x, y\rangle|^2 = (\sum x_i y_i)^2 \leq \|x\|^2 \cdot \|y\|^2$$

Under what conditions will equality hold?

A.4 Gram-Schmidt Orthogonalization and Triangularization

1. Let $\{\vec{x}_1, \vec{x}_2, \ldots, \vec{x}_N\}$ be a set of vectors. Show how to find a set of orthonormal vectors $\{\vec{y}_1, \vec{y}_2, \ldots, \vec{y}_M, M \leq N\}$, spanning the same space as the $\{\vec{x}_i\}$. (The space spanned by the $\{\vec{x}_i\}$ is the space of all linear combinations of the $\{\vec{x}_i\}$.) *Hint:* Let $\vec{y}_1 = \vec{x}_1 / \|\vec{x}_1\|, \vec{y}_2 = \vec{x}_2 - \alpha \vec{y}_1 / \|\vec{x}_2 - \alpha \vec{y}_1\|$, where α is chosen so that $\langle \vec{y}_2, \vec{y}_1 \rangle = 0$. Consider the geometric interpretation of this procedure, which is known as the *Gram-Schmidt technique.*

2. Using the above, show that if A is $m \times n$, there exists an $n \times n$ upper triangular matrix R (i.e., one with zero entries below the main diagonal) and an $m \times n$ matrix Q with orthonormal columns (i.e., columns that are orthogonal and have unit length) such that $A = QR$. *Hint:* Apply the Gram-Schmidt technique to the columns of A and write the results in matrix notation. This is an important exercise. Work out a numerical example for yourself, taking, say, four different 3×1 vectors.

Alternative procedures for orthogonalization and triangularization—modified Gram-Schmidt, Householder, and Givens—are described, for example, in [12]–[19] and, more briefly, in [2, pp. 276–282]; they generally have better numerical properties than the Gram-Schmidt method (however also see the remarks in [17, p. 9.26]).

A.5 Complex Numbers and 3-Vectors as Matrices

1. Verify that complex numbers can be represented by matrices via the correspondence

$$a + ib \sim \begin{bmatrix} a & -b \\ b & a \end{bmatrix}$$

with the usual rules for matrix manipulation. This representation is useful in doing complex arithmetic on a digital computer.

2. Consider the association of 3-vectors with matrices, say,

$$\vec{x} = x_1\vec{i} + x_2\vec{j} + x_3\vec{k} \sim \begin{bmatrix} 0 & x_3 & x_2 \\ -x_3 & 0 & x_1 \\ -x_2 & -x_1 & 0 \end{bmatrix} = X$$

where $\{\vec{i}, \vec{j}, \vec{k}\}$ are the orthogonal unit vectors in 3-space. Show that the Lie product rule (cf. Exercise A.1), $[X, Y] = XY - YX$, gives us the matrix associated with the "vector product" of $\vec{x}$ and $\vec{y}$. This representation is useful for calculations in electromagnetic theory, inertial navigation, etc.

A.6   Toeplitz Matrices

A matrix is Toeplitz if its (i, j)th entry depends only on the value of $i - j$. Thus such a matrix is "constant along the diagonals."

Note that a lower (upper) triangular Toeplitz matrix is completely specified by the elements of the first column (row). Among many other things, Toeplitz matrices provide a convenient representation for polynomial multiplication.

1. Thus, show that if $a(s)b(s) = c(s)$, with $a(s)$ of degree n and $b(s)$ of degree m, then $\mathbf{T}(a)b = c$, where (in an obvious notation) $b' = [b_0 \cdots b_m]$, $c' = [c_0 \cdots c_{m+n}]$ and $\mathbf{T}(a)$ is an $(m + n + 1) \times (m + 1)$ lower triangular Toeplitz matrix with first column $[a_0 \cdots a_n\ 0 \cdots 0]'$.
2. Show that we can also write $c = \mathbf{T}(b)a$, where $\mathbf{T}(b)$ is an $(m + n + 1) \times (n + 1)$ matrix with first column $[b_0 \cdots b_m\ 0 \cdots 0]'$.
3. What can you say about the commutativity of Toeplitz matrices from the fact that polynomials commute?

[Further algebraic properties are discussed in the review papers of T. Kailath, A. Vieira and M. Morf, *SIAM Review*, January 1978, and F. Gustavson and D. Yun, *IEEE Trans. Circuits and Systems*, September 1979.]

2   SOME DETERMINANT FORMULAS

We assume that the reader is familiar with the elementary properties of determinants. We recall only that the determinant can be evaluated by using *Laplace's expansion*,

$$\det A = \sum_j a_{ij}\gamma_{ij} \qquad \text{for any } i = 1, 2, \ldots, n$$

where γ_{ij} denotes the *cofactor* corresponding to a_{ij} and is equal to $(-1)^{i+j}$

det M_{ij}, where M_{ij} is the $(n-1) \times (n-1)$ matrix obtained by deleting the ith row and jth column of A. Det M_{ij} is called the ijth *minor* of the matrix; the *leading* or *principal* minors are obtained when $i = j$.

An important consequence of Laplace's expansion is that the *determinant of a triangular matrix is equal to the product of the diagonal elements*.

To assist in the computation of det A, we often use the above fact and some combination of the following results. The operations described in 1–3 below are called *elementary (row or column) operations*.

1. If any column (or row) of A is multiplied by a scalar c and the resulting matrix is denoted by $\tilde{A}$, then det $\tilde{A} = c$ det A. Hence it is easy to deduce that det $[cA] = c^n$ det A.
2. If $\tilde{A}$ is the matrix obtained from A by interchanging any two rows (or columns) of A, then det $\tilde{A} = -\det A$.
3. If $\tilde{A}$ is obtained from A by adding a multiple of any one row (or column) to another, then det $\tilde{A} = \det A$.
4. det $A' = \det A$ (transposition does not affect the determinant).
5. *Determinant of a product:* If A and B are any two *square* matrices, then det $AB = \det A \det B = \det BA$.

A.7   Vandermonde Determinants

Given $\{\lambda_1, \ldots, \lambda_n\}$, the Vandermonde matrix is one that has ith row $[1 \; \lambda_i \; \cdots \; \lambda_i^{n-1}]$. By direct evaluation (or by row operations), show that the 3×3 Vandermonde determinant is given by $(\lambda_3 - \lambda_1)(\lambda_3 - \lambda_2)(\lambda_2 - \lambda_1)$. Then show that the $n \times n$ Vandermonde determinant is equal to $\prod (\lambda_j - \lambda_i)$, where the product is taken over all (i, j) such that $1 \leq i < j \leq n$.

A.8   Cramer's Rule

There is a famous formula, useful chiefly for theoretical analyses, that gives the inverse of a matrix as $A^{-1} = \text{Adj } A / \det A$, where Adj A is the adjugate matrix of A, defined by (note the transpose)

$$\text{Adj } A = [\gamma_{ij}]'$$

and γ_{ij} is, as noted earlier, the cofactor of a_{ij}.

Now suppose we have to solve $Ax = y$. Show that $x = [x_1 \; \cdots \; x_n]'$ can be computed as $x_i = \det A_i / \det A$, $i = 1, \ldots, n$, where $A_i = $ the matrix A with the ith column replaced by y.

A.9   The Binet-Cauchy Formula [5]

If A is an $m \times n$ matrix, let $|A|_{i,k}^r = $ the $r \times r$ minor formed by selecting r rows, say $\{i_1, \ldots, i_r\}$, and r columns, say $\{k_1, \ldots, k_r\}$, of A, where r is any integer such that $r \leq \min(m, n)$. Then if B is any $n \times p$ matrix, the Binet-

Cauchy formula says that

$$|AB|_{i,k}^r = \sum_l |A|_{i,l}^r |B|_{l,k}^r$$

where l ranges over all sets of r integers, and $1 \le r \le \min(m, n, p)$.

1. Show that if A and B are *both square*, then det $AB = $ det A det B $=$ det BA.
2. If A is $m \times n$ and B is $n \times m$, $m < n$, find an expression for det AB.

3   BLOCK MATRICES AND THEIR DETERMINANTS

We shall often encounter matrices whose elements are themselves matrices. No additional special rules are necessary for such matrices: We just have to be sure that we are not violating any rules for ordinary matrices.

A.10   Multiplication of Block Matrices

Specify the dimensions that will make the following formula valid when A is $n \times m$ and H is $p \times q$:

$$\begin{bmatrix} A & B \\ C & D \end{bmatrix}\begin{bmatrix} E & F \\ G & H \end{bmatrix} = \begin{bmatrix} AE + BG & AF + BH \\ CE + DG & CF + DH \end{bmatrix}$$

A.11   Determinants of Block Matrices

1. Show that when A and B are square

$$\det \begin{bmatrix} A & 0 \\ C & B \end{bmatrix} = \det A \det B$$

Hint: If A or B is singular, the identity is trivial. Otherwise, observe that

$$\begin{bmatrix} A & 0 \\ C & B \end{bmatrix} = \begin{bmatrix} A & 0 \\ 0 & I \end{bmatrix}\begin{bmatrix} I & 0 \\ 0 & B \end{bmatrix}\begin{bmatrix} I & 0 \\ B^{-1}C & I \end{bmatrix}$$

2. Hence, if A is nonsingular, show that

$$\det \begin{bmatrix} A & D \\ C & B \end{bmatrix} = \det A \det [B - CA^{-1}D]$$

What if we only know that B is nonsingular?

A.12   A Useful Identity for Determinants of Products

We have stated that det $AB = $ det BA if A and B are square but that this does not hold for nonsquare A and B. However, show that if A is $n \times m$ and B is $m \times n$, then det $[I_n - AB] = $ det $[I_m - BA]$, where I_p is the $p \times p$ identity matrix. The special case when $m = 1$, so that A is a column and B is a row, is especially useful. *Hint:* Apply elementary row and column operations and Exercise A.11 to the block matrix with first row $[I_n \ A]$ and second row $[B \ I_m]$.

A.13   Useful Representations for Transfer Functions

1. Show that if A is $n \times n$, b is $n \times 1$, c is $1 \times n$, and d is a scalar, then

$$\det (sI - A)[c(sI - A)^{-1}b + d] = \det \begin{bmatrix} sI - A & b \\ \hline -c & d \end{bmatrix}$$

The two sides are sometimes said to be *reciprocal forms* (see [10, p. 162]).

2. If $G(s) = c(sI - A)^{-1}b$, show that we can write

$$G(s) = \frac{\det (sI - A + bc) - \det (sI - A)}{\det (sI - A)}$$

3. Let $G(s) = D + C(sI_n - A)^{-1}B$ and $S(g) = A + B(gI_m - D)^{-1}C$ be $m \times m$ and $n \times n$ matrix functions, and define $\Delta(g, s) = \det [gI_m - G(s)]$ and $\nabla(s, g) = \det [sI_n - S(g)]$. Show that

a. $\det (sI_n - A) \, \Delta(g, s) = \det \begin{bmatrix} sI_n - A & -B \\ \hline -C & gI_m - D \end{bmatrix}$

b. $\det (gI_m - D) \, \nabla(s, g) = \det (sI_n - A) \, \Delta (g, s)$

[Reference: A. MacFarlane and N. Karcanias, in *Proceedings of the Seventh IFAC World Congress, Preprints 3*, Helsinki, Aug. 1978, pp. 1771–1779.]

4   SOME REMARKS ON LINEAR EQUATIONS

Many questions in system theory reduce ultimately to the study of the linear equation

$$\begin{matrix} A & \cdot & x & = & y \\ m \times n & & n \times 1 & & m \times 1 \end{matrix}$$

for an unknown vector x. Depending on the values of the matrices A and y, this equation may have a unique solution, or many solutions, or no solution.

If we recall that Ax can be regarded as a linear combination of the columns of A, with weights equal to the corresponding components of x, then it is clear that the equations will have a solution if and only if the right-hand side y is some linear combination of the columns of A. In this case, the equations are said to be *consistent*, and y is said to lie in the *range space* of A, denoted $\mathfrak{R}(A)$. The word *space* is used because of the following *closure* property: If y_1 and y_2 are in the range of A (i.e., if they are linear combinations of the columns of A), then any linear combination of them, $c_1 y_1 + c_2 y_2$, where c_1 and c_2 are scalars, will also be in the range of A.

If the equations are consistent, when can we have more than one solution? Note that if x_1 and x_2 are any two solutions (i.e., $Ax_1 = y$ and $Ax_2 = y$), then

$$A(x_1 - x_2) = 0$$

It is then clear that x_1 will be different from x_2, and we shall have more than one solution if and only if $A(x_1 - x_2) = 0$ has a nontrivial solution, i.e., a solution not all of whose components are zero. The *null-space* of A, denoted $\mathfrak{N}(A)$, is the space of all solutions of the equation

$$Ax = 0$$

and this space will be empty if and only if no nontrivial combination of the columns of A can be zero.† Any set of columns with this property is said to be a *linearly independent* set, and any matrix whose columns are linearly independent will be said to have *full column rank*. The null-space of A will be nonempty (i.e., will contain something besides the zero column) if and only if A does not have full column rank. It should now be clear that the equations $Ax = y$ will have a unique solution if and only if A has full column rank.

An $m \times n$ matrix will be said to have *column rank* α if at most α columns of A are linearly independent.

So far we have talked only about the columns of A and the equation $Ax = y$. Clearly, similar statements can be made about the rows of A and the equation

$$xA = y$$

(where again the context is used to set the dimensions). Thus the *row rank* of A is the maximum number of linearly independent rows of A.

The special case of square matrices ($m = n$) is particularly important. The properties of determinants show that a square, full- (column or row)

---

†The null-space of A is often also called the *kernel* of A and written ker $\{A\}$.

rank matrix can never have a zero determinant and therefore will be *non-singular*. A nonsingular matrix is always invertible, so that for nonsingular A the equation $Ax = y$ will always be consistent for any y: a solution is $A^{-1}y$. Moreover, the solution will be unique because $Ax = 0$ will imply $x = A^{-1}0 = 0$.

On the other hand, if A is square and not of full rank, its determinant will be zero and A will be singular. Now the null-space will not be empty, but whether or not we have a solution will depend on whether or not y lies in $\mathfrak{R}(A)$.

A.14   A Simple Linear Equation

Let

$$A = \begin{bmatrix} 1 & 1 & 1 \\ 2 & 1 & 1 \\ 4 & 1 & 1 \end{bmatrix}$$

1. Does the equation $Ax = y$ have a solution for $y = [2\ 3\ 5]'$?
2. If there is more than one solution, give a simple way of describing the family of all possible solutions.
3. Repeat the above questions for $y = [2\ 3\ 6]'$.
4. How many linearly independent columns does A have? How many linearly independent rows?
5. Repeat all the above questions for the matrix obtained from A by changing the $(2, 2)$ entry from 1 to 2.

A.15   Range Spaces and Null Spaces

1. Suppose x is a vector in $\mathfrak{N}(A)$, the null-space of A. Prove that x is orthogonal to every vector in $\mathfrak{R}(A')$, the (column) range space of A'.
2. Prove the converse: if x is orthogonal to every vector in $\mathfrak{R}(A')$, then x is in $\mathfrak{N}(A)$.
3. Make and explore similar statements about $\mathfrak{R}(A)$ and $\mathfrak{N}(A')$.

There are many other results in the theory of linear equations that we shall not enter into here, though some more geometric aspects are also explored in Sec. 6.2.

Elementary Operations and Matrix Factorizations.   The question of ways to solve the equation $Ax = y$ is a vast one (see, e.g., [1]–[3], [12]–[19]), but in all methods it is useful to be able to "simplify" the given matrix A by performing some elementary, reversible operations on it. As might already be familiar (e.g., from knowledge of determinants), there are three basic *elementary row operations*: (1) multiplying a row by a constant, (2) interchanging two rows, and (3) adding a multiple of one row to another.

We can also analogously define *elementary column operations*. These elementary operations are clearly reversible, and combinations of them can be used to induce various desirable properties for a matrix.

A.16   Elementary Matrices

We know that premultiplying a matrix A on the left describes certain operations on the rows of A.

1. Find the matrices that describe the above elementary row operations on A. These will be called *elementary matrices*.
2. Show that multiplication on the right by these elementary matrices will correspond to the elementary column operations on A.
3. Suppose A is $m \times n$ and has row rank $\alpha < m$. Show that by elementary row operations we can transform A to a matrix with its first α rows linearly independent and the remaining $m - \alpha$ rows identically zero. Why is it that we cannot make more than $m - \alpha$ rows identically zero?

A.17   LDU Decompositions

1. If the first α rows of an $m \times n$ matrix A are linearly independent, where α is the row rank of A, show that we can always find a sequence of elementary operations that will transform A to an upper triangular matrix.

    Hence, show that such a matrix A can always be written in the form $A = LDU$, where L is an $m \times \alpha$ lower triangular matrix *with diagonal elements equal to unity*, U is an $\alpha \times n$ upper triangular matrix *with unity diagonal entries*, and D is an $\alpha \times \alpha$ diagonal matrix. (Note, of course, that the D can be absorbed into L or U to obtain an LU factorization.)
2. Assume that $m = n = \alpha$. Show that if there are two factorizations $A = L_1 D_1 U_1 = L_2 D_2 U_2$, then we must have $L_1 = L_2$, $D_1 = D_2$, $U_1 = U_2$. That is, the LDU factorization (or decomposition) is unique. *Hint*: $L_2^{-1} L_1 = D_2 U_2 U_1^{-1} D_1^{-1}$.
3. If A is $n \times n$, of full rank, and if $A = LDU$, what can you say about the LDU-*factorization* of A_m, the matrix formed by taking the first m rows and first m columns of A, $1 \le m \le n$.

5   SOME RESULTS ON RANK

A perhaps surprising, but fundamental, result of the theory of linear equations is that for any matrix A,

$$\text{the column rank} = \text{the row rank}$$

Proofs can be found in [1, Chap. 5] and [2, Chap. 2]; a short elegant proof is given by H. Liebeck (*Am. Math. Mon.*, Vol. 73, p. 1114, 1966). It is also true that these ranks are equal to the so-called *determinantal rank* of A, which is α if all submatrices formed from $\alpha + 1$ rows and $\alpha + 1$ columns of A are singular and at least one $\alpha \times \alpha$ submatrix is nonsingular.

In future, therefore, we can just talk of the rank of a matrix A, which we shall write as $\rho(A)$.

A.18   Rank of Products

1. Prove that $\rho(AB) \le \min \{\rho(A), \rho(B)\}$ unless B is nonsingular, in which case $\rho(AB) = \rho(A)$.
2. If A is $m \times n$, $m > n$ and of full rank, prove that $A'A$ is nonsingular. *Hint:* Consider $x'A'Ax$.
3. *Sylvester's inequality* [1, Sec. 5.5]. If A is $m \times n$ and B is $n \times p$, show that $\rho(A) + \rho(B) - n \le \rho(AB) \le \min \{\rho(A), \rho(B)\}$.

A.19   Matrices of Rank α

1. If u is $n \times 1$ and v is $m \times 1$, show that the matrix uv' has rank 1. Show, conversely, that any matrix of rank 1 can be expressed as the product of a column matrix and a row matrix.
2. Show that in general $\rho(A + B) \le \rho(A) + \rho(B)$.
3. Show that a matrix A has rank α if and only if it can be written in the form

$$A = \sum_{1}^{\alpha} u_i v_i'$$

where $\{u_i\}$ and $\{v_i\}$ are sets of independent column vectors.

It is important to note that the actual determination of the rank of a matrix is a notoriously dangerous numerical problem; the rank is very sensitive to small changes in the element values, a very simple example being the matrix diag $\{1, \epsilon\}$. The QR factorization of Exercise A.4 is often used for rank determination, though singular value criteria (cf. Sec. 15, p. 667) are being increasingly advocated. For some perspective on these choices, see the remarks in [17, p. 11.23].

6   SOME FORMULAS ON INVERSES

A.20   A Matrix Inverse

Let A be a nonsingular matrix. Let u and v be column matrices, and assume that $A + uv'$ is nonsingular. Verify by using the definition of the inverse of a matrix that

$$(A + uv')^{-1} = A^{-1} - \frac{(A^{-1}u)(v'A^{-1})}{1 + v'A^{-1}u}$$

The point of this formula is that if A^{-1} is known, the inverse of A augmented by a rank-1 matrix can be obtained by a simple modification of the known A^{-1}.

A.21   A Generalization—The Modified Matrices Formula

A and C are nonsingular $m \times m$ and $n \times n$ matrices, respectively. Verify, by using the definition of the matrix inverse, that

$$(A + BCD)^{-1} = A^{-1} - A^{-1}B(DA^{-1}B + C^{-1})^{-1}DA^{-1}.$$

This general formula has many applications in system theory, especially in the form

$$[I + C(sI - A)^{-1}B]^{-1} = I - C(sI - A + BC)^{-1}B.$$

Its exact origin is hard to trace; it is usually attributed to Woodbury (1950), but according to Householder [15], it may be in the work of Schur. In any case, it is a very useful result. See also Gill et al, *Math. Comp.*, **27**, pp. 1051–1077, 1975, for methods of modifying LDU factorizations.

A.22   Inverse of Block Matrices

1. Show, when A^{-1} and B^{-1} exist, that

$$\begin{bmatrix} A & 0 \\ C & B \end{bmatrix}^{-1} = \begin{bmatrix} A^{-1} & 0 \\ -B^{-1}CA^{-1} & B^{-1} \end{bmatrix}$$

and

$$\begin{bmatrix} A & D \\ 0 & B \end{bmatrix}^{-1} = \begin{bmatrix} A^{-1} & -A^{-1}DB^{-1} \\ 0 & B^{-1} \end{bmatrix}$$

2. If A^{-1} exists, show that

$$\begin{bmatrix} A & D \\ C & B \end{bmatrix}^{-1} = \begin{bmatrix} A^{-1} + E\Delta^{-1}F & -E\Delta^{-1} \\ -\Delta^{-1}F & \Delta^{-1} \end{bmatrix}$$

where $\Delta = B - CA^{-1}D$, $E = A^{-1}D$, and $F = CA^{-1}$. Show that if B^{-1} exists, the (1, 1) block element of the inverse can also be written as $[A - DB^{-1}C]^{-1}$. Δ is known as the *Schur complement* of A.

7   CHARACTERISTIC POLYNOMIALS AND RESOLVENTS

For any square $n \times n$ matrix A, the polynomial

$$a(s) \triangleq \det (sI - A) = s^n + a_1 s^{n-1} + a_2 s^{n-2} + \cdots + a_{n-1}s + a_n$$

is called the *characteristic polynomial* of A. The n roots of $a(s)$ [i.e., the solutions of the *characteristic equation* $a(s) = 0$] are called the *eigenvalues* of A. Thus we can write

$$a(s) = (s - \lambda_1)(s - \lambda_2) \cdots (s - \lambda_n)$$

where the $\{\lambda_i\}$ are the eigenvalues of A. The existence of n roots is, of course, due to the fact that we allow the coefficients and the roots of $a(s)$ to be complex numbers.

The matrix $(sI - A)^{-1}$ is known as the *resolvent* of A.

A.23   Some Resolvent Identities

1. Verify (by multiplying both sides by $sI - A$) that

$\text{Adj}(sI - A)$

$$= [Is^{n-1} + (A + a_1 I)s^{n-2} + \cdots + (A^{n-1} + a_1 A^{n-2} + \cdots + a_{n-1} I)]$$

$$= [A^{n-1} + (s + a_1)A^{n-2} + \cdots + (s^{n-1} + a_1 s^{n-2} + \cdots + a_{n-1})I]$$

(Note that these expressions differ only by the substitution $sI \leftrightarrow A$. The real reason for this can be seen from an alternative, deeper approach to this whole topic, based on some results from polynomial matrix theory—see Exercise 6.3-7.)

2. Show that the coefficients, say S_i, of the powers s^{n-i} in the first expression can be recursively computed as

$$S_1 = I, \quad S_2 = S_1 A + a_1 I, \quad S_3 = S_2 A + a_2 I, \quad \cdots,$$
$$S_n = S_{n-1} A + a_{n-1} I, \quad 0 = S_n A + a_n I$$

These formulas show that the adjugate matrix can be recursively determined from knowledge of A and the coefficients of $a(s)$.

3. Show that the above formulas can be used to obtain a recursive method of simultaneously calculating the coefficients $\{a_i\}$ of the characteristic polynomial and the matrices $\{S_i\}$ that enter into the numerator of $(sI - A)^{-1}$ via the identities

$$a_1 = -\text{tr } A \qquad\qquad S_1 = I$$
$$a_2 = -\tfrac{1}{2}\text{tr }(S_2 A) \qquad\qquad S_2 = S_1 A + a_1 I$$
$$a_3 = -\tfrac{1}{3}\text{tr }(S_3 A) \qquad\qquad S_3 = S_2 A + a_2 I$$
$$\vdots \qquad\qquad\qquad\qquad \vdots$$
$$a_{n-1} = -\frac{1}{n-1}\text{tr }(S_{n-1} A) \qquad S_n = S_{n-1} A + a_{n-1} I$$
$$a_n = -\frac{1}{n}\text{tr }(S_n A) \qquad\qquad 0 = S_n A + a_n I$$

Here tr A denotes *trace of A*—see Exercise A.24 for a definition. These formulas were independently derived by various authors and are often known as the *Leverrier-Souriau-Faddeeva-Frame formulas*.

A.24 Trace and Determinant

1. Show that $\det A = \prod_{i=1}^{n} \lambda_i = (-1)^n a_n$, where $a(s) = \prod_{1}^{n} (s - \lambda_i)$; i.e., the $\{\lambda_i\}$ are the eigenvalues of A.
2. Show that

    $$\text{trace } [A] = \text{sum of the elements on the main diagonal of } A$$

    $$= \sum_{i=1}^{n} \lambda_i = -a_1$$

3. Show that $a_k = (-1)^k \sum \lambda_{i_1} \lambda_{i_2} \ldots \lambda_{i_k}$, where the sum ranges over $\{1 \leq i_1 < i_2 < \cdots < i_k \leq n\}$.
4. Show that $\det (I + \epsilon A) = 1 + \epsilon \text{ tr } A +$ terms in higher powers of ϵ.

A.25 Traces of Products

1. Show that if A and B are $n \times n$ matrices, then AB and BA have the same characteristic polynomial even if $AB \neq BA$.
2. Show that tr $AB =$ tr BA whenever both products are meaningful.

A.26 Rank-1 Matrices

1. If a matrix A has rank 1, show that $(sI - A)^{-1} = [I + A(s - \text{tr } A)^{-1}] s^{-1}$ and $a(s) = \det (sI - A) = s^{n-1}(s - \lambda)$, where λ is the only nonzero eigenvalue of A.
2. Hence, show that if a matrix A has rank 1, $\det (I + A) = 1 + \text{tr } (A)$. Relate this to the result of Exercise A.12.

8 THE CAYLEY-HAMILTON THEOREM

The *Cayley-Hamilton theorem* states that every square matrix A satisfies its characteristic equation. That is, if A is an $n \times n$ square matrix with characteristic polynomial $a(s) = s^n + a_1 s^{n-1} + \cdots + a_{n-1} s + a_n$, then

$$a(A) \triangleq A^n + a_1 A^{n-1} + \cdots + a_{n-1} A + a_n I = 0$$

A.27 A Proof of the Cayley-Hamilton Theorem

Establish the theorem by using the recursions developed in connection with the resolvent expansion (Exercise A.23).

A.28   The Cayley-Hamilton Theorem for Diagonal Matrices

Let Λ be a diagonal matrix with diagonal entries $\{\lambda_1, \ldots, \lambda_n\}$, $\Lambda =$ diag $\{\lambda_i, i = 1, \ldots, n\}$. Prove that $a(\Lambda) = 0$, and show that this result also holds for matrices of the form $A = P\Lambda P^{-1}$. The extension to arbitrary matrices can be made by a limiting procedure [4].

Several different proofs of this theorem are known. Another will be found in Exercise 6.3-6, and system-theoretical proofs are outlined in Exercises 2.3-31 and 2.5-10.

A.29   The Minimal Polynomial of A

While the Cayley-Hamilton theorem shows that $a(A) = 0$, there may be polynomials $\delta(s)$ of lower degree than $a(s)$ such that $\delta(A) = 0$. The monic polynomial of lowest degree for which $\mu(A) = 0$ is called the *minimal polynomial* of A. Find the characteristic and minimal polynomials of

$$A_1 = \begin{bmatrix} \lambda & & \bigcirc \\ & \lambda & \\ \bigcirc & & \lambda \end{bmatrix}, \qquad A_2 = \begin{bmatrix} \lambda & 1 & 0 \\ 0 & \lambda & 0 \\ 0 & 0 & \lambda \end{bmatrix}, \qquad A_3 = \begin{bmatrix} \lambda & 1 & 0 \\ 0 & \lambda & 1 \\ 0 & 0 & \lambda \end{bmatrix}$$

9   COMPANION MATRICES

A matrix A_c of the form

$$A_c = \begin{bmatrix} -a_1 & -a_2 & \cdots & -a_n \\ 1 & 0 & \cdots & 0 \\ 0 & 1 & 0 & \cdots & 0 \\ \vdots & & \ddots & \vdots \\ 0 & & \cdots & 1 & 0 \end{bmatrix}$$

is said to be a *companion matrix*, or a matrix companion to the polynomial $s^n + a_1 s^{n-1} + \cdots + a_{n-1} s + a_n$. This name was suggested by MacDuffee [6] in 1933 for a reason that will be clear from Exercise A.30.

There can be four forms of companion matrices depending on whether the $\{a_i\}$ occur in the first or last row or first or last column. All four forms are useful in system theory, and we shall often refer to them as top-, bottom-, left- or right-companion matrices.

A.30   Characteristic Polynomial of a Companion Matrix

1. Show by column operations that $a(s) = \det (sI - A_c) = s^n + a_1 s^{n-1} + \cdots + a_{n-1}s + a_n$. This illustrates in part the reason for the name *companion matrix*.
2. Show that the rank of $\lambda_i I - A_c$, where λ_i is an eigenvalue of A_c, is equal to $n - 1$.

A.31   Shifting Property of Companion Matrices

If we let e_i' denote the row vector $[0\ 0\ \cdots\ 0\ 1\ 0\ \cdots\ 0]$ with the entry 1 in the ith place, then show that $e_i' A_c = e_{i-1}'$ for $2 \leq i \leq n$ and $e_1' A_c = [-a_1\ -a_2\ \cdots\ -a_n]$. Note here that if we had chosen the companion matrix with last column $-[a_n\ \cdots\ a_1]'$, then we would have obtained the shift property in the form $Ae_i = e_{i+1}$, $i = 1, 2, \ldots, n - 1$.

A.32   The Inverse of a Companion Matrix Is Companion

Let A be a top-companion matrix with first row $-[a_1\ \cdots\ a_n]$.

1. Show that A is nonsingular if and only if $a_n \neq 0$.
2. Show that if A is nonsingular, then A^{-1} is a bottom-companion matrix with last row $-[1\ a_1\ \cdots\ a_{n-1}]/a_n$.
3. How are the eigenvalues of A^{-1} related to those of A?

(Reference: L. Brand, *Am. Math. Mon.*, Vol. 71, pp. 629–634, 1964; Vol. 75, pp. 146–152, 1968.)

A.33   Some Companion Matrix Identities

1. Show that if $b_c' = [1\ 0\ \cdots\ 0]$, then $(sI - A_c)^{-1}b_c = [s^{n-1}\ s^{n-2}\ \cdots\ s\ 1]'/a(s)$. [*Remark*: This result can be shown directly or by observing that the ith entry is given by $e_i'(sI - A_c)^{-1}b_c$, where $e_i' = $ the ith row of I, and then using the identity proved in Exercise A.13.]
2. Let $g(s) = g_1 s^{n-1} + \cdots + g_{n-1}s + g_n$. Show that $g(A_c) = [(A_c^{n-1})'c'\ \cdots\ A_c'c'\ c']'$, where $c = [g_1\ g_2\ \cdots\ g_n]$. *Hint*: Show that $e_n'g(A_c) = c$, and then use the shifting property (Exercise A.31) to show that $e_{n-1}'g(A_c) = cA_c$, etc.
3. Show that we can write

$$
\text{Adj}\,(sI - A_c) = \begin{bmatrix} s^{n-1} \\ s^{n-2} \\ \cdot \\ \cdot \\ \cdot \\ s \\ 1 \end{bmatrix} [1 \quad s \quad \cdots \quad s^{n-1}] a_-' - \begin{bmatrix} 0 & 1 & s & \cdots & s^{n-2} \\ & 0 & 1 & & \cdot \\ & & 0 & & \cdot \\ & & & \cdot & s \\ & \bigcirc & & \cdot & 1 \\ & & & & 0 \end{bmatrix} a(s)
$$

where $\mathcal{C}_- =$ a lower-triangular Toeplitz matrix with $[1 \; a_1 \; \cdots \; a_n]'$ as its first column.

4. Hence, show that for any row and column vectors $c = [c_1, \ldots, c_n]$, $b' = [b_1, \ldots, b_n]$ we can write $c(sI - A_c)^{-1}b = [\gamma(s)\beta(s) \bmod a(s)]/ a(s)$ with $\gamma(s) = c_1 s^{n-1} + \cdots + c_n$, $\beta(s) = \beta_n s^{n-1} + \cdots + \beta_1$, where the $\{\beta_i\}$ are defined as $\beta = \mathcal{C}_-' b$.

10   EIGENVECTORS AND EIGENVALUES

A vector $p \neq 0$ for which

$$Ap = \lambda p$$

is said to be an *eigenvector* (of the square matrix A) associated with the eigenvalue λ (λ is a scalar). To reconcile this with our earlier discussion of eigenvalues, note that by rearranging terms we have

$$(\lambda I - A)p = 0$$

But $p \neq 0$, and thus

$$\det(\lambda I - A) = 0$$

so that λ is an eigenvalue of A as defined earlier.

Note that if p is an eigenvector of A, so is αp for any constant α. Therefore, the eigenvectors can be normalized to have unit length; i.e., $p'p = 1$. We shall always assume this to have been done.

A.34   Existence of Eigenvectors

Show that a matrix has at least one eigenvector for every distinct eigenvalue. Find an example of a 2×2 matrix that has only one eigenvector. Find a matrix that has an infinite number of eigenvectors.

A.35   Eigenvectors of a Companion Matrix

Let A_c be a top-companion matrix with $-[a_1 \; \cdots \; a_n]$ as its top row.

1. Show that $p = [\lambda^{n-1} \; \cdots \; \lambda \; 1]'$ is an eigenvector of A_c corresponding to the eigenvalue λ.
2. Show that $q = [1 \; \lambda \; \cdots \; \lambda^{n-1}]\mathcal{C}_-'$, where $\mathcal{C}_-$ is as defined in Exercise A.33, is a left eigenvector of A_c associated with λ, i.e., that $qA_c = \lambda q$.

A.36   Eigenvalues of Matrix Polynomials

1. If $f(A)$ is a polynomial in A and p an eigenvector of A associated with the eigenvalue λ, then $f(A)p = f(\lambda)p$, and thus $f(\lambda)$ is an eigenvalue of $f(A)$ and p the corresponding eigenvector.

2. If A is lower triangular, i.e., $a_{ij} = 0$, $i < j$, show that the eigenvalues of A^r are equal to the rth powers of the diagonal elements of A.

A.37   Nilpotent Matrices

A matrix A such that $A^r = 0$ for some $r > 1$ is said to be nilpotent. Show that all the eigenvalues of A must be zero.

A.38   Eigenvectors and the Adjugate Matrix [6, p. 86]

If λ is an eigenvalue of A, show that the nonzero columns of the adjugate $a(\lambda)(\lambda I - A)^{-1}$ are eigenvectors of A (cf. the formulas in Exercise A.23).

A.39   Products of Square Matrices; Commuting Matrices

1. If A and B are square matrices, show that AB and BA have the same eigenvalues. What can you say if A is $m \times n$ and B is $n \times m$?
2. Show that if the $n \times n$ matrices A and B have a common set of n linearly independent eigenvectors, then $AB = BA$.
3. Give an example to show that the converse is not always true. [A sufficient condition is that both A and B have distinct eigenvalues or at least n distinct (normalized) eigenvectors.]

A.40   A Matrix Equation

Show that the matrix equation $AX + XB = -C$, where A, B, C are given matrices of compatible dimensions, has a unique solution if and only if A and $-B$ have no common eigenvalues. *Hint*: Assume that A and B have distinct eigenvalues. The more general case can be treated by a limiting procedure [4] (also see Exercise A.41). *Note*: It can be shown also that if $\text{Re}[\lambda_i(A)] + \text{Re}[\lambda_j(B)] < 0$, all i, j, then the unique solution can be written as $X = \int_0^\infty e^{At} C e^{Bt}\,dt$, where $e^{At} \triangleq$ the matrix exponential of $At \triangleq I + At + A^2 t^2 / 2! + \cdots$.

A.41   Kronecker Products

If A is an $m \times m$ matrix with eigenvalues $\{\lambda_1, \ldots, \lambda_m\}$ and B is an $n \times n$ matrix with eigenvalues $\{\mu_1, \ldots, \mu_n\}$, show that

1. The eigenvalues of $A \otimes B$, the Kronecker product of A and B (cf. Exercise A.1), are $\{\lambda_i \mu_j\}$.
2. Det $A \otimes B = (\det A)^n (\det B)^m$.
3. The equation $AX + XB = C$ can be written as $(A \otimes I_n + I_m \otimes B)\chi = \gamma$, where $\{\chi, \gamma\}$ are column vectors of mn elements formed from the rows of X, C (respectively) taken in order; i.e., $\chi = [x_{11} \ x_{12} \ \cdots \ x_{1n} \ \cdots \ x_{mn}]'$.

Use 1 and 3 to show that the matrix equation has a unique solution if and only if A and $-B$ have no eigenvalues in common.

4. Show that the matrix equation $P - FPF' = Q$, where F and P are given $n \times n$ matrices, can be written as $(I - F \otimes F)\chi = \gamma$ where χ and γ are obtained from P and Q by "stacking" the rows as in 3 above.

[For a review of Kronecker products in system theory, see a paper by J. W. Brewer (*IEEE Trans. on Circuits and Systems,* **CAS-25**, pp. 772–781, Sept. 1978) and the references therein.]

11   SPECTRAL DECOMPOSITIONS AND MATRIX DIAGONALIZATION

A.42   Distinct Eigenvalues Imply Linearly Independent Eigenvectors

Suppose an $n \times n$ matrix A has m distinct eigenvalues, which we shall label as $\{\lambda_1, \ldots, \lambda_m\}$, with $\{p_1, \ldots, p_m\}$ being the corresponding eigenvectors. Show that the $\{p_i\}$ are linearly independent, i.e., that there exists no set $\{c_1, \ldots, c_m\}$ with not all the $\{c_i\}$ being zero such that $\sum_1^m c_i p_i = 0$. *Hint*: Suppose that $\sum_1^m c_i p_i = 0$, with $c_1 \neq 0$. Apply $(A - \lambda_2 I) \cdots (A - \lambda_m I)$ to both sides, and show that this implies $p_1 = 0$, which is impossible!

A.43   Symmetric Matrices

1. Show that if $A = A'$ all its eigenvalues are real and that eigenvectors $\{p_i, p_j\}$ associated with distinct eigenvalues $\lambda_i \neq \lambda_j$ are orthogonal to each other, i.e., $p_i' p_j = 0$, $i \neq j$.
2. For any $n \times n$ matrix A such that $A = A'$, show that we can always find n orthogonal eigenvectors whether or not A has distinct eigenvalues.
3. A matrix A is said to be normal if it satisfies $AA' = A'A$. Show that an $n \times n$ normal matrix has a (complete) set of n linearly independent eigenvectors.

A.44   Right and Left Eigenvectors

Suppose the $n \times n$ matrix A has n eigenvalues $\{\lambda_i\}$ with n linearly independent associated eigenvectors $\{p_i\}$. Let $P = [p_1, \ldots p_n]$, $Q = P^{-1}$, and let $\{q_i'\}$ be the rows of Q.

1. Why does P^{-1} exist?
2. Show that

$$q_j' p_i = \delta_{ij} = \begin{cases} 1, & i = j \\ 0, & i \neq j \end{cases}$$

3. Show that $A' q_i = \lambda_i q_i$ or equivalently that $q_i' A = \lambda_i q_i'$. The $\{q_i'\}$ are often called the *left* (or *row*) eigenvectors of A in contrast to the *right*

(or *column*) eigenvectors $\{p_i\}$. However, note that only one of the sets $\{p_i\}, \{q_i\}$ can be taken as *normalized* (to unit length) eigenvectors.

A.45 Spectral Decomposition

Let A, P, Q be as in Exercise A.44. Show that we can write

1. $A = \sum \lambda_i p_i q_i' = P\Lambda Q$, where $\Lambda = \text{diag}\{\lambda_1, \ldots, \lambda_n\}$.
2. $A^2 = \sum \lambda_i^2 p_i q_i' = P\Lambda^2 Q$.
3. $f(A) = \sum f(\lambda_i) p_i q_i' = Pf(\Lambda)Q$, where $f(A)$ is any polynomial in A.

A.46 Resolution of the Identity (cf. Sec. 1.3)

Let A, P, Q be as in Exercise A.44, and let $R_i = p_i q_i'$. Show that

1. $R_i R_j = R_i \delta_{ij}$.
2. $I = \sum_1^n R_i$.
3. $A = \sum_1^n \lambda_i R_i = \sum \lambda_i p_i q_i'$, $(sI - A)^{-1} = \sum_1^n R_i/(s - \lambda_i)$.

12 SIMILARITY TRANSFORMATIONS AND TRIANGULAR FORMS

In the previous section we encountered transformations of the form

$$A = T\bar{A}T^{-1}, \qquad \bar{A} = T^{-1}AT$$

For arbitrary T, det $T \neq 0$, such transformations are known as *similarity* transformations. (We refer to Sec. 5.2 for a useful geometric interpretation of these transformations.) When A has a *complete set of eigenvectors* (i.e., when A is $n \times n$ and has n linearly independent eigenvectors), we see from Exercise A.45 that, by choosing T to be the so-called *modal matrix* formed by these n eigenvectors, $\bar{A} = T^{-1}AT$ is diagonal. The diagonal form is clearly quite a simple form of matrix and is useful for various computational and theoretical purposes. Unfortunately not all matrices can be diagonalized by similarity transformations: Matrices of this type do not have a complete set of eigenvectors and are said to be *defective*. A general study of such matrices is rather complicated, but in the next section we shall outline some aspects that are relevant to this course. First, however, we note a rather different type of result which is often as useful as the reduction to diagonal form and is more general.

A.47 Triangular or Schur Form for Arbitrary Matrices

While not all matrices can be diagonalized, show that any matrix can be reduced to triangular form by a similarity transformation. *Hint*: Any matrix A has at least one eigenvector, say p_1. Let $T_1 = [p_1 \ g_2 \ \cdots \ g_n]$, where the $\{g_i\}$ are any set of vectors that will ensure det $T_1 \neq 0$. Now note that the first column of $A_1 = T_1^{-1}AT_1$ will be zero except for the top entry. Repeat the procedure for the (generally full) lower $(n-1) \times (n-1)$ submatrix of A_1.

A.48 An Application of the Schur Form

If A has eigenvalues $\{\lambda_1, \ldots, \lambda_n\}$, not necessarily distinct, show that $f(A)$ has eigenvalues $\{f(\lambda_1), \ldots, f(\lambda_n)\}$, where $f(\lambda)$ is any polynomial in λ.

13 DEFECTIVE MATRICES AND JORDAN FORMS

A.49 Examples of Defective Matrices

Find eigenvalues and eigenvectors of the following matrices.

1.
$$\begin{bmatrix} \lambda & 1 & & & & \\ & \lambda & 1 & & \bigcirc & \\ & & \cdot & \cdot & & \\ & & & \cdot & \cdot & \\ & \bigcirc & & & \cdot & 1 \\ & & & & & \lambda \end{bmatrix}$$

2.
$$\begin{bmatrix} \lambda & & & & \\ 1 & \lambda & & \bigcirc & \\ & \cdot & \cdot & & \\ & & \cdot & \cdot & \\ \bigcirc & & & 1 & \lambda \end{bmatrix}$$

3.
$$\begin{bmatrix} \lambda & 1 & 0 & 0 \\ & \lambda & 0 & 0 \\ & & \lambda & 1 \\ \bigcirc & & & \lambda \end{bmatrix}$$

4.
$$\begin{bmatrix} \lambda & 0 & 0 & 0 \\ & \lambda & 0 & 0 \\ & & \lambda & 1 \\ \bigcirc & & & \lambda \end{bmatrix}$$

A.50 Instability of Defective Matrices

Find the eigenvalues and eigenvectors of the matrices

$$\begin{bmatrix} 1 & 0 \\ 0 & 1 \end{bmatrix}, \quad \begin{bmatrix} 1 & \epsilon \\ 0 & 1 \end{bmatrix}$$

Note that the second matrix is always defective no matter how small ϵ is. (It can be proved that given any matrix A there exists another matrix B all of whose elements are smaller in magnitude than any preassigned number such that $A + B$ is nondefective. However, we note that this fact must be used with care.)

A.51 Generalized Eigenvectors

Suppose A is a 3×3 matrix with eigenvalues $\lambda_1, \lambda_2, \lambda_2$. Let A_ϵ be a perturbation of A, say, $A_\epsilon = A + \epsilon B$, $|\epsilon| \ll 1$, $B =$ arbitrary, such that A_ϵ has distinct eigenvalues $\lambda_1, \lambda_2, \lambda_3 = \lambda_2 + \delta$, $|\delta| \ll 1$, and corresponding eigenvectors $\{p_1, p_2\}$ and $p_3 = p_2 + \delta z$, say. Show that as $\epsilon \rightarrow 0$

1. $(A - \lambda_i I)p_i = 0,\ i = 1, 2.$
2. $(A - \lambda_2 I)z = p_2.$
3. $\{p_1, p_2, z\}$ are linearly independent.

The vector z is called a *generalized eigenvector* of A.

A.52   Jordan Forms

Let A, z, p_1, p_2 be as in Exercise A.51, and let $P = [p_1, p_2, z]$. Show that

$$\bar{A} = P^{-1} A P = \begin{bmatrix} \lambda_1 & 0 & 0 \\ 0 & \lambda_2 & 1 \\ 0 & 0 & \lambda_2 \end{bmatrix}$$

By using suitable sets of eigenvectors and generalized eigenvectors, it can be shown that any matrix can be reduced by a similarity transformation to a *modified-diagonal* or *Jordan* form,

$$J = \begin{bmatrix} J_1 & & & \\ & J_2 & & \bigcirc \\ & & \ddots & \\ \bigcirc & & & J_s \end{bmatrix}, \quad J_i = \begin{bmatrix} \lambda_i & * & & \\ & \lambda_i & * & \bigcirc \\ & & \ddots & \\ & \bigcirc & & \ddots & * \\ & & & & \lambda_i \end{bmatrix}_{(n_i \times n_i)}$$

where $\{\lambda_1, \ldots, \lambda_s\}$ are the eigenvalues of A, n_i being the number of times λ_i occurs, and the $*$ on the superdiagonal are either 1 or 0 depending on further characteristics of the matrix A (see Exercise A.49). There are procedures for obtaining more information on the $\{*\}$ without computing generalized eigenvectors—see [6, pp. 73–74].

We should remark that except when J is diagonal its numerical determination can be unstable (see Exercise A.50). For this and other reasons we shall generally avoid Jordan forms in this book. However, it should also be noted that when the eigenvalues are somehow known (e.g., on physical grounds), then much important information associated with the Jordan form can often still be accurately determined (see [20] and [21]).

A.53   *Eigenvectors of a Companion Matrix*

1. If $\alpha(s) \triangleq \det(sI - A_c)$, where A_c is as in Exercise A.35, has exactly $m\ (\leq n)$ distinct roots λ_i, then show that A_c has exactly m linearly independent eigenvectors of the form $p'_i = [\lambda_i^{n-1}, \lambda_i^{n-2}, \ldots, \lambda_i^2, \lambda_i, 1]$, $i = 1, 2, \ldots, m$, each associated, respectively, with the eigenvalues $\lambda_1, \lambda_2, \ldots, \lambda_m$.

2. When A_c has an eigenvalue λ with multiplicity k [i.e., $(s - \lambda)^k$ is a factor of $\alpha(s)$ but $(s - \lambda)^{k+1}$ is not], show that there are k generalized eigenvectors $\{p_i\}$ given by

$$p_1' = [\lambda^{n-1}, \lambda^{n-2}, \dots, \lambda^2, \lambda, 1]$$
$$p_2' = [(n - 1)\lambda^{n-2}, \dots, 3\lambda^2, 2\lambda, 1, 0]$$

.
.
.

$$p_k' = \left[\binom{n-1}{k-1}\lambda^{n-k}, \dots, 0, 0, 0, 0 \right]$$

associated with the eigenvalue λ.

14   POSITIVE-DEFINITE MATRICES

The matrix A is *positive semidefinite* (p.s.d.) if (the quadratic form) $x'Ax \geq 0$, all x. If equality holds only when $x \equiv 0$, we say that A is *positive definite* (p.d.). Note that because $2x'Ax = x'(A + A')x + x'(A - A')x = x'(A + A')x$, we usually assume that A is symmetric.

A.54   Eigenvalue Test

Show that a symmetric matrix A is p.s.d. (p.d.) if and only if all its eigenvalues are nonnegative (positive).

A.55   Principal Minors

Show that a symmetric matrix A is *positive definite* if and only if the *leading* principal minors (the determinants of the $k \times k$ matrices in the top left-hand corner of A, $k = 1, 2, \dots, n$) are all *positive*.

On the other hand, we have to check that *all* the principal minors are nonnegative to test if a matrix is p.s.d.

A.56   Factorization

Prove that A is p.s.d. if and only if it can be written in the factored form $A = TT'$ for some matrix T, known as a *square root* of A. *Hint*: Use Exercise A.45.

15   SINGULAR VALUES OF A MATRIX

A.57   Singular-Value Decomposition

Let A be a rectangular $p \times m$ matrix, with $p > m$. Show that we can write $A = U\Sigma V'$, where U is a $p \times p$ orthogonal matrix of the eigenvectors of AA', V is an $m \times m$ orthogonal matrix of the eigenvectors of $A'A$, and Σ is a diagonal matrix whose entries are the nonnegative square roots of the eigen-values $\{\sigma_i, i = 1, \ldots, r\}$ of $A'A$ (r is the rank of A). The $\{\sigma_i\}$ are known as the *singular values of A* (and are usually arranged in descending order). The corresponding decomposition of A is known as the *singular-value decomposition of A*.

A.58   An Example

Find the eigenvalues and the singular values of a square matrix with 1s on the first superdiagonal and 0s everywhere else.

A.59   Eigenvalues vs. Singular Values (Kahan)

Consider a lower triangular matrix with -1s on the diagonal and $+1$s everywhere else. Show that the smallest singular value behaves as 2^{-n}, so that for large n the matrix is almost "singular" even though all its eigenvalues are clearly nonzero.

A.60   Bounds on Eigenvalues

Let $\sigma_{\min}$ and $\sigma_{\max}$ be the smallest and largest singular values of a matrix A. Let $\{\lambda_i\}$ denote the eigenvalues of A. Show that $\sigma_{\min} \leq \min_i |\lambda_i| \leq \max_i |\lambda_i| \leq \sigma_{\max}$.

Singular-value decompositions have been found to be very useful in numerical analysis and correspondingly should play an important role in numerical calculations in system theory. They are discussed in [1]–[3] and [12]–[19], while [22] gives a review from a systems point of view (see also [17, p. 11.23] for a discussion of "several limitations that have not been stressed sufficiently in the literature").

REFERENCES

1. B. NOBLE, *Applied Linear Algebra*, Prentice-Hall, Englewood Cliffs, N.J., 1969. Covers a large number of topics with unusual attention to computational questions. (2nd rev. edition, 1977).

2. G. STRANG, *Linear Algebra and Its Applications*, Academic Press, New York, 1976. Emphasizes core topics, with excellent perspective; stress on discussions and explanations rather than on formal deductive proofs.

3. C. LANCZOS, *Applied Analysis*, Prentice-Hall, Englewood Cliffs, N.J., 1950, and *Linear Differential Operators*, Van Nostrand Reinhold, New York, 1959. Both books have excellent but somewhat unconventional chapters on matrix theory, with particular reference to the solution of linear equations.

4. R. E. BELLMAN, *Matrix Analysis*, 2nd ed., McGraw-Hill, New York, 1968, Skillful presentation of selected topics; many annotated references for advanced results.

5. F. R. GANTMAKHER, *Theory of Matrices*, Chelsea Publishing Co., New York, 1959, 2 vols. Comprehensive; contains ideas that are still the basis of new research in linear system theory.

6. C. C. MACDUFFEE, *The Theory of Matrices*, Springer, Berlin, 1933; reprinted by Chelsea Publishing Co., New York, 1950. Good source for original references, with some unique results on matrix polynomials.

7. I. A. GLAZMAN and JU. I. LJUBIC, *Finite-Dimensional Linear Analysis—A Systematic Presentation in Problem Form*, M.I.T. Press, Cambridge, Mass., 1974 (Russian original, 1968). A book for browsing and self-study along the lines of this appendix; Chapters II and III are the relevant ones for our readers.

8. S. BARNETT, *Matrices in Control Theory*, Van Nostrand Reinhold, New York, 1972. Contains several results from the system theory literature.

9. A. ARCHBOLD, *Algebra*, Sir Isaac Pitman & Sons Ltd., London, 1964. A much reprinted English textbook containing a large number of old and new results, with several worked examples.

10. J. H. M. WEDDERBURN, *Lectures on Matrices, Am. Math. Soc.*, 1934; Dover ed., New York, 1964. Special topics; interesting historical notes and an extensive bibliography of early papers.

11. H. W. TURNBULL and A. C. AITKEN, *An Introduction to the Theory of Canonical Matrices*, Blackie, Glasgow, 1932; Dover reprint, New York, 1962.

Books and Papers on Numerical Analysis

12. G. W. STEWART, *Introduction to Matrix Computations*, Academic Press, New York, 1973.

13. G. FORSYTHE and C. B. MOLER, *Computer Solution of Linear Algebraic Systems*, Prentice-Hall, Englewood Cliffs, N.J., 1967.

14. G. E. FORSYTHE, M. A. MALCOLM, and C. B. MOLER, *Computer Methods for Mathematical Computations*, Prentice-Hall, Englewood Cliffs, N.J., 1977.

15. A. S. HOUSEHOLDER, *The Theory of Matrices in Numerical Analysis*, Ginn-Blaisdell, Waltham, Mass., 1964. See also *Lectures on Numerical Algebra*, Mathematical Association of America, 1972.

16. J. H. WILKINSON, *The Algebraic Eigenvalue Problem*, Oxford University Press, London, 1965.

17. J. J. DONGARRA ET AL., *LINPACK User's Guide, SIAM*, Philadelphia, PA. 1979. A collection of machine independent, fully portable, and very efficient Fortran subroutines for analyzing and solving various systems of simultaneous linear equations.

18. B. S. GARBOW ET AL., "Matrix Eigensystem Routines—Eispack Guide Extensions," *Lecture Notes in Computer Science*, Vol. 51, Springer-Verlag, New York, Inc., New York, 1977. A systematized collection of FORTRAN IV subroutines for computing eigenvalues (and singular values) and eigenvectors of important classes of matrices.

19. J. H. WILKINSON and C. REINSCH, EDS., "Handbook for Automatic Computation," *Linear Algebra*, Vol. II, Springer-Verlag New York, Inc., New York, 1971. A collection of well-tested programs written in ALGOL.

20. G. H. GOLUB and J. H. WILKINSON, "Ill-Conditioned Eigensystems and the Computation of the Jordan Canonical Form," *SIAM Rev.*, vol. 18, pp. 578–619, 1976.

21. G. W. STEWART, "Error and Perturbation Bounds for Subspaces Associated with Certain Eigenvalue Problems," *SIAM Rev.*, vol. 15, pp. 727–764, October 1973. See also *ibid.*, vol. 19, pp. 634–662, 1977.

22. B. C. MOORE, "Singular Value Analysis of Linear Systems," in *Proc. 1978 IEEE Conference on Decision and Control,* San Diego, Ca, Jan. 8–11, 1979. *IEEE Trans. Automat. Contr.,* 1980.

23. G. H. GOLUB, V. C. KLEMA, and G. W. STEWART, "Rank Degeneracy and Least-Squares Problems," *Tech. Rept. STAN-CS-76-559,* Computer Science Dept., Stanford University, Stanford, Calif., August 1976.

24. P. VAN DOOREN, "The Generalized Eigenstructure Problem and Applications in Linear Systems," Ph.D. Dissertation, Dept. of Applied Math., Katholieke Universiteit Leuven, Leuven, Belgium, June 1979.

NAME INDEX

SUBJECT INDEX